CAPILLARY ELECTROPHORESIS AND MICROCHIP CAPILLARY ELECTROPHORESIS

CAPILLARY ELECTROPHORESIS AND MICROCHIP CAPILLARY ELECTROPHORESIS

Principles, Applications, and Limitations

Edited by

CARLOS D. GARCÍA
KARIN Y. CHUMBIMUNI-TORRES
Department of Chemistry
The University of Texas at San Antonio
San Antonio, TX, USA

EMANUEL CARRILHO
Institute of Chemistry
University of São Paulo
São Paulo, Brazil

WILEY

Published by John Wiley & Sons, Inc., Hoboken, New Jersey.
Published simultaneously in Canada.

Library of Congress Cataloging-in-Publication Data:

Capillary electrophoresis and microchip capillary electrophoresis : principles, applications, and limitations / edited by Carlos D. García, Ph.D., University of Texas at San Antonio; Karin Y. Chumbimuni-Torres, Ph.D., University of Texas at San Antonio; Emanuel Carrilho, Ph.D., University of Sao Paulo.
pages cm
Includes bibliographical references and index.
ISBN 978-0-470-57217-7 (cloth)
1. Capillary electrophoresis. 2. Microtechnique. I. García, Carlos D., 1972- editor of compilation. II. Chumbimuni-Torres, Karin Y., editor of compilation. III. Carrilho, Emanuel, 1965- editor of compilation.
TP248.25.C37C365 2013
502.8'2–dc23

2012031794

ISBN: 9780470572177

10 9 8 7 6 5 4 3 2 1

CONTENTS

PREFACE

Capillary electrophoresis (CE) is a powerful tool for analytical separations, but in particular, it excels in the separation of biomolecules. In comparison with traditional analytical techniques, CE offers speed, small sample and solvent consumption, low cost, and the possibility of miniaturization. As a consequence, the initial concept describing the separation of proteins in an open tube has evolved into an array of techniques and methodologies that can be competitively applied into the chemical, physical, environmental, clinical, and forensic fields. On the basis of the most current information and the state of the art related to separations by CE and microchip capillary electrophoresis, this book aims to provide a critical perspective related not only to the advantages but also to the limitations of the available techniques. The chapters of this book provide information related to fundamental aspects of electrophoretically driven separations, instrumentation, sampling techniques, separation modes, detection systems, optimization strategies for method development, as well as timely applications. Emphasis was placed in applications in which CE has objectively demonstrated advantages with respect to other analytical approaches. The book not only provides relevant information to researchers actively working in the field but also represents an instructive platform to introduce the topic in educational settings.

Dr. Carlos D. García
Dr. Karin Y. Chumbimuni-Torres
Dr. Emanuel Carrilho

ACKNOWLEDGMENTS

We would like to acknowledge the invaluable contributions and patience from the authors of each chapter of this book. We take this opportunity to highlight their commitment to share their time, experience, insightful projections, and critical views of the field, which we believe is the major aspect of this project. We would also like to express gratitude to John Wiley & Sons, Inc. and especially to Bob Esposito and Michael Leventhal for their help during the development of the book.

Dr. Carlos D. García
Dr. Karin Y. Chumbimuni-Torres
Dr. Emanuel Carrilho

CONTRIBUTORS

Isabel Álvarez-Martos, Department of Physical and Analytical Chemistry, University of Oviedo, Asturias, Spain

Toni Ann Riveros, Department of Chemistry and Biochemistry, California State University, Los Angeles, CA, USA

Fernando Armani Aguiar, Department of Pharmaceutical Sciences, University of São Paulo, São Paulo, Brazil

Lucas Blanes, Centre for Forensic Science, University of Technology, Sydney, Australia

Pierina Sueli Bonato, Department of Physics and Chemistry, University of São Paulo, São Paulo, Brazil

Hector Carmona, Department of Chemistry, California State University, Los Angeles, CA, USA

Emanuel Carrilho, Institute of Chemistry, University of São Paulo, São Paulo, Brazil

Karin Y. Chumbimuni-Torres, Department of Chemistry, The University of Texas at San Antonio, San Antonio, TX, USA

Wendell Karlos Tomazelli Coltro, Institute of Chemistry, Federal University of Goias, Goias, Brazil

Agustín Costa-García, Department of Physical and Analytical Chemistry, University of Oviedo, Asturias, Spain

Anderson Rodrigo Moraes de Oliveira, Departamento de Química, Universidade de São Paulo, São Paulo, Brazil

Philip Doble, Faculty of Science Store, NSW, Australia

Chaitanya Dongre, Integrated Optical Micro Systems (IOMS) Group, MESA Institute for Nanotechnology, University of Twente, Enschede, The Netherlands

Jessica L. Felhofer, Department of Chemistry, The University of Texas at San Antonio, San Antonio, TX, USA

M. Teresa Fernández-Abedul, Department of Physical and Analytical Chemistry, University of Oviedo, Asturias, Spain

José Alberto Fracassi da Silva, Department of Analytical Chemistry, Chemistry Institute, State University of Campinas, São Paulo, Brazil

Francisco Javier García Alonso, Department of Organic and Inorganic Chemistry, University of Oviedo, Asturias, Spain

Carlos D. García, Department of Chemistry, The University of Texas at San Antonio, San Antonio, TX, USA

Zdeněk Glatz, Department of Biochemistry, Central European Institute of Technology, Masaryk University, Brno, Czech Republic

Frank A. Gomez, Department of Chemistry and Biochemistry, California State University, Los Angeles, CA, USA

Gabrielle G. Haby, Department of Chemistry, The University of Texas at San Antonio, San Antonio, TX, USA

Josef Havel, Department of Chemistry and Department Physical Electronics, Masaryk University, Brno, Czech Republic

Xiaoping He, BioTherapeutics Pharmaceutical Sciences, Pfizer Worldwide Research and Development, Chesterfield, MO, USA

Yan He, BioTherapeutics Pharmaceutical Sciences, Pfizer Worldwide Research and Development, Chesterfield, MO, USA

Charles S. Henry, Department of Chemistry, Colorado State University, Fort Collins, Co, USA

Hugo J. W. M. Hoekstra, Integrated Optical Micro Systems (IOMS) Group, MESA Institute for Nanotechnology, University of Twente, Enschede, The Netherlands

Jos Hoogmartens, Laboratory for Pharmaceutical Analysis, KU Leuven, Leuven, Belgium

Nobuhiko Iki, Graduate School of Environmental Studies, Tohoku University, Sendai, Japan

Mihkel Kaljurand, Department of Science, Institute of Chemistry, Tallinn University of Technology, Tallinn, Estonia

Petr Kubáň, Institute of Chemistry, Tallinn University of Technology, Tallinn, Estonia

Qian Liang, School of Mechanical and Aerospace Engineering, Nanyang Technological University, Singapore

Xiaojun Liu, Department of Chemistry and Biochemistry, California State University, Los Angeles, CA, USA

Roger Lo, Department of Chemical Engineering, California State University, Long Beach, CA, USA

Claudimir Lucio do Lago, Department of Fundamental Chemistry, University of São Paulo, São Paulo, Brazil

Cristiane Masetto de Gaitani, Department of Pharmaceutical Sciences, University of São Paulo, São Paulo, Brazil

Yehia Mechref, Department of Chemistry and Biochemistry, Texas Tech University, Lubbock, TX, USA

Germán A. Messina, INQUISAL, Department of Analytical Chemistry, National University of San Luis, San Luis, Argentina

Jianmin Miao, School of Mechanical and Aerospace Engineering, Nanyang Technological University, Singapore

Jim Mo, BioTherapeutics Pharmaceutical Sciences, Pfizer Worldwide Research and Development, Chesterfield, MO, USA

Maria F. Mora, Jet Propulsion Laboratory, California Institute of Technology, Pasadena, CA, USA

Scott D. Noblitt, 1872 Campus Delivery, Fort Collins, CO, USA

Roberto A. Olsina, INQUISAL, Department of Analytical Chemistry, National University of San Luis, San Luis, Argentina

Jochen Pauwels, Laboratory for Pharmaceutical Analysis, KU Leuven, Leuven, Belgium

Eladia María Peña-Méndez, Department of Analytical Chemistry, University of La Laguna, Tenerife, Spain

Dosil Pereira de Jesus, Analytical Department, Institute of Chemistry, State University of Campinas, São Paulo, Brazil

Bruno Perlatti, Chemistry Department, Federal University of São Carlos, São Paulo, Brazil

Markus Pollnau, Integrated Optical Micro Systems (IOMS) Group, MESA Institute for Nanotechnology, University of Twente, Enschede, The Netherlands

Roman Řemínek, Department of Biochemistry, Masaryk University, Brno, Czech Republic

Alberto Rojas-Hernández, Kampus Bohunice, Kamenice, Brno, Czech Republic

Claude Roux, Centre for Forensic Science, University of Technology, Sydney, Australia

Margaret Ruesch, BioTherapeutics Pharmaceutical Sciences, Pfizer Worldwide Research and Development, Chesterfield, MO, USA

Renata Mayumi Saito, Instituto de Quimica, Universidade de São Paulo, São Paulo, Brazil

Marisol Salgado, Department of Chemistry, California State University, Los Angeles, CA, USA

Andrus Seiman, Department of Chemistry, Tallinn University of Technology, Tallinn, Estonia

Patricia W. Stege, INQUISAL, Department of Analytical Chemistry, National University of San Luis, San Luis, Argentina

Amanda M. Stockton, Jet Propulsion Laboratory, California Institute of Technology, Pasadena, CA, USA

Hanno Stutz, Department of Molecular Biology, University of Salzburg, Salzburg, Austria

Toru Takahashi, Graduate School of Engineering, University of Fukui, Fukui, Japan

Ann Van Schepdael, Laboratory for Pharmaceutical Analysis, KU Leuven, Leuven, Belgium

Xu Wang, Laboratory for Pharmaceutical Analysis, KU Leuven, Leuven, Belgium

Robert Weinberger, CE Technologies, Inc., Chappaqua, NY, USA

Blánaid White, School of Chemical Sciences, Dublin City University, Dublin, Ireland

Peter A. Willis, Jet Propulsion Laboratory, California Institute of Technology, Pasadena, CA, USA

Chun Yang, School of Mechanical and Aerospace Engineering, Nanyang Technological University, Singapore

1

CRITICAL EVALUATION OF THE USE OF SURFACTANTS IN CAPILLARY ELECTROPHORESIS

Jessica L. Felhofer, Karin Y. Chumbimuni-Torres, Maria F. Mora, Gabrielle G. Haby, and Carlos D. García

1.1 INTRODUCTION

Surfactants are amphiphilic molecules consisting of a hydrophobic tail and a polar head group (Figure 1.1a) [1]. This unique structure imparts to surfactants a dual affinity for polar and nonpolar solvents, therefore decreasing the surface tension of the medium in which they are dissolved and/or interfacial tension with other phases.

One of the most interesting properties of surfactants is their ability to form aggregates in solution. The concentration at which this aggregation process starts is called the critical micellar concentration (CMC) [2,3]. Because the aggregation process encompasses a delicate balance between hydrophobic attraction and electrostatic repulsion, the CMC is dependent on the structure of the surfactant, the composition of the solution, and the temperature. Probably the most common of these aggregates are spherical structures known as micelles (Figure 1.1b). In some cases, surfactants can also aggregate in solution to form vesicles (Figure 1.1c), which are spherical or ellipsoidal particles formed by a bilayer wrapping around to enclose a volume of solution [1]. Single-chain surfactants tend to form micelles and double-chained surfactants vesicles. The aggregation behavior can be predicted by calculating the packing factor (P) using Equation 1.1:

$$P = \frac{V_c}{l_c a_h} \tag{1.1}$$

where V_c and l_c are the volume and length of the hydrophobic region of the surfactant, respectively, and a_h is the electrostatic cross-sectional area of the head group [1]. When the packing factor is lower than one-third, the surfactant molecule is cone-shaped, and tends to form spherical micelles. When the packing factor is between one-third and one-half, the molecules have a truncated cone shape and tend to form either spherical or cylindrical micelles [4]. When the packing factor is between one-half and one, the surfactant is cylindrical in shape and the formation of a bilayer is favored. In all cases, the shape and number of molecules in the aggregates are affected by factors such as pH, ionic strength, temperature, presence of organic solvents, and salts [5].

Surfactants can also spontaneously aggregate at interfaces, such as the solid–liquid interface that exists between solution and the capillary wall in capillary electrophoresis (CE). In this case, the aggregation/adsorption can occur even at concentrations below the CMC [6–10]. In general, the adsorption of surfactant to solid surfaces comprises the transport of the surfactant molecule toward the interface, followed by interaction with the sorbent surface [11]. Surfactants could then attach at, or detach from, the sorbent surface producing two fluxes—one forward and one backward. The relative contribution of each of these fluxes to the overall adsorption process depends on both the attraction exerted by the surfactant to the surface and the solvent–sorbent surface interactions.

The aggregates of surfactants that form on surfaces are similar to those observed in solution, and include spheres/hemispheres, cylinders/hemicylinders, bilayers (which can be interdigitated), and others such as multilayers (Figure 1.2) [12–14]. Just as in solution, aggregation

Capillary Electrophoresis and Microchip Capillary Electrophoresis: Principles, Applications, and Limitations, First Edition.
Edited by Carlos D. García, Karin Y. Chumbimuni-Torres, and Emanuel Carrilho.

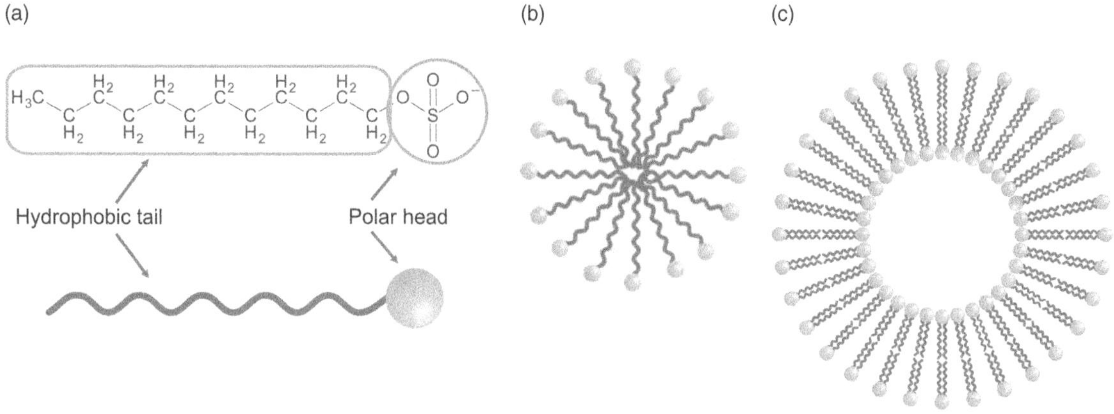

FIGURE 1.1 (a) Structure of a typical surfactant, outlining polar and nonpolar parts, and representations of (b) a micelle, and (c) a vesicle.

tendencies at interfaces are dependent on the packing factor and solution conditions. Additionally, hydrophobic and electrostatic interactions between the surface and the surfactant, as well as lateral surfactant–surfactant interactions, can influence the balance of forces defining the adsorption and aggregation [9,15].

Based on the nature of the head group, surfactants are classified as anionic, cationic, neutral, and zwitterionic. Examples of each class are shown in Figure 1.3. As a compromise between solubility and aggregation properties, the most commonly used anionic surfactants in CE contain linear chains (of 10–14 carbons) and highly polar head groups such as sulfate, carboxylate, sulfonate, or phosphate. Among them, sodium dodecyl sulfate (SDS) is probably the most used surfactant and has been successfully applied to improve the separation of a wide range of analytes [16–22]. Several cationic surfactants have also been used in CE. Like cetyltrimethylammonium bromide (CTAB) and tetradecyltrimethylammonium bromide (TDAB), the most common cationic surfactants used in CE are quaternary ammonium salts. Later described in more detail, neutral surfactants such as Tween 20, Brij 35, and Triton X-100 have also been used in CE. Zwitterionic surfactants contain both anionic and cationic groups (typically an ammonium salt) and the overall charge and effect can be regulated by solution pH. Palmityl sulfobetaine and 3-[(3-cholamidopropyl)dimethylammonio]-2-hydroxy-1-propanesulfonate (CHAPSO) are two examples of zwitterionic surfactants commonly used in CE.

Although outside the aforementioned classification based on charge, other surfactants used in CE include gemini

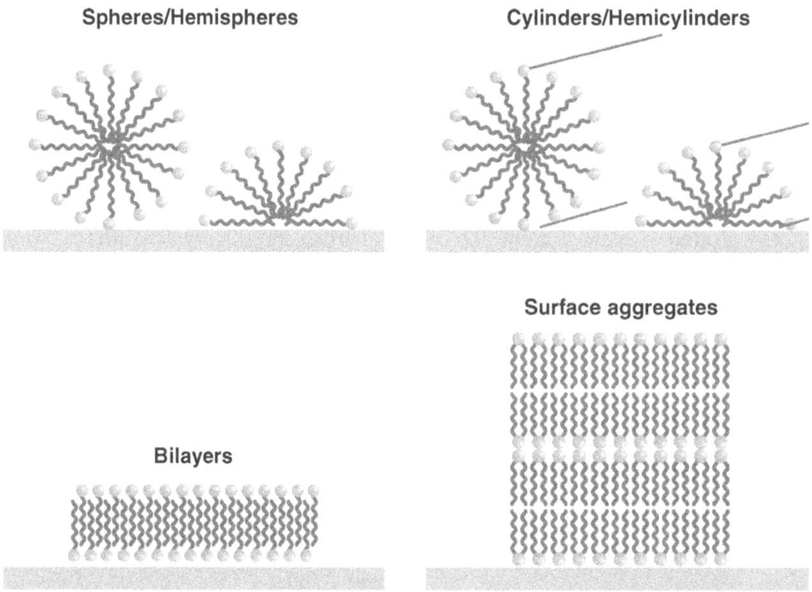

FIGURE 1.2 Various types of surfactant aggregates at the solid–liquid interface. Adapted from References 12 and 14.

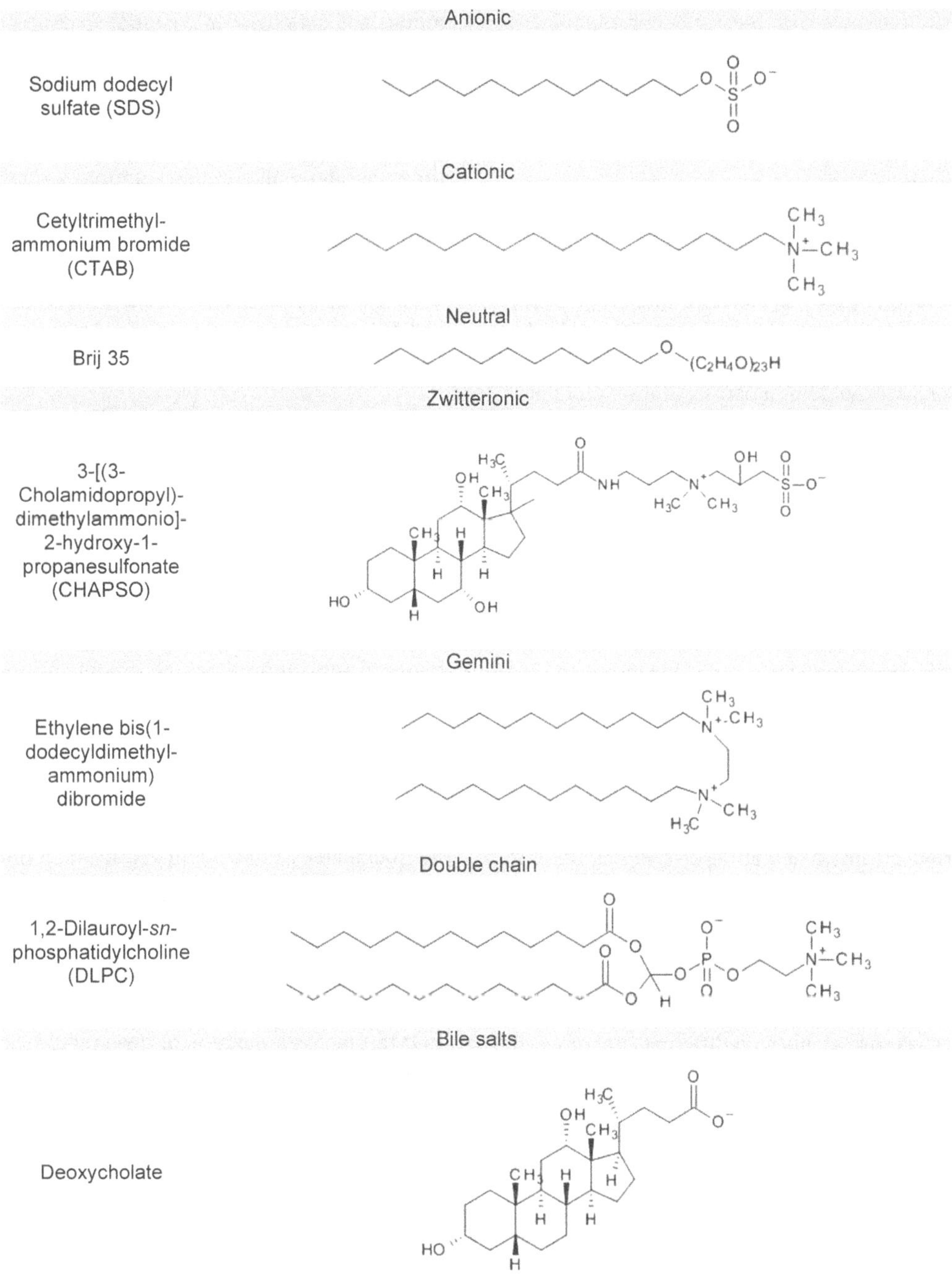

FIGURE 1.3 Examples of various classes of surfactants.

surfactants, double-chain surfactants, bile salts, and polymeric surfactants (Figure 1.3). Gemini surfactants are a relatively new class of amphiphilic compounds, consisting of two identical surfactant moieties connected by an alkyl spacer [23–25]. These molecules have attracted attention because of their high surface activity in addition to their particular aggregation structures, formed as a consequence of their unusual geometries, derived from the nature of the spacer group [26,27]. Advantageously, the CMC of gemini surfactants can be up to 100 times lower than the corresponding monomeric surfactant units [27]. Double-chain surfactants, such as 1,2-dilauroyl-*sn*-phosphatidylcholine (DLPC), are mainly derived from phospholipids. Bile salts are biological surfactants consisting of a large, rigid, and planar hydrophobic steroid moiety with two or three hydroxyl groups and a carboxylate [28]. Examples of bile salts include sodium cholate and sodium deoxycholate. Polymeric surfactants are structurally similar to other surfactants except that the hydrophobic tails in the micelles are covalently linked [29,30]. As a result, there is no dynamic equilibrium of the polymer with the surfactant monomers, and the CMC is effectively zero.

1.2 SURFACTANTS FOR WALL COATINGS

Coating the walls of a capillary with surfactants can improve the performance of CE in two major ways: first, the electroosmotic flow (EOF) can be controlled and stabilized, leading to efficient separations and reproducible migration times; second, adsorption of sample components (proteins, analytes, and the sample matrix) onto the capillary walls can be significantly decreased, yielding good separation efficiency (approaching 1–2 million plates m^{-1}) and excellent recovery values (in some cases reaching 100%) [11,31–34]. These improvements are especially relevant for microchannels used in microchip-CE, which are often fabricated using polymeric substrates, such as poly(dimethylsiloxane) (PDMS) [35–38]. In most cases, the addition of surfactants can help minimize the negative impact of hydrophobic interactions between the sample and the capillary walls, while enabling facile generation and regeneration of the coating, low cost, applicability in a wide range of buffer conditions, and minimal interference with the detection system [31,34]. In addition, surfactants are particularly convenient as dynamic wall coatings for several practical reasons. First, surfactants spontaneously adsorb to most surfaces, requiring no instrumentation or specific skills to prepare the coating; second, the surfactant concentration can easily be controlled; third, the coating, in equilibrium with excess surfactant in the buffer, is continuously replenished; fourth, most surfactants can be removed by simply rinsing the surface with water or the separation electrolyte [39–42]; and lastly, numerous surfactants with different structures are commercially available, simplifying the experimental design [43,44].

As mentioned earlier and further described in Chapter 6, coating the walls of a capillary with surfactants can improve the performance of CE by enabling control of the EOF and by minimizing the adsorption of sample components to the capillary walls. Although these goals are often not mutually exclusive, each of these is herein described separately.

1.2.1 Controlling the Electroosmotic Flow

The EOF is a key parameter in CE because it has a crucial effect on both the efficiency and reproducibility of the separations. The EOF is generated as a consequence of the surface charge of the capillary wall and the applied separation potential. In capillaries made with glass, the presence of ionized silanol groups ($pK_a \sim 4$) provides negative charges on the surface [45]. When in contact with an electrolyte, the deprotonated silanol groups induce a buildup of counterions, followed by a diffuse double layer of anions and cations. The potential spanning the diffuse layer to the capillary wall is called the zeta potential, ζ, and is dependent on the surface charge density of the capillary wall. Upon application of normal polarity potential, solvated cations in the diffuse layer migrate toward the cathode, dragging the bulk of the solution through the capillary, toward the detector. Equation 1.2 shows the relationship between parameters that define the velocity of the EOF (ν_{EOF}): the dielectric constant (ε), zeta potential (ζ), the viscosity of the solution (η), and the applied electric field (E) [15,46].

$$\nu_{EOF} = \frac{\varepsilon\zeta}{\eta}E \qquad (1.2)$$

Since the zeta potential depends on the charge density of the capillary surface, modifications of the capillary walls by surfactants, polymers, and other charged compounds have the potential to affect (increase, decrease, suppress, or reverse) the EOF [19,47–52]. Most often, a precise control of the EOF can aid in achieving an adequate balance between separation time and efficiency.

Stabilization of the EOF is an important target in CE, especially in microchips fabricated from PDMS. Monomers from the bulk of the polymer can migrate to the surface of the channel, changing the surface charge density and resulting in EOF drift. To stabilize the EOF, the charge of the microchannel can be stabilized by including surfactants in the running buffer, providing a dynamic coating. Surfactants spontaneously adsorb to PDMS, via hydrophobic interactions between the surface and the tail of the surfactant, exposing the charged head group to the solution. In this regard, Mora et al. investigated the change in the EOF of PDMS films as a function of the structure and concentration of anionic alkyl surfactants. The selected alkyl tails ranged from eight to fourteen carbon atoms long, and the head groups included sulfates, phosphates, and carboxylates. This allowed a systematic evaluation of the hydrophobic contribution of the surfactant's tail and the electrostatic contribution of the head group to the adsorption behavior. It was shown that all surfactants produced a significant increase in the EOF and that the affinity of each surfactant for the PDMS surface correlates to the corresponding CMC value [15]. Other groups have also demonstrated the role of SDS for effective modification and stabilization of EOF in PDMS microchannels [53–55]. Anionic surfactants are not effective coatings for silica mainly due to unfavorable electrostatic interactions [56,57]. On the other hand, electrostatic interactions with cationic surfactants are widely used to manipulate surface chemistry of silica capillaries [16,17]. When cationic surfactants are adsorbed (head-to-surface), silanol groups become neutralized, resulting in reduction or complete elimination of the EOF (for concentrations leading to the formation of a monolayer of surfactant). If the concentration is further increased, a bilayer is formed (positive head groups interact with the solution) and the EOF is reversed. Thus, the proper choice of substrate as well as surfactant type and concentration allows a simple way to control the magnitude and the direction of the EOF [58,59]. Among other examples of this approach, it is worth mentioning the work from Liu et al. who reversed the EOF on PDMS by using a didodecyldimethylammonium bromide (DDAB) dynamic coating, enabling the analysis of ATP [60].

Mixtures of cationic and anionic surfactants have been used to improve the stabilities of the coatings. For example, a mixture of CTAB with SDS formed more stable coatings in fused-silica capillaries than CTAB alone. This mixture created a reversed EOF that remained stable for over 80 min, even after the removal of the surfactants from the buffer [40].

In cases where the presence of a significant EOF is not desirable, zwitterionic surfactants, such as CHAPSO and palmityl sulfobetaine, can be used to effectively neutralize the charge of the capillary surface [61–64]. The EOF of silica has been suppressed up to 90% with this approach, while providing high-efficiency separations of basic proteins [65].

Gemini surfactants such as hexyl-α,ω-bis(dodecyldimethylammonium bromide) are also used to control the EOF. This surfactant is able to adopt a wide variety of structures and gives surface coverage of up to 92% at 5.0 mM concentration [66]. Dynamic coatings of the cationic gemini surfactant ethylene bis (1-dodecyldimethylammonium) dibromide have also been used to control the EOF. Although this surfactant can reverse the EOF even at concentrations lower than 0.01 mM, the magnitude of the resulting EOF is affected by the surfactant concentration, pH, ionic strength, and secondary electrolytes added to the buffer [67].

Lipid vesicles have been employed as coating precursors to obtain semipermanent lipid bilayers on silica capillaries. For example, after acid/base preconditioning of the capillary, a positively charged film can be formed by exposing the capillary to a solution containing 1 mM 1,2-dioleoyl 1-3-trymethylammonium propane solution for 30 min. In this case, a reverse and stable EOF (measured in 40 mM acetate, pH = 5.0) can be achieved [68]. Anionic phospholipids such as 1,2-dimyristoyl-*sn*-glycero-3-[phosphor-*rac*-(1-glycerol)] (DMPG), on the other hand, can be used to generate a weakly cathodic EOF. At neutral pH, this coating is difficult to obtain due to charge repulsion with the negatively charged surface of the capillary; however, at pH < 4, a 5:95 DMPG:1,2-dimyristoyl-*sn*-glycero-3-phosphocholine mixture can be used to suppress the EOF [69,70].

1.2.2 Preventing Adsorption to the Capillary

One of the major problems encountered in CE is the adsorption of sample matrix components, including proteins and analytes, to the capillary wall. In general, the main consequences of this adsorption process are increased peak broadening, poor reproducibility, and low recovery. Adsorption can occur by electrostatic interactions between sample components and the capillary surface, as in the case of positively charged proteins or amines. In other cases, hydrophobic interactions are the main driving force of the adsorption, as in the case of the analysis of proteins [71,72]. Dynamic coatings of anionic surfactants, mostly SDS, and neutral surfactants, such as Tween 20, Brij 35, and Triton X-100, have been widely used to reduce hydrophobicity and prevent unwanted adsorption to capillary walls and microchannels, especially those made of PDMS [19–22,53–55]. Cationic and zwitterionic surfactants have also been used for CE coatings [18,61–64]. The surfactant *n*-dodecyl-β-D-maltosine, an alkyl polyglucoside that belongs to a family of mild nonionic surfactants, adsorbs to PDMS surfaces forming an uncharged hydrophilic monolayer [73] that reduces hydrophobic and electrostatic interactions between protein and the surface. Since alkyl polyglucosides do not affect the native structure of many proteins, they can be added to the separation buffer without significantly affecting other variables. Alternatively, semipermanent coatings by surfactants can also be used [74–76]. Among others, gemini surfactants, double-chain surfactants, bile salts, and polymeric surfactants form stable semipermanent coatings [77] that remain adsorbed even after rinsing a capillary with buffer, thereby avoiding undesired interactions between surfactant and analytes. Examples of semipermanent coatings include cationic polymers such as polybrene [78] and neutral polymers such as hydroxyethylcellulose and poly(vinyl alcohol) [79]. Pluronic, a triblock copolymer of poly(ethylene oxide)–poly(propylene oxide)–poly(ethylene oxide), can be spontaneously adsorbed to a variety of hydrophobic polymeric materials via the hydrophobic poly(propylene oxide) moiety [80]. Hellmich et al. [81] investigated the coating of PDMS with Pluronic, which resulted in 85% reduction of serum protein adsorption relative to native PDMS. In one study, nonionic surfactants such as Triton X-100 and Tween 20 proved less effective than zwitterionic surfactants, such as palmityl sulfobetaine. A solution of >10% nonionic Triton X-100 was required to reduce protein adsorption by 90%, versus only 0.3% of the zwitterionic palmityl sulfobetaine. For zwitterionic surfactants, the inhibition of protein adsorption increases as the length of the carbon chain increases and the CMC decreases [32]. To further improve the stability of semipermanent surfactant coatings, a layer-by-layer adsorption of oppositely charged surfactants has shown to be effective. For example, DDAB and a gemini surfactant (18-6-18) were used for coating along with SDS. With an increase in the concentration of SDS in the coating, the stability dramatically increases due to the enhanced packing capabilities. These coatings showed excellent stability, efficiency, reproducibility, and high recovery of proteins, indicating powerful suppression of adsorption of sample components to the capillary walls [82].

For a more efficient prevention of adsorption, the geometry of the surfactant monomer should be cylindrical so that aggregates form a bilayer [39]. If the surfactant geometry is conical, such as with single-chained surfactants such as CTAB, the surfactants tend to aggregate on the capillary wall as spherical hemi-micelles, which can provide incomplete surface coverage, therefore decreasing the efficiency of the coating [83]. Two-tailed surfactants such as DDAB and dioctadecyldimethylammonium bromide possess the cylindrical geometry to support a bilayer on flat surfaces [39,84,85]. A mixture of these two surfactants allowed protein recoveries of 85–100% in comparison to 0–81% for CTAB [4]. Melanson et al. produced a semipermanent coating with the double-chained surfactant DDAB by rinsing a capillary for less than 20 min at a concentration of 0.1 mM [34,39]. After that, the separation of four basic proteins, ribonuclease A, cytochrome C, α-chymotrypsinogen, and lysozyme, was achieved with recoveries ranging from 92 to 100%. DDAB has also been employed to reduce the adsorption of fluorescent dyes on PDMS microchannels [85].

To further improve the stability of surfactant coatings to inhibit protein adsorption, polymerized surfactants have been used [86]. The surfactant 1,2-dioleoyl-*sn*-glycero-3-phosphocholine was used to coat a glass capillary, which was then heated in a solution containing the free radical initiator 2,2′-azobis(2-methylpropionitrile) for oligomerization. With this modified capillary, a separation of cationic and anionic proteins was carried out at pH 7.4 with good efficiencies. Protein recoveries for cationic proteins were improved ($97 \pm 6\%$) with respect to a capillary dynamically coated with DLPC ($75 \pm 6\%$) [87]. Highly stable cross-linked phospholipid bilayers have also been used as coatings on silica to improve protein and peptide separations. Recently, 1,2′-bis[10-(2′,4′-hexadienoyloxy)decanoyl]-*sn*-glycero-3-phosphocholine was used to prepare planar, self-assembled phosphorylcholine phospholipid bilayers with radical polymerization [88]. Protein separations for cationic and anionic mixtures were carried out, and as a result of increased resolution, more protein peaks were observed, including one for α-chymotrypsinogenin, which could not be detected after separation in a bare capillary. These films allowed continuous protein separations for up to 10 h per day with no reduction in the separation performance.

1.3 SURFACTANTS AS BUFFER ADDITIVES

The components of the running buffer directly influence the quality of separations by CE. The buffer type and concentration, pH, ionic strength, organic solvent concentration, and type and concentration of any additives affect the zeta potential and, thus, can be used to control the EOF. Additionally, these factors can be adjusted to tune the resolution. In this section, we describe methods that use surfactant additives to impart selectivity in CE, namely, micellar electrokinetic chromatography, microemulsion electrokinetic chromatography, and nonaqueous CE with added surfactants.

1.3.1 Micellar Electrokinetic Chromatography

The basis of CE separations is the difference in the charge to size ratio of analytes. The introduction of micellar electrokinetic chromatography (MEKC) in 1984 by Terabe et al. [89] expanded the versatility of CE by enabling the separation of neutral analytes in addition to charged ones. To perform MEKC, surfactants must be included in the running buffer at a concentration higher than the CMC. Micelles form a charged pseudostationary phase in solution, creating a hybrid system of electrophoresis and chromatography upon application of the separation potential. In the presence of an electric field, the micelles move with or against the EOF, depending on the charge of the head groups of the surfactants (described later). During electromigration of the micelles, analytes in solution can interact with the micelles through hydrophobic and electrostatic interactions with the core of the micelle and the surface of the micelle, respectively. Thus, the differential analyte/micelle interactions/partitions affect the migration time of the analytes and determine the resolution/selectivity of MEKC. (As a side note, analytes can also associate with surfactant monomers that are in equilibrium with micelles. These analyte–monomer associates have different partitioning behavior into micelles than free analytes have [90].) Neutral compounds are separated mainly by hydrophobicity. Less hydrophobic compounds partition to a lesser extent into the hydrophobic core of the micelles and consequently have a relatively fast migration time. In contrast, more hydrophobic compounds spend more time within the pseudostationary phase and thus have longer migration times. Charged analytes are separated based on a combination of electromigration and chromatography. The analyte charge and size influence its electrophoretic mobility. Additionally, any repulsion between the analyte and the micelles (when both have like charges) or attraction/ion-pairing effects (when both have opposite charges) influence separation [91]. Other factors such as shape, dipolar moment, and hydrogen bond interactions between the analytes and the surfactant may play fundamental roles in the separation [92]. One of the most important advantages of MEKC is that the selectivity can be manipulated by using different types of surfactants—anionic, cationic, zwitterionic, nonionic, and chiral—or even mixtures of surfactants. Also, the chain length and branching of the surfactants tails can be varied to change the physical nature of the micelles. Organic solvents can also be used to adjust the resolution in MEKC, although the amount has to be controlled as the solvents may compromise the integrity of the micelles [93–95]. It is also worth mentioning that the fundamental and theoretical aspects of MEKC, such as calculations of retention factors and resolution, have been described extensively in many books, chapters, and reviews [96–105]. Applications of MEKC for the analysis of pharmaceutical, forensic, food, environmental, and clinical samples abound in the literature, so the reader is referred to recent reviews for further information [30,106–116].

The first micelles used for MEKC comprised the anionic surfactant SDS [89]. To date, it still remains one of the most widely used surfactants for MEKC separations due to high stability, relatively low Krafft point, low UV absorbance, high solubilizing capability, and the availability of high-quality reagent [102]. Upon application of the normal polarity electric field (anode at inlet to cathode at outlet), the anionic pseudostationary phase migrates toward the anode/inlet (Figure 1.4). However, under neutral or basic conditions, the capillary walls are negatively charged and thus the direction of the EOF is toward the detector/cathode. Although the pseudostationary phase is attracted toward the inlet (allowing separation of incorporated analytes), the EOF drives the bulk solution toward the detector. Very hydrophobic analytes such as aromatic compounds and steroids can be difficult to resolve with SDS-based MEKC because rather than having differential partitioning in and out of the micelle, the analytes are believed to completely reside in the hydrophobic core. To remedy this problem, less hydrophobic surfactants, such as the bile salts sodium cholate, sodium deoxycholate, sodium taurocholate, and sodium taurodeoxycholate, can be added, resulting in significant improvements in resolution [115]. Other nonconventional pseudostationary phases can be used to optimize MEKC separations as well. Cationic surfactants such as CTAB provide different selectivity than anionic surfactants, although a downfall is a smaller migration window (relative to SDS) [115]. Zwitterionic surfactants are not commonly used in MEKC, but they have been used to improve the separation of proteins and peptides [30,92,115]. Nonionic surfactants, such as Brij 35, Tween, and alkoglucosides, are useful for improving the separation of charged analytes [92]. They are not effective in separating neutral analytes because they have no mobility, although it is possible to form *in situ* charged micelles by complexation with ions such as borate [30]. A major advantage of nonionic surfactants is that they do not contribute to the ionic strength of the running buffer, and therefore they do not contribute to the electrical current [117]. For this reason, nonionic surfactants are used to form mixed micelles with ionic surfactants to improve resolution and optimize selectivity while avoiding increases in Joule heating [92,115]. Mixing surfactants offers powerful optimization capabilities as mixed micelles have tunable hydrophobicity and surface charge density, which alters analyte–micelle interactions [92]. Another advantage of including nonionic surfactants in mixed micelles is that the electrophoretic mobility of the micelles and incorporated analytes can be controlled without significantly increasing the electroosmotic velocity, which can be useful for extending the elution window of a separation [118]. Ahuja et al. achieved an infinite elution range in MEKC using a mixture of Brij 35 and SDS for the micelles [119]. Although countless combinations of surfactant types have been used, mixed micelles typically consist of a nonionic surfactant and an anionic surfactant, usually SDS. Chiral separations, important for pharmaceutical and biomedical fields, can be achieved by using chiral surfactants for MEKC. These surfactants include naturally occurring bile salts, saponins, digitonin, and semisynthetic surfactants derived from naturally occurring sugars and amino acids [28,120–122]. These surfactants achieve enantioseparations of analytes by stereoselective recognition interactions such as hydrophobic, steric, and electrostatic interactions [123]. Chiral surfactants have also

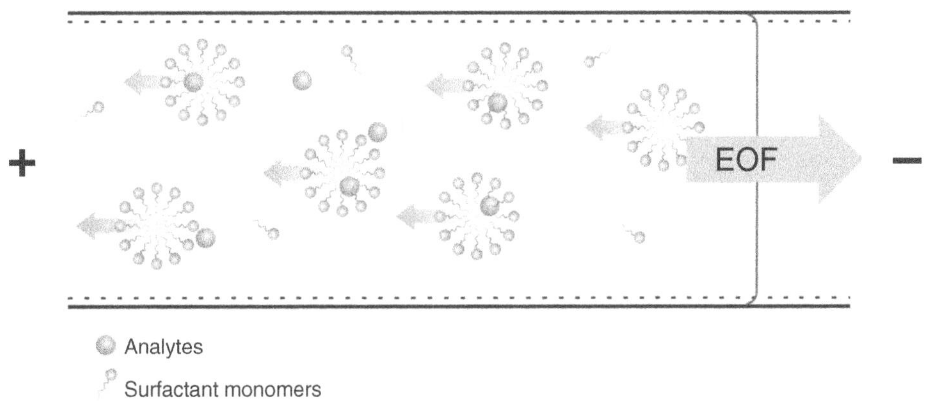

FIGURE 1.4 MEKC separation by anionic surfactants under neutral or basic conditions.

been combined with other types of surfactants to form mixed micelles, allowing for tunable selectivity in enantioseparations [30,124].

1.3.2 Microemulsion Electrokinetic Chromatography

Another surfactant-based CE mode is microemulsion electrokinetic chromatography (MEEKC), which was first reported in 1991 by Watarai [125]. In MEEKC, the separation is carried out in a microemulsion: an optically transparent mixture of immiscible liquids, stabilized by a monolayer of surfactants. Microemulsions can consist of nanometer-scale oil droplets dispersed in a bulk phase of water (o/w) or water droplets dispersed in a bulk phase of oil, the most commonly used type for CE being o/w. In that case, the charged head groups of the surfactants (usually anionic) face outside of the oil droplet toward the aqueous phase while the lipophilic tails are buried inside (Figure 1.5), reducing the surface tension between the two phases. To further stabilize the nanodroplets, a nonionic, amphiphilic co-surfactant (butan-1-ol, in Figure 1.5) can also be added to the solution. The co-surfactant has a polar, but uncharged head group, which decreases the repulsion between the primary surfactant head groups by decreasing the charge density.

Unique ratios of water, oil, surfactant, and co-surfactant are required to form a stable microemulsion [126]. As an example, 0.8% organic solvent, 3.3% SDS, 6.6% cosurfactant, and 89.3% aqueous buffer were used to form an o/w microemulsion for MEEKC [127]. The aqueous phase typically comprises borate or phosphate buffer. Common oils for emulsions include octane or heptane. Others oils that have been evaluated for performance in MEEKC include diethyl ether, cyclohexane, chloroform, methylene chloride, and amyl alcohol. Each type slightly affects selectivity and migration times of analytes. Again, the most frequently used surfactant is SDS, although sodium cholate (bile salts) and CTAB have been used. Separation with the neutral surfactant Triton X-100 has been demonstrated, although its use for separation of uncharged analytes is very limited.

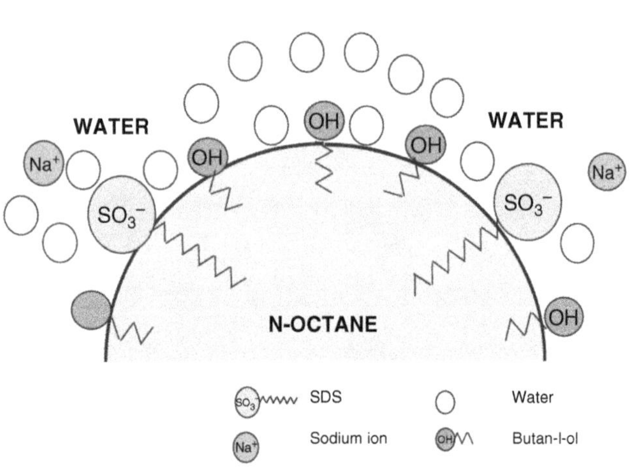

FIGURE 1.5 Schematic representation of an o/w microemulsion droplet. Extracted from Ref. 128.

The type of surfactant can be selected to tune the size and charge of the emulsion droplets, the magnitude and direction of the EOF, and ion-pairing or repulsion effects between analytes and the pseudophase. A longer length of hydrophobic tail can increase the stability of the microemulsion by reducing the polydispersity of the emulsion. The concentration of surfactant used should greatly exceed the CMC to stabilize the emulsions to a greater extent (e.g., 110 mM SDS produces an emulsion that is stable for over several months). The concentration of nonionic co-surfactants (typically a short-chain alcohol, of which the most commonly used is butan-1-ol) can affect migration times due to changes in viscosity, and consequently EOF. Also, the amount of co-surfactant used can influence the size of the droplets by changing the charge density on the surface of the droplets [91,128]. Altria et al. [129] and Hansen [127,130] have investigated the effects of type and concentration of surfactants, pH, and ionic strength of aqueous phase, type of oil used, method of emulsion preparation, as well as many other factors, including the addition of organic solvents, urea, cyclodextrins, and ion-pairing reagents. The background and principles of MEEKC methodology and descriptions of operating parameters have been discussed in several reviews [91,126,128,129,131–134].

The separation principles of MEEKC are the same as those of MEKC (described earlier), and some groups have shown that selectivity and efficiency are essentially the same in MEEKC as in solvent-modified MEKC [135,136]. However, some researchers claim that MEEKC has several advantages. Among them, authors highlight that oil droplets are larger and less rigid than standard micelles, allowing compounds with extremely high hydrophobicity to partition into the droplet. The high solubilizing characteristics of microemulsions allow dissolution of a wider variety of compounds and complex sample matrices. Thus, less sample preparation is required before analysis. The solubilizing power of microemulsions also prevents matrix precipitation during analysis, avoiding capillary deterioration. Another potential advantage of MEEKC over MEKC is the widened migration time window, due to the presence of organic solvents in solution. Since its inception in 1991, MEEKC has been demonstrated for numerous analyses of water-soluble, water-insoluble, neutral, and charged analytes, including proteins, agrochemicals, pharmaceuticals, natural products, dyes, cosmetics, foods, and biological and environmental samples. Chiral separations have been accomplished with MEEKC by including an enantioselective component either as a primary surfactant, co-surfactant alcohol, or oil (e.g., (*R*)- and (*S*)-dodecoxycarbonylvaline, (*R*)- and (*S*)-2-hexanol, or D- and L-diethyl tartrate, respectively) in the microemulsion [128,137]. Another important application of MEEKC is the determination of partition coefficients (log P values) of hydrophobic pharmaceutical compounds. Several groups have thoroughly reviewed these applications of MEEKC [91,113,128,132,133,138–141]. As a side note, emulsions have a variety of other applications aside from CE, including oil recovery, delivery of pharmaceuticals and cosmetics, nanoparticle preparation, liquid–liquid extraction, and high-performance liquid chromatography [91,134,140].

1.3.3 Nonaqueous Capillary Electrophoresis with Added Surfactants

In 1984, Walbroehl and Jorgenson demonstrated that CE can actually be carried out in pure organic solvents [142], a separation mode termed nonaqueous capillary electrophoresis (NACE). NACE is particularly advantageous for separations of hydrophobic analytes. The increase in solubility and the decrease in aggregation of analytes achieved under nonaqueous conditions lead to a better resolution and higher throughput [143]. NACE has several other advantages. Compared to water, organic solvents have a wider range of physical and chemical characteristics such as dielectric constants, polarity, viscosity, and density [144]. Thus, the physicochemical differences between analytes can be maximized, improving the separation of closely related compounds [145]. The wide range of dielectric constants and autoprotolytic behavior in organic solvents permits a wide range of compounds to be ionized through versatile acid/base chemistry [146–148]. Organic solvents can increase separation time by decreasing the EOF through compression of the electrical double layer and a consequent decrease in the zeta potential [143]. Organic solvents reduce adsorption of analytes to the capillary surface, avoiding changes in EOF [143]. Because lower currents are generated in nonaqueous media (compared to aqueous solutions with the same ionic strength), a higher separation potential can be applied, further amplifying the physicochemical differences between analytes in the sample [147,149,150]. Thus, by selecting the type of organic solvent (and ratio, if a mixture of organic solvents is used), one can tune the separation resolution, selectivity, efficiency, and analysis time [144]. Finally, the use of volatile solvents for NACE supports the compatibility of CE and mass spectrometry. Many reviews [145,151–158] are also available detailing the fundamental physicochemical aspects, advantages, and applications of NACE, as well as critical evaluations [159] of the claims of NACE proponents.

Just as in aqueous CE, NACE separates analytes based on charge to size ratio, and therefore cannot separate non-ionized species. However, surfactants have been used as additives in NACE (Table 1.1) to accomplish separations of neutral compounds. Analytes can be associated with charged monomers of surfactants for separation, as Lu et al. [143] and Li and Fritz [144] demonstrated for the separation of tamoxifen metabolites and polycyclic aromatic hydrocarbons, respectively. The majority of NACE separations have included a high enough concentration of surfactants to form aggregates (as a pseudophase) in a separation mode called nonaqueous micellar electrokinetic chromatography (NAMEKC). Typically, micelles do not form in organic solvents because the hydrophobic interactions that cause aggregation are weaker than in water and also CMCs of surfactants are higher than in water [156]. However, there are organic solvents compatible with micelle formation, particularly those with high dielectric constants, such as formamide [149,160–162]. Table 1.1 lists other organic solvents and the corresponding surfactants that are compatible with NAMEKC. The analyte–surfactant interactions in NACE include solvophobic interactions, electrostatic interactions (ion–ion, ion–dipole, and dipole–dipole interactions), and donor–acceptor interactions [147,162]. The chemical nature and concentration of the organic solvent and surfactant can be adjusted to tune such interactions. Nonpolar solvents with low dielectric constants support ion–pair interactions that can be used for chiral separations [163]. NAMEKC is particularly useful for highly lipophilic racemates with limited solubility in water. Although cyclodextrins are widely used for chiral separations, Table 1.1 also shows examples where surfactant chiral selectors have been used.

1.4 SURFACTANTS FOR ANALYTE PRECONCENTRATION

The limited dimensions of a capillary constrain the volume of sample that can be injected and also provide only a short path length for UV absorption detection (the most standard detection method for CE). Consequently, sensitivity can be poor and limits of detection relatively high. Electrochemical or fluorescence detection is a viable option with superior sensitivity; however, they sometimes require time-consuming derivatization steps. A convenient and popular alternative to overcome such difficulties is to develop online preconcentration strategies. The goal is to compress the width of analyte bands in the capillary so that a larger volume of sample can be injected without sacrificing separation efficiency. Two main approaches have been used that take advantage of the partitioning of analytes into stationary or pseudostationary phases: stacking, which manipulates the electrophoretic mobility of analytes, and extraction [176,177]. The myriad of strategies in both classes has been thoroughly discussed in many reviews [176–196]. Stacking techniques include field-amplified sample stacking/normal stacking mode, large-volume sample stacking/-stacking with matrix removal, field-enhanced sample injection, transient isotachophoresis, electrokinetic supercharging, and dynamic pH junction. Membrane filtration and

TABLE 1.1 NACE Separations Utilizing Surfactants

Organic Solvents	Surfactants	Analytes	References
Acetonitrile	Camphorsulfonic acid, Tween 20	Basic chiral drugs: atenolol, bisoprolol, bunitrolol, metroprolol, pindolol, propanolol, salbutamol, ephedrine, epinephrine, cisapride, and synthetic impurities	164
	Tetrahexylammonium perchlorate	Polycyclic aromatic hydrocarbons	146
	Planar organic cations, for example, tropylium tetrafluoroborate or 2,4,6-triphenylpyrylium tetrafluoroborate	Polycyclic aromatic hydrocarbons	147
Acetonitrile and methanol	Sodium cholate	*Trans*- and *cis*-resveratrol	165,166
	Sodium cholate, SDS	Linoleic acid oxidation products	167
	Trimethyloctadecylammonium bromide	Aromatic compounds	168
	Ammonium acetate, tetrabutylammonium bromide, tetrabutylammonium hydrogen sulfate, and tetrapentylammonium bromide	Tamoxifen and four phase I metabolites	169
Methanol	Brij 35	Porphyrin acids	170, 171
	Sodium cholate	Cryptotanshinone, tanshinone IIA, and tanshinone I	172
	SDS	Tamoxifen metabolites	143
	Sodium tetradecyl sulfate	Polycyclic aromatic hydrocarbons	144
Methanol and formamide	Sodium cholate	3,4-Methylenedioxymethamphetamine	173
Formamide	SDS	Dimethyl phthalate, diethyl phthalate, dibutyl phthalate	174
	SDS, diethylhexyl sodium sulfosuccinate, taurodeoxycholic acid sodium salt	*p*-Arylacetophones	175
Dimethyl formamide, dimethyl acetamide	Sodium caprylate, sodium laurate, sodium palmitate	Four tetracyclines	149
Propylene carbonate	Tetraalkylammonium ions, long-chain trimethylammonium ions	Phenanthrene, β-naphthol, methylparaben, ethylparaben, propylparaben, and vitamin K_1	93

solid-phase extraction are also used for preconcentration. The aim of this section is to focus on the preconcentration strategies that rely on the use of surfactants, namely sweeping, transient trapping, analyte focusing by micelle collapse, micelle to solvent stacking, combinations of methods, and offline cloud point extractions.

1.4.1 Sweeping

Sweeping is a powerful technique that was first demonstrated in 1998, yielding a 5000-fold preconcentration factor [197]. The theoretical aspects of the technique and experimental corroboration were discussed in detail soon after [198]. Briefly, sweeping is carried out by including surfactants above the CMC as a pseudostationary phase in the separation buffer, but not in the sample solution. After conditioning the capillary by rinsing with the separation buffer, the sample plug, devoid of micelles, is injected. Preconcentration occurs as analytes partition into the pseudostationary phase as it penetrates the sample plug upon electrophoresis [197]. Then, the analytes, which are typically preconcentrated up to a factor of several thousand [178], can be separated by MEKC. Figure 1.6 gives a schematic drawing of this process. Sweeping can be used for the preconcentration of neutral as well as charged analytes due to hydrophobic interactions between the neutral analytes and the surfactant tails and also through electrostatic interactions between oppositely charged analytes and surfactants [195]. In fact, the degree of preconcentration is proportional to the strength of the interactions between analytes and the micelles [198]. Sweeping can be carried out using anionic, cationic, nonionic, zwitterionic, and mixed micelles. Excellent reviews detailing sweeping with micelles and providing pertinent applications of the technique (e.g., to determine pesticides, phenols, illicit drugs, pharmaceuticals, and herbal medicines) have been published [178,185,191]. Sweeping has also been carried out as a preconcentration method hyphenated to MEEKC for

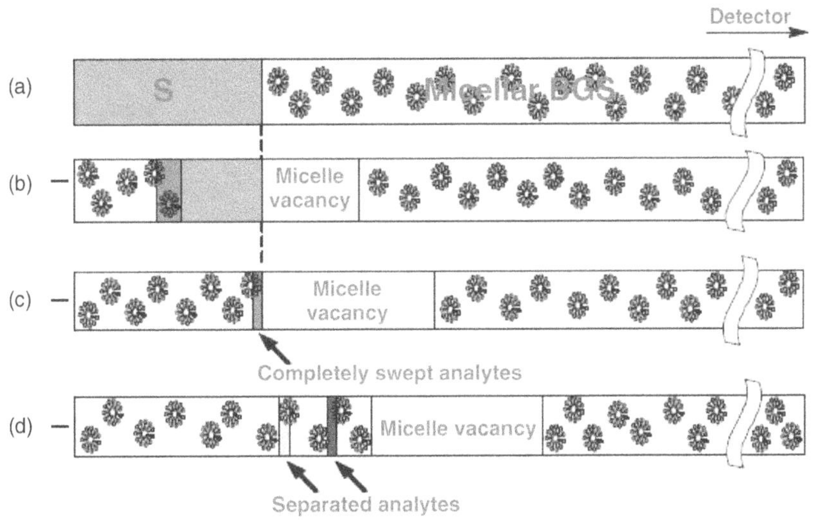

FIGURE 1.6 Mechanism of sweeping preconcentration using negatively charged surfactants and zero EOF conditions. (a) A sample plug (S), devoid of micelles, is pressure injected into a capillary that has been conditioned with a background solution (BGS) that contains micelles as a pseudostationary phase. (b) On application of the separation potential, the BGS with micelles penetrates the sample plug and begins to sweep/preconcentrate the analytes. (c) The sample plug is completely swept and the analyte zone has been compressed. (d) Preconcentrated analytes are then separated by MEKC. Extracted from Ref. 178.

analytes such as tobacco alkaloids [199], catechins [200], and phenolic compounds [201], with up to 238,000-fold increases in detection sensitivity [201].

1.4.2 Transient Trapping

Sueyoshi et al. aimed to perform mass spectrometry (MS) detection after MEKC separation on a microchip. As described later, surfactants can decrease the sensitivity of MS detection. To avoid introducing surfactants to the detector, and considering the short separation channel length on a microchip, the group used a partial filling technique in which plugs of a micellar solution and the sample solution were injected into a channel filled with separation buffer devoid of micelles [202]. As shown in Figure 1.7, upon application of separation potential, the sample plug migrates through the micellar plug and separates by MEKC. When the separated analytes reach the end of the micellar plug, they are released into the separation buffer devoid of surfactants, where they continue to migrate to the detector. The micellar plug never reaches the detector, avoiding incompatibility issues with MS. Along with allowing the compatibility of MEKC with MS, this partial filling technique was observed to actually preconcentrate the sample plug. In 2008, Otsuka's group investigated the effect further and proposed that the preconcentration occurs due to a transient trapping mechanism [203]. When a short micellar plug is injected into the separation channel before the sample plug, analytes are trapped and released in a concentrated sample zone at the interface between the two plugs, similar to the mechanism in sweeping. In sweeping, a large volume of sample can be injected and preconcentrated into a narrow zone to improve the detection limits. However, in microchips, the short channel length limits the injection volume. Thus, the transient trapping mechanism is a convenient preconcentration option for microchips [203].

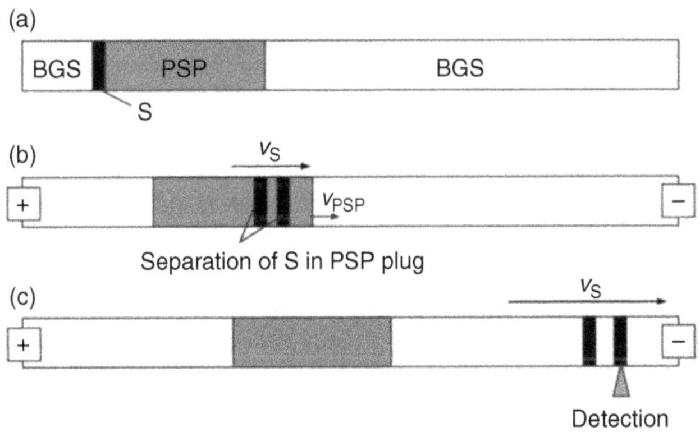

FIGURE 1.7 Schematic of transient trapping preconcentration that occurs during the partial-filling technique. (a) In a capillary filled with background solution (BGS) devoid of micelles, a short pseudostationary phase micellar plug (PSP) is injected, followed by a short plug of sample (S). (b) On application of separation potential, the sample plug is preconcentrated by transient trapping and then separated into analyte bands, which migrate faster (ν_S) than the PSP plug (ν_{PSP}). (c) The analyte bands are introduced into the detector without interference from the PSP plug. Extracted from Ref. 202.

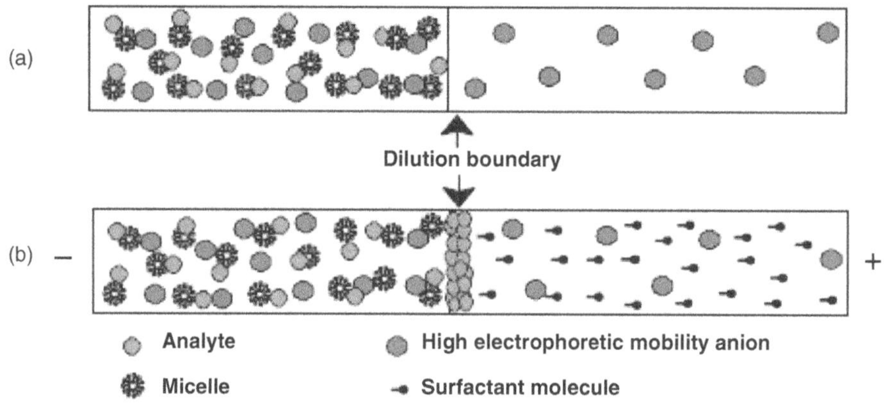

FIGURE 1.8 The basis of preconcentration in AFMC. (a) The sample plug contains analytes, anionic micelles, and an additional anion with high electrophoretic mobility. (b) On application of separation potential, the micelles in the sample plug, containing the analytes, migrate toward the separation buffer and on reaching the dilution boundary, collapse into monomers, preconcentrating and releasing analytes. Extracted from Ref. 178.

1.4.3 Analyte Focusing by Micelle Collapse

Quirino and Haddad introduced analyte focusing by micelle collapse (AFMC) in 2008 for the online preconcentration of neutral analytes [204]. As shown in Figure 1.8, to perform AFMC, a solution containing anionic micelles and an additional anion that has high electrophoretic mobility is used to transport the analyte molecules via the hydrophobic core of the micelles. As the micelles migrate due to electrophoresis, they encounter the separation buffer, which contains no micelles and has a lower conductivity than the sample solution. The micelles become diluted and collapse into monomers, releasing and accumulating their contents into a micellar dilution zone. By manipulating solution conditions in the capillary, the sample enrichment of AFMC can be combined with the separation capabilities of MEKC to preconcentrate and then separate neutral analytes. The initial proof of concept of AFMC MEKC resulted in a preconcentration factor of 160–200 for the analysis of the steroids cortisone, hydrocortisone, and prednisolone [204]. The optimal conditions for producing the best peak height enhancements were shown to be a concentration of surfactant in the sample solution just above the CMC and a minimum conductivity ratio (separation solution to sample solution) needed to collapse the micelles [205]. AFMC was demonstrated to be compatible with MS detection since the preconcentrated and separated analyte zones contain surfactant monomers rather than micelles, which do not reduce the sensitivity of MS to such a high degree [206].

1.4.4 Micelle to Solvent Stacking

Micelle to solvent stacking was recently developed for the online preconcentration of small organic cations [207]. In this technique, a capillary is conditioned with a separation buffer that contains an organic solvent. An anionic micellar sample solution, which carries the organic cations, is then injected and the separation potential applied. The electrophoretic mobility of the cations when inside the micelles is directed toward the detector. At the interface between the sample solution and the separation buffer, the cations have less affinity for the micelles due to the presence of the organic modifier, and are released. The cations then show reversed electrophoretic mobility (with respect to the micelles) directed toward the capillary inlet, causing the cations to accumulate at the boundary. After all of the micelles have passed through this interface, the analytes are focused into a narrow zone. Guidote and Quirino [208] developed a model for micelle to solvent stacking and also demonstrated the preconcentration of organic anions by using cationic micelles, for which 10-fold peak height enhancement was achieved.

1.4.5 Combinations of Preconcentration Methods

To exploit sample preconcentration to maximum capabilities, some groups have combined complementary preconcentration steps before analysis. Table 1.2 gives examples of combinations of preconcentration methods that use surfactants in at least one step. As can be seen from the table, sweeping is compatible with a wide variety of other preconcentration methods.

1.4.6 Cloud Point Extraction

Although cloud point extraction (CPE) is not a technique that has been integrated online, it is worth mentioning as a preconcentration method amenable to CE. By using the temperature-dependent phase separation behavior of aqueous surfactant solutions, cloud point preconcentration is a

TABLE 1.2 Hyphenated Preconcentration Methods that Require the Use of Surfactants

Techniques	Analytes	Preconcentration Factor	References
Dynamic pH junction + sweeping	Flavin derivatives	1,200	164
	Androgens, corticosteroids, estrogens	30	209
	Pyrrolizidine alkaloids	24–90	210
Field-amplified sample stacking + sweeping	Hypolipidemic drugs	80	211
Full capillary sample stacking + sweeping	Derivatized amino acids	400	212
Hollow fiber-based liquid-phase microextraction + sweeping	*Strychnos* alkaloids	35–50	213
Homogeneous liquid–liquid extraction + sweeping	Polycyclic aromatic hydrocarbons	100,000	214
Micelle to solvent stacking + sweeping	Beta blocker and tricyclic antidepressant drugs	20–50	215
Selective electrokinetic injection with a water plug + sweeping	Phenols	96,000–238,000	201
	Flavonoids	45–194	216
Selective exhaustive injection + sweeping	Laudanosine, 1-napthylamine	550,000–900,000	217
	Aromatic amines	39,000–146,000	218
	Carboxylic acids, dansyl amino acids, naphthalenedisulfonic acids	1,000–6,000	219
	Environmental pollutants	100	220
	Lysergic acid diethylamine, *iso*-lysergic acid diethylamide, lysergic acid *N*,*N*-methylpropylamide	100,000	221
	Lysergic acid diethylamine, *iso*-lysergic acid diethylamide, lysergic acid *N*,*N*-methylpropylamide	100,000	222
	Herbicides (paraquat, diquat, difenzoquat)	50,000	223
	Phenoxy acid herbicides	100,000	224
	Lysergic acid diethylamide	100,000	225
	Reserpine	2,500–3,800	226
	Ephedra alkaloids in herbal extracts	10,000	227
	Corticosterone, 17-hydroxycorticosterone	1,500	228
	Methamphetamine, ketamine, morphine, codeine	6,000 for methamphetamine	229
	Amphetamine, methamphetamine, *p*-hydroxymethamphetamine	Not given; LOD 15–20 ng mL^{-1}	230
	Amphetamine, methamphetamine, methylenedioxymethamphetamine	1,000	231
	Fluorescein, 5-carboxyfluorescein	4,000–4,500	232
	Morphine, codeine, ketamine, methamphetamine	1,000	233
	Amphetamine, methamphetamine, methylenedioxymethamphetamine	2,500–10,000	234
	Tobacco alkaloids	180–540	199
	Morphine and its metabolites	2,500	235
	Serotonin reuptake inhibitors	57,000–120,000	236
	Cocaine and its metabolites	1,750–39,600	237
Single-drop microextraction + sweeping	Fluorescein, 6-carboxyfluorescein	28,000–32,000	238
Solid-phase extraction + sweeping	Testosterone, progesterone, testosterone propionate	700–1,100	239

green alternative to liquid–liquid extractions that use organic solvents. When an aqueous solution of certain nonionic or zwitterionic surfactants is brought to a temperature called the cloud point (specific to the surfactant type and solution conditions), the solution separates into two phases—one layer is clear, and the other is turbid. The clear, surfactant-dilute layer is composed of micelles, and the turbid surfactant-rich layer is composed of larger aggregates that are able to scatter light. The difference in density between each phase drives the separation. Analytes that are solubilized by the micelles in the total volume of solution become extracted into the surfactant-rich layer upon reaching the cloud point, and are effectively preconcentrated since the volume of the turbid layer is much smaller than that of the clear layer. The surfactant-rich layer with preconcentrated analytes can then be separated from the clear layer by filtration or centrifugation and then analyzed. Carabias-Martinez et al. and Quina and Hinze have reviewed the properties, behavior, and experimental considerations for the cloud point extraction method [240,241]. Some applications of this preconcentration method before CE separation include the extraction and analysis of triazine herbicides [242], Cu(II) and Co(II) [243], Pt and Pd [244], lead [245], mercury species [246], phenol and nitrophenol [247], and auxins [248]. In most of these cases, preconcentration factors of 1 to 2 orders of magnitude were achieved.

1.5 SURFACTANTS AND DETECTION IN CE

The use of surfactants is compatible with most optical detection systems, including UV–Vis absorbance, fluorescence, and thermo-optical absorbance (thermal lensing microscopy) [249,250], provided that the optical properties of the surfactants do not interfere with those of the analytes. With fluorescence detection, the separation efficiency is particularly important to maximize as labels can decrease the structural differentials of analytes, and excess labels and their impurities must be separated from analytes [115]. As described next, in MS and electrochemical detection, surfactants have larger impacts—they can either diminish or improve detection capabilities.

1.5.1 Mass Spectrometry

Coupling CE with MS yields a powerful analytical tool with excellent separation capabilities combined with identification and characterization of analytes. This is a relatively straightforward integration because, in comparison with traditional separation techniques such as gas chromatography and liquid chromatography, CE has the advantage of handling very low sample volumes and extremely low flow rates, which simplifies the interface. Typically, electrospray ionization (ESI) is the interface used to couple CE with MS because it efficiently creates gas-phase ions from nonvolatile solutes. NACE is optimal for hyphenation with MS because the high volatility and relatively low current generated by organic solvents are ideal for ESI [145]. The ionization efficiency and, consequently, detection sensitivity are higher in organic solvents versus aqueous solutions due to lower surface tension of the electrospray droplet and more rapid solvent evaporation. Also, the lower surface tension of organic solvents allows for using lower electrospray potentials, which decreases the likelihood of electric discharges, and thus stabilizes the electrospray [143].

As described throughout the chapter, surfactants are widely used to enhance CE performance. However, their presence in the background electrolyte can greatly impair compatibility with MS detection. Surfactants are nonvolatile, have high surface tension, and decrease ESI efficiency. Rundlett and Armstrong explained that the high concentration of nonvolatile surfactants at the liquid–vapor interface (at the Taylor cone) inhibits the amount and efficiency of droplet formation [251]. Over time, the ion source can become fouled by a build-up of surfactants. Another problem with integration of CE and MS is that surfactants suppress analyte ion signals and increase noise, reducing the sensitivity of the system [252]. Rundlett and Armstrong also stated that Coulombic interactions between oppositely charged surfactant and solute ions suppress analyte ionization by inhibiting the charged analyte ions from transferring from the liquid to the gas phase [251]. The high ionic strength of the background electrolyte resulting from the inclusion of surfactants can also lead to high currents during separation, which then leads to unstable electrospray conditions [253].

Overcoming the compatibility issues of surfactants with MS has been an aim since the 1990s [254], and strategies continue to be sought after today [255]. The compatibility of surfactant-based separations with MS can be increased by using more ESI-friendly surfactants such as high-molecular-weight/polymeric surfactants or semivolatile surfactants, by avoiding the entrance of surfactants into the ESI interface, or by using alternate ionization strategies before MS. The basis of each strategy will be described, but for more specific examples, the reader is referred to a review that has thoroughly tabulated applications [255].

As mentioned earlier, polymeric or high-molecular-weight surfactants are large molecules that aggregate into micelles at an effective CMC of zero, while still retaining the ability to solubilize analytes and provide a pseudostationary phase for MEKC. Because very low concentrations are required for separation, noise generated by ionized surfactants is minimized. The ions that do form from high-molecular-weight surfactants fall out of the mass range covered by the spectrometer, leading to increased signal to noise ratios (compared to low-molecular-weight surfactants) [255]. Again, Rundlett and Armstrong suggested that high-molecular-weight surfactants are not very surface active and hence do not inhibit electrospray efficiency to the same degree as lower-molecular-weight surfactants [251]. An additional advantage of polymeric surfactants is that they can be used in conjunction with high amounts of organic modifier to tune chromatographic selectivity without disintegration of the micellar structure [256]. Some examples of such surfactants include butyl acrylate–butyl methacrylate–methacrylic acid copolymer [257], poly(sodium undecylenic sulfate) [258] and a suite of chiral molecular micelles—poly(sodium *N*-undecanoyl-L-valinate) [259], poly(sodium *N*-undecenoxy carbonyl-L-leucinate) [260–262], poly(sodium *N*-undecenoxy carbonyl-L,L-leucyl-valinate) [263], poly(sodium *N*-undecenoyl-L,L-leucyl-valinate) [264], poly(sodium *N*-undecenoyl-L-leucine sulfate), poly(sodium *N*-undecenoyl-L-valine sulfate), and poly(sodium *N*-undecenoyl-L-isoleucine sulfate) [265]—investigated by the Shamsi group for enantioseparations.

Using semivolatile fluorinated surfactants has been shown to decrease the build-up of surfactant concentration at the liquid–gas interface, avoiding decreases in electrospray efficiency. Ishihama et al. demonstrated that decreases in sensitivity caused by more commonly used MEKC surfactants are not experienced when using perfluorooctanoic acid (PFOA) and perfluorooctanesulfonic acid [266]. Petersson et al. demonstrated that signal suppression was not a problem even when using 100 mM PFOA. This group also showed that the separation performance of PFOA was comparable to that of SDS, and that only the selectivity was different [267]. Van Biesen and Bottaro also demonstrated a successful MEKC–ESI–MS analysis using ammonium perfluorooctanoate as a surfactant [268]. Though promising, semivolatile surfactants have surprisingly not been thoroughly explored as a simple solution to the surfactant–ESI compatibility issue.

Much focus has been directed toward methods in which surfactant introduction into the ESI interface is totally avoided, namely partial-filling and reverse-migration MEKC. As described previously, in partial-filling MEKC, the separation capillary is filled with background electrolyte. Then, a small plug of micellar solution is injected, followed by the sample plug. On application of the separation potential, analytes migrate through the micellar zone and separate based on MEKC principles. After passing through this zone, the analytes then sequentially migrate through the micelle-free background electrolyte to the detector. The separation potential is terminated promptly after detection of the analytes to prevent surfactants from entering the ESI–MS system [269]. Drawbacks of the partial filling technique include lower separation efficiencies, lower resolution, and lower peak capacity, when compared to traditional MEKC [270]. The differences in electric field strength and in viscosity between the micellar plug and the background electrolyte cause laminar flows, leading to band broadening. Additionally, the reproducibility of the method is questionable.

In reverse-migration MEKC, the capillary is filled with a micellar background electrolyte. The pH is adjusted so that there is a low EOF toward the detector on application of the separation potential. Meanwhile, during the separation, the micelles move toward the capillary inlet, avoiding introduction into the ESI–MS interface. Typically, negatively charged surfactants are used in a low pH buffer. In these cases, the applicability of the technique is limited to positively charged analytes since negatively charged analytes will not have an overall mobility toward the detector. With a high enough EOF and a relatively low affinity for the micelles, neutral analytes can also be detected [255]. As mentioned, specific applications of both the reverse-migration and the partial-filling techniques have thoroughly been reviewed [255].

Alternatives to ESI, atmospheric pressure chemical ionization (APCI) and atmospheric pressure photoionization (APPI) are not inhibited by the presence of nonvolatile buffer constituents, including surfactants [271–275]. This is because in both the methods, the sample is vaporized in a heated nebulizer before being ionized; therefore, nonvolatile buffer constituents are left behind and do not decrease ionization efficiency as in ESI [274,276]. Mol et al. showed that even up to 60 mM SDS does not cause suppression of analyte signals, excess background noise, nor affects the stability of the APPI interface performance [277]. Himmelsbach et al. [278,279] and Schappler et al. [280] carried out MEEKC separations, which use even higher concentrations of surfactants than MEKC, with no impact on APPI-MS performance. APCI and APPI have the disadvantage of lower sensitivity [272,273,275]; however, coupling these ionization techniques with CE has only recently been attempted and further research will likely lead to improved interfaces.

1.5.2 Electrochemical Detection

Electrochemical detection (ECD) techniques such as amperometry and conductivity are commonly coupled to CE and microchip-CE. An integrated electrode offers a liquid–solid interface for surfactants in solution to interact with, as described in Section 1.2. An important characteristic of an electrode surface in particular is that the potential applied to the electrode determines the surface charge. This, in turn, can induce changes in the structure of the aggregations adsorbed to the surface [13,281–283]. For example, Burgess et al. showed that with small-to-moderate absolute charge density, SDS surfactant monomers aggregate into hemicylindrical stripes on the electrode surface. If the applied potential is increased to equal or superequivalent to the charge of the surfactants, these aggregations undergo a phase transition into a more condensed state (the surface concentration of SDS doubles), probably in an interdigitated bilayer structure (Figure 1.2) [283]. The transition between the hemicylindrical aggregations and the condensed state was shown to be reversible [281]. In contrast to the adverse effects on MS detection, surfactants actually have been shown to enhance the performance of electrodes [284].

One mechanism of enhancement occurs when adsorbed surfactants electrostatically preconcentrate analytes onto the surface of the electrode. For instance, when the positively charged surfactant CTAB was adsorbed to a glassy carbon electrode, the concentration of negatively charged dinitrophenols on the electrode was increased via electrostatic attraction and the reduction peak currents were enhanced, increasing detection sensitivity [285]. Similarly, redox peak currents were increased due to enhanced accumulation of sodium nitroprusside on a CTAB-coated electrode [286]. With a negatively charged SDS layer hydrophobically adsorbed to the surface of a carbon paste electrode, the response of dopamine was enhanced while at the same time, a negatively charged interferant (ascorbic acid) was repelled [287]. Many other instances of this electrostatic preconcentration onto electrodes can be found in the literature [286,288–298]. Hu et al. found that a Nafion-modified electrode, which contains $—SO_3^-$ groups, adsorbed positively charged CTAB through ion exchange. This surfactant allowed for preconcentration of estradiol, estrone, and estriol by hydrophobic interactions, improving the electrochemical reaction between analyte and electrode and increasing sensitivity [299]. Other hydrophobic analytes such as diethylstilbestrol have been detected with improved sensitivity as a result of surfactants [12,300]. To systematically study the effects of surfactants on signal enhancement ECD, Ding et al. analyzed six phenolic compounds in the presence of four different surfactants. The hypothesis presented was that the electrochemical method used—pulsed amperometric detection (PAD)—promotes the formation of hemimicellar aggregates on electrode surfaces [282,283]. After computationally calculating the partial charges and dipole moments, it was determined that the analytes that have a surfactant-like structure have enhanced analytical signals, due to ease of incorporation into the hemimicelles [301]. In all of these examples, the aggregation of surfactants on the electrode surface resulted in enhanced redox peak currents of analytes by facilitating electron exchange. The high concentration of the analyte and proximity to the electrode surface increases the electron transfer rate, decreasing the overvoltage of the electrode [287,288]. Additionally, the surfactant microstructures may increase the stability of electrogenerated radicals, intermediates, and products [12,284,302–305].

In conjunction with the use of surfactants, PAD has also exhibited improvements in the electrochemical response of carbohydrates [54,306], biomarkers [307], metabolites and biomolecules (glucose, penicillin, phenol, homovanillic acid) [38,306], and antioxidants [308]. PAD shows a greater enhancement by the presence of surfactants than DC amperometry. The higher electrode potentials likely allow a greater amount of surfactant to adsorb to the electrode surface. These effects can be seen even when the concentration of surfactant is below the CMC value [282,283]. PAD may also show better response because of the repeated cycling of applied potentials that can induce repeated adsorption and desorption of surfactant aggregates. This allows enhanced response at the detection by providing a clean electrode surface with no oxidation products as well as fresh surfactant aggregates capable of preconcentration [309–311]. Surfactants have been demonstrated to stabilize electrodes during the analysis of highly selfpassivating analytes such as phenol, tyramine, tyrosine, serotonin, tryptophan, and 3,4-dihydroxyphenyl acetic acid [312,313]. Surfactants are believed to prevent electrode fouling by competitive adsorption with excessive oxidation products in addition to electrostatic repulsion of these products from the electrode surface. Similarly, surfactants have been shown to suppress protein interference in ECD by competitive adsorption. This has allowed the analysis of cadmium and lead in the presence of albumin and lysozyme and lead and copper in milk powder [314–316]. In summary, the combination of surfactant-modified CE separations with surfactant-enhanced ECD is simple, yet highly advantageous. This has proven particularly important in microchip-CE–PAD as surfactants can control/stabilize the EOF of the polymeric substrates of the microchips, while at the same time improving the detection.

1.6 CONCLUSIONS

This chapter summarizes different approaches that take advantage of the unique structure and properties of surfactants to improve analysis in CE. As described, surfactants can be used to coat the capillary surface, allowing a simple way to control EOF and minimize adsorption of analytes and sample matrix to the capillary wall. When used as buffer additives, surfactants can improve the separation of hydrophobic analytes, allowing the development of a wide variety of separation modes including MEKC, MEEKC, and NAMEKC. Surfactants can also be used to improve the sensitivity of CE by enabling preconcentration of analytes by sweeping, transient trapping, focusing by micelle collapse, micelle-to-solvent stacking, and cloud point extraction. Last but not least, the performance and selectivity of some detection modes can be improved by the addition of surfactants. The addition of surfactants is one of the most versatile alternatives to further improve the performance of CE. Theoretical aspects, advantages, and guidelines to rationally design CE methods that employ various surfactants have been herein described. The use of surfactants in CE, however, seems to be mostly limited by cost and commercial availability. Consequently, most researchers seem inclined to develop applications based on the more traditional surfactants (such as SDS or CTAB). It is also interesting to see that in most cases, reports describe the use of surfactants only for specific tasks in the analytical procedure without necessarily considering the effects on multiple steps of the analysis. In summary, we believe that the use of surfactants in CE will continue to grow, despite the aforementioned limitations. As new types of surfactants become available to researchers, innovative applications will continue to push the limits of CE in terms of performance and ability to handle real samples.

REFERENCES

1. Lucy, C.A., MacDonald, A.M., and Gulcev, M.D. (2008) *J. Chromatogr. A 1184*(1–2), 81–105.
2. Otzen, D. (2011) *Biochim. Biophys. Acta Proteins Proteomics 1814*(5), 562–591.
3. Starov, V., Zhdanov, V., and Kovalchuk, N.M. (2010) *Colloids Surf. A. 354*(1–3), 268–278.
4. Baryla, N.E. et al. (2001) *Anal. Chem. 73*, 4558–4565.
5. Rosholm, K., González-Pérez, A., and Mouritsen, O. (2010) *Colloid Polym. Sci. 288*(14), 1351–1357.
6. Wanless, E.J. and Ducker, W.A. (1996) *J. Phys. Chem. 100*(8), 3207–3214.
7. Wang, W. and Kwak, J.C.T. (1999) *Colloids Surf. A. 156*, 95–110.
8. Diamant, H., Ariel, G., and Andelman, D. (2001) *Colloids Surf. A. 183 185*, 259 276.
9. Atkin, R. et al. (2003) *Adv. Colloid Interface Sci. 103*, 219–304.
10. Paria, S. and Khilar, K.C. (2004) *Adv. Colloid Interface Sci. 110*, 75–95.
11. Mora, M.F. et al. (2008) *Anal. Lett. 41*(2), 312–334.
12. Rusling, J.F. (1991) *Acc. Chem. Res. 24*(3), 75–81.
13. Hu, K. and Bard, A.J. (1997) *Langmuir 13*(20), 5418–5425.
14. Choi, K.S., McFarland, E.W., and Stucky, G.D. (2003) *Adv. Mater. 15*(23), 2018–2021.
15. Mora, M.F., Giacomelli, C.E., and Garcia, C.D. (2007) *Anal. Chem. 79*, 6675–6681.
16. Melanson, J.E., Baryla, N.E., and Lucy, C.A. (2001) *Trends Anal. Chem. 20*(6–7), 365–374.
17. Lucy, C.A. and Underhill, R.S. (1996) *Anal. Chem. 68*(2), 300–305.
18. Ye, M. et al. (1999) *J. Chromatogr. A 855*(1), 137–145.
19. Towns, J.K. and Regnier, F.E. (1991) *Anal. Chem. 63*(11), 1126–1132.
20. Yowell, G.G., Fazio, S.D., and Vivilecchia, R.V. (1993) *J. Chromatogr. A 652*(1), 215–224.
21. Antczak, A., Ramstad, T., and Johnson, T. (2006) *Chromatographia 64*(1), 57–64.
22. Paterson, G.R., Hill, J.P., and Otter, D.E. (1995) *J. Chromatogr. A 700*(1–2), 105–110.
23. Manet, S. et al. (2010). *Langmuir 26*(13), 10645–10656.
24. Akbay, C. et al. (2010) *J. Chromatogr. A 1217*(32), 5279–5287.
25. Liu, Q. et al. (2009) *J. Sep. Sci. 32*(23–24), 4148–4154.
26. Danino, D., Talmon, Y., and Zana, R. (1995) *Langmuir 11*(5), 1448–1456.
27. Zana, R., Benrraou, M., and Rueff, R. (1991) *Langmuir 7*(6), 1072–1075.
28. Herrero-Martínez, J.M. et al. (2001) *Electrophoresis 22*, 526–534.
29. Haddadian, F., Shamsi, S.A., and Warner, I.M. (1999) *Electrophoresis 20*(15–16), 3011–3026.
30. Molina, M. and Silva, M. (2002) *Electrophoresis 23*(22–23), 3907–3921.
31. Mazzeo, J.R. and Krull, I.S. (2002) *Biotechniques 10*(5), 638–645.
32. Doherty, E.A.S. et al. (2003) *Electrophoresis 24*(1–2), 34–54.
33. Dolnik, V. (2004) *Electrophoresis 25*, 3589–3601.
34. Lucy, C.A., Baryla, N.E., and Yeung, K.K.-C. (2004) Surfactant-based methods for prevention of protein adsorption in capillary electrophoresis, In: Strege, M.A. and Lagu, A.L. (eds), *Capillary Electrophoresis of Proteins and Peptides*, Springer-Verlag, New York, NY, pp. 1–14.
35. Zhou, J., Ellis, A.V., and Voelcker, N.H. (2010) *Electrophoresis 31*(1), 2–16.
36. Belder, D. and Ludwig, M. (2003) *Electrophoresis 24*(21), 3595–3606.
37. Olah, A., Hillborg, H., and Vancso, G.J. (2005) *Appl. Surf. Sci. 239*, 410–423.
38. García, C.D. et al. (2005) *Electrophoresis 26*(3), 703–709.
39. Melanson, J.E., Baryla, N.E., and Lucy, C.A. (2000) *Anal. Chem. 72*(17), 4110–4114.
40. Wang, C.Z. and Lucy, C.A. (2004) *Electrophoresis 25*, 825–832.
41. Yassine, M.M. and Lucy, C.A. (2004) *Anal. Chem. 76*, 2983–2990.
42. Yassine, M.M. and Lucy, C.A. (2005) *Anal. Chem. 77*, 620–625.
43. Righetti, P.G. et al. (2001) *Electrophoresis 22*(4), 603–611.
44. Mohanty, A. and Dey, J. (2005) *J. Chromatogr. A 1070*, 185–192.
45. Gilman, S.D. and Chapman, P.J. (2006) Measuring electroosmotic flow in microchips and capillaries, In: Henry, C.S. (ed.), *Microchip Capillary Electrophoresis: Methods and Protocols*, Humana Press Inc., Totowa, NJ, pp. 187–201.
46. Foley, J.P. and Zhou, M.X. (2006) *Anal. Chem. 78*, 1849–1858.
47. Madabhushi, R.S. (1998) *Electrophoresis 19*(2), 224–230.
48. Albarghouthi, M.N., Stein, T.M., and Barron, A.E. (2003) *Electrophoresis 24*, 1166–1175.
49. Kleemiss, M.H., Gilges, M., and Schomburgh, G. (1993) *Electrophoresis 14*(5–6), 515–522.
50. Bean, S.R. and Lookhart, G.L. (1998) *Electrophoresis 19*(18), 3190–3198.
51. Gong, B.Y. and Ho, J.W. (1997) *Electrophoresis 18*(5), 732–735.
52. Baryla, N.E. and Lucy, C.A. (2000) *Anal. Chem. 72*, 2280–2284.
53. Bao, N. et al. (2005) *J. Chromatogr. A 1099*, 203–206.
54. García, C.D. and Henry, C.S. (2005) *Electroanalysis 17*(3), 223–230.
55. Kirby, B.J. and Hasselbrink, E.F.J. (2004) *Electrophoresis 25*, 203–213.
56. Berthod, A., Girard, I., and Gonnet, C. (1986) *Anal. Chem. 58*, 1356–1358.
57. Thibaut, A. et al. (2000) *Langmuir 16*, 9192–9198.

58. Tavares, M.F., Colombara, R., and Massaro, S. (1997) *J. Chromatogr. A 772*, 171–178.
59. Quirino, J.P. and Terabe, S. (2000) *Electrophoresis 21*, 355–359.
60. Liu, B.F. et al. (2005) *Anal. Chem. 77*, 573–578.
61. MacDonald, A.M., Sheppard, M.A.W., and Lucy, C.A. (2005) *Electrophoresis 26*(23), 4421–4428.
62. Chiesl, T.N. et al. (2009) *Anal. Chem. 81*(7), 2537–2544.
63. Hadley, M.R., Harrison, M.W., and Hutt, A.J. (2003) *Electrophoresis 24*(15), 2508–2513.
64. Wei, W. and Ju, H. (2005) *Electrophoresis 26*(3), 586–592.
65. Yeung, K.K.-C. and Lucy, C.A. (1997) *Anal. Chem. 69*, 3435–3441.
66. Cao, M. et al. (2006) *J. Colloid Interface Sci. 300*, 519–525.
67. Liu, Q. et al. (2007) *Electrophoresis 28*, 2275–2282.
68. Bonoli, M., Varjo, S.J., Wiedmer, S.K.R., and Riekkola, M. L. (2006) *J. Chromatogr. A 1119*, 163–169.
69. Zhang, H.X. et al. (2005) *Anal. Chem. 77*(18), 6078–6084.
70. Jurcic, K., Nesbitt, C.A., and Yeung, K.K.C. (2006) *J. Chromatogr. A 1134*, 317–325.
71. Kurrat, R., Prenosil, J.E., and Ramsden, J.J. (1997) *J. Colloid Interface Sci. 185*(1), 1–8.
72. Van Oss, C.J., Docoslis, A., and Giese, R.F. (2001) *Colloids Surf. B 22*, 285–300.
73. Huang, B. et al. (2005) *Lab Chip. 5*, 1005–1007.
74. Castelleti, L. et al. (2000) *J. Chromatogr. A 894*, 281–289.
75. Mohabbati, S. and Westerlund, D. (2006) *J. Chromatogr. A 1121*, 32–39.
76. Sebastiano, R. et al. (2000) *J. Chromatogr. A 894*, 53–61.
77. Stutz, H. (2009) *Electrophoresis 30*, 2032–2061.
78. Cordova, E., Gao, J., and Whitesides, G.M. (1997) *Anal. Chem. 69*, 1370–1379.
79. Horvath, J. and Dolnik, V. (2001) *Electrophoresis 22*(4), 644–655.
80. Wong, I. and Ho, C.-M. (2009) *Microfluid. Nanofluid. 7*, 291–306.
81. Hellmich, W. et al. (2005) *Langmuir 21*, 7551–7557.
82. Liu, Q., Yang, Y., and Yao, S. (2008) *J. Chromatogr. A 1187*, 260–266.
83. Baryla, N.E. et al. (2001) *Anal. Chem. 73*, 4558–4565.
84. Cunliffe, J.M., Baryla, N.E., and Lucy, C.A. (2002) *Anal. Chem. 74*(4), 776–783.
85. Han, B. et al. (2009) *Talanta 79*, 959–962.
86. Ross, E.E. et al. (2003) *Langmuir 19*, 1766–1774.
87. Wang, J. et al. (2005) *J. Chromatogr. A 1091*(1–2), 177–182.
88. Mansfield, E., Ross, E.E., and Aspinwall, C.A. (2007) *Anal. Chem. 79*, 3135–3141.
89. Terabe, S. et al. (1984) *Anal. Chem. 56*(1), 111–113.
90. Téllez, A., Weiss, V.U., and Kenndler, E. (2008) *Electrophoresis 29*(18), 3916–3923.
91. Altria, K.D., (2000) *J. Chromatogr. A 892*(1–2), 171–186.
92. Muijselaar, P.G., Otsuka, K., and Terabe, S. (1997) *J. Chromatogr. A 780*(1–2), 41–61.
93. Tjørnelund, J. and Hansen, S.H. (1997) *J. Chromatogr. A 792* (1–2), 475–482.
94. Tonin, F.G. et al. (2005) *Electrophoresis 26*(17), 3387–3396.
95. Silva, C.A. et al. (2007) *Electrophoresis 28*(20), 3722–3730.
96. Terabe, S. (1992) *J. Pharm. Biomed. Anal 10*(10–12), 705–715.
97. Khaledi, M.G. (1998) Micellar electrokinetic chromatography, In: Khaledi, M.G. (ed.), *High Performance Capillary Electrophoresis*, Wiley, New York.
98. Otsuka, K. and Terabe, S. (1998) *Bull. Chem. Soc. Jpn. 71* (11), 2465–2481.
99. Pyell, U. (ed.) (2007) *Electrokinetic Chromatography: Theory, Instrumentation and Applications.* Wiley, Chichester, UK.
100. Terabe, S. (1992) *Micellar Electrokinetic Chromatography*, Beckman, Fullerton.
101. Terabe, S. (2004) *Anal. Chem. 76*(13), 240 A–246 A.
102. Terabe, S. (2008) Micellar electrokinetic chromatography, In: Landers, J.P. (ed.), *Handbook of Capillary and Microchip Electrophoresis and Associated Microtechniques*, CRC Press, New York.
103. Terabe, S., Chen, N., and Otsuka, K. (1994) *Adv. Electrophoresis 7*, 87–153.
104. Vindevogel, J. and Sandra, P. (1992) Introduction to micellar electrokinetic chromatography. In: Bertsch, W. et al. (eds), *Chromatographic Methods*, Hüthig, Heidelberg.
105. Terabe, S. (2008) *Chem. Rec. 8*(5), 291–301.
106. Pappas, T.J., Gayton-Ely, M., and Holland, L.A. (2005) *Electrophoresis. 26*(4–5), 719–734.
107. Song, L. et al. (1997) *J. Chromatogr. A 780*(1–2), 297–328.
108. Iadarola, P. et al. (2008) *Electrophoresis 29*(1), 224–236.
109. Viglio, S. et al. (2010) *Electrophoresis 31*(1), 93–104.
110. Watanabe, T. and Terabe, S. (2000) *J. Chromatogr. A 880*(1–2), 295–301.
111. Strege, M.A. and Lagu, A.L. (1997) *J. Chromatogr. A 780*(1–2), 285–296.
112. Haddad, P.R. et al. (1997) *J. Chromatogr. A 780*(1–2), 329–341.
113. Nishi, H. (1997) *J. Chromatogr. A 780*(1–2), 243–264.
114. Otsuka, K. and Terabe, S. (2000) *J. Chromatogr. A 875*(1–2), 163–178.
115. Silva, M. (2007) *Electrophoresis 28*(1–2), 174–192.
116. Silva, M. (2009) *Electrophoresis 30*(1), 50–64.
117. Maier, V. et al. (2009) *J. Chromatogr. A 1216*(20), 4492–4498.
118. Bumgarner, J.G. and Khaledi, M.G. (1996) *J. Chromatogr. A 738*(2), 275–283.
119. Ahuja, E.S. et al. (1995) *Anal. Chem. 67*(1), 26–33.
120. Terabe, S., Otsuka, K., and Nishi, H. (1994) *J. Chromatogr. A 666*(1–2), 295–319.
121. Nishi, H. et al. (1990) *J. Chromatogr. A 498*, 313–323.

122. Navazo, A.L.C., Marina, M.L., and Gonzalez, M.J. (1998) *Electrophoresis 19*(12), 2113–2118.
123. Camilleri, P. (1997) *Electrophoresis 18*(12–13), 2322–2330.
124. Herrero-Martínez, J.M., Simó-Alfonso, E.F., and Ramis-Ramos, G. (2003) *Electrophoresis 24*, 681–686.
125. Watarai, H. (1991) *Chem. Lett. 20*(3), 391–394.
126. Altria, K.D., Mahuzier, P.-E., and Clark, B.J. (2003) *Electrophoresis 24*(3), 315–324.
127. Pedersen-Bjergaard, S., Gabel-Jensen, C. and Honoré Hansen, S. (2000) *J. Chromatogr. A 897*(1–2), 375–381.
128. Ryan, R. et al. (2009) *Electrophoresis 30*(1), 65–82.
129. Altria, K., Clark, B., and Mahuzier, P. (2000) *Chromatographia 52*(11), 758–768.
130. Gabel-Jensen, C., Hansen, S.H., and Pedersen-Bjergaard, S. (2001) *Electrophoresis 22*(7), 1330–1336.
131. Klampfl, C.W. (2003) *Electrophoresis 24*(10), 1537–1543.
132. Marsh, A. et al. (2004) *Electrophoresis 25*(23–24), 3970–3980.
133. McEvoy, E. et al. (2007) *Electrophoresis 28*(1–2), 193–207.
134. Watarai, H. (1997) *J. Chromatogr. A 780*(1–2), 93–102.
135. Hansen, S.H. et al. (2001) *Trends Anal. Chem. 20*(11), 614–619.
136. Honoré Hansen, S., Gabel-Jensen, C., and Pedersen-Bjergaard, S. (2001) *J. Sep. Sci. 24*(8), 643–650.
137. Kahle, K.A. and Foley, J.P. (2007) *Electrophoresis 28*(15), 2503–2526.
138. Altria, K.D. (1999) *J. Chromatogr. A 844*(1–2), 371–386.
139. Hansen, S.H. (2003) *Electrophoresis 24*(22–23), 3900–3907.
140. Huie, C.W. (2006) *Electrophoresis 27*(1), 60–75.
141. Miola, M.F., Snowden, M.J., and Altria, K.D. (1998) *J. Pharm. Biomed. Anal. 18*(4–5), 785–797.
142. Walbroehl, Y. and Jorgenson, J.W. (1984) *J. Chromatogr. A 315*, 135–143.
143. Lu, W. et al. (1996) *Anal. Chem. 68*(4), 668–674.
144. Li, J. and Fritz, J.S. (1999) *Electrophoresis 20*(1), 84–91.
145. Bjørnsdottir, I., Tjørnelund, J., and Hansen, S.H. (1998) *Electrophoresis 19*(12), 2179–2186.
146. Walbroehl, Y. and Jorgenson, J.W. (1986) *Anal. Chem. 58*(2), 479–481.
147. Miller, J.L., Khaledi, M.G., and Shea, D. (1997) *Anal. Chem. 69*(6), 1223–1229.
148. Cantu, M.D., Hillebrand, S., and Carrilho, E. (2005) *J. Chromatogr. A 1068*(1), 99–105.
149. Nozal, L. et al. (2006) *Anal. Chim. Acta. 560*(1–2), 69–76.
150. Sahota, R.S. and Khaledi, M.G. (1994) *Anal. Chem. 66*(7), 1141–1146.
151. Chankvetadze, B. and Blaschke, G. (2000) *Electrophoresis 21*(18), 4159–4178.
152. Geiser, L. and Veuthey, J.L. (2007) *Electrophoresis 28*(1–2), 45–57.
153. Geiser, L. and Veuthey, J.L. (2009) *Electrophoresis 30*(1), 36–49.
154. Porras, S.P., Riekkola, M.-L., and Kenndler, E. (2003) *Electrophoresis 24*(10), 1485–1498.
155. Riekkola, M.-L. (2002) *Electrophoresis 23*(22–23), 3865–3883.
156. Riekkola, M.-L. et al. (2000) *J. Chromatogr. A 892*(1–2), 155–170.
157. Steiner, F. and Hassel, M. (2000) *Electrophoresis 21*(18), 3994–4016.
158. Wang, F. and Khaledi, M.G. (2000) *J. Chromatogr. A 875*(1–2), 277–293.
159. Porras, S.P. and Kenndler, E. (2005) *Electrophoresis 26*(17), 3203–3220.
160. Akhter, M.S. (1999) *Colloids Surf. A 150*(1–3), 25–30.
161. Akhter, M.S. and Alawi, S.M. (2000) *Colloids Surf. A 175*(3), 311–320.
162. Bowser, M.T., Kranack, A.R., and Chen, D.D.Y. (1998) *Trends Anal. Chem. 17*(7), 424–434.
163. Bjørnsdottir, I., HonoréHansen, S., and Terabe, S. (1996) *J. Chromatogr. A 745*(1–2), 37–44.
164. Britz-McKibbin, P., Otsuka, K., and Terabe, S. (2002) *Anal. Chem. 74*(15), 3736–3743.
165. Chen, Y.-H., Chung, Y.-L. and Lin, C.-H. (2002) *J. Chromatogr. A 943*(2), 287–294.
166. Lin, C.-H. and Chen, Y.-H. (2001) *Electrophoresis 22*(12), 2574–2579.
167. Jussila, M. et al. (1999) *Electrophoresis 20*(1), 111–117.
168. Bellini, M.S., Deyl, Z., and Mikík, I. (1998) *Forensic Sci. Int. 92*(2–3), 185–199.
169. Ng, C.L., Lee, H.K., and Li, S.F.Y. (1994) *J. Liq. Chromatogr. 17*(18), 3847–3857.
170. Bowser, M.T., Sternberg, E.D., and Chen, D.D.Y. (1997) *Electrophoresis 18*(1), 82–91.
171. Bowser, M.T., Sternberg, E.D., and Chen, D.D.Y. (1996) *Anal. Biochem. 241*(2), 143–150.
172. Chen, A. et al. (2005) *J. Pharm. Biomed. Anal. 37*(4), 811–816.
173. Fang, C. et al. (2002) *Forensic Sci. Int. 125*(2–3), 142–148.
174. Huang, R. et al. (2006) *Chin. J. Chromatogr. 24*(6), 597–600.
175. Lin, J.-M. et al. (2002) *Electrophoresis 23*(3), 421–425.
176. Breadmore, M.C. (2007) *Electrophoresis 28*(1–2), 254–281.
177. Breadmore, M.C. et al. (2009) *Electrophoresis 30*(1), 230–248.
178. Aranas, A., Guidote, A., and Quirino, J. (2009) *Anal. Bioanal. Chem. 394*(1), 175–185.
179. Beckers, J.L. and Boček, P. (2000) *Electrophoresis 21*(14), 2747–2767.
180. Britz-McKibbin, P. and Terabe, S. (2003) *J. Chromatogr. A 1000*(1–2), 917–934.
181. Breadmore, M. and Haddad, P. (2001) *Electrophoresis 22* (12), 2464–2489.
182. Chien, R.L. (2003) *Electrophoresis 24*(3), 486–497.
183. Guzman, N.A. and Stubbs, R.J. (2001) *Electrophoresis 22*(17), 3602–3628.
184. Kim, J.-B. et al. (2001) *J. Chromatogr. A 916*(1–2), 123–130.

185. Kim, J.-B. and Terabe, S. (2003) *J. Pharm. Biomed. Anal. 30*(6), 1625–1643.
186. Lin, C.H. and Kaneta, T. (2004) *Electrophoresis 25*(23–24), 4058–4073.
187. Malá, Z. et al. (2007) *Electrophoresis 28*(1–2), 243–253.
188. Monton, M.R.N. and Terabe, S. (2006) *J. Chromatogr. B 841*(1–2), 88–95.
189. Osbourn, D.M., Weiss, D.J., and Lunte, C.E. (2000) *Electrophoresis 21*(14), 2768–2779.
190. Puig, P. et al. (2008) *Anal. Chim. Acta. 616*(1), 1–18.
191. Quirino, J.P., Kim, J.-B., and Terabe, S. (2002) *J. Chromatogr. A 965*(1–2), 357–373.
192. Quirino, J.P. and Terabe, S. (2000) *J. Chromatogr. A 902*(1), 119–135.
193. Sentellas, S., Puignou, L., and Galceran, M.T. (2002) *J. Sep. Sci. 25*(15–17), 975–987.
194. Shihabi, Z.K. (2000) *J. Chromatogr. A 902*(1), 107–117.
195. Simpson, S.L.Jr., Quirino, J.P., and Terabe, S. (2008) *J. Chromatogr. A 1184*(1–2), 504–541.
196. Urbánek, M., Křivánková, L., and Boček, P. (2003) *Electrophoresis 24*(3), 466–485.
197. Quirino, J.P. and Terabe, S. (1998) *Science 282*(5388), 465–468.
198. Quirino, J.P. and Terabe, S. (1999) *Anal. Chem. 71*(8), 1638–1644.
199. Huang, H.-Y. and Hsieh, S.-H. (2007) *J. Chromatogr. A 1164*(1–2), 313–319.
200. Huang, H.-Y. et al. (2007) *Electrophoresis 28*(11), 1735–1743.
201. Huang, H.Y., Lien, W.C., and Huang, I.Y. (2006) *Electrophoresis 27*(16), 3202–3209.
202. Sueyoshi, K. et al. (2006) *Meas. Sci. Technol. 17*(12), 3154–3161.
203. Sueyoshi, K., Kitagawa, F., and Otsuka, K. (2008) *Anal. Chem. 80*(4), 1255–1262.
204. Quirino, J.P. and Haddad, P.R. (2008) *Anal. Chem. 80*(17), 6824–6829.
205. Quirino, J.P., (2008) *J. Chromatogr. A 1214*(1–2), 171–177.
206. Quirino, J.P. and Haddad, P.R. (2009) *Electrophoresis 30*(10), 1670–1674.
207. Quirino, J.P. (2009) *J. Chromatogr. A 1216*(2), 294–299.
208. Guidote, A.M.Jr and Quirino, J.P. (2010) *J. Chromatogr. A 1217*(40), 6290–6295.
209. Britz-McKibbin, P. et al. (2003) *J. Chromatogr. A 1013*(1–2), 65–76.
210. Yu, L. and Li, S.F.Y. (2005) *Electrophoresis 26*(22), 4360–4367.
211. Dawod, M. et al. (2010) *J. Chromatogr. A 1217*(3), 386–393.
212. Shih, C.M. and Lin, C.H. (2005) *Electrophoresis 26*(18), 3495–3499.
213. Wang, C. et al. (2007) *J. Chromatogr. A 1143*(1–2), 270–275.
214. Takagai, Y. and Igarashi, S. (2002) *Anal. Bioanal. Chem. 373*, 87–92.
215. Quirino, J.P. (2010) *J. Chromatogr. A 1217*, 7776–7780.
216. Zhu, J. et al. (2007) *J. Chromatogr. A 1166*(1–2), 191–200.
217. Quirino, J.P. and Terabe, S. (2000) *Anal. Chem. 72*(5), 1023–1030.
218. Quirino, J.P. et al. (2000) *Electrophoresis 21*(14), 2899–2903.
219. Kim, J.-B., Otsuka, K., and Terabe, S. (2001) *J. Chromatogr. A 932*(1–2), 129–137.
220. Isoo, K., Otsuka, K., and Terabe, S. (2001) *Electrophoresis 22*(16), 3426–3432.
221. Fang, C., Liu, J.-T., and Lin, C.-H. (2002) *J. Chromatogr. B 775*(1), 37–47.
222. Fang, C., Liu, J.-T., and Lin, C.-H. (2002) *Talanta 58*(4), 691–699.
223. Nuñez, O. et al. (2002) *J. Chromatogr. A 961*(1), 65–75.
224. Zhu, L., Tu, C., and Lee, H.K. (2002) *Anal. Chem. 74*(22), 5820–5825.
225. Fang, C. et al. (2003) *Electrophoresis 24*(6), 1031–1037.
226. Tsai, C.H., Huang, H.M., and Lin, C.H. (2003) *Electrophoresis 24*(17), 3083–3088.
227. Chiang, H.Y. and Sheu, S.J. (2004) *Electrophoresis 25*(4–5), 670–676.
228. Chen, M.-C., Chou, S.-H., and Lin, C.-H. (2004) *J. Chromatogr. B 801*(2), 347–353.
229. Lin, Y.-H. et al. (2006) *J. Chromatogr. A 1130*(2), 281–286.
230. Cheng, H.L. et al. (2006) *Electrophoresis 27*(23), 4711–4716.
231. Meng, P. et al. (2006) *Electrophoresis 27*(16), 3210–3217.
232. Gong, M. et al. (2006) *Anal. Chem. 78*(17), 6035–6042.
233. Lin, Y.-H. et al. (2007) *J. Chromatogr. A 1145*(1–2), 234–240.
234. Fang, N. et al. (2007) *Analyst 132*(2), 127–134.
235. Lin, Y.H. et al. (2008) *Electrophoresis 29*(11), 2340–2347.
236. Su, H.-L. and Hsieh, Y.-Z. (2008) *J. Chromatogr. A 1209*(1–2), 253–259.
237. Su, H.L. et al. (2008) *Electrophoresis 29*(20), 4270–4276.
238. Liang, G., Choi, K., and Chung, D.S. (2009) *Electrophoresis 30*(11), 1953–1957.
239. Liu, F.-K. (2008) *J. Chromatogr. A 1215*(1–2), 194–202.
240. Carabias-Martínez, R. et al. (2000) *J. Chromatogr. A 902*(1), 251–265.
241. Quina, F.H. and Hinze, W.L. (1999) *Ind. Eng. Chem. Res. 38*(11), 4150–4168.
242. Carabias-Martínez, R. et al. (1999) *Anal. Chem. 71*(13), 2468–2474.
243. Tang, A.-N., Jiang, D.-Q., and Yan, X.-P. (2004) *Anal. Chim. Acta. 507*(2), 199–204.
244. Cerutti, S. et al. (2005) *Electrophoresis 26*(18), 3500–3506.
245. Luconi, M.O. et al. (2006) *J. Hazard. Mater. 128*(2–3), 240–246.
246. Yin, X.-B. (2007) *J. Chromatogr. A 1154*(1–2), 437–443.
247. Wei, W., Yin, X.-B., and He, X.-W. (2008) *J. Chromatogr. A 1202*(2), 212–215.
248. Yin, X.-B., Guo, J.-M., and Wei, W. (2010) *J. Chromatogr. A 1217*(8), 1399–1406.

249. Ragozina, N.Y. et al. (2004) *Anal. Chem. 76*(13), 3804–3809.
250. Waldron, K.C. and Dovichi, N.J. (1992) *Anal. Chem. 64*(13), 1396–1399.
251. Rundlett, K.L. and Armstrong, D.W. (1996) *Anal. Chem. 68*(19), 3493–3497.
252. Tomlinson, A.J., Benson, L.M., and Naylor, S. (1994) *J. Capill. Electrophor. 2*, 127–135.
253. Lu, W., Yang, G., and Cole, R.B. (1995) *Electrophoresis 16* (1), 487–492.
254. Yang, L. and Lee, C.S. (1997) *J. Chromatogr. A 780*(1–2), 207–218.
255. Somsen, G.W., Mol, R., and de Jong, G.J. (2010) *J. Chromatogr. A 1217*(25), 3978–3991.
256. Shamsi, S.A., Akbay, C., and Warner, I.M. (1998) *Anal. Chem. 70*(14), 3078–3083.
257. Ozaki, H. et al. (1995) *J. Chromatogr. A 716*(1–2), 69–79.
258. Lu, W. et al. (1998) *Electrophoresis 19*(12), 2193–2199.
259. Shamsi, S.A. (2001) *Anal. Chem. 73*(21), 5103–5108.
260. Akbay, C., Rizvi, S.A.A., and Shamsi, S.A. (2005) *Anal. Chem. 77*(6), 1672–1683.
261. Hou, J. et al. (2007) *Electrophoresis 28*(9), 1352–1363.
262. Hou, J., Zheng, J., and Shamsi, S.A. (2007) *Electrophoresis 28*(9), 1426–1434.
263. Hou, J. et al. (2006) *Electrophoresis 27*(5–6), 1263–1275.
264. Hou, J., Zheng, J., and Shamsi, S.A. (2007) *J. Chromatogr. A 1159*(1–2), 208–216.
265. Rizvi, S.A.A. et al. (2006) *Anal. Chem. 79*(3), 879–898.
266. Ishihama, Y., Katayama, H., and Asakawa, N. (2000) *Anal. Biochem. 287*(1), 45–54.
267. Petersson, P., Jörntén-Karlsson, M., and Stålebro, M. (2003) *Electrophoresis 24*(6), 999–1007.
268. Van Biesen, G. and Bottaro, C.S. (2006) *Electrophoresis 27*(22), 4456–4468.
269. Terabe, S. et al. (1995) In: *Proceedings of the 7th International Symposium on High Performance Capillary Electrophoresis*, Wurzburg, Germany.
270. Nelson, W.M. and Lee, C.S. (1996) *Anal. Chem. 68*(18), 3265–3269.
271. Hommerson, P. et al. (2008) *J. Chromatogr. A 1204*(2), 197–203.
272. Hommerson, P. et al. (2009) *J. Am. Soc. Mass. Spectrom 20*(7), 1311–1318.
273. Mol, R., de Jong, G.J., and Somsen, G.W. (2005) *Electrophoresis 26*(1), 146–154.
274. Somsen, G., Mol, R., and de Jong, G. (2006) *Anal. Bioanal. Chem. 384*(1), 31–33.
275. Takada, Y., Sakairi, M., and Koizumi, H. (1995) *Rapid Commun. Mass Spectrom. 9*(6), 488–490.
276. Schappler, J. et al. (2007) *Electrophoresis 28*(17), 3078–3087.
277. Mol, R., de Jong, G.J., and Somsen, G.W. (2005) *Anal. Chem. 77*(16), 5277–5282.
278. Himmelsbach, M. et al. (2007) *J. Chromatogr. A 1159*(1–2), 58–62.
279. Himmelsbach, M. et al. (2007) *Anal. Chem. 79*(4), 1564–1568.
280. Schappler, J. et al. (2008) *Electrophoresis 29*(1), 11–19.
281. Burgess, I. et al. (1999) *Langmuir 15*(8), 2607–2616.
282. Zamlynny, V. et al. (2000) *Langmuir 16*(25), 9861–9870.
283. Burgess, I. et al. (2001) *Langmuir 17*(11), 3355–3367.
284. Vittal, R., Gomathi, H., and Kim, K.-J. (2006) *Adv. Colloid Interface Sci. 119*(1), 55–68.
285. Wang, X.-G. et al. (2006) *Electrochim. Acta. 52*(2), 589–594.
286. Yang, X., Wang, F., and Hu, S. (2007) *Colloids Surf. B 54*(1), 60–66.
287. Zheng, J. and Zhou, X. (2007) *Bioelectrochemistry 70*(2), 408–415.
288. He, Q. et al. (2004) *Colloids Surf. B 35*(2), 93–98.
289. Hernández, M. et al. (2007) *Anal. Chim. Acta. 597*(2), 245–256.
290. Svancara, I., Foret, P., and Vytras, K. (2004) *Talanta 64*(4), 844–852.
291. Xie, P. et al. (2006) *Colloids Surf. B 48*(1), 17–23.
292. Stadiober, M. et al. (1996) *Talanta 43*(11), 1915–1924.
293. Hu, S., Yan, Y., and Zhao, Z. (1991) *Anal. Chim. Acta. 248*(1), 103–108.
294. Yi, H. et al. (2001) *Talanta 55*(6), 1205–1210.
295. Hu, C. and Hu, S. (2004) *Electrochim. Acta. 49*(3), 405–412.
296. Wang, F., Fei, J., and Hu, S. (2004) *Colloids Surf. B 39*(1–2), 95–101.
297. Hu, C., Dang, X., and Hu, S. (2004) *J. Electroanal. Chem. 572*(1), 161–171.
298. Stadlober, M., Kalcher, K., and Raber, G. (1997) *Anal. Chim. Acta. 350*(3), 319–328.
299. Hu, S. et al. (2002) *Anal. Chim. Acta. 464*(2), 209–216.
300. Zhang, S., Wu, K., and Hu, S. (2002) *Talanta 58*(4), 747–754.
301. Ding, Y. et al. (2007) *Analyst 132*(10), 997–1004.
302. Proske, G.E.O. (1952) *Anal. Chem. 24*(11), 1834–1836.
303. Wallingford, R.A. and Ewing, A.G. (1988) *Anal. Chem. 60*(3), 258–263.
304. Wen, X.-L., Jia, Y.-H., and Liu, Z.-L. (1999) *Talanta 50*(5), 1027–1033.
305. McIntire, G.L. et al. (1982) *J. Phys. Chem. 86*(14), 2632–2640.
306. García, C.D. and Henry, C.S. (2004) *Anal. Chim. Acta. 508*(1), 1–9.
307. Garcia, C.D. and Henry, C.S. (2004) *Analyst 129*(7), 579–584.
308. Ding, Y., Mora, M.F., and Garcia, C.D. (2006) *Anal. Chim. Acta. 561*(1–2), 126–132.
309. Hompesch, R.W. et al. (2005) *Analyst 130*(5), 694–700.
310. Ding, Y. and Garcia, C.D. (2006) *Analyst 131*(2), 208–214.

311. Ding, Y. and Garcia, C.D. (2006) *Electroanalysis 18*(22), 2202–2209.

312. Hoyer, B. and Jensen, N. (2006) *Electrochem. Commun. 8*(2), 323–328.

313. Hoyer, B. and Jensen, N. (2007) *J. Electroanal. Chem. 601*(1–2), 153–160.

314. Hoyer, B. and Jensen, N. (2003) *Electrochem. Commun. 5*(3), 257–261.

315. Hoyer, B. and Jensen, N. (2004) *Analyst 129*(8), 751–754.

316. Hoyer, B. and Jensen, N. (2005) *Electroanalysis 17*(22), 2037–2042.

2

SAMPLE STACKING: A VERSATILE APPROACH FOR ANALYTE ENRICHMENT IN CE AND MICROCHIP-CE

Bruno Perlatti, Emanuel Carrilho, and Fernando Armani Aguiar

2.1 INTRODUCTION

In recent years, capillary electrophoresis (CE) has become an alternative and complementary technique to chromatographic analysis [1–3], being applied to a wide range of compounds [4–7], especially those that are hard to analyze by gas chromatography (GC) or high-performance liquid chromatography (HPLC) [8].

Capillary electrophoresis offers numerous advantages over HPLC, including higher separation efficiency, shorter analysis time, lower sample and solvent consumption, and a lower relative cost [2,5,9]. Despite such benefits, CE has some limitations, including the difficulty of analyzing dilute samples. This is particularly important for the determination of compounds in biological matrices [5,10]. This shortcoming, which significantly affects the sensitivity [11], can be attributed to two main factors: (i) the limited optical path length, generally in the range 50–100 μm (the inner diameter of the capillary column), and (ii) the small sample volume injected.

In the first case, as in HPLC, most of the commercial instruments are provided (by default) with UV detectors. In such instruments, quantification is performed following the Lambert–Beer law, according to Equation 2.1:

$$A = \varepsilon bc \tag{2.1}$$

where A is the absorbance (a.u.), ε is the molar absorptivity (L mol^{-1} cm^{-1}), b is the optical path length (cm), and c is the molar concentration (mol L^{-1}).

Owing to the short optical path and the small sample volume injected (usually in the nanoliter range), the sample must be concentrated to the maximum in order to improve the signal-to-noise (S/N) ratio and therefore reduce the detection limits [8,12].

To overcome the drawbacks of reduced optical path, different strategies have been used. Although bubble-shaped cells [13,14] and Z-shaped cells [15,16] can provide significant improvements in signal, these strategies can also induce losses in resolution [2,17]. Another strategy to improve sensitivity in CE includes the use of laser-induced fluorescence (LIF) [16], mass spectrometry [18–20], and electrochemical detection (ED) [16,21], to name a few.

The second issue that adds to the disadvantages is the small volume of sample injected into the capillary. Due to the small dimensions of the CE column, injecting large volumes of sample typically promotes peak distortion and compromises the separation [2,8].

With the aim of increasing the concentration of samples, distinct techniques have been developed. Among those derived from chromatographic methods, in which partitioning of a sample occurs between stationary phases, it is worth mentioning solid-phase extraction (SPE), liquid–liquid extraction (LLE), and their evolutions [22–25]. Swartz and Merion [26] reported a twofold enhancement in sensitivity in electrophoretic analysis using an *online* preconcentration method. They used commercial capillaries with an internal polymeric coating of 1 mm, held together by two glass tapes. After an acetonitrile (ACN) preconditioning of the packed material, followed by a water and buffer purge, sample was injected electrokinetically, using a mixture of water and ACN to elute the sample by electromigration. After the sample elution, electrophoretic separation proceeded normally [26].

Capillary Electrophoresis and Microchip Capillary Electrophoresis: Principles, Applications, and Limitations, First Edition.
Edited by Carlos D. García, Karin Y. Chumbimuni-Torres, and Emanuel Carrilho.

Beattie et al. [27] investigated the potential of solid-phase online preconcentration in the separation of the two isoforms of metallothionein (MT). In this study, a C_{18}-packed preconcentration column was used, installed at the capillary inlet. Samples were injected hydrodynamically for 5 min and eluted with H_2O–ACN (2:1), followed by electrophoretic separation using phosphate buffer (pH 7). The authors reported a 700-fold enhancement in sensitivity [27].

Xie et al. [25] developed a method using liquid–liquid–liquid microextraction (LLLME) coupled with capillary zone electrophoresis (CZE) for preconcentration of samples. For such experiments, four *N*-methylcarbamates were used as analytes in order to evaluate the enrichment efficiency. After the microextraction step, the analytes were stacked with the help of injected hydroxide anions (OH^-) that neutralized the background solution (BGS) anions promoting an enhanced field zone. Using this approach, the authors achieved a near 1100-fold enrichment compared with normal CZE analysis [25].

Although these techniques were useful for increasing the concentration of analytes, aided in sample cleanup, and improved sensitivity up to 1000-fold, they also showed some problems such as peak distortion, efficiency loss, peak broadening, electroosmotic flow (EOF) disturbance, and complex capillary setups, which restrict their application in online systems [10,28,29].

Techniques in which the preconcentration occurs by means of electrophoretic methods, based on the manipulation of the electrophoretic mobility of the analytes in different buffered conditions in the same analysis—sometimes in different regions of the capillary—have also been used and have become a suitable alternative to partition-based preconcentration strategies.

In 1950, Haglund and Tiselius [30] used for the first time in CZE, in a glass column, field-enhanced injection in which the sample was prepared in diluted buffer with the same composition as that of the run buffer. In this setup, ions moved with high velocity through the sample plug, under the influence of a high electric field. When analyte ions reach the sample–buffer interface, they encountered a much lower electric field and their mobility was drastically reduced. An increase in the intensity of the applied electric field at the injection point was observed when high voltage was applied, leading to a large increase in the number of injected ions [30].

In 1979, Mikkers et al. [31] reported a sample stacking using a BGS with higher conductivity compared with the sample. Moring et al. [32], in 1989, reported an increase of about 10-fold in sample detectability with CE using sample stacking. These preconcentration methods based on physicochemical phenomena that involved the manipulation of the electrophoretic velocity of analyte and running electrolyte, together with simple injection procedures, are classified as electric field amplification (enhanced field) and/or electrophoretic velocity [33].

2.2 ISOTACHOPHORESIS

In isotachophoresis (ITP), analytes are separated by successive zones, which are disposed in a descending electrophoretic mobility order. The sample is introduced at the buffer interface, consisting of a leading electrolyte (LE) and a terminating electrolyte (TE). To accomplish the separation, the difference in the electrophoretic mobilities of cations or anions must be intermediate to those of the leading and terminal electrolytes. It must be known, however, that either cations or anions can be concentrated by ITP because analytes have to be disposed between a leading and a terminating electrolyte that migrates in the same direction as that of the analytes (coions) [17].

To perform a separation, the leading electrolyte must possess ions whose effective electrophoretic mobility is higher than the mobility of the sample ions, and the terminating electrolyte must have ions with a lower effective electrophoretic mobility than all ions in the capillary. When high voltage is applied, it creates a potential gradient inside the capillary, where sample ions are separated in bands according to their electrophoretic mobility. The highest-mobility ion zone in the leading electrolyte shows a higher conductivity and thus a weaker electric field strength. If a sample ion penetrates into a neighboring zone, it will face a different electric field and will be accelerated or slowed in such a way as to return to its own zone. In this way, sample zones are ordered between both electrolytes until each zone contains only one kind of ion. All zones migrate at the same speed of the terminating electrolyte toward the corresponding electrode [2,16,33].

ITP has numerous advantages; for example, it allows the possibility of large sample injection volumes, up to 50% of the total capillary length, without compromising the separation. This fact allows the lowering of the limits of detection (LOD) in direct comparison with a CZE separation [17,34].

The use of ITP as an online preconcentration strategy was reported initially in 1992 by Foret et al. in protein analysis [35]. There are two ways of executing an ITP–CZE method. In the first mode (cITP–CZE), two online coupled capillaries are used—the first capillary being used for the ITP preconcentration step and the second carrying a normal CZE procedure. In the second mode (tITP–CZE), both ITP preconcentration and CZE separation are performed at the same capillary [36].

ITP combined with CZE can be accomplished by two common strategies:

1. **B_L–S–B_T:** This approach uses a buffer system in which the first buffer injected into the capillary has ions with higher effective electrophoretic mobility than the sample ions ($\mu_L > \mu_S$), with B_L being the leading buffer. Following the sample injection (S), a second run buffer is introduced, possessing a lower

mobility than the sample ($\mu_S > \mu_T$), named the terminating electrolyte (B_T). After application of high voltage, ions are focused in well-defined bands, causing concentration. The B_T is then replaced with the running buffer, in this case B_L, performing a formal CZE separation.

2. **B_T–S–B_L:** In this configuration, first the B_T is introduced, but as its mobility is lower than that of the sample ions, a convenient leading ion (L) must be added to the sample. As the ITP process follows, the leading ion zone will start to vanish because of electromigration scattering and electrophoretic separation by conventional CZE is carried out.

Krivánková et al. [37], using a combination of cITP and CZE in fluorinated ethylene–propylene (FEP) capillaries, obtained a 100-fold decrease in LOD in the determination of hyaluronic acid in serum samples compared with analysis by ITP or CZE alone, showing that this was an appropriate method for the analysis of complex matrices.

Praus [38] used a combination of ITP and CE (cITP–CZE) in the analysis of chloride anions in commercial water samples. Capillaries were made of FEP, measuring 160 mm × 0.8 mm i.d. and the analytical capillary 160 mm × 0.3 mm i.d. With this method, the author obtained LOD and LOQ values lower than the regulatory requirements [38].

Jastrzebská et al. [39] described the determination of sodium triphosphate in meat and derivatives using cITP–CZE, obtaining LOD and LOQ of 0.80 mg P_2O_5 dm^{-3} and 2.69 mg P_2O_5 dm^{-3}, respectively. They used a separation capillary tube of polytetrafluoroethylene (PTFE) (160 mm; 0.8 mm i.d.) connected to an analytical PTFE capillary (160 mm; 0.3 mm i.d.). This method presented better statistical parameters than cITP and cITP–cITP due to its better repeatability and reproducibility [39].

The great virtue of ITP–CZE coupling was its ability both to explore the best of each method and to inject large volumes of sample without losses in efficiency [39]. The good detectability and precision, the low solvent consumption, and the low cost of analysis make this method an alternative to other separation methods such as ion chromatography (IC) and CZE. However, cITP–CZE needs special instrumentation, which restricted its use. Other applications have also been described [16,17,40].

Transient isotachophoresis (tITP) is a relatively simple strategy with easy automation and is applicable to most commercial equipment. Analysis of bromide, iodide, nitrate, and oxalic acid in biological fluid samples can also be carried out using a combination of tITP and CZE. Similar results to those given by the CZE method without tITP were obtained with a simple ultracentrifugation and dilution of serum and urine, respectively, for sample cleanup. This work showed that the samples did not need to have a complex matrix pretreatment, and that all target ions were detected in a single run, with a relative short analysis time [41].

Xu et al. [42] described the use of tITP–CZE for the determination of inorganic ions in saliva samples. This method of analysis allowed the authors to achieve an LOD of bromide and iodide 50 times lower than that with CZE without tITP, detecting also nitrate and thiocyanate in concentrations of 4.6 and 3.8 $\mu g\,L^{-1}$, respectively, in 1:1000 dilutions of saliva [42].

2.3 CHROMATOGRAPHY-BASED SAMPLE STACKING

Usually when the sample is injected in CE, the maximum amount of analyte is limited to a maximum injection time or length due to dispersion and band broadening. Using a capillary equipped with a chromatography-based preconcentration chamber, it is feasible to load multiple times the column volume and then concentrate the analytes with minimal adverse effect on the separation process [29].

The system most used is preconcentration with low-specificity chromatography. In this system, online sample preconcentration strategies based on solid-phase extraction (SPE) can avoid tedious manipulation and sample contamination, thus becoming of great interest when the amount of sample available is an issue. It can be used with distinct stationary phases, fixed by membranes or fritted disks. Although the latter is more common, it presents some problem of hydrodynamic flow arising from the disturbance of the electroosmotic flow because of back-pressure built up by the retainers [29]. A regular SPE cartridge inserted in the inlet of the capillary is presented in Figure 2.1.

Another low-specificity chromatography concentration strategy, developed to overcome the problems encountered with SPE cartridges, is membrane preconcentration (mPC) using cartridges impregnated with an organic solvent or stationary phase. The cartridge is fitted at the inlet of a normal CE capillary, and the setup does not experience the

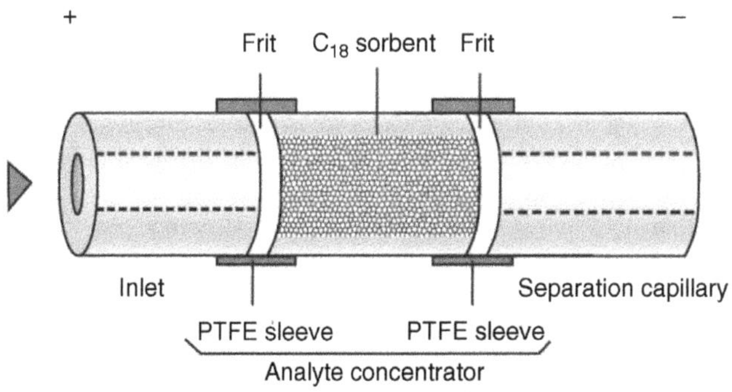

FIGURE 2.1 Representation of a C_{18} solid-phase extraction preconcentration chamber fixed at the capillary inlet using two frits to retain the stationary phase. The two capillaries, the frits, and the preconcentration chamber are held together by PTFE sleeves. Adapted from Ref. 29 with permission.

back-pressure caused by frits as in conventional SPE [43]. The membrane can act as a cleanup and concentration device at the same time, retaining target analytes while eluting biological matrix components or chemical reagents [44]. This method allows the injection of more than 100 μL of analyte solution [45].

By using liquid–liquid extraction techniques, chromatographic concentration of target analytes can also be achieved. In electroextraction (EE), voltage is used to extract charged analytes from organic solvents, while counterflow pressure is applied to avoid solvent injection, as shown in Figure 2.2 [46]. The driving force of the concentration strategy is the very high electric field present in aqueous organic solvents such as ethyl acetate in comparison with the aqueous buffer in the rest of the capillary. In this enhanced field environment, charged analytes of opposite applied polarity are strongly directed to the inlet of the capillary [47].

Supported liquid membranes (SLMs) to enhance cleanup and separation of analytes from the sample matrix are widely used, with distinct setups depending on the kind of target compound. The most basic method consists of two phases of aqueous buffer with different pH values in which the analyte has different charge. Both solutions are inserted in the concentration chamber in counterflow, having a porous membrane impregnated with organic solvent separating the two liquid phases, but allowing the migration of neutral species. As neutral species diffuse through the membrane to the acceptor buffer, they immediately become charged, staying in the acceptor phase, becoming concentrated as more of the neutral analytes are absorbed [48]. Several setups can be obtained, using inexpensive products such as an improvised PFTE membrane impregnated with water-immiscible solvents [49] for application of miniaturized extraction and electrical forces to enhance exchange content [50].

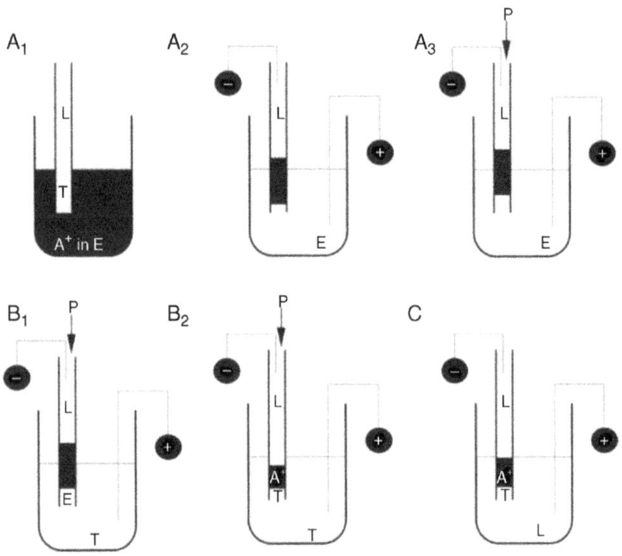

FIGURE 2.2 Representation of an EE–ITP–CZE preconcentration experiment in which the analyte is injected electrokinetically through application of voltage in the sample dissolved in organic solvent. (A_1) The capillary filled with leading buffer (L) and terminating buffer (T) is placed in the ethyl acetate (E) solution containing the cationic analyte molecules (A^+). (A_2) Electroextraction (EE) is performed by applying a voltage. (A_3) During EE, using a pressure-induced counterflow (P) in the direction of the arrow, the sample solution is electroextracted. (B_1) The analytes are focused by means of pressure-induced counterflow ITP, after the capillary has been placed in the terminating buffer vial. (B_2) During ITP, the remaining ethyl acetate plug is pushed out of the capillary, while reaching the steady state. (C) The capillary is placed in the leading buffer vial and CZE starts. Adapted from Ref. 47 with permission.

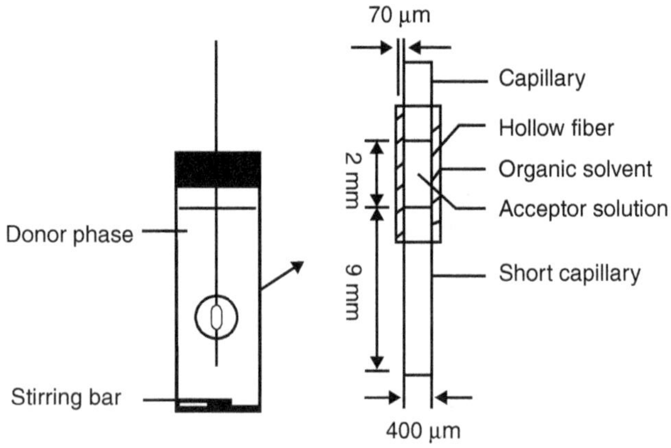

FIGURE 2.3 Schematic representation of a liquid–liquid–liquid microextraction using an impregnated hollow fiber to extract the analytes into the separation capillary. Adapted from Ref. 25 with permission.

Use of an impregnated hollow fiber followed by a back-extraction can further enhance separation, generating a liquid–liquid–liquid microextraction (LLLME) setup, as presented in Figure 2.3 [25]. In this way, a hollow fiber impregnated with an organic solvent is fixed at the junction of two capillaries, promoting the extraction of neutral analytes from the aqueous sample matrix to the organic layer. A second aqueous buffer, in contact with the hollow fiber from the inside of the CE capillary having a pH that will cause the analyte to be protonated, will back-extract the analyte in its charged form, preventing it from being released again through the fiber. The fiber may be used to promote size exclusion in addition to solubility blocking.

2.4 METHODS BASED ON ELECTROPHORETIC MOBILITY AND VELOCITY MANIPULATION (ELECTROPHORETIC METHODS)

Capillary electrophoresis is a very useful technique for the analysis and separation of many compounds, ionic or ionizable, owing to its high separation efficiency. The short optical path length of the capillaries, however, makes detection based on concentration very poor, becoming an inappropriate technique for many practical applications. To overcome this limitation, sample stacking was developed, compressing analytes into narrower bands, thus improving detection limits. Some methods, based on the difference in electrophoretic mobilities of ions, can stack the sample inside the capillary before the separation [33,51].

The concentration (stacking) techniques use discontinuous buffer systems to vary charge and electric field strength, causing a change in the electrophoretic velocity of analytes, concentrating the sample into narrower bands. The buffer discontinuity can be achieved by simple changes in the experimental setup, such as changing the sample conductivity or pH, distinct from the background electrolyte [8,17,51,52].

2.4.1 Field-Enhanced Sample Stacking (FESS)

Field-enhanced sample stacking, or more commonly field-amplified sample stacking (FASS), although the use of this nomenclature is not to be recommended because the term "field amplified" is not philologically a suitable term for electric field [16], is the simplest electrophoretic preconcentration strategy employing hydrodynamic injection, applying pressure at the inlet or vacuum at the outlet. It allows a larger than normal injection volume in CE without ruining the separation efficiency, lowering the detection limit 10–100-fold [2,16,31,33,53].

In FESS, both the BGS and sample solvent have the same composition, but the sample matrix is prepared in a diluted solution that is, with lower electric conductivity than the run buffer. After the conditioning of the capillary with buffer and injection of a large sample plug, voltage is applied, and the region with low ion concentration will present a higher resistivity (ρ) than the rest of the capillary, according to Equation 2.2:

$$R = \rho \frac{L}{A} \tag{2.2}$$

where R is the resistance of the medium, L is the capillary length, and A is the capillary cross-section. As L and A do not change during the application of electric current, an increase in ρ will increase the resistance of the solution in the sample plug. This will cause a rise in the electric field at the sample zone, following Equation 2.3:

$$E = Ri \tag{2.3}$$

where E is the electric field strength and i is the electric current, kept constant, which in turn will cause the analytes to undergo an increase in their electrophoretic velocity (v_{ep}), since it depends on the electric field strength according to Equation 2.4:

$$v_{ep} = \mu_{ep} E \tag{2.4}$$

where μ_{ep} is the electrophoretic mobility of the analyte. When ions reach the interface between the sample zone (high E) and the run buffer (low E), they will be slowed, causing shrinkage of the sample plug, providing a sample zone with higher concentration than before. The electric field becomes homogeneous and the sample is separated following normal CE principles.

Assuming that the capillary diameter is constant and homogeneous, the ratio between the electric field of the sample plug (E_s) and the electric field of the rest of the capillary (E_c) is quantified by

$$\frac{E_s}{E_c} = \gamma \tag{2.5}$$

where γ is the enrichment factor. As the electric field is directly proportional to resistivity and inversely proportional to the buffer concentration, we have

$$\frac{C_{ss}}{C_a} = \gamma \tag{2.6}$$

where C_{ss} is the concentration of analyte after sample stacking and C_a is the concentration of the analyte in the sample plug injected. Equation 2.6 shows that the analyte concentration will increase by a factor γ, with reduction of band length. According to this equation, stacking efficiency would always be higher in samples dissolved in water than in those dissolved in diluted buffer solutions, but this is not always observed. Some band broadening arises from the difference in electroosmotic velocities between zones of low and high electric field, generating an electroosmotic pressure at the interfaces of both zones, resulting in a laminar flow [3,16,53].

The use of field-enhanced sample injection (FESI) to enhance detection has been demonstrated by several authors and recently by Danger et al. [54], who used FESI to preconcentrate α,α-dialkylated amino acids, obtaining an enrichment efficiency of 40-fold, using ACN to dilute the sample. The use of ACN increased the matrix resistance and consequently an increase in ion velocity was observed.

2.4.2 Field-Enhanced Sample Injection (FESI)

Introduced in 1991 by Chien and Burgi [12], FESI (sometimes also referred to as FASI, although this term is also not philologically suitable as mentioned in Section 2.1) is a preconcentration technique with the same basic principle as FESS. The sample is dissolved in a low-conductivity buffer or even water and analyzed using a higher-conductivity solution as BGS. The difference lies in the mode of injection and focusing of ions. In FESI, electrokinetic injection is used and the focusing occurs at the time of injection.

The low conductivity of the sample solution increases the quantity of ions introduced in the capillary, caused by the increase in the velocity of the ions, as in FESS. In this way, ions are focused in a narrow zone in the BGS, which has a high conductivity, reducing the velocity of the ions. Cations and anions can be separated by simply switching polarity. For this to happen, the EOF must be kept low or even absent because a high EOF would allow only ions with mobility in the same direction as the EOF to be introduced [12,55]. Simply injecting a water plug before sample injection can resolve this problem. As the water plug has a lower electroosmotic velocity than the sample ions and also a lower conductivity than the BGS, there is an enhancement in the injection of the sample ions, even in the presence of EOF [55].

This technique has been used in the analysis of acrylamide in food products [56]; in this case, acrylamide was derivatized with 2-mercaptobenzoic acid and an LOD near $1\,\mu g\,mL^{-1}$ was achieved. Another example of the use of this method was reported by Law et al. [57], who separated basic proteins from water using a poly(vinyl alcohol) (PVA)-coated capillary and laser-induced fluorescence (LIF). They obtained an enrichment factor of ~1000-fold based on the field-enhanced sample injection [57]. Other examples have been described [58–61].

FESI is an important preconcentration technique, having several factors that can affect its performance, for example, (i) the size of the water plug—it cannot be too short so as to avoid sample overload and not too long as it can cause peak broadening and loss of efficiency separation due to the excessive time to remove the water plug, (ii) the length of the injection time—an excessively long injection time leads to band broadening due to EOF pressure, generating laminar flow, whereas a short plug leads to a decrease in enrichment factor due to incomplete use of stacking capacity [2,12,55,58].

2.4.3 Large-Volume Sample Stacking (LVSS)

Injection of large amounts of sample in sample stacking is limited because, as previously discussed, band broadening due to incomplete stacking can occur and also the sample enrichment factor is not always exactly proportional to the ratio between sample and buffer electric fields [53,55,62]. Facing this impasse, Chien and Burgi [62] developed a method to perform sample stacking after the injection of a large volume of sample.

In general, large-volume sample stacking starts with dissolution of the sample in water or diluted buffer and hydrodynamic injection of sample filling from one-third up to the entire capillary. The preconcentration is mainly based on the difference between the electrophoretic velocities of ions in sample and the BGS electroosmotic velocity. After sample injection, in this example the anionic analytes, negative polarity is applied. Ions are accelerated due to the low resistivity of the sample solution toward the oppositely charged electrode. Sample anions migrate toward the positive polarity electrode and stack at the interface of the injection zone and BGS. While EOF is directed to the negative pole, it increases the overall migration speed in order to remove excess sample solution and neutral analytes. Cations are removed from the capillary because they also migrate to the BGS in the inlet vial. When sample solution has been almost completely removed from the capillary, noted by the current reaching 97–99% of the normal current under separation, the polarity is switched and analytes are rapidly separated and eluted. To analyze cations, both the EOF and current polarity must be switched, in such a way that only cations or anions can be separated in a single run by LVSS [16,62,63].

Another way of performing the method without the need to switch the polarity is to carry it out in a low EOF and with high-mobility ions, as they migrate in the direction opposite to the EOF. Quirino and Terabe [64], using this technique, analyzed bromide, nitrate, and bromate, reaching an enrichment of 100-fold compared with normal hydrodynamic injection without stacking.

Chen and Naidu [65] analyzed the formation of an anionic complex of Fe^{2+} and Fe^{3+} with 2,6-pyridinedicarboxylic acid. Analytes were separated by CZE using direct UV detection at 214 nm. The detection sensitivity was increased over 50-fold compared with that obtained with conventional hydrodynamic injection [65]. Other examples of the use of LVSS with and without polarity switching have been reported [63,66–71].

2.4.4 Dynamic pH Junction

All electrophoretic preconcentration strategies described so far, FESS, FESI, and LVSS, are carried out in sample matrices of low ionic strength. However, this represents a great limitation for complex matrix samples, such as biological fluids, which normally have a high content of salts [72]. This high sample ionic strength leads to spreading of analytes in the BGS, due to a higher ion velocity [73].

To overcome this drawback without the need for further sample dilution or even the use of dialysis or other preliminary extraction for removal of excess salt, several authors have reported the use of techniques in which the stacking of analyte is based on changes in the pH of the sample and BGS.

In general terms, pH values of both the sample and the BGS are selected in order to increase the mobility of analytes, based on their pK_a to weak acids or bases or on their isoelectric point (pI) to zwitterionic compounds such as peptides or proteins [74].

In the analysis of anions, the analyte is solubilized in an acidic matrix, where it is maintained in its protonated form. The solution is injected into a capillary already filled with basic BGS. Upon application of a positive polarity mode, the acidic sample zone is gradually titrated by hydroxide anions from the basic buffer. When the analytes come in contact with the titrated zone, they are deprotonated and acquire charge, tending to move backwards to the anode. When they reach the sample zone that has not yet been titrated, they are again protonated, losing their charge and moving only under the influence of the EOF, and in this way all samples become stacked at the boundary of the titrated and nontitrated zones [16,74,75].

One of the first applications of this technique used a buffer system with different pH values for the analysis and separation of zwitterionic peptides. The sample was diluted in a solution with a pH higher than the isoelectric point of the peptides, leaving them negatively charged. The sample was then injected into the capillary already filled with low-pH BGS. Using this procedure, LODs lower than 1 ng mL^{-1} were obtained [76].

Dynamic pH junction has been used to improve the detection of a wide range of compounds in CE analysis [74,77,78]. More recently, Su and Yu [79] used this technique for the preconcentration of the recombinant peptide Arg–Gly–Asp–hirudin in urine samples. They managed to achieve an ~100-fold increase in the sample enrichment factor, without pretreatment of the complex sample.

Another way of performing this stacking technique, also using pH variation in solutions, named “pH-mediated stacking” by Arnett and Lunte [80], has been well described by Zhao and Lunte [73]. They demonstrated the benefit of using stacking in the analysis of four organic acids. Under optimal conditions, they obtained sample enrichment factors of ~100-fold compared with normal CE with hydrodynamic injection [73].

The mechanism of the method proposed by Zhao and Lunte is based on dissolution of analytes in a concentrated Ringer solution[1] of a chosen pH. They used a run buffer consisting of a salt of a weak base, such as ammonium (NH_4^+); however, the mobility of both the analyte and EOF must be in the same direction, both running toward the detector end. Sample can be inserted in the capillary through hydrodynamic or electrokinetic injection. Next, a strong base is injected, in such a way that hydroxide anions, having a greater mobility, can pass through the sample plug, neutralizing the NH_4^+ in the BGS. A neutral low-conductivity zone is created inside the capillary, where the enhancement of field and stacking of analytes occur before separation by CZE. Other examples of applications have been described [80–83].

In addition to the pH values of the sample and BGS, other factors such as ionic strength, time, and mode of sample injection can have a great influence on the sample enrichment factor. Arnett and Lunte [80] demonstrated that the combination of electrokinetic injections, for both the sample and the NaOH solution, was efficient because a smaller capillary length was used to perform the stacking before separation, improving resolution and peak shape.

[1] Ringer solution: 155 mM NaCl, 5.5 mM KCl, and 2.3 mM $CaCl_3$, pH 7.4.

2.5 SAMPLE STACKING IN PSEUDO-STATIONARY PHASES

Online preconcentration methods can also be applied for the separation of neutral analytes using similar procedures in micellar electrokinetic chromatography (MEKC). Although more details can be found in Chapter 1, the technique involves the use of a surfactant above its critical micelle concentration (CMC) as the pseudo-stationary phase (PS), in which there is a dynamic equilibrium of molecules of surfactant between its individual and aggregated structures. The micelle interacts with neutral and charged analytes to some extent, and is this degree of interaction that promotes the separation process, thus turning neutral analytes indirectly sensitive to applied voltage [84]. Thus, separation of neutral molecules is achieved, based on the partition coefficients of analytes between the mobile phase and the PS and the electrophoretic migration of micelles to the capillary.

As in other types of CE, a major drawback, especially when related to the use of UV detection, is the small injection volume and short optical path length, resulting in poor sensitivity. To overcome this obstacle, online preconcentration strategies were developed in order to improve resolution and sensitivity, described in several reviews focused on fundamentals and applications [8,16,84–87]. There are several experiments that can enhance detection, based on either hydrodynamic insertion of longer sample volumes, using micelles to stack the analytes in thin bands, or types of preconcentration that involve the formation of micelles at the sample inlet and the use of voltage to inject and compress the analytes electrokinetically into the capillary [87].

The mostly used techniques can be split into two groups, namely (i) stacking using a field-enhanced zone to promote a variation of the micelles' speed—which can be further separated into hydrodynamic injection and electrokinetic injection—and (ii) sweeping—which is applicable when there is a homogeneous electric field throughout the capillary [8].

2.5.1 Field-Enhanced Sample Stacking

Stacking using the difference in mobilities of charged molecules between two regions in the capillary, which presents different conductivity, can be achieved using several experimental setups. The concentration step occurs at the boundary between these two phases, where the differential mobility causes the micelles to stop, focusing the molecules that have been carried away. The stacking may occur inside the capillary after the hydrodynamic injection or conducted at the time of injection with electrokinetic injection. Table 2.1 describes the field-enhanced sample stacking strategies and some of their characteristics.

TABLE 2.1 Conditions of Sample Stacking Modes Over Discontinuous Electric Field in MEKC

Mode[a]	Sample Matrix	Water Plug[b]	pH (BGS)	EOF[c]	Polarity
Hydrodynamic injection					
NSM	Low conductivity (nonmicellar)	N	Basic or neutral	Cathodic (strong)	+
REPSM	Low conductivity (nonmicellar)	N	Basic or neutral	Cathodic (strong)	$- \rightarrow +$[d]
SRMM	Low conductivity (nonmicellar)	N	Acidic	Cathodic (weak)	−
SRW	Low conductivity (micellar)	Y	Acidic	Cathodic (weak)	−
Electrokinetic injection					
FESI	Low conductivity (micellar)	Y	Basic or neutral	Cathodic (strong)	$- \rightarrow +$[e]
FESI–RMM	Low conductivity (micellar)	Y	Acidic	Cathodic (weak)	−

[a]Each mode is explained in details in sections 2.5.2 and 2.5.3.
[b]A water plug is injected into the capillary at the inlet end after conditioning the capillary with BGS.
[c]EOF is suppressed under the acidic conditions.
[d]Polarity is to be switched when current reaches close to 97% of the original value.
[e]Electrokinetic injection with negative polarity, then separation voltage is applied at positive polarity.
Source: Adapted from Ref. 84.

2.5.2 Hydrodynamic Injection Techniques

Hydrodynamic injection can be performed by applying positive pressure at the sample inlet vial or negative pressure at the outlet vial, with the former being the most commonly used. There are several ways to achieve the stacking after the pressure injection of solutions into the capillary, as described next.

2.5.2.1 Normal Stacking Mode (NSM) The normal stacking mode (NSM) is the simplest stacking technique involving micelles, using the variation of the migration speed of the pseudo-stationary phase in two different environments to promote the stacking of analytes at the boundary between the sample zone and the BGS [88]. The stacking allows the capillary to be filled with a larger sample plug than usual, dissolved in a low-conductivity matrix. The capillary is already conditioned with BGS containing an appropriate concentration of surfactant. The sample inlet is replaced with a vial containing BGS; with the application of a positive separation voltage, the anionic micelles move toward the anode, entering the sample plug, and start to move faster, influenced by the lower-conductivity zone. As the micelles move and reach the other side of the sample zone, they slow down, bringing the analytes together into the boundary between the sample zone and the BGS. After stacking has occurred, separation proceeds as a normal MEKC. The NSM with cationic surfactants is also possible; the only difference being the use of the opposite voltage to generate a reverse EOF.

2.5.2.2 Reverse Electrode Polarity Stacking Mode (REPSM) In the reverse electrode polarity stacking mode (REPSM), a great advantage is that the plug inserted may be much longer than that used in the NSM because the band broadening effect is minimized by removal of sample matrix. The whole stacking method involves three distinct steps: filling of the capillary with sample, the stacking process together with matrix removal, and normal separation [89]. In this mode, the capillary is filled with alkaline or neutral BGS (with EOF higher than micelle mobility)-containing micelles, followed by insertion of a large sample plug dissolved in a low-conductivity matrix. The sample inlet is replaced by a vial containing BGS, and voltage is applied in the negative polarity mode. At the same time, the EOF pushes the sample matrix out of the capillary and the micelles in the cathodic reservoir start to migrate in the opposite direction, passing through the sample matrix with enhanced speed due to the stronger electric field, and stack the analytes at the boundary between the sample and the BGS, where they are forced to slow by the increase of conductivity. The current starts to increase as the sample matrix is removed, and it is monitored until it reaches 97–99%, indicating that almost all of the sample plug was pumped out of the capillary, leaving the analytes stacked. At this time, the polarity is switched, and the separation is performed as in regular MEKC. This mode has the advantage of allowing the introduction of a large volume of sample in the capillary. In turn, it can be very difficult to maintain reproducibility unless current control is applied very rigorously [90].

2.5.2.3 Stacking with Reverse Migrating Micelles (SRMM) The use of stacking with reverse migrating micelles (SRMM) requires an acidic BGS, making the mobility of micelles higher than the EOF. In this way, the separation does not involve the switching of polarity during separation [91]. Briefly, a large volume of sample in a low-conductivity matrix is inserted in the capillary already filled with acidic buffer solution. As the polarity is switched to the negative mode, the micelles from the cathodic vial migrate toward the anode, passing through the sample and pushing the analytes with them until they reach the boundary between the sample and the buffer zones, slowing and stacking the analytes. At the same time, the weak EOF pushes the sample matrix slowly out of the capillary, filling it with the buffer solution at the anode. After focusing at the band, the micelles continue to separate the analytes according to their partition coefficients. The EOF removes the entire sample matrix from the capillary, and the separation proceeds as normal MEKC. The advantage of SRMM is that the focusing step and the removal of sample matrix occur without the need for careful monitoring and switching of current, avoiding possible sample loss due to excess matrix removal. The low speed is also a positive point for analytes with low or medium partition coefficients [92].

2.5.2.4 Stacking Using Reverse Migrating Micelles and a Water Plug (SRW) In stacking using reverse migrating micelles and a water plug (SRW), as in other reverse migrating micelle methods, the EOF is suppressed by the use of an acidic buffer or coating of the capillary, so the micelle mobility becomes higher than the EOF [93]. Sample is prepared in a low-conductivity matrix, containing surfactant in a concentration that is slightly higher than the CMC, thus improving the solubility capacity of the sample matrix. In this way, more concentrated solutions of slightly soluble analytes can be analyzed. The modification of this method compared with SRMM is the insertion of a water plug after the buffer and before the sample plug. This water plug is responsible for the low-conductivity zone and higher electric field, promoting stacking [94]. The capillary is filled with acidic BGS, a water plug, and the sample plug, which can be fairly large. The voltage is turned on in the negative polarity mode; the micelles from the cathodic vial and the sample plug start to migrate to the anode, and stacking is promoted at the interface between the water plug and the acidic BGS, due to the great difference in conductivity. The slow EOF keeps pushing the sample and the water plug out of the capillary to the inlet, while the focusing of the analytes in the focused band starts to change, forming bands as the analytes are pushed through the capillary by the micelles.

2.5.2.5 High-Conductivity Sample Stacking (HCSS) In high-conductivity sample stacking (HCSS), described for the first time by Palmer et al. [95], a high concentration of salt is used in the sample matrix in order to increase the conductivity of the solution. Using a micellar buffer with weaker ionic strength than the sample matrix will cause the discontinuity of the conductivity inside the capillary to promote acceleration and stacking of micelles at the interface of the two solutions [96]. Subsequent stacking of analytes and destacking of micelles are proposed as some of the mechanisms involved in the stacking process [97]. To promote stacking, the parameters given in Equation 2.7 are necessary:

$$\mu_{\text{sample}} E_{\text{sample}} < \mu_{\text{EV}} E_{\text{EV}} \tag{2.7}$$

where μ_{sample} is the electrophoretic mobility of the sample matrix coion, μ_{EV} is the mobility of the electrokinetic vector, in this case the micelles, and E_{sample} and E_{EV} are the electric field strength in the sample and separation solution, respectively. To have efficient stacking, the mobility of the coion present in the sample matrix must be higher than the mobility of the micelle, in order to prevent the micelles from entering the high-conductivity zone because $\mu_{\text{sample}} E_{\text{sample}}$ is always greater than $\mu_{\text{EV}} E_{\text{sample}}$. Stacking of micelles occurs first which in turn promotes the indirect stacking of neutral analytes as the high-conductivity zone moves toward the anode [98].

First, the capillary is filled with neutral or basic BGS in order to achieve a fast EOF. Then, the sample dissolved in a high-conductivity matrix is inserted using hydrodynamic injection. The voltage is turned on in the positive polarity mode and the micelles migrate at high speed from the cathode vial, slowing and stacking at the boundary between the low-conductivity BGS and the high-conductivity sample matrix, while the EOF pushes the sample plug. The higher-mobility coion present in the sample matrix starts to migrate toward the anode, leaving the neutral analytes exposed to the stacked micelles. Separation is then driven by the EOF to the detector as a normal MEKC. As a more specific example, Figure 2.4 illustrates high-conductivity stacking using

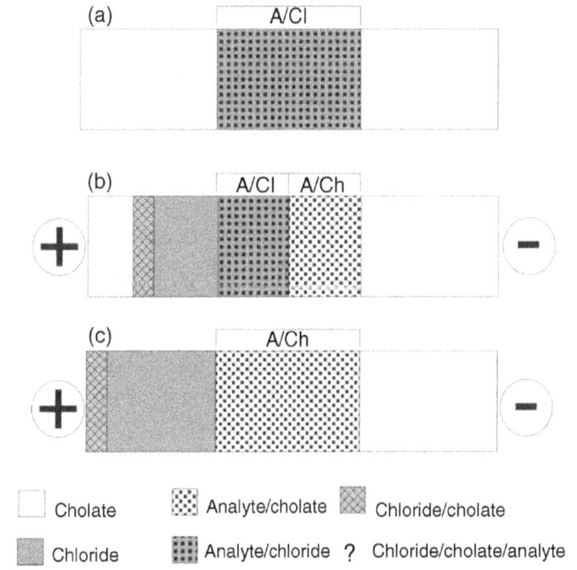

FIGURE 2.4 A schematic illustration of the movement of anionic species under high-salt sample stacking conditions. (a) The high-salt sample matrix is illustrated in a dark shade in the middle of a capillary. The micellar separation buffer are the regions on either side of the sample zone. Neutral analytes are depicted as open circles. (b) Electrophoresis is initiated with the anode to the left. The movement of the anodic components is shown. Chloride moves to the left, and diffuses into the cholate on the left side. On the right side of the chloride region, the lower-concentration cholate will stack without overtaking the higher-mobility chloride region. (c) Neutral analytes partition into the micelles emerging from the chloride region into the cholate region. The question mark points out the inexistence of a region containing simultaneously analyte, chloride, and cholate. Adapted from Ref. 98 with permission.

sodium chloride as the salt to increase ionic strength of sample and sodium cholate as the micelle.

2.5.3 Electrokinetic Injection Techniques

Field-enhanced sample injection is a common technique used for the concentration of ionic analytes, using an applied voltage at the inlet vial to introduce ionic species electrokinetically inside the capillary. As neutral analytes are not affected by this applied field in a normal environment, it is necessary to add surfactant above its CMC in the sample, forming micelles that interact with analytes now in a charged environment, being transported inside the capillary when polarity opposite to that of the micelles is applied.

2.5.3.1 Field-Enhanced Sample Injection (FESI–MEKC) In field-enhanced sample injection (FESI), a capillary is filled with buffer (neutral pH) with EOF greater than the micelle mobility, and just after the buffer a large plug of water is introduced. The sample dissolved in a low-conductivity matrix together with surfactant above the CMC is positioned at the inlet and negative polarity is applied. The EOF pushes the water plug out of the capillary, but the analytes dissolved in the anionic surfactant enter the capillary driven by the negative net charge of the analyte involved by micelles. The enhanced electric field of the water plug that generates electrophoretic mobility higher than the EOF brings the micelles together with analytes to the boundary between the water plug and the buffer, where they slow down because of the abrupt change in the electric field and focus into a narrow band. The current is monitored until it reaches 97–99% of the total, indicating that the water plug has been almost entirely removed from the capillary by the EOF. Then, voltage is turned off, and a BGS vial replaces the sample vial and polarity is switched to the positive mode. With voltage turned on again, separation proceeds driven by the EOF, as in the regular MEKC [99].

2.5.3.2 Field-Enhanced Sample Injection with Reverse Migrating Micelles (FESI–RMM) Field-enhanced sample injection with reverse migrating micelles (FESI–RMM) differs from the former method in that the buffer used here is acidic, rendering a slower EOF than the electrophoretic mobility of the micelles [100]. The capillary is filled with acidic buffer and a large water plug. The sample is dissolved in a low-conductivity matrix with surfactant above the CMC placed in the inlet, and voltage is applied in the negative polarity mode, forcing the water out of the capillary inlet through a small but existing EOF; the micelles migrate together with the analytes through the sample plug toward the anode, stacking on the boundary between the water and the acidic buffer. The current should be monitored until it reaches 97–99% of the total current, indicating the removal of almost the entire water plug. At this time, the voltage is turned off, the sample vial is replaced with a vial containing acidic buffer, and the voltage is restarted in the same polarity mode. Separation then proceeds as in normal MEKC.

2.5.4 Sweeping

Sweeping is the most promising stacking method in MEKC, based on the phenomena of interaction and transportation of analytes from a sample matrix devoid of PS by micelles contained in the BGS; it was developed by Quirino and Terabe in 1998 [101]. The technique can be performed in conditions where the EOF is stronger than the micelle mobility, named sweeping with reverse migrating micelles (RM-sweep) or sweeping with normal migrating micelles (NM-sweep) in EOF-independent conditions, using an acidic buffer or coated capillary [102].

The sample is usually prepared in the same matrix as the buffer, without surfactant. In this way a homogeneous electric field in the entire capillary solution causes the micelles to move with constant velocity. When there is a heterogeneous environment, micelles have different mobilities and hence the stacking follows other parameters, as in high-salt sample stacking [86,103].

The stronger is the interaction between micelles and analytes—evaluated by the retention factor k—the better is the separation and the detection achieved. According to this statement, one can think of the length of the swept zone of analytes represented as [104]

$$l_{\text{swept}} = l_{\text{inj}}\left(\frac{1}{1+k}\right) \tag{2.8}$$

From Equation 2.8, it can be deduced that a high k would result in bands more concentrated and sharper, improving resolution and detection. For an experiment with an acidic run buffer [84], the capillary is filled with BGS and then sample in matrix is dissolved in buffer or another solution with a conductivity similar to that of the buffer. With the voltage applied in the negative polarity mode, the micelles migrate through the sample plug and pull the analytes together, forming well-defined bands according to their partition coefficients. As the micelle mobility is higher than the EOF, separation proceeds in the same way as in SRMM.

The use of nonionic surfactants is also possible, but only for separation of charged analytes in an EOF-suppressed background. As the micelles do not have electrophoretic mobility, the sample region passes through a micellar BGS, where separation of charged analytes occurs through different partition coefficients between the nonionic micelle and charged analytes [105].

There is some discussion about whether high-salt MEKC stacking is a variation of sweeping [96,104,106]. To date, there are several pieces of evidence distinguishing each one;

however, further studies are still necessary to elucidate proper explanations and mechanisms of both methods [16].

2.5.5 Combined Techniques

It is possible to use more than one preconcentration strategy at a time to enhance the detection sensitivity further in CE. Not all techniques are capable of being hyphenated; usually a method is used to enhance electrokinetic injection and another to compress the analytes into thin bands.

2.5.5.1 Dynamic pH Junction: Sweeping The use of a dynamic pH junction to concentrate weakly ionic analytes that have their mobility changed as a function of pH can be enhanced by a sweeping step. First, the sample devoid of PS and with pH different from that of the micellar BGS is inserted in the capillary. The experiment setup is designed to promote neutralization of analytes at the junction, ceasing their mobility. As the difference in pH experienced at the junction of the sample and BGS influences the mobility of analytes, they become stacked, and further concentration is achieved by interaction with micelles from the BGS that sweep the sample plug.

2.5.5.2 Selective Exhaustive Injection (SEI) Selective exhaustive injection (SEI) is an electrokinetic injection performed in the FESI mode for a long period (up to 1000 s); hence, it might even deplete the sample vial, injecting virtually all of the vial contents into the capillary. As this technique is based on electrokinetic injection, only one kind of charged analyte can be injected at a time. When performed for cation injection, it is named cation-selective exhaustive injection (CSEI) [107], and for anions it is called anion-selective exhaustive injection (ASEI) [108]. Although it produces a large injection volume with this operation, the long time involved causes band broadening and so the use of another technique is needed. SEI can be used in the concentration steps of SRMM to insert more ions into the sample zone while the sample plug is being removed, thus increasing the enhancement factor [109], or to promote a concentrated electrokinetic injection before sweeping, where the highest enhancement factors reported in CE were obtained, almost 10^6-fold [110].

2.5.6 New Techniques

The ability to work not only with voltage but also with properties of micelles and electrolytes in samples and run buffers has turned attention to new stacking techniques. Analyte focusing by micelle collapse (AFMC) was first described by Quirino and Haddad in 2008 [111], and it is based on the capture, accumulation, transport, and release of analytes by anionic micelles contained in the sample matrix, together with a high electrophoretic mobility coion. This is achieved through micellar electrokinetic dilution, in which the surfactant micelles are continuously diluted and collapsed into a liquid-phase zone that follows the Kohlrausch regulating function (KRF) [111].

Another technique of online concentration evolved while tests of AFMC having different electrolytes in the composition were being performed. High-mobility coions may cause a micelle-mediated isotachophoresis (MM-ITP) on one side of the capillary, thus enhancing stacking of analytes, especially those with high retention factors. As both techniques involve similar experimental conditions, care must be taken as both methods can produce two peaks caused by different preconcentration strategies acting at each end of the sample plug, degrading the detection enhancement [112].

A different approach by Quirino in 2009 [113] introduced a novel stacking mechanism called micelle to solvent stacking (MSS), based on the changing of the electrophoretic mobility of weakly ionic analytes diluted in the micellar samples at the boundary with a run buffer with organic modifier, creating a micelle at the solvent-stacking boundary zone between the two phases. The presence of a trapping solution between the sample and the BGS, and also a nonionic surfactant in buffer, was tested as a method for optimizing the stacking efficiency [114].

Table 2.2 presents examples of most of the methods cited in this chapter, their detection mode, the kind of analyte screened, the limits of detection obtained, and the selective enhancement factor (SEF) related to peak area, height, or LOD.

2.6 STACKING TECHNIQUES IN MICROCHIPS

The development of microfabricated devices (microchips) has attracted growing attention during the last decade. Miniaturization of analytical systems allows the realization of chemical analysis using smaller quantities of samples, faster separation times, and, at the same time, reduced usage of reagents, solvents, and power [131–134].

In addition to the benefits of CE, already described, these devices also allow one to analyze/resolve all kinds of analytes, due to their applicability in different separation modes. Nevertheless, they also bring some of the intrinsic problems of CE such as the small optical path length and the difficulty of performing analyses of small-sized diluted samples [134,135]. Due to these drawbacks and the difficulties in performing modifications in the transversal section of the microchannels, sample preconcentration techniques may provide ways to overcome the obstacles and therefore concentrate analytes into small volumes, improving detection limits [134,136,137]. However, the adaptation of CE to microchip CE (MCE) is not that simple. The first problem is the exact control of sample loading, stacking, and separation procedures that demand precise changes in pressure and voltage to avoid problems such as diffusion [138,139].

TABLE 2.2 Examples Described in the Literature for Sample Stacking of Analytes Using MEKC

Technique	Mode	Analyte	LOD	Stacking Enhancement Factor (SEF)	References
NSM	CE–UV	Melatonin and related compounds	0.04 $\mu g\ mL^{-1}$	52.8[a]	102
NSM	CE–DAD	Fungicides	2 $\mu g\ mL^{-1}$	5–6[a], 5[b]	115
NSM	CE–UV	Antibiotics	0.3 $\mu g\ mL^{-1}$	5[a], 10[b]	116
NSM	CE–UV	2,3,7,8-TCDD and 2,3,7,8-TCDF	0.05; 0.34 $\mu g\ mL^{-1}$	330[a]; 65[b]	117
REPSM	CE–UV	Sildenafil citrate	17 ng mL^{-1}	11	90
REPSM	CE–DAD	Pesticides	0.05–1.7 $\mu g\ mL^{-1}$	—	118
REPSM	CE–DAD	Au nanoparticles	—	380	119
SRMM	CE-DAD	Herbicides	10–20 ppb	>50[a]	92
SRMM	CE–DAD	Pesticides	2.7–35 $\mu g\ L^{-1}$	5.5–15[c]; 2.7–35[b]	94
SRMM	CE–UV	Melatonin and related compounds	0.03 $\mu g\ mL^{-1}$	8.6[a]	102
SRMM	CE–DAD	Nonsteroidal anti-inflammatory drugs	7–25 $\mu g\ L^{-1}$	146–300[a]; 63–154[c]; 65–143[b]	109
SRMM	CE–UV	Antibiotic	0.3 $\mu g\ mL^{-1}$	4.3[a]; 6.3[b]	116
SRMM	CE–DAD	Triazole fungicides	1.2–4.0 $\mu g\ mL^{-1}$	10[a]	120
SRW	CE–UV	Dioxin compounds	0.1 ppm	200[a]	121
SRW	CE–DAD	Pesticides	5.3–46 $\mu g\ L^{-1}$	2.9–15[c]; 1.6–7.4[b]	94
SRW	CE–UV	Alkylphenols	0.009–0.09 $\mu g\ mL^{-1}$	121–193[a];71–211[b]	122
FESI	CE–DAD	Alkylphenols	—	5.4–65[a]; 4.1–18.1[c]	99
FESI	CE–DAD	Estrogens	—	60–106[a]; 14–23[c]	99
FESI	CE–DAD	Trihexyphenidyl	0.92 ng mL^{-1}	490[b]	123
FESI–RMM	CE–DAD	Alkylphenol ketones	—	2–20[a]; 2–18[c]	100
FESI–RMM	CE–DAD	Phenol derivatives	—	3–50[a]; 3–43[c]	100
FESI–RMM	CE–DAD	Steroids	—	70–106[a]; 41–78[c]	100
FESI–RMM	CE–DAD	Nonsteroidal anti-inflammatory drugs	40–75 $\mu g\ L^{-1}$	37–112[a]; 23–63[c]; 20–60[b]	109
HCSS	CE–LIF	Alkylphosphonic acids	12–17 ppb	—	124
HCSS	CE–DAD	Herbicides	0.13–2.73 $\mu g\ L^{-1}$	—	125
HCSS	CE–DAD	3-Nitrotyrosine	0.1 μM	1 (Without pretreatment)	126
Sweeping	CE–DAD	Pesticides	3.6–43 $\mu g\ L^{-1}$	2.6–19[c]; 2.5–18[b]	94
Sweeping	CE–UV	Saikosaponins	0.02–0.11 $\mu g\ mL^{-1}$	300[b]	127
Sweeping	CE–DAD	Fungicides	0.4 $\mu g\ mL^{-1}$	34–38[a]; 25[b]	115
Sweeping	CE–UV	Melatonin and related compounds	0.059 $\mu g\ mL^{-1}$ (NM–sweep); 0.036 $\mu g\ mL^{-1}$ (RM–sweep)	41.8 (NM–sweep); 11.9 (RM–sweep)	102
Sweeping	CE–UV	Steroids	2.5–3 $\mu g\ L^{-1}$	100[a]	128
Sweeping	CE–DAD	Hypolipidemic drugs	50 $\mu g\ L^{-1}$	60[b]	129
Sweeping	CE–DAD	Triazole fungicides	0.1–0.2 $\mu g\ mL^{-1}$	62–67[a]	120
SRMM–ASEI	CE–DAD	Nonsteroidal anti-inflammatory drugs	515 $\mu g\ L^{-1}$	103–270[a]; 105–263[c]; 99–250[b]	109
FESS–sweeping	CE–DAD	Hypolipidemic drugs	36 $\mu g\ L^{-1}$	80[c]	129
Dynamic pH-sweep	CE–LIF	Flavin derivatives	4.0 pM	1,200	130
CSEI–sweeping	CE–UV	Drugs of abuse	5–15 ng mL^{-1}	6,000	107
ASEI–sweeping	CE–DAD	Phenoxy acidic herbicides	100 pg mL^{-1}	100,000[c]	107
CSEI–sweeping	CE–DAD	Laudanosine and 1-naphthylamine	4.1–8.0 ppt	550,000–900,000	110
AFMC	CE–DAD	Hypolipidemic drugs	38 $\mu g\ L^{-1}$	83[c]	129
MSS	CE–ESI–MS	β-Blockers	0.01–0.03 $\mu g\ mL^{-1}$	15–16[b]	113
MSS	CE–UV	Alkaloids	16–26 ng mL^{-1}	113–123[a]	114

[a]Peak area.
[b]LOD.
[c]Peak height.

Jacobson and Ramsey [140] first reported the use of microchip sample stacking. They used FESS to obtain a stacking enhancement factor of 10–30-fold in the analysis of fluorescent dyes. FESS is a widely used technique in MCE, and it is based on the fact that ions that migrate from a low-conductivity solution to a high-conductivity solution have their speed drastically reduced at the limit of the two solutions, being stacked at the interface, as already discussed [31].

In 2001, Yang and Chien performed the analysis of two fluorescent dyes, also employing microchip FESS [141]. To promote EOF suppression, they used an internal microchannel wall coating of poly(allyl glycidyl ether *N*,*N*-dimethylacrylamide) (pDMA), and to prevent any residual EOF a small pressure was applied at the sample inlet; in this way, they were able to keep the concentration boundary at the injection channel. With this design, the authors demonstrated an enhancement of 100-fold in electrochemical signal, with a theoretical relative conductivity (TRC[2]) of 200. The experimental TRC, however, was about 140 due to impurities present in the BGS.

Zhang et al. [142] also proposed the analysis of mono- and trisulfated glycosaminoglycans, using a simple cross-shaped microchip. The microchip was made of poly(methyl methacrylate) (PMMA), having the internal walls covered with a dynamic coating of methylcellulose, to achieve EOF suppression and to avoid adsorption of analytes on the microchannel wall. The filling of channels with BGS was performed with the aid of a syringe, and sample was introduced electrokinetically by the application of 400 V at the sample inlet, while other reservoirs were grounded. During separation, a low voltage was applied between the sample well and at the sample waste well, while the BGS well was grounded and a high voltage was applied at the BGS waste reservoir.

An interesting fact reported by Zhang et al. was that even though pH did not interfere with resolution, it influenced peak height, probably due to modification of the molecules' π-bonding, thus changing the fluorescent properties, giving a higher signal at lower pH values. With this experimental setup, they achieved LODs ranging from 2.6×10^{-8} to $6.6 \times 10^{-8}\,\mathrm{mol\,L^{-1}}$ and efficiencies of 1×10^{6} plates per meter.

The sample stacking techniques described here used small injection volumes, bearing in mind that large volumes might reduce the separation efficiency, leading to poor resolution [62]. As a result, many researchers have looked for ways to overcome this problem.

In 2001, Lichtenberg et al. [135] developed a microchip with a long injection channel so that a large volume could be injected, thereby increasing the number of molecules to be stacked. The stacking and separation process starts with filling of the preconcentration microchannel by applying a voltage between the sample well and the "stacking waste" well and monitoring the current to verify when the microchannel is full. Next, the electric field is inverted, applying a voltage between the "stacking buffer" well and "sample waste" well, so that analyte ions can be dragged and stacked on the separation channel. Subsequently an electric field is applied between the buffer and buffer waste wells to complete the separation. It is noted in this experiment that the longer the preconcentration channel, the more the analytes can be stacked, which increases the sensitivity. Thus, the authors demonstrated a 65-fold increase in concentration.

More recently, Kawai et al. [143] performed ATP-derivatized oligosaccharides analysis with a single-channel microchip coated with PVA. The experiment is based on the large sample injection into the separation channel. However, because this procedure was performed in a single channel, it might generate low resolution and reproducibility due to the short separation channel, but this was not the case. After filling the channel with a low ionic strength sample by pressure, a voltage was applied, making the analytes stack at the sample–BGS interface by migration of BGS ions in the opposite direction to the analyte ions. After removal of most of the sample solution, the electrophoretic mobility of the analytes overcomes the EOF mobility generated by the low concentration and high pH of the sample solution, allowing the ions now to migrate, following normal separation, toward the detector [143]. In the analysis of oligomers of glucose, the authors noted that the longer the analyte, the lower was its mobility, which increased the efficiency of enrichment; this may be explained by the lower diffusion of the analytes, which corroborates the results reported by Jeong et al. [144]. Through this technique, which they named large-volume sample stacking with an electroosmotic flow pump (LVSEP)–microchip zone electrophoresis (MCZE), the authors achieved an enhancement of 2200–2900-fold in detection compared with that obtained with normal MCZE [143].

Although FESI techniques are very useful for preconcentration in microchips, they are vulnerable to the ionic strength of the sample solution, making it difficult to perform them in complex sample matrices such as biological fluids and high-salt concentration samples such as seawater. tITP–MCE has been shown to be a very interesting alternative to be applied to these kinds of matrixes [134,144,145].

In tITP, sample is introduced between the leading electrolyte (LE) and the terminating electrolyte (TE), where the mobilities of the analytes (A) are lower than that of the LE and higher than that of the TE. Under these conditions, sample ions are kept in narrow zones between the buffers and separated into successive and adjacent zones.

In 1998, Walker et al. reported the first use of ITP in MCE, where they performed the separation of the herbicides paraquat and diquat in glass microchips with Raman spectroscopic detection, obtaining LODs of 2.3×10^{-7} M [146].

[2] TRC is the conductivities of BGSs with high and low concentration ratios, measured separately in the microchip.

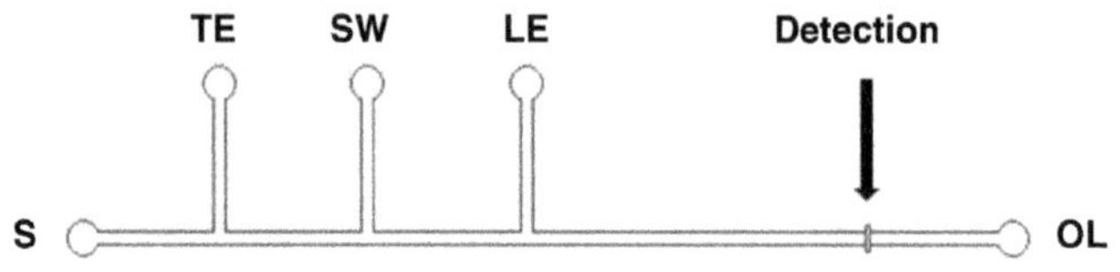

FIGURE 2.5 Design of a microchip CE for sample stacking. Buffer wells are sample (S), terminating electrolyte (TE), sample waste (SW), leading electrolyte (LE), and outlet (OL). The channels and wells SW, LE, and OL are filled with LE buffer; S and TE are filled with sample and TE buffer. Adapted from Ref. 136 with permission.

Under optimal conditions, Wainright et al. [136] obtained an enhancement of 400-fold in the detection of 13 eTags (fluorescent biological tags) of an untreated cell lysate solution. The design of the microchip used, shown in Figure 2.5 , allowed a reduction of errors caused by the electrokinetic injection [136]. They also noted that when the pH of the TE was increased, the stacking process occurred faster. However, as reported by Jeong et al. [144] and corroborated by Jung et al. [147], only the LE concentration interfered with the concentration of the sample stacking zone, as governed by Equation 2.9:

$$C_i = C_L \frac{\mu_i(\mu_L + \mu_{LC})}{\mu_L(\mu_i + \mu_{LC})} \tag{2.9}$$

where C_i is the concentration of sample in the stacking zone, C_L is the concentration of the LE, and μ_i, μ_L, and μ_{LC} are the mobilities of the analyte, LE, and coion, respectively.

Using a solution of HCl as LE and *N*-tris(hydroxymethyl) methyl-3-aminopropanesulfonic acid (TAPS) as TE, Jeong et al. [144] achieved a threefold improvement in the fluorescent peak height of fluorescein and 2,7-dichlorofluorescein; with HCl concentration in the range 100–250 mM, an increase of 200–500-fold in concentration was observed.

Wang et al. [148] reported the analysis of BSA (bovine serum albumin) and its immunocomplexes by tITP–MCGE with mAB (monoclonal antibody) in a simple cross-designed microchip of PMMA. They achieved an increment of more than 2000-fold in the concentration of the BSA immunocomplexes. The enhancement in the concentration of BSA diluted in water exceeded 20,000-fold [148].

Another microchip preconcentration technique using changes of analyte velocities is isoelectric focusing (IEF), where analytes are conducted to zones with the same pH of their isoelectric point (where the molecular analyte charge is zero) using a complex ampholyte solution. At this point of neutrality, analytes lose their mobility, being focused [149–151].

2.7 CONCLUDING REMARKS

This chapter described preconcentration strategies based on ionic analytes. For neutral analytes, sweeping-MEKC is fairly effective, and in 2001, Sera et al. [152] achieved enhancements of up to 1500-fold in fluorescence signal.

Another way to perform the preconcentration is to use chromatographic techniques, such as SPE), in which the analyte is separated between a liquid phase and the extraction material (solid phase), following the partition mechanism, depending on the extraction material used. Augustin et al. [153] reported the use of monoliths to perform SPE–MCE, obtaining a stacking enhancement factor of 200-fold in polycyclic aromatic hydrocarbons (PAHs). Following the reasoning of separation and concentration of analytes through partition mechanisms, Tokeshi et al. [154] reported a two-phase microextraction system using a microchip; here Fe^{2+} complexes were extracted from an aqueous solution using chloroform, in such a way that 7.7–180 zmol of the Fe^{2+} was detected in the integrated microextraction system.

The above-mentioned techniques come from the adaptation of CE to MCE; however, there are other processes that have been developed only for microchip preconcentration: temperature-gradient focusing (TGF) [155,156] and nanofluids filtration (NF) [157] are examples of developed methods used solely in microchips, in which enrichment factors of about 10^6-fold can be obtained.

The recent development of and progress in microchip techniques are largely due to the new technological applications, such as in the clinical analysis field, environmental analysis, and quality control (areas that need multiple and fast sample analyses) and chemical sciences [158]. This technology has numerous advantages, including reduced sample consumption and short analysis times. However, these microchip systems provide small sample detection volumes that can result in low sensitivity. Therefore, online sample preconcentration will remain a useful strategy to improve detection limits in microanalysis.

One of the major drawbacks of CE is the lack of sensitivity, especially with the most common and widely used UV absorbance detector. This is due to the limited sample loading capacity of capillaries, owing to their small diameter and length, which limits the total sample loading and the detection window path lengths [8,11,12]. To overcome such obstacles, useful methods to enhance detection limits in all modes of CE and MCE have been developed over the past two decades to promote efficient sample stacking inside the capillary prior to analysis. These strategies include chromatographic and electrophoretic approaches such as focusing agents (e.g., SPE, ITP, and FESS) [17,29,31,33]; it is also possible to use combinations of these focusing principles to maximize sample enrichment (e.g., selective exhaustive injection followed by SRRM or sweeping) [109,110]. With these strategies, enhancements of up to five orders of magnitude could be achieved compared with similar analyses without sample stacking [110]. Thus, it is possible to perform electrophoretic separations employing the UV–Vis absorption detection, commonly found in most commercial devices using these strategies, in cases where such applications would be impossible due to the low concentration of analytes.

Online preconcentration also allows sample enrichment in microchannels with detection limit improvements of as much as 10^7-fold, as seen in recent developments in microchip electrophoresis and analysis: a solid format that continues to grow and has shown excellent output in many areas [143,148,155–157].

The use of state-of-the-art online strategies to increase sensitivity in CE and MCE has become an analytical technique that is particularly attractive for application in different areas, such as biochemical, environmental, pharmaceutical, and toxicological analyses. Mastering such methods can be a powerful tool to improve even further the great scope of different electrophoretic methods.

REFERENCES

1. Mayer, B.X. (2001) *J. Chromatogr. A 907*, 21–37.
2. Sentellas, S., Puignou, L., and Galceran, M.T. (2002) *J. Sep. Sci. 25*, 975–987.
3. Jun, C., Fangjun, H., and Ping, C. (2007) *Chin. J. Chromatogr. 25*, 183–188.
4. Tagliaro, F., Turrina, S., and Smith, F.P. (1996) *Forensic Sci. Int. 77*, 211–229.
5. Bessanova, E.A., Kartsova, L.A., and Shmukov, A.U. (2007) *J. Chromatogr. A 1150*, 332–338.
6. King, J.T. and Desai, U.R. (2008) *Anal. Biochem. 380*, 229–234.
7. Fonslow, B.R. and Yates, J.R., III (2009) *J. Sep. Sci. 32*, 1175–1188.
8. Osbourn, D.M., Weiss, D.J., and Lunte, C.E. (2000) *Electrophoresis 21*, 2768–2779.
9. Morzunova, T.G. (2006) *Pharm. Chem. J. 40*, 158–169.
10. Britz-McKibbin, P. and Terabe, S. (2003) *J. Chromatogr. A 1000*, 917–934.
11. Martínez, D., Borrull, F., and Calull, M. (1997) *J. Chromatogr. A 788*, 185–193.
12. Chien, R.L. and Burgi, D.S. (1991) *J. Chromatogr. 559*, 141–152.
13. Tsuda, T., Sweedler, J.V., and Zare, R.N. (1990) *Anal. Chem. 62*, 2149–2152.
14. Liu, Z., Otsuka, K., and Terabe, S. (2002) *J. Chromatogr. A 961*, 285–291.
15. Moring, S.E., Reel, R.T., and van Soest, R.E.J. (1993) *Anal. Chem. 85*, 3454–3459.
16. Simpson, S.L.Jr., Quirino, J.P., and Terabe, S. (2008) *J. Chromatogr. A 1184*, 504–541.
17. Lin, C.H. and Kaneta, T. (2004) *Electrophoresis 25*, 4058–4073.
18. Pacifici, R., Pichini, S., Altieri, I., Caronna, A., Passa, A.R., and Zuccaro, P. (1995) *J. Chromatogr. B 664*, 329–334.
19. Figeys, D. and Aebersold, R. (1998) *Electrophoresis 19*, 885–892.
20. Nieddu, M., Boatto, G., Pirisi, M.A., and Dessì, G. (2010) *Rapid Commun. Mass Spectrom. 24*, 2357–2362.
21. Mason, J.L., Ashmore, S.P., and Aitkenhead, A.R. (1991) *J. Chromatogr. B 570*, 191–197.
22. Platoff, G.E. and Gere, J.A. (1991) *Forensic Sci. Rev. 3*, 117–133.
23. Lord, H.L. and Pawliszyn, J. (1997) *Anal. Chem. 69*, 3899–3906.
24. Bishop, S.C., Lerch, M., and McCord, B. (2004) *Forensic Sci. Int. 141*, 7–15.
25. Xie, H.Y., He, Y.Z., Gan, W.E., Fu, G.N., Li, L., Han, F., and Gao, Y. (2009) *J. Chromatogr. A 1216*, 3353–3359.
26. Swartz, M. and Merion, M. (1993) *J. Chromatogr. 632*, 209–213.
27. Beattie, J.H., Self, R., and Richards, M.P. (1995) *Electrophoresis 33*22–328.
28. Tomlinson, A.J., Benson, L.M., Guzman, N.A., and Naylor, S.J. (1996) *J. Chromatogr. A 744*, 3–15.
29. Saavedra, L. and Barbas, C. (2007) *J. Biochem. Biophys. Methods 70*, 289–297.
30. Haglund, H. and Tiselius, A. (1950) *Acta Chem. Scand. 4*, 957–962.
31. Mikkers, F.E.P., Everaerts, F.M., and Verheggen, Th.P.E.M. (1979) *J. Chromatogr. 169*, 1–10.
32. Moring, S.E., Colburn, J., Grossman, P.D., and Lauer, H.H. (1989) *LC-GC 8*, 26–34.
33. de Moraes, M.L.L., Moraes, S.L., Pereira, E.A., and Tavares, M.F.M. (2009) *Quim. Nova 32*, 1041–1046.
34. Wang, C.C., McCann, W.P., and Beale, S.C. (1996) *J. Chromatogr. B 676*, 19–28.
35. Foret, F., Szoko, E., and Karger, B.L. (1992) *J. Chromatogr. 608*, 3–12.
36. Waterval, J.C., Krabbe, H., Teeuwsen, J., Bult, A., Lingeman, H., and Underberg, W.J.M. (1999) *Electrophoresis 20*, 2909–2916.
37. Krivánková, L., Pantucková, P., and Bocek, P. (1999) *J. Chromatogr. A 838*, 55–70.
38. Praus, P. (2004) *Talanta 62*, 977–982.
39. Jastrzebská, A. (2006) *Talanta 69*, 1018–1024.
40. Jung, B.G., Zhu, Y.G., and Santiago, J.G. (2007) *Anal. Chem. 79*, 345–349.
41. Hirokawa, T., Yoshioka, M., Okamoto, H., Timerbaev, A.R., and Blaschke, G. (2004) *J. Chromatogr. B 811*, 165–170.
42. Xu, Z., Doi, T., Timerbaev, A.R., and Hirokawa, T. (2008) *Talanta 77*, 278–281.
43. Tomlinson, A.J., Benson, L.M., Jameson, S., Johnson, D.H., and Naylor, S. (1997) *J. Am. Soc. Mass Spectrom. 8*, 15–24.
44. Hempel, G. (2000) *Electrophoresis 21*, 691–698.
45. Kurian, E., Prendergast, F.G., Tomlinson, A.J., Holmes, M. W., and Naylor, S. (1997) *J. Am. Soc. Mass Spectrom. 8*, 8–14.
46. Pedersen-Bjergaard, S., Rasmussen, K.E., and Halvorsen, T. G. (2000) *J. Chromatogr. A 902*, 91–105.

47. van der Vlis, E., Mazereeuw, M., Tjaden, U.R., and van der Greef, H.I.J. (1995) *J. Chromatogr. A 712*, 227–234.
48. Stroink, T., Paarlberg, E., Waterval, J.C.M., Bult, A., and Underberg, W.J.M. (2001) *Electrophoresis 22*, 2374–2383.
49. Almeda, S., Nozal, L., Arce, L., and Valcárcel, M. (2007) *Anal. Chim. Acta 587*, 97–103.
50. Xu, L., Hauser, P.C., and Lee, H.K. (2008) *J. Chromatogr. A 1214*, 17–22.
51. Shihabi, Z.K. (2000) *J. Chromatogr. A 902*, 107–117.
52. Malá, Z., Slampová, A., Gebauer, P., and Bocek, P. (2009) *Electrophoresis 30*, 215–229.
53. Burgi, D.S. and Chien, R.L. (1991) *Anal. Chem. 63*, 2042–2047.
54. Danger, G., Pascal, R., and Cottet, H. (2009) *J. Chromatogr. A 1216*, 5748–5754.
55. Chien, R.L. (1991) *Anal. Chem. 63*, 2866–2869.
56. Bermudo, E., Núñez, O., Moyano, E., Puignou, L., and Galceran, M.T. (2007) *J. Chromatogr. A 1159*, 225–232.
57. Law, W.S., Zhao, J.H., and Li, S.F.Y. (2005) *Electrophoresis 26*, 3486–3494.
58. Zhang, L.Y. and Sun, M.X. (2005) *J. Chromatogr. A 1100*, 230–235.
59. Fang, H., Zeng, Z., and Liu, L. (2006) *Anal. Chem. 78*, 6043–6049.
60. Ma, M., Zhang, H.S., Xiao, L.Y., Xiao, L., Wang, P., Cui, H. R., and Wang, H. (2007) *Electrophoresis 28*, 4091–4100.
61. Wu, Y.W., Liu, J.F., Xiao, T.X., Han, D.Y., Zhang, H.L., and Pan, J.C. (2009) *Electrophoresis 30*, 668–673.
62. Chien, R.L. and Burgi, D.S. (1992) *Anal. Chem. 64*, 1046–1050.
63. Albert, M., Debusschere, L., Demesmay, C., and Rocca, J.L. (1997) *J. Chromatogr. A 757*, 281–289.
64. Quirino, J.P. and Terabe, S. (1999) *J. Chromatogr. A 850*, 339–344.
65. Chen, Z. and Naidu, R. (2004) *J. Chromatogr. A 1023*, 151–157.
66. Albert, M., Debusschere, L., Demesmay, C., and Rocca, J.L. (1997) *J. Chromatogr. A 757*, 291–296.
67. Yang, Y., Liu, Q., Tao, W., Nie, L., and Yao, S. (2007) *J. Sep. Sci. 30*, 3296–3301.
68. Puig, P., Tempels, F.W.A., Somsen, G.W., de Jong, G.J., Borrull, F., Aguilar, C., and Calull, M. (2008) *Electrophoresis 29*, 1339–1346.
69. Zhu, Z.F., Yan, N., Zhou, X., Zhou, L., and Chen, X. (2009) *J. Sep. Sci. 32*, 3481–3488.
70. Kim, J., Chun, M.S., Choi, K., and Chung, D.S. (2009) *Electrophoresis 30*, 1046–1051.
71. See, H.H., Hauser, P.C., Ibrahim, W.A.W., and Sanagi, M.M. (2010) *Electrophoresis 31*, 575–582.
72. Britz-McKibbin, P. and Chen, D.D.Y. (2000) *Anal. Chem. 72*, 1242–1252.
73. Zhao, Y. and Lunte, C.E. (1999) *Anal. Chem. 18*, 3985–3991.
74. Britz-McKibbin, P. and Terabe, S. (2003) *J. Chromatogr. A 1000*, 917–934.
75. Kim, J.B., Britz-McKibbin, P., Hirokawa, T., and Terabe, S. (2003) *Anal. Chem. 75*, 3986–3993.
76. Aebersold, R. and Morrison, D.H. (1990) *J. Chromatogr. 516*, 79–88.
77. Jaafar, J., Irwan, Z., Ahamad, R., Terabe, S., Ikegami, T., and Tanaka, N. (2007) *J. Sep. Sci. 30*, 391–398.
78. Du, Y., Jia, L., Liu, H., and Xing, D. (2007) *Anal. Lett. 40*, 2005–2015.
79. Su, S. and Yu, Y. (2009) *J. Chromatogr. A 1216*, 1490–1495.
80. Arnett, S.D. and Lunte, C.E. (2003) *Electrophoresis 24*, 1745–1752.
81. Hoque, M.E., Arnett, S.D., and Lunte, C.E. (2005) *J. Chromatogr. B 827*, 51–57.
82. Arnett, S.D. and Lunte, C.E. (2007) *Electrophoresis 28*, 3786–3793.
83. Bykova, L. and Holland, L.A. (2008) *Electrophoresis 29*, 3794–3800.
84. Kim, J.-B. and Terabe, S. (2003) *J. Pharm. Biomed. Anal. 30*, 1625–1643.
85. Quirino, J.P. and Terabe, S. (1999) *J. Chromatogr. A 856*, 465–482.
86. Aranas, A.T., Guidote Jr., A.M., and Quirino, J.P. (2009) *Anal. Bioanal. Chem. 394*, 175–185.
87. de Moraes, L.M.M., Tavares, M.F.M., and Pereira, E.A. (2010) *Quim. Nova 33*(2), 466–470.
88. Quirino, J.P. and Terabe, S. (1997) *J. Chromatogr. A 781*, 119–128.
89. Quirino, J.P. and Terabe, S. (1997) *J. Chromatogr. A 791*, 255–267.
90. Nevado, J.J.B., Flores, J.R., Peñalvo, G.C., and Fariñas, N.R. (2002) *J. Chromatogr. A 953*, 279–286.
91. Quirino, J.P. and Terabe, S. (1998) *Anal. Chem. 70*, 149–157.
92. Quirino, J.P., Inoue, N., and Terabe, S. (2000) *J. Chromatogr. A 892*, 187–194.
93. Quirino, J.P., Otsuka, K., and Terabe, S. (1998) *J. Chromatogr. B 714*, 29–38.
94. da Silva, C.L., de Lima, E.C., and Tavares, M.F.M. (2003) *J. Chromatogr. A 1014*, 109–116.
95. Palmer, J.F., Munro, N.J., and Landers, J.P. (1999) *Anal. Chem. 71*, 1679–1687.
96. Palmer, J.F. and Landers, J.P. (2000) *Anal. Chem. 72*, 1941–1943.
97. Quirino, J.P., Terabe, S., and Bocek, P. (2000) *Anal. Chem. 72*, 1934–1940.
98. Palmer, J.F. (2004) *J. Chromatogr. A 1036* 95–100.
99. Quirino, J.P. and Terabe, S. (1998) *J. Chromatogr. A 798*, 251–257.
100. Quirino, J.P. and Terabe, S. (1998) *Anal. Chem. 70*, 1893–1901.
101. Quirino, J.P. and Terabe, S. (1998) *Science 282*, 465–468.
102. Musijowski, J., Poboży, E., and Trojanowicz, M. (2006) *J. Chromatogr. A 1104*, 337–345.
103. Quirino, J.P. and Terabe, S. (1999) *J. High Resolut. Chromatogr. 22*(7), 367–372.
104. Quirino, J.P. and Terabe, S. (1999) *Anal. Chem. 71*, 1638–1644.
105. Monton, M.R.N., Quirino, J.P., Otsuka, K., and Terabe, S. (2001) *J. Chromatogr. A 939*, 99–108.

106. Giordano, B.C., Newman, C.I.D., Federowicz, P.M., Collins, G.E., and Burgi, D.S. (2007) *Anal. Chem. 79*, 6287–6294.
107. Lin, Y.-H., Li, J.-H., Ko, W.-K., and Wu, S.-M. (2006) *J. Chromatogr. A 1130*, 281–286.
108. Zhu, L., Tu, C., and Lee, H.K. (2002) *Anal. Chem. 74*, 5820–5825.
109. Macià, A., Borrull, F., Calull, M., and Aguilar, C. (2006) *J. Chromatogr. A 1117*, 234–245.
110. Quirino, J.P. and Terabe, S. (2000) *Anal. Chem. 72*, 1023–1030.
111. Quirino, J.P. and Haddad, P.R. (2008) *Anal. Chem. 80*, 6824–6829.
112. Quirino, J.P. (2008) *J. Chromatogr. A 1214*, 171–177.
113. Quirino, J.P. (2009) *J. Chromatogr. A 1216*, 294–299.
114. Liu, L., Deng, X., and Chen, X. (2010) *J. Chromatogr. A 1217*, 175–178.
115. Ibrahim, W.A.W., Hermawan, D., and Sanagi, M.M. (2007) *J. Chromatogr. A 1170*, 107–113.
116. Michalska, K., Pajchel, G., and Tyski, S. (2009) *J. Chromatogr. A 1216*, 2934–2942.
117. Ibrahim, W.A.W., Sanagi, M.M., and Liew, S. (2009) *Malaysian J. Anal. Sci. 13*, 44–51.
118. Ravelo-Pérez, L.M., Hernández-Borges, J., Borges-Miquel, T.M., and Rodríguez-Delgado, M.A. (2008) *Food Chem. 111*, 764–770.
119. Liu, F.-K. (2009) *J. Chromatogr. A 1216*, 2554–2559.
120. Ibrahim, W.A.W., Hermawan, D., Sanagi, M.M., and Aboul-Enein, H.Y. (2010) *Chromatographia 71*, 305–309.
121. Otsuka, K., Hayashibara, H., Yamauchi, S., Quirino, J.P., and Terabe, S. (1999) *J. Chromatogr. A 853*, 413–420.
122. Li, X., Chu, S., Fu, S., MA, L., Liu, X., and Xu, X. (2005) *Chromatographia 61*, 161–166.
123. Li, H., Wang, P.-H., Li, C., Wang, H., and Zhang, H.-S. (2008) *Microchem. J. 89*, 34–41.
124. Jiang, J. and Lucy, C.A. (2002) *J. Chromatogr. A 966*, 239–244.
125. Carabias-Martínez, R., Rodríguez-Gonzalo, E., Revilla-Ruiz, P., and Domínguez-Álvarez, J. (2003) *J. Chromatogr. A 990*, 291–302.
126. Hai-Ling, L., Hui-Bi, X., and Zhong-Hong, G. (2007) *Chin. J. Anal. Chem. 35*, 1087–1091.
127. Sheng, P.-W., Huang, W.-Y., and Sheu, S.-J. (2003) *J. Food Drug Anal. 11*, 209–213.
128. Kartsova, L.A. and Bessonova, E.A. (2007) *J. Anal. Chem. 62*, 68–75.
129. Dawod, M., Breadmore, M.C., Guijt, R.M., and Haddad, P.R. (2010) *J. Chromatogr. A 1217*, 386–393.
130. Britz-McKibbin, P., Otsuka, K., and Terabe, S. (2002) *Anal. Chem. 74*, 3736–3743.
131. Grass, B., Neyer, A., Johnck, M., Siepe, D., Eisenbeiss, F., Weber, G., and Hergenroder, R. (2001) *Sens. Actuators B 72*, 249–258.
132. Rìos, A., Escarpa, A., González, M.A., and Crevillén, A.G. (2006) *Trends Anal. Chem. 25*, 467–479.
133. Schappler, J., Veuthey, J.L., and Rudaz, S. (2008) In: S. Ahuja and M.I. Jimidar (eds), *Capillary Electrophoresis Methods for Pharmaceutical Analysis*, Vol. 9, 1st ed., Academic Press, San Diego, CA, pp. 477–521.
134. Sueyoshi, K, Kitagawa, F., and Otsuka, K. (2008) *J. Sep. Sci. 31*, 2650–2666.
135. Lichtenberg, J., Verpoorte, E., and Rooij, N.F. (2001) *Electrophoresis 22*, 258–271.
136. Wainright, A., Williams, S.J., Ciambrone, G., Xue, Q., Wei, J., and Harris, D. (2002) *J. Chromatogr. A 979*, 69–80.
137. van Midwoud, P.M. and Verpoorte, E. (2008) In: J.P. Landers (ed.), *Handbook of Capillary and Microchip Electrophoresis and Associated Microtechniques*, 3rd ed., CRC Press, Boca Raton, FL, pp. 1375–1419.
138. Zhang, L. and Yin, X.F. (2006) *J. Chromatogr. A 1137*, 243–248.
139. Gong, M., Wehmeyer, K.R., Limbach, P.A., Arias, F., and Heineman, W.R. (2006) *Anal. Chem. 78*, 3730–3737.
140. Jacobson, S.C. and Ramsey, J.M. (1995) *Electrophoresis 16*, 481–486.
141. Yang, H. and Chien, R.L. (2001) *J. Chromatogr. A 924*, 155–163.
142. Zhang, Y., Ping, G., Zhu, B., Kaji, N., Tokeshi, M., and Baba, Y. (2007) *Electrophoresis 28*, 414–421.
143. Kawai, T., Sueyoshi, K., Kitagawa, F., and Otsuka, K. (2010) *Anal. Chem. 82*, 6504–6511.
144. Jeong, Y., Choi, K., Kang, M.K., Chun, K., and Chung, D.S. (2005) *Sens. Actuators B. 104*, 269–275.
145. Liu, D., Shi, M., Huang, H., Long, Z., Zhou, X., Qin, J., and Lin, B. (2006) *J. Chromatogr. B 844*, 32–38.
146. Walker, III, P.A., Morris, M.D., Burns, M.A., and Johnson, B. N. (1998) *Anal. Chem. 70*, 3766–3769.
147. Jung, B., Bharadwaj, R., and Santiago, J.G. (2006) *Anal. Chem. 78*, 2319–2327.
148. Wang, J., Zhang, Y., Mohamadi, M.R., Kaji, N., Tokeshi, M., and Baba, Y. (2009) *Electrophoresis 30*, 3250–3256.
149. Cui, H., Horiuchi, K., Dutta, P., and Ivory, C.F. (2005) *Anal. Chem. 77*, 1303–1309.
150. Thormann, W., Caslavska, J., and Mosher, R.A. (2007) *J. Chromatogr. A 1155*, 154–163.
151. Ren, K., Liang, Q., Yao, B., Luo, G., Wang, L., Gao, Y., Wang, Y., and Qiu, Y. (2007) *Lab Chip 7*, 1574–1580.
152. Sera, Y., Matsubara, N., Otsuka, K., and Terabe, S. (2001) *Electrophoresis 22*, 3509–3513.
153. Augustin, V., Proczek, G., Dugay, J., Descroix, S., and Hennion, M.-C. (2007) *J. Sep. Sci. 30*, 2858–2865.
154. Tokeshi, M., Minagawa, T., and Kitamori, T. (2000) *Anal. Chem. 72*, 1711–1714.
155. Sommer, G.J., Kim, S.M., Littrell, R.J., and Hasselbrink, E.F. (2007) *Lab Chip 7*, 898–907.
156. Matsui, T., Franzke, J., Manz, A., and Janasek, D. (2007) *Electrophoresis 28*, 4606–4611.
157. Wang, Y.C., Stevens, A.L., and Han, J. (2005) *Anal. Chem. 77*, 4293–4299.
158. Honda, S., Suzuki, S., and Taga, A. (2003) *J. Pharm. Biomed. Anal. 30*, 1689–1714.

3

SAMPLING AND QUANTITATIVE ANALYSIS IN CAPILLARY ELECTROPHORESIS

Petr Kubáň, Andrus Seiman, and Mihkel Kaljurand

3.1 INTRODUCTION

Capillary electrophoresis (CE) techniques have now been around for about three decades following the seminal publication by Jorgenson and Lukacs in 1981 [1]. In spite of the very high expectations and predictions for a bright future, CE has not entirely fulfilled the hopes that were initially predicted. The reasons for the current status of CE are manyfold: low sensitivity, low reproducibility, and sample matrix dependence are among the most frequently cited drawbacks. It is perhaps the low reproducibility that is one of the most serious problems that CE is being faced with.

Undoubtedly, there are research areas where high separation efficiencies, short analysis times, and extremely small sample and electrolyte consumption present a clear benefit compared with liquid chromatography (LC) and outweigh the poorer reproducibility of CE. Chiral analysis by CE is a well-established technique; minimal amounts of expensive chiral selectors are required for the separations, making method development and analysis extremely cost-effective. CE can be effectively used for the screening and characterization (determination of pK_a, solubility, etc.) of various compounds, particularly biomolecules, biopharmaceuticals, and proteins. DNA sequencing is another field where fast capillary gel electrophoretic separations have been of clear benefit. Capillary isoelectric focusing (cIEF) is being applied to the characterization of biomolecules, and gel electrophoresis in capillaries is far superior to the planar gel-based techniques such as sodium dodecyl sulfate polyacrylamide gel electrophoresis (SDS-PAGE) or isoelectric focusing (IEF).

Modern instrumentation has, to some extent, eliminated some of the early variability issues as the suppliers of CE equipment have responded to the need for improved robustness and reliability. The recently upgraded CE instruments now include various hardware and software improvements that have helped to resolve some reproducibility issues. There is also a trend toward the development of well-controlled and validated chemistry/capillary kits to improve the performance, for example, in the analysis of inorganic anions and metal cations. The emergence of more powerful, automated instruments makes CE more accessible than ever. There are some signs that the future of this technique may be somewhat brighter, as CE is again attracting increased attention, as confirmed in the report from a recent panel discussion.[1] For instance, the recent acetonitrile (ACN) shortage influenced the field of analytical separations and alternative, ACN-free, methods had to be quickly developed. The wider usage of environmentally friendly solutions and wider acceptance of green chemistry by analysts may also help to break this impasse.

Nevertheless, in general, the acceptance of CE by the industry has been slow. Analytical chemists have learned that CE is highly sensitive to parameter changes and thus not very reproducible and robust. In fact, CE has not, to any large extent, replaced the current LC techniques used for routine analysis in accredited laboratories, nor has it replaced to a significant level any reference methods published by regulatory agencies worldwide. For instance, only

[1] http://www.chromatographyonline.com/lcgc/Articles/Technology-Forum-Capillary-Electrophoresis/ArticleStandard/Article/detail/735500?contextCategoryId=42946.

Capillary Electrophoresis and Microchip Capillary Electrophoresis: Principles, Applications, and Limitations, First Edition.
Edited by Carlos D. García, Karin Y. Chumbimuni-Torres, and Emanuel Carrilho.

gel electrophoretic methods are listed in the AOAC International *Official Methods of Analysis* handbook's current edition.[2]

It may be the lack of familiarity with CE fundamentals that can lead to methods being developed that are not robust, perform poorly, and give CE a negative reputation. However, other factors, such as poor sampling precision, changes in electroosmotic flow (EOF) induced by temperature and viscosity changes, adsorption of sample compounds on the capillaries, various inherent sampling and detection biases, improper buffering of the electrolytes, decomposition due to electrolysis, and so on, contribute to the overall poor performance of CE compared with its LC counterparts. It is quite evident now that CE will not replace high-performance liquid chromatography (HPLC) or ion chromatography (IC) as it is complementary/orthogonal to these techniques. Currently, CE still needs better support from instrument manufacturers (to build more robust and reliable instrumentation), consumables vendors (to continue to devise kits and reagents), and scientists themselves (to develop novel methodologies and applications). Without this type of investment, CE cannot expand, but without expansion, no-one wants to invest.

Reproducible sample introduction in any analytical system is of great importance, as injection, being the initial step in the separation process, probably represents the largest error in a well-controlled method. Therefore, the main focus of this chapter will be on the various sample injection techniques and devices used in CE. Sampling in general and sampling devices in particular differ significantly from those used in LC techniques. Proper understanding of the sampling fundamentals is important for achieving good reproducibility and consequently good accuracy of quantitative analysis. Different sampling methods that are currently used by CE practitioners using home-made and commercial instruments will be overviewed first. A significant part will be devoted to alternative sampling devices and techniques that have been developed over time as a response to the reproducibility issues. The advantages and disadvantages of the available injection techniques and devices will be critically assessed and the ways that can be used for correcting some of the injection-related problems will also be outlined. Some aspects of the sampling related to quantitative analysis will also be briefly discussed. Qualitative analysis, for example, compound identification, will not be of particular concern here; however, some issues that are related to both qualitative and quantitative analyses will also be overviewed.

3.2 INJECTION TECHNIQUES IN CE

One of the major differences between CE and LC separation techniques is the way in which the sample is introduced into the separation capillary. Another difference relates to the total volumetric amount of sample injected. As the total volume of a typical CE separation capillary is only a few microliters, the sample volumes must be very small, otherwise the high separation efficiency of CE will be compromised. Typically, a sample volume that represents ~1% of the total capillary volume is injected, resulting in low nanoliter volumes of samples introduced [2]. Small injection volumes are often referred to as one of the advantages of CE over the corresponding LC techniques (HPLC, IC). This is indeed true, but only in specific applications, where the sample amount is limited. Some explicit examples include the analysis of individual rain drops [3,4], individual cell analysis [5], and the analysis of biological fluids. There is, however, a significant downside when decreasing the sample volume injected, as the requirements for sampling device precision increase dramatically. In addition, effects such as molecular diffusion, surface tension, and liquid evaporation play much more significant roles and should not be neglected.

There are two commonly used injection techniques in CE: hydrodynamic (HD) and electrokinetic (EK) injections. They have been used since the early days of CE and have changed very little since then, although their performance has not been excellent. Currently, the most often reported reproducibility values, expressed as the relative standard deviation (RSD) of measured peak areas, achieved with manually operated HD or EK injection are no better than 5–10%. They can be improved to ~1–3% RSD using fully automated injection devices in commercial instruments. Comparing these values with the routinely achieved performance of valve-based injectors in HPLC or IC, in which the values typically do not exceed 1% RSD, it is apparent that there is still some ample space for improvement. Moreover, the reported reproducibility values for CE are somewhat inconsistent throughout the literature. In some studies, for instance in a study conducted by Kunkel et al. [6], an excellent precision of most of the commercial instruments (RSD of peak areas below 0.5%) was reported. In other interlaboratory CE validation studies, the precision was between 1% and 2.6% [7–9]. These values are contradicted by other similar studies, in which large discrepancies were found, with reported RSD values as high as 7% [10] or 17.7% [11] using the same, fully automated commercial instruments. From the authors' own experience, the best achievable figures lie somewhere in the range 2–3% RSD. It should be noted that the sample injection is not the only factor that contributes to the overall lower reproducibility of CE, as there are other factors that have been mentioned previously. If not properly understood, however, the injection process itself may make a significant contribution to the overall poor reproducibility and, more importantly, lead to poor accuracy in quantitative analysis. The two major injection methods used in CE and some alternative

[2] http://www.eoma.aoac.org/.

approaches will be critically assessed and compared in the following sections.

3.2.1 Hydrodynamic Sample Injection

3.2.1.1 Principle Hydrodynamic sample injection is perhaps one of the most widely used injection techniques in both home-made and commercial CE instruments. The volume of the sample introduced by HD injection relates directly to the pressure differential applied to the capillary end immersed in the sample solution, as shown in the following equation [12]:

$$V = \frac{\Delta p \pi r^4 t_{\text{inj}}}{8\eta L} \tag{3.1}$$

where V is the injection volume, Δp is the pressure differential applied, r is the capillary inner radius, t_{inj} is the injection time, η is the sample solution viscosity, and L is the total capillary length. The amount of a particular ion, i, injected can then be calculated by substituting Q_i/C_i for V:

$$Q_i = \frac{\Delta p \pi r^4 C_i t_{\text{inj}}}{8\eta L} \tag{3.2}$$

where Q_i is the amount of a particular ion i (in moles) and C_i is the concentration of that ion in the sample.

It follows from Equation 3.2 that the amount of a particular analyte, Q_i, injected hydrodynamically from the sample will be directly related to its concentration in the sample (C_i) and will not depend on its electrophoretic mobility, but only on various other injection parameters that can be kept constant, in most cases. A representative sample fraction that reflects a true composition of the original sample will be injected by the HD injection technique. This characteristic of HD injection will have some important consequences for quantitative analysis, as discussed later.

There are various modes of HD injection, such as gravity injection, pressure injection, and vacuum injection, that differ only in the way in which the pressure differential, Δp, is achieved. Among the available methods, gravity injection (also called siphoning injection) is still the most popular in home-made devices. It is a very simple means of sample introduction—no injection valve or additional pressure-based devices are required. The separation capillary filled with the background electrolyte (BGE) serves as a column of liquid. When one capillary end (immersed in the solution) is elevated to a specific height, Δh, above the other capillary end, HD flow results. The volumetric flow rate, F, can be calculated from the Hagen–Poiseuille equation:

$$F = \frac{\Delta h \rho g \pi r^4}{8\eta L} \tag{3.3}$$

where Δh is the height difference, ρ is the liquid density, and g is the acceleration due to gravity. The injection volume is then the product of volumetric flow rate and the time during which the capillary end is elevated. The injection in this mode can be accomplished by simple manual manipulation of the capillary or using an in-house-made carousel. By knowing the height difference, injection duration, and other capillary and solution parameters, the exact volume injected can be calculated. Gravity injection is depicted schematically in Figure 3.1a. Note that except for the purpose of the injection, the siphoning effects should be avoided as any HD flow disturbs the plug-like profile of the separated analyte zones during the CE separation process. Keeping the liquid levels in the vials at exactly the same level is one of the requirements for highly efficient CE separation to be achieved.

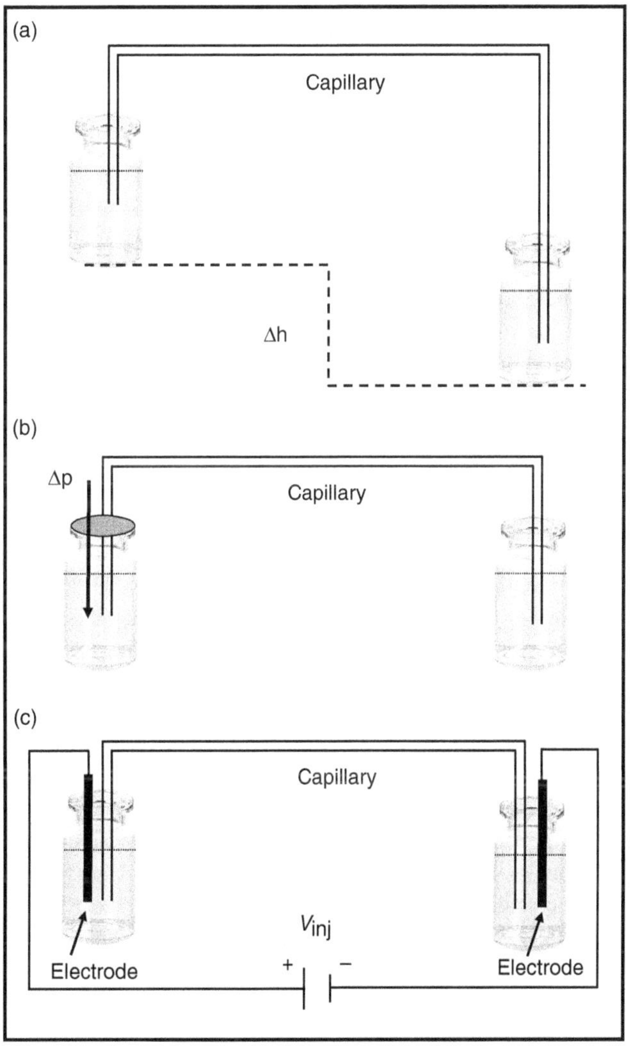

FIGURE 3.1 Schematic of HD and EK injection modes in CE. (a) Hydrodynamic injection using gravity. (b) Hydrodynamic injection using pressure. (c) EK injection.

Due to its simplicity, the gravity injection mode was popular with the early versions of commercial CE equipment and was used, for instance, in the Waters Quanta 4000 CE instrument introduced in 1990. It used an automated carousel platform that allowed vial replacement and elevating one end of the separation capillary relative to the other end. The model has been discontinued, however, and CE manufacturers have replaced gravity injection by various pressure-based injection systems. In pressure-based injection (see Figure 3.1b), the pressure difference (Δp) from Equation 3.1 is accomplished by high-precision air pumps and a set of regulation valves.

The reason for this replacement of commercial instrumentation may have been the belief that the applied pressure can be more precisely controlled than simple siphoning of the liquid, leading to better injection reproducibility. The current versions of the instruments include rather sophisticated means to minimize the errors in the injected sample amounts caused by pressure variations. The performance of the instruments has indeed improved over the years, but at the cost of increased complexity. The general platform, that is, a rotating carousel with lift mechanisms, still needs to be present in order to replace the sample and electrolyte vials. Malfunctioning of the lift systems and also the air pressure pumps in commercial instruments is often reported by users, and in this respect there is still room for improvements to be addressed by the suppliers.

The third type of HD injection, the vacuum injection, as opposed to pressure injection, has only been rarely used in a few commercial instruments (for instance, in the older models of the Applied Biosystems CE apparatus). It is typically not offered as a standard feature in commercial instruments, nor is widely applied in research laboratories.

3.2.1.2 Advantages and Performance HD injection, as used in CE, is relatively straightforward and simple. The injection, similarly to any valve injection in HPLC–IC, introduces a representative sample aliquot into the separation capillary, the content of which is identical with the content of the analytes in the original sample. The volume of the introduced sample is fairly constant, unless samples differing significantly in viscosity are injected or the temperature changes appreciably between sample injections. Provided that the viscosity of the standards and samples is fairly constant, external calibration for analyte quantitation can be easily performed. The repeatability of HD injection for manually injecting samples has not been very good. Typically, 5–10% RSDs of peak areas are reported; these, however, can be significantly improved to about 1–3% by using automated injection systems that are present in all commercial instruments.

3.2.1.3 Disadvantages There are relatively few disadvantages of the HD injection method in general. The viscosity problem is commonly mentioned in analyses of real samples that contain large amount of matrix compounds, for instance, high concentrations of inorganic salts and proteins. In this case, correction for sample viscosity must be applied or standard addition/internal standardization should be performed rather than simple external calibration with standards prepared in deionized water. However, it seems that rather than the differences in viscosity of the samples, other instrumental parameters contribute to the poor precision of the sample introduction. In home-made instruments, using siphoning injection, the injection timing accuracy and repeatability of all steps performed during the sampling sequence, such as lifting, vial replacement, and capillary immersion into vials, are difficult to control. A repeatability similar to that given by robotic/automated CE samplers can hardly be achieved. On the other hand, in automated instruments, other factors may be limiting. For instance, the variability of the vials, vial types, quality of the vial septa, and so on have been of concern [13]. It has been found that although the automated instruments can inject reproducibly from the same sample vial, they cannot compensate completely for differences between the vials. Other parameters that have been studied and shown to have an effect on the injection variability include, for instance, the quality of the capillary tip at the injection side [14], the total injection volume and the corresponding extraneous/ubiquitous injection contribution [15,16], and so on. These effects, unfortunately, apply in both the manually operated and fully automated injection systems.

Nevertheless, if the variability of the measured peak areas is due only to the variation of the injection volume, it should be possible to apply a simple correction, for instance, measure the injection plug length. It has been shown that the injection plug length can be assessed indirectly, for instance, by monitoring the separation current [17]. Another possibility that seems fairly straightforward and logical is to use the EOF peak to assess the introduced injection plug volume. Additionally, a simple and inexpensive optical- or conductivity-based monitoring device could be mounted at the injection end of the separation capillary for this purpose. Provided that the exact length of the injected sample plug can be reliably assessed, the HD injection technique should be ideally suited for CE.

3.2.2 Electrokinetic Sample Injection

3.2.2.1 Principle Electrokinetic sample injection is the other injection technique that can be used in CE. EK injection, similarly to the gravity-based HD injection, does not require any additional sampling device. The capillary injection end is immersed in the sample vial together with one electrode and a high voltage is applied across the capillary. The injection time is typically short (5–10 s) unless longer injection times, for instance in sample stacking, are used. The capillary is then reimmersed in the BGE vial and the separation is commenced. A schematic of EK injection is shown in Figure 3.1c.

EK injection thus relies on the EOF as a sample pumping mechanism. The amount of a particular ion injected also depends on its own electrophoretic mobility, μ_i, and on the conductivity of the sample. These fundamental differences make the amount of ions injected by EK injection more difficult to assess accurately than with HD injection. The following equation expresses the amount of a particular ion injected in the EK injection mode:

$$Qi = \frac{[\mu_i(k_e/k_s) + \mu_{EOF}]\pi r^2 C_i t_{inj} V_{inj}}{L} \tag{3.4}$$

where Q_i is the injected amount of a particular ion, μ_i is the ion's own clectrophoretic mobility, μ_{EOF} is the mobility of the EOF, k_e/k_s is the ratio of the conductivities of the BGE and the sample, and V_{inj} is the applied injection voltage. Unlike in HD injection, the amount of ion injected with EK injection is a function of its own electrophoretic mobility in addition to the conductivity of the sample, and the quantitation becomes more difficult than with HD injection.

3.2.2.2 Advantages and Performance The experimental setup for EK injection is very simple, as only existing CE equipment is utilized. The sensitivity of CE analysis is typically better with EK injection than with HD injection, as the sample is concentrated at the sample/BGE boundary due to the stacking effects described in a later section. The reproducibility of EK injection is, however, slightly worse than that of a corresponding HD injection; typically values between 2% and 5% are reported for both manual and automated injection techniques.

3.2.2.3 Disadvantages From Equation 3.4 that relates the injected sample amount to the injection conditions and sample properties, the difficulties in quantitative analysis may be apparent. EK injection suffers from two inherent biases, namely, the "mobility bias" and the "sample composition bias." The "mobility bias" results in larger amounts of faster-migrating ions being injected relative to the slower-migrating ions [18] (this represented by μ_i in Equation 3.4). This means that the injected sample aliquot inside the CE capillary does not correspond exactly to the original composition in the sample vial. The "mobility bias" can, under some circumstances, be ignored because it is cancelled out by the "detection bias," as has been demonstrated by several research groups [19,20].

The second bias, the "sample composition bias" [18] has, however, some more serious consequences, especially in quantitative analysis— in fact it does not allow external calibration to be used, unless extensive corrections are applied. Unlike in HD injection, the total amount of a particular sample injected (and consequently the total amount of all ions) will depend on the sample resistivity (conductivity), as seen in the k_e/k_s term in Equation 3.4. This implies that if two samples with the same concentration of a particular ion have different conductivities, the total amount of that ion injected will differ between those samples. This will render the external calibration completely inadequate, as the measured peak areas will be dependent on the sample conductivity, which is not known, unless it is measured in a separate experiment. Although several methods for correcting this bias have been suggested, for instance in Ref. 21, the accuracy of quantitative analysis with EK injection has not been improved to values better than 5% RSD. The need for extensive corrections that have to be applied in quantitative analysis limits the use of EK injection, unless the sensitivity requirements or instrumental setup clearly favor its use. EK injection has therefore been applied mostly when quantitation is not important, as a proof of concept in newly developed CE methods, or in the evaluation of various sampling schemes and devices. It has also been used extensively in automated flow injection sampling devices.

3.2.3 Bias-Free Electrokinetic Injection

The "mobility bias" that is observed in EK injection due to the different migration velocities of the analyte ions present in the sample can be resolved by using an EOF pump for sample aspiration [22,23]. This approach basically eliminates the ions' own mobility in Equation 3.4, as solely the EOF drags the sample portion into the separation capillary. The construction of the "bias-free" sampling device is, however, nontrivial. A schematic of such a device is shown in Figure 3.2.

A small fracture is created directly on the separation capillary a few centimeters from the injection end. The fractured capillary portion is inserted into the vial with the BGE. Three Pt electrodes are used. The first Pt electrode (Pt1) is inserted in the outlet vial together with the detection end of the separation capillary. The second Pt electrode (Pt2) is inserted in the BGE vial with the fractured capillary part. The third Pt electrode (Pt3) is inserted in the vial containing the injection capillary end. By connecting the electric circuit using the first and second Pt electrodes, a voltage-free region is created between the fracture and the injection capillary end. Provided that the EOF is directed toward the detection end, it will act as a pump and eventually drag the liquid from the injection end inside the separation capillary. When the injection end is inserted in the sample solution this will result in a sample portion being dragged inside the capillary, it is free from the "mobility bias." The EOF-based injection from sample solution is precisely timed, and subsequently the voltage is switched from the second to the third Pt electrode and the CE separation is started. The electrode

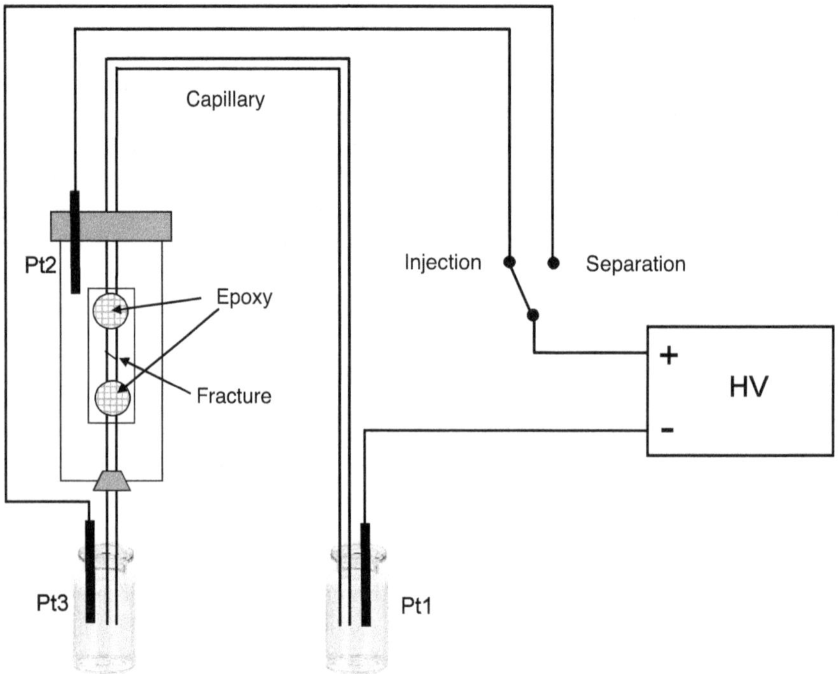

FIGURE 3.2 Schematic of the "bias-free" sampling device. Switching the high voltage between the second and third electrodes is used for sample injection and separation, respectively.

switching after the injection is necessary, otherwise buffer ions from the vial with the fractured capillary part will be preferentially introduced into the separation capillary through the fracture, resulting in large baseline drifts. The authors called the developed injection technique "electroosmotic syringe" as the short capillary segment acts as the needle and the long segment as the syringe barrel. It has been demonstrated by Pu et al. [23] that the "electroosmotic syringe" is free from "mobility bias" and compares well with HD injection when samples are diluted to some extent with the BGE. The effectiveness of this bias-free injection in removing the second bias, for example, the "sample composition bias," has not, however, been demonstrated.

In 1998, Wei et al. [24] showed that "bias-free" EK injection using the on-column fracture may not work exactly as shown previously, especially for samples prepared in a low-conductivity matrix [deionized (DI) water]. They observed that on injecting positively charged, neutral, and negatively charged analytes from DI water, some unexpected peak area deviations were observed. Compared with HD injection, the peak area of the positively charged ion decreased, whereas the peak area of the negatively charged analyte increased. The observed deviations were exactly opposite to the "mobility bias" in EK injection. Ions with lower mobilities were over-represented in the injected sample portion. Wei et al. theoretically modeled the "bias-free" injection device and showed that the short, voltage-free capillary end (extending from the fracture to the injection end) behaves as an isolated metal sphere. The portion of the BGE in the capillary will therefore possess a certain capacitance and will be charged during the injection process, resulting in the observed discrimination injection. After the charging process has been completed, the potential of the sample solution will be equal to that of the running buffer and no more discrimination injection will occur. To avoid the charging phenomenon, the authors suggested using very small injection voltages. Using too low voltages is, however, impractical since the injection would take too long. Instead a relatively slow ramping of the applied voltage ($\sim$0.2 kV s^{-1}) during the initial injection stage was shown to have a similar effect as low-voltage injection and the observed discrimination effect was nearly eliminated. Although no results were presented, the authors claimed that the method works efficiently also with samples having different conductivities.

3.2.4 Extraneous Sample Introduction Accompanying Injections in CE

One of the possible reasons for the low reproducibility of gravity and manual EK injections may be the difficulty in repeating the procedure of sample/electrolyte vial replacement and capillary elevation in exactly the same manner. In 1985, Grushka and McCormick [15] observed that when performing gravity injection in CE, a small portion of the sample may be irreproducibly introduced into the capillary as a result of capillary immersion and removal from the sample/BGE vial. This would also apply to EK injection. In

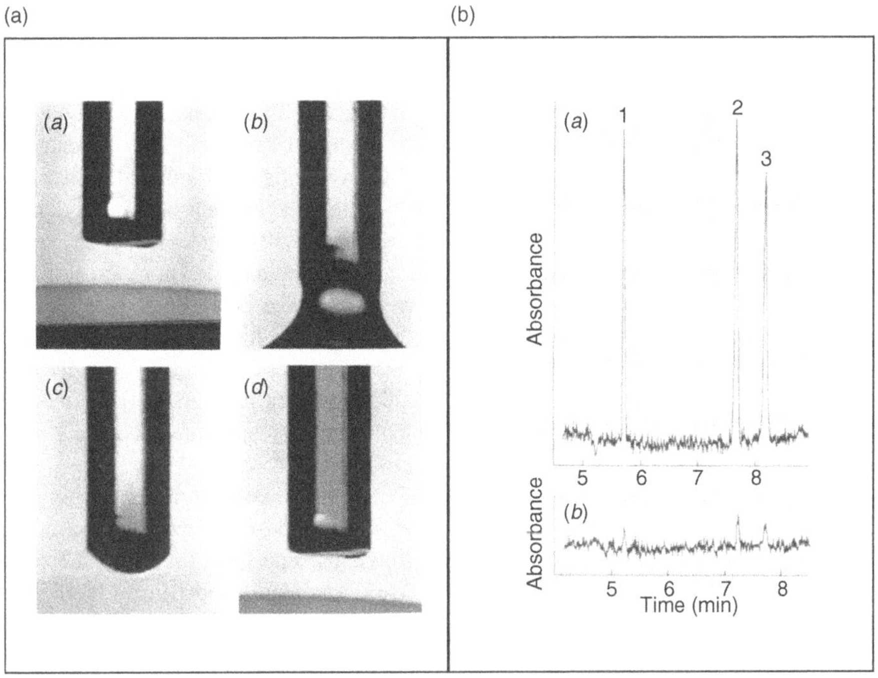

FIGURE 3.3 (a) Images showing the extraneous injection. (*a*) Capillary tip before immersion into the sample; (*b*) before removal from the sample; (*c*) liquid drop on the capillary tip immediately after removal from the sample; (*d*) introduction of the sample droplet into the capillary 10 s after removal. (b) Electropherogram comparing spontaneous extraneous injection in an (*a*) untreated and (*b*) etched capillary. Separation of 0.5 mM (1) dansyllysine, (2) dansyltryptophan, and (3) dansylalanine. Separation capillary, 75 pm i.d. × 62 cm (27 cm to detector); buffer, 15 mM phosphate, pH 8.9. Reprinted with permission from Ref. 25. Copyright (1994) American Chemical Society.

1994, Fishmann et al. [25] used a CCD camera to monitor the exact process that occurs when the capillary is inserted and withdrawn from the sample solution. The images revealed that when the capillary is withdrawn from the solution, a liquid droplet of the sample is formed on the capillary tip and contributes to the injected amount. This can be seen in Figure 3.3a, panel (*c*).

They also observed that holding the capillary in the air after the injection for an extended period of time results in the droplet entering the capillary and contributing to the total amount of sample injected. When the capillary tip is reimmersed in the BGE vial before CE separation is started, the rest of the sample droplet that has not penetrated inside the capillary is washed away. In another study, Dose and Guiochon [16] theoretically simulated the process occurring during the sample injection in CE. Except for extraneous injection, diffusion due to the presence of a concentration gradient at the capillary tip can contribute to the amount of the analyte injected. They calculated that up to 10% of the typical amount injected may arise due to the spontaneous diffusion of analytes upon insertion of the capillary tip into the sample vial.

To minimize extraneous (ubiquitous) injection in CE, the droplet formed upon removal of the capillary tip from the sample should be minimized. This can be achieved by decreasing the surface area of the capillary tip, for example, by etching the outer capillary wall or by using thin-walled capillaries. Modification of the surface properties of the injection end of the FS capillary is another alternative that has not, however, been very reproducible. The sample volume injected should be as high as possible so that the extraneous injection contribution does not account for a significant proportion of the total sample volume. It has been shown that up to a 12-fold decrease in ubiquitous injection volume could be achieved when the capillary was etched and other precautions were fulfilled (see Figure 3.3b).

If accurately controlled, extraneous injection could be used for spontaneous sample injection by fluid displacement, as shown by Fishmann et al. [26]. The capillary end was just shortly immersed in the sample solution, keeping the liquid levels of the sample and outlet reservoir even, so that there was no HD flow contributing to the injected amount. After the short immersion, the capillary end was removed from the sample, leaving a liquid droplet on the capillary tip. The capillary was then held in the saturated vapor atmosphere for a defined period until the droplet was completely aspirated inside the capillary due to the pressure difference occurring across its curved surface. A well-defined, very small sample volume (~3.5 nL) was injected with reasonable reproducibility (better than 6% RSD). As the injection volumes are very small, several injection/sample parameters need to be precisely controlled to achieve highly reproducible sample injection. These include surface tension of the sample solution, relative humidity of the surrounding atmosphere, and the quality of the capillary tip, in addition to the immersion angle. The extraneous injection may be useful when extremely small amounts of sample are available for quantitation and no substitute injection techniques can be used.

As the above-described extraneous injection cannot be completely avoided, some general precautions are necessary for reproducible manual injection in CE in general. The capillary end needs to be carefully cut to produce a straight and smooth tip and the delay time between the removal from the sample and insertion into the BGE should be kept constant. Some of these parameters are better controlled in automated injection systems and are possibly the reason for their superior performance over manual injections. It could be argued that the injection devices in which the manipulation of the separation capillary during injection is eliminated, such as rotary valve injection or sample splitters, described later, should provide superior performance.

3.2.5 Sample Stacking

3.2.5.1 Principle Sample stacking can be regarded as an extension of both HD and EK injection techniques in CE and can be utilized to improve the low concentration sensitivity of CE. The stacking phenomenon in CE was discovered as early as 1979 by Mikkers et al. [27], who noticed that the sensitivity of CE analysis was enhanced for EK injected samples prepared in a matrix having low conductivity relative to the BGE. They concluded that the stacking phenomenon is caused by a sudden change in analyte velocity at the sample/BGE boundary due to the difference in electric field in the sample zone and the rest of the capillary. Various modes of sample stacking have been developed. They utilize, for instance, the difference in conductivity or pH of the sample/BGE solutions, stacking in the presence of ACN and high salt concentration, stacking of neutral analytes in the micellar electrokinetic chromatography (MEKC) separation mode, sample sweeping, or transient isotachophoresis, to name just a few.

The normal stacking mode utilizes the conductivity differences between the sample and the BGE. Chien et al. [28–30] explained theoretically and demonstrated practically the phenomenon of the normal mode [field amplified sample stacking (FASS)]. The sample is prepared in a matrix (typically DI water or dilute BGE) with conductivity lower than that of the BGE. Part of or even the entire separation capillary is initially filled with this low-conductivity sample, typically done by a long HD injection. Upon application of a high voltage, the analyte ions in the sample zone experience a high electric field and move quickly to the sample/BGE boundary, where they are slowed and stack (concentrate) because the electric field in the BGE zone is much lower. One of the important aspects of the normal sample stacking mode that has to be considered is the large volume of the sample matrix (usually water or dilute BGE) that is present in the separation capillary after the stacking is completed. This can cause a mismatch in the local EOF velocities and induce laminar flow inside the capillary. Note that the laminar flow has a deleterious effect on the CE separation efficiencies and the remaining sample matrix should be removed. Various mechanisms for sample matrix removal have been developed; these include, for instance, polarity switching [30], using an EOF modifier as a pump [31], suppression of EOF by pH adjustment [32,33], and application of pressure [34]. In general, FASS is an efficient concentration technique and analytes can be preconcentrated 10–100-fold.

As the preconcentration factors that can be achieved with normal-mode FASS with HD sample introduction are only moderate and are limited by the volume of the separation capillary, Chien and Burgi [35,36] suggested an alternative to FASS, called head column field amplified sample stacking (HC-FASS). This technique relies on the introduction of a short sample plug prepared in a low-conductivity matrix or a plug of DI water into the separation capillary and stacking from the entire volume of the sample vial using the EK injection mode. Obviously, this technique offers several advantages over the FASS method. First, since the sample/DI water plug introduced into the capillary is relatively short in comparison with the total capillary length/volume, the laminar flow, caused by the mismatch of the EOF velocities, is small, and removal of the sample matrix is not necessary if relatively short stacking times and voltages are applied. Second, the stacking occurs from the whole sample volume in the vial, so it is not limited by the capillary volume, and higher preconcentration factors than in FASS can be achieved. Zhang and Thormann [37,38] demonstrated up to a 1000-fold concentration of positively charged hydrophobic compounds in their model system after fine-tuning the sample/electrolyte composition (by adding an organic solvent and small amounts of H^+) and optimizing the sampling time/voltage. They also found that it was necessary to introduce a short water plug before the stacking started, and to reduce the product of the HV and injection time, to prevent excessive band broadening.

Kuldvee and Kaljurand [39] were the first to demonstrate HC–FASS from a flow by using a home-made pneumatic sampling device coupled to the CE system. Stacking from a flow presents an additional benefit, as the volume of the sample can be further enhanced compared with the volumetric limitations of the HC–FASS from a vial. In the flow mode, the sample reservoir is pressurized and a small fraction of the sample is forced to enter the capillary through a combined EK–HD injection mechanism. It is similar, in principle, to HC–FASS except that the sample plug introduction and stacking take place at the same time. The analytes are continuously concentrated at the sample-/electrolyte boundary region. A 100-fold concentration factor was achieved for benzyltriethylammonium (BTEA) and benzyltributylammonium (BTBA) cations in a 10 s stacking mode at 18 kV compared with traditional EK injection. The time limitation of the stacking step was mainly due to the excess of sample introduced by HD forces during the stacking process. The excess of the sample matrix entering the capillary could be efficiently removed by applying counterpressure during or after the stacking step [34]. In another study, Kubáň et al. [40] achieved up to 2000-fold preconcentration by HC–FASS from the flow by careful adjustment of the separation electrolyte, minimizing the movement of the electrolyte solution inside the capillary, and thereby improving the stacking efficiency.

The stacking of samples with high salt content has not been possible with the two aforementioned modes because the condition of low conductivity of the sample cannot be fulfilled. pH-mediated sample stacking [41] represents an alternative where the stacking of analytes is not based on the difference in conductivities of the respective zones, but is caused by a change in the analyte charge upon a pH change. This can typically be applied to zwitterionic compounds because their charge and presence as anions or cations are dependent on pH. In 1995, Shihabi [42] presented an alternative stacking method that can be used in the presence of high salt content. Acetonitrile is added to the sample solution at a specific concentration (66%). Originally, ACN was used to precipitate the proteins from the sample; however, it has been found that a suitable concentration ratio of ACN and the inorganic salts can produce preconcentration of selected analytes (e.g., metabolites in urine samples) up to 20-fold. The mechanism for the observed stacking has not been fully understood. Most probably, the inorganic salts that have a high electrophoretic mobility migrate quickly toward the sample/BGE boundary leaving behind a low-conductivity zone of analytes. The analytes experience eventually a high electric field in this newly created zone and move rapidly toward the salt zone and stack behind it.

The problem of stacking the neutral analytes has also been of interest and several ingenious approaches for this class of analytes were developed [43–48]. When neutral analytes are preconcentrated, charged reagents are used to facilitate the stacking. Various compounds that are able to form micelles have been used for both sample stacking and separation in MEKC mode. The major idea here is the accumulation of the analytes in the micellar phase. The efficiency of stacking utilizing the micelles depends primarily on the affinity of the analyte to the micellar phase. The stacking then occurs when the micelles with the analytes concentrate at the sample/BGE boundary. The micelles can also be used in a so-called sweeping mechanism that is defined as picking and accumulating the analytes by the micellar phase. The two approaches can even be combined, as shown by Quirino and Terabe [49], who presented a two-step approach to concentrate neutral chargeable analytes. The sample is first accumulated inside the capillary by field-enhanced sample injection (FESI). The broad zone of the analytes is then swept into a narrow zone by a micellar pseudo-stationary phase and further separated in MEKC mode. Up to 10^6-fold preconcentration factors have been reported with this method.

Transient isotachophoresis (tITP) can also be used for sample stacking. In tITP, leading or terminating electrolytes are present in the sample/BGE that provides suitable conditions for analyte band narrowing due to the ITP process. The goal here is to employ the ITP only briefly during the injection process for analyte enrichment. Discontinuous BGE conditions only exist transiently for a short time, followed by the re-establishment of the continuous BGE conditions. The stacked analytes are then separated in a normal CZE separation mode.

In most of the stacking techniques, the process takes place at the liquid–liquid interface. Sjögren and Dasgupta [50] have shown that it is possible to perform stacking even without liquid contact. Shortly before the capillary tip is returned back from the sample to the electrolyte vial, at a distance of ~1.5 cm above the BGE vial, a high voltage is applied. This basically forces the liquid residing at the capillary tip to flow slowly inside the capillary and at the same time the analytes (anions) stack at the sample/BGE boundary. The air stacking should not be too long, as a bubble may be introduced to the capillary, so the maximum stacking times were about 2.5 s. The authors observed that the peak area increased slightly compared with the case when the capillary tip was returned to the electrolyte vial without the air stacking and also, in some cases (slow anions), an increase in plate numbers.

Some of the major trends in sample stacking and stacking modes were presented briefly above, but various other mechanisms and techniques have been developed in recent years for a wide range of analytes of interest. Discussions on the details, theoretical description, and the up-to-date status of sample stacking can be found in a few recent excellent reviews [51–54].

3.2.5.2 Advantages and Performance The major advantage of sample stacking is the enhancement of concentration sensitivity. Various modes of sample stacking perform differently with respect to what concentration factors can be achieved. The normal stacking mode (FASS) provides only moderate preconcentration factors between 10 and 100, because the stacking takes place from a limited volume of the sample present in the separation capillary. For instance, for a 50 μm i.d. length capillary half-filled with the sample solution, the volume of the sample is only 500 nL. The stacking using ACN also provides only moderate preconcentration factors comparable to FASS; however, due to the different stacking mechanism, it can be used with samples that contain high salt concentrations. On the other hand, HC–FASS that utilizes the introduction of a short sample plug into the capillary and EK stacking from the whole volume of the sample vial can achieve preconcentration factors of up to several thousand. Stacking from the flow presents an interesting modification of HC–FASS, where theoretically the sample volume is unlimited. Preconcetration of neutral analytes using field-amplified sample injection or sweeping can preconcentrate analytes up to several thousand-fold, depending on the affinity of a particular analyte to the micellar phase. The combination of stacking and sweeping mechanisms provides even higher preconcentration factors that can reach a 10^6-fold increase in sensitivity.

3.2.5.3 Disadvantages The sample stacking is a very efficient technique to overcome the low sensitivity of CE. The whole stacking process requires, however, precise control of the applied conditions. Some stacking techniques require manual operation and monitoring of the electric current during the sample plug removal (in case the plug is removed) to achieve reproducible results. Other matrix removal techniques may only be applicable to a specific sample/matrix composition. Large variability of the remaining matrix plug inside the capillary can have serious effects on migration time reproducibility as well as on the overall performance. Additionally, the sample stacking is not very efficient when the sample contains medium to large amounts of fast migrating ions (such as inorganic anions or cations), as these ions will be stacked most efficiently and their peaks will become very large, causing co-migration and other effects on the remaining peaks of interest. Stacking of neutral analytes can be performed from samples with high salt content; however, as discussed before, the preconcentration factors are influenced by the affinity of the analyte and the micelles, so the concept may not be universally applicable to all neutral analytes. The same applies for the ACN stacking method, which only works with very specific concentration ratios of ACN and salts.

Quantitation with the sample stacking technique may be problematic as the sample matrix will affect the amount of analytes stacked. In most of the published literature, the major focus was on the increase in concentration sensitivity. Only rarely has the quantitation of the analytes in the samples been performed. Thus, sample stacking is so far predominantly used to detect trace amounts of compounds with a lack of subsequent quantitative analysis. Appropriate validation of quantitative analysis with sample stacking will be an asset in future investigations.

3.2.6 Alternative Batch Sample Injection Techniques

Besides the most commonly used gravity/pressure-based HD or EK injection, several alternative sampling methods and devices have been developed, driven by the unsatisfactory results obtained with manually operated HD or EK injection and, in some cases, due to the need to develop a case-specific sampling application, such as sampling in portable instruments and sampling of miniature volumes.

Only the "batch-type" alternative sampling techniques will be discussed here, as the automated flow-based approaches will be treated separately in a later section.

3.2.6.1 Rotary-Type Injectors for CE In 1989, Tsuda et al. [55] modified a rotary-type LC injector for CE sample injection. To achieve compatibility with CE, its parts were constructed from ceramics and PTFE. The selected materials eliminated the bubble formation frequently observed with an identical injector made of stainless steel. The repeatability of the injection was shown to be below 2.2%, which represents relatively good value compared with manual HD injection. The advantage of using the rotary-type injector is that exactly the same sample plug is injected regardless of the sample viscosity; thus, one of the few drawbacks of the HD pressure-based injection can be eliminated. Second, the separation capillary is not moved during the injection and the extraneous injection should be eliminated. The developed injector prototype had a fairly large sample loop volume (350 nL). Such a volume would be too large to be used with 50 or 75 μm FS capillaries, as the volume of such capillaries ranges between 1 and 2 μL. The CE separation with rotary valve injection was thus performed in a 200 μm i.d. tube made of PTFE and only a single analyte was analyzed. In 2003, Iizuka et al. [56] published an extension of the work on the development of a rotary-type injector for CE. A new injector prototype was constructed having a much smaller injection volume. A 3 mm long FS capillary (h) was embedded into a precisely machined hole inside a rotor part (e) of the injector using epoxy glue. A schematic of this injector is shown in Figure 3.4.

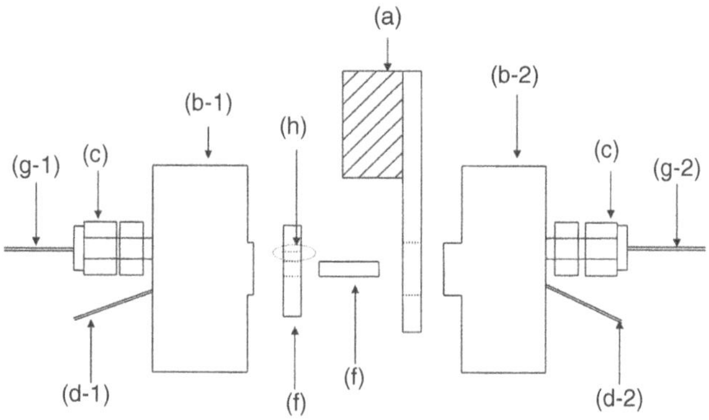

FIGURE 3.4 Schematic diagram of proposed nanoinjector for CE. (a) Knob screws; (b-1) inlet stator (PI); (b-2) outlet stator (PI); (c) 5/16 in union nut; (d) fused-silica capillary tubing (7–100 cm) for sample inlet (d-1) and vent (d-2); (e) rotor (PI); (f) central pin; (g-1) inlet capillary tubing; (g-2) capillary column; (h) sample loop. Reprinted with permission from Ref. 56. Copyright (1994) American Chemical Society.

The injection volume could be varied by selecting the i.d. of the 3 mm capillary and was between 6 and 24 nL for 50–100 μm capillaries. These injection volumes are compatible with CE separation capillaries. The repeatability of a peak area using a single analyte injection and CE separation was about 3%, which is relatively good considering the small injection volumes. This rotary-type injector represents a very useful sampling device because it is fairly robust and it could be used as the best alternative to HD sample injection, especially if the injection precision could be improved. Because the reported values include both the injector and CE separation errors, it would be valuable to measure the actual performance of this injection valve as a stand-alone device.

3.2.6.2 Hydrodynamic Sample Splitting as Injection Method for CE Splitting of a liquid stream is a common method to achieve very low flow rates in microbore LC techniques, such as capillary LC or IC. The HD splitter is a simple device that divides an incoming flow of high flow rate into two outgoing channels, one being the waste channel and the other the split-flow channel. The desired flow rate can be achieved by careful selection of the diameter and length ratios of these two channels. Typically, most of the total flow is diverted into the waste channel, while only a very small flow passes through the split-flow channel. The splitting principle can also be used for sample injection in CE; a defined volume of sample (typically several tens to hundreds of microliters) can be introduced into the splitter. The sample amount that enters the separation capillary is an infinitesimal part of the total sample plug passing by the capillary end at the "splitting" point [57]. The sample injection is followed by an identical injection of BGE to remove the sample left in the splitter interface. During this BGE flush, an additional plug of liquid is introduced into the separation capillary; thus, both the sample and the BGE volumes injected have to be precisely controlled. In 2004, Tsukagoshi et al. [58] described a splitter device constructed from two T-pieces made from PTFE connected in series. Sample injections were accomplished by using a syringe pump that delivered the sample at a defined flow rate (typically 100 mL h^{-1}) for 10 s. The sample injection was followed by a BGE flush of the same duration (10 s) and flow rate. The calculated injection volume was about 7 nL, and the repeatability was 2.7%. The reported repeatability value was the RSD of peak areas after CE separation, demonstrating the overall performance of the system, including the variations of the injection volume and the CE separation.

A similar principle of splitting the sample stream is also used in the so-called cross samplers. Cross samplers are built according to the ideas used in microchip electrophoresis. A simple cross sampler consists of two crossing channels. Two pieces of a separation capillary with a small gap in between are inserted into one channel (separation channel) while the sample introduction channel is perpendicular to the separation capillaries. The other two ends of the separation capillaries are placed in the BGE vials. Only the longer part of the separation capillary is used for separating the analytes, while the shorter part is used only for BGE supply. Injection is performed by hydrodynamically pushing the sample solution through the sample introduction channel. The sample amount pushed past the crossing point is split into three streams, two of which form a sample plug in the separation capillary and the third stream exits through the waste channel. Similarly to sample stream splitters, the size of the sample plug is determined by the dimensions of the separation channel and the waste channel. Unlike in the sample stream splitter, there is no need to flush the cross sampler with the BGE after the sample injection. The EOF will carry only the sample plug located in the separation capillary toward the detector. Various designs of sample splitters and cross samplers are shown in Figure 3.5.

Kulp et al. [60] used cross samplers mechanically manufactured out of poly(methyl methacrylate) (PMMA) to

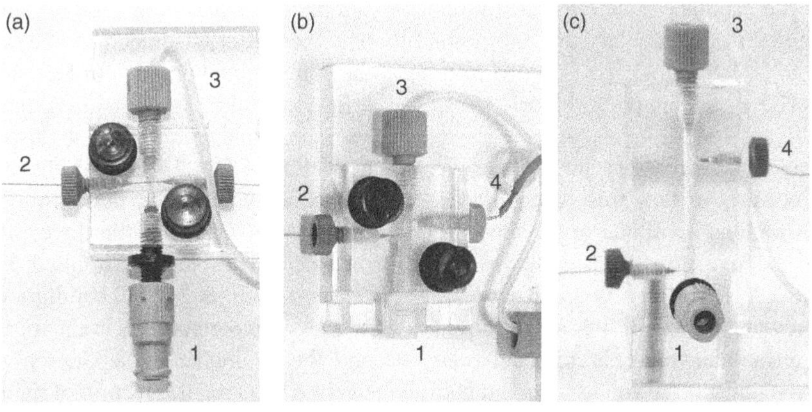

FIGURE 3.5 Various designs of (a) cross sampler and (b, c) sample splitters. (1) Sample inlet; (2) separation capillary; (3) sample outlet; and (4) Pt electrode. Reprinted from Ref. 59 with permission from Elsevier.

monitor the oxidation reaction of L-ascorbic acid by hydrogen peroxide. The sample introduction channel with a diameter of 0.4 mm was filled with two pieces of fused-silica capillary (0.75 μm i.d.) with lengths of 18 and 50 cm. The distance between two ends of the separation capillary was about 30 μm. The dimensions of the two pieces of sample introduction channel made of Teflon tubing were i.d. 0.3 mm and lengths of 9 and 20 cm. The minimum pressure necessary for the proper functioning of the sampler was about 98 kPa. The same group used this type of cross sampler with a portable CE instrument where automated pressurized injection was replaced by manual injection using plastic syringes [61]. When performing the injections manually, it was very difficult to control the velocity of the liquid flow, and therefore the reproducibility of peak areas was poor, the RSD being ~50%. The use of an internal standard improved the reproducibility to a satisfactory level (RSD <8%). On comparing the cross sampler and the simple splitter using the manual injection mode, the cross sampler was found to perform slightly better [59].

The sample splitters and cross samplers are interesting alternative injection devices that could be useful in portable CE instruments because their operation is very simple. In cross samplers, compared with sample splitters, one operational step is removed that could eventually lead to an overall better performance. On the other hand, precise positioning of the separation capillaries in the sampler cross-section and achievement of a tiny gap between them are not trivial and some improvements to the currently available devices are still needed.

3.2.6.3 Electrokinetic Sample Splitting as Injection Method for CE The splitting principle described above is also applicable in electrokinetically driven injections as similar splitting principles also apply to electric currents. Instead of using the HD pressure to drive the sample and the BGE past the splitting point in a splitter, the flow is achieved by means of electroosmosis and electrophoresis. The sample portion introduced into the capillary is determined by the ratio of the currents applied to the capillary and the drain tube [62,63].

Analogously to the HD cross samplers, an EK cross sampler developed by Evans [64] uses electroosmosis to generate the sample liquid flow necessary for the sample injection. The sampler consists of two crossed capillaries with equal lengths of 30 cm and a volume in the crossing region of ~30 nL. For the injection, the sample vial is grounded while a negative high voltage is applied to the sample outlet vial. To prevent the flow of the sample liquid into the separation channel, a small voltage is applied to both BGE vials. It was discovered that if the voltage is about half of the voltage applied to the sample vial, the flow through the separation capillary is maintained close to zero. During the separation, the situation is opposite: the BGE vial is grounded, while a high voltage is applied to the BGE waste vial and a smaller voltage to the sample vials. Switching between sample injection and separation mode is achieved by a simple high-voltage (HV) switch. The reproducibility of five consecutive runs was between 1.7% and 4.1% for peak heights and 2.3–10.7% for peak areas. The lower reproducibility of peak areas was attributed to partial overlap between two peaks. Compared with HD splitting devices, the EK sample splitting is likely to be prone to suffer from the biases accompanying EK injection. No detailed study on these effects was provided by the authors as the devices were only presented as a proof of concept.

3.2.6.4 Dual-Opposite End Injection in CE The concept of dual-opposite end injection (DOI) represents a unique approach applicable for sample injection in CE only that involves the injection of two sample aliquots from the opposite ends of the separation capillary. This approach was first presented by Kubáň and Karlberg [65] in 1998, followed by a very similar concept by Padarauskas et al. [66]. In CE, the charge balance must be maintained during the injection and separation steps and for each number of positive charges (carried by the sample and/or electrolyte molecules) that are transported toward the cathode there must be the same number of negative charges transported to the anode. It is thus intuitive that during a separation run, both anionic and cationic species migrate through the capillary in opposite directions, provided that the EOF is suppressed or very low. In a typical CE system with one end injection, only a specific group of ions is detected (either cations or anions), although the migration of analytes with opposite charge exists. If a sample is injected from both ends of the separation capillary, both anions and cations can be detected. Typically, the detector is placed in the center of the separation capillary, but can be moved closer to one end according to the separation selectivity needs. Several interesting aspects of the dual-opposite end injection approach are worth detailed description. First, the injection of the sample plugs into both capillary ends can be accomplished by either EK or HD injection, and there is no difference from the conventional practice of HD or EK injection. When HD injection of the sample is used, however, the injection of a second sample would expel the first injected sample portion and either HD injection of the electrolyte solution plug immediately after the first sample portion or application of a short HV pulse while the capillary end is immersed in the electrolyte solution is applied. Then the second sample is injected from the second capillary end. As a result, the two injected sample portions are maintained inside the capillary and the simultaneous separation can be started. Alternatively, a sequential injection of the two sample plugs can be used with a time delay. This approach can resolve the problems with co-migration of cationic and anionic analytes, as they may arrive in the detector at the same time.

The concept of DOI–CE has become fairly popular recently, as demonstrated by a growing number of applications that include the analysis of small inorganic anions and cations [67–71], large organic ions [72–75], and neutral analytes in DOI EK chromatography [76], and has been even applied in microchip analysis [77] and in flow injection samplers [78], as described later. The principles and applications of this technique have also been reviewed recently [79,80]. The performance in terms of precision and sensitivity of the DOI–CE techniques is comparable to the that achieved with single HD or EK injections. The advantage of DOI–CE is in the ability to analyze cationic and anionic analytes simultaneously with potentially higher sample throughput.

3.3 MICROMACHINED/MICROCHIP INJECTION DEVICES

Micromachined devices for sample injection in CE occurred recently as a consequence of the development of microfluidic devices and microchip electrophoresis. The micromachined/microchip injection devices could be advantageous as very low dead volumes between the sampling point and the injection end of the separation capillary can be achieved. This is indeed true when such devices are fabricated on the same platform with the microchip separation system. The microchip CE separation systems are, however, outside the scope of this chapter as they are a rather specific topic. Here, the focus is only on the micromachined/microchip devices that are connected to an ordinary fused-silica CE capillary for sample injection. Ideally, the volume between the injection port of such a device and the injection end of the separation capillary should be zero. This is hardly achievable; however, very low volumes in the picoliter to low nanoliter range should be possible, although difficult, to achieve. This may be one of the reasons why the examples of such devices are limited. A micromachined device for sample injection in CE was described by van der Moolen et al. [81]. The device is a T-shaped channel machined in silicon. The leg of this device is glued directly to the separation capillary, and the two arms are connected to the sample and buffer reservoirs. The construction principle resembles the pneumatic CE sampler described by Kaljurand et al. [82]; however, the injection in this device is only EK. The device was used for correlation CE, where repeated pseudorandom injections of sample and buffer were made with 5 s sequences and the correlogram was recorded and processed. The authors showed that the repeatability of the EK injection was excellent, about 1%, and they also showed that for low sample concentrations the improvement in signal-to-noise ratio (SNR was about eightfold. In 1999, Bings et al. [83] used a drilling procedure to generate extremely low dead volume couplings between a microfluidic chip and an external capillary. A 200 μm tungsten carbide drill with a flat tip was used to eliminate the dead volume, as even a 0.7 nL dead volume had a significant effect on separation efficiencies in the CE separation capillary.

It has been demonstrated that the injection from a microchip device into a CE separation capillary is extremely delicate and low dead volume connections are critical for transferring separated solutions off the chip. Here, the CE separation capillary served merely as a connection between the microchip platform and the electrospray mass spectrometer (ESMS) detector, and was later integrated in the microchip–CE–ESMS system [84]. In principle, a similar design was later used by several other researchers for constructing low-dead-volume connectors to couple CE and MS. A few other examples in which a microchip platform is used for sample injection into a fused-silica CE capillary have recently been summarized in a review by Fang and Fang [85], although the approaches presented fall rather into the category of automated flow sampling devices presented later.

3.3.1 Droplet Sampler Based on Digital Microfluidics

Digital microfluidics (DMF) uses discrete unit-volume droplets that are moved on the DMF platform. Typically, 1 unit of fluid is moved over 1 unit of distance. This facilitates the use of microfluidic biochip design approaches such as cell-based assays. Because each droplet can be controlled independently, these systems can be reconfigured (reprogrammed) to change their functionality during the concurrent execution of a set of bioassays.

DMF is an attractive platform for biological applications, which often require the use of expensive or rare reagents and small amounts of samples. DMF has been applied for solving various biological problems such as enzymatic and immunoassays, proteomics, and DNA analysis. Applications such as cell-sorting and cell-based assays, polymerase chain reaction (PCR), and pyrosequencing have been reported. These and several others were described by Miller and Wheeler [86]. As DMF devices are not yet very common, a few words will be devoted to their preparation and their actual function. Various techniques for droplet manipulation using surface acoustic waves, optoelectrowetting, electrowetting-on-dielectric (EWOD), and so on have been demonstrated recently. EWOD is a phenomenon in which an electric field changes the contact angle and thus the wetting behavior of a polarizable and/or conductive liquid droplet in contact with a hydrophobic, insulated electrode. The insulating film is frequently made of two layers of Paralyene C and Teflon-AF. The application of a voltage to a series of adjacent electrodes (that can be turned on or off) can be used to manipulate droplets because the hemispherical shape of the droplet becomes asymmetric and the internal tension of

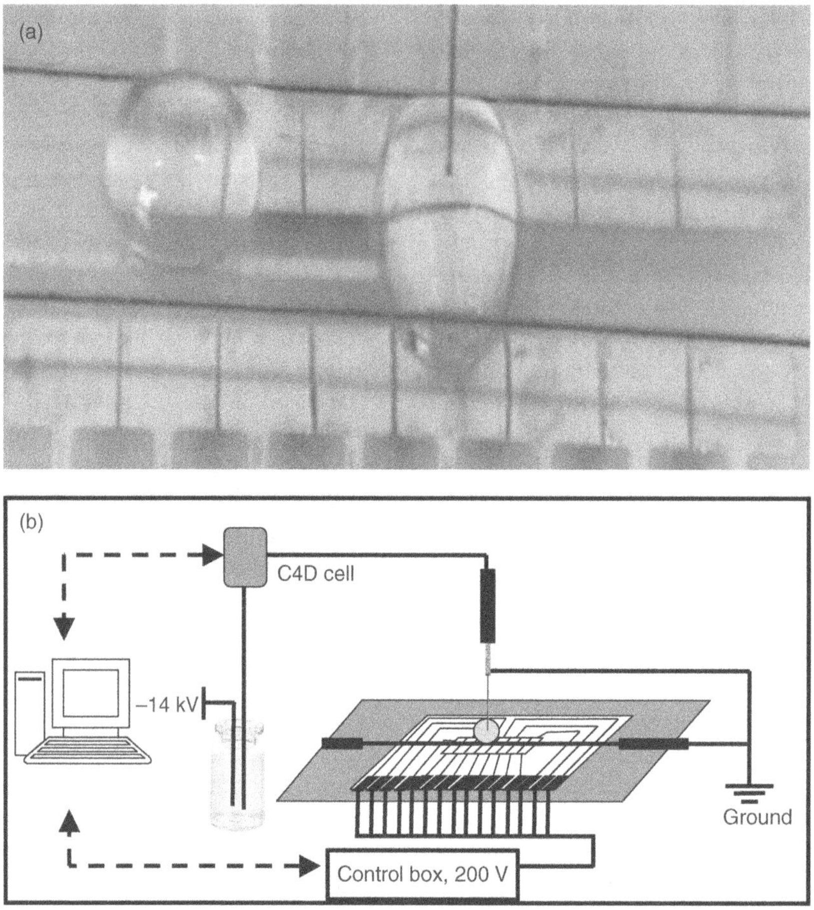

FIGURE 3.6 (a) Droplet deformation due to the application of electric potential to one of the electrodes in an open setup. Droplet movement results from a change in the geometric shape of the droplet. Left, no voltage applied; right, voltage applied across the two adjacent electrodes. (b) Schematic of a digital microfluidic sampler for CE. Reprinted with permission from Ref. 88. Copyright (2009) American Chemical Society.

the droplet actuates its movement [87]. This effect is illustrated in Figure 3.6a.

DMF makes possible a completely new quality of sampling in CE—programmability of sample preprocessing and manipulation. Abdelgawad et al. [89] recently described a hybrid microfluidic device in which a sample was delivered to the separation channel by a DMF device. In another study, a common CE device with a contactless conductivity detector was interfaced to DMF platforms [88]. The DMF sample injection was performed by transporting sample and buffer droplets in succession below the end of the CE capillary inlet, immersing the capillary in the sample/buffer droplet, and performing CE separation by applying a high voltage between the (grounded) buffer droplet and the CE outlet reservoir. Movements of the sample and buffer droplets were controlled by a computer sending electrical pulses to the electrodes covered by common food wrap (see Figure 3.6b). Using a DMF sampler, CE separation of a mixture of vitamins was achieved [88].

3.3.2 Wire Loop Injection

Dasgupta and Surowiec [90,91] described yet another alternative device for injection in CE by using a wire loop deployed at the injection end of the separation capillary. Various geometric configurations of the loop samplers are depicted in Figure 3.7. When the capillary tip with an attached Pt wire loop is immersed in the sample solution and then slowly removed, a thin film of sample with a well-defined volume will reside in the loop.

This volume has been shown to be independent of the sample surface tension and, to a large extent, also independent of the sample viscosity. The whole sample volume from the loop could be reproducibly aspirated into the injection capillary end by applying precisely controlled negative pressure. By carefully adjusting the pressure to about 3.5 kPa, the introduction of air bubbles was avoided. The injection volume was ~13 nL and a single analyte injection gave a reproducibility of peak areas better than 2% RSD,

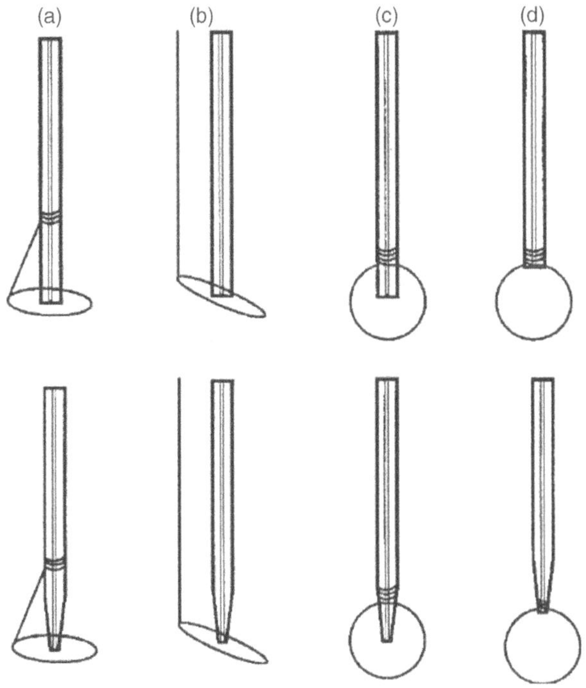

FIGURE 3.7 Various modes of fabricating small loops at the tip of a capillary. The top row shows capillaries with square-cut ends and the bottom row shows the same configuration with a tapered tip capillary. Reprinted with permission from Ref. 91. Copyright (1994) American Chemical Society.

while multianalyte determination using CE system resulted in RSD values between 2.8% and 3.7%. The loop injection device represents an interesting alternative for HD injection that is independent of the sample viscosity but, as pointed out by the authors, it is difficult to reproduce exactly the same sample loop. The commercial construction of this sampler may be therefore difficult. Also, the loop injection device requires precise control of both the speed of capillary tip withdrawal from the sample and the pressure that is applied to aspirate the loop content into the capillary. The precision and repeatability are comparable to those of other HD injection devices. This sampling device may be useful when samples with significantly different viscosities need to be quantitatively analyzed. In a later publication [91], the same authors showed that the sample volume from the sample loop can be introduced into the separation capillary by electromigration, resulting in a dramatic decrease in the "mobility bias" typically encountered in EK injection. An additional benefit of the whole loop EK injection was the increase in the linear dynamic range of the calibration curve to over three orders of magnitude. It was also shown that during EK injection the applied field geometry is very important. A decrease in the "mobility bias" was also observed in EK injection from a small hemispherical depression having a volume of 10 μL. Fuller and Sweedler [92] studied the effect of EK injection by sampling from micro- and nanovials (volume less than 1 μL, typically 10–340 nL), and severe problems were encountered due to such low sample volumes. Analyte depletion due to the electrolysis, formation of new species (redox reactions), and extraneous sample injection at 0 kV (probably due to the surface tension when using nanovials) were observed. There were also great differences when different sample vial volumes were used. The positioning of the capillary tip and electrode was also found to have a large effect on the amount of sample injected. It was concluded that it is very difficult to inject samples reproducibly from such small volumes, and that quantitative analysis may be almost impossible.

3.4 AUTOMATED FLOW SAMPLE INJECTION AND HYPHENATED SYSTEMS

3.4.1 Introduction

Automated *batch injection* of samples has been successfully used in commercial CE equipment and a considerable improvement of the injection precision has been achieved compared with manually operated systems. Alternatively, automated *flow injection* in CE can be used because automation and also coupling of CE to other analytical techniques are much more easily achieved in a flow than a batch mode. Flow sampling methods can provide some degree of simplification, as the sophisticated mechanical carousel and lifting devices are replaced by easy-to-construct flow interfaces. In flow sampling systems, the capillary is usually not moved for sample injection; thus, the problems related to capillary manipulations (for instance, extraneous injection) are eliminated. In principle, flow samplers rely on the injection of a portion of a sample from a flowing stream that passes by the injection capillary end. The sample can be delivered to the capillary injection end as a continuous stream or as a discrete plug by using low-pressure pumps and valves, as known from flow injection analysis (FIA) [93]. Often, EK injection is used as it is the simplest technique that does not require any modification of the instrument and is easily implemented for flow sampling. Typically, repeated injection sequences of samples, as in FIA, can be made without the need to interrupt the separation voltage, thus increasing the sample throughput. Automated flow samplers are also very suitable for special techniques such as correlation CE, as will be discussed later.

The flow sampling devices for CE have not yet been commercialized, although their general performance is comparable to that of automated batch-type sampling devices. This may be because the vendors of CE equipment do not fully understand the principles behind FIA and the possible benefits that it could bring as a sampling and pretreatment method for CE. On the other hand, research interest in flow sampling devices has accelerated in recent decades. Various types of devices and principles have been developed and are reviewed in several excellent papers [94–100]. It is not possible to encompass all flow sample techniques and devices here, so only a brief overview of historical aspects with some important milestones, advantages, drawbacks, and some performance data will be given. Further information can be found in the previously mentioned reviews.

Historically, some of the previously described sampling devices, for instance, the electric sample splitter developed by Deml et al. [62], can be regarded as a simple flow sampling device. Probably the work of Liu and Dasgupta [101] in 1993, in which a 100 nL volume of sample was injected via a microvalve into a 75 μm separation capillary performing repeated multianalyte determinations, can be considered the first real example of flow sampler in CE. Repeated injections of sample without the voltage interruption were shown for the first time. An online coupled dialysis system presented by Hogan et al. [102] in 1994 is an example of coupling of two analytical techniques using a flow interface. A pneumatic sampler was constructed by Kaljurand et al. [82] in 1995. It relies on quick pressure actuations of the sample and BGE vessels connected to the separation capillary through a T-shaped channel via a set of automatically operated solenoid valves. The sampler was shown to be very well suited for monitoring purposes, as the composition of the reaction (sample) vessel can be repeatedly injected into the CE capillary [60,103,104].

The continued interest in flow (or flow injection) sampling for CE was stimulated by two publications by Kubáň et al. [105] and Fang et al. [106] in 1997, who independently presented a design of a flow injection–CE (FI–CE or FIA–CE) interface. These interfaces are shown in Figure 3.8.

Since then, there has been a surge of interest in flow sampling systems and different modified versions thereof were developed. These include a, falling droplet interface [107], microfabricated CE interface [108], and coupling of sequential injection analysis to CE [109,110]. Additionally, various hyphenated systems for sample treatment, derivatization, preconcentration, dialysis, gas diffusion, ion exchange, and monitoring were also developed.

3.4.2 Advantages and Performance

The overall performance of the developed flow devices for injection in CE shows relatively good precision (as defined by the RSD of peak areas), and typically RSD values around 1–3% were achieved. These values compare well with the values reported for automated batch-type commercial instruments and again represent a significant improvement over manually performed injections. It can be therefore concluded that the automated operation of the sampling devices improves significantly the overall precision of the CE analysis. In flow samplers, the good reproducibility obtained may be due to the following reasons: first, the injection precision of injector valves used in flow systems typically reaches the precision obtained with other liquid

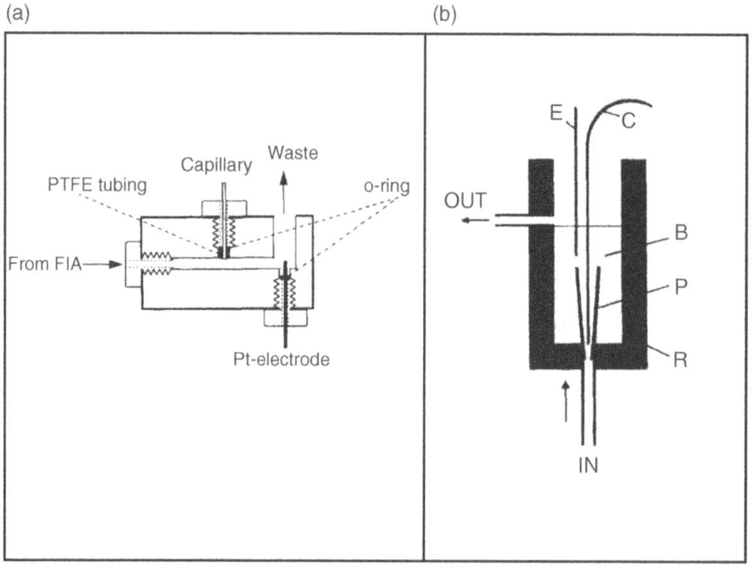

FIGURE 3.8 Schematic diagram of the two original on-line FI–CE interfaces. (a) The interface developed by Kubáň et al. (b) The interface developed by Fang et al. R, Plastic reservoir body; E, platinum electrode; C, capillary column; OUT, waste solution outlet; IN, inlet; P, conical pipette tip; and B, buffer electrolyte solution (with sample waste). Reprinted from Refs. 105,106 with permission from Elsevier.

chromatographic techniques, for example, RSD values below 1% are common. Second, in most of the flow injection devices, the separation capillary is not moved, thus eliminating some of the previously mentioned problems, including, but not limited to, extraneous injection, diffusion, and introduction of bubbles. There is one more feature of the FI–CE devices that may have an effect on the stability of the CE separations. In CE analysis, it should be good practice to include a voltage conditioning step prior to each injection and CE analysis. Ross [111] showed that by including a short (2 min) voltage preconditioning step prior to each CE analysis, the reproducibility of the migration times and peak areas could be greatly improved. This practice is often overlooked in CE. In FI–CE systems, the high voltage is always on during the separation and sequential injections, so the capillary is being continuously electro-conditioned during the injection sequence. This may lead to an improved performance of the system. Because the repeated injections are often accomplished within a very narrow time interval, the chance that the EOF will significantly change during the analysis is minimized. For instance, Kubáň et al. [105] showed that 20 repeated injections and separations of three inorganic anions could be accomplished within 10 min in one run. Each injection was performed with a 20 s interval. It is apparent that during the 10 min time span the EOF will likely remain constant and its effect on peak areas and migration times will be negligible. Excellent repeatability below 2% was achieved. In a separate experiment, the injection precision of the FIA injector was evaluated as 1.6%; hence the contribution of the CE separation to the overall error was negligible. A similar performance is frequently obtained in other coupled FI–CE systems. Apparently, such a short injection interval is not applicable for samples with a large number of analytes, but even in such a case, the sample throughput can be significantly increased, as the peak overlapping technique [105] can be used.

3.4.3 Disadvantages

In FI–CE systems, the major drawback is their limited applicability to quantitative analysis. Because in most automated/hyphenated systems the injection is EK, the previously discussed biases apply. A simple external calibration is not possible (although it could be very easily performed in an automated fashion). Applying various corrections to improve the results of quantitative analysis did not improve the overall accuracy of the system to below 5–6%. Although there is a significant improvement in repeatability of injection and also the possibility of automated sample preconcentration/pretreatment, the bottleneck in flow injection systems combined with CE is the lack of reliable quantitative analysis. Most of the publications on flow samplers do not deal with quantitation and, more importantly, analysis and quantitation of real samples have yet to be developed.

3.5 COMPUTERIZED SAMPLING AND DATA ANALYSIS

With the spur in the development of information technologies in recent decades, the separation techniques and consequently also CE may benefit from the extreme power that computer technology currently provides. Nowadays, computers can easily perform extensive multitasking, ranging from automatically controlling the injection sequences, recording various parameters of the separation system, evaluating and processing the measured data on the fly, to eventually applying sophisticated integration and evaluation protocols to the measured data.

The ability to process large data files and multiple inputs has recently raised interest in performing multiple separations in parallel or in series. Multiplexing of separation capillaries (several capillaries in parallel), however, puts an increased demand on the detection device; of our interest here will be the introduction of multiple injections in series. If the injections are performed in a sequence, as for instance in FI–CE systems, the total analysis time will increase proportionally to the number of injections performed and the overall procedure will then be somewhat time consuming. Rather than that, it is possible to perform injections in a rapid sequence without the need to wait for the analytes to separate and register the overlapping signals of several injections. Later, mathematical procedures can be used to correlate the obtained detector signal with a complex injection sequence. Such signal averaging typically results in a single electropherogram with improved SNR. This is due to the fact that summing up signals is proportional to the number of signals whereas summing up noise is proportional to the square of the number of signals. Techniques employing this kind of multiplex advantage to achieve better sensitivity are known as cross-correlation CE [112,113] or Hadamard transform CE [114]. Cross-correlation CE relies on the injection of multiple, discrete sample plugs into the separation capillary in a well-defined injection sequence, and treating the detector signal as a sum of single injection responses. From the measured detector signal it is possible to calculate an average electropherogram of single injections if the multiple injections have been performed according to so-called pseudo-random binary sequence (PRBS) or any other binary sequence that contains no repetitions. This condition needs to be satisfied to process and correlate the detector signal and injection sequence mathematically. PRBS generators construct injection sequences with length of 2^{n-1}, where n is a natural number. The maximum theoretical SNR improvement factor can be calculated as

$$\mathrm{SNR} = \frac{2^{n-1}}{\sqrt{2^{n-1}}} \approx \frac{2^{n-1}}{\sqrt{2^{n}}} = \frac{\sqrt{2^{n-1}}}{2} \qquad (3.5)$$

For instance, if $n = 10$, the length of the injection sequence is 511 and the maximum theoretical achievable SNR improvement factor is ~11. Although the multiplex advantage seems very attractive, especially when the number of injections is significantly increased, the SNR improvement is always lower than the theoretically calculated maximum. To implement multiplex techniques, extremely high precision of sample injection is required, as reduced injection precision leads to a number of detrimental effects, eventually reducing SNR enhancements. Low sample injection repeatability is then the source of artifacts such as synthetic low-frequency noise (correlation noise) or artificial peaks that carry no chemical information (ghost peaks).

Cross-correlation CE can be successfully applied with various detection techniques such as UV [81,112,113], laser-induced fluorescence (LIF) [114], or even amperometric detection [115]. LIF performed slightly better than UV detection in cross-correlation CE experiments . SNR improvements in LIF reported so far have been up to 18-fold [116], compared with 6.5 in UV detection [117]. This is explained by the smaller sample concentrations used with LIF detection, resulting in a much more homogeneous electric field across the capillary than with a UV detector. Consequently, maintaining the homogeneity of the separation medium with higher sample concentrations is more problematic. In extreme cases of sample stacking conditions, cross-correlation CE is impossible [118].

As correlation CE works best when the high voltage is not interrupted during the PRBS injection sequence, preferably nonconventional samplers, such as a pneumatic autosampler [113], microchip with HV switch [112], or photolytic optical gating [114], are used. HV interruptions could become a possible source of uncertainty as it always takes some time after switching the HV power supply on to reach its final value. The same applies when the HV is switched off. This error could be tolerated in the case of a single injection, but is multiplied during cross-correlation CE experiments. This is probably why only one example of the use of commercial CE equipment with an autosampler without any modifications has been reported recently by Seiman and Reijenga [119]. Two- to threefold SNR improvements were still achieved with only slightly longer analysis times compared with normal single injection experiments. In addition to lower precision due to HV interruptions during the sample injection, several other limitations specific to commercial devices have been reported. For instance, the length of the injection sequence was limited because the instrument was not capable of registering the signal while it was in the sample injection mode.

The cross-correlation technique is theoretically an excellent way to improve the low sensitivity that often accompanies CE analysis. In practice, the lack of injection precision and problems associated with data registration still limit the SNR improvement. Moreover, in the case of complex samples, this effect is multiplied as every sample component produces extra amount of low-frequency correlation noise. Therefore, cross-correlation experiments reported so far have not been applied for real sample analysis and improvement in this area is needed.

3.6 SAMPLING IN PORTABLE CE INSTRUMENTATION

Capillary electrophoretic methods offer great potential for portability, compared with other liquid separation methods, such as LC or IC, since no highly sophisticated high-pressure pumping systems are required. The HV power supply necessary for the separation can be provided by an inexpensive, lightweight, DC–DC converter to which the input power can be delivered by a battery source, because the electric current values in CE systems are very low (typically 10 μA or less). With the development of contactless conductometric detection or LED-based detectors, the power and size requirements of the portable CE instruments have decreased further and it is now possible to operate the whole instruments for extended periods using a battery source only.

In the last decade, the development of portable CE instruments has seen a remarkable surge [61,120–128]. Portable CE instruments were also recently reviewed [129]. The most common detection mode is C^4D, although other detection concepts including LED-based detectors [127], potential gradient detectors [123,126], and potentiometric and amperometric detectors [120,121] have been presented. There is even a portable CE instrument commercially available.[3]

As the portable instruments are mainly intended to be used in the outdoor environment, maintaining highly reproducible operating conditions comparable to the analytical laboratory is difficult. Outside temperature and humidity may differ and field sampling is also more prone to various sources of contamination. As a consequence, the portable systems should provide an increased degree of robustness toward these effects. Additionally, sampling devices should be robust yet reproducible to allow operation under difficult conditions. For instance, portable instruments are often used to detect the presence of warfare agents that are highly toxic and require a protective suit and gloves to be worn. The sampling in such a case should be as simple as possible; handling microliter sample vials with rubber gloves is not an easy task. Perhaps, the best option for sampling under these conditions is to use syringes. The injection should preferably be done so that the capillary movement is limited. Ideally, the two capillary ends should be in a fixed position during

[3] http://www.ce-resources.com/.

the sample injection and analysis. The injection in most of the presented portable instruments is, however, still accomplished by conventional injection procedures, either by EK injection [120–128] or by applying pressure [61,124–128]. Some of the portable CE instruments include a simple sampling carousel, similar to those present in the commercial instrument that is operated manually.

The overall performance of the portable instruments, expressed mostly as RSD of peak areas, is inferior to those of similar laboratory-operated instruments. Typical reported values are usually only slightly better than 10% RSD. Attempts to simplify the design of the samplers in portable CE instrument were reported by Kaljurand's group [59,61]. The tedious manual HD of EK injection was replaced by flow-splitting devices of various geometries. However, even with this simplified injection based on flow splitting, the reproducibility of peak areas was no better than 9% RSD.

Concerning the application of portable instruments in quantitative analysis, it has been rather limited, as only in five of the 11 published papers was a quantitative analysis performed. Currently, the performance of portable CE instruments still deserves improvement as a few potential drawbacks need to be resolved. There is a need to develop robust yet simple injection techniques, providing increased precision. A proper correction for temperature changes without the use of a thermostating device would also be desirable. Reliable and accurate quantitation methods still need to be developed as well. For quantitative analysis in CE, various techniques have been suggested for improving the overall performance, some of which may be tested in portable instruments. A short overview of the quantitative aspects of CE follows.

3.7 QUANTITATIVE ANALYSIS IN CE

3.7.1 Introduction

It is not an overstatement to say that quantitation of the analytes separated in samples is the ultimate goal of any separation technique used by the analytical chemists. Accurate estimation of the analyte content in a sample from the measured data depends on many parameters of the technique. A prerequisite for such an accurate assessment is a reproducible injection of a sample aliquot that is a representative portion of the analyzed sample. As has been discussed before, unlike with valve injection in LC or IC, the injection methods in CE vary greatly with regard to the representativeness of the sample portion injected into the CE separation capillary. HD injection is preferred as in most cases no bias in the injection step is introduced, in contrast to EK injection that suffers from multiple biases. If an accurate and representative sample injection is performed, the measured peak areas (rather than peak heights) in the electropherogram should reflect the concentrations of each individual analyte in the sample. When estimating concentrations of the analytes in the sample from their respective peak areas, an additional difficulty exists in CE due to the so-called detection bias. In chromatography, all the separated analytes move through the detector at the same speed—the speed of the mobile phase. In CE, in contrast, the separated analytes move through the detector with different speeds. As a result of this "detection bias," if two analyte zones with identical bandwidth move through the detector at different speeds, their peak widths and consequently peak areas will differ because the slower-migrating zone will spend considerably more time in the detector. This will result in an increase in the peak area of the latter. The "detection bias" is not a problem provided that the migration times are stable; the plot of the peak areas versus concentration of analytes yields a linear dependence. However, once the migration velocity of the analyte zone is changed, that is, by small changes in the EOF, adsorption of the sample compounds on the capillary wall, or temperature changes, the strictly linear relationship will no longer be valid and large deviations can be observed. Fortunately, a simple correction technique, suggested by Huang et al. [130], can be used for correcting the "detection bias." When the peak area of each individual analyte is divided by its migration time, the corrected peak area is obtained that should be free from any variation induced by differences in the migration velocities. However, it should be noted that this correction does not account for other sources of peak broadening not related to the change in migration velocity. For instance, if adsorption of the analyte on the capillary walls occurs, it often results in excessive peak tailing that cannot be corrected for by this simple method. If the precision of the injected amount is low, the migration time correction will not be effective either. The following are some basic means for standardization of the measured data with both the HD and the EK injection modes.

3.7.2 Quantitative Analysis with HD Injection

Quantitative analysis with HD injection is mostly straightforward and easy. An external calibration with HD injection results in linear calibration plots with excellent RSDs, provided that the standards and sample are prepared in the same solvent. DI water or diluted separation electrolyte can be used for sample and standard preparation. In HD injection, the practice of dissolve the sample in BGE known from LC is also used. If repeatability problems due to the injection are suspected, an internal standard can be added, which may help to correct these deviations. An internal standard should not be present in the sample, should have migration velocities similar to those of the analytes but not co-migrate with any of the analytes, and should be added in a sufficient concentration so that it produces a large peak, the area of which is unaffected by the integration errors. The internal standard should also be stable in the sample and BGE system and should not interact with the sample or BGE components. Dose and Guiochon [131] suggested that using one internal standard with HD injection can considerably improve the precision. In their work, raw peak areas produced RSDs of ~5%. When a simple correction using a single internal standard was performed by dividing the respective peak areas with the peak area of the internal standard, the precision was improved to ~1%. The internal standard method is generally accepted as an efficient method in CE and has been used in numerous studies on the validation of developed CE separation methods.

If the prerequisite of exactly the same sample and standard matrix is not fulfilled, the external calibration may still work. Other possibilities exist that can be used when the sample and standard have significantly different matrices. Diluting the sample with the corresponding BGE (in this case the sensitivity is compromised) is a viable option. If dilution does not produce at least a somewhat similar matrix composition, the use of a standard addition method is recommended. This applies particularly if the viscosities of the sample and standards differ or if the viscosity differs among the samples. This means, however, that the number of experiments that need to be performed is doubled or even tripled. Care must also be taken with various other parameters using HD injection, such as the amount of extraneous injection [15], the sample plug length, accurate timing of the capillary movement between the sample and the electrolyte vials, and positioning of the capillary end and Pt electrode in the vials. All these parameters may have an influence on the amount injected and quantitative analysis. As a general recommendation, the internal standard method should be used when possible.

3.7.3 Quantitative Analysis with EK Injection

In EK injection, quantitative analysis is much more troublesome, as the two commonly recognized biases influence the amount of each ion injected. These biases are the "mobility bias" and the "sample conductivity bias," as mentioned previously. The first type of bias should not pose any significant problem in quantitative analysis, as has been demonstrated by several groups [19,20]. Under some circumstances, the mobility bias and the detection bias can even cancel each other. This may be intuitive because, on the one hand, a higher amount of a fast-migrating analyte is introduced during the injection compared with the slow-migrating analyte due to the migration bias. This effect can be compensated for by using multiplication of the peak area by its migration time [18]. On the other hand, due to the detection bias, the peak areas of the slowly migrating analytes will be over-represented. For correction of the detection bias, the division of the peak area with the corresponding migration time is used. It turns out that these two corrections have an exactly opposite effect, so they can cancel each other—no correction is necessary. Independently, van der Moolen et al. [19] and Qi et al. [20] demonstrated that this is indeed true, provided that the BGE and sample have the same pH and conductivity.

The second bias, which is caused by differences in sample and/or standard solution conductivities, however, is not easy to correct for. An appropriate method has not yet been found that would be able to decrease the variations due to the different sample conductivities to the same level as the internal standard method in HD injection. In some early attempts to correct for EK injection, Dose and Guiochon [131] described a correction method that uses two internal standards. The standards are chosen so that their peaks bracket the analyte peaks in the electropherogram. Variations in injection reproducibility could be corrected to about 1% using EK injection; however, this applied only for samples with similar conductivity. Lee and Yeung [21] applied another correction by measuring the integrated separation current during the analysis. For samples with similar conductivities, excellent precision can be obtained. When samples differed in conductivity, a slight modification of the correction equation was used. Satisfactory results were obtained except for the samples with the lowest conductivity. They concluded that while the applied correction will not work in all cases, the previously described method of two internal standards suggested by Dose and Guichon will work even in the case of different sample conductivities with reproducibility better than 5%. In the pursuit of finding another suitable, universal, calibration method with EK injection that would provide reasonable accuracy, most research failed to improve the results below the "magic" value of 5%. These methods include the following: internal standardization using one or two internal standards, correction and normalization of peak areas by measuring the sample conductivity (cluster coherent potential approximation, CCPA) [132], use of the so-called COPA method [133], in which correction factors (the measured PA in the sample matrix is divided with the PA of the same analyte in the calibration matrix) were measured for 12 different matrices, use of artificial neural network corrections [134], use of transformation of the time-based scale of the measured electropherograms to the mobility scale, and correction of the peak area by migration time correction [135,136]. Gaspar and Gabor [137] showed that when working with low concentrations of analytes in dilute solutions, some degree of linearity is obtained and the effect of the second bias can be neglected, with RSDs better than 6%. Nevertheless, the number of studies that involve quantitative analysis of samples with EK injection and particularly those that deal with samples with differing matrix compositions and conductivities are relatively few and generally RSD values range between 5% and 6%.

3.7.4 Validation of the Developed CE Methods

Validation of the developed CE methods is closely related to the previously discussed issues on precision and accuracy in quantitative analysis. In general, validation in CE includes a separate validation of the CE instrument and a validation of the CE method. The general requirements were described by Ross [138]. The validation of the CE instrument includes validation of the stability of the applied voltage, temperature, and function of the injection device and detector. The validation of the developed method then includes selectivity, peak purity and identity, migration time and peak area precision, detector linearity, and limit of detection. These are some general considerations that should be addressed before submission to the regulatory agency. In general, CE and corresponding developed methods have been validated, to a large extent, for analysis in the pharmaceutical industry. Watzig et al. [139] published an extensive review with strategies for the validation of CE methods for pharmaceutical and biological applications. Altria [140,141] discussed the application and validation of CE in main component assays, stoichiometric determinations, and vitamin analysis, and also the analysis of drug-related impurities and chiral separations. Currently, however, the number of validated CE methods applied by industry, licensing authorities, and the international pharmacopoeias, for example, the European Pharmacopoeia (EP), are limited to specific areas only, such as the analysis of selected amino acids, peptides, and protein compounds.

3.7.5 Computer Data Treatment in Quantitative Analysis

Quantitative analysis relies as much on precise sample injection and highly repetitive separations as on the use of accurate data treatment for postprocessing of the electropherograms. Peak integration may constitute a significant source of irreproducibility of the peak areas in CE analysis. Precise quantification and good reproducibility of peak areas are dependent on the correct determination of peak boundaries. Even small errors in locating these boundaries can cause significant changes in peak areas. Standard procedures for integration of peak areas consist of several operations performed in the following order [142]:

(i) first the baseline is corrected—any drift is removed, the baseline is smoothed if necessary;
(ii) the beginning and the end of the peak are located;
(iii) the peak area is integrated;
(iv) very small peaks are discarded as potential noise artifacts.

A wide range of baseline correction algorithms have been developed. Various tactics from simple accelerated multiple-pass moving average filter [143], improved iterative polynomial fitting with automatic threshold [144], use of discrete wavelet transform [145], and many other methods have been developed for baseline correction. Less attention has been paid to developing algorithms for locating correct start and end points of the peak. Schirm and Wätzig [146] proposed an algorithm for finding peak boundaries that alternatively expand the initial peak by one data point at a time to the left or the right. As a result, peak area is increased. The process stops when the increase is smaller than the threshold, which is set by the standard deviation of the baseline noise. Together with their baseline correction algorithm using smoothing spline functions, an RSD reduction by ~50% was achieved.

Commercially available integration software follows point-by-point the earlier described procedures. Most data can be successfully integrated using this type of software, but peak tailing or other irregularities in peak shapes could become a great challenge, as it can be difficult to recognize a difference between the baseline shift and the tailing peak.

Faller and Engelhardt studied errors introduced by integration software in CE [147]. They used an identical data set, transferred it to different commercially available CE integration software, and compared the reproducibilities of peak areas. All the tested software packages were able to determine correctly the true concentration of an unknown solution within the limits of the standard deviation. However, the new generation of software designed especially for CE does it with slightly better reproducibility than older software originally designed for data acquisition of HPLC systems. Two additional laboratory-written integration algorithms were used by the authors that can cope better with the tailing peaks that occurred in some of the CE analyses. A normal practice in quantitative CE analysis is that the total peak area is divided by the migration time at the peak maximum. It is obvious that in the case of strongly tailing and very wide triangular-shaped peaks the migration time at the peak maximum is not representative of the whole peak. Therefore, one of the developed laboratory-written algorithms divides the value of every single point by the corresponding migration time before calculating the peak area. The principle and comparison of conventional migration time correction and the improved algorithms are shown in Figure 3.9. For the strongly tailing peaks, the algorithm with the migration time correction gives better RSD values, whereas RSD values for symmetrical peaks are about the same for both algorithms.

The integration algorithms have improved over the years of CE practice and now include various ingenious algorithms that can cope better with the particularities of the analytes separated by CE. These include corrections for migration times, peak areas, and peak shapes. The newly developed software should be preferred as it provides more accurate results.

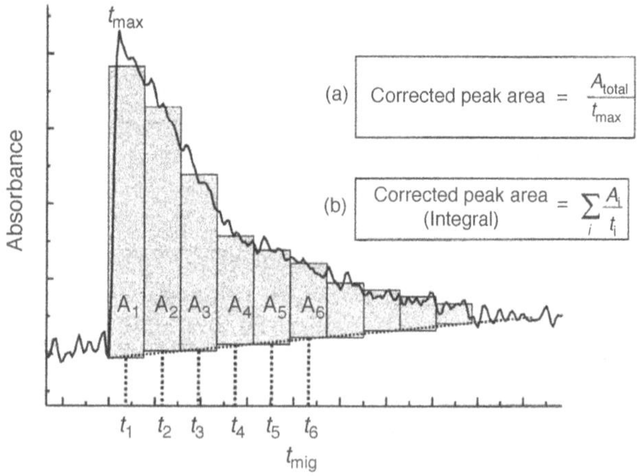

FIGURE 3.9 (a) Comparison of conventional migration time correction and (b) the improved algorithm. Reprinted from Ref. 147 with permission from Elsevier.

3.8 CONCLUSIONS

Sampling in CE is a delicate procedure that requires a proper understanding of the underlying principles. Use of an appropriate sampling technique is of vital importance for subsequent quantitative analysis. Of the two available injection principles, HD injection should preferably be used, as it is less prone to sample-to-sample variations, is free from any sampling bias, and provides higher precision. Injection devices in which the capillary ends are not moved should be preferred as these are typically more robust and do not suffer from additional sampling errors that accompany capillary movements. Rotary injection valves with small injection volumes offer several advantages as sampling bias, and also the influence of varying properties between different samples, are eliminated to the greatest extent. The capillary end(s) are not moved but stationary. They should be easily transferable to the portable instrumentation and should work reproducibly regardless of the operator skills and other conditions. A good alternative for HD sample injection may be sample splitters or cross samplers because their operation can be fully automated, thus eliminating the currently experienced limitations due to manual usage.

An analytical performance of CE comparable to that of LC or IC can only be obtained if a large number of parameters are carefully controlled. The use of improved injection devices with high precision can help, but other procedures should be carefully followed. They include, for instance, precise cutting of the capillary end, washing of the capillary between the runs, thermostating the CE system, and applying voltage conditioning between runs. With the availability of powerful computers able to perform multitasking, advantage should be taken of monitoring various measurable parameters during the injection, separation, and detection. For instance, migration times, EOF migration time, EOF peak width and area, actual current and voltage applied during the analysis, temperature measurements, and so on should all be monitored. They can help process the final electropherograms by applying various correction procedures to yield improved reproducibility of the overall separation process. There is, therefore, a need to develop not only robust enough injectors but also universally applicable procedures for data analysis and postanalysis data treatment.

Only if consistent results with reproducibility ~1% RSD or better can be achieved for analysis of standards and real samples could CE replace other techniques currently used in routine applications and be validated as a standard analytical method.

Quantitative analysis in CE relates strongly to the proper usage and understanding of the sample injection principles. It still requires some attention before it can be used with confidence in analyzing samples other that those prepared in the laboratory by dilution of well-behaved standards in DI water. A large number of validation tests that will prove that the newly developed CE methods are appropriate for the analysis of real samples still need to be performed. Without proper validation, CE will hardly become accepted as a method of choice.

REFERENCES

1. Jorgenson, J.W. and Lukacs, K.D. (1981) *Anal. Chem. 53*, 1298–1302.
2. Altria, K.D. (2002) *LC-GC Eur. 9*, 1–5.
3. Bächmann, K., Steeg, K.H., Groh, T., Röder, A., Haumann, I., and Boden, J. (1992) *Int. J. Environ. Anal. Chem. 49*, 87–101.
4. Tenberken, B. and Bächmann, K. (1996) *J. Chromatogr. A 755*, 121–126.
5. Stuart, J.N. and Sweedler, J.V. (2003) *Anal. Bioanal. Chem. 375*, 28–29.
6. Kunkel, A., Degenhardt, M., Schirm, B., and Watzig, H. (1997) *J. Chromatogr. A 768*, 17–27.
7. Altria, K.D., Harden, R.C., Hart, M., Hevizi, J., Hailey, P.A., Makwana, J.V., and Portsmouth, M.J. (1993) *J. Chromatogr. 641*, 147–153.
8. Altria, K.D., Clayton, N.G., Hart, M., Hevizi, J., Hailey, P.A., Makwana, J.V., and Portsmouth, M.J. (1994) *Chromatographia 39*, 180–184.
9. Altria, K.D., Clayton, N.G., Harden, R.C., Makwana, J.V., and Portsmouth, M.J. (1995) *Chromatographia 40*, 47–50.
10. Mopper, B. and Sciacchitano, C.J. (1997) *J. Capillary Electrophor. 4*, 73–76.
11. Boone, C.M., Franke, J.P., de Zeeuw, R.A., and Ensing, K. (2000) *Electrophoresis 21*, 1545–1551.
12. Tsuda, T., Nomura, K., and Nakagawa, J. (1983) *J. Chromatogr. 264*, 385–392.

13. Schaeper, J.P. and Sepaniak, M.J. (2000) *Electrophoresis 21*, 1421–1429.
14. Cohen, N. and Grushka, E. (1994) *J. Chromatogr. A 684*, 323–328.
15. Grushka, E.M. and McCormick, R.M. (1989) *J. Chromatogr. A 471*, 421–428.
16. Dose, E.V. and Guiochon, G. (1992) *Anal. Chem. 64*, 123–128.
17. Erny, G.L. and Cifuentes, A. (2006) *Electrophoresis 27*, 4166–4173.
18. Huang, X., Gordon, M.J., and Zare, R.N. (1988) *Anal. Chem. 60*, 375–377.
19. Van der Moolen, J.N., Boelens, H.F.M., Poppe, H., and Smit, H.C. (1996) *J. Chromatogr. A 744*, 103–113.
20. Qi, S., Huang, A., and Sun, Y. (1996) *Anal. Chem. 68*, 1342–1346.
21. Lee, T.T. and Yeung, E.S. (1992) *Anal. Chem. 64*, 1226.
22. Linhhares, M.C. and Kissinger, P.T. (1991) *Anal. Chem. 63*, 2076–2078.
23. Pu, Q.-S. and Fang, Z.-L. (1999) *Anal. Chim. Acta 398*, 65–74.
24. Wei, H., Ang, K.C., and Li, S.F.Y. (1998) *Anal. Chem. 70*, 2248–2253.
25. Fishman, H.A., Amudi, N.M., Lee, T.T., Scheller, R.H., and Zare, R.N. (1994) *Anal. Chem. 66*, 2318–2329.
26. Fishmann, H.A., Scheller, R.H., and Zare, R.N. (1994) *J. Chromatogr. A 680*, 99–107.
27. Mikkers, F.E.P., Everaerts, F.M., and Verheggenm, Th.P.E.M. (1979) *J. Chromatogr. A 169*, 11–20.
28. Chien, R.-L. and Helmer, J.C. (1991) *Anal. Chem. 63*, 1354–1361.
29. Burgi, D.S. and Chien, R.-L. (1991) *Anal. Chem. 63*, 2042–2047.
30. Chien, R.-L. and Burgi, D.S. (1992) *Anal. Chem. 64*, 1046–1050.
31. Burgi, D.L. (1993) *Anal. Chem. 65*, 3726–3729.
32. He, Y. and Lee, H.K. (1999) *Anal. Chem. 71*, 995–1001.
33. Zhu, L.Y. and Lee, H.K. (2001) *Anal. Chem. 73*, 3065–3072.
34. Kuldvee, R., Kubáň, P., Vunder, K., and Kaljurand, M. (2000) *Electrophoresis 21*, 2879–2885.
35. Chien, R.-L. and Burgi, D.S. (1991) *J. Chromatogr. 559*, 141–152.
36. Chien, R.-L. and Burgi, D.S. (1991) *J. Chromatogr. 559*, 153–161.
37. Zhang, C.-X. and Thormann, W. (1996) *Anal. Chem. 68*, 2523–2532.
38. Zhang, C.-X. and Thormann, W. (1998) *Anal. Chem. 70*, 540–548.
39. Kuldvee, R. and Kaljurand, M. (1998) *Anal. Chem. 70*, 3695–3698.
40. Kubáň, P., Berg, M., Garcia, C., and Karlberg, B. (2001) *J. Chromatogr. A 912*, 163–170.
41. Aebersold, R. and Morrison, H. (1990) *J. Chromatogr. 516*, 79–88.
42. Shihabi, Z.K. (1995) *J. Capillary Electrophor. 2*, 267–271.
43. Liu, Z., Sam, P., Sirimanne, S.R., McLure, P.C., Grainger, J., and Patterson, D.G.Jr. (1994) *J. Chromatogr. A 673*, 125–132.
44. Nielsen, K.R. and Foley, J.P. (1994) *J. Chromatogr. A 686*, 283–291.
45. Quirino, J.P. and Terabe, S. (1997) *J. Capillary Electrophor. 4*, 233–245.
46. Quirino, J.P. and Terabe, S. (1997) *J. Chromatogr. A 781*, 119–128.
47. Quirino, J.P. and Terabe, S. (1998) *Anal. Chem. 70*, 149–157.
48. Palmer, J., Munro, N.J., and Landers, J.P. (1999) *Anal. Chem. 71*, 1679–1687.
49. Quirino, J. and Terabe, S. (2000) *Anal. Chem. 72*, 1023–1030.
50. Sjogren, A. and Dasgupta, P.K. (1996) *Anal. Chem. 68*, 1933–1940.
51. Quirino, J.P. and Terabe, S. (2000) *J. Chromatogr. A 902*, 119–135.
52. Shihabi, Z.K. (2000) *J. Chromatogr. A 902*, 107–117.
53. Beckers, J.L. and Boček, P. (2000) *Electrophoresis 21*, 2747–2767.
54. Urbánek, M., Křivánková, L., and Boček, P. (2003) *Electrophoresis 24*, 466–485.
55. Tsuda, T., Mizuno, T., and Junichi, A. (1987) *Anal. Chem. 59*, 799–800.
56. Iizuka, E., Tsuda, T., Mumesue, M., and Samizo, S. (2003) *Anal. Chem. 75*, 3929–3933.
57. Tehrani, J., Macomber, R., and Day, L. (1991) *J. High Resolut. Chromatogr. 14*, 10–14.
58. Tsukagoshi, K., Nakahama, K., and Nakajima, R. (2004) *Anal. Sci. 20*, 379–381.
59. Makarõtševa, N., Seiman, A., Vaher, M., and Kaljurand, M. (2010) *Proceedia Chem. 2*, 20–25.
60. Kulp, M., Vaher, M., and Kaljurand, M. (2005) *J. Chromatogr. A 1100*, 126–129.
61. Seiman, A., Jaanus, M., Vaher, M., and Kaljurand, M. (2009) *Electrophoresis 30*, 507–514.
62. Deml, M., Foret, F., and Boček, P. (1985) *J. Chromatogr. 320*, 159–165.
63. Tsuda, T. and Zare, R.N. (1991) *J. Chromatogr. 559*, 103–110.
64. Evans, C.E. (1997) *Anal. Chem. 69*, 2952–2954.
65. Kubáň, P. and Karlberg, B. (1998) *Anal. Chem. 70*, 360–365.
66. Padarauskas, A., Olsauskaite, V., and Schwedt, G. (1998) *J. Chromatogr. A 800*, 369–375.
67. Padarauskas, A., Olsauskaite, V., and Paliulionyte, V. (1998) *J. Chromatogr. A 829*, 359–365.
68. Padarauskas, A., Paliulionyte, V., and Pranaityte, B. (2001) *Anal. Chem. 73*, 267–271.
69. Unterholzner, V., Macka, M., Haddad, P.R., and Zemann, A. (2002) *Analyst 127*, 715–718.
70. Kubáň, P., Karlberg, B., Kubáň, P., and Kubáň, V. (2002) *J. Chromatogr. A 964*, 227–241.

71. Kubáň, P., Kubáň, P., and Kubáň, V. (2002) *Electrophoresis 23*, 3725–3734.
72. Durkin, D. and Foley, J.P. (2000) *Electrophoresis 21*, 1997–2009.
73. Priego-Capote, F. and de Castro, M.D.L. (2005) *Electrophoresis 26*, 2283–2292.
74. Nehme, R., Lascaux, A., Delepee, R., Berengere, C., and Morin, P. (2010) *Anal. Chim. Acta 663*, 190–197.
75. Hou, X.L., Deng, D.L., Wu, X., Lv, Y., and Zhang, J.Y. (2010) *J. Chromatogr. A 1217*, 5622–5627.
76. Zhou, M.X. and Foley, J.P. (2004) *Electrophoresis 25*, 653–663.
77. Wang, J., Chen, G., Muck, A., and Collins, G.E. (2003) *Electrophoresis 24*, 3728–3734.
78. Kubáň, P. Kubáň, P., Hauser, P.C., and Kubáň, V. (2004) *Electrophoresis 25*, 35–42.
79. Padarauskas, A. (2001) *Rev. Anal. Chem. 20*, 271–301.
80. Priego-Capote, F. and de Castro, M.D.L. (2004) *Electrophoresis 25*, 4074–4085.
81. Van der Moolen, J.N., Poppe, H., and Smit, H.C. (1997) *Anal. Chem. 69*, 4220–4225.
82. Kaljurand, M., Ebber, A., and Somer, T. (1995) *J. High Resolut. Chromatogr. 18*, 263–265.
83. Bings, N.H., Wang, C., Skinner, C.D., Colyer, C.L., Thibault, P., and Harrison, D.J. (1999) *Anal. Chem. 71*, 3292–3296.
84. Li, J., Thibault, P., Bings, N.H., Skinner, C.D., Wang, C., Colyer, C., and Harrison, J. (1999) *Anal. Chem. 71*, 3036–3045.
85. Fang, Z.L., and Fang, Q., (2001) *Fresen. J. Anal. Chem. 370*, 978–983.
86. Miller, E. M. and Wheeler, A.R. (2009) *Anal. Bioanal. Chem. 393*, 419–426.
87. Berthier, J. (2008) *Microdrops and Digital Microfluidics*, William Andrew, Norwich, NY, p. 435.
88. Gorbatšova, J., Jaanus, M., and Kaljurand, M. (2009) *Anal. Chem. 81*, 8590–8595.
89. Abdelgawad, M., Watson, M.W.L, and Wheeler, A.R. (2009) *Lab Chip 9*, 1046–1051.
90. Dasgupta, P.K. and Surowiec, K. (1996) *Anal. Chem. 68*, 1164–1168.
91. Dasgupta, P.K. and Surowiec, K. (1996) *Anal. Chem. 68*, 4291–4299.
92. Fuller, R.R. and Sweedler, J.V. (1999) *Anal. Chem. 71*, 4014–4022.
93. Růžička, J. and Hansen, E.H.H. (1988) *Flow Injection Analysis*, Wiley, New York.
94. Kubáň, P. and Karlberg, B. (1998) *Trends Anal. Chem. 17*, 34–41.
95. Fang, Z.L. (1999) *Anal. Chim. Acta 400*, 233–247.
96. Fang, Z.L., Chen, H.W., Fang, Q., and Pu, Q.S. (2000) *Anal. Sci. 16*, 197–203.
97. Kuldvee, R. and Kaljurand, M. (1999) *Crit. Rev. Anal. Chem. 29*, 29–68.
98. Theodoris, G.A., Zacharis, C.K., and Voulgaropoulos, A.N. (2007) *J. Biochem. Biophys. Methods 70*, 243–252.
99. Santos, B., Simonet, B.M., Rios, A., and Valcarcel, M. (2006) *Trends Anal. Chem. 25*, 968–976.
100. Kubáň, P. and Karlberg, B. (2009) *Anal. Chim. Acta 648*, 129–145.
101. Liu, S. and Dasgupta, P.K. (1993) *Anal. Chim. Acta 283*, 739–745.
102. Hogan, B.L., Lunte, S.M., Stobaugh, J.F., and Lunte, C.E. (1994) *Anal. Chem. 66*, 596–602.
103. Kulp, M., Vassiljeva, I., Vilu, R., and Kaljurand, M. (2002) *J. Sep. Sci. 25*, 1129–1135.
104. Ehala, S., Vassiljeva, I., Kuldvee, R., Vilu, R., and Kaljurand, M., (2001) *Fresen. J. Anal. Chem. 371*, 168–173.
105. Kubáň, P., Engström, A., Olsson, J.C., Thorsen, G., Tryzell, R. and Karlberg, B. (1997) *Anal. Chim. Acta 337*, 117–124.
106. Fang, Z.L., Liu, Z.S., and Shen, Q. (1997) *Anal. Chim. Acta 346*, 135–143.
107. Liu, H.H. and Dasgupta, P.K. (1997) *Anal. Chem. 69*, 1211–1216.
108. Fang, Q., Wang, F.-R., Wang, S.-L., Liu, S.-S., Xu, S.-K., and Fang, Z.-L. (1999) *Anal. Chim. Acta 390*, 27–37.
109. Wu, C.H., Scampavia, L., and Ruzicka, J. (2002) *Analyst 127*, 898–905.
110. Wu, C.H., Scampavia, L., and Ruzicka, J. (2003) *Analyst 128*, 1123–1130.
111. Ross, G.A. (1995) *J. Chromatogr. A 718*, 444–447.
112. Van der Moolen, J.N., Louwerse, D.J., Poppe, H., and Smit, H.C. (1995) *Chromatographia 40*, 368–374.
113. Kuldvee, R., Kaljurand, M., and Smit, H.C. (1997) *J. Chromatogr. A 789*, 247–257.
114. Kaneta, T., Yamaguchi, Y., and Imasaka, T. (1999) *Anal. Chem. 71*, 5444–5446.
115. Guchardi, R. and Schwarz, M.A. (2005) *Electrophoresis 26*, 3151–3159.
116. Kaneta, T., Kosai, K., and Imasaka, T. (2002) *Anal. Chem. 74*, 2257–2260.
117. Hata, K., Kaneta, T., and Imasaka, T. (2004) *Anal. Chem. 76*, 4421–25.
118. Seiman, A., Kaljurand, M., and Ebber, A. (2007) *Anal. Chim. Acta 589*, 71–75.
119. Seiman, A. and Reijenga, J. (2010) *Proceedia Chem. 2*, 59–66.
120. Kappes, T. and Hauser, P.C. (1998) *Anal. Commun. 35*, 325–329.
121. Kappes, T., Schnierle, P., and Hauser, P.C. (1999) *Anal. Chim. Acta 393*, 77–82.
122. Kappes, T., Galliker, B., Schwarz, M.A., and Hauser, P.C. (2001) *Trends Anal. Chem. 20*, 133–139.
123. Feng, H.-T., Wei, H.-P., and Li, S.F.Y. (2004) *Electrophoresis 25*, 909–913.
124. Xu, Y., Qin, W., and Li, S.F.Y. (2005) *Electrophoresis 26*, 517–523.
125. Kubáň, P., Nguyen, H.T.A., Macka, M., Haddad, P.R., and Hauser, P.C. (2007) *Electroanalysis 19*, 2059–2065.

126. Xu, Y., Wang, W., and Li, S.F.Y. (2007) *Electrophoresis 28*, 1530–1539.

127. Hutchinson, J.P., Evenhuis, C.J., Johns, C., Kazarian, A.A., Breadmore, M.C., Macka, M., Hilder, E.F., Guijt, R.M., Dicinoski, G.W., and Haddad, P.R. (2007) *Anal. Chem. 79*, 7005–7013.

128. Hutchinson, J.P., Johns, C., Breadmore, M.C., Hilder, E., Guijt, R.M., Lennard, C., Dicinoski, G., and Haddad, P.R. (2008) *Electrophoresis 29*, 4593–4602.

129. Ryvolová, M., Preisler, J., Brabazon, D., and Macka, M. (2010) *Trends Anal. Chem. 29*, 339–353.

130. Huang, X.H., Coleman, W.F., and Zare, R.N. (1989) *J. Chromatogr. 480*, 95–110.

131. Dose, E.V. and Guiochon, G.A. (1991) *Anal. Chem. 63*, 1154–1158.

132. Kubáň, P., Tennberg, K., Tryzell, R., and Karlberg, B. (1998) *J. Chromatogr. A 808*, 219–227.

133. Leube, J. and Roeckel, O. (1994) *Anal. Chem. 66*, 1090–1096.

134. Polášková, P., Bocaz, G., Li, H., and Havel, J. (2002) *J. Chromatogr. A 979*, 59–67.

135. Schmitt-Kopplin, P., Garmash, A.V., Kudryavtsev, A.V., Menzinger, F., Perminova, I.V., Hetkorn, N., Freitag, D., Petrosyan, V.S., and Kettrup, A. *Electrophoresis* 2001, *22*, 77–87.

136. Ikuta, N., Yamada, Y., Yoshiyama, T., and Hirokawa, T. (2000) *J. Chromatogr. A 894*, 111–117.

137. Gáspar, A. and Gábor, L. (2005) *J. Chromatogr. A 1091*, 163–168.

138. Ross, G.A. (1997) *Accred. Qual. Assur. 2*, 275–284.

139. Watzig, H., Degenhardt, M., and Kunkel, A. (1998) *Electrophoresis 19*, 2695–2752.

140. Altria, K.D., Kelly, M.A., and Clark, B.J. (1998) *Trends Anal. Chem. 17*, 204–214.

141. Altria, K.D., Kelly, M.A., and Clark, B.J. (1998) *Trends Anal. Chem. 17*, 214–226.

142. Dyson, N. (1990) *Chromatographic Integration Methods*, Royal Society of Chemistry, Cambridge.

143. Solis, M., Rex, A., Campiglia, D.A., and Sojo, P. (2007) *Electrophoresis 28*, 1181–1188.

144. Gan, F., Ruan, G., and Mo, J. (2006) *Chemom. Intell. Lab. Syst. 82*, 59–65.

145. Liu, B.-F., Sera, Y., Matsubara, N., Otsuka, K., and Terabe. S. (2003) *Electrophoresis 24*, 3260–3265.

146. Schirm, B. and Wätzig, H. (1998) *Chromatographia 48*, 331–346.

147. Faller, T. and Engelhardt. H. (1999) *J. Chromatogr. A 853*, 83–94.

4

PRACTICAL CONSIDERATIONS FOR THE DESIGN AND IMPLEMENTATION OF HIGH-VOLTAGE POWER SUPPLIES FOR CAPILLARY AND MICROCHIP CAPILLARY ELECTROPHORESIS

Lucas Blanes, Wendell Karlos Tomazelli Coltro, Renata Mayumi Saito, Claudimir Lucio do Lago, Claude Roux, and Philip Doble

4.1 INTRODUCTION

4.1.1 High-Voltage Fundamentals

Sample injection and separation are two crucial aspects in capillary electrophoresis (CE) and microchip capillary electrophoresis (MCE). Separation of the analytes occurs due to the influence of an applied potential difference between electrodes placed at the ends of the capillary or channel. As a consequence, the components present in the plug of the sample are driven toward the detector with a velocity proportional to the electroosmotic flow (EOF), the charge-to-size ratio, the electrolyte conditions, the temperature, and the magnitude of the applied potential.

In CE, the velocity of a species (v) is directly proportional to the voltage applied (V), according to Equation 4.1 [1]. Thus, highest voltages are usually preferable to reduce the analysis time.

$$v = \mu E = \mu \frac{V}{L} \tag{4.1}$$

where μ is the mobility of the species, E is the electric field, V is the applied voltage, and L is the capillary total length.

Moreover, the voltage applied has an important role in the resolution and efficiency of separation. The efficiency of the electrophoretic system is measured by the number of theoretical plates N, which also depends on the nature of the species involved (analyte and BGE (background electrolyte)) and the applied potential (Equation 4.2) [1].

$$N = \frac{(\mu + \mu_{\text{osm}})V}{2D}\frac{l}{L} \tag{4.2}$$

where μ is the mobility of the species, μ_{osm} is the electroosmotic mobility, l and L are the effective and total capillary length, respectively, V is the applied voltage, and D is the diffusion coefficient.

Although N gives some insight into the behavior of the separation system, the ultimate parameter is resolution (R_s), which can be predicted by (Equation 4.3)

$$R_s = 0.177(\mu_1 - \mu_2)\sqrt{\frac{V}{D_{\text{avg}}(\bar{\mu} + \mu_{\text{osm}})}} \tag{4.3}$$

where μ_1 and μ_2 are the electrophoretic mobility of the two analytes involved, $\bar{\mu}$ is the average electrophoretic mobility, V is the applied voltage, μ_{osm} is the electroosmotic mobility, and D_{avg} is the average diffusion coefficient.

Therefore, for a given electrolyte, increasing the applied voltage, and consequently the electric field, results in higher resolution and efficiency. However, under high electric fields, significant Joule heating can be generated reducing the separation efficiency and, consequently, resolution. According to Ohm's law, for an ideal resistor, there is a

Capillary Electrophoresis and Microchip Capillary Electrophoresis: Principles, Applications, and Limitations, First Edition.
Edited by Carlos D. García, Karin Y. Chumbimuni-Torres, and Emanuel Carrilho.

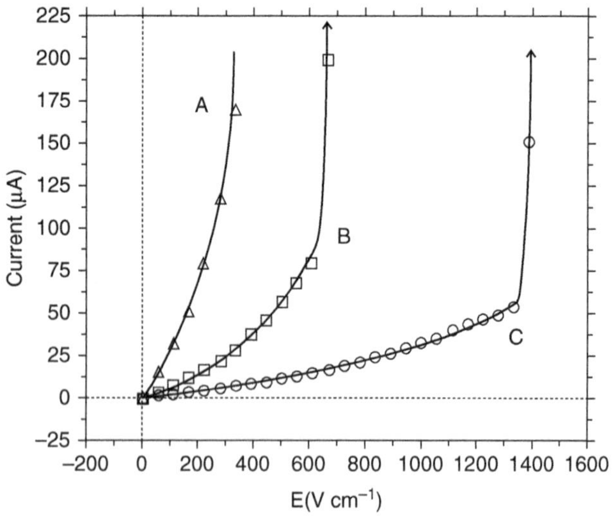

FIGURE 4.1 Ohm's law plot of current versus electric field for natural air convection on a capillary column length with internal diameters of (a) 200, (b) 100, and (c) 50 μm. Buffer: 0.1 M Tris, 0.25 M boric acid, and 7 M urea (pH 7.6). Reproduced with permission from Ref. 2.

linear dependence between the current and voltage. As can be observed in Figure 4.1, the current varies linearly with the applied voltage until a certain threshold value. Beyond this threshold, there is an abrupt increase in the current, which indicates the point of the thermal breakdown where the Ohm's law is no longer valid and the rate of heat generation is greater than its dissipation. This phenomenon may be recognized by the variations of resistance that increases the conductivity of the solution due to the increase in temperature. In this region, it is common for the electrolyte to boil, generating bubbles that ultimately open the circuit. As depicted in Figure 4.1, the rate of heat dissipation depends on the capillary inner diameter.

There are mainly two modes of operation usually employed in CE: constant voltage and constant current. Some authors show that the use of constant current provides better separation than using constant voltage. The majority of the literature makes use of high-voltage power supplies (HVPS) operating in constant voltage mode. This practice is most likely attributed to (i) the previous experience of the researchers and/ (ii) the instrumental limitations of the HVPS that do not necessarily allow the operation in the constant current mode.

Moreover, the parameters necessary to derive the mobility of the species using constant voltage mode are easier to obtain and more precise compared to the constant current mode. Equations 4.1 and 4.4 show how to calculate the mobility of species (μ) using constant voltage and constant current, respectively:

$$\mu = \frac{l\kappa A}{ti} \tag{4.4}$$

where μ is the mobility of the species, l is the capillary effective length, t is the migration time, κ is the conductivity of the solution, i is the current, and A is the capillary cross-sectional area.

High voltage (HV) is also used to perform electrokinetic injection. This method of injection is very popular due to its instrumental simplicity, and is the main injection mode in MCE. The main disadvantage is that the sample plug injected electrokinetically is not representative of the real sample composition.

In CE separations, HV can be applied directly or using a voltage ramp. A voltage ramp is often used in the beginning of an electrophoretic run in order to improve the separation efficiency.

For example, Palonen et al. improved the number of theoretical plates for the separation of selected organic acids by more than threefold employing a voltage ramp of 45 s with nonaqueous solutions and a separation voltage of 60 kV, compared to a run with no voltage ramp [3]. This effect was theoretically studied by Xuan et al. [4]. Using a model that considered temporal and spatial variations of the non-uniform temperature and flow fields, they concluded that the application of an initial voltage ramp delays the development of Joule heating-induced pressure-driven flows, significantly reducing thermal dispersion and improving the separation efficiency.

4.1.2 Electroosmotic Flow Control

Control of the electroosmotic flow is essential for all CE and MCE analyses. Most EOF control strategies are based on manipulation of electrolyte composition and chemical modification of the surface of the capillary. An alternative instrumental approach is based on applying a radial electric field. This electric field can be generated by applying an electrical potential difference perpendicular to the separation voltage; so the radial voltage (V_{rad}) is the difference between the outer potential (V_{out}) and the inner potential (V_{in}), where the inner potential corresponds to the separation voltage. This allows manipulation of the ζ potential and consequently the velocity and direction of the EOF.

$$V_{rad} = V_{out} - V_{in} \tag{4.5}$$

Lee et al. described the first use of radial electric field to control the EOF. They showed that the EOF could be enhanced, decreased, eliminated, and even reversed by simply using a radial electric field to control the ζ potential at the aqueous/inner capillary interface [5].

The instrumental setup may use two or three HVPS. Hayes and Ewing used three HVPS, one to control the inner voltage and two to control the outer voltage [6]. The capillary was also covered with a conductive sheath of Nafion. As can be seen in Figure 4.2, the radial voltage was constant

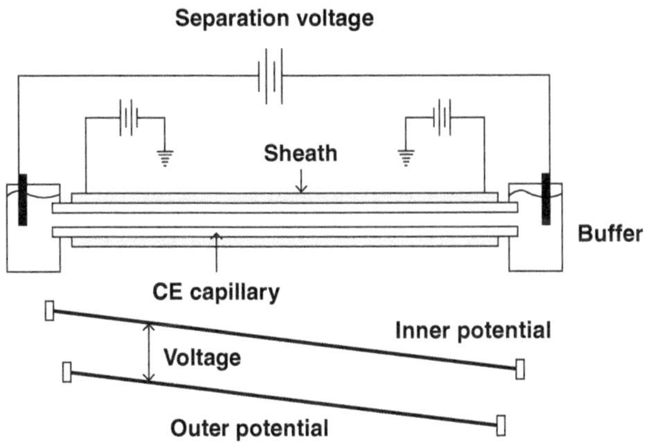

FIGURE 4.2 Scheme of the experimental apparatus used to control EOF by applied radial voltage. Adapted from Ref. 6.

because the inner and outer potential decayed in parallel, according to Equation 4.5.

Culbertson and Jorgenson designed a simpler and safer apparatus with two HVPS (Figure 4.3b) [7]. The two HVPS were used to vary the inner voltage while the capillary was covered with a silver paint and grounded to establish 0 V of outer voltage. It was also possible to use aluminum foil instead of the silver paint. According to Hayes et al. [8], the effects of the radial voltage over the EOF were observed up to 4% of the capillary covered with the conductivity coating.

In the setup described by Culbertson, the radial voltage varied linearly along the length of the capillary because the outer voltage was constant while the inner voltage varied. The profile of voltage along the capillary is graphically illustrated in Figure 4.3a. Once the radial voltage varies along the capillary, the authors considered the average radial voltage (ΔV_{rad}) for their studies, which can be obtained by

$$\Delta V_{\text{rad}} = V_{\text{out}} - \bar{V}_{\text{in}} = -\frac{V_1 + V_2}{2} \tag{4.6}$$

$$V_{\text{out}} = 0 \tag{4.7}$$

$$\bar{V}_{\text{in}} = \frac{V_1 + V_2}{2} \tag{4.8}$$

$$\Delta V_{\text{rad}} = V_{\text{out}} - \bar{V}_{\text{in}} = -\frac{V_1 + V_2}{2} \tag{4.9}$$

where ΔV_{rad} is the average inner voltage and V_1 and V_2 are the potentials applied by the two HVPS positioned at the two extremities of the capillary.

A specific feature of Culbertson's assembly is the possibility to control the radial voltage while keeping the separation voltage constant, since the two HVPS change at the same time. For example, if one applies +10 and −10 kV at V_1 and V_2, respectively, the separation and average radial potential will be +20 kV and 0 V, respectively. On the other

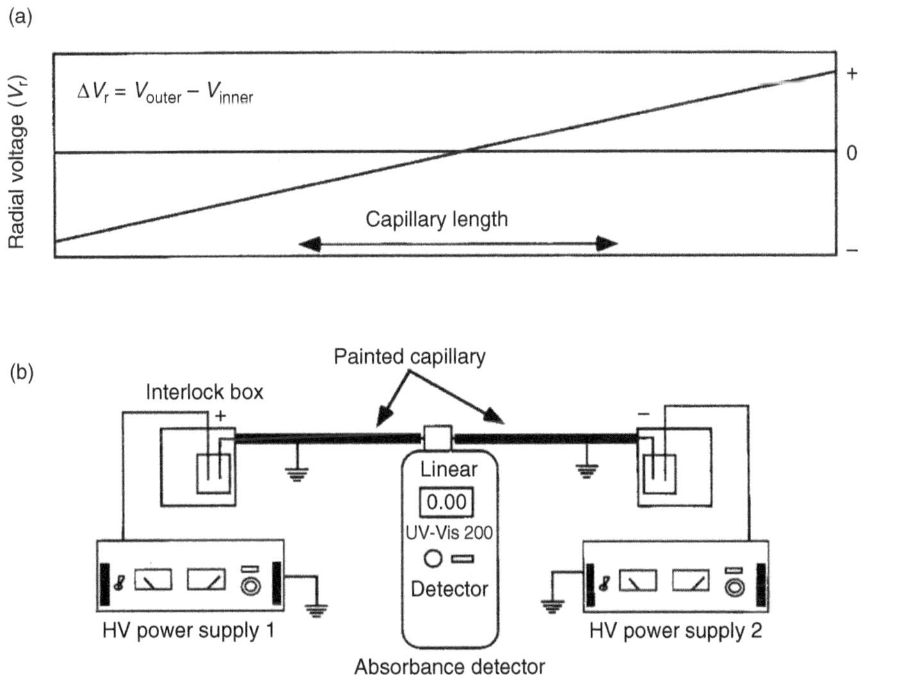

FIGURE 4.3 (a) Radial voltage as a function of location along the length of the capillary. (b) Instrumental diagram of the apparatus used to control EOF by applied radial voltage. Adapted from Ref. 7.

hand, if the voltage of V_1 and V_2 is changed to −15 and −5 kV, respectively, the separation voltage will remain the same, but the average radial voltage will become −5 kV.

The effects of different ways of application of radial electric field over the magnitude and orientation of EOF were studied by Aturki and coworkers [9]. The authors demonstrated that the EOF decreased as the radial electric field varied from negative to positive values. Furthermore, they concluded that the EOF mobility at constant or variable radial voltage with the average value close to the constant value was approximately the same.

Some studies show that the magnitude of the effect of radial electric field over the EOF depends on the pH. According to Ewing and coworkers [10], the EOF is influenced by the surface charge density, which has two additive components: (i) the dissociation of silanol groups and (ii) the radial electric field. At low pH, the contribution of silanol groups dissociation to the total surface charge density is less significant. Thus, the surface charge density is mostly governed by the radial electric field and the variations in EOF due to the applied radial voltage are greater.

An example of application of radial electric field to control the EOF to increase the resident time of leucine and isoleucine inside the capillary and hence increase the resolution of the peaks is shown in Figure 4.4. In this example, the detector was positioned in the middle of the column. The applied average radial voltage was −5 kV ($V_1 = +20$ kV; $V_2 = -10$ kV), and the EOF mobility was $0.86 \times 10^{-4}\,\mathrm{cm^2\,V^{-1}\,s^{-1}}$. The initial separation is shown in Figure 4.4a. After the peaks had passed the first time through the detection window, the radial voltage was reversed and increased to +15 kV ($V_1 = 0$ kV; $V_2 = -30$ kV), resulting in a change in the EOF mobility to $-1.82 \times 10^{-4}\,\mathrm{cm^2\,V^{-1}\,s^{-1}}$ and thus reversing the direction of migration of EOF and the analytes. After the analytes migrated back through the detection window, the radial voltage was restored back to −5 kV, similar to the first radial voltage applied, changing again the direction of EOF and moving the analytes forward, recording a new electropherogram (Figure 4.4b). Finally, the process of moving back and forward, similar to that used to record in Figure 4.4b, was employed to record the last electropherogram (Figure 4.4c). The result was an increase in resolution of peaks with each successive back and forward pass.

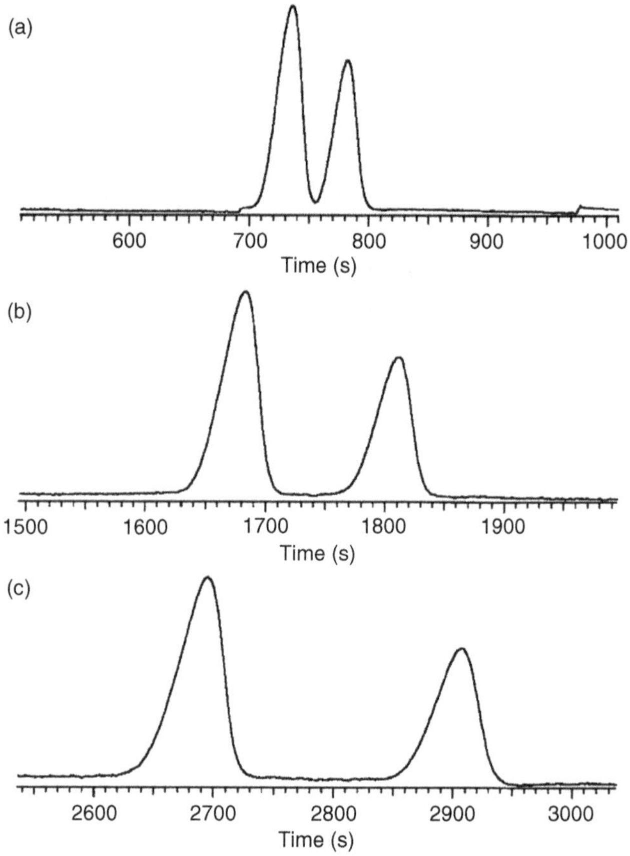

FIGURE 4.4 Separation of 0.20 mM DNS–leucine and 0.25 mM DNS–isoleucine using radial field control. (a) First forward pass. (b) Second forward pass. (c) Third forward pass. Buffer: 5 mM sodium citrate (pH 3.1). Electric field: $380\,\mathrm{V\,cm^{-1}}$. Adapted from Ref. 7.

4.1.3 Technical Aspects

In the constant voltage mode, a HVPS should ideally maintain the voltage at the constant value independent of the current consumption. Similarly, in the constant current mode, the power supply should maintain the current at a constant value, independent of the voltage. However, in practice, commercially available HVPS only come close to these ideals. One of the problems commonly found in HVPS is its output impedance. In an ideal case, this impedance should be equal to zero, which implies that the measured voltage is in fact the actual voltage applied to the electrodes.

However, sometimes manufacturers add an inner resistor at the output to prevent a permanent damage caused by a short circuit. Thus, the measured voltage is greater than the actual one, and the greater the consumed current, the greater the voltage error.

4.1.4 Construction of Bipolar HVPS from Unipolar HVPS

The basic idea for construction of a bipolar HVPS from a unipolar HVPS is depicted in Figure 4.5. Two voltage sources and two resistors can be used to generate intermediate voltage values between the limits established by individual values of the sources (Figure 4.5a). The resulting output voltage is given by

$$V_{\mathrm{out}} = \frac{R_2}{(R_1 + R_2)} V_1 + \frac{R_1}{(R_1 + R_2)} V_2 \tag{4.10}$$

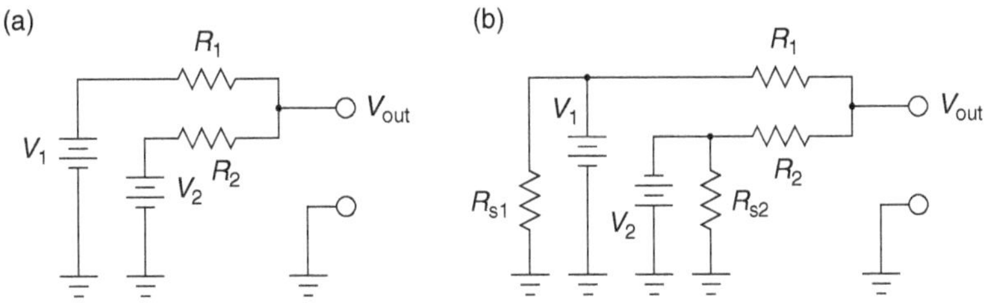

FIGURE 4.5 Schemes for combining HVPS. The generic model shown in (a) allows one obtaining any positive voltage between V_1 and V_2 by choosing the correct values of resistors. The model shown in (b) is a bipolar setup starting from two unipolar HVPS. The shunt resistors R_{S1} and R_{S2} are necessary when the HVPS have asymmetric capacity of source and sink current.

Given that R_1 and R_2 are not, simultaneously, equal to zero, it is possible to get any intermediate value between the limits established by V_1 and V_2. However, when one source is positive and the other one is negative (Figure 4.5b), the resulting voltage can be set to a positive voltage, a negative voltage, or even zero.

This general scheme has some drawbacks, the first being the need for two HVPS. One should take into account, however, that unipolar HVPS are, in general, smaller, lighter, and cheaper than the bipolar ones.

The second drawback is the current consumption because an additional current will flow through resistors R_1 and R_2. Obviously, the greater the resistance, the smaller the current is and the greater the output impedance is. Thus, there is a trade-off between power dissipation and precision of the output voltage.

The third drawback lies in the asymmetric impedance of the output of an actual HVPS. For example, a negative HVPS is more efficient when it is providing electrons than removing them from the external electrical circuitry. This situation is potentially hazardous when two HVPS with the same polarity are used to set an intermediate voltage, which occurs very often in microchip CE systems. The user must consult the datasheet provided by the manufacturer prior to building the double-HVPS. Nevertheless, one strategy to minimize this problem is to insert an additional resistor in parallel with each HVPS (Figure 4.5b). This resistor provides a permanent channel for the current to flow from or to the power source as required by the manufacturer. Again, the drawback is the noticeable increase in the total current drawn from the HVPS.

In addition, it should be noted that the voltages involved in these assemblies (HVPS) are generally higher than the limiting safe voltage of the resistors produced for low-voltage electronic devices.

Thus, in order to prevent leakage of high voltage, which can cause serious harm to people and equipment, components and assemblies produced specifically for HV values should be used.

For the sake of exemplification, let us take the circuit from Figure 4.5a with the following values: $V_1 = 30\,\text{kV}$, $V_2 = 10\,\text{kV}$, $R_1 = R_2 = 100\,\text{M}\Omega$. The setup will behave as a 20 kV HVPS with output impedance of $50\,\text{M}\Omega$, which corresponds to R_1 and R_2 in parallel. This output impedance allows a maximum current of $400\,\mu\text{A}$, which is enough for most applications. However, one must consider that, independent of this current, an additional overhead of $100\,\mu\text{A}$ is constantly flowing from V_1 to V_2. Thus, for example, if the electrophoretic system is consuming $50\,\mu\text{A}$, V_1 is supplying $150\,\mu\text{A}$ and V_2 is sinking $50\,\mu\text{A}$. The question is whether V_2 is able to sink such a current. If not, a shunt resistor (similar to R_{S2}) should be added to the circuit. For example, a shunt resistor of $150\,\text{M}\Omega$ would consume $67\,\mu\text{A}$ from V_2. Thus, by adding R_{S2} to the circuit, V_2 source is $17\,\mu\text{A}$ instead of sink $50\,\mu\text{A}$.

4.1.5 Safety Considerations

Although the voltages supplied by the HVPS used for CE and MCE are very high (~kV), the current is in general too low to cause death (~μA). However, HV electric shock can cause serious injuries and may be painful. Leaks of HV can damage internal electronic circuits and is one of the main noise sources. Implementation of hardware and software security switches, light/sound alerts, and adequate mechanical protection are fundamental when working with HVPS.

4.1.6 HVPS Commercially Available

Some of the HVPS models can cost from tens to thousands of dollars. Companies such as Glassman (http://www.glassmanhv.com/), Spellman (http://www.spellmanhv.com/), UltraVolt (http://www.ultravolt.com/), EMCO (http://www.emcohighvoltage.com/), VMI (http://www.highvoltagepowersupplies.com/), Matsusada Precision Inc. (http://www.matsusada.com/), CE Resources (http://www.ce-resources.com/), and Elektro Automatik (http://www.elektroautomatik.de/) offer a variety of options in terms of technical specifications and have been described in a recent publication [11].

Matsusada produces a HVPS (model K12-30R) that is compatible with CE. This model is very complete for off-the-shelf integration and is operated with a regulated input voltage of +24 V. The output voltage and current range from 0 to ±30 kV and ±400 μA, respectively. Important features as the current and voltage monitoring, output voltage control, and TTL reverse polarity control are present in this HVPS. Another HVPS with very similar characteristics that can be used to build CE equipments is the module Spellman CZE2000 (from 0 to ±30 kV, ±300 μA). The only problem with these HVPS is the relatively high volume and weight.

The model CZE1000R 30 kV, 300 μA is a dual polarity HVPS "rack mountable" from Spellman. Rack mountable HVPS are a good choice for laboratory experiments with HV or to build simplified CE instruments [12]. Matsusada also produces several racks and benchtop models that can be used for this purpose. "Rack" and "benchtop" HVPS are good choices to carry out experiments at the laboratory. On the other hand, modular HVPS are the best options to build CE instruments.

EMCO produces miniature/compact DC/DC converters. Several models produced by this company are a good choice to build portable and battery-operated CE and MCE equipment [13]. Recently, EMCO reported "plug and play" USB converters with complete software control and accessories. With a footprint of ~6 × 9 cm^2, this HVPS is apparently a good and simple choice for microchip experiments in desktop or workbench areas. The model USB20P provides 0 to +2 kV, and the model USB20N ranges from 0 to −2 kV, 400 μA.

Although a wide range of HVPS and DC/DC converters are commercially available, the researcher needs to consider a range of issues such as the HVPS characteristics, the electronic and mechanic resources available in the laboratory, and also the personal knowledge in electronics before to build CE instrument or MCE controller.

4.1.7 Practical Considerations

The HVPS model normally provides important information about the characteristic of the HVPS. However, each manufacturer has a different and specific code to specify each product. Some models can use direct AC input but more frequently require a DC power source (normally 5–24 V) for operation. The simplest way to control the output voltage of a HVPS is using a potentiometer; however, the use of electronic interfaces to control this device using a computer is recommended. The term "HVPS" is normally associated with a source that produces more than 5 kV. Lower voltages than this are referred to as "DC/DC" converters. Essentially a HVPS is a DC/HVDC converter if supplied by a DC converter or an AC/HVDC converter if supplied by an AC voltage. The main difference is that normally a HVPS is instrumentally more complex and complete than a DC/HVDC converter. HVPS normally works under high frequencies that vary around 10–100 kHz. The use of HVPS with shield protection is also desirable to avoid interferences. It is common to find in the literature nomenclature errors, models discontinued, or the absence of the HVPS used in the experiments. It is essential to read the datasheet of the HVPS or DC/DC converter to choose the best option to the specific application.

4.1.8 Alternative Sources of HV

In 2006, da Silva and Frasatto [14] described the use of a conventional CRT (cathode ray tube) monitor as a HVPS for CE. This alternative HVPS was successfully used to build inexpensive homemade CE equipment coupled with a capacitive contactless conductivity detector (C^{4}D). Monitors use a flyback to generate HV, which is employed to accelerate the electrons to the CRT. This specific flyback can produce a 23 kV HV with a maximum output current of about 180 μA. The authors estimated the ripple voltage of about 303 V.

Figure 4.6 depicts the instrumental setup using a CRT monitor as HVPS for CE (Figure 4.6a). The stability of the output HV and a typical electropherogram recorded with C^{4}D showing a separation of inorganic cations (K^+, Na^+, and Li^+) are presented in Figure 4.6b and c, respectively. The relative standard deviation (RSD) values for peak area and migration times were below 2.5 and 1.0%, respectively.

4.1.9 HVPS Controllers for MCE

MCE offers several advantages over conventional CE, including short analysis time, reduced sample consumption, and low cost per analysis [15]. However, the MCE experiments are often performed with conventional HVPS, which limits the portability of microanalytical systems. When an electrophoretic separation is carried out in microfluidic channels, the applied potentials are lower than those required for CE because the separation channel is shorter (around 1–6 cm). The most important factor is to maintain the electric field strength (E) (V cm^{-1}). Compared to the conventional CE, the other remarkable feature of MCE is in regard to the sample introduction process. In CE, the sample plug is introduced by electrokinetic or hydrodynamic injection.

In MCE, the injection is almost exclusively performed by electrokinetic transport when gated (time-based) or pinched (volume-based) injection modes are used. When gated or pinched injection modes are used, different voltage values should be applied to the solution reservoirs. This is required to prevent sample leakage and also to inject small sample plugs, thus focusing high separation efficiency. One of the limitations to implement these systems is the requirement of more than one conventional HVPS. This feature is expensive and is not always readily accessible to many researchers.

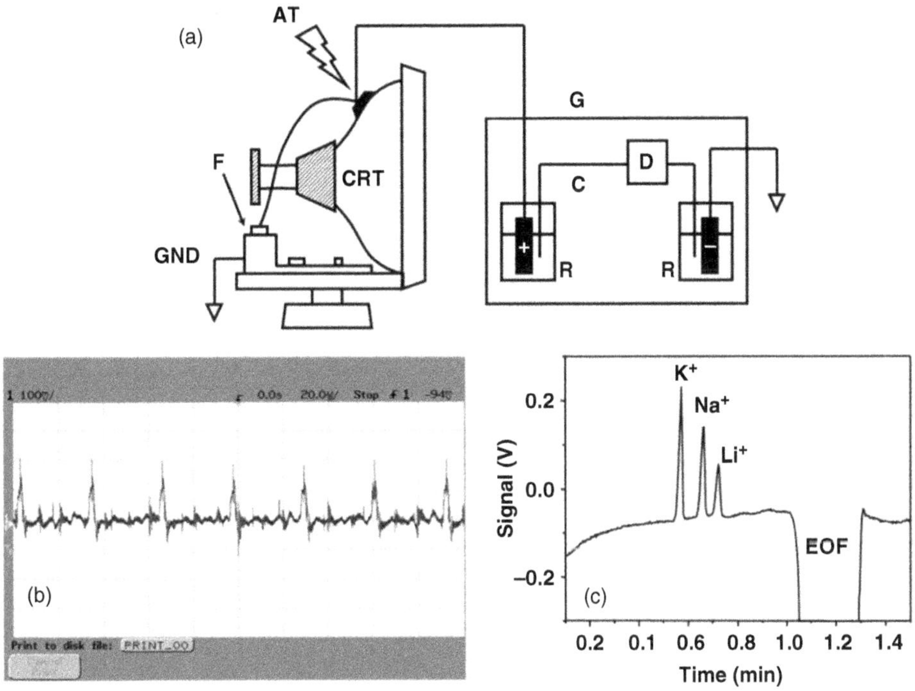

FIGURE 4.6 (a) Representation of the experimental setup using a CRT monitor as HVPS for CE. (b) Stability of the HV applied by the CRT monitor. (c) Typical electropherogram showing the separation of inorganic cations using the proposed instrumentation. Reprinted with permission from Ref. 14.

Specifically for MCE, HVPS with high accuracy and reliability in a low potential range (0–3 kV) is required and the typical length of a microchannel is about 10-fold lower than that of a capillary. For this reason, a few research groups have reported instrumental advances related to the development of miniaturized or compact HVPS sequencers to control these devices [16–21]. Two pioneering reports were published by Jackson et al. [16] and Garcia et al. [17].

Jackson et al. [16] described a miniaturized, battery-powered HVPS and interface circuits designed for MCE. The circuit was designed using the small (4 in. × 6 in. × 1 in.) and lightweight (0.35 kg) EMCO modules, which are rated at +1.2 and −1.2 kV. The EMCO modules are ideal for portable battery-powered operation and their use associated with microfluidic platforms is highly compatible. Due to the low required power, the modules can operate continuously for 15 h without recharging.

Garcia et al. [17] reported the construction of a three-channel battery-operated HVPS for MCE. The authors used two positive and one negative UltraVolt DC–DC converters (maximum output voltage of 4 kV), a microprocessor-controlled timer, a battery, and a transformer to recharge the battery and feed the high-voltage relays. Using this HVPS, the authors performed either pinched or gated injection by changing the external connections between the DC–DC converter outputs and the timer inputs. The inclusion of a microprocessor-controlled timer simplified the analysis and allowed triggering of the detector simultaneously with the separation process.

The instrumentation presented in the literature is based on the use of small DC/HVDC converters, especially those produced by EMCO or UltraVolt companies. These converters are normally powered by single or multiple batteries as well as by universal serial bus (USB) ports and controlled by personal computers. Instrumental advances in HV have allowed meeting the power requirements at the microchip scale, that is, decoupling from the conventional benchtop instrumentation.

Figure 4.7 shows an example of a PDMS microchip powered by a compact HVPS.

Researchers who want to start with microchip experiments but do not have good knowledge of electronics may opt to use a commercial benchtop high-voltage sequencer (http://www.labsmith.com) to perform their experiments simply and safely.

4.2 HIGH-VOLTAGE MEASUREMENT

High voltages can be measured using special voltmeters or through indirect methods. As HV voltmeters are normally not accessible to CE users/developers, the easiest way to measure high voltage in the range 30 kV is stepping down the voltage using potential dividers. In this method,

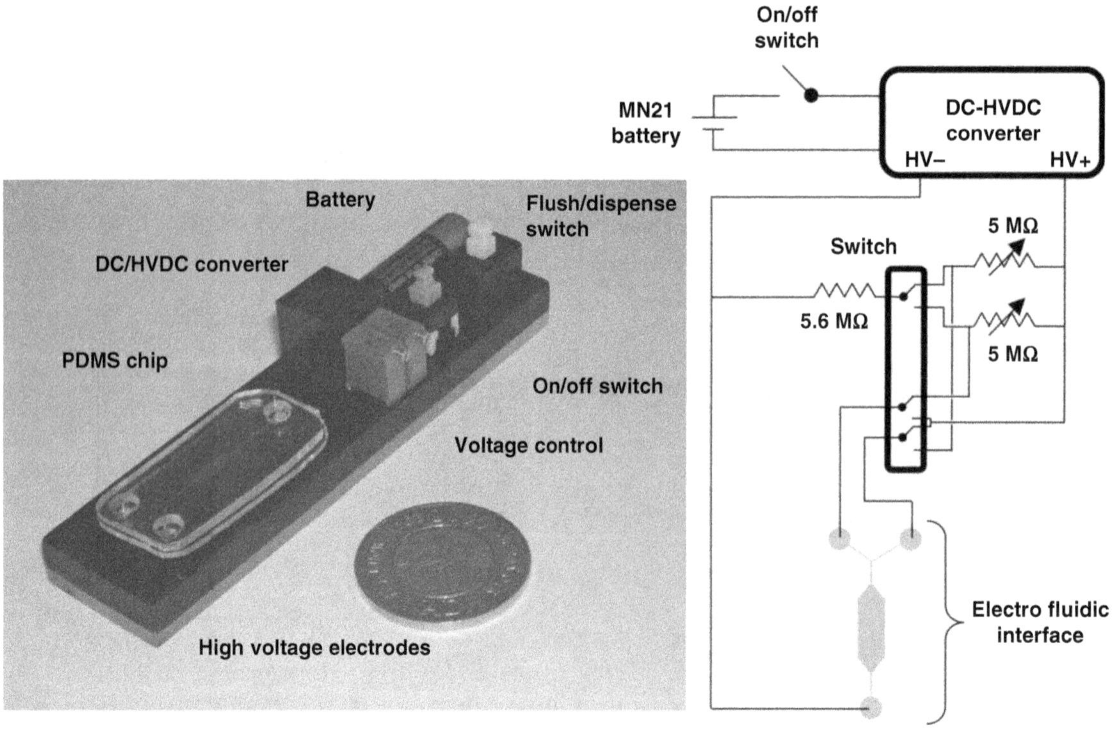

FIGURE 4.7 Representation of a PDMS chip in operation with the miniaturized high-voltage integrated power supply. Reprinted with permission from Ref. 22.

a high-resistance potential divider is connected between the high-voltage output and the ground, and a definite fraction of the total voltage is measured by means of a low-voltage voltmeter connected between the divider and the ground.

One problem with such an approach is the lack of high-voltage resistors on the market. Common resistors are rated for only a few hundred volts. One possible solution is to prepare a thread of small resistors. For example, 25 resistors of 15 MΩ soldered in series will result in an equivalent resistor of 375 MΩ and the voltage across each one is only 1200 V when 30 kV is being measured. Although much smaller, 1200 V is still above the rated voltage of commercial resistors. The final solution is to insert the long thread of resistor in, for example, polyester resin, which improves the dielectric breakdown voltage of the setup. This high-voltage resistor in series with a 37.5 kΩ resistor allows measuring, over this small resistor, voltages in the range 0–3 V, when 0–30 kV voltages are generated at the HVPS.

4.3 CONCLUDING REMARKS

The more popular the electrophoresis equipment, the easier the construction of a new apparatus by the developer, since an appropriate model of HVPS is easily accessible. This tendency is particularly true for MCE developers who were able to find an increasing number of possibilities in recent years. However, there is always room for new implementations, which require knowledge about the bases of the circuitry and alertness about the intrinsic risk of working with high voltages.

REFERENCES

1. Harris, D.C. (2008) *Quantitative Chemical Analysis*, 7th ed., W. H. Freeman and Company, New York, pp. 588–627.
2. Nelson, R.J. et al. (1989) *J. Chromatogr. 480*, 111–127.
3. Palonen, S., Jussila, M., and Riekkola, M.L. (2005) *J. Chromatogr. A 1068*(1), 107–114.
4. Xuan, X.C., Hu, G.Q., and Li, D.Q. (2006) *Electrophoresis 27* (16), 3171–3180.
5. Lee, C.S., Blanchard, W.C., and Wu, C.T. (1990) *Anal. Chem. 62*(14), 1550–1552.
6. Hayes, M.A. and Ewing, A.G. (1992) *Anal. Chem. 64*(5), 512–516.
7. Culbertson, C.T. and Jorgenson, J.W. (1999) *J. Microcolumn Sep. 11*(3), 167–174.
8. Hayes, M.A., Kheterpal, I., and Ewing, A.G. (1993) *Anal. Chem. 65*(15), 2010–2013.
9. Sazelova, P. et al. (2007) *Electrophoresis 28*(5), 756–766.
10. Hayes, M.A., Kheterpal, I., and Ewing, A.G. (1993) *Anal. Chem. 65*(1), 27–31.

11. Felhofer, J.L., Blanes, L., and Garcia, C.D. (2010) *Electrophoresis 31*(15), 2469–2486.
12. Tanyanyiwa, J. and Hauser, P.C. (2004) *Electrophoresis 25* (17), 3010–3016.
13. Ryvolova, M. et al. (2010) *Trends Anal. Chem. 29*(4), 339–353.
14. Fracassi da Silva, J.A. and Frasatto, L. (2006) *Quim. Nova 29* (6), 1377–1380.
15. Arora, A. et al. (2010) *Anal. Chem. 82*(12), 4830–4847.
16. Jackson, D.J. et al. (2003) *Anal. Chem. 75*(14), 3643–3649.
17. Garcia, C.D. et al. (2003) *Lab Chip 3*(4), 324–328.
18. Collins, G.E. et al. (2004) *Lab Chip 4*(4), 408–411.
19. Shadpour, H. et al. (2007) *Anal. Chem. 79*(3), 870–878.
20. Berg, C. et al. (2008) *Electrophoresis 29*(24), 4914–4921.
21. Li, Q.L. et al. (2009) *Sens. Actuators B 136*(1), 265–274.
22. Erickson, D., Sinton, D., and Li, D.Q. (2004) *Lab Chip 4*(2), 87–90.

5

ARTIFICIAL NEURAL NETWORKS IN CAPILLARY ELECTROPHORESIS

JOSEF HAVEL, ELADIA MARÍA PEÑA-MÉNDEZ, AND ALBERTO ROJAS-HERNÁNDEZ

5.1 INTRODUCTION

Over the past quarter of a century, capillary electrophoresis (CE) has become an important analytical technique offering advantages of speed, reproducibility, and high resolution. It is an excellent tool in bio-analytics, requires a small quantity of sample and separation is done mostly in aqueous buffers without any organic solvents, and so on. However, separation is influenced by many factors such as variance of electroosmotic flow (EOF), adsorption of charged species on capillary walls, Joule heating, pH, buffer concentration, and many others [1–3]. Thus, to find the optimal conditions in a CE method development is often necessary and an important prerequisite.

Chemometrics and computational methods are penetrating all scientific disciplines including CE. They help to increase effectiveness by optimizing the performance of CE methods. Optimization, modeling, and/or prediction of separation conditions might shorten the method development, which can often be a tedious and time-consuming process, and facilitate or improve the performance of applications. Even if several efficient and advanced optimization procedures have been developed during recent decades, sometimes, unfortunately, the classical experimental approach based on the "one-variable-at-a-time" and/or the so-called *relaxation method* or single variable approach (SVA) is still used [4]. SVA might work in specific cases but usually requires too much experimental work. Instead, multivariate methods should be used.

The aim of this chapter is to demonstrate the limits of traditional SVA and to stress the advantages of the multiparametric approach, where experimental design (ED) should be adopted to design the experiments. Even if response surface modeling can be applied to evaluate multivariate data of EDs, such data can still be modeled with advantage using artificial intelligence methods, namely by artificial neural networks (ANNs) [5–8]. ANN belongs to a group of nonlinear computational methods which mimic, to certain extent, the function of a human brain. In an increasing number of fields of analytical chemistry, these methods have found practical use, and they are already used in electrophoresis.

Neural network computing can be classified as "soft modeling." It is robust and can expose hidden relationships in a large data set. Applications of ANN in electrophoresis are quite wide and they will be reviewed and discussed in this chapter. However, an exhaustive review is not the purpose here. Rather, we would like to elucidate the fundamental principles and, on selected examples, to demonstrate efficacy, beauty, and advantages of the ANN applications in electromigration methods.

5.2 OPTIMIZATION IN CE: FROM SINGLE VARIABLE APPROACH TOWARD ARTIFICIAL NEURAL NETWORKS

Optimization in electromigration methods including CE, micellar electrokinetic chromatography (MEKC), or capillary electro chromatography (CEC) implies to find such values of the parameters under which "best" separation is obtained, that is, speed, resolution, sensitivity, and so on.

Capillary Electrophoresis and Microchip Capillary Electrophoresis: Principles, Applications, and Limitations, First Edition.
Edited by Carlos D. García, Karin Y. Chumbimuni-Torres, and Emanuel Carrilho.

Optimization is achieved via a search for the optimal (minimal or maximal) value of an objective function (OF) that usually depends on several variables (factors). The OF can be, for example, resolution, sensitivity, reproducibility, and so on. Prediction of the OF can be principally achieved by the so-called *hard* or *soft* models. The phenomenological or "hard" model requires a thorough knowledge of the processes that take place during the separation, that is, exact physicochemical description and also the determination of corresponding physicochemical constants. An example of such an approach is the computer program CELET of Havel et al. where the exact mathematical formula of effective mobility, μ_{eff}, is given [9]. The effective mobility of the metal ion M under the presence of a ligand L, where ML, ML_2, ..., ML_n complexes are formed is given by Equation 5.1

$$\mu_{\text{eff}} = \frac{\sum_{i=0}^{n} \mu_{\text{ML}_i}\beta_i[L]^i}{\sum_{i=0}^{n} \beta_i[L]^i} = \frac{\mu_{\text{M}}\beta_0 + \mu_{\text{ML}}\beta_1[L] + \mu_{\text{ML}_2}\beta_2[L]^2 + \cdots + \mu_{\text{ML}_n}\beta_n[L]^n}{\beta_0 + \beta_1[L] + \beta_2[L]^2 + \cdots + \beta_n[L]^n} \tag{5.1}$$

where μ_{ML} are ionic mobilities and β_n, stability constants of individual ML_n complexes. The CELET program has been used for capillary electrophoretic estimation of sulfate stability constants of metal ions and optimized determination of alkali and alkaline earth metals in water [10], for capillary electrophoretic determination of stability constants of alkali and alkaline earth metal cations with a crown ether, a sulfated β-cyclodextrin, and a zwitterionic buffer, and so on [10,11]. A nice example of how knowledge of stability constants might help to develop a CE method is provided in a paper by Pirogov and Havel [12]. In this work, the stability constants of palladium chloro complexes $Pd(II)Cl_n$, ($n = 2$–4) and their mobilities were determined at ambient and elevated temperatures. Studying the influence of pH, concentration of chloride in the background electrolyte, temperature, cetyltrimetyl ammonium concentration, and other factors on the capillary zone electrophoresis (CZE) separation of the platinum group metal ions, it was possible to optimize the method. High efficiency of the separation of Os(IV), Pt(IV), Ir(III), Pd(II), Au(III), Rh(III) with reduction of the analysis time down to 5–6 min, under the optimal conditions, was achieved at an elevated temperature. Simultaneous determination of Pt(II) and Pt(IV), or Os(IV) and Os(III) has also been demonstrated and in the case of Pd(II) it was shown that the increase in the efficiency with temperature can be explained by changes to the stability constants of chloro complexes. The determination of platinum group metals was made and compared to that of inductively coupled plasma atomic emission spectrometry while the detection limits achieved were in the range 0.1–0.6 ppm.

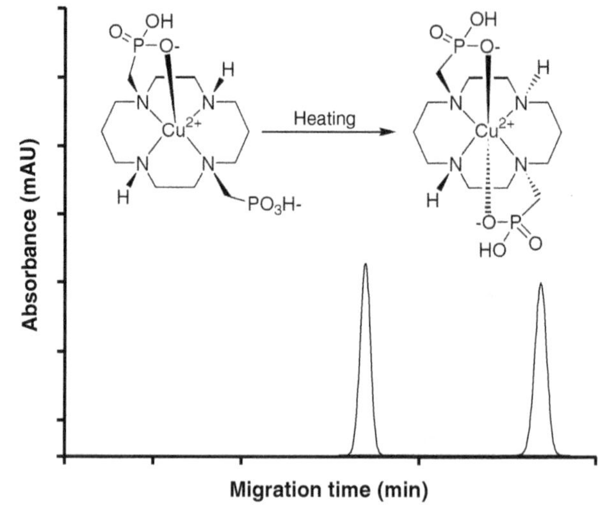

FIGURE 5.1 Separation of Cu^{2+} complexes of H_4dipon as achieved in 75 mM disodium tetraborate aqueous solution (pH 9.5).

Another, even more complicated system was that one of copper(II) and 1,8-bis(methylphosphonate) derivative of cyclam studied by Lubal et al. [13]. In this work, the behavior of penta- and hexa-coordinated species of two structural isomers of the copper(II) complex of 1,4,8,11-tetraazacyclotetradecane-1,8-bis(methylphosphonic acid) (H_4dipon) was studied by CE. Understanding complicated kinetics, structure, and stability constant values enabled separation and determination of both the isomers (with the same molecular weight and charge) in 75 mM disodium tetraborate aqueous solution (pH 9.5) as background electrolyte while simultaneously determining the effective mobilities. The separation method was applied to the study of the reaction mechanism of the isomerization and rate constants were also determined. An example of the separation is shown in Figure 5.1.

In conclusion, knowing all physicochemical constants, ionic mobilities, complex equilibrium constants and/or kinetics, and so on, the optimal conditions can be predicted. However, such an approach is tedious and laborious as many physicochemical constants must be determined.

On the other hand, it will be shown in the following sections that the experimental approach can be used with the application of "soft" modeling. Such models do not require knowledge of the processes involved in the techniques and the number of experiments can be significantly reduced especially if a multivariate approach is used. Several extensive reviews dealing with optimization methods in electromigration have been published [14–18]. Traditionally, the optimal conditions are searched for experimentally and the most common procedure applied here is, alas, a SVA (Section 5.2.1).

5.2.1 Limitations of "Traditional" Single Variable Approach

The experimental approach in optimization of electromigration methods is often based on the SVA. It can be simple but sometimes quite complex due to the numerous parameters to be manipulated. Typical parameters (factors) to be optimized are pH, buffer concentration, organic solvent content in the background electrolyte, temperature, separation voltage, and so on, all of which should be incorporated in the optimized strategy in order to achieve an adequate separation of complex mixtures within an acceptable analysis time. Traditionally, alas, the parameters are changed step by step, varying one factor at a time while keeping all the others constant. Because all the remaining parameters except that one under examination are "relaxing," the method is also called "the relaxation method." This approach works successfully if the number of parameters is low or there is just one parameter to be optimized. However, this relaxation method might generate a high number of experiments; it is time-consuming (and might be expensive) and noneffective. The performance of the approach in the case of two factors is demonstrated in Figure 5.2a.

Keeping the Factor 1 constant one can start at various values for this factor. Starting, for example, from START one can, in some 5–6 experiments, find MINIMUM – 1. Changing the direction and keeping now Factor 2 constant one can change Factor 1 to reach MINIMUM – 2, and so on. Assuming that five points should be measured for each iteration, a total of ~50 experiments are needed to reach global minimum. Often, the first factor is changed repeatedly and the second one just once. True, global minimum is then not reached.

Even if the strategy can get successful results in localizing the minimum in some cases, it may happen that the property of the response surface does not allow convergence and the experiments will be stacked in a local pseudo-minimum and global minimum will not be reached. Therefore, it is important to start the experiments at several different levels of the Factor 1 (Figure 5.2b).

In conclusion, in spite of their simplicity, the relaxation method and SVA might be slow or even nonconverging. On the other hand, multivariate approaches (i.e., ED, modeling, ANN) have shown numerous advantages when replacing these "trial and error methods." Such advantages are a reduction in the number of experiments and reduced time of analysis.

5.2.2 Multivariate Approach with Experimental Design and Response Surface Modeling

5.2.2.1 Experimental Design The design of experiments is an efficient procedure for planning experimental work concerning the electrophoretic system to be optimized, in which several variables are involved. In contrast to SVA, the data obtained is multidimensional and it can be analyzed to yield

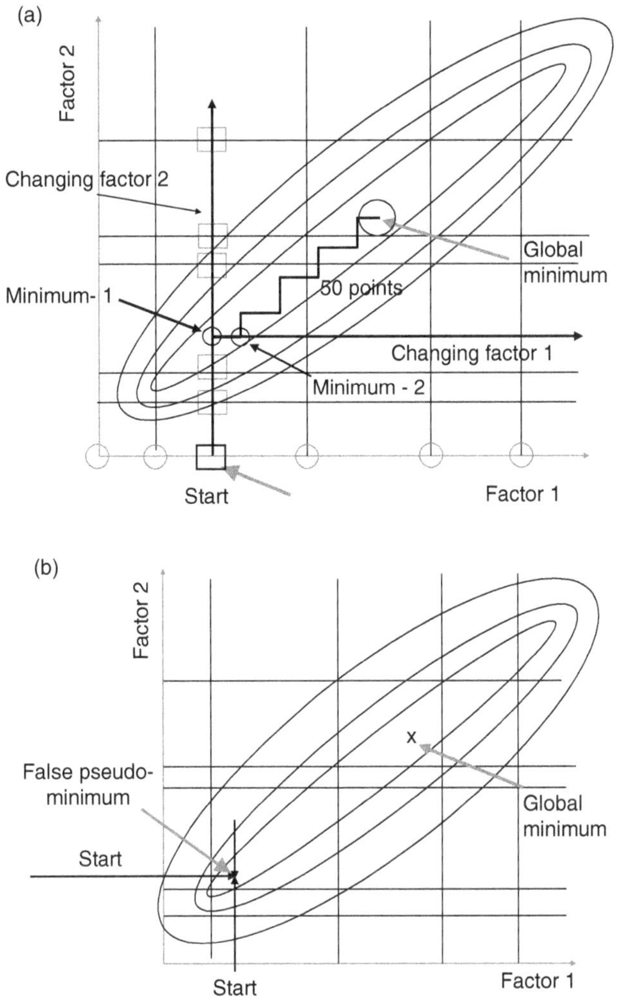

FIGURE 5.2 Performance of the relaxation method in case of two factors. (a) Slow convergence and (b) unconvergence.

valid and objective conclusions. In the context of optimization, ED has been defined by Massart et al. as the process "to determine in an efficient way the set of conditions that are required to obtain a product or process with desirable, often optimal, characteristics" [19]. In other words, the aim of ED is to determine a set of conditions (factors) that will be changed in a controlled way and to follow their effects in order to find such set of conditions that minimize (or maximize) performance of a given process. To select the level of factors is the first step of ED. Conventionally, the levels are marked as − (minus) for the low level, and as + (plus) for the high level of the variable (factor).

The "response(s)" are the characteristics of the process to be optimized for a given set of variables. Choice of the proper ED depends on the objectives of the experiment (estimation of factor effects and to model the dependence of the responses on the factors) and the number of factors of interest. Generally, ED may be considered as the methodology of how to conduct and plan experiments in order to extract the maximum amount of information in the fewest number of runs (experiments) [20].

The term "screening design" is used here, as well. It refers to an initial experimental plan that is intended to find the fewest significant factors from a list of many potential ones. Alternatively, we refer to a design as a screening design if its primary purpose is to identify significant main effects, rather than interaction effects, the latter being assumed one order of magnitude less important. Different designs are central composite designs (CCDs), 3-level factorial design, Box–Behnken designs (B–BD), and D-optimal designs. The application of them will depend on the problem [21].

An example of the simplest ED in the case of two factors is a two-level full factorial design (FFD) with a central point, as given in Figure 5.3a, where the total number of points is just five.

Instead of changing Factor 1 step-by-step and then Factor 2, as is done in SVA, during the experiments done according to ED, all factors are changed at the same time for low and high levels of the parameters. In this way, there is a high probability that in such a two-dimensional space (Factors 1 and 2) we will find the optimal conditions. The central point in the middle (often marked as (0, 0)) has high importance as it can account for nonlinear behavior. Often, ED is wrongly declared as an optimization method. This is not true at all; ED is just a way of how to distribute the experiments in a two- or multidimensional space of parameters (factors), analogously, in the case of three parameters, nine experiments are required in a two-level factorial design with a central point, as shown in Figure 5.3b.

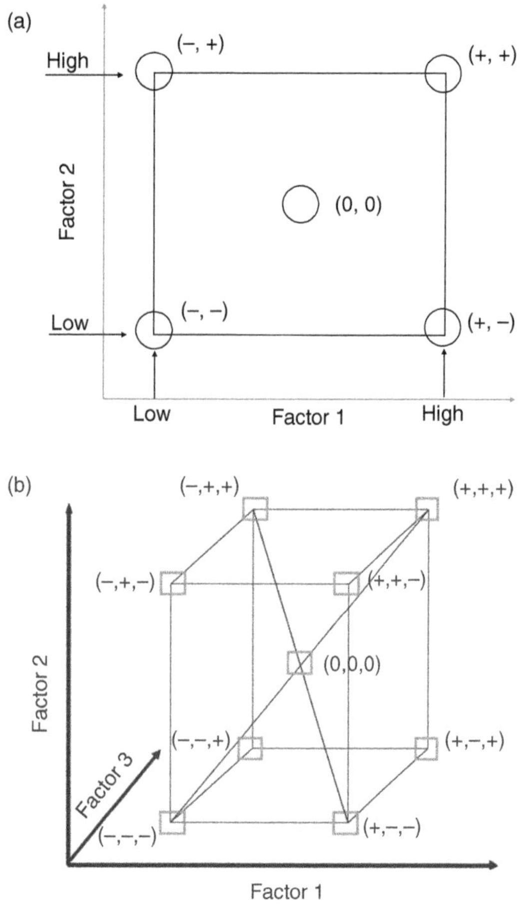

FIGURE 5.3 (a) Two-level factorial design with central point for two factors and (b) scheme of a two-level FFD with central point.

The points in a two-level factorial experiment are often abbreviated with strings of + and − signs. The strings have as many symbols as factors, and their values dictate the level of each factor: conventionally, − for the low level, and + for the high level of the variable.

Increasing the number of factors significantly increases the number of experiments. If the number of experiments will be dramatically high then a fraction of these experiments might be carried out. These designs are called fractional factorial designs. There are situations where it is only important to estimate the main effects. In such a case the Plackett–Burman design (P–BD) can be used. The interest to model the interactions between the factors requires empirical models. The situation when highly nonlinear curved relationships are to be modeled, at least three levels of each factor have to be considered. Such designs are CCD and B–BD; they are also known as response surface designs [22]. The so-called Box–Wilson CCD contains an imbedded factorial or fractional factorial design with center points that is augmented with a group of points called *stars*. These star points allow estimation of the surface curvature.

5.2.2.2 Response Surface Modeling As already mentioned earlier, ED is NOT an optimization routine. In order to perform optimization using the advantage of multivariate approach, it is necessary to have some OF and to use an approach where the minimum/maximum is searched for. The most simple first step in the optimization procedure is to define the OF and objective criteria (OC). As in CE or MEKC, or generally in analytical chemistry, the analytical performance characteristics can be considered as important OC: calibration function, determination of detection and determination limits, precision, specificity and selectivity, and so on. The OF measures the misfit between the data and the corresponding modeling result. The performance characteristics should be considered simultaneously, that is, more than one factor has to be considered at the same time. The OF defined after aggregation of performance characteristics, *considering q individual criteria, f_i, is given by* Equation 5.2

$$\mathrm{OF} = w_1 f_1 + w_2 f_2 + \cdots + w_q f_q \tag{5.2}$$

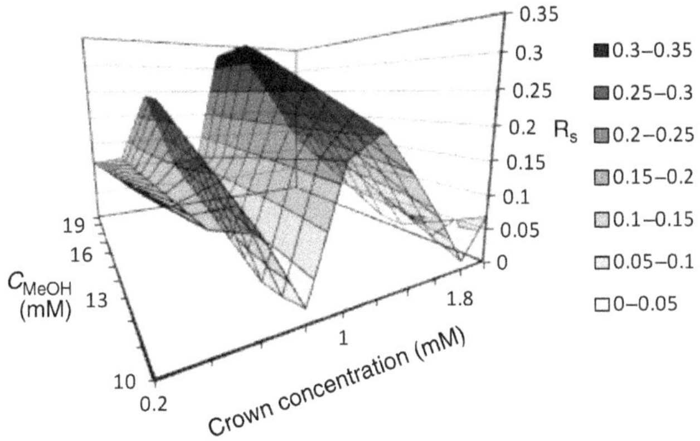

FIGURE 5.4 Simulated three-dimensional response resolution surface. Rs means resolution. Adapted from Yang et al. [23].

where w_q is the weight of the objective criterion f_q. An example of OF is the chromatographic response function.

After screening of the data the next step is to draw a map of the experimental domain given by the significant variables (factors) and their ranges. This map is usually obtained using a quadratic polynomial model. Studying the determination of the mono-, di-, and trivalent cations by CE, Yang et al. applied CCD and modeling techniques for the optimization using an empirical model as given by Equation 5.3 [23].

$$\mu_{M_i} = k_{0i} + k_{1i}[\mathrm{HL}] + k_{2i}[\mathrm{Crown}] + k_{3i}[\mathrm{MeOH}] + k_{22i}[\mathrm{Crown}]^2 \tag{5.3}$$

where k's are empirical constants used to calculate electrophoretic mobilities. Equation 5.3 relates the ion mobility (μ_{Mi}) of the ith metal ion (Mi) with the complexing agent concentration [HL], 18-crown-6 concentration [Crown], and methanol concentration [MeOH]. The simulated surface of resolution Rs changes, which represents the response surface for the separation conditions of three different cations and is shown in Figure 5.4.

5.3 ARTIFICIAL NEURAL NETWORKS IN ELECTROMIGRATION METHODS

ANN—as a part of artificial intelligence methods—are playing an irreplaceable role in modern analytical chemistry and in separation methods as well [20]. Because of the importance of ANN, the basic abbreviated theory of ANN will be given here. We consider this useful as according to our experience, the use of ANN is often hindered by a fundamental misunderstanding. As for applications of ANN in CE and separation methods we refer to the literature and references given below [6,24–29].

5.3.1 Introduction—Basic Principles of ANN

The theory of different networks has been reviewed by Zupan and Gasteiger [30,31]. Aleksander and Morton defined ANN computing as "the study of networks of adaptable nodes, which through a process of learning from task examples, store experimental knowledge and make it available for use" [32].

The use of ANN for data evaluation is often claimed to constitute the so-called *soft model*, that is, the model able to represent behavior of the system when the exact description is missing or is too complex. Neural networks consist of arrays of simple activation units linked by weighted connections (Figure 5.5).

The basic processing unit in ANN is a node or a simulated neuron. ANN can consist of multiple layers of neurons

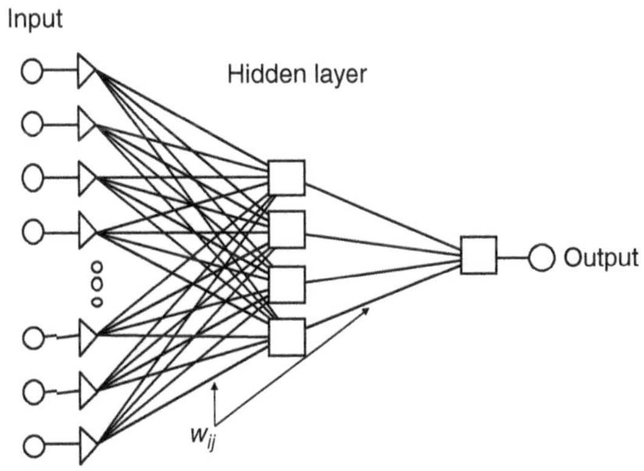

FIGURE 5.5 Scheme of a three-layer ANN architecture. w_{ij} are weights assigned to each connection.

arranged so that each neuron in one layer is connected with each neuron in the next layer. Most often, a multilayered forward-fed neural network using as a learning scheme the algorithm of the back propagation (BP) of errors and the generalized "delta rule" is used for the adjustment of the connection weights (so-called *BP networks*). ANN modeling is in contrast, to so called *hard models* in chemistry, where there are formulae and equations, and the system is exactly described by these equations and the values of physicochemical constants.

The number of nodes in the input, hidden, or output layers is defined by the problem being solved. The input layer contains the experimental information and/or values of parameters; the output layer delivers the response sought for property value, for example, the value of resolution. The hidden layer encodes and organizes the information received from the input layer, and delivers it to the output layer. The number of nodes in the hidden layer, which somewhat play the role of intermediate variables, may be considered as an adjustable parameter. Thus each neuron has a series of weighted inputs, w_{ij}, which may be either output from other neurons or input from external sources. Each neuron calculates a sum of the weighted inputs and transforms it by a transfer function (Equation 5.4),

$$\delta_j = \frac{1}{1 + e^{((1-x)/\gamma)}} \tag{5.4}$$

where δ_j is the output from the jth neuron connected to ith neurons in the previous layer and γ is the gain determining the slope of the sigmoid transfer function, and x is given by Equation 5.5

$$x = \sum_{i=1}^{n} w_{ij} o_i + \theta_j \tag{5.5}$$

where w_{ij} represents the weight applied to the connection from neuron i to neuron j, o_i is the output from the ith neuron in the previous layer, and θ_j is a bias term.

BP networks operate in a supervised learning mode. In the first step (training), known data are given to the network (a series of coupled experimental parameter-response values constituting the learning set). Using the BP algorithm, the network iteratively adjusts connection weights w_{ij} and biases t_i (starting from initial random values) so it can calculate output values satisfactorily matching the observed values. In the BP training algorithm, this adjustment is carried out comparing the calculated value (target t_{ij}) and the desired output (o_i) of the network by means of calculation of the total sum of squares (t_{ss}) of the deviation ($t_{ij} - o_{ij}$) for the n patterns of the learning set, to get the value of the root mean square (RMS) minimal (Equation 5.6):

$$\text{RMS} = \sum_{i=1}^{n} (t_{ij} - o_{ij})^2 \min \tag{5.6}$$

The weights are then adjusted for all the interconnections between the output layer and the hidden layer. Likewise, the weights are adjusted for the interconnections between the hidden layer and the input layer, as well.

The practical use of ANN consists of several steps: *training*, *verification*, and *prediction*. In the *training* phase for the optimal ANN structure the training data set is used. Weight coefficients and biases are searched until the value RMS is minimal, that is, calculated values t_{ij} are close to experimentally observed o_{ij} values. In the case that the training phase is successful, the ANN model should then be verified with a test data set. Finally, if also the *verification* is successful, the ANN can be used for *prediction* of the optimal values of parameters. The prediction should be done within the range of input data used for the process of learning as extrapolation might be difficult or problematic.

5.3.2 Optimization Using a Combination of ED and ANN

It was shown in Section 5.2.2 that data obtained by ED can be elaborated by response surface modeling using, for example, polynomial or a general least squares method. However, it has been shown by Havel et al. that the combination of ED with artificial ANN modeling, ED–ANN, has many advantages over that [6]. This combination of ED–ANN was proposed in CE for the first time by Havel et al. and later on in MEKC [26,27].

In addition to these papers and many other applications since Havel et al. [6], it has been demonstrated that the proposed ED–ANN method can be used with great advantage as a new optimization algorithm and not only in electrophoresis [6]. For example, the optimization of an analytical method can be speeded up considerably using a multivariate approach applying suitable ED and soft modeling with ANN. The schema is shown in Figure 5.6.

It might be seen from Figure 5.6 that, initially, two levels FFD with central point (points 1–5) are used. The corresponding OF is measured for each of the points 1–5 and then such data are used for ANN prediction of point 6. For predicted values of parameters corresponding to the coordinates of point 6 experiments are done. Eventually, another design around the point 6 can be done and those corresponding experiments, as well. All new experiments done are included in the database and so no information is lost. A flow chart of this ED–ANN algorithm is shown in Figure 5.7. It is recommended to first conduct some screening experiments to estimate boundaries, that is, low and high levels of parameters. The suitable ED is made and the OF is then modeled using ANN. More and more experiments are done

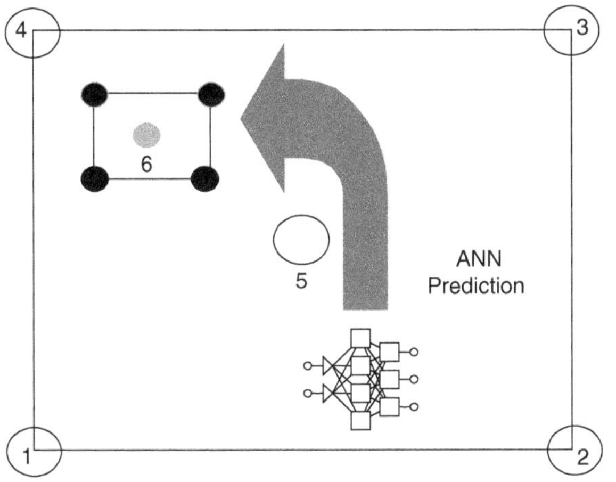

FIGURE 5.6 Scheme of experimental design combined with ANN.

and ANN is applied again and again. This iterative process is repeated until satisfactory results are obtained.

5.3.2.1 Testing of ED–ANN Algorithm The ED–ANN algorithm (Figure 5.7) was tested here using Rosenbrock's function, which is a nonconvex function and it is used as a standard performance test for the performance of optimization algorithms, introduced by Rosenbrock [33]. It is also well known as Rosenbrock's valley or Rosenbrock's "banana" function.

The function is defined by Equation 5.7

$$f(x,y) = (1-x)^2 + 100(y-x)^2 \tag{5.7}$$

The global minimum, $(x, y) = (1,1)$ where $f(x, y) = 0$, is inside a long, narrow, parabolic-shaped flat valley. To find the valley is trivial. To converge to the global minimum, however, is difficult and many optimization algorithms or methods fail.

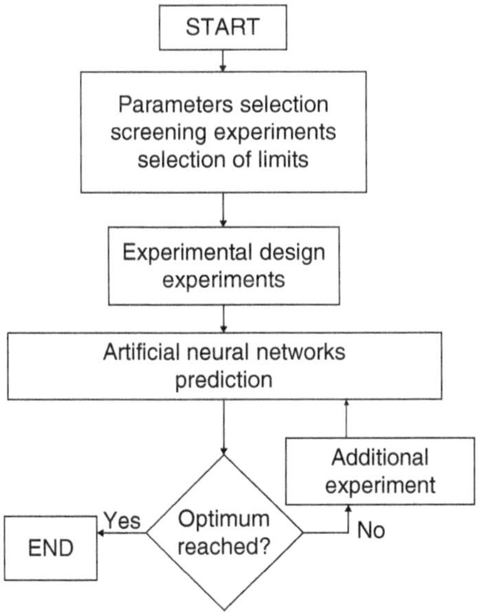

FIGURE 5.7 Flow chart of ED–ANN method.

In the beginning of the procedure, a two-level two-factor FFD with central point was chosen as starting ED (Figure 5.8). The procedure of minimum searching using the ED–ANN method is schematically shown. The points 1–5 represent starting ED. Points 6, 7, and so on were consecutively obtained by ANN prediction (in consecutive iterations). The ANN architecture (2:4:1) (number of neurons in the inputs: number of hidden nodes: output, respectively) was used. Further searching was stopped if the function value of the next point was higher than the function value for the previous point or if sufficiently close to the minimum.

The starting set was chosen so that their dislocation corresponded to the two-level two-factor FFD with a central point. Repeatedly, the neural network is applied and trained, and the prediction of the point in which the function value is the lowest is done. This point can be regarded as the function minimum in *i*th iteration. The comparison of true and found minima was then done and this difference was a criterion of the iteration success. Many excellent minimizing algorithms have failed; however, it is shown here that the ED–ANN approach converges well even for such a difficult surface such as this Rosenbrock's banana function.

5.3.2.2 Practical Applications of ED–ANN The procedure combining ED and ANN does not require either prior knowledge of physicochemical constants or an understanding of the separation mechanisms, and the "best" conditions of the analytical process can be predicted from a limited number of experiments. For example, the empirical model given by Equation 5.3 and CCD for the optimization of the separation of several metals was described in Section 5.2.2.

Another extensive application is that of Farková et al. in which the ED–ANN approach was also used extensively [24]. In this paper, the ANN modeling was applied extensively and it was shown that all capillary electrophoretic peaks existing in electrophoresis, including Gaussian, Weibull, Lorentz, and Gaussian-exponential functions, can be precisely simulated using ANN. The ED–ANN method was applied to demonstrate the optimization of mono-, di-, and trivalent cations as a function of 18-crown-6 and methanol concentrations. Once the simulation of the resolution was concluded, the simulated values of the resolution were rearranged to represent two inputs, [Crown] and [Methanol] concentrations, and one output (resolution, Rs). The optimal performance was found for the architecture 2:7:1 and, after training the ANNs, a very good fit was obtained more easily than using response surface modeling.

Since the first paper, the ED–ANN methodology has successfully been applied in several cases, for example, calibration curves, use of ED–ANN for improvement of pharmaceutical analysis, enhancement of precision in the

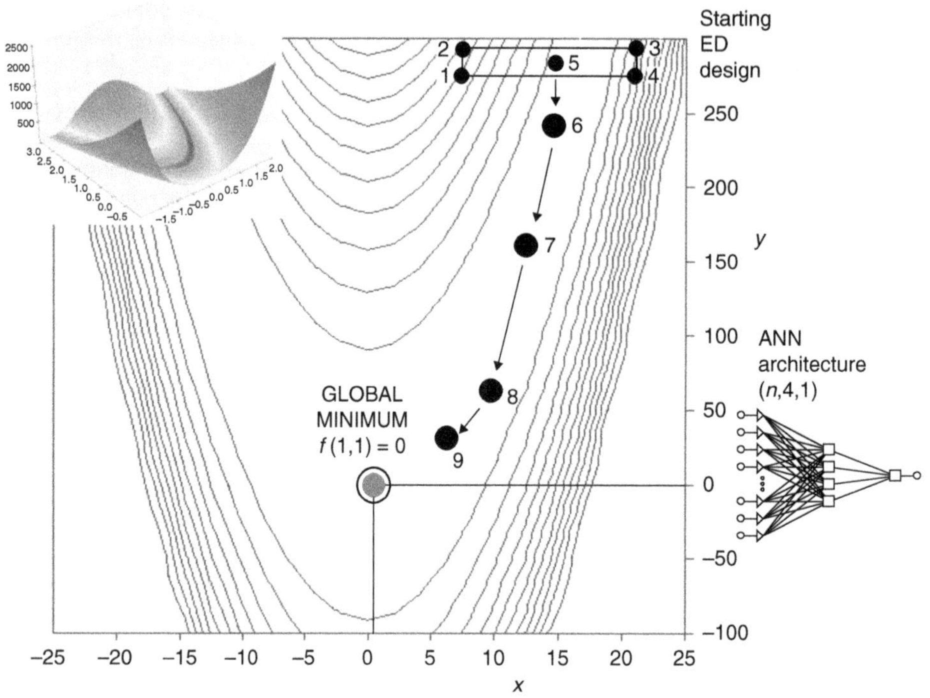

FIGURE 5.8 Testing the ED–ANN algorithm using the Rosenbrock's function.

analysis of medicines by CE using ANN, ANN for computer-aided modeling and optimization in MEKC, quantification in overlapped CE peaks [28,34,35], and including extensive applications in chiral CE separations [36].

5.3.3 Quantitative CE Analysis and Determination from Overlapped Peaks

CE is used extensively for quantitative drug analysis as an alternative and complementary technique to HPLC with the convenience of a shorter-analysis time [37–40]. Precision of CE analysis, however, is often not sufficient and represents, to certain extent, a limitation for the use of CE in drug quantitative analysis.

5.3.3.1 Evaluation of Calibration Plots in CE Using ANN to Increase Precision of Analysis Traditionally, calibration curves are used in quantitative capillary electrophoretic analysis, plotting either peak height or peak area versus concentration of the analyte (drug, pharmaceutical product, metal, etc.). Calibration curves are mostly evaluated using linear least squares method applying the linear model:

$$Y = ax + b \tag{5.8}$$

where a is slope and b is the intercept, Figure 5.9. Finding the parameters a and b in Equation 5.8, the estimation (prediction) of the concentration of the analyte (drug) is done using the regression equation.

Several inter-laboratory studies with different mathematical and experimental efforts, including the use of internal standards or markers, have been published in order to eliminate the influence of an EOF, to improve the injection repeatability and reach more precise CE analysis [41,42]. Normalization of electropherograms using markers added to the analyzed solution is also one of the beneficial aspects of the recommended procedure, which is included in an available commercial CE instrumentation of Agilent Technology software. A survey of various possibilities of how to improve precision of CE analysis is reviewed by Mayer [43].

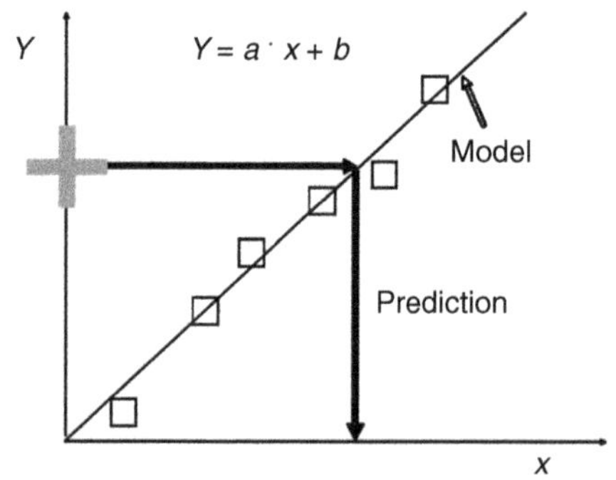

FIGURE 5.9 "Classical" linear calibration plot used for the quantitative analysis in CE. Y is either peak height or peak area, x is concentration of the analyte, and $Y = ax + b$ is regression equation (MODEL). Using this model the prediction of the concentration is done from the sample analysis (marked as +).

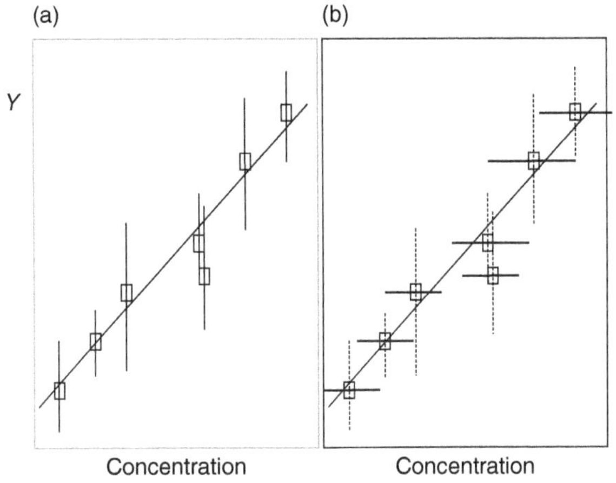

FIGURE 5.10 Schematic calibration plots. (a) Errors in Y are assumed only; (b) values on both Y and x axes are subject of random errors.

Using the linear regression equation, the prediction of the analyte concentration can be done from the response of the analyte on the sample Y_{sample}. (+ symbol on Y axis, Figure 5.9). However, the classical formula of standard linear least squares (SLS) implicitly assumes that only dependent variable (Y) is the subject of random errors. But this is generally not true. In reality, all practical measurements of both Y and x axes values are subject to error (Figure 5.10a and b) and the application of SLS introduces an error in the analysis and in principle is not correct.

The solution to the problem is offered by the bi-variate linear least squares (BLS) method, which is the generic name for a technique used for fitting bi-variate data, where the variables in both axes are subject to random errors.

In conclusion, generally, the bi-variate regression method should always be used for regression; however, in BLS, the formulae are much more complicated than those for SLS and they are given as follows. The method consists of minimizing the sum of the weighted residuals, S, expressed in equation

$$S = \sum_{i=1}^{n} \frac{(y_i - \hat{y}_i)^2}{w_i^2} = (n-2)\hat{s}^2 \tag{5.9}$$

where w_i is the weighting factor, $\hat{y}_i$ is the estimated value for the y predicted, and $\hat{s}^2$ is the estimated value of error, measured in terms of variance, for the set of n experimental data points (x_i, y_i). This method uses the variance of residuals ($s_{\text{E}_i}^2$), which can be expressed using the Taylor series, as a weighting factor even when the covariance between variables for each data pair is not equal to zero, cf. Equation 5.10:

$$s_{\text{E}_i}^2 = w_i = s_{y_i}^2 + \hat{b}^2 s_{x_i}^2 - 2\hat{b}cov(x_i, y_i) \tag{5.10}$$

where $s_{x_i}^2$ and $s_{y_i}^2$ stand for the variance of each (x_i, y_i) data point. By minimizing the sum of the weighted residuals in relation to the slope and the intercept, two nonlinear equations are obtained and, by including the partial derivatives of the squared residuals, the equations can be written in matrix form as Equations 5.11 and 5.12:

$$R\widehat{b} = g \tag{5.11}$$

$$\begin{bmatrix} \sum_{i=1}^{n} \frac{1}{s_{\text{E}_i}^2} & \sum_{i=1}^{n} \frac{x_i}{s_{\text{E}_i}^2} \\ \sum_{i=1}^{n} \frac{x_i}{s_{\text{E}_i}^2} & \sum_{i=1}^{n} \frac{x_i^2}{s_{\text{E}_i}^2} \end{bmatrix} \times \begin{bmatrix} \hat{a} \\ \hat{b} \end{bmatrix} = \begin{bmatrix} \sum_{i=1}^{n} \left(\frac{y_i}{s_{\text{E}_i}^2} + 1/2 \left(\frac{R_i}{s_{\text{E}_i}^2} \right)^2 \frac{\partial s_{\text{E}_i}^2}{\partial \hat{a}} \right) \\ \sum_{i=1}^{n} \left(\frac{x_i y_i}{s_{\text{E}_i}^2} + 1/2 \left(\frac{R_i}{s_{\text{E}_i}^2} \right)^2 \frac{\partial s_{\text{E}_i}^2}{\partial \hat{a}} \right) \end{bmatrix} \tag{5.12}$$

The slope and the intercept values, which are components of vector $\hat{b}$ in Equations 5.11 and 5.12, can be determined by carrying out an iterative process on the following matrix form given in Equation 5.13

$$\hat{b} = R^{-1}g \tag{5.13}$$

Using this method, assuming that the straight line model is the correct one, the variance–covariance matrix of the calibration straight line coefficients, B, is obtained by multiplying the final matrix R^{-1} by the experimental error s^2. As the experimental error is unknown, the estimated value, s^2, expressed in Equation 5.9 should be used. It should be pointed out that if $s_{\text{E}_i}^2 = 1$ (if all errors are due to the experimental measurement in the ordinate axis), then the expressions obtained are the same as for SLS. Even if the computer program for BLS was published, it is not commercially available [44].

A program has been developed and written (Havel et al., unpublished results) and also the use of ANN has been examined to solve the problem and resolve the situation when Y and x are subject to random errors. Because simultaneously in CE there is also a variation of EOF, markers have also been applied. As for calibration models in CE, the previously mentioned studies incorporating neural networks based on ANN and multilayer perceptron (ANN–MLP) have been used to examine UV–Vis electrophoretic data to determine concentration levels of different drugs and peptides [27,28]. ANN has been shown to give better results than the partial least squares (PLS) method and principal components regression (PCR) in common regression analysis. This can be explained by its ability to model nonlinear relationships in the data. The developed neural networks approach demonstrates that instead of these relatively complicated

BLS formulae, the use of ANN represents a real contribution to improving the quantitative analysis when only SLS is applied. Thus, ANN is able to account for errors in both x- and y-axes, for nonlinear response of detector and nonlinearity of calibration curves, eventually [27,28].

In addition, the applications develop and improve the results of the analysis when simultaneously used with markers in the analysis in order to compensate for the variation of EOF [27,45]. For example, the improvement in the analysis can be seen in the results for the determination of antiviral agents used for prophylaxis and treatment of influenza A such as memantine (3,5-dimethyl-1-adamantanamine) and rimantadine (α-methyl-1-adamantanemethylamine). Moreover, another example of the applicability of the method is the analysis of rutin (3-[[6-*O*-(6-deoxy-α-L-mannopyranosyl)-β-glucopyranosyl]oxy]-2-(3,4-dihydroxyphenyl)-5,7-dihydroxy-4*H*-1-benzopyran-4-one), a flavonoid *O*-glycoside derivative of a quercetin (important drug used in the treatment of capillary bleeding and increased capillary fragility). While even if SLS and BLS give similar results, ANN yields lower prediction errors than traditional linear least squares methods and the precision of analysis was found to be in the range 0.5–1.5% RSD.

5.3.3.2 ANN in Quantitative CE Analysis from Overlapped Peaks Often, in spite of great efforts to reach baseline separation or optimal separation, the baseline separation is not obtained. In this case, we can still use the CE results anyway but the de-convolution of overlapped peaks should be applied. Even if some companies are offering corresponding software, this is however a rather tedious procedure as the function describing the peak profile must be known or it should be determined. It is rather easy in the case of Gaussian profiles but this is a rare case in CE. We have demonstrated that deconvolution is not needed and it can be eliminated using ANN-supported multivariate calibration. It has been demonstrated for the first time by Havel et al. Simply, the same approach and advantages of ANN as mentioned earlier were applied in multianalyte determination from overlapping peaks from CE without the necessity to do any deconvolution [45]. This procedure has been deeply studied and further developed. For example, quantitative analysis from overlapped peaks using a combination of ED–ANN was also applied to chiral CE separations [36]. Zhang et al. studied the use of ANN networks based on genetic input selection for the quantification of overlapped electrophoretic peaks and also considered the use of ANN based on principal component analysis input selection for the quantification in overlapped CE peaks [46].

The combination of classical classification and regression approaches for the quantification of overlapped electrophoretic peaks by using neural computing was proved to be a promising tool for solving highly overlapped peaks with acceptable errors [47].

In conclusion, multicomponent analysis in CE is possible even from highly overlapped peaks, and using ANN-based multivariate calibration approach, knowledge of the peak profile function is not needed. In addition, variation of EOF and changes of peak profiles during the separation because of dispersion effects can be taken into account simultaneously even without knowing the source of the errors. It could be said that the community of users of CE is still not completely realizing the great possibilities and power of ANNs.

5.3.4 ANN in CEC and MEKC

CEC is a separation technique that combines the separation efficiency of CE with the selectivity and sample capacity of liquid chromatography (LC). CEC was introduced by Pretorius et al. [48] and, in 1981, Jorgensen and Lucas extended the concept of CEC to use packed capillary columns [49]. Since that time CEC has developed rapidly as an exciting alternative separation technique, especially due to the recent high demand for new miniaturized separation methods, representing a promising alternative to LC [50,51]. CEC is an attractive technique especially for the separation of enantiomeric substances [3,14,52–54].

MEKC may be understood as a hybrid technique of reversed phase liquid chromatography (RPLC) with CZE. Viewing the micelles as a pseudo-stationary phase, the theoretical modeling of MEKC mainly requires the understanding of the interaction of noncharged species with charged micelles [14,15,55]. It is not difficult to realize that theoretical modeling for the optimization of separation methods began with the optimization concepts of LC, which is explaining the efforts from different authors to use classical chromatographic equations and parameters to optimize MEKC separations [56].

Optimization by the SVA is accomplished when a specific problem should be solved empirically. In these cases, it is achieved by the optimization of separation by surfactant, modifier concentration, pH, or operation voltage variations. Some examples related to the application of SVA for optimization in MEKC may be found [57–59].

Classical multivariate analysis such as ANOVA and ED were used in the optimization of MEKC parameters. However, the multivariate approach does not always solve the identification of the significant factors and predict optimum experimental conditions [55]. A classical chemometrics approach has been followed by several authors [60–63]. They used soft and robust models—as well as response surface plots—in order to predict resolution (Rs) and some other properties as migration time (t_m), capacity factors, peak width at half height ($w_{1/2}$), or the number of the peaks to optimize separation systems, as an OF controlled by experimental variables as pH and surfactant or buffer concentration.

ED has widely been used to simultaneously optimize separation parameters leading to higher resolution in relatively short analysis times [64,65]. The only work related to theoretical hard modeling (as far as we know) was developed by Pačes et al. [66]. The main problem when applying the results of this work to optimization conditions arises from the numeric nature of mathematical solutions given. In this case, the application of chemometrics or empirical approaches is obligatory.

The use of the controlled weighted centroid variable size simplex algorithm does not depend on the number of factors, the complexity, or their interactions. This advantage was used to optimize the separation of four local anesthetics by MEKC, only 11 steps being necessary for the optimization [67]. As multilevel design is rather costly in the number of experiments, it is necessary, therefore, to decide carefully the star factors.

As for the optimization procedure, the simplex method can be applied in combination with other ED. FFD has been used to follow the effects of factors like voltage, temperature β-CD concentration, and pH for the separation of some endogenous urinary compounds by MEKC. Once the interactions and effects of the factors had been studied, a simplex optimization was carried out around the values of the factors used previously; a number of peaks were followed during the optimization [68].

There are a number of applications in which ANN is involved. Hilder et al. presented systematic investigations on the combination of both of the above-mentioned methodologies using ED and ANN for optimization, allowing the CEC separation of eight inorganic anions employing an anion-exchange stationary phase in less than 2.5 min [69]. Using ANN this separation system was modeled and the results obtained were used to determine the optimum conditions (9 mM perchlorate and −10 kV) for the separation of eight inorganic anions. A more rapid separation of these analytes was only obtained by applying a higher voltage.

The combination of ED–ANN in MEKC responded to the necessity of getting better optimization with lower number of experiments in a shorter time. Examples of the use of the ED–ANN approach for nonlinear regression in CE are many, and involve different electrophoretic modes. For instance, Havel et al. and Breadmore et al. derived and validated a soft model when two electrolyte parameters (concentration of methanol and surfactant) were varied [34,70]. That model allowed the description and prediction of the observed mobilities of anionic metal complexes formed using two ligands: bis(2-hydroxyethyl) dithiocarbamate (HEDTC) and *trans*-1,2-diamino-*N*,*N*,*N*′,*N*′-cyclohexanetetraacetic acid (CDTA). Nonlinear regression was used to derive parameters for the model from experimental data and these parameters were used to predict the effective mobilities of the analytes to within 3.1% RSD. The experimental data has been examined using ANN. The network was shown to

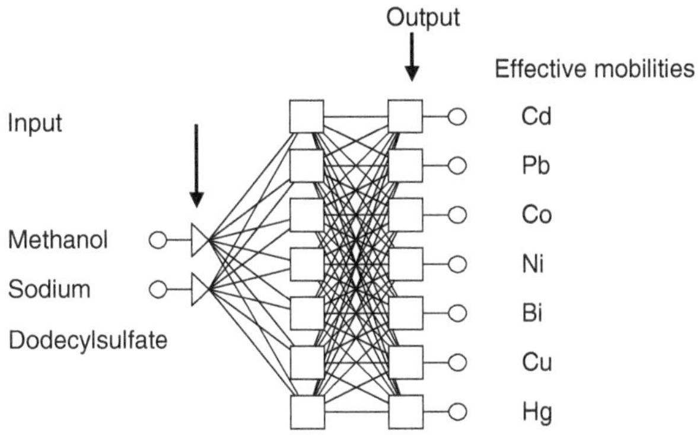

FIGURE 5.11 The ANN architecture for modeling the separation of anionic metal complexes in MEKC.

predict the effective mobilities, in which no explicit mathematical model was needed and, using a suitable ANN architecture (Figure 5.11), an excellent model provided higher accuracy than in nonlinear regression.

In parallel, the combination of ED–ANN gave the researchers the possibility of substantially reducing the number of experiments and that the optimized conditions could be found using just nine points from the ED (Figure 5.12). Using a two-level factorial design (making a total of nine experimental points) enabled the reduction of the average error to <4%, which is within the error of reproducibility of CE. Finally, optimum separation conditions were obtained for ~8% (v/v) methanol and 10 mM SDS using nine experimental points (Figure 5.12).

In conclusion, this is a new optimization procedure based on a combination of ED–ANN that exhibits the general advantage of "soft model" modeling, that is, there is no necessity to formulate an explicit mathematical model in order to provide an accurate result, even in the presence of

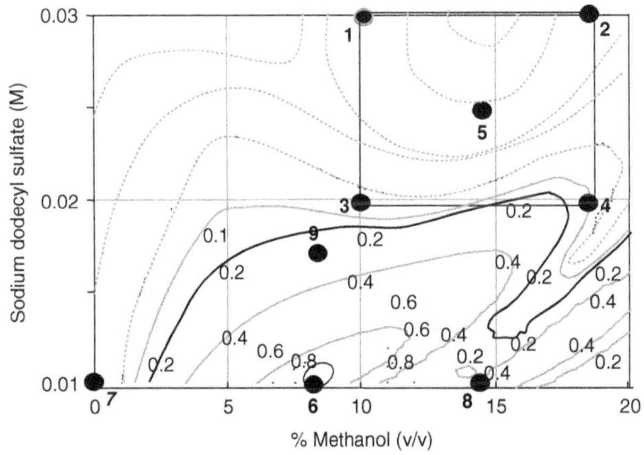

FIGURE 5.12 The map of the response surface as obtained by artificial neural networks during computer-aided modeling and optimization in micellar electrokinetic chromatography and ED designs. Adapted from Havel et al. [34].

noisy data. Nevertheless this should be an incentive for the application of ANN for the optimization of CEC and MEKC separations.

5.3.5 ANN for Peptides Modeling

Recently, because of the highly important PROTEOME project of mankind in which the aim is to determine the structure of perhaps up to 10 million proteins, CE for peptide analysis is almost irreplaceable. As hundreds of peptides are coming from systematic enzymatic cleavage of proteins, some estimates for peptides mobilities are required to facilitate the analysis. Thus, intensive efforts are also concentrated on the modeling of oligopeptides mobilities which, in the first place, may be due to the studies of fractionation of proteins. First of all, some efforts have been undertaken to understand the relationships between shape, charge, and size of oligopeptides with regard to mobilities by hard modeling or by empirical and statistical modeling [71–78]. Also, some work has been reported specifically using ANN to predict the mobility of peptides [79–81]. Finally, there is some research addressed to achieve some quantitative structure–property relationship (QSPR) studies of peptides with CZE parameters by ANN [82–86].

Peptide mobility modeling is quite a wide task. Therefore, we would only like to briefly mention here that ANN can be used for this purpose. Principally, having a sufficiently large database, structure descriptors of the compounds can be related to various properties such as biological activity and physicochemical properties, including peptides electrophoretic mobilities, and so on.

In this section, only one example is commented on. Recently, Jalali-Herawi et al. studied the prediction of electrophoretic mobilities of peptides in CZE by QSPR using the Offord model (i.e., $\mu \equiv Q/M^{2/3}$ for the charge-over-mass term for the QSPR equation) and ANN [83]. Mobilities from 125 peptides (varying in size from 2 to 14 amino acids) were measured at pH 2.5 and it was shown that peptides mobilities are best correlated with a charge-over-size ratio. The peptides were characterized by three molecular descriptors, QM, Es'c, and MR, where QM is charge-over-size ratio, Es'c means corrected steric substituent constant, and multiple regression (MR) means molecular refractivity. The values of molecular descriptors QM, Es'c, and MR were used as the input and peptide mobilities as the output. We have used part of the data from Jalali-Herawi et al. (Table 1, in Ref. 83), and ANN was applied using Trajan software [85]. The results were proved to be in agreement with the paper and that a simple ANN architecture (3,4,1) is able not only to model the mobilities of peptides but also to make predictions. The structure of the ANN is shown in Figure 5.13, and the correlation between the calculated and experimental mobilities (cm^2 V^{-1} s^{-1}) are shown in Figure 5.14.

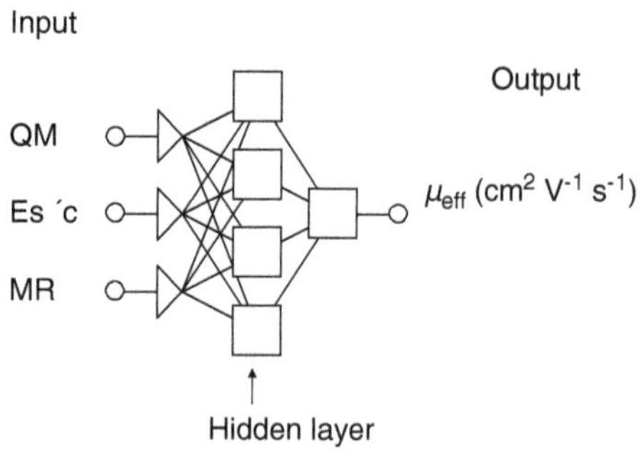

FIGURE 5.13 Structure of the ANN architecture for prediction of peptides mobilities.

5.3.6 Classification and Fingerprinting

To perform the classification in a multivariate system, traditionally, pattern recognition methods based on principal component analysis, factor analysis, discriminant analysis, and so on are required, where prior to classification, data preprocessing is essential for enhanced performance of the classification. In general, supervised learning techniques in chemometrics use a calibration or training set in order to build the model or to obtain the classification. In the case of classification methods, the classes to which the objects of the training set belong to are *a priori* known. This knowledge is then used to obtain the classification by means of a given technique. Neural computing is able to estimate the final probabilities, establish classification rules, and then perform statistical analysis [87,88]. If a proper mathematical model that describes a data set is already known, a neural network is unlikely to be needed, but when the rules that underlie the data are only partially known or not known at all, a neural network may discover interesting relationships as it ploughs through the database. Neural networks are able to learn complex behavior and are highly capable of undertaking a remarkable variety of tasks. Moreover, MLP offer a form of nonlinear mapping that is adaptable to pattern recognition assignments present in both regression and classification, and they have been used to identify fingerprints.

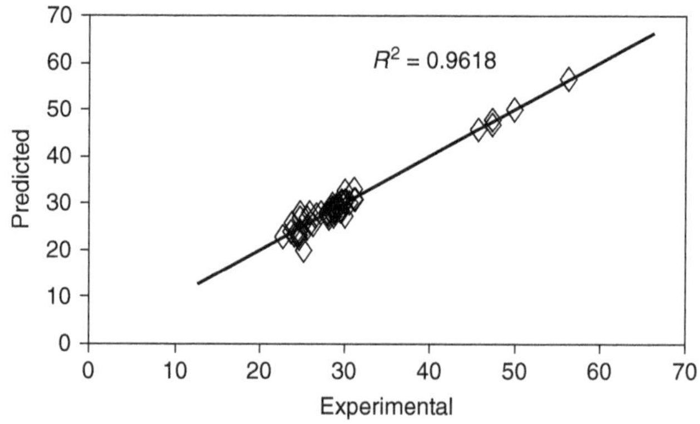

FIGURE 5.14 Correlation of experimental and predicted mobilities. Part of data from Ref. 83.

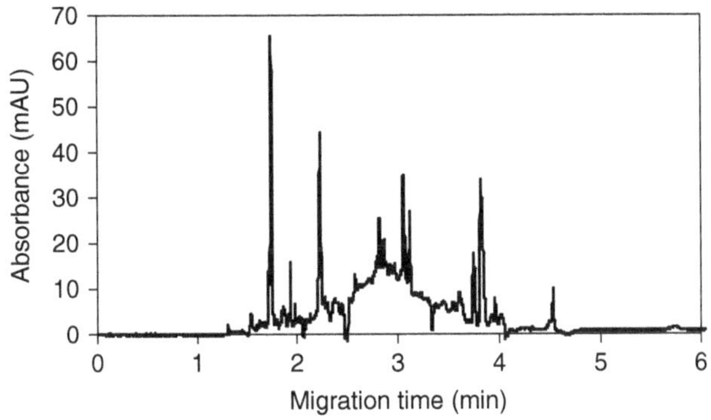

FIGURE 5.15 Example of electrophoretic analysis. Fingerprint of extract from wine.

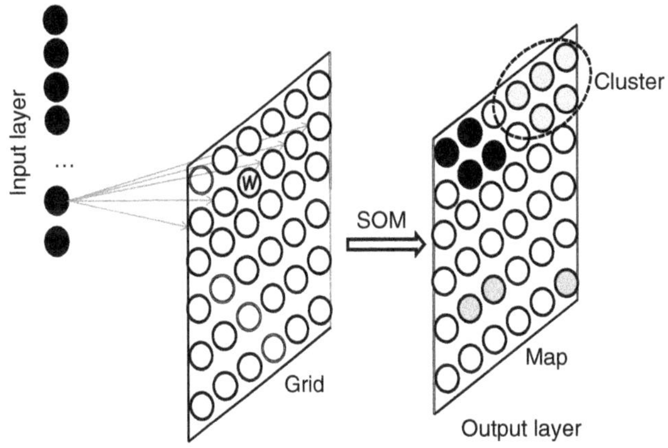

FIGURE 5.16 Self-organizing map (*W* winner).

Classification can be done, based on some selected variables, from the electropherograms and/or using electrophoretic profiles (fingerprints). An electrophoretic profile consists of a set of peak locations (migration time) and areas/heights. As a consequence fingerprints of compounds present in a sample can be evaluated to determine the presence of detectable signals (unknown or known) [89–92]. A typical example of electrophoretic analysis (Figure 5.15), which is significant as an early real-world application of the methods, is the assessment of wine identification.

As has been mentioned earlier, neural networks are nonlinear models which makes them easily adaptable in modeling complex relationships; having the capability of learning input–output relationships from a training data set and this feature is used to carry out correlation studies, and thus establishing neural networks as a promising alternative to various classical multivariate classification methods. Although many types of neural networks can be used for classification purposes, in CE back-propagation multilayer perceptron (BP-MLP) has been implemented for different problems (environmental, medical, pharmaceutical, food chemistry, etc.) [45,93].

The appropriately trained and validated MLP neural network has demonstrated efficient classification of the electrophoretic data and enabled the processing of spectra, hence proving their utilization for the classification and processing of routine data [93].

Pattern recognition is a difficult problem and involves different steps. The Kohonen neural network differs considerably from the feed forward, BP ANN. The Kohonen neural network differs in both how it is trained and how it recalls a pattern. Fundamentally, the Kohonen neural network modifies itself in response to input patterns (self-organization) and the result is reached using so-called competitive learning. The Kohonen neural network contains only an input and an output layer in the architecture, Figure 5.16. The self-organizing map (SOM) creates topologically ordered mappings between input data and processing elements of the map. SOM is usually represented as a two-dimensional sheet (also other dimensions can be used) of processing elements (Figure 5.16). Each processing element has its own weight vector and the learning of SOM is based on the adaptation of these weight vectors [30].

Using capillary electrophoresis mass spectrometry (CE-MS) and capillary electrophoresis diode-array detection (CE-DAD), unsupervised clustering of comprehensive metabolic profiles using a neural network based on Kohonen's SOM for evaluation of electrophoretic fingerprints allows the classification of biochemical pathways activated by a light and dark cycle through the dynamic changes in the level of 56 basic metabolites in plant foliage (*Oryza sativa* L. ssp. *japonica*). The use of SOM facilitates an understanding of large-scale interactions among components in biological systems.

Support vector machines (SVMs) are a set of related supervised learning methods that analyze data and recognize patterns and are used for classification and regression analysis. MLP networks provide an alternative training method for radial basis function and MLP classifiers by solving a quadratic programming problem with linear constraints. SVMs provide a nonspecific mechanism where three basic steps are defined: (i) the basic mathematical derivation for linearly separable classes, (ii) the extension to the nonlinearly separable case with the use of kernel functions, and (iii) the presentation of the generalized solution with the incorporation of the trade-off parameter [94,95]. SVM models are considered by many to have a close relationship to MLP networks and provide an alternative training method for these algorithms [96]. By introducing the concept of similarity between data defines completely pattern classes [97]. There are not many applications of SVM to capillary electrophoretic data [98–102]. As an example of an application, Mao et al. demonstrated the utilization of SVM for bladder cancer recognition using CE-mass spectrometry in pattern analysis of urinary nucleosides [100].

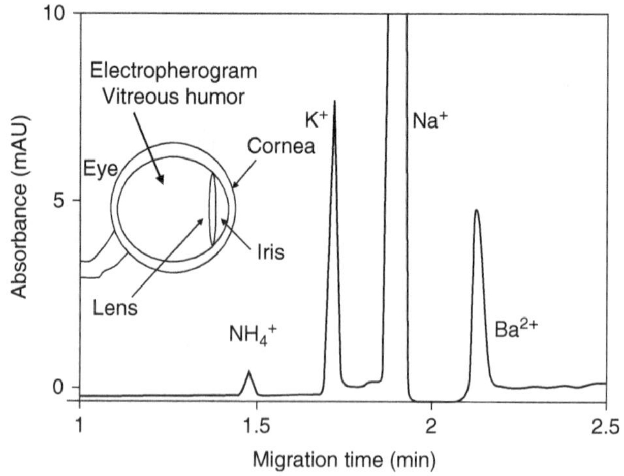

FIGURE 5.17 Capillary electrophoresis in human death investigations. Example of electropherogram of human vitreous humor.

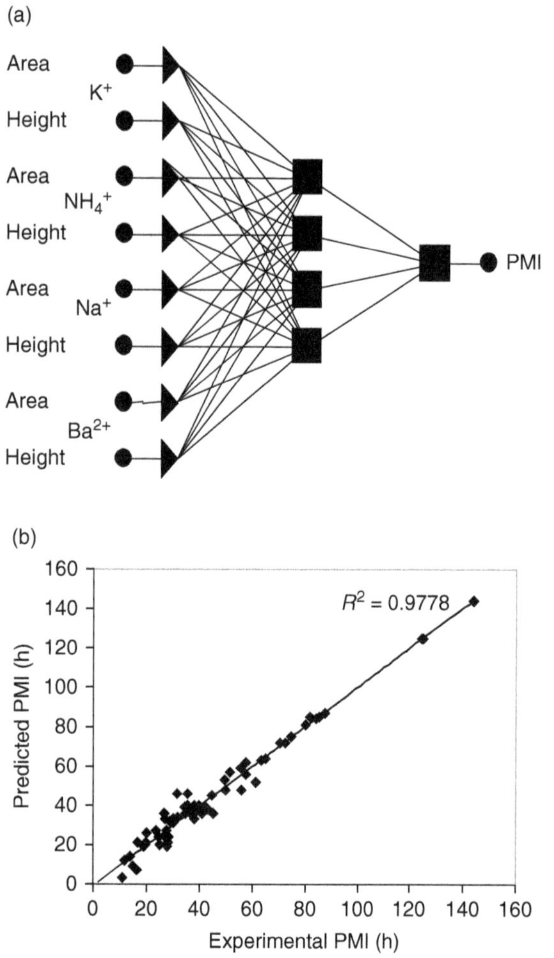

FIGURE 5.18 (a) ANN used of the PMI study; (b) correlation of predicted PMI by ANN versus real experimental PMI.

5.3.7 Other Applications

There are numerous applications of ANN in electromigration methods. As mentioned before, a complete overview is not the declared aim of this chapter. From many possibilities a curious application of the use of CE in forensic analysis is discussed here. The methodology includes several items: the use of ANN for method development, multivariate calibration and quantitative determination while EOF variation, viscosity changes of a biological sample, and random error distribution in measured signals, and x values are simultaneously present.

Neural computing can also be successfully applied to correlate the outcomes of CE in human death investigations (Figure 5.17).

For instance, BP multilayer perceptron (MLP) was used for modeling the relationship and determination of electrolytes K^+, NH_4^+, Na^+, and Ba^{2+} concentration (mainly sodium) in vitreous humor, Figure 5.18a [103]. In human death investigations, vitreous humor has long been considered as an important tool for the estimation of the postmortem interval (PMI). Because of the variation in vitreous humor viscosity that causes variations in EOF, Ba^{2+} ions are added to a sample to mark the changes in EOF. In comparison to the traditional LLS method, the optimal BP-MLP model showed good linear correlation ($R^2 = 0.98$) with a higher accuracy of PMI estimation (Figure 5.18b).

5.4 CONCLUSIONS

ANNs have found an important position in electromigration methods. Especially, when combined with suitable ED, the ED–ANN combination is perhaps the most effective and suitable approach to be used for method development and/or optimization. The use of SVA when changing just one variable at a time should be avoided as it requires more experiments; it is laborious and might not be converging. Multivariate chemometrics techniques should be applied and the use of ANN for response surface modeling shows many advantages. The ANN can be used even when separation principles are not known and so the laborious determination of physicochemical parameters can be avoided.

Summarizing, ANN can be used in electromigration methods for

- modeling of separation processes;
- optimization in any of electromigration methods and for efficient method development in free zone electrophoresis, CEC, or MEKC;
- improvement of precision and accuracy in analyte determination from calibration curves –taking into account all kind of random errors, variation in EOF, and other side effects;

- multicomponent analysis from overlapping peaks where deconvolution can be eliminated.

Basic principles of ED, ANN, and a combination of ED and ANN (ED–ANN) approaches are given as well as a critical overview of other ANN applications.

ACKNOWLEDGMENTS

Support from Ministry of Education, Youth and Sports of the Czech Republic (Projects MSM, 0021622411, 0021627501 and Czech Science Foundation (Projects No. 104/08/0229, 202/07/1669) are gratefully acknowledged. This research has also been supported by the project R&D center for low-cost plasma and nanotechnology surface modifications CZ.1.05/2.1.00/03.0086 funding by European Regional Development Fund and EU project SUDSOE FP7-INCO-2011-6.

E.M. P-M. would like to thank the University of La Laguna (ULL) and Erasmus exchange program between ULL and Masaryk University for partial financial support. Also support from PROPMEP-SEP (SIATA Network), Mexico, is recognized.

REFERENCES

1. Khaledi, M.G., Smith, S.C., and Strasters, J. K. (1991) *Anal. Chem. 63*, 1820–1830.
2. Altria, K.D. and McLean, R. (1998) *J. Pharm. Biomed. Anal. 18*, 807–813.
3. Gübitz, G. and Schmid, M.G. (2000) *Enantiomer 5*, 5–11.
4. Hanrahan, G., Montes, R., and Gomez, F.A. (2008) *Anal. Bioanal. Chem. 390*, 169–179.
5. Gasteiger, J. and Zupan, J. (1993) *Angew. Chem. 32*, 503–527.
6. Havel, J., Peña-Méndez, E.M., Rojas-Hernández, A., Doucet, J.-P., and Panaye, A. (1998) *J. Chromatogr. A 793*, 317–329.
7. Marengo, E., Gianotti, V., Angioi, S., and Gennaro, M.C. (2004) *J. Chromatogr. A 1029*, 57–65.
8. Hanrahan, G. (2010) *Anal. Chem. 82*, 4307–4313.
9. Havel, J. and Janoš, P. (1997) *J. Chromatogr. A 786*(2), 321–331.
10. Havel, J., Janoš, P., and Jandik, P. (1996) *J. Chromatogr. A 745*, 127–134.
11. Muzikář, M., Havel, J., and Macka, M. (2002) *Electrophoresis 23*, 1796–1802.
12. Pirogov, A.V. and Havel, J. (1997) *J. Chromatogr. A 772*, 347–355.
13. Lubal, P., Maleček, J., Hermann, P., Kotek, J., and Havel, J. (2006) *Polyhedron 25*, 1884–1892.
14. Corstjens, H., Billiet, H.A.H., Frank, J., and Luyben, K.Ch.A. M. (1995) *J. Chromatogr. A 715*, 1–11.
15. Khaledi, M.G. (1997) *J. Chromatogr. A 780*, 3–40.
16. Beale, S.C. (1998) *Anal. Chem. 70*, 279–300.
17. Siouffi, A. M. and Phan-Tan-Luu, R. (2000) *J. Chromatogr. A 892*, 75–106.
18. Gübitz, G.;and Schmid, M.G. (2001) *Biopharm. Drug Dispos. 22*, 291–336.
19. Massart, D.L., Vandeginste, B.G.M., Buydens, L.M.C., De Jong, S., Lewi, P.J., and Smeyers-Verbeke, J. (1997) *Handbook of Chemometrics and Qualimetrics: Part A*, Elsevier, Amsterdam.
20. Otto, M. (2007) *Chemometrics. Statistics and Computer Application in Analytical Chemistry*, 2nd ed., Wiley-VCH, Weinheim, Germany.
21. Phan-Tan-Luu, R. and Cela, R. (2009) In: Sarabia, L., Phan-Tan-Luu, R., and Leardi, R. (eds), *Comprehensive Chemometrics: Chemical and Biochemical Data Analysis*, Vol. *1*, Elsevier, Slovenia.
22. Montgomery, D.C. (1997) *Introduction to Statistical Quality Control*, 3rd ed., Wiley, New York.
23. Yang, Q., Smeyers-Verbeke, J., Wu, W., Khots, M.S., and Massart, D.L. (1994) *J. Chromatogr. A 688*, 339–349.
24. Farková, M., Peña-Méndez, E. M., and Havel, J. (1999) *J. Chromatogr. A 848*, 365–374.
25. Dohnal, V., Farková, M., and Havel, J. (1999) *Chirality 11*, 616–621.
26. Havel, J., Madden, J. E., and Haddad, P. R. (1999) *Chromatographia 49*, 481–488.
27. Polášková, P., Bocaz, G., Li, H., and Havel, J. (2002) *J. Chromatogr. A 979*, 59–67.
28. Li, H., Zhang, Y.X., Polášková, P., and Havel, J. (2002) *Acta Chim. Sinica 60*(7), 1264–1268.
29. Havel, J., Vaňhara, J., Peña-Méndez, E.M., and Báez, M. (2010) In: *Redes Neuronales en Ciencia, Artificial Neural Networks in Science*, Universidad de La Laguna, Tenerife, Spain, Language: English, 54 pp. Depósito Legal TF-1750/2009, Edicion Septiembre. http://www.arte-c-c.com.
30. Zupan, J. and Gasteiger, J. (1993) In: *Neural Networks for Chemists: An Introduction*, Wiley-VCH, Weinheim, Germany.
31. Zupan, J. and Gasteiger, J. (1991) *Anal. Chim. Acta 248*, 1–30.
32. Aleksander, I. and Morton, H. (1995) *An Introduction to Neural Computing*, 2nd ed., Chapman and Hall. London.
33. Rosenbrock, H. H. (1960) *Comput. J. 3*, 175–184.
34. Havel, J., Breadmore, M., Macka, M., and Haddad, P.R. (1999) *J. Chromatogr. A 850*, 345–353.
35. Zhang, Y., Li, H., Hou, A., and Havel, J. (2005) *Talanta 65* (1), 118–128.
36. Dohnal, V., Li, H., Farková, M., and Havel, J. (2002) *Chirality 14*, 509–518.
37. Altria, K.D., Goodall, D.M., and Rogan, M.M. (1994) *Electrophoresis 15*, 824–827.
38. Wynia, G.S., Windhorst, G., Post, P.C., and Maris, F.A. (1997) *J. Chromatogr. A 773*, 339–350.
39. Altria, K.D., Kelly, M.A., and Clark, B.J. (1998) *Trends Anal. Chem. 17*, 204–226.

40. Shihabi, Z.K. (1998) *J. Chromatogr. A 807*, 25–30.
41. Schaeper, J.P. and Sepaniak, M.J. (2000) *Electrophoresis 21*, 1421–1429.
42. Ikuta, N., Yamada, Y., Yoshiyama, T., and Hirokawa, T. (2000) *J. Chromatogr. A 894*, 11–17.
43. Mayer, B.X. (2001) *J. Chromatogr. A 907*, 21–37.
44. York, D. (1966) *Can. J. Phys. 44*, 1079–1086.
45. Bocaz-Beneventi, G., Latorre, R., Farková, M., and Havel J. (2002) *Anal. Chim. Acta 452*(1), 47–63.
46. Zhang, Y.X., Li, H., Hou, A.X., and Havel, J. *Chemometr. Intell. Lab. Syst.* (2006) *82*, 165–175.
47. Hervás, C., Gutierrez, P.A., Silva, M., and Serrano, J.M.J. (2007) *Chemometrics 21*, 567–577.
48. Pretorius, V., Hopkins, B.J., and Schieke, J.D. (1974) *J. Chromatogr. 99*, 23–30.
49. Jorgenson, J.W. and Lukacs, K.D. (1981) *J. Chromatogr. 218*, 209–216.
50. Knox, J.H. and Boughtflower, R. (2000) *Trends Anal. Chem. 19*(11), 643–647.
51. Dittmann, M.M., Wienand, K., Bek, F., and Rozing, G.P. (1995) *LC/GC 13*, 802–814.
52. Deng, F Y., Zhang, F J., Tsuda, F T., Yu, F P.H., Boulton, A.A., and Cassidy, F R.M. (1998) *Anal. Chem. 70*(21), 4586–4593.
53. Wistuba, D. and Schurig, V. (2000) *Electrophoresis 21*, 4136–4158.
54. Hilder, E.F., Svec, F., and Fréchet, J.M.J. (2004) *J. Chromatogr. A 1044*, 3–22.
55. He, J. and Shamsi, S.A. (2009) *J. Chromatogr. A 1216*, 845–856.
56. Garcia-Alvarez-Coque, M.C., Torres-Lapasio, J. R., and Baeza-Baeza, J.J. (1996) *Anal. Chim. Acta 324*, 163–173.
57. Agnew-Heard, K. M., Sanchez, P.M., Shamsi, S.A., and Wagner, I. M. (1997) *Anal. Chem. 69*, 958–864.
58. Luo, J.W., Chen, H.W., and He, Q.H. (2001) *Chromatographia 53*, 295–300.
59. Lemus-Gallego, J.M. and Pérez Arroyo, J. (2003) *J. Sep. Sci. 26*, 947–952.
60. van Zomeren, P.V., Hilhorst, M.J., Coenegracht, P.M.J., and de Jong, G.J. (2000) *J. Chromatogr. A 867*, 247–259.
61. Kang, J., Schepdael, A., Roets, E., and Hoogmartens, J. (2001) *Electrophoresis 22*, 2588–2592.
62. Smet, E., Staelens, L., Heyden, Y.V., Baeyens, W.R.G., Aboul-Enein, H.Y., van der Weken, G., and García-Campaña, A.M. (2001) *Chirality 13*, 556–567.
63. Hillaert, S., De Beer, T.R.M., De Beer, J.O., and Van den Bossche, W. (2003) *J. Chromatogr. A 984*, 135–146.
64. Mangelings, D., Maftouh, M., Massart, D.L., and Vander Heyden, Y.J. (2004) *Pharm. Biomed. Anal.* 35789–35799.
65. Mangelings, D., Perrin, C., Massart, D.L., Maftouh, M., Eeltink, S., Kok, W.Th., Schoenmakers, P.J., and Vander Heyden, Y. (2004) *Anal. Chim. Acta 509*, 11–19.
66. Pačes, M., Kosek, J., Marek, M., Tallarek, U., and Seidel-Morgenstern, A. (2003) *Electrophoresis 24*, 380–389.
67. Wang, H., Gu, J.-L., Hu, H.-F., Ding, T.-H., and Fu, R.-N. (1998) *Fresen. J. Anal. Chem. 361*, 419–423.
68. Alfazema, L.N., Howells, S., and Perret, D. (2000) *Electrophoresis 21*, 2503–2058.
69. Hilder, E.F., Klampfl, C.W., and Haddad, P.R. (2000) *J. Chromatogr. A 890*(2), 337–345.
70. Breadmore, M.C., Macka, M., and Haddad, P.R. (1999) *Anal. Chem. 71*(9), 1826–1833.
71. Castagnola, M., Rossetti, D.V., Corda, M., Pellegrini, M., Misiti, F., Olianas, A., Giardina, B., and Messana, I. (1998) *Electrophoresis 19*, 2273–2277.
72. Piaggio, M.V., Peirotti, M.B., and Deiber, J.A. (2006) *Electrophoresis 27*, 4631–4647.
73. Piaggio, M.V., Peirotti, M.B., and Deiber, J.A. (2007) *Electrophoresis 28*, 2223–2234.
74. Piaggio, M.V., Peirotti, M.B., and Deiber, J.A. (2009) *Electrophoresis 30*, 2328–2336.
75. Piaggio, M.V., Peirotti, M.B., and Deiber, J.A. (2010) *J. Sep. Sci. 33*, 2423–2429.
76. Janini, G.M., Metral, C.J., Issaq, H.J., and Muschik, G.M. (1999) *J. Chromatogr. A 848*, 417–433.
77. Janini, G.M., Metral, C.J., and Issaq, H.J. (2001) *J. Chromatogr. A 924*, 291–306.
78. Solínová, V., Kasicka, V., Sázelová, P., Barth, T., and Miksík, I. (2007) *J. Chromatogr. A 1155*, 146–153.
79. Yu, K. and Cheng, Y. (2007) *Talanta 71*, 676–682.
80. Olajos, M., Chován, T., Mittermayr, S., Kenesei, T., Hajos, P., Molnár, I., Darvas, F., and Guttman, A. (2008) *J. Liquid Chromatogr. Rel. Technol. 31*, 2348–2362.
81. Mittermayr, S., Chovan, T., and Guttman, A. (2009) *Electrophoresis 30*, 890–896.
82. Li, Q., Dong, L., Jia, R., Chen, X., Hu, Z., and Fan, B. (2002) *Comput. Chem. 26*, 245–251.
83. Jalali-Heravi, M., Shen, Y., Hassanisadi, M., and Khaledi, M.G. (2005) *Electrophoresis 26*, 1874–1885.
84. Ma, W., Luan, F., Zhang, H., Zhang, X., Liu, M., Hua, Z., and Fan, B. (2006) *Analyst 131*, 1254–1260.
85. Trajan Neural Network Simulator, Release 3.0 D, Copyright © Trajan Software Ltd., Trajan House, 68 Lesbury Close, Chester-le-Street, Co. Durham, DH2 3SR, UK.
86. Liu, K.P., Xia, B.B., and Zhang, X.Y. (2008) *J. Liquid Chromatogr. Rel. Technol. 31*, 1808–1822.
87. Richard, M.D. and Lippmann, R. (1991) *Neural Comput. 3*, 461–483.
88. Zhang, G.P. (2000) *IEEE Trans. Syst. Man Cybern. C Appl. Rev. 30*(4), 451–462.
89. Zhang, Q.-F. and Cheung, H.-Y. (2011) *Phytochem. Anal. 22*, 18–25.
90. Garrido-Delgado, R., López-Vidal, S., Arce, L., and Valcárcel, M. (2009) *J. Sep. Sci. 32*, 3809–3816.
91. Kruzlicova, D., Mocák, J., Balla, B., Petka, J., Farková, M., and Havel, J. (2009) *Food Chem. 112*(4), 1046–1052.
92. Cartwright, H.M. (2009) Artificial neural networks in biology and chemistry—the evolution of a new analytical tool.

In: Livingstone, D.J. (ed.), *Artificial Neural Networks Methods and Applications*, Vol. *458*, Humana Press, pp. 1–13.

93. Pazourek, J., Gajdošová, D., Spanilá, M. Farková, M., Novotná, K., and Havel, J. (2005) *J. Chromatogr. A 1081* (1), 48–54.
94. Zhang, Y.X. (2007) *Talanta 73*, 68–75.
95. Xua, Y., Zomera, S., and Brereton, R.G. (2006) *Crit. Rev. Anal. Chem. 36*, 177–188.
96. Zürbig, P. and Mischak, H. (2008) *Contrib. Nephrol. Basel, Karger 160*, 107–126.
97. Malrey, L. and Taeho, J. (2006) *Lect. Notes Contr. Inform. Sci. 344*, 1056–1067.
98. Noble, W.S. (2006) *Nat. Biotechnol. 24*(12), 1565–1567.
99. Rogers, S., Girolami, M., Krebs, R., and Mischak. H. (2005) In: Singh, S. et al. (eds), *Disease Classification from Capillary Electrophoresis: Mass Spectrometry*, ICAPR 2005, LNCS 3686, pp. 183–191.
100. Mao, Y., Zhao, X., Wang, S., and Cheng, Y. (2007) *Anal. Chim. Acta 598*, 34–40.
101. Drube, J., Schiffer, E., Mischak, H., Kemper, M.J., Neuhaus, T., Pape, L., Lichtinghagen, R., and Ehrich, J.H.H. (2009) *Nephrol. Dial. Transplant. 24*, 2161–2169.
102. Haubitz, M., Good, D.M., Woywodt, A., Haller, H., Rupprecht, H., Theodorescu, D., Dakna, M. Coon, J.J., and Mischak, H. (2009) *Proteomics 8*, 2296–2307.
103. Bocaz-Beneventi, G., Tagliaro, F., Bortolotti, F., Manetto, G., and Havel J. (2002) *Int. J. Legal Med. 116*(1), 5–11.

6

IMPROVING THE SEPARATION IN MICROCHIP ELECTROPHORESIS BY SURFACE MODIFICATION

M. Teresa Fernández-Abedul, Isabel Álvarez-Martos, Francisco Javier García Alonso, and Agustín Costa-García

6.1 INTRODUCTION

Analytical Chemistry is a well-established discipline but also a very vivid and young one. The analytical process aims to solve solving problems by translating the socio-economic or scientific-technological statement into an analytical problem related to some of the four known questions: What and how much? How structured? How bound? and/or How distributed? [1]. The process followed to find the answers is well structured and usually starts with a sampling step and continues with the sample preparation or pretreatment step. Once the sample has been conditioned for measurement, an optimized methodology is followed to obtain the analytical signal that will give the chemical information, commonly an electronic signal. This will help in understanding and solving the analytical problem. In this context, the main goals of Analytical Chemistry are to obtain more (bio) chemical information with better quality, increasingly using less materials, time, cost, and risk for personnel and environment. This is summarized in three basic and interconnected trends, namely, automation, simplification, and miniaturization [2]. The last one is a basic trend shared with many technical and scientific areas. One of the best examples is the development of analytical devices with extremely small size, including operations such as sample injection, separation (i.e., capillary electrophoresis, CE), and detection [3]. This milestone started a new period that involves an intensive research line in microfluidics (manipulation of fluids in channels with dimensions of tens of micrometers), in order to achieve important technological improvements [4,5] in addition to relevant applications [6,7] that lead to real "lab-on-a-chip" (LOC) devices. The use of flow systems had produced an important advance in Analytical Chemistry, since many methodologies that were performed in discrete steps could be automated [i.e., flow injection analysis (FIA) methodologies]. Now, miniaturization has opened up many possibilities of transferring methods already matured or developing new ones. The latest developments with regard to technology, standard operations, and applications have been reviewed recently [8]. This field, also referred as micro total analysis systems (μTAS), is highly interdisciplinary and has served as a focal point to bring together research fields from different disciplines. On the other hand, achieving real μTAS or LOC devices involves integration of the multiple steps performed in conventional analysis systems. However, not all analytical steps are equally incorporated and the incidence level of miniaturization is not homogeneous. Concerning separation techniques, CE has demonstrated to be a robust technique with relevance in projects of paramount importance [9] and an invaluable help to pathologists and doctors in the evaluation of patients' status [10]. The fusion of the two principles (μTAS devices and CE), which is denoted microchip electrophoresis (ME) (at the beginning it was named microchip capillary electrophoresis, MCE), is extremely promising, with many different unexplored paths. Research in this field is active and new developments are being continuously reported [11]. However, limitations are still there and solutions need to be found

Capillary Electrophoresis and Microchip Capillary Electrophoresis: Principles, Applications, and Limitations, First Edition.
Edited by Carlos D. García, Karin Y. Chumbimuni-Torres, and Emanuel Carrilho.

for continuous improvement of the methodologies. Enhancement of sensitivity and selectivity is among the main goals.

6.2 STRATEGIES FOR IMPROVING SEPARATION

The determination of closely related analytes, with very similar structures, is still a challenge. Quantitation of analytes in complex matrices has required the use of different strategies in order to differentiate the analyte from a mixture of interfering substances. Complex pretreatments are sometimes performed in order to execute this separation. In some cases, even when pretreatments are performed, similar molecules are present in the sample. Moreover, it has to be taken into account that two related molecules can both be relevant analytes and they have to be in the same sample for simultaneous determination. This is the field of separation techniques, using different approaches for separating molecules. Requirements are increasing continuously and. with time, more similar molecules must be separated.

6.2.1 Selection of an Adequate Technique: ME

Strategies for resolving signals coming from related analytes are a field of active research. For example, resolution of peaks by means of voltammetric techniques is commonly performed through electrode modification. Polymeric films with electrocatalytic activity have been combined with carbon electrodes for the voltammetric resolution of dopamine (DA) [12] or epinephrine (EP) [13] in the presence of ascorbic and uric acids. However, to distinguish between chemical species that are involved in diffusion-controlled, one-electron electrolysis processes, it is required that their E° values differ by at least 0.118 V [14], and therefore combination with separation techniques is usually required. Even with the use of these techniques, the challenge is still there. Strategies to improve resolution in CE, including the use of nonaqueous capillary electrophoresis (NACE) or the use of isoelectric buffers or ionic liquids (ILs) as additives, have been reported [15].

An important example of these aims (improvement of the separation between similar molecules) is the resolution of very structurally related molecules such as chiral analytes [16,17], in which a special strategy has to be employed [18]. A high-speed ME method for the enantiomeric separation of (*R*,*S*)-naproxen using methyl-β-cyclodextrin was compared with CE methodology [19]. Poly(ethylene glycol)-functionalized polymeric microchips were fabricated, and evaluated for the electrophoretic chiral separation of 10 different D,L-amino acid pairs with the addition of β-cyclodextrin as chiral selector in the running buffer [20]. A different strategy for enantioseparation was the *in situ* molecular imprinting of the microchannel wall for separation of model enantiomers Boc-D-Trp and Boc-L-Trp, detected amperometrically [21].

6.2.2 Microchannel Design

Referring to their dimensions, CE (diameter) and microchip electrophoresis (microchannel width and height) work on the same scale, both on the order of micrometers. However, a great difference in dimensions is found in the length: capillaries are usually between 50 and 100 cm, whereas separation microchannels are around 5 cm, 10 times smaller. It has been shown that resolution in free zone electrophoresis is directly proportional to the square root of separation length [22]. Therefore, increasing the separation channel length will lead to an increase in resolution between closely eluting analytes. Nevertheless, the microfabrication procedure is not usually considered for very long microchips because of the equipment employed [ultraviolet (UV) insolation or laser equipment, spin coaters, etc.]. Consequently, the increase in length is made through serpentine microchannels with incorporated turns [23]. However, it has to be kept in mind that this design increases analyte dispersion [24] and requires an optimized design [25].

6.2.3 Selection of an Appropriate ME Material

A different possibility that ME offers (and actually it is one of its major advantages) is the employment of different materials. Owing to the limited separation channel length, the resolution obtainable in ME is in practice often lower than that in classical CE. Accordingly, the optimization of resolution is of great importance. In conventional CE, fused silica is nearly exclusively used as capillary material. However, quite diverse materials can be employed for microchip fabrication. The similarity between glass (microchip) and fused silica (capillary) made this material one of the first developed in the ME field together with some examples of silicon and quartz. However, the properties of polymers were quickly highlighted and microchips were manufactured in various materials with technologies different from those employed in conventional CE, as is extensively commented upon throughout the literature. As examples, polymeric materials such as polydimethylsiloxane (PDMS) [26,27], poly(methyl methacrylate) (PMMA) [28,29], polycarbonate (PC) [30], and poly(ethylene terephthalate) (PET) [31,32] are among the most employed for ME devices and also those consisting of cyclic olefin polymers such as Zeonex/Zeonor [33,34] and thermoplastic olefinic polymer of amorphous structure (TOPAS) [35] have been employed. The silicone elastomer PDMS is very popular for the ease of fabrication that allows speed in creating new fluidic designs and complex multilayer channel networks, but incompatibility with most nonpolar solvents [36] and surface instability have created the need for analogous materials that address these issues. One such polymer is thermoset polyester (TPE) [37,38], which can be shaped by a replica molding process similar to PDMS, allowing for rapid prototyping of fluidic designs, but with a surface stability and solvent resistance similar to those of glass. The use of polyester toner has also been reported [39], in addition to the use of EPON SU-8 resin [40,41], typically employed as sacrificial material in photolithographic processes, or elastomeric polyurethane (PU) [42]. Moreover, microchip materials such as low-temperature co-fired ceramics [43] or paper, an inexpensive material, easy to use, and specifically designed for developing countries [44], can be employed.

More recently, a new commercial hybrid ceramic polymer, Ormocomp, was introduced for the fabrication of microfluidic separation chips. This organic–inorganic polymer structure natively resists biofouling; even so, it can be used in intact protein analysis without prior surface modification [45]. A different strategy can be employed for generating materials that allow the fabrication of microfluidic devices: synthesis of block copolymers. As an example, a polydimethylsiloxane–poly(ethylene oxide) (PDMS–PEO) vinyl-terminated block copolymer has been synthesized *via* a simple hydrosilylation reaction between hydride-terminated PDMS and PEO divinyl ether [46]. This prepolymer can be subsequently cross-linked into an elastomer in a second hydrosilylation reaction involving a methylhydrosiloxane–dimethylsiloxane copolymer. The presence of the PEO block in the prepolymer chain results in a much more hydrophilic material following cross-linking. The length of the PDMS–PEO prepolymer chain and the multifunctional hydride cross-linker chains dictate the durability of the elastomeric material. Another possibility, which is commented upon in the section Surface Modification, is the use of materials obtained by copolymerization, that is, an acrylic copolymer such as poly(glycidyl methacrylate)-*co*-(methyl methacrylate) (PGMAMMA) [47] or poly(ethylene glycol) (PEG)-functionalized acrylic copolymer [48,49].

Apart from the consideration of the microchip material, it has to be taken into account that hybrid microchips are commonly employed. Microfabrication of channels is usually made in one of the plates of the microchip and the other plate can be different (e.g., PDMS–glass and silicon–PDMS). Differences with microchips in which both plates are made of the same material are very possible and have to be considered.

Comparison of the analytical performance of electrophoresis microchannels fabricated in different materials has been reported. In this way, the performance of microchips fabricated with different technologies in quartz, glass, PMMA, and PDMS has been studied [50]. Characteristics of PDMS and glass microchips have been evaluated in the separation of peptides [51]. Similarly, different PDMS and glass microchips and also polyester toner chips micromachined by a direct printing process using an office laser printer have been compared [52], in addition to the surface chemistry and optical properties of SU-8 and glass microchips [53].

The fact that ME devices have been manufactured in all these materials, very different in properties and characteristics, suggests the great potential of this methodology. As will be commented upon later, ME is still under development and researchers are still looking for the material with the best performance. The main goal in this area is the optimization of the separations, which relates to the surface of the microchannel, which in turns depends on the material and also on the treatments of the surface. For example, the surface of native PDMS does not contain ionizable surface groups to produce a significant zeta potential when it is in contact with solution. Electrokinetic pumping, the method by which solution flow is induced within the microchannels, relies on the electroosmotic force generated as a result of an electrical double layer. The lack of surface ionizable groups prevents PDMS from generating significant electroosmotic flow (EOF) [54]. Therefore, separations are affected by the microchannel material responsible for EOF and the state of its surface.

Apart from the improvement of the separation in terms of resolution between adjacent peaks, solving problems such as microchannel hydrophobicity is pursued. Such hydrophobicity makes it more difficult to wet the channels and easily form air bubbles. Moreover, adsorption of molecules on the surface that even penetrate spontaneously into the polymer matrix (e.g., PDMS) can occur and the EOF can be unstable and poorly controlled [54,55]. Therefore, it can be said that, in general, the main motivation for finding an adequate material is the control of the EOF in order to improve the precision of migration times but also to perform a robust injection. Another important goal is the reduction of analyte–wall interactions.

6.2.4 Optimization of the Working Conditions

It is known that molecules migrate under the electric field following two different electrokinetic effects: electrophoretic and EOF. Apart from the material, the background electrolyte (BGE), which affect the apparent mobility of molecules, especially the pH of the running buffer, are dominant factors in the separation of analytes [56]. Moreover, operational conditions such as separation voltage, injection time, and injection format have to be carefully optimized. However, when the differences in molecules are very small and also dependent on the material, the use of appropriate conditions is not sufficient. In these cases, modification of the microchannel surface is a possible strategy for improving the separations.

6.2.5 Surface Modification

The field of microfluidics is moving very fast on the research front, but the number of real-world, routine applications is still limited. One probable reason is that the very richness of the field also increases the already large number of challenges that must be solved before a robust practical system can be achieved. As commented upon in a review on surface treatment and characterization [57], surface interactions and treatments are among these challenges, being, in contrast to what happens in CE, in their infancy. Another limitation is the moderate number of surface characterization methods that can be applied inside a capillary or microchannel. Microfluidic systems are often assembled by the bonding of flat substrates. It is therefore often possible to transpose to microfluidics the developments in the chemical physics of surfaces. It has to be taken into account that the devices have a large surface area-to-volume ratio and, therefore, the state of the surface becomes very important. Changes in the surface will undoubtedly have a positive/negative influence on separations.

The number of microchip materials and surface chemistries prohibit an exhaustive account of methodologies here and, moreover, have been adequately dealt with elsewhere. Revisions on the surface modification in ME have been reported [58–65]. Since modification of chips was driven by different aims, this is sometimes the criterion for making a classification of the modification procedures: modulation of the EOF, solid-phase extraction, suppression of protein adsorption, electrophoretic separations, or design of biochips/arrays [61]. On the other hand, in other revisions [58,60], classification into dynamic coating with surface-active compounds that presents physical adsorption, and permanent coating with molecules that are covalently bound through functional groups, is made. Subclassification attending to the material, glass/quartz or plastic devices is possible. Several approaches for surface modification of glass substrates have been transferred from classical capillaries in CE to ME, but various different approaches were generated for the new materials employed in ME. Concretely, most of the developments are made for polymers [18,58,61,62], which are cheaper and employ simpler fabrication procedures, PDMS being the most popular [58,62,66]. In reported reviews, the special case of wall [60] / permanent [62] coatings for protein separations is considered. In this chapter, the most important modifiers employed in ME as a strategy for improving the resolution are commented upon. As a summary, in Figure 6.1 all the considerations that have to be taken into account when a surface modification is going to be performed and the criteria that can be followed for classifying the modification procedures are schematized.

6.2.5.1 Surface Micro- and Nanostructuring Microchannel surfaces can be modified by different means. One of these is to construct on the microchannels special micro- and nanostructures that differentiate the passage of similar molecules. Electrophoresis over nanopatterned surfaces has some analogies with gel electrophoresis, taking into account that the gel is replaced by nanofabricated structures. Depending on the dimensions, they can act as a sieving matrix. The main advantage of such structures is the uniformity of spacing and size of the obstacles and their

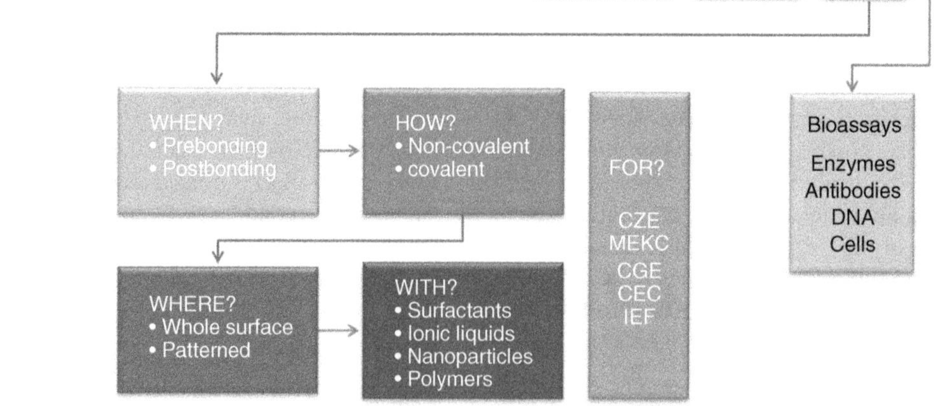

FIGURE 6.1 Classification of surface modification procedures according to diverse criteria.

robustness for repeated use [67]. Surface morphology or surface roughness of the microchannels undoubtedly has an effect on separation quality; the influence of surface microstructures on the separation efficiency in glass ME has been evaluated for the model system DA–EP, molecules that differ only in a hydroxyl group [68]. However, even though these structures have been applied to the separation of small analytes, the main application is found for large molecules, such as DNA. Thus, nanochannels with cross-sections as small as $10\,\text{nm} \times 50\,\text{nm}$ allow the elongation of very long biopolymers by space confinement. They can be gently elongated and analyzed [69]. Special networks such as nanoslit arrays were imprinted in PMMA, cyclic olefin copolymer (COC), and PC fluidic devices using a nanoimprinting tool for studying the double-stranded DNA electrokinetic transport mobility [70]. Nanoscaled structures inside microchannels such as nanopillar, nanoball, and nanofibers contribute to obtaining higher performances in DNA separations [71]. In these cases, the flow passes through the different motifs and, since the size of analytes approaches that of the structures, transport is affected and separation can be improved. Actually, a rectangular array of micropillars was used to replace gels [polyacrylamide (PA) or agarose] to perform electrokinetic separation, since the microarray mimicked a highly diluted gel [72]. Instead of using a regular array of posts or holes, size separation of DNA molecules can be achieved by using anodic porous alumina as a separation medium [73]. The bed of the microfluidic channel was made of a porous alumina membrane having nanoscale pores on its surface. Larger molecules eluted first as the small ones are much more frequently trapped by the nanopores. This is a field that is probably going to increase with the development of nanotechnology, and promising work on the modification with ordered structures of controlled dimensions is expected.

6.2.5.2 *Employment of Energy Sources* When dealing with surface modifications, two main options are possible: physical or chemical. It is known that exposing materials to various energy sources (i.e., oxygen plasmas, UV light, and corona discharges) can alter surface properties. Plasma is a mixture of electrons and ions with high energy. This modification employs gases such as oxygen, nitrogen, and hydrogen, which dissociate and react with the substrate surface, creating chemical functional groups. UV treatment can also be employed, with the advantage that it facilitates much deeper modification of the surface. On the other hand, chemical vapor deposition (CVD) is a chemical process used to produce a thin film on a substrate surface by means of the deposition of gaseous molecules, which react chemically on the surface. The sequence of events is typically sublimation, pyrolysis, and deposition. PDMS is the material in which these treatments were mostly employed [66], exposure to oxygen plasma being, by far, the approach most often employed, as will be shown later in this chapter, in Table 6.5.

Surface modification of PDMS by Tesla coil oxidation can be utilized to yield a hydrophilic surface that is capable of possessing a significant zeta potential at high pH values. The Si—CH_3 groups on the PDMS surface can be modified through oxidation to Si—OH [74]. Three consequences can be envisaged. First, the surface is much more hydrophilic, enabling facile filling of the micron-sized channels with aqueous buffers. Second, the surface now closely resembles glass in terms of functionality (Si—OH) and, as a result, a strong EOF, very similar to glass, results at pH values above the pK_a of the silanol group [75]. Third, the oxidized surface is able to form a permanent bond when placed in conformal contact with a similarly oxidized piece of PDMS. Nevertheless, the modification is only temporary because the hydrophilic surface begins to revert back to its original hydrophobic form, exhibiting a 75% decrease in flow in less than 24 h. The process can be slowed somewhat by contacting with a buffer solution, but finally it reverts to its original form [76]. However, when the PDMS surface is derivatized with 3-aminopropyltriethylsilane (APTES) following oxidation, the modification is stable for more than 10 days [77].

The goal of this chapter, however, is to present another possibility for surface modification, namely by using chemical reagents. This does not mean that a choice has to be made between the two possibilities (physical or chemical modification) because in most cases (see Table 6.5) the physical treatment is employed as a pretreatment that prepares the surface for further chemical modification.

6.2.5.3 *Chemical Surface Modification* There is a wide range of reports on the use of surface chemistry in order to add new functionalities. This would advance the introduction of more steps in the same device, which will approximate the concept of LOC devices. However, different objectives can be pursued with the addition of new functional groups. On the one hand, improvements in the separation in ME devices are one of the objectives, which normally is related to EOF control and elimination of analyte adsorption. This objective is the aim of this chapter and is considered in the following sections. However, on the other hand, high immobilization capacity and low nonspecific binding are aimed at for the development of bioassays. Summarizing, the purposes of modifying the surface of microfluidic devices fall into two main categories: improvement of the surface properties for normal functioning of the devices (this includes EOF control that greatly affects the separation efficiency and precision of analysis, control of the adsorption of biological components, modification of the hydrophobic/hydrophilic character of some materials, etc.) and functionalization of the surface to tailor the device to a specific use [78]. In addition, in this section different considerations on when, where, or how the modification is made are discussed, in addition to the electrophoresis mode that is commonly employed.

Application to Bioassays. Surface chemistry procedures can be shared with those employed in bioassays such as arrays for DNA or protein determination [79]. In conventional solid-phase immunoassays (and bioassays in general), a long incubation time is a result of inefficient mass transport by molecular diffusion from the solution to the solid surface, although the immunoreaction itself is a rapid process [80]. The use of microfluidic devices is very appropriate for this type of analysis. However, because of the large surface-to-volume ratio of the microchannels, controlling the surface properties becomes one of the most critical issues in developing those assays. Conventional passive adsorption of bioreagents on the surface often causes protein denaturation and reduces functional sites or activities. Therefore, developing an efficient surface modification method to enhance the binding efficiency and activity is critical in bioassay development. Poly(ethylenimine) (PEI, MW 75,000 Da), which produces amine functionalization, allows binding of antibodies *via* glutaraldehyde interaction, with 10 times more active antibody bound to the surface. The results were superior to those obtained with other amino-bearing chemicals such as poly(allylamine hydrochloride) (PAH), hexamethylenediamine (HMD), and 1,3-diaminopropane (DAP). The amine-bearing polymers give a better response than small diamine molecules because the spacer function of polymers can preserve most of the biological activity of the bound protein molecules [81]. Antibodies directed toward human cardiac troponin I were immobilized on the internal surface of the PDMS channel on which a protein G layer that enhances proper orientation of the antibodies had been generated by silanization [82]. This was done by injection of 7-octenyltrichlorosilane and further oxidation of the vinyl groups into carboxylic groups. *N*-(3-Dimethylaminopropyl)-*N*-ethylcarbodiimide (EDC) and *N*-hydroxysulfosuccinimide (sulfo-NHS) were injected into the channel to activate the carboxyl groups and protein G was anchored via chemical reaction between its free amino groups and the NHS-activated carboxyl groups. However, not only PDMS is employed as a substrate material for immunoassay development:, immobilization of antibodies on PMMA microchannels has also been effected using PEI containing abundant NH_2 groups to immobilize the monoclonal antibodies covalently for the determination of α-fetoprotein (AFP) [83].

Combination of pillar arrays and grafting with hyperbranched polyglycerols (HPGs) on an amine-modified PDMS was applied for the selective capture of positively charged proteins (avidin). The charge density on grafted HPG was optimized to minimize the nonspecific protein adsorption and increase the selective capture [84].

Surface modifications can be performed with the aim of carrying out bioassays, and in this field not only immunoassays are considered but also DNA hybridization assays. In this case, the final objective is the immobilization of the affinity reagent. Chemical modification of PDMS by curing a mixture of undecylenic acid (UDA) in PDMS prepolymer on a gold-coated glass slide was performed. This gold slide had been previously pretreated with a self-assembled hydrophilic monolayer of 3-mercaptopropionic acid (MPA). During curing of UDA–PDMS, the hydrophilic UDA carboxyl moieties diffused toward the hydrophilic MPA carboxyl moieties on the gold surface. Once completely cured, the PDMS is peeled off the gold substrate, exposing the interfacial carboxyl groups, which are available for subsequent attachment of 5′-amino-terminated DNA oligonucleotides via amide linkages [85]. Further, DNA extraction from samples can be performed by a procedure based on the electrostatic interaction between amine groups (from APTES) or 3-[2-(2-aminoethylamino)ethylamino]propyltrimethoxysilane (AEEA) on a solid support (silicon wafer with a microchannel that after functionalization is enclosed with PDMS) [86].

It is clear that for the development of bioassays, immobilization of reagents is often required. The absence of functional groups makes the immobilization of reactive coatings with excellent adhesion difficult when deposited in thin films. For this type of applications, glass is not very appropriate because of its intrinsic stiffness and its incompatibility with soft materials for making valves and actuators. Poly(*p*-xylylenecarboxylic acid pentafluorophenol ester-*co*-*p*-xylylene) (PPX-PPF) was deposited on the luminal surface of PDMS. The high chemical reactivity of their functional groups supported conversion with biological ligands, proteins, and cells [87]. PDMS microchannels with a poly(*N*-isopropylacrylamide) (PNIPAAm)-grafted surface were employed for cell-based assays. PNIPAAm is a type of thermoresponsive smart polymer that exhibits a reversible phase transition at around 32 °C (lower critical solution temperature, LCST). At temperatures below its LCST, the polymer chains swell and become hydrophilic, whereas at temperatures above the chains collapse and become hydrophobic [88]. When PNIPAAm is grafted on to different substrates, cells can adhere and proliferate at temperatures above the LCST. The adhering cells can be detached from the surfaces by lowering the temperature to below the LCST without the help of trypsin digestion. The most attractive properties of PDMS are biological compatibility and gas permeability, and therefore it has been chosen for cell culture [89] and cell-based assays [90].

Proteolysis is a key process for protein sequencing in proteome research. High-quality alternatives for conventional in-solution digestion of proteins that is time consuming are needed. Immobilizing protease enzymes on the channel walls of microchips for further coupling to mass spectrometry (MS) produced a layer that is more stable and highly resistant to environmental changes. Moreover, it provides molecular-level interactions between the immobilized enzymes and the flowing protein substrates. Trypsin is a proteolytic enzyme that has been immobilized on the channel walls of microchips by sol–gel encapsulation [91], covalent linking [58], or layer-by-layer (LBL) procedures with multilayer assembly of chitosan and hyaluronic acid [92], or poly(diallyldimethylammonium chloride) (PDDA) and nanozeolite crystals [93]. In these cases, as in the majority, microchannel surface modification has been reported but special and interesting cases of a different type of ME-modifications can be found. Fibers made from various substances are versatile materials that can be employed as platforms for protein digestion. A piece of trypsin-immobilized glass fiber was inserted into the channel of a PMMA microchip to fabricate a core-changeable microfluidic bioreactor. The reactor could be regenerated by changing the core, which consisted of a piece of glass fiber and a layer of trypsin immobilized on a silica coating [94]. The reactor was prepared by a sol–gel method with tetraethoxysilane (TEOS) and APTES as precursors, offering a porous surface to accommodate trypsin. It was immobilized on the primary amino group-containing coating (from APTES) with the aid of glutaraldehyde. The original waste reservoir was cut off to leave the channel outlet for inserting the fiber core in such a way that 3 mm protruded outside the channel outlet, facilitating its removal.

When is the Surface Modification Made? It is said that the next generation of microchips will incorporate more diverse surface functionalities in already complex geometries, requiring advanced methods for patterning the internal surfaces of the microchannels. One of the issues that have to be taken into account is that the modification of the microchannel surface is far less amenable to the patterning methods employed for planar surfaces. Two different approaches, prebonding and postbonding modification, are possible. The first one is effective for selected systems; however, in many cases the bonding conditions can damage the surface treatment. The second one has the advantage of using highly effective bonding methods. Moreover, modification can be carried out immediately prior to microchip use, avoiding aging of the surface chemistry in the time between fabrication and application [66]. This type of modification also circumvents potential alignment problems, where chemical patterns are required at specific locations within the microchannel network and offer the reuse of expensive or difficult-to-fabricate microfluidic devices by chemically stripping under flow. Postbonding modification methodologies such as laminar flow and capillarity, photolithography, use of microplasmas, and electrochemical biolithography have recently been reviewed [95]. In most of the cases commented upon here, postbonding procedures with introduction of reagents *via* flow through the channels, either by capillarity or by external pressure, and further reaction/adsorption at the channel walls are performed. Alternatively, and especially with covalent coatings, this is made in one of the plates that is further sealed with the other microchip plate.

Where does the Modification Take Place? Another aspect to be considered is the homogeneity of the modified microchannel surface. Most of the procedures achieved homogeneous modification but exclusive modification of discrete regions will be required in the near future, where microchannel walls are no longer simply guides for fluid flow but important instruments in the functionality of microchips. Here, greater spatial control over surface modification to perform more complex designs is needed in addition to the development of inexpensive polymers for disposable biochips [95]. In this case, it should be noted that prebonding patterning with inorganic materials (metals, metal oxides, etc.) can facilitate postbonding patterning of less robust materials (organic self-assembled monolayers, proteins, DNA, etc.). Metallic materials can act as electrodes that provide a suitable platform for postbonding surface biomodification by electrochemical methodologies [96,97]. Moreover, photopatterning of the channels by UV-mediated grafting is possible, either before assembling the microchip plates [98] or directly in the microchannel [99], as will be commented on in the section dedicated to polymer modifications.

How is the Modification Performed? In most of the reports and related reviews, modification is classified in dynamic and static or noncovalent and covalent coatings. The first is the easiest way for surface modification and usually surface-active or other components such as polymers or nanostructures are included in running buffers. Alternatively, they can be added to the sample reservoir or employed first in a preliminary treatment. Since a not very strongly adsorbed layer is formed on the wall surface, this is sometimes referred to as a pseudostationary phase (PSP). The benefit of this approach is that uncomplicated procedures are employed apart from the fact that the "one-time use" approach allows continuous regeneration and absence of a stationary phase carry-over effect. The dynamic coating therefore has the same basis of disposable devices, something simple that can be employed each time for improving the resolution. Among the aims of dynamic modifiers is to control/change the EOF, whose velocity is given by the equation [100]:

$$\mu_{eo} = (\epsilon_0/4\pi\eta)E\xi$$

where ε_0 is the dielectric constant; η is the solution viscosity, E is the electric field strength in V cm^{-1}, and ξ is the zeta potential that depends on the surface charge. Therefore, the dynamic modifier can act by altering the viscosity or the surface charge, both with effects on the EOF.

Low cost and ease of performing procedures are achieved simply by physical adsorption of molecules on a surface. However, dynamic coatings are degraded with electrophoresis or cause problems when coupled with some types of detection such as MS [101]. Therefore, another possibility is a permanent or covalent coating that is generated when a chemical compound is covalently bound to functional groups of the surface. In this case, and by comparison with chromatographic methods, a stationary phase is created. This subject is one of the focuses of research because improving resolution is still among the problems to be solved in electrophoresis, and there are many papers related to capillary coatings and surface modifications in recent reviews on CE [102].

What Electrophoresis Mode is the Modification Made for? Referring to the electrophoretic modes, most of the treatments are performed for capillary zone electrophoresis (CZE) in microchips but treatments are also advantageous for other modes, such as isoelectric focusing (IEF). Immobilization of hydroxypropylcellulose (HPC) and poly(dimethylacrylamide-*co*-allyl glycidyl ether) (PDMA-AGE) on glass and PDMS microchips demonstrated that a coating was needed to avoid pH gradient drift [103]. Both coatings were efficient on glass microchips, but only the second one allowed focusing of p*I* markers on PDMS microchips. A double coating using PDMA-AGE and methylcellulose (MC)–Tween-20 was employed for the separation of β-peptides in cerebrospinal fluid by gel electrophoresis in PDMS microchips [104].

Gel electrophoresis is one of the most commonly employed modes because of its applicability to DNA separation and new gels are continuously appearing in order to improve resolution, such as the cholesterol-bearing pullulan nanogels [105]. The formation of the gel is due to weak hydrophobic interactions between cholesterols above 30 mg mL^{-1} and can be easily deformed by external forces. Thus, the loading process of the nanogels into the microchannels becomes easier. Gel electrophoresis has frequently been combined with the use of surfactants (e.g., sodium dodecyl sulfate polyacrylamide gel electrophoresis, SDS-PAGE), and also surfactants are the basis of the micellar electrokinetic chromatography (MEKC) mode, due to the formation of micelles that allows the separation of neutral molecules and the improvement for ionic analytes because of the differential interaction with micelles. Although surfactants will be considered in this chapter, these modes will not be exhaustively considered. Some special cases that refer to surface modification will be commented upon.

Referring to the CE modes, electrochromatography is also sometimes involved where the use of stationary or PSP is concerned. Electrochromatography is a technique that combines the efficiency of electrophoresis with the selectivity of chromatography. In capillary electrochromatography (CEC), three different types are generally distinguished, based on the capillary column used: packed, open-tubular, or monolithic. Porous polymer monolithic materials, which emerged in the early 1990s [106], have become popular owing to their ease of preparation. The first syntheses of organic monoliths were carried out in glass chips [107], but different materials, such as cyclic olefin copolymers (COCs) that are compatible with a broad range of chemicals and solvents, can be employed. As an example, a lauryl methacrylate monolith was synthesized into a COC microdevice for reversed-phase electrochromatography [108]. A fourth type of CEC can also be distinguished, which uses a PSP, micelles being the first PSP to be used in the MEKC technique. Other macromolecular entities such as liposomes, proteins, cyclodextrins, dendrimers, and nanoparticles (NPs) can be employed. In this case, they can be considered as modifiers. The use of NPs will be particularly considered in this chapter. Frontiers are not rigid and therefore a PSP–CEC methodology with micelles can be considered MEKC or a NP-PSP as a dynamic coating for CZE. In the same manner, a surface modification for improving separation such as a static coating for CZE can be considered as a CEC. It depends on the focus: if the coating produces a differential interaction of the analytes, a CEC methodology can be considered; if the coating just modifies the surface for avoiding adsorption or controlling the EOF for improving resolution, a CZE mode with modifiers can be considered.

6.3 CHEMICAL MODIFIERS

Chemical modifiers are a group of substances with varying properties such as size, charge, functionalities, and so on that affect the performance of electrophoresis, especially in the case of ME, where the area-to-volume ratio is very high. The use of organic solvents has been extended to the use of electrolyte solutions prepared from pure organic solvents or their mixtures in NACE [109]. Organic solvents can solubilize compounds hardly soluble in water and thus afford the possibility of separating such compounds by CE. Moreover, their physical and chemical properties can offer large changes in separation factor/resolution, analysis time, selectivity, and reduced electrophoretic currents. Therefore, changing organic solvents, usually amines (e.g., diethanolamine) or alcohols (e.g., methanol), which are the most common, or varying the proportions allows a simple tuning of separation.

Special consideration will be given to other types of modifiers: surfactants, ionic liquids, nanostructures, and polymers. Special modification procedures such as sol–gel protocols make modification possible with different types of chemical modifiers, but they are scarce. Sol–gel methods can be employed to fabricate PMMA electrophoresis microchips with a hydrophilic channel wall. TEOS is injected into the channel and is allowed to diffuse into the surface layer for 24 h. After removing the extra material, the channel is filled with acidic solution for 3 h, flushed with water, and maintained in an oven to obtain a sol–gel-modified PMMA microchip [110]. In this chapter, however, only the four types commented on at the beginning of the paragraph will be considered.

In Figure 6.2, we present histograms with the number of publications in which surfactants, ionic liquids, and NPs appeared in relation to CE or ME in the period from January 2000 to April 2011. Several differences can be seen. First, the use of surfactants as modifiers is a well-established methodology. Even though a slight increase over the years can be noted, they have been employed from the beginning of the decade. Actually, SDS has long been used in well-known methodologies such as SDS-PAGE and MEKC. Further, there has been an increase in use in recent years, not only for CE but also for ME.

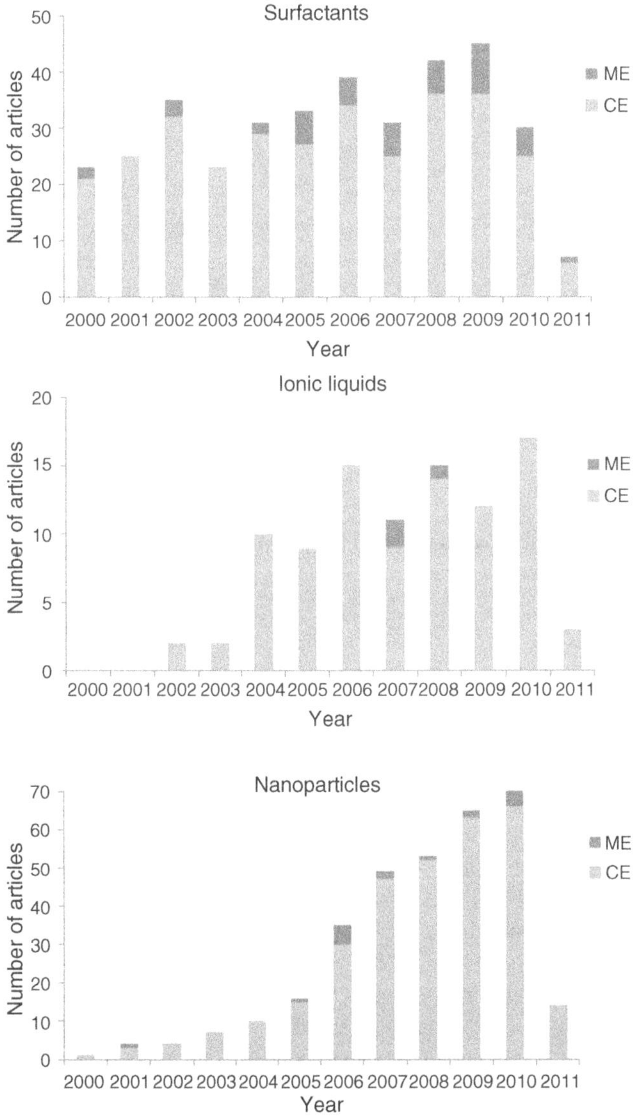

FIGURE 6.2 Number of publications in which surfactants, ionic liquids, and nanoparticles appeared related to CE or ME applications. ISI Web of Knowledge, April 2011.

The words "surfactants" and "capillary electrophoresis"/"microchip electrophoresis" were employed for the Internet search. It should be noted that when "SDS" was employed instead of surfactants, for CE, the number of articles was at least doubled (e.g., it changed from 21 to 42, from 27 to 53, or from 25 to 57 for the years 2000, 2005, and 2010, respectively). In the case of ME devices, the search was refined with the terms "SDS," "CTAB," "Triton X-100," or "Tween 20" and the results were checked to avoid duplication. The search was also performed with "microchip capillary electrophoresis" for all the modifiers, since this term was more used at the beginning of the decade. The results were also cross-checked to avoid repetitions.

The case of ILs and NPs is totally different from that of surfactants. In both cases, the research in this field, in combination with CE, originated in 2002 and 2000, respectively, and an increase (continuous in the case of NPs) can be seen. The situation for ME is slightly different. In the case of NPs, the first work of the decade appears in 2001 and a continuous increase is seen from 2005 to date. However, in the case of ILs, first works appeared in 2007 and 2008, and no further research is reported to date. The words employed in the search were "ionic liquids" and "capillary electrophoresis"/"microchip electrophoresis" as well as "nanoparticles" and "capillary electrophoresis"/"microchip electrophoresis." Refinement was made for ME with "background additive" for ILs and "nanotubes"/"nanostructures" for NPs.

Therefore, it can be concluded that surfactants are widely employed for electrophoresis methodologies and an increase in their use is expected. The field of application of ILs is still unexplored in ME, and in the case of NPs, an almost exponential increase is suspected. The number of publications is significant for 2010: the number of publications on NPs overtook that for surfactants. Moreover, new materials are continuously being incorporated (e.g., graphene, NPs of smaller sizes, and new functionalizations) and future possibilities are still unknown.

In the following sections, these modifiers will be considered in more detail. It has to be taken into account that some procedures can employ more than one type of modifier. An explanatory scheme with the different types of modifiers is presented in Figure 6.3, together with the classification of the modifiers that have been employed in ME. In further sections, each modifier will be considered in depth. This chapter does not try to provide exhaustive information, but explains the more representative strategies. Some more information can be found in the tables and in the reviews that are cited in the text.

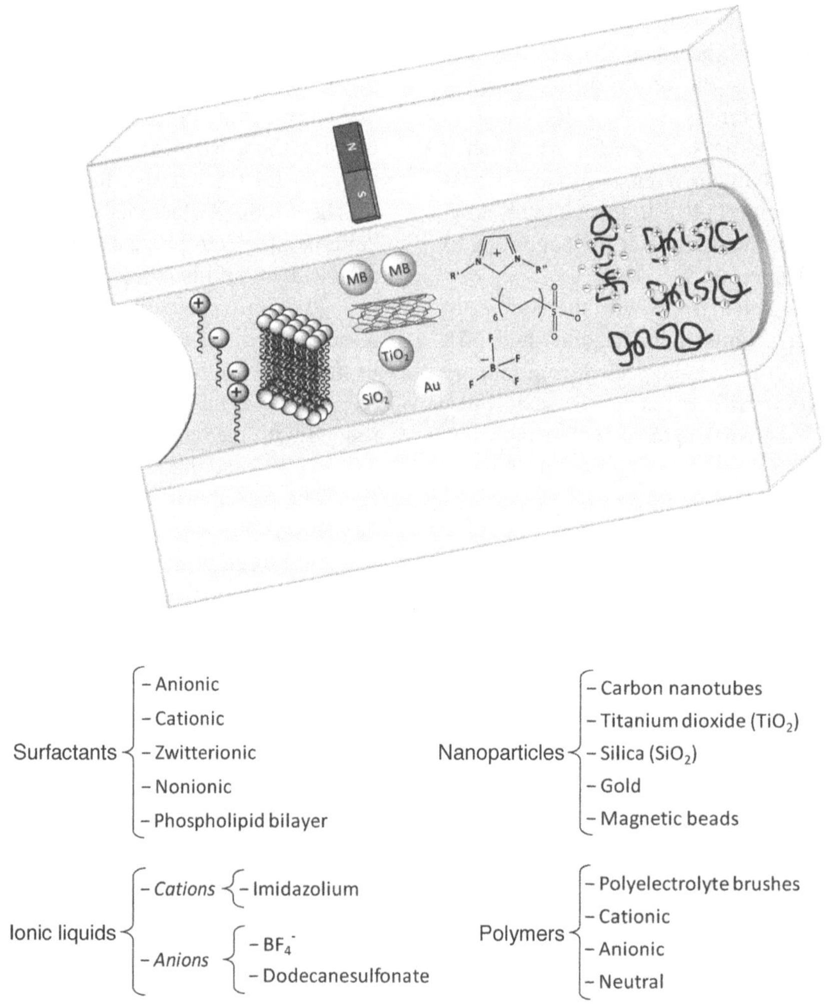

FIGURE 6.3 Scheme of the different types of modifiers that have been employed in combination with ME.

6.3.1 Surfactants

Surfactants are considered among the small molecules that can be used as modifiers. It is known that the use of small molecules has some effect on the separation. Owing to the nonpolymeric nature of these additives, any physically adsorbed layer on the surface is likely to be fairly thin [63]. Salts, amines, and surfactants have been reported as additives for improving protein separations by capillary and chip electrophoresis [63]. Surfactant is an abbreviation for surface-active agent, which literally means "active on a surface." The surface can be established between a solid and a liquid, between air and a liquid, or between two immiscible liquids. The primary property of a surfactant in a solution is that the concentration of the surfactant is higher on the surface than in the bulk of the liquid. Thus, the surfactant concentrates on the surface, where it is active. Surfactants consist of a hydrophilic group (anionic, cationic, nonionic or amphoteric) and a hydrophobic group, hydrocarbon chains in the majority and usually linear because of the demand for biodegradability (commonly between C_{10} and C_{16}). The hydrophilic group determines the main differences between the majority of surfactants, the hydrophobic group being the largest part of the molecule. As noted earlier, one of the drawbacks of hydrophobic materials (e.g., PDMS) is that they adsorb organic solvents and hydrophobic analytes due to their hydrophobic surface, causing fouling of the material and poor separation efficiency. Surfactant-based dynamic modification is a favorable method to avoid these adsorptions owing to its versatility and low cost.

Surfactants can be classified according to their ionic (cationic, anionic, zwitterionic) or nonionic character, SDS (anionic) and cetyltrimethylammonium bromide (CTAB) (cationic) being the most commonly employed. Although usually a surfactant is associated with small molecules, nonionic polymers such as Tween 20 (polysorbate surfactant with three hydroxyl groups *per* molecule) and Triton X-100 (nonionic surfactant with a hydrophilic PEO group) are also widely employed. Other less common surfactants can be combined for the better performance of ME devices. *n*-Dodecyl β-D-maltoside (DDM) is an alkylpolyglucoside that belongs to a family of very mild nonionic surfactants that adsorbs strongly on hydrophobic surfaces forming a monolayer, which causes the surface to become hydrophilic and nonionic, thus reducing the interaction between proteins and PDMS surface [111].

A review that considers the perspectives of surface treatment in electrophoresis and lab-on-chips [57] differentiates between thin and thick films when preparing coatings. The most common method for the build-up of thin films is Langmuir films (Langmuir–Blodgett films by transferring the first to a solid substrate). They are prepared from amphiphilic molecules scattered on the surface of a liquid, in such a way that the hydrophilic headgroup positions at the water/film interface while the hydrophobic end group sticks out from the film/solid interface. These films are mostly bonded through weak hydrogen bonds and van der Waals interactions. Self-assembled monolayers (SAMs) relate to the formation of ordered molecular assemblies that are formed spontaneously by the adsorption of a surfactant with a specific affinity of its headgroup to a substrate [112].

Single-chain surfactants, such as those noted above, have been widely employed but, in recent years, double-chain surfactants, which are also called lipid or supported lipid bilayer membranes, have been increasingly employed. On the other hand, biomimicking phospholipid membranes and liposomes are utilized in many liquid chromatography (LC) and CE techniques, either as coatings or as carriers [113,114]. Lipids can group to form different aggregates: liposomes, micelles, bicelles, and discoidal micelles. Adding PEG (PEGylation) provides structural stabilization and also prevents protein adsorption on the lipid layer [115]. Phospholipid disks based on 1-palmitoyl-2-oleyl-*sn*-glycero-phosphatidylcholine and 1,2-distearoyl-*sn*-glycero-3-phosphoethanolamine-*N*-[methoxy-(polyethylene glycol)] with a molar mass of 3000 were employed as a dynamic coating for SU-8 microfluidic chips [116]. The effects on surface charge and nonspecific protein adsorption was compared with those of two common surfactants, anionic SDS and neutral Tween-20, employed as buffer additives. The effect on the EOF was more prominent than that of SDS and similar to that of Tween-20. This indicates that the neutral PEG chains are able to shield the negative charges of lipids, suggesting that the disks maintain their discoid form upon immobilization on microchannel surfaces. A similar shield against nonspecific adsorption was obtained.

A negatively charged phospholipid polymer biointerface containing 2-methacryloyloxyethylphosphorylcholine (MPC), *n*-butyl methacrylate (BMA), potassium 3-methacryloyloxypropylsulfonate (PMPS), and 3-methacryloxypropyltrimethoxysilane (MPTMSi) moieties was synthesized to introduce such phosphorylcholine segments and also surface charges on to a silica-based microchannel. Along with another uncharged copolymer, poly(MPC-*co*-MPTMSi), the regulation of the surface charge density can be realized by adjusting the initial concentrations, providing a feasible approach to controlling the EOF, especially in systems with biological applications requiring neutral buffer conditions [117].

In Table 6.1, ME studies that employed surfactants are reported. The type of surfactant, substrate material, analyte, type of detection, and some comments are included.

As can be deduced from the references, the rows follow a chronological order. It should be noted that of the ten oldest references, eight employed SDS, and of the most recent ten, only two used this surfactant. Referring to the material, initially most of the microchips in these studies were made of PDMS, later glass was the material of choice, and nowadays there is a coexistence of glass and polymers. Fluorescence is the most common principle of detection, followed by the use of electroanalytical methodologies (amperometry and conductometry).

6.3.2 Ionic Liquids

Room-temperature ILs are compounds formed of organic (commonly nitrogen-containing) cations and inorganic or organic anions that are either liquid at room temperature or whose melting points are slightly higher than ambient temperature. Any salt with a melting point below 100 °C is considered to be an IL [154], whereas those with higher melting points are referred to as molten salts. Although ILs were first reported in 1914 [155], it was not until 2000 that they were widely exploited. Applications are increasing: from large potential window electrolytes in electrochemical devices, media for enzyme catalysis, solvents in organic synthesis or extraction processes to sensor layers or modifiers in analytical separations. Their application in analytical chemistry [156], especially in separating analytes [157], is merited because of their unique properties (hydrophobicity, water solubility, viscosity, conductivity, acidity, etc.). They can be tuned depending on the anions (usually chloride, hexafluorophosphate, tetrafluoroborate, etc.) and cations (heterocyclic aromatic ions, quaternary ammonium, phosphonium, etc.). Since ILs are not used as pure solvents, but rather diluted in aqueous solutions, they are just salts, and therefore the influence of the cations and anions has to be considered, although too often the properties of the cations are taken as the properties of the IL itself [157].

In this context, ILs, initially based on tetraalkylammonium ions, have been employed as modifiers in CE. An excellent review on the combination of IL and CE [158] focuses on two different approaches: application of CE for

TABLE 6.1 Studies Related to ME with Employment of Surfactants

Surfactants	ME Material	Analyte	Detection	Comments	Reference
Phospholipid disks	SU-8, PDMS, glass	BSA	Fluorescence	Reduce adsorption	117
Zwitterionic surfactants	PDMS	Perchlorate	Contact conductivity detection	Separate perchlorate from common water anions	118
SDS	PMMA	Proteins	LIF	2D separation: SDS μ-CGE with μ-MEEKC	119
Tween-20	PDMS	Peptides	Fluorescence	MGE	104
DTAC and DDM	PMMA	N-linked glycans	LIF	Reduce adsorption	120
Acid-labile surfactant and SDS	Glass	Proteins	Fluorescence	Surfactant effects on proteins separation	121
SDS	Glass	Small-chain aldehydes	Amperometry	Sensitive and rapid determination of harmful atmosphere aldehydes	122
SDS	PDMS–glass	Biogenic amines	LIF	MEKC	123
SDS	Glass	Nitrate ester explosives	Amperometry	Multichannel microchip	124
Zwitterionic surfactant	Glass	PB-amines and amino acids	Fluorescence	MEKC	125
SDS + MeOH	Glass	Serotonin, tryptophan, propranolol, and acetaminophen	UV absorbance and UV fluorescence	MEKC	126
Zwitterionic surfactant	PDMS	Atmospheric aerosol	Conductivity	MEKC	127
SDS	Glass	Lipoproteins	LIF	Rapid diagnostics	128
SDS	PMMA	Proteins	LIF	CGE + MEKC	129
SDS	Glass	Proteins	LIF	Size-based protein separation	130
SDS	Glass	Endotoxins	LIF	MGE	131
SDS	PMMA	Proteins	LED	MGE	132
SE_8S, $SD_{10}S$, $SD_{12}S$, and $ST_{14}S$	PDMS	Phenols	PAD	Adsorption behavior	133
SDS + MeOH	Glass	Nitrated benzodiazepines	LIF	MEKC	134
SDS	Glass	Virus	LED	SEC	135
SDS	PDMS	NDA-amino acids	Fluorescence	Increased reproducibility	136
SDS + PAMAM	Glass	Cationic neurotransmitters	Amperometry	CZE	137
SDS	Glass	Rhodamine derivatives	LIF	MEKC	138
CTAB	Glass	Anionic species	Conductivity	Reverse the EOF	139
SDS	PDMS–PDMS, PDMS–glass	Neuropetides	Fluorescence	On-chip electrokinetic sample stacking	140
Tween 20	PDMS	Amino acids	Amperometry	Reduce adsorption	141
SDS	PDMS	Amino acids	LIF	MEKC	142
SDS, PA, DOCh	PDMS	Glucose, penicillin, phenol, homovanillic acid	PAD	Decrease analysis time and increase electrochemical response	143
CTAB	Glass	Inorganic arsenic	Fluorescence	Reverse the EOF	144
SDS	PDMS	Carbohydrates	iPAD	Improve separation and detector response	145
SDS	Glass	NBD-DL-amino acids	LIF	Enantiomeric separation	146
Triton X-100	PDMS–glass	Rhodamine B	LIF	Reduce adsorption	147
SDS	PDMS	Glucose and glucosamine	PAD	Signal enhancement and flow stabilization	148
SDS + α-CD	PMMA	Amino acids	LIF	Serpentine ME	149
SS	Glass	Lipoproteins	LIF	Rapid diagnostics	150
SDS	PMMA	Proteins	Conductivity	MEKC	151
SDS	PDMS		Fluorescence	CZE	152
SDS + γ-CD	Glass	FITC-amino acids	Fluorescence	MEKC	153

CD, cyclodextrin; CGE, capillary gel electrophoresis; DTAC, *n*-dodecyltrimethylammonium chloride; iPAD, indirect pulsed amperometric detection; LED, light-emitting diodes; LIF, laser-induced fluorescence; μCGE, microcapillary gel electrophoresis; μMEEKC, microemulsion electrokinetic chromatography; PAD, pulsed amperometric detection; $SD_{10}S$, sodium decyl sulfate; SDS (equivalent to $SD_{12}S$), sodium dodecyl sulfate; SE_8S, sodium 2-ethylhexyl sulfate; $ST_{14}S$, sodium tetradecyl sulfate; SEC, size-exclusion chromatography.

the analysis of ILs and application of ILs in CE. In the later context, ILs are considered for sample treatment before CE analysis or for CE resolution enhancement. The high capacity to interact/solvate analytes, the high tendency to be adsorbed on silanol groups of the capillary wall and change the EOF providing dynamic coatings, the change in the ionic strength, conductivity, and viscosity of the BGE, or the formation of a stationary (CEC) or PSP because of the capacity of some ILs to form micelles (MEKC) being some of the reasons for the improvement. Although ILs have been employed in combination with CE, their use in ME is very scarce (see Figure 6.2). Referring to CE, most of the studies employed imidazolium cations, 1-ethyl-3-methyl [159,160] as the most common followed by 1-butyl-3-methyl [161,162] substituted. (*S*)-(−)-2-Hydroxymethyl-1,1-dimethylpyrrolidinium was also employed as a cation [163]. In this case, there is suppression of the magnitude of the EOF and a gradual change in its direction. Therefore, this IL was employed as an alternative to cationic surfactants for the determination of tricyclic antidepressants.

Concerning the anions, most of them are BF_4^- apart from bis(trifluoromethanesulfonyl)imide (NTf2) [164] or acetate, fluoroacetate, heptafluorobutanoate [109], or dodecyl sulfate [165]. ILs have been employed in NACE since alkylimidazolium- and alkylammonium-based salts are well soluble in organic solvents and can be successfully used as BGE components [109]. Different molecules such as nicotinic acid and structural isomers [160], anthraquinones [161], benzodiazepines [164], and the amino acids proline and hydroxyproline [165] have been separated. However, the differences between the materials employed in the two methodologies (CE and ME) make a thorough study necessary.

Only a few reports have appeared on ILs combined with ME devices and most of them were for the analysis of proteins and peptides. In Table 6.2, studies reported for IL combined with ME devices (glass, PDMS, and hybrid PDMS–glass) are presented. In all cases, the detection method was laser-induced fluorescence (LIF).

The effect of ILs used as supporting electrolytes was studied for Rhodamine B and fluorescently labeled streptavidin and IgGs. A hybrid coating with IL (1-ethyl-3-methylimidazolium tetrafluoroborate, EMIm BF_4) [which gave better results than BMIm BF_4 (1-butyl-3-methylimidazolium tetrafluoroborate] and BMIm PF_6 (hexafluorophosphate), which possess butyl radicals) and the nonionic surfactant Triton X-100, produced a more efficient modification [166]. This IL, which possesses high conductivity, low viscosity, miscibility with water, and good solvating properties, was also employed in combination with an anionic surfactant (SDS) [167]. EMIm BF_4 was directly employed as the supporting electrolyte in a PDMS–glass microchip without adding other buffers to adjust its pH. The role of SDS is that cationic dyes (Pyronin Y in this case) can form weak fluorescent dimers with SDS, whereas when proteins are added to the dye–surfactant system, some dye dimers turn into monomers, resulting in fluorescence enhancement of the dye–surfactant system. On the other hand, IL was applied to overcome protein adsorption in the microchannels. Phosphate-buffered saline (PBS) and EMIm BF_4 were compared at similar concentrations and a lower background current and shorter migration time (due to an increase in the EOF) were achieved for the IL. It not only had the ability to separate hydrophobic analytes while maintaining an adequate background current, but also produced smaller Joule heating in the microchannels, which induced more rapid sample analysis and better reproducibility. The combination of IL and surfactant was designed in the same molecule, with the aim of combining the advantages of the two. In this way, 1-butyl-3-methylimidazolium dodecanesulfonate (BAS) was applied to microchip MEKC in PDMS material, obtaining an eightfold increase in the EOF [168].

The most common IL (EMIm BF_4) exhibited great potential in the glass microchip electrophoretic separation of complex optical isomers such as peptides, when employed as the working electrolyte with chiral selectors included [169]. A chiral IL, (*S*)-[3-(chloro-2-hydroxypropyl)trimethylammonium] [bis((trifluoromethyl)sulfonyl)amide] was synthesized for use as both co-electrolyte and chiral selector [170], although it has not yet been employed in ME.

The influence of ILs on the separation of analytes with different ME materials has not been thoroughly studied as well as the influence of different ion compositions, including synthetic ILs. In Figure 6.4, the electropherograms obtained for two model analytes, DA and EP, in a native glass microchip and in a dynamically IL-coated glass ME are presented. The improvement in the separation is demonstrated through the baseline resolution of the peaks.

TABLE 6.2 Studies Related to ME with Employment of Ionic Liquids

Ionic Liquid	ME Material	Analyte	Detection	Comments	Reference
BAS	PDMS	Cy3-extravidin and Cy3-IgG	LIF	MEKC	166
[EMIm] BF_4 + Triton X-100	PDMS	TRITC-IgG, Cy3-IgG, FITC-IgG	LIF	Studies with Rhodamine B	167
[EMIm] BF_4	PDMS–glass	BSA, bovine hemoglobin, cytochrome *c*, and trypsin	LIF	Employment of SDS	168
[EMIm] BF_4	Glass	NBD-peptides	LIF	Chiral separation	169

TRITC, tetramethyl rhodamine isothiocyanate; NBD, nucleotide-binding domain.

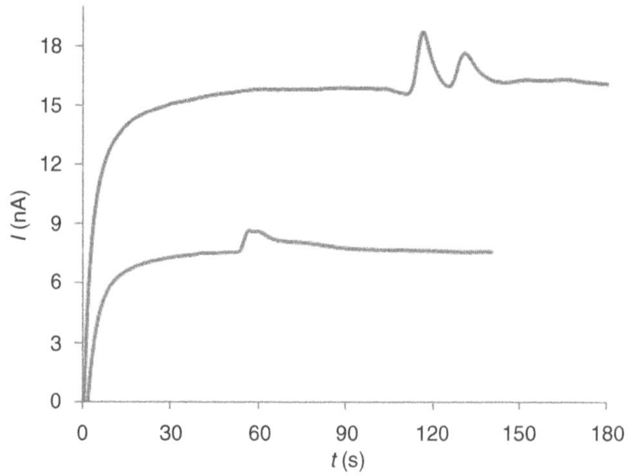

FIGURE 6.4 Electropherograms recorded in a solution of dopamine (100 μM) and epinephrine (100 μM): top curve, with IL; bottom curve, without IL. V_{sep}, 1500 V; V_{inj}, 500 V; t_{inj}, 1 s; working electrode (WE), Au wire, 100 μm diameter; E_d, 0.8 V; ME material, soda-lime glass; L_{sep}, 55 mm; BGE, 25 mM MES-His, pH 5.5; IL (in case of use; 20 mM [BMIm] HSO_4^-); detection electrolyte, 0.1 M H_2SO_4 (Álvarez-Martos et al., unpublished results).

6.3.3 Nanoparticles

NPs are having a significant impact in many scientific fields with the advance and promotion of nanotechnology. They are particles with a size in the nanometer range, that is, up to 1000 nm, and due to their large surface-to-volume ratio they have been extensively studied in recent years in separation science. Therefore, NPs are also applied to capillary and microchip electrochromatography since they can be used as stationary phases or PSPs. NPs need to fulfill some important requirements to be suitable in PSP-CE: be able to form stable suspensions in a wide range of electrolytes, provide the desired selectivity in the interaction with electrolytes, be charged, and show equal velocity to prevent peak broadening, show small mass transfer resistance, not disturb detection, and be small so as to provide a high surface area to improve sample capacity [171]. The use of polymer, gold, and silica NPs in addition to fullerenes and carbon nanotubes (CNTs) in CE has been reviewed [172,173]. The first attempt to use NPs as PSPs in CEC was described in 1989 [174]. Even though they have many advantages, the main problem is the negative effect of NPs on UV–Vis detection due to the light-scattering properties of the NPs. Another difficulty in NP-based ME is the necessity to form stable suspensions in aqueous solutions, which often can be solved by dynamic or covalent modification of the NPs. Referring to protocols, they can be used as additives in the running buffer (PSPs) or they can act as modifiers adsorbed on the channel surface. In the former case, they are suspended in the electrolyte solution and continuously pumped by the EOF.

In Table 6.3, studies related to the use of NPs in combination with ME are presented, indicating the analytes, detection system, and ME material employed.

A potential alternative for modifying the surface properties of PDMS is to use a sol–gel process to form nanometer-sized silica particles throughout a polymerized PDMS

TABLE 6.3 Studies Related to ME with Employment of Nanoparticles

Nanoparticles	ME Material	Analyte	Detection	Comments	Reference
Porous liquid crystalline lipid-based NPs	COP	GFP-proteins	LIF	Reduce adsorption	197
Au	PDMS	FITC-labeled myoglobin	LIF	Reduce adsorption	175
Au	Glass	Endocrine disruptors	Amperometry	Improve separation and increase sensitivity	176
Magnetic beads	PDMS	Peptides	LIF	Binding constants determination	177
Titanium dioxide (TiO_2)	PDMS	Amino acids	Indirect amperometry	LBL assembly technique	187
SWCNTs	Glass	Proteins	Fluorescence	SWCNTs–PEG–acrylate hydrogel composite	178
Au	Glass	DNA	Amperometry	MGE	179
Au	PDMS	Dopamine, epinephrine, arginine, and histidine	Amperometry	EOF-switchable ME	180
Silica	PDMS	Neurotransmitters	Amperometry	LBL assembly technique	181
Au	PDMS	Neurotransmitters	Amperometry	LBL assembly technique	190
Au	PDMS	Neurotransmitters and pollutants	Amperometry	LBL assembly technique	182
PEGylated latex	PMMA	dsDNA (10 bp to 2 kbp)	LED	Improved DNA separations	183
Au	Glass	Aminophenols	Amperometry	LBL	188
Silica (SiO_2)	PMMA	dsDNA (100 bp to 1.5 kbp)	LED	Improved DNA separations	184

COP, cyclic olefin polymer.

matrix. To form the particles, an alkoxysilane precursor such as tetraethyl orthosilicate (TEOS), which is soluble in the polymerized PDMS, is first hydrolyzed. Particles (~10 nm) are then formed through condensation of the hydrolyzed silanes. The condensation reaction is generally catalyzed using either an acid or a base. The process should result in the generation of permanent free silanol groups at the surface of the PDMS, thereby increasing the EOF and wettability. Particle formation in the bulk PDMS should also allow for irreversible sealing of PDMS. The silanol groups on these particles should allow a wide range of surface modifications developed for glass to be used on these modified devices. This process has been extensively studied [185] and NPs have been characterized [186].

Although metallic NPs made of titanium dioxide have been reported for modifying a PDMS microfluidic channel surface by sequentially immobilizing PDDA and NPs by a layer-by-layer technique [187] (see the electropherograms in Figure 6.5), gold NPs are without any doubt the most commonly employed. The first use of gold NPs (10 nm) was for the separation and amperometric detection of aminophenols [188]. Gold NPs form stable (red to purple) solutions with most buffer solutions used in electrophoresis. Owing to their sizes, ease of synthesis, and properties, they have been tested as additives in CE. The selectivity and efficiency increased compared with uncoated microchannels.

Gold NPs can be synthesized *in situ* by infusing $HAuCl_4$ solution into channels and incubating at 37 °C for 48 h [189]. Once they have been formed, a stable bovine serum albumin (BSA) blocking layer can be prepared simply by incubation.

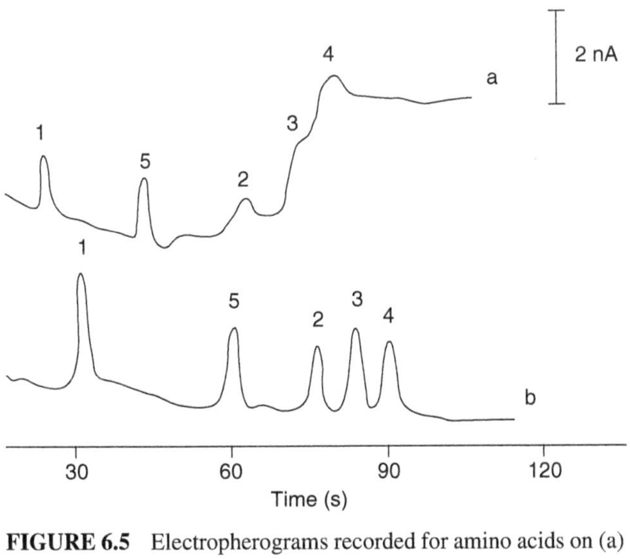

FIGURE 6.5 Electropherograms recorded for amino acids on (a) native, and (b) PDDA–TiO_2 NP-coated PDMS microchips. Experimental parameters: BGE, 5.0 mM, pH 9.2 STB; V_{inj}, 800 V; t_{inj}, 5 s; V_{sep}, 1200 V; E_d, 0 V. (1) Arginine; (2) phenylalanine; (3) serine; (4) threonine; (5) EOF. Reprinted from Ref. 187 with permission, © John Wiley & Sons, Inc.

The PDMS–AuNP–BSA-treated microchannel could successfully suppress protein adsorption, as demonstrated by the lack of fluorescence in a coated channel with incubated fluorescein isothiocyanate (FITC)-labeled myoglobin. Contact angle measurements revealed that the coated surface was hydrophilic.

Modification of the microchannel can be achieved by a layer-by-layer (LBL) assembly technique, employed for taking advantage of electrostatic interactions. Polyelectrolytes (PEs) are commonly employed since charges vary from one layer to the other. A simple procedure is usually employed: in this case, a PE was used as the first coating and then citrate-stabilized gold NPs were adsorbed [190]. The procedure is as simple as pumping the running buffer containing the PE through the microchip for polymer adsorption. Then, the microchip is flushed with the buffer and then with gold NP solution for a fixed time. Improvement in the separation of neurotransmitters that were detected amperometrically was achieved. Gold NPs were also employed as bifunctional linkers to immobilize cysteine on the surface. After coating a PDMS microchannel with PDDA, gold NPs were immobilized to chemisorbed cysteine *via* their –SH functionality. The unique feature of the chemisorbed amino acid is that, depending on the solution pH, the surface can have excess positive charge (low pH), no net charge (isoelecric point), or excess negative charge (high pH). As a result, the EOF can be switched by using running buffer with different pH [191].

Two different strategies were employed using LBL procedures for the separation of neurotransmitters and environmental pollutants: cationic PE (chitosan), gold NPs, and albumin in the first case, and lysozyme and albumin in the second one [182]. The adsorption properties of the microchip material (PDMS) were due to proteins. Therefore, adsorption of small molecules is avoided and separation improved. In this way, even when NPs are employed, the surface-active compounds are proteins. Gold NPs were also employed in MEKC that uses SDS as a micelle-forming agent for the separation of neutral analytes [192]. In this case, the purpose of the NPs is to provide additional interaction sites with which the solutes can interact. The presence of AuNPs in the running buffer can significantly affect the apparent mobility of the solute and also change the electroosmotic mobility of the running buffer.

Apart from small analytes, the use of NPs has been combined with DNA analysis. Gel electrophoresis sometimes caused trapping of the molecules. Gold NPs were immobilized following a more complicated procedure, consisting of a three-layer procedure: coating with PVP, PEO, and PEO–gold NPs (citrate stabilized) that produced a sharper peak profile and improvement in resolution [193]. Multilayer deposition minimized unspecific adsorption of DNA [194].

Polymeric NPs can also be prepared. PEG-coated polystyrene latex NPs were prepared by polymerization of styrene in the presence of PEG. Improved separation is achieved when PEG is present on the NP surface. The smaller the size, the better the DNA separation when mixed with hydroxylpropylmethylcellulose buffer [195]. This conventional polymer was mixed with silica NPs, which are supposed to cause repulsion in DNA (polyanionic) and accelerate it, resulting in an increase in resolution and speed [196].

A lipid-based liquid crystalline NP suspension (average diameter 70 nm) was prepared by one-step procedure based on lipid self-assembly and applied to the separation of green fluorescent protein (GFP) mutants differing in only one amino acid [197]. An aqueous NP suspension was formed when soy phosphatidylcholine, glycerol dioleate, polysorbate 80, oleic acid, and ethanol were added to water and stirred for 48 h.

Magnetic beads (MBs) are a material with many possibilities in analytical chemistry. Magnetic particles offer an additional advantage over other solid surfaces and conventional particles: having embedded magnetic entities, they can be easily manipulated using permanent magnets or electromagnets, independently of normal microfluidic or biological processes [198]. The main advantage is the reusability, since magnetic particles are released by blocking/removing the magnetic field. Moreover, the configuration can easily be changed by functionalization. In most cases, they are used for performing bioassays [199], rather than for improving the separation. In this way, in-line extraction has been performed with C_{18} functionalized magnetic particles (silica-coated iron oxide particles) on a disposable PMMA microfluidic device [200]. The use of magnetic NPs as obstacles in the channel by applying a magnetic field allows the separation of large dsDNAs [201]. The NP suspension becomes self-organized into a fixed array when a constant homogeneous magnetic field is applied.

Increasing developments in technology have produced carbon nanoparticles (CNPs). However, their use among the rest of NPs is not widespread, mainly where microchips are concerned. An overview of the use of CNPs as PSPs in electrokinetic chromatography (EKC) has been reported [202]. A variety of carbon nanostructures including fullerenes, nano-onions, nanodiamonds, nanotubes, nanohorns, peapods, nanofibers, and nanotube rings have been reported. CNTs are the carbon nanostructures most widely used recently in analytical chemistry [203,204], including their use as a separation carrier in CE [205]. Although Wiles and Abrahamson [206] observed a thick mat of fine fibers of graphite anodes with diameters ranging from 4 to 100 nm composed of graphitic layers with a hollow core, the discovery of CNTs was attributed to Iijima [207], who prepared carbon structures consisting of needle-like tubes, each needle comprising coaxial tubes of graphitic sheets. On each tube, the carbon-atom hexagons are arranged in a helical manner about the needle axis, resulting in high-aspect (length to diameter) ratio materials with outstanding electrical, mechanical, chemical, and thermal properties. They can be present in different forms: multiple-walled carbon nanotubes (MWCNTs) (in "hollow-tube," "herringbone," or "bamboo" morphological variations) and single-walled carbon nanotubes (SWCNTs); metallic (armchair) and semiconducting (zigzag or chiral); closed- or open-ended CNTs; and different functionalizations are possible. Therefore, even when they have a very simple chemical composition and atomic-bond configuration, they exhibit the most extreme diversity in structure and in turn in properties and behavior.

Although some reports can be found on CE combining the use of CNTs and surfactants [208], or composites with polymers [209], its application in ME is still scarce. A special case is the use of CNTs (shortened carboxylic SWCNTs) that were bound after chemical oxidation under ultrasonication to BSA and used as a chiral detector for tryptophan [210]. The separation is based on the fact that BSA has different binding affinities to D- or L-tryptophan. This stationary phase was adsorbed taking advantage of the adsorption properties of proteins. Similarly to most of the cases related to modification by NPs, the detection system is amperometry.

ILs have been shown to have thermophysical properties that justify the replacement of several of the chemical processes now under exploitation because they can be considered as green solvents. Dissolving NPs (MWCNTs) in ILs forms "bucky gels" or ionanofluids, which have recently been shown to have thermal conductivity enhancements ranging from 5% to 35% for [(HMIm) BF_4], [(BMIm) PF_6], [(HMIm) PF_6], [(BMIm) CF_3SO_3], [(BMPyrr) $(CF_3SO_2)_2N$] (1-hexyl-3-methylimidazolium, HMIm; 1-butyl-3-methylpyrrolidinium, BMPyrr) with MWCNTs [211]. Considering that solid materials, namely metals or CNTs, have a thermal conductivity at room temperature several orders of magnitude higher than that of fluids, it has been shown that the thermal conductivity of fluids containing suspended particles could be significantly higher than that of the base fluids. The term nanofluid [212], in this case, designates a new class of heat transfer fluid formed by the dispersion of nanometer-sized solid particles, rods, or tubes in traditional heat transfer fluids. This is supposedly based on nanocluster formation. Combination with ME methodologies is a promising and unexplored methodology.

Multifunctional NPs are exciting nanomaterials with promising applications: (bio)sensing, (bio)assays, catalysis, and separations based on their magnetic, optical, and electrochemical properties [213]. In Figure 6.6, a scheme of the possibilities of these NPs is shown.

6.3.4 Polymers

Polymers cover a type of molecules that are different in size, dispersity, charge, and so on, which is very appropriate for

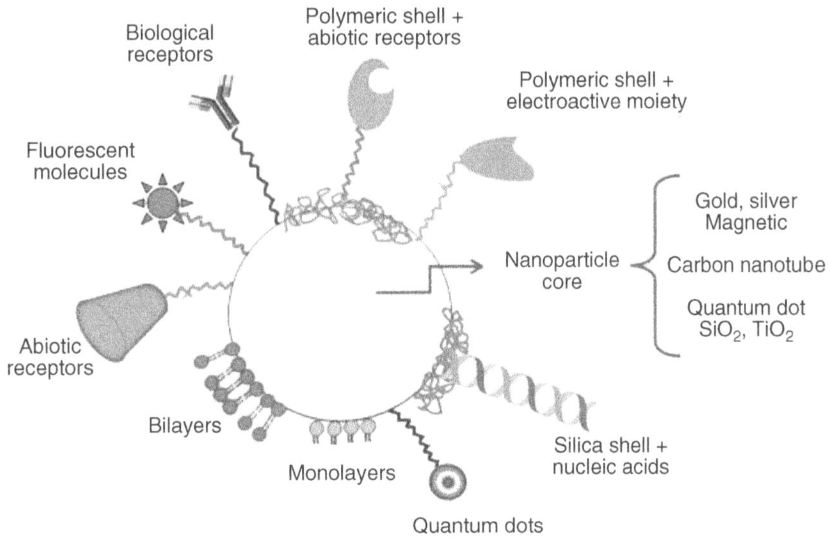

FIGURE 6.6 Multifunctional nanoparticles that schematized all the possibilities of use. Reprinted from Ref. 213 with permission from Elsevier.

surface modification due to, among other features, its versatility. In this section, the more representative cases of different strategies will be presented. Some examples have already been commented on because they can be employed in combination with other types of chemical modifiers in order to improve jointly the characteristics of the analytical methodology. In many cases, polymers are applied as preconcentration matrices. The applications of monoliths and related porous polymer gels in microfluidic devices have been reviewed [214]. Monolithic columns can be formed within PMMA microchannels by *in situ* photopolymerization, for performing preconcentration through retention/elution mechanisms [215], and chambers for including monolithic disks were prepared at the junction between the injection and separation channels for adsorption and elution of catecholamines [216]. In many cases, they have been employed for improving resolution, as dynamic or static coatings. They have been mainly applied in combination with CZE methodologies, but they also improve microchip gel electrophoresis (MGE) (constituting the gel or the viscous solution), MEKC (accompanying the micellar medium or as a pretreatment), or even IEF methodologies.

Many polymers, commonly derived from cellulose, have been employed for improving separations, usually for DNA or protein analysis. DNA analysis has moved from the slow and labor-intensive slab gels to the fast and automated capillary arrays and, more recently, to the even faster and potentially integrated microfluidic devices that can be fabricated for running several assays in parallel. Depending on the concentration and the molecular weight, the viscosity varies in addition to the pore size. Polymers can act as a sieving matrix and different macromolecule sizes can be separated. Surface modification of polymer microchips (PMMA and COC) with HPC dissolved in a matrix of the copolymer [poly(methyl methacrylate-8.5-methacrylic acid)] in a spin-coated thin film on the surface has been proposed [217]. On the other hand, linear polyacrylamide (LPA) [218] has been commonly employed as a sieving matrix, not only the concentration but also the molecular weight being important parameters in the resolution of DNA fragments (e.g., differences are observed between a 4% w/v >5.5 MDa LPA or a mixed system consisting of 3% w/w 10 MDa LPA and 1% w/w 50 kDa LPA). Easy handling for matrix preparation, loading into channels and replacement after runs, and consistency in batch quality are parameters that have to be taken into consideration. To suppress the EOF of plastic (polyolefinic) channels, a coating solution of polydimethylacrylamide–diethylacrylamide was filled to allow noncovalent coating before loading with a denaturing sieving matrix.

The first step to improve separation is the optimization of the instrumental conditions, mainly buffer pH, concentration, and composition, and also the separation voltage. When separation is still not adequate, surface modification and polymers can be an alternative. In Table 6.4, some terms employed in surface modification with polymers are reported.

The use of additives as a dynamic coating is the first simple solution. Three major classes of dynamic coating modifiers, amines, surfactants, and neutral polymers, have been widely adopted in conventional CE [219]. This is mainly due to the possibility of forming hydrogen bonds through hydroxyl and amine groups and also hydrophobic interactions through (hydrocarbon chains) with the surface. Among candidate dynamic coating additives, amines and surfactants have attracted perhaps the greatest attention because very small amounts of such additives can significantly affect the EOF and suppress analyte interaction. Such minute amounts of surfactant do not alter the properties (pH

TABLE 6.4 Terms Employed in ME Surface Modification with Polymers

Layer-by-layer (LBL)	Formation of a film by depositing alternating layers of opposite charge	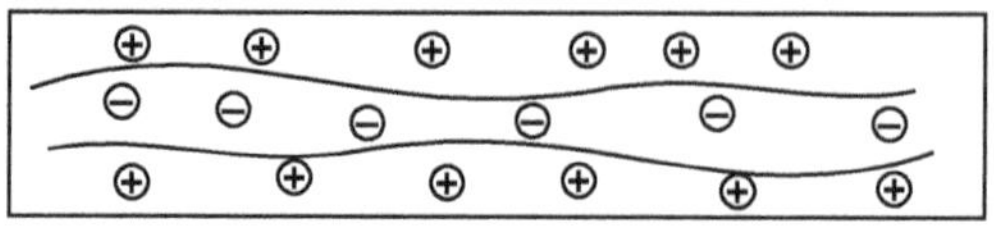
Click chemistry	Usually, it is a reaction between a terminal alkyne and an organic azide to yield a triazole. It was given that name because of its reliability	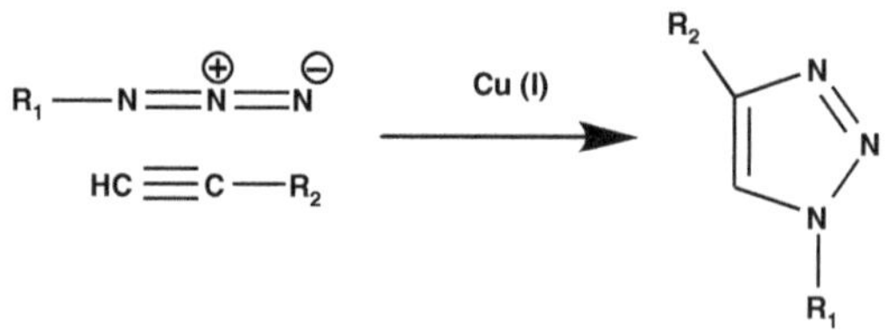
ATRP	A living radical polymerization that uses a metal catalyst, usually copper	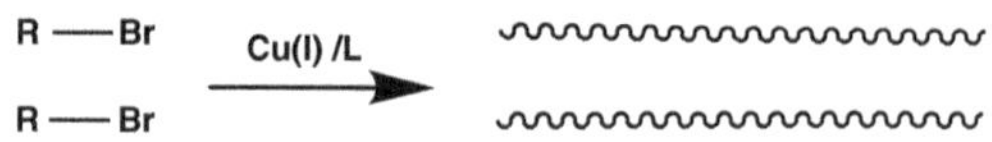
Grafting to	Consists in attaching a functionalized polymer to an active site on the surface	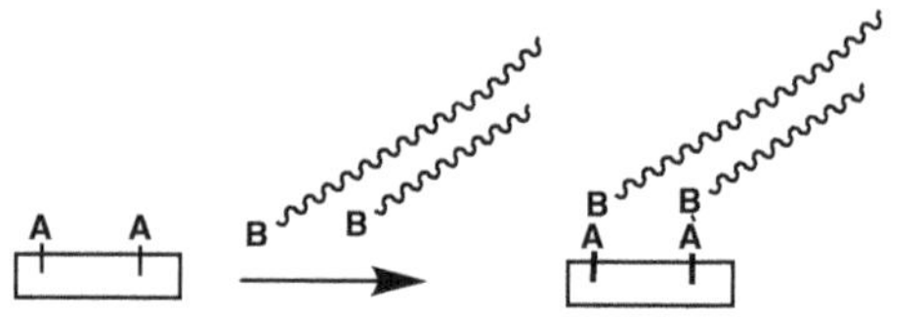
Grafting from	A procedure according to which a polymerization initiator or a monomer is covalently bonded to the surface and then the polymer is conveniently grown from the initiator or the monomer	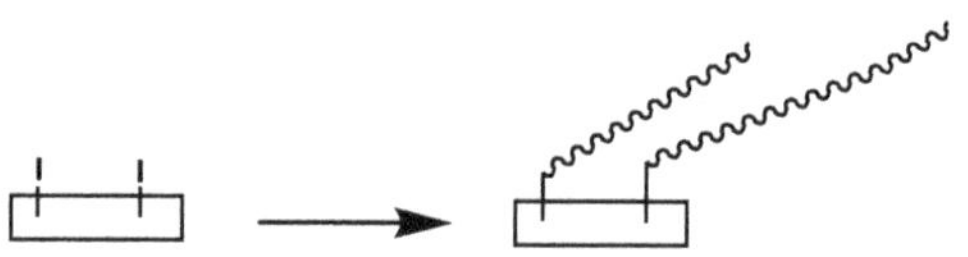

and conductivity) of the BGE and therefore do not adversely affect the speed of separation. Amines such as ethylamine, diethylamine, triethylamine, and triethanolamine, and surfactants such as SDS, dodecyltrimethylammonium chloride (DTAC), and CTAB, did not show any observable effects on suppressing oligosaccharide adsorption under acidic or alkaline conditions [220]. In this case, also PA, well known for its excellent capillary wall coating and sieving properties in conventional CE, only very slightly reduced adsorption. Neutral polymers with hydroxyl groups such as PEG, hydroxyethylcellulose (HEC), hydroxypropylmethylcellulose (HPMC), and MC clearly suppressed adsorption and improved separation. Among them, MC is preferred because it gave a better performance and less tailing than HPC. The low molecular weight of HEC and the very small number of hydroxyl groups in PEG produced poorer separations. Therefore, as commented on before, molecular weight, hydrophilicity/hydrophobicity, and concentration of polymer are important parameters to be considered. In many cases, neutral polymers are chosen because they do not impart a charge to the surface to give extra EOF. Neutral polymers with polyhydroxyl groups such as PEG, HPMC, MC, and HEC have been employed together with amines and surfactants for the improvement of derivatized oligosaccharide separation on PMMA microchips. There are special cases where the alternatives have not been widely exploited for ME, for example, dendrimers. The first use of dendrimers as PSPs in CE separations was described in 1992. One of the important conclusions is that separation was influenced by the size and charge of the dendrimers and by the composition of the electrolyte [221]. In Figure 6.7, electropherograms showing the improvement in the resolution of several neurotransmitters and cationic metabolites are shown.

As commented on previously, polymeric coatings on solid surfaces can be performed following two main strategies: physisorption and covalent attachment. For polymer physisorption, a polymer or a block copolymer can be adsorbed *via* interaction of the polymer or one of the blocks with suitable surfaces. The technique of LBL has been developed as a versatile method to functionalize surfaces. The process is based on the sequential deposition of interactive polymers from their solutions in such a way that a film is formed by deposition of alternating layers of opposite charge.

Chitosan is a biopolymer derived from chitin, made primarily of repeating units of glucosamine, among which amino groups have been used in the immobilization of proteins and other molecules. Furthermore, chitosan is positively charged in mildly acidic aqueous solutions, and its charge density is high. Therefore, it can be used as a polycation in an LBL assembly system, on which various polyanions can be adsorbed [92].

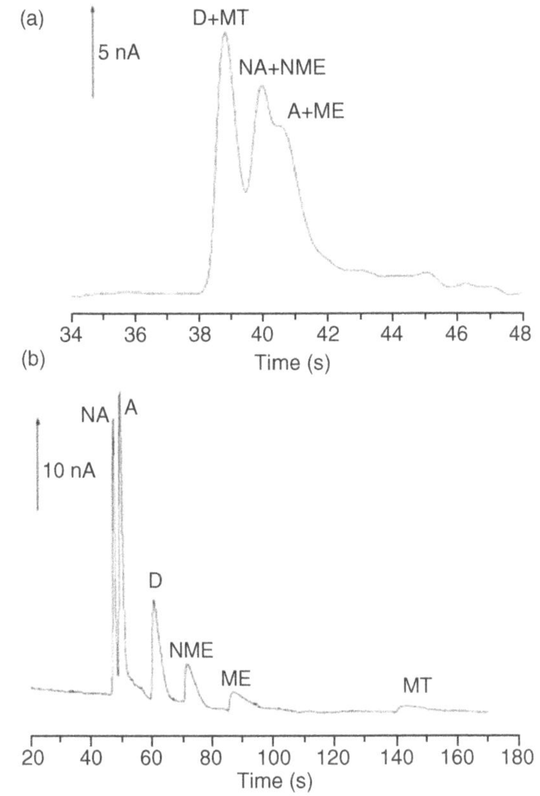

FIGURE 6.7 Electrophoretic separation of catecholamines and cationic metabolites on the microchip. BGE, 5 mM borate–phosphate buffer (pH 7) containing SDS and PAMAM dendrimer; V_{sep}, 3 kV; V_{inj}, 1 kV (3 s); Au electrode; E_d, 1400 mV. Reprinted from Ref. 137 with permission from Elsevier.

PE coatings have an important effect on the velocity and direction of the EOF as well as on the separation efficiency for PDMS-ME. Successive multiple ionic layer coatings indicated that the EOF direction reversed with each additional deposited polymer layer (first anionic and then cationic or *vice versa*), but did not vary significantly with layer number or polymer structure [222]. The maximum separation efficiency was achieved for coatings of polybrene (cationic) and dextran sulfate (anionic) polymers after deposition of six layers. Complexation between oppositely charged polyions occurs through electrostatic interactions and is a means to control the structure and order of assemblies.

Another possibility is the generation of PE brushes. When long linear PE chains are grafted densely to a solid surface, a PE brush results [223]. One of the main features of PE brushes is the strong confinement of the counterions within the brush layer. There are large differences compared with monolayers of end-grafted uncharged macromolecules. Two classes of PE brushes are possible [223]: if strongly dissociating PE chains such as poly(styrenesulfonic acid) (PSS) are grafted to surfaces, a quenched PE brush results. The degree of dissociation of ionic monomer units within these systems is independent of the pH. On the other hand, an annealed PE brush is obtained if weak PEs such as poly (acrylic acid) (PAA) are affixed to surfaces. Here, the state of charging of the chains depends directly on the local pH, which depends both on pH and on the ionic strength of the solutions. In both cases, salt concentration is important.

For a covalent attachment, two approaches have been envisaged: "grafting to" or "grafting from." According to the first, a functionalized polymer is attached to an active site on the surface; mostly the functional group is allocated at one of the polymer ends. In the second procedure, a monomer-/polymerization initiator is covalently bonded to the surface and then the polymer is conveniently grown from the monomer/initiator.

The advantages of the "grafting to" procedure is that free polymers that are going to be attached to the surface by formation of a chemical bond between the end groups of the polymer and the solid [224] can be synthesized and characterized prior to attachment. Neutral brushes can be converted to PE brushes by chemical treatment, often requiring rough conditions. "Grafting from" techniques present a viable alternative for the synthesis of dense PE brushes [225]. A monolayer of initiators is attached to the surface of the solid substrate and a surface-initiated polymerization is started that is not hampered by kinetic or thermodynamic barriers. In principle, any chain reaction that leads to long polymer chains can be used to create brushes. Block copolymers consisting of polymers with hydrophobic and PE blocks can also be employed [226], in addition to grafting of several different functional polymers on to a solid substrate *via* functional end groups, which results in mixed polymer brushes [227] e.g. PE and non-PE, and also two oppositely charged PEs are possible. Depending on the pH of the system, the charge of these brushes could be switched. PE block copolymers can be used to tune the nanostructure assembly by changing the charged block or the degree of ionization, adding salts, and changing the solution conditions [228].

Some examples for both approaches are considered in the following paragraphs. Prebonding or postbonding procedures can be performed (see the section When Is the Surface Modification Made?). If the latter is chosen, the most common procedure is introduction of the reagent solutions, with incubation there for a given time, or pumping them through the microchannel at a fixed flow rate. In both cases, washing the channel with deionized water (or another solvent such as ethanol depending on the functionalization and material of the channel) is carried out for a while. Long times are usually required, but they are decreased when flow is employed. Conditioning with the BGE is always mandatory.

Covalent modification can be performed in simple one-step procedures or more complex approaches. An example of covalent immobilization is presented in Figure 6.8. In this

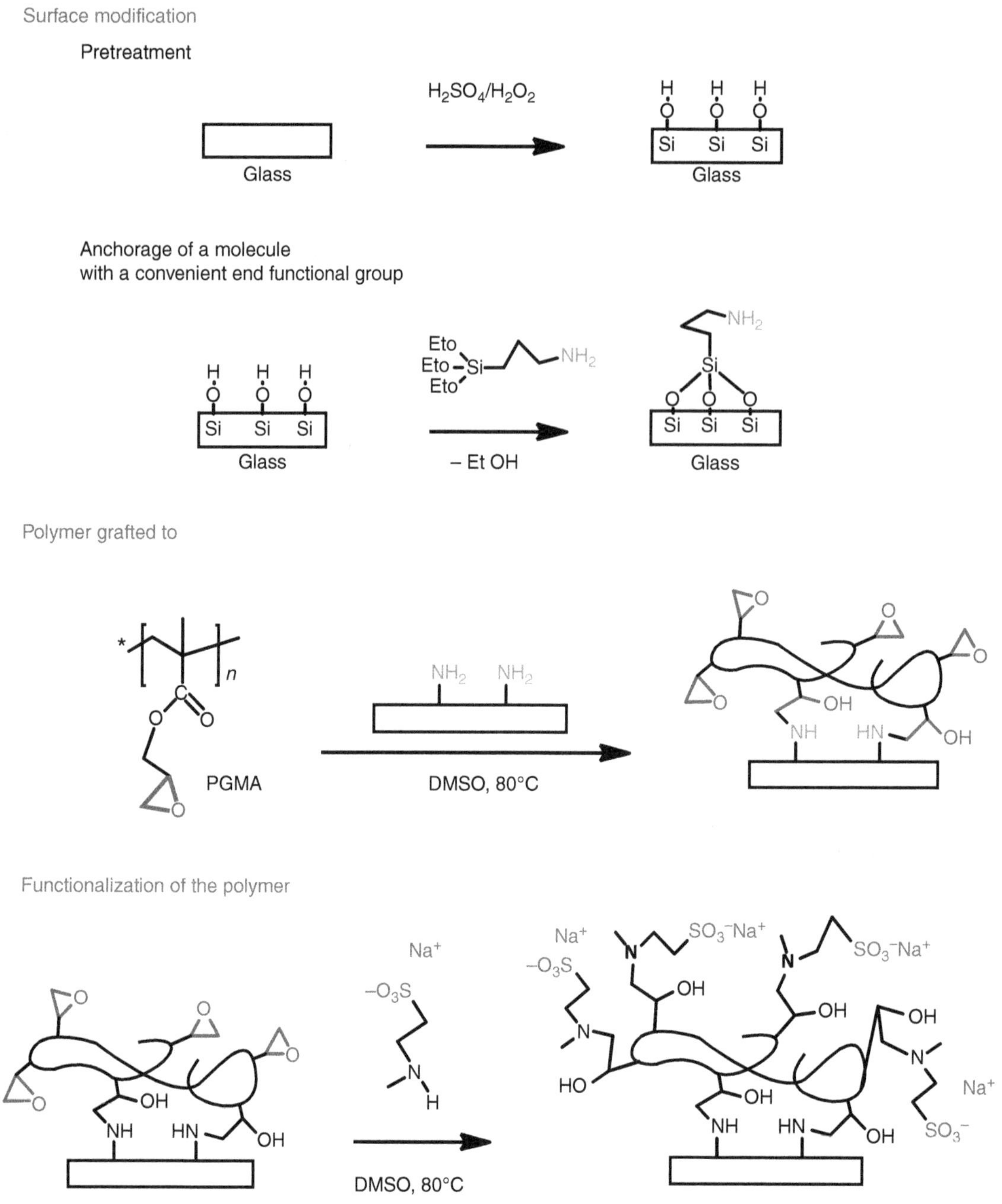

FIGURE 6.8 Successive steps for immobilization and functionalization of polymers on a microchip surface. Alonso-García et al. (unpublished results).

case, a glass microchannel is modified by a "grafting to" procedure. A pretreatment for enhancement of silanol groups is performed by immersion of the glass surface in a piranha solution [H_2O_2 (30%)–H_2SO_4 (18 M) 1:3 v/v]. Then, the surface is covered with primary amines by reaction of the glass silanols with 3-aminopropyl(triethoxy)silane (APTES), $(EtO)_3SiCH_2CH_2CH_2NH_2$, liberating EtOH. Later poly(glycidyl methacrylate) (PGMA) is immobilized on the glass surface by opening some of the epoxy rings in the polymer by the NH_2 groups previously anchored to the surface. Finally, the attached PGMA further reacts with the sodium salt of the methyltaurine, coating the surface with a negatively charged polymer (Alonso-García et al., unpublished results).

In Table 6.5, the different steps for permanent modification of the ME surface with polymers are summarized. Although there are particular cases, a general procedure can be envisaged. Taking into account the objective, namely to anchor a polymer to a substrate, the steps can be divided into two. The first one consists in preparing the surface for

TABLE 6.5 Studies Related to Covalent Immobilization of Polymers in ME

Surface Modification			Bonded Polymer		Analyte	Reference
Substrate	Pretreatment, Generated Group	Group (Molecule Anchorage)	Graft To	Graft From		
PDMS	O_2 plasma, Si—OH	—NH_2 (APTES)	Oxidized dextran		Neuro transmitters peptides, proteins	238
	O_2 plasma, Si—OH	—NH_2 (APTES)	NSS-mPEG		Amino acids	101
	O_2 plasma, Si—OH	—NH_2 (APTES)	P(AAM-*co*-GMA), P(DAM-*co*-GMA), (PVP-*g*-GMA), (PVA-*g*-GMA)[a]		Peptides, proteins	248
	O_2 plasma, Si—OH	—N_3 (CPTCS, NaN_3)	PEG-alkyne		Amino acids, proteins	241
	O_2 plasma, Si—OH	Epoxy (GPTMS)	PEG-NH_2		Amino acids, proteins	236
	O_2 plasma, Si—OH	—NH_2 (APDMES)	PGA		Amino acids	237
		(a) 3-Chloropropyltrichlorosilane (b) NaN_3 for "click chemistry"	Alkyne-PEG		Proteins	241
	O_2 plasma, Si—OH	1-Trichlorosilyl-2-(*m,p*-chloromethylphenyl)ethane		PA (ATRP)	Proteins	252
	O_2 plasma, generating Si—OH	MPTS		PAAm[b] $(NH_4)_2SO_5$	Protein	229
		Benzophenone[c]		PAA, PA-*g*-PEG	Peptides	99
	O_2 plasma, generating Si—OH			AMPS, C_{18}-AMPS, PSS, VSA, PAA, StMA-AMPS	Peptides	244
	UV			PAA, PAAm, etc.	Peptides	55
P(GMA-MMA)	O_2 plasma, C(O)OH, C—OH	2-Bromoisobutyryl bromide		PMMA-*g*-PEG (ATRP)	Amino acids, peptides, proteins	249
	Air plasma or an acid	2-Bromoisobutyryl bromide		PMMA-*g*-PEG (ATRP)	Peptides, proteins	246
PMMA	Basic solution		PEG-NH_2		Amino acids, proteins	235
	$LiNHCH_2CH_2NH_2$ UV, etylendiamine (en), EDC	Methacrylic acid, EDC		PA	DNA samples	243
	O_2 plasma, C(O)OH, C—OH	2-Bromoisobutyryl bromide[d]		PMMA-*g*-PEG (ATRP)	Peptides, proteins	245
	UV–O_3 or HNO_3		HPMC or PVA		DNA ladder	230
TPE	UV radiation	2-Bromoisobutyryl bromide		PMMA-*g*-PEG (ATRP)	Amino acids, peptides, proteins	250
P(MMA-MAA)			HPC		Proteins	217
Glass	NaOH	MPTS		PAA	DNA sequencing	232
	H_2O_2, NH_3, H_2O	MET[d]		PAAm	PCR products	232
	Acetone, isopropyl alcohol, and deionized water	(a) Deposition of 11-bromoundecyltrichlorosilane (BUTS) (b) NaN_3 in DMF for "click chemistry"	Alkyne (a) PAMAM (b) PEOX (c) PEI (d) PEG			240

[a]Remaining epoxy groups were blocked by 0.1 M aminoethanol.
[b]In addition, methylcellulose was dynamically coated on the PAAm.
[c]Cerium(IV) catalyzed.
[d]Other coating methods are described in this chapter.
3-CPTCS, chloropropyltrichlorosilane; VSA, poly(vinylsulfonic acid); PA, polyacrylamide (equivalent to PAAm); PA-*g*-PEG, polyacrylate-graft-polyethylene glycol; PCR, polymerase chain reaction; PMMA-*g*-PEG, poly(methyl methacrylate)-graft-poly(ethylene glycol); P(MMA-MAA), poly(methyl methacrylate-*co*-methacrylic acid); P(AAM-*co*-GMA), poly(acrylamide-*co*-glycidyl methacrylate); P(DAM-*co*-GMA), poly(dimethylacrylamide-*co*-glycidyl methacrylate); PVP-*g*-GMA, polyvinylpyrrolidone-graft-poly(glycidyl methacrylate); PVA-*g*-GMA, poly(vinyl alcohol)-graft-poly(glycidyl methacrylate); StMA-AMPS, poly(stearyl methacrylate-*co*-acrylamido-2-methyl-1-propanesulfonic acid); MET, 3-methacryloxypropyltrimethoxysilane (equivalent to MPTMSi and MPTS).

receiving the (pre)polymer and can be subdivided into a pretreatment of the surface and later anchorage of a convenient molecule. The pretreatment aims to clean and activate the surface for later performing the immobilization of a molecule with an appropriate end functional group. The pretreatment, which can be chemical (piranha solution, basic solutions, etc.) or physical (UV irradiation, oxygen plasma, etc.), tries to generate active groups such as —Si—OH groups in PDMS or —C(O)OH groups in PMMA. On the other hand, the pretreatment is very important because the oxidized surface must have enough surface silanols to allow a polymer film to be of sufficient density to protect the substrate from unspecific analyte adsorption.

Concerning the immobilization of the functional molecule, this differs depending on the chosen conjugation procedure and can vary from amine, azide, epoxy, vinyl, carboxylate, hydroxyl, and so on. A different possibility is to anchor a polymerization initiator in case an atom transfer radical polymerization (ATRP) is performed. Since the possibilities vary depending on the material employed, classification is made taking into account the material: glass or polymer [PDMS, PMMA, or less common polymers such as P(GMA-MMA), P(MMA-MAA), or TPE].

The second step in the covalent immobilization of a polymeric modifier is bonding the polymer. As commented on earlier, two different approaches can be performed. A "grafting to" procedure involves the attachment of the polymer to the functional group in the substrate surface. Different polymers have been reported: dextran, PEG, PEI, and so on, or proteins, mainly depending on the analytes under separation. In many cases, polymers are also functionalized in order to perform the covalent bonding with the counterpart in the surface (e.g., PE can be amine, alkyne, or NSS-functionalized).

The "grafting from" procedure is mainly executed by polymerization of a monomer that modifies the substrate surface or by the ATRP procedure. This is a controlled polymerization in which the polymer chains grow at a more constant rate than in traditional chain polymerization and their lengths remain very similar (very low polydispersity index). This is due to the use of a transition metal-based catalyst (usually copper) that provides an equilibrium between active (propagating) and inactive (dormant) forms of the polymer, which suppresses side reactions. Many functional groups for use as monomer or initiator are possible.

Surface bonding of PEG or PEO is a common strategy for retarding the nonspecific adsorption of proteins and other biological species. PEG is an uncharged and nontoxic polymer. Proteins are macromolecules bearing hydrophobic, hydrophilic, and ionic chemical groups; they are complex natural organic PEs. There are several reasons for the adsorption resistance of PEO surfaces in water. It was proposed that polymer chains have the ability to extend into an aqueous medium and thus create a steric repulsion for other adsorbates [233] apart from the ability of ethylene oxide units to form helices [234]; much more hydrogen bonds are available resulting in a more hydrophilic surface. On the other hand, PEG is generally recognized as the benchmark for the resistance of nonspecific adsorption of proteins due to its biocompatibility, low toxicity, and excellent protein resistance. The hydroxyl group of PEG is readily modified to produce a functional end group. For the determination of anionic proteins, amino-terminal PEG is covalently immobilized by reaction between the acyl carbon of PMMA and the primary amino group in PEG-NH_2 dissolved in a basic aqueous solution [235]. In both cases, a one-step immobilization procedure is performed.

In a two-step grafting method, PEG-NH_2 was used for an environmentally friendly PDMS modification. The PDMS was epoxy functionalized through reaction of the silanol groups on the surface with 3-glycidoxypropyltrimethoxysilane (GPTMS). Significantly improved electrophoretic performance of the modified microchips, with a noticeable EOF suppression and resistance to nonspecific protein adsorption for more than 30 days, was obtained [236]. NSS–mPEG {*O*-[(*N*-succinimidyl)succinyl]-*O*′-methyl–poly(ethylene glycol)} was grafted on to an amine-functionalized (with APTES reagent) surface. In this way, hydrophilic polymer brushes were obtained on a PDMS microchannel for improvement of amino acid separation [101], as shown in Figure 6.9.

Novel covalent strategies are being continuously developed. Although PEG is one of the more common polymers because of its hydrophilic characteristics, different polymers have been covalently grafted on microchip surfaces. Silanization with 3-aminopropyldimethylethoxysilane (APDMES) of a plasma oxidized PDMS surface produced amine groups on the surface. Further, amide bonds (through EDC and NHS) were formed with polyglutamic acid (PGA) [237] for the separation of four amino acid derivatives and homocysteine. Dextran, a glucose polymer, was selectively oxidized to aldehyde groups with sodium periodate and subsequently grafted on to amine-functionalized PDMS (through APTES reagent) *via* a Schiff base reaction (between an aldehyde and an amine producing an imine) [238]. The coated PDMS surface efficiently prevented biomolecules from adsorption and suppressed EOF. Improvements in the separation of different analytes are shown in Figure 6.10.

When choosing a surface modification procedure, the total procedure has to be kept in mind: consumption of time, working conditions, and so on. For example, oxygen-free conditions that can be required for radical polymerization are difficult to handle during *in situ* modification of PDMS microchips. Moreover, nonpolar organic solvents can severely damage the materials, similarly to what happens with high temperatures when polymeric microchips are employed. Special chemistries such as "click" chemistry are sometimes involved.

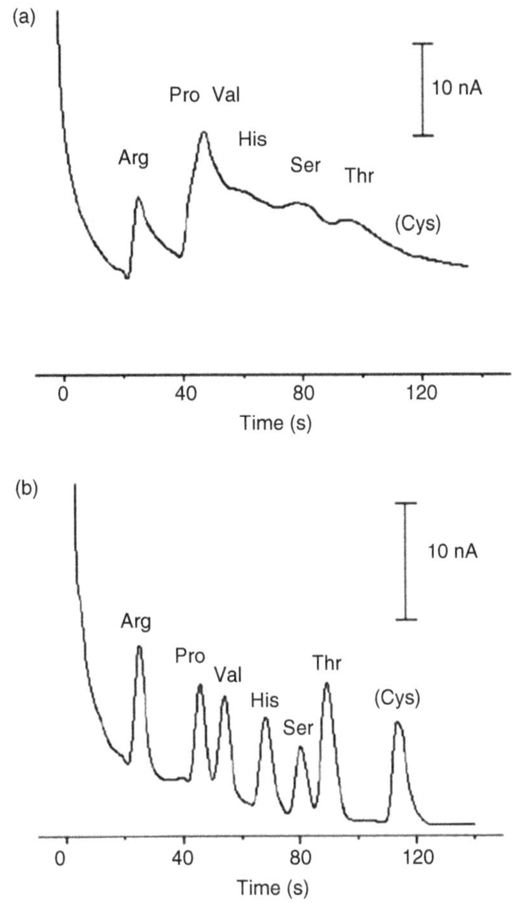

FIGURE 6.9 Electropherograms for amino acids: (a) on the native PDMS–PDMS microchip; (b) on the mPEG-coated PDMS–PDMS microchip. Conditions: BGE: 5.0 mM STB; V_{inj}: 600 V for 3 s; V_{sep}: 1600 V. Reprinted from Ref. 101 with permission from Elsevier.

"Click" chemistry is a type of highly efficient surface modification method with high specificity, nearly quantitative with mild ambient conditions in both organic and aqueous solutions, suitable for *in situ* reactions [239]. Among these, the copper-catalyzed Huisgen cycloaddition of alkynes and azides is considered particularly useful. Covalent attachment of preformed terminally functionalized linear and dendritic polymers [polyamidoamine (PAMAM), linear *m*-PEG, PEI, and poly(2-ethyloxazoline) (PEOX)] were synthesized with terminal alkynes for "click" modification of glass microfluidic channels [240]. A "click" chemistry-based surface modification strategy was also developed for PDMS microchips to enhance the separation performance for both amino acids and proteins. Alkyne-PEG was synthesized by a conventional procedure and then "click" grafted to azido-PDMS. EOF regulation and stability of the modification were evident [241].

As can be seen in Table 6.5, proteins and peptides together with amino acids are the most common analytes. Most of the materials used in the production of analytical microchips have the inconvenience of the strong adsorption of hydrophobic molecules. Proteins are particularly prone to surface adsorption, and therefore strategies for obtaining hydrophilic microchannels are being continuously developed.

Recently, "grafting from" procedures have advanced considerably for the covalent immobilization of polymers. Immobilization of a monomer in an aminated PMMA surface [through either a chemical [242] (lithiated diamine) or photochemical (UV modification) procedure] was performed prior to surface coating with an LPA. The aminated surface is used to covalently anchor methacrylic acid, which was used as a scaffold to produce LPAs on the surface through radical polymerization of acrylamide. It was applied to single-base resolution of ssDNA [243].

In the more usual procedure, only disassembled channels could be coated by UV-mediated graft polymerization. Ce-catalyzed polymerization has also been reported [244]. In this case, vinylsulfonic acid, acrylic acid, 2-acrylamido-2-methylpropanesulfonic acid (AMPS), 4-styrenesulfonic acid, and stearyl methacrylate were used for successful modification of the surface of PDMS by cerium(IV)-catalyzed polymerization. AMPS-coated PDMS channels were shown to give a reproducible separation of a synthetic peptide mixture for over 1 month. Referring to UV-mediated polymerization, by preadsorbing a photoinitiator (benzophenone) on the surface (PDMS), the rate of polymer formation at the surface was greatly accelerated compared with that in solution. Thus, a gel did not form in the lumen of enclosed microchannels. After addition of a variety of monomer solutions [acrylic acid, poly(ethylene glycol), monomethoxyl acrylate, or poly(ethylene glycol) diacrylate] and illumination with UV light, a stable coating was produced. The placement of an opaque mask over a portion of the channel permitted photopatterning of the microchannels. By using an appropriate mixture of monomers combined with masks, it should be possible to fabricate PDMS devices with distinct surface properties in different regions or channels [99]. In this case, two different problems need to be solved at the same time: the difficulty in applying the coating in an intact microchannel and the fact that many coatings that can be accomplished through the use of laminar flow are limited to patterning.

UV-mediated grafting has been applied but with channels that had to be assembled after coating [99]. When applied to an intact channel, it forms a gel rather than a surface coating. In addition, the small amount of free monomer within the channel is rapidly consumed by gel formation, leaving little to no monomer available for polymerization on the surface. The gel also restricts diffusion of any residual monomer to the surface of the channels. Unless the channel can be disassembled and reassembled, the gel forms a plug in the channel, restricting movement. By adsorbing the

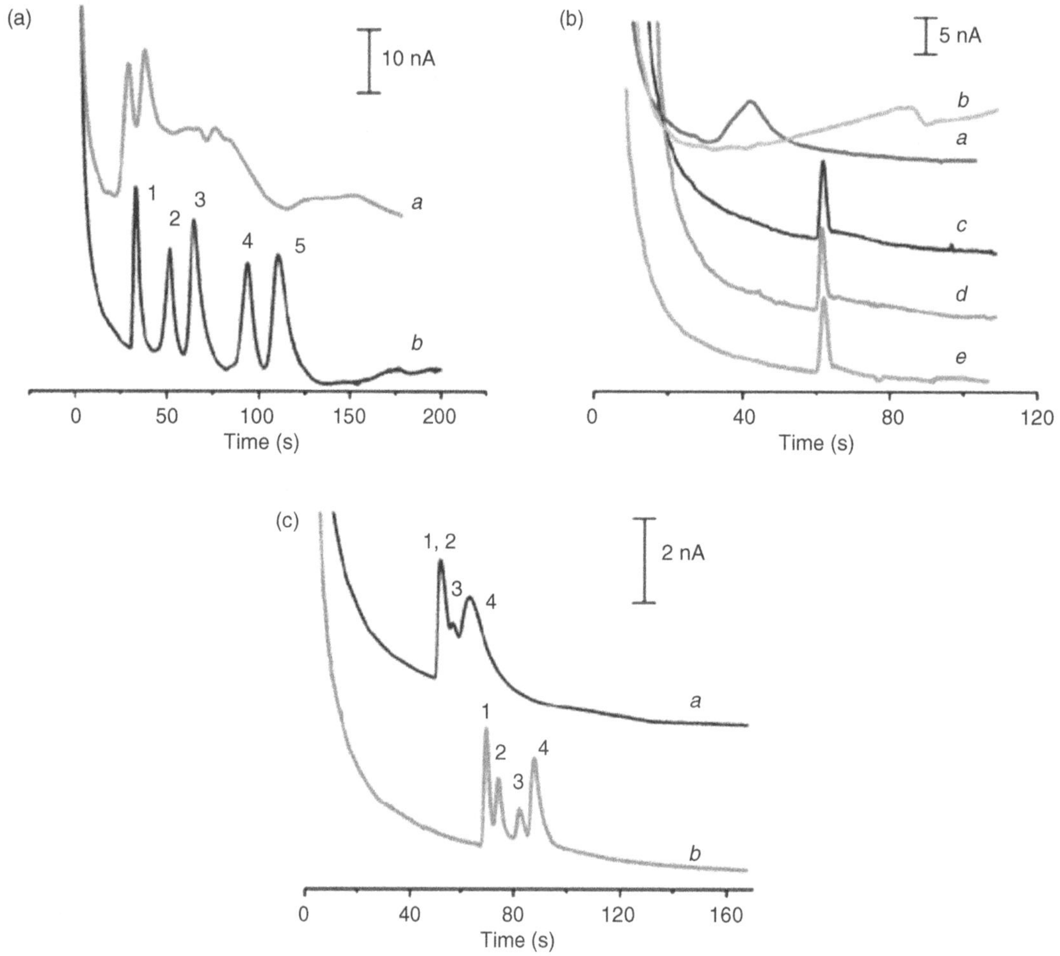

FIGURE 6.10 Electropherograms for peptides (a), protein (b), and neurotransmitters (c) on native and oxidized-dextran coated PDMS microchips. (a) (1) Thr–Arg–Lys; (2) Lys–Gly–Ser; (3) Ile–His–Pro; (4) Val–Ala–Ser; (5) Ile–Leu–Met. Conditions: BGE, 5.0 mM STB (pH 9.5); V_{inj}, +700 V for 3 s; V_{sep}, +1600 V; WE, copper microelectrode; E_d, +0.6 V. (b) Glucose oxidase (GOx); *a*, the first injection on the native PDMS microchip; *b*, the third injection on the native PDMS microchip; *c–e*, the first, third, and fifth injections on the coated PDMS microchip. Conditions: BGE, 30 mM NaOH; V_{inj}, +600 V for 3 s; V_{sep}, +1100 V; WE, copper microelectrode, E_d, +0.6 V. (c) *a*, on the native PDMS microchip; *b*, on the oxidized dextran-coated PDMS microchip; (1) 5-HT (5-hydroxytryptamine); (2) DA; (3) EP; (4) DBA (dobutamine). Conditions: BGE, 40 mM PB (pH 7.0); V_{inj}, +600 V for 3 s; V_{sep}, +1000 V; WE, single carbon fiber electrode; E_d, +1.40 V. Reprinted from Ref. 238 with permission from Elsevier.

photoinitiator on the surface and removing free photoinitiator prior to graft polymerization, an accelerated rate of formation of the polymer on the surface relative to that in solution during the grafting step occurs.

On the other hand, the most common procedure for radical polymerization is ATRP, sometimes called surface initiated atom transfer radical polymerization (SI-ATRP), in which an initiator of the polymerization is anchored on the surface. This is one of the methods employed for covering surfaces with hydrophilic polymers on microchannel surfaces. Different from traditional radical polymerization, ATRP is a transition metal-catalyzed free-radical living polymerization method, which produces well-defined polymers. In this case, an initiator is anchored in the surface and then the monomer is polymerized. An example is the anchorage of 2-bromoisobutyryl bromide to the hydroxyl groups of oxidized PMMA, and polymerization of a PEG-functionalized monomer (methyl ether methacrylate) [245].

The same initiator and monomer are employed with this procedure but using a surface-reactive acrylic polymer, PGMAMMA. The advantage of this approach is that this copolymer presents epoxy groups on the surface that can be activated by air plasma treatment, hydrolysis, or aminolysis

[246]. Actually, tentacle-type polymer chains with epoxy groups were fabricated for open-tubular CEC but no test was made in ME format. The fabrication procedure of the stationary phase included pretreatment of the capillary inner wall, silanization, and glycidyl methacrylate (GMA)-grafted polymerization. The epoxy group-containing monomer GMA is the commonly used precursor monomer. The prepared GMA-grafted stationary phase could be easily modified with different chromatographic ligands by the ring-opening reaction of the epoxy groups [247]. A procedure for anchoring epoxy groups, following two different paths, copolymerization of GMA and acrylamide–dimethylacrylamide or graft polymerization of GMA on a pretreated, amine-functionalized (*via* APTES) PDMS microdevice, has been reported for resisting nonspecific protein adsorption [248]. It has to be pointed out that SU-8, an epoxy-based polymer (glycidyl ether of bisphenol A), mainly used as a negative tone photoresist, is employed for ME fabrication (http://www.micruxfluidic.com) and shows promising possibilities for surface chemistry.

During the ATRP modification processes described earlier, the total surface of both plates was grafted with a PEG layer, which affected the thermal bonding strength of the resultant microchip. ATRP grafting is performed on the surfaces of top and bottom plates before bonding them together thermally to form the enclosed microchannels, but they can delaminate spontaneously during use, due to the weak bonding strength of the PEG-functionalized surfaces. Furthermore, the underlying substrate surface polymers were prevented from interacting because of the more hydrophilic PEG-functionalized surface layer. Modification of only the microchannel surface is also possible if the initiator is anchored on the whole surface of both plates before bonding them together to form the enclosed microchannel, instead of bonding before introducing initiator into the channels. In this way, the reaction is faster and more efficient, avoiding swelling and deforming the microchannel and also its clogging with precipitate generated in the reagent solutions. After thermal bonding of the plates, an aqueous solution containing monomer (PEG-MEMA), catalyst, and ligand, prepared in an oxygen-free environment, was pumped into the microchannels to perform the in-channel ATRP modification [249].

ATRP involves activation of the surface, immobilization of the initiator, and subsequent grafting of the chosen polymer. The length of polymer tethered to the initiator can be controlled readily. In-channel ATRP grafting of a thin film of PEG on the surface of TPE microchannels was performed using 2-bromoisobutyryl bromide as initiator. Immobilization occurred on TPE microdevices in a water-free environment. Further, grafting solution (Cu, PEG-MEMA) was pumped through the TPE microchannels at a sufficiently slow flow rate for the reaction conditions to be close enough still to allow ATRP functionalization. Since ATRP is performed after TPE microdevices have been bonded together, it does not interfere with device fabrication [250].

ATRP of acrylamide on PDMS tries to achieve a lasting hydrophilicity due to the growth of PA chains from oxidized PDMS surface. Their immiscibility ought to promote stability of the hydrophilic surface. A 20-fold improvement in resisting irreversible adsorption of lysozyme, compared with bare PDMS and a 10-fold improvement compared with bare glass was reported. Moreover, unlike oxidized PDMS, which reverts fully to being hydrophobic after 2 days, the surface treated by ATRP of acrylamide was shown to remain hydrophilic for at least 1 month [251]. Similarly, a thin PA layer on a PDMS microchannel surface was formed by ATRP by using 1-trichlorosilyl-2-(*m*,*p*-chloromethylphenyl) ethane as initiator [252]. Because of the hydrophilic nature of the PA-coated silica surface, the aqueous solution of reagents was able to fill the channels by capillary force within 0.5 h. This is very important because pressure must be used to fill hydrophobic channels with aqueous buffers and this is hampered by bubble formation due to the permeability of PDMS [253]. The surface can be oxidized to become somewhat hydrophilic due to surface silanol and carboxylic acid groups, allowing filling to occur by capillary action, but hydrophobicity can be recovered.

6.4 CONCLUSIONS

ME devices are extremely promising tools with many analytical possibilities and applications, but they are still under development. Important advances from their beginning in all the areas involved in ME—new materials, more appropriate designs, integration of different steps, improvement of detection methodologies, and so on—have been reported. However, modern analytical chemistry has important aims, such as the simultaneous determination of closely related analytes that are still a challenge. Therefore, there are wide research activities on ME resolution improvement. With this objective, the state of the device surface is of paramount importance and different methodologies have been reported for obtaining the most suitable one. Micro- and nanostructuring, employment of energy sources, and chemical modification are reported methodologies. They can be combined and, in many cases, a physical treatment (oxygen plasma and UV radiation are among the most common) is employed as a means of conditioning the surface before a chemical modification is made. The main objective is to improve separation through control of the EOF and suppression of analyte adsorption, but a totally different one can be found in the development of bioassays.

Procedures have to consider when (before or after bonding the plates that constitute the microchip), where (in the whole surface or in discrete patterned locations), how (through covalent or noncovalent bonds), and which modifiers to use. Different modes of electrophoresis are available (CZE, MEKC, CGE, CEC, etc.).

Referring to chemical modifiers, the use of surfactants is well established. However, ILs and NPs are at the beginning of their application and further promising results are envisaged. The possibilities for polymers are wide, and therefore much research is found in this area. As in the case of the rest of the modifiers, a dynamic or permanent coating can be performed. The former is a very simple procedure but more unstable layers are obtained. Covalent coating often requires organic solvents or high temperatures, and the procedures are more tedious. However, the possibilities are varied and very appropriate for nondisposable devices.

Consideration of the material is of paramount importance. Similarity among glass (microchip) and fused silica (capillaries) has made this material the first choice. Different materials are employed taking advantage of the possibilities of polymers and most of the modification strategies have been applied to polymeric materials, probably due to their more hydrophobic nature. However, one of the major drawbacks when developing modification chemistries is that the procedure employed for one particular type of polymer may not necessarily transfer to another substrate with the same degree of success [59]. Research is still needed and interesting options, many of them still unexplored, are available: surface tuning through the properties of block copolymers, use of branched molecules, patterning of surfaces, exploitation of the possibilities of ILs, application of new multifunctional NPs, and so on.

Surface chemistry is of great importance in ME, especially in highly miniaturized devices, due to the high surface area-to-volume ratio. With reduction of the size of the channels, the properties of channel walls are becoming probably more important than originally thought. The properties are commonly explained on the basis of the structures of the material employed, but some additives (heat stabilizers, plasticizers, antioxidants, UV stabilizers, etc.) can affect the surface of the microchannel and the effect has not been thoroughly studied. Moreover, although the surface science community has developed many protocols, they have not yet been exploited deeply. The field of surface chemistry is not yet well known by the analytical chemistry community, and multidisciplinary efforts have to continue in order to obtain real applications. Coated capillaries are commonly applied in CE and they are commercially available. However, in the case of ME, the basic research has to focus on commercial applications.

ACKNOWLEDGMENTS

This work was supported by FICYT under project FC-11-PC10-30. Isabel Álvarez Martos thanks the MICINN for the award of a PhD grant.

REFERENCES

1. Ortner, H.M. (1992) *Fresen. J. Anal. Chem. 343*, 825–826.
2. Valcarcel, M. (1997) *Trends Anal. Chem. 16*, 124–131.
3. Manz, A., Graber, N., and Widmer, H.M. (1990) *Sens. Actuators B 1*, 244–248.
4. Vazquez, M., Frankenfeld, C., Coltro, W.K.T., Carrilho, E., Diamond, D., and Lunte, S.M. (2010) *Analyst 135*, 96–103.
5. Fernández-la-Villa, A., Pozo-Ayuso, D.F., and Castaño-Álvarez, M. (2010) *Electrophoresis 31*, 2641–2649.
6. Kovachev, N., Canals, A., and Escarpa, A. (2010) *Anal. Chem. 82*, 2925–2931.
7. Oita, I., Halewyck, H., Thys, B., Rombaut, B., van der Heyden, Y., and Mangelings, D. (2010) *Anal. Bioanal. Chem. 398*, 239–264.
8. Arora, A., Simone, G., Salieb-Beugelaar, G.B., Kim, J.T., and Manz, A. (2010) *Anal. Chem. 82*, 4830–4847.
9. Zubritsky, E. (2002) *Anal. Chem. 74*, 22A–26A.
10. Kraly, J.R., Jones, M.R., Gomez, D.G., Dickerson, J.A., Harwood, M.M., Eggertson, M., Paulson, T.G., Sanchez, C.A., Odze, R., Feng, Z., Reid, B.J., and Dovichi, N.J. (2006) *Anal. Chem. 78*5977–5986.
11. Kenyon, S.M., Meighan, M.M., and Hayes, M.A. (2011) *Electrophoresis 32*, 482–493.
12. Chandra, U., Swamy, B.E.K., Gilbert, O., and Sherigara, B.S. (2010) *Electrochim. Acta 55*, 7166–7174.
13. Ensafi, A.A., Rezaei, B., Zare, S.Z.M., and Taei, M. (2010) *Sens. Actuators B 150*, 321–329.
14. Kissinger, P.T. and Heineman, W.R. (1984) *Laboratory Techniques in Electroanalytical Chemistry*, 2nd ed., Marcel Dekker, New York.
15. Varenne, A. and Descroix, S. (2008) *Anal. Chim. Acta 628*, 9–23.
16. Scharwz, M.A. and Hauser, P.C. (2001) *J. Chromatogr. A 198*, 225–232.
17. Nagl, S., Schulze, P., Ludwig, M., and Belder, D. (2009) *Electrophoresis 30*, 2765–2772.
18. Ward, T.J. and Ward, K.D. (2010) *Anal. Chem. 82*, 4712–4722.
19. Guihen, E., Hogan, A., and Glennon, J. (2009) *Chirality 21*, 292–298.
20. Sun, X., Li, D., and Lee, M. (2009) *Anal. Chem. 81*, 6278–6284.
21. Qu, P., Lei, J., Ouyang, R., and Ju, H. (2009) *Anal. Chem. 81*, 9651–9656.
22. Weinberger, R. (2000) *Practical Capillary Electrophoresis*, Academic Press, San Diego, CA.
23. Bowen, A.L. and Martin, R.S. (2009) *Electrophoresis 30*, 3347–3354.
24. Zubritsky, E. (2000) *Anal. Chem. 72*, 687A–690A.
25. Ramsey, J.D., Jacobson, C.T., Culbertson, C.T., and Ramsey, J.M. (2003) *Anal. Chem. 75*, 3758–3764.
26. Li, M.W., Huynh, B.H., Hulvey, M.K., Lunte, S.M., and Martin, R.S. (2006) *Anal. Chem. 78*, 1042–1051.

27. Castaño-Álvarez, M., Fernández-la-Villa, A., Pozo-Ayuso, D. F., Fernández-Abedul, M.T., and Costa-García, A. (2009) *Electrophoresis 30*, 3372–3380.

28. Muck, A., Wang, J., Jacobs, M., Chen, G., Chatrathi, M.P., Jurka, V., Výborný, Z., Spillman, S.D., Sridharan, G., and Schoning, M.J. (2004) *Anal. Chem. 76*, 2290–2297.

29. Chen, Y., Zhang, L., and Chen, G. (2008) *Electrophoresis 29*, 1801–1814.

30. Wang, Y., Chen, H., He, Q., and Soper, S.A. (2008) *Electrophoresis 29*, 1881–1888.

31. Liu, A.-L., He, F.-Y., Hu, Y.-L., and Xia, X.-H. (2006) *Talanta 68*, 1303–1308.

32. Bai, X.X., Roussel, C., Jensen, H., and Girault, H.H. (2004) *Electrophoresis 25*, 931–935.

33. Kameoka, J., Craighead, H.G., Zhang, H., and Henion, J. (2001) *Anal. Chem. 73*, 1935–1941.

34. Stachowiak, T.B., Rohr, T., Hilder, E.F., Peterson, D.S., Yi, M., Svec, F., and Fréchet, J.M.J. (2003) *Electrophoresis 24*, 3689–3693.

35. Castaño-Álvarez, M., Fernández-Abedul, M.T., and Costa-García, A. (2005) *Electrophoresis 26*, 3160–3168.

36. Lee, J.N., Park, C., and Whitesides, G.M. (2003) *Anal. Chem. 75*, 6544–6554.

37. Vickers, J.A., Dressen, B.M., Weston, M.C., Boonsong, K., Chailapakul, O., Cropek, D.M., and Henry, C.S. (2007) *Electrophoresis 28*, 1123–1129.

38. Fiorini, G.S., Lorenz, R.M., Kuo, J.S., and Chiu, D.T. (2004) *Anal. Chem. 76*, 4697–4704.

39. Coltro, W.K.T., da Silva, J.A.F., da Silva, H.D.T., Richter, E. M., Burlan, R., Angnes, L., do Lago, C.L., Mazo, L.H., and Carrilho, E. (2004) *Electrophoresis 25*, 3832–3839.

40. Castaño-Álvarez, M., Fernández-Abedul, M.T., Costa-García, A., Agirregabiria, M., Fernández, L.J., Ruano-López, J.M., and Barredo-Presa, B. (2009) *Talanta 80*, 24–30.

41. Agirregabiria, M., Blanco, F.J., Berganzo, J., Fullaondo, A., Subyaga, A.M., Mayora, K., and Ruano-López, J.M. (2006) *Electrophoresis 27*, 3627–3634.

42. Piccin, E., Coltro, W.K.T., da Silva, J.A.F., Neto, S.C., Mazo, L. H., and Carrilho, E. (2007) *J. Chromatogr. A 1173*, 151–158.

43. Fercher, G., Haller, A., Smetan, W., and Vellekoop, M.J. (2010) *Analyst 135*, 965–970.

44. Martinez, A.W., Philipps, S.T., Whitesides, G.M., and Carrrilho, E. (2010) *Anal. Chem. 82*, 3–10.

45. Silkanen, T., Aura, S., Heikkila, L., Kotiaho, T., Franssila, S., and Kostiainen, R. (2010) *Anal. Chem. 82*, 3874–3882.

46. Klasner, S.A., Metto, E.C., Roman, G.T., and Culbertson, C. T. (2009) *Langmuir 25*, 10390–10396.

47. Liu, J.K., Sun, X.F., and Lee, M.L. (2005) *Anal. Chem. 77*, 6280–6287.

48. Kim, P., Jeong, H.F., Khademhosseini, A., and Suh, K.Y. (2006) *Lab Chip 6*, 1432–1437.

49. Liu, J.K., Sun, X.F., and Lee, M.L. (2007) *Anal. Chem. 79*, 1926–1931.

50. Kim, M.-S., Cho, S., II, Lee, K.-N., and Kim, Y.-K. (2005) *Sens. Actuators B 107*, 818–824.

51. Lacher, N.A., de Rooij, N.F., Verpoorte, E., and Lunte, S.M. (2003) *J. Chromatogr. A 1004*, 225–235.

52. Coltro, W.K.T., Lunte, S.M., and Carrilho, E. (2008) *Electrophoresis 29*, 4928–4937.

53. Sikanen, T., Heikkila, L., Tuomikoski, S., Ketola, R.A., Kostiainen, R., Franssila, S., and Kotiaho, T. (2007) *Anal. Chem. 79*, 6255–6263.

54. Duffy, D.C., McDonald, J.C., Schueller, O.J.A., and Whitesides, G.M. (1998) *Anal. Chem. 70*, 4974–4984.

55. Hu, S., Ren, X., Backman, M., Sims, C.E., Li, G.P., and Li, Allbritton, N. (2002) *Anal. Chem. 74*, 4117–4123.

56. Yu, H., He, F.-Y., Lu, Y., Hu, Y.-L., Zhong, H.-Y., and Xia, X.-H. (2008) *Talanta 75*, 43–48.

57. Pallandre, A., de Lambert, B., Attia, R., Jonas, A.M., and Viovy, J.-L. (2006) *Electrophoresis 27*, 584–610.

58. Fan, H.Z. and Chen, G. (2007) *Proteomics 7*, 3445–3449.

59. Belder, D. and Ludwig, M. (2003) *Electrophoresis 24*, 3595–3606.

60. Dolnik, V. (2004) *Electrophoresis 25*, 3589–3601.

61. Muck, A. and Svatos, A. (2007) *Talanta 74*, 333–341.

62. Makamba, H., Kim, J.H., Lim, K., Park, N., and Hahn, J.H. (2003) *Electrophoresis 24*, 3607–3619.

63. Doherty, W.A.S., Meagher, R.J., Albarghouthi, M.N., and Barron, A.E. (2003) *Electrophoresis 24*, 34–54.

64. Chen, Y., Zhang, L., and Chen, G. (2008) *Electrophoresis 29*, 1801–1814.

65. Liu, J. and Lee, M.L. (2006) *Electrophoresis 27*, 3533–3546.

66. Zhou, J., Ellis, A.V., and Voelcker, N.H. (2010) *Electrophoresis 31*, 2–16.

67. Kist, T.B.L. and Mandaji, M. (2004) *Electrophoresis 25*, 3492–3497.

68. Lu, Y., Hu, Y.-L., and Xia, X.-H. (2009) *Talanta 79*, 1270–1275.

69. Cao, H., Yu, Z., Wang, J., Tegenfeldt, J.O., Austin, R.H., Chen, E., Wu, W., and Chou, S.Y. (2002) *Appl. Phys. Lett. 81*, 174–176.

70. Chantiwas, R., Hupert, M.L., Pullagurla, S.R., Balamurugan, S., Tamarit-Lopez, J., Park, S., Datta, P., Goettert, J., Cho, Y. K., and Soper, S.A. (2010) *Lab Chip 10*, 3255–3264.

71. Kaji, N., Okamoto, Y., Tokeshi, M., and Baba, Y. (2010) *Chem. Soc. Rev. 39*, 948–956.

72. Li, H.Y., Dauriac, V., Thibert, V., Senechal, H., Peltre, G., Zhang, X.X., and Descroix, S. (2010) *Lab Chip 10*, 2597–2604.

73. Sano, T., Iguchi, N., Iida, K., Sakamoto, T., Baba, M., and Kawaura, H. (2003) *Appl. Phys. Lett. 83*, 4438–4440.

74. McDonald, J.C., Duffy, D.C., Anderson, J.R., Chiu, D.T., Wu, H., Schueller, O.J.A., and Whitesides, G.M. (2000) *Electrophoresis 21*, 27–40.

75. Wang, B., Abdulali-Kanji, Z., Dodwell, E., Horton, J.H., and Oleschuk, R.D. (2003) *Electrophoresis 24*, 1442–1450.

76. Murakami, T., Kuroda, S., and Osawa, Z. (1998) *J. Colloid Interface Sci. 202*, 37–44.
77. Wang, B., Chen, L., Abdulali-Kanji, Z., Horton, J.H., and Oleschuk, R.D. (2003) *Langmuir 19*, 9792–9798.
78. Jones, B.J. and Hayes, M.A. (2006) Henry, C.S. (ed.) *Microchip Capillary Electrophoresis, Methods and Protocols*, Humana Press, Totowa, NJ, pp. 49–56.
79. Preininger, C, Sauer, U., Kern, W., and Dayteg, J. (2004) *Anal. Chem. 76*, 6130–6136.
80. Rossier, J.S. and Girault, H.H. (2001) *Lab Chip 1*, 153–157.
81. Bai, Y., Koh, C.G., Boreman, M., Juang, Y.-J., Tang, I.-C., Lee, L.J., and Yang, S.-T. (2006) *Langmuir 22*, 9458–9467.
82. Ko, S., Kim, B., Jo, S.-S., Oh, S.Y., and Park, J.-K. (2007) *Biosens. Bioelectron. 23*, 51–59.
83. Liu, Y., Wang, H., Huang, J., Yang, J., Liu, B., and Yang, P. (2009) *Anal. Chim. Acta 650*, 77–82.
84. Yen, P.-Y., Rossi, N.A.A., Kizhkkedathu, J.N., and Mu, C.A. (2010) *Microfluid. Nanofluid. 9*, 199–209.
85. Zhou, J.W., Voelcker, N.H., and Ellis, A.V. (2010) *Biomicrofluidics 4*, 046504.
86. Nakagawa, T., Tanaka, T., Niwa, D., Osaka, T., Takeyama, H., and Matsunaga, T. (2005) *J. Biotechnol. 116*, 105–111.
87. Lahann, J., Balcells, M., Lu, H., Rodon, T., Jensen, K.F., and Langer, R. (2003) *Anal. Chem. 75*, 2117–2122.
88. Gil, E.S. and Hudson, S.A. (2004) *Prog. Polym. Sci. 29*, 1173–1222.
89. Dai, W., Zheng, Y., Luo, K.Q., and Wu, H. (2010) *Biomicrofluidics 4*, 024101.
90. Ye, N.N., Win, J.H., Shi, W.W., Liu, X., and Lin, B.C. (2007) *Lab Chip 7*, 1696–1704.
91. Qu, H.Y., Wang, H.T., Huang, Y., Zhong, W., Lu, H.J., Kong, J.L., Yang, P.Y., and Liu, B.H. (2004) *Anal. Chem. 76*, 6426–6433.
92. Liu, Y., Lu, H., Zhong, W., Song, P., Kong, J., Yang, P., Girault, H.H., and Liu, B. (2006) *Anal. Chem. 78*, 801–808.
93. Ji, J., Zhang, Y., Zhou, X., Kong, J., Tang, Y., and Liu, B. (2008) *Anal. Chem. 80*, 2457–2463.
94. Liu, T., Wang, S., and Chen, G. (2009) *Talanta 77*, 1767–1773.
95. Priest, C. (2010) *Biomicrofluidics 4*, 032206.
96. Luo, X., Lewandowski, A.T., Yi, H., Payne, G.F., Ghodssi, R., Bentley, W.E., and Rubloff, G.W. (2008) *Lab Chip 8*, 420–430.
97. Kaji, H., Hashimoto, M., and Nishizawa, M. (2006) *Anal. Chem. 78*, 5469–5473.
98. Hu, S., Ren, X., Bachman, M., Sims, C.E., Li, G.P., and Allbritton, N.L. (2003) *Electrophoresis 524*, 3679–3688.
99. Hu, S., Ren, X., Bachman, M., Sims, C.E., Li, G.P., and Allbritton, N.L. (2004) *Anal. Chem. 76*, 1865–1870.
100. Camilleri, P. (1998) *Capillary Electrophoresis, Theory and Practice*, 2nd ed., CRC Press, Boca Raton, FL, p. 103.
101. Wang, A.-J., Feng, J.-J., and Fan, J. (2008) *J. Chromatogr. A 1192*, 173–179.
102. Frost, N.W., Jing, M., and Bowser, M.T. (2010) *Anal. Chem. 82*, 4682–4698.
103. Poitevin, M., Shakalisava, Y., Miserere, S., Peltre, G., Viovy, J.L., and Descroix, S. (2009) *Electrophoresis 30*, 4256–4263.
104. Mohamadi, M.R., Svobodova, Z., Verpillot, R., Esselmann, H., Wiltfang, J., Otto, M., Taverna, M., Bilkova, Z., and Viovy, J.L. (2010) *Anal. Chem. 82*, 7611–7617.
105. Kondo, K., Kaji, N., Toita, S., Okamoto, Y., Tokeshi, M., Akiyoshi, K., and Baba, Y. (2010) *Biomicrofluidics 4*, 032210.
106. Hjerten, S., Kiao, J.-L., and Zhang, R. (1989) *J. Chromatogr. 473*, 273–275.
107. Ngola, S.M., Fintschenko, Y., Choi, W.Y., and Shepodd, T.J. (2001) *Anal. Chem. 73*, 849–856.
108. Landner, Y., Crétier, G., and Faure, K. (2010) *J. Chromatogr. A 1217*, 8001–8008.
109. Vaher, M. and Kohel, M. (2005) *J. Chromatogr. A 1068*, 83–88.
110. Chen, G., Xu, X.J., Lin, J.Y.H., and Wang, J. (2007) *Chem. Eur. J. 13*, 6461–6467.
111. Wang, H., Wang, D., Wang, J., Wang, H., Gu, J., Han, Ch., Jin, Q., Xu, B., He, Ch., Cao, L., Wang, Y., and Zhao, J. (2009) *J. Chromatogr. A 1216*, 6343–6347.
112. Schreiber, F. (2000) *Prog. Surf. Sci. 65*, 151–256.
113. Wiedmer, S.K., Jussila, M.S., and Riekkola, M.L. (2004) *Trends Anal. Chem. 23*, 562–582.
114. Gomez-Hens, A. and Fernandez-Romero, J.M. (2006) *Trends Anal. Chem. 25*, 167–178.
115. Vermette, P. and Meagher, L. (2003) *Colloids Surf. B 28*, 153–198.
116. Sikanen, T., Wiedmer, S.K., Heikkilä, L., Franssila, S., Kostiainen, R., and Kotiaho, T. (2010) *Electrophoresis 31*, 2566–2574.
117. Xu, Y., Takai, M., Konno, T., and Ishihara, K. (2007) *Lab Chip 7*, 199–206.
118. Gertsch, J.C., Noblitt, S.D., Cropek, D.M., and Henry, C.S. (2010) *Anal. Chem. 82*, 3426–3429.
119. Osiri, J.K., Shadpour, H., and Soper, S.A. (2010) *Anal. Bioanal. Chem. 398*, 489–498.
120. Dang, F., Haseqawa, T., Biju, V., Ishikawa, M., Kaji, N., Yasui, T., and Baba, Y. (2009) *Langmuir 25*, 9296–9301.
121. Root, B.E., Zhang, B., and Barron, A.E. (2009) *Electrophoresis 30*, 2117–2122.
122. Dossi, N., Susmel, S., Toniolo, R., Pizzariello, A., and Bontempelli, G. (2009) *Electrophoresis 30*, 3465–3471.
123. Shi, M., Huang, Y., Li, X.T., and Zhao, S.L. (2009) *Chromatographia 70*, 1651–1657.
124. Piccin, E., Dossi, N., Cagan, A., Carrilho, E., and Wang, J. (2009) *Analyst 134*, 528–532.
125. Chiesl, T.N., Chu, W.K., Stockton, A.M., Amashukeli, X., Gruthaner, F., and Mathies, R.A. (2009) *Anal. Chem. 81*, 2537–2544.
126. Ohlsson, P.D., Ordeig, O., Mogensen, K.B., and Kutter, J.P. (2009) *Electrophoresis 30*, 4172–4178.
127. Noblitt, S.D., Schwandner, F.M., Hering, S.V., Collet, J.L., Jr., and Henry, C.S. (2009) *J. Chromatogr. A 1216*, 1503–1510.

128. Wang, H., Wang, H.M., Jin, Q.H., Cong, H., Zhuang, G.S., Zhao, J.L., Sun, C.L., Song, H.W., and Wang, W. (2008) *Electrophoresis 29*, 1932–1941.

129. Osiri, J.K., Shadpour, H., Park, S., Snowden, B.C., Chen, Z. Y., and Soper, S.A. (2008) *Electrophoresis 29*, 4984–4992.

130. Okada, H., Kaji, N., Tokeshi, M., and Baba, Y. (2008) *Electrophoresis 29*, 2533–2538.

131. Kilár, A., Farkas, V., Kovács, K., Kocsis, B., and Kilár, F. (2008) *Electrophoresis 29*, 1713–1722.

132. Okada, H., Kaji, N., Tokeshi, M., and Baba, Y. (2008) *Anal. Sci. 24*, 321–325.

133. Ding, Y., Mora, M.F., Merrill, G.N., and Garcia, C.D. (2007) *Analyst 132*, 997–1004.

134. Bishop, S.C., Lerch, M., and McCord, B.R. (2007) *J. Chromatogr. A 1154*, 481–484.

135. Kolivoska, V., Weiss, V.U., Kremser, L., Gas, B., Blaas, D., and Kenndler, E. (2007) *Electrophoresis 28*, 4734–4740.

136. Li, M.W. and Martin, R.S. (2007) *Electrophoresis 28*, 2478–2488.

137. Vlcková, M. and Schwarz, M.A. (2007) *J. Chromatogr. A 1142*, 214–221.

138. Sueyoshi, K., Nagai, H., Wakida, S., Nishii, J., Kitagawa, F., and Otsuka, K. (2006) *Meas. Sci. Technol. 17*, 3154–3161.

139. Vrouwe, E.X., Luttge, R., Olthuis, W., and van der Berg, A. (2006) *J. Chromatogr. A 1102*, 287–293.

140. Mourzina, Y., Kalyagin, D., Steffen, A., and Offenhäusser, A. (2006) *Talanta 70*, 489–498.

141. Wang, A.J., Xu, J.J., and Chen, H.Y. (2006) *Anal. Chim. Acta 569*, 188–194.

142. Roman, G.T., McDaniel, K., and Culbertson, C.T. (2006) *Analyst 131*, 194–201.

143. García, C.D., Dressen, B.M., Henderson, A., and Henry, C.S. (2005) *Electrophoresis 26*, 703–709.

144. Li, F., Wang, D.D., Yan, X.P., Su, R.G., and Lin, J.M. (2005) *J. Chromatogr. A 1081*, 232–237.

145. García, C.D. and Henry, C.S. (2005) *Electroanalysis 17*, 223–230.

146. Nakajima, H., Kawata, K., Shen, H., Nakagama, T., and Uchiyama, K. (2005) *Anal. Sci. 21*, 67–71.

147. Kang, J., Yan, J., Liu, J., Qiu, H., Yin, X.B., Yang, X., and Wang, E. (2005) *Talanta 66*, 1018–1024.

148. García, C.D. and Henry, C.S. (2003) *Anal. Chem. 75*, 4778–4783.

149. Charvátová, J., Deyl, Z., Klevar, M., Miksík, I., and Eckhardt, A. (2004) *J. Chromatogr. B 800*, 83–89.

150. Ceriotti, L., Shibata, T., Folmer, B., Weiller, B.H., Roberts, M.A., de Rooij, N.F., and Verpoorte, E. (2002) *Electrophoresis 23*, 3615–3622.

151. Galloway, M., Stryjewski, W., Henry, A., Ford, S.M., Llopis, S., McCarley, R.L., and Soper, S.A. (2002) *Anal. Chem. 74*, 2407–2415.

152. Ocvirk, G., Munroe, M., Tang, T., Oleschuk, R., Westra, K., and Harrison, D.J. (2000) *Electrophoresis 21*, 107–115.

153. Rodríguez, I., Jin, L.J., and Li, S.F. (2000) *Electrophoresis 21*, 211–219.

154. Anderson, J.L., Armstrong, D.W., and Wei, G.T. (2006) *Anal. Chem. 78*, 2892–2902.

155. Walden, P. (1914) *Bull. Acad. Sci. St. Petersburg* 405–422.

156. Liu, J.-f., Jönsson, J.A., and Jiang, G.-B. (2005) *Trends Anal. Chem. 24*, 20–27.

157. Berthod, A., Ruiz-Angel, M., and Carda-Broch, S. (2008) *J. Chromatogr. A 1184*, 6–18.

158. López-Pastor, M., Simonet, B.M., Lendl, B., and Valcárcel, M. (2008) *Electrophoresis 29*, 94–107.

159. Jiang, T.-F., Gu, Y.-L., Liang, B., Li, J.-B., Shi, Y.-P., and Ou, Q.-Y. (2003) *Anal. Chim. Acta 479*, 249–254.

160. Marszatt, M.P., Markuszewski, M.J., and Kaliszan, R. (2006) *J. Pharm. Biomed. Anal. 41*, 329–332.

161. Tian, K., Wang, Y., Chen, Y., Chen, X., and Hu, Z. (2007) *Talanta 72*, 587–593.

162. Corradini, D., Nicoletti, I., and Bonn, G.K. (2009) *Electrophoresis 30*, 1969–1976.

163. Maier, V., Horakova, J., Petr, J., Drahonovsky, D., and Sevcik, J. (2006) *J. Chromatogr. A 1103*, 337–343.

164. Su, H.-L., Kao, W.-C., Lin, K.-W., Lee, C.-y., and Hsieh, Y.-Z. (2010) *J. Chromatogr. A 1217*, 2973–2979.

165. Xu, Y.H., Fang, L.Y., and Wang, E.K. (2009) *Electrophoresis 30*, 365–371.

166. Xu, Y.H., Jiang, H., and Wang, E.K. (2007) *Electrophoresis 28*, 4597–4605.

167. Xu, Y., Li, J., and Wang, E. (2008) *Electrophoresis 29*, 1852–1858.

168. Xu, Y.H., Li, J., and Wang, E.K. (2008) *J. Chromatogr. A 1207*, 175–180.

169. Zen, H.-L., Shen, H., Nakagama, T., and Uchiyama, K. (2007) *Electrophoresis 28*, 4590–4596.

170. Tran, C.D. and Mejac, I. (2008) *J. Chromatogr. A 1204*, 204–209.

171. Göttlicher, B. and Bächmann, K. (1997) *J. Chromatogr. A 780*, 63–73.

172. Nilsson, C. and Nilsson, S. (2006) *Electrophoresis 27*, 76–83.

173. Nilsson, C., Birnbaum, S., and Nilsson, S. (2007) *J. Chromatogr. A 1168*, 212–224.

174. Wallingford, R.A. and Ewing, A.G. (1989) *Adv. Chromatogr. 29*, 1–76.

175. Fan, D.H., Yuan, S.W., and Shen, Y.M. (2010) *Colloids Surf. B 75*, 608–611.

176. Noh, H.B., Lee, K.S., Lim, B.S., Kim, S.J., and Shim, Y.B. (2010) *Electrophoresis 31*, 3053–3060.

177. Liu, X.J. and Gomez, F. (2009) *Anal. Bioanal. Chem. 393*, 615–621.

178. Makamba, H., Huang, J.W., Chen, H.H., and Chen, S.H. (2008) *Electrophoresis 29*, 2458–2465.

179. Shiddiky, M.J. and Shim, Y.B. (2007) *Anal. Chem. 79*, 3724–3733.

180. Wang, W., Zhao, L., Zhou, F., Zhu, J.J., and Zhang, J.R. (2007) *Talanta 73*, 534–539.

181. Wang, W., Zhao, L., Zhang, J.R., Wang, X.M., Zhu, J.J., and Chen, H.Y. (2006) *J. Chromatogr. A 1136*, 111–117.

182. Wang, A.J., Xu, J.J., and Chen, H.Y. (2006) *J. Chromatogr. A 1107*, 257–264.
183. Tabuchi, M., Katsuyama, Y., Nogami, K., Nagata, H., Wakuda, K., Fujimoto, M., Nagasaki, Y., Yoshikawa, K., Kataoka, K., and Baba, Y. (2005) *Lab Chip 5*, 199–204.
184. Ohno, T., Matsuda, T., Suzuki, H., and Fujimoto, M. (2006) *Adv. Powder Technol. 17*, 167–179.
185. Mark, J.E., Jiang, C.-Y., and Tang, M.-Y. (1984) *Macromolecules 17*, 2613–2616.
186. Roman, G.T., Hlaus, T., Bass, K.J., Seelhammer, T.G., and Culbertson, C.T. (2005) *Anal. Chem. 77*, 1414–1422.
187. Qiu, J.-D., Wang, L., Liang, R.-P., and Wang, J.-W. (2009) *Electrophoresis 30*, 3472–3479.
188. Pumera, M., Wang, J., Grushka, W., and Polsky, R. (2001) *Anal. Chem. 73*, 5625–5628.
189. Fan, D.H., Yuan, S.W., and Shen, Y.M. (2010) *Colloids Surf. B Biointerfaces 75*, 608–611.
190. Wang, A.J., Xu, J.J., Zhang, W., and Chen, H.Y. (2006) *Talanta 69*, 210–215.
191. Wang, W., Zhao, L., Zhoyu, F., Zhu, J.-J., and Zhang, J.-R. (2007) *Talanta 73*, 534–539.
192. Noh, H.B., Lee, K.S., Lim, B.S., Kim, S.J., and Shim, Y.B. (2010) *Electrophoresis 31*, 3053–3060.
193. Lin, Y.W., Huang, M.J., and Chang, H.T. (2003) *J. Chromatogr. A 1014*, 47–55.
194. Lin, Y.W. and Chang, H.T. (2005) *J. Chromatogr. A 1073*, 191–199.
195. Tabuchi, M., Ueda, M., Kaji, N., Yamasaki, Y., Nagasaki, Y., Yoshikawa, K., Kataoka, K., and Baba, Y. (2004) *Nat. Biotechnol. 22*, 337–340.
196. Tabuchi, M., Katsuyama, Y., Nogami, K., Nagata, H., Wakuda, K., Fujimoto, M., Nagasaki, Y., Yoshikawa, K., Kataoka, K., and Baba, Y. (2005) *Lab Chip 5*, 199–204.
197. Ohlsson, P., Ordeig, O., Nilsson, C., Harwigsson, I., Kutter, J.P., and Nilsson, S. (2010) *Electrophoresis 31*, 3696–3702.
198. Pamme, N. (2006) *Lab Chip 6*, 24–38.
199. Ambrosi, A., Guix, A.M., and Merkoci, A. (2011) *Electrophoresis 32*, 861–869.
200. Tnnico, Y.H. and Remcho, V.T. (2010) *Electrophoresis 31*, 2548–2557.
201. Doyle, P.S., Bibette, J., Bancaud, A., and Viovy, J.L. (2002) *Science 295*, 2237–2237.
202. Moliner-Martinez, Y., Cárdenas, S., Simonet, B.M., and Valcarcel, M. (2009) *Electrophoresis 30*, 169–175.
203. Fernández-Abedul, M.T. and Costa-García, A. (2008) *Anal. Bioanal. Chem. 390*, 293–298.
204. Trojanowicz, M. (2006) *Trends Anal. Chem. 25*, 480–489.
205. Wang, Z., Luo, G., Chen, J., Xiao, S., and Wang, Y. (2003) *Electrophoresis 24*, 4181–4188.
206. Wiles, P.G. and Abrahamson, J. (1978) *Carbon 16*, 341–349.
207. Iijima, S. (1991) *Nature 354*, 56–58.
208. Cao, J., Dun, W.L., and Qu, H.B. (2011) *Electrophoresis 32*, 408–413.
209. Chen, J.L. and Hsieh, K.H. (2010) *Electrophoresis 31*, 3937–3948.
210. Weng, X.X., Bi, H.Y., Liu, B.H., and Kong, J.L. (2006) *Electrophoresis 27*, 3129–3135.
211. Nieto de Castro, C.A., Lourenco, M.J.V., Ribeiro, A.P.C., Langa, E., Vieira, S.I.C., Goodrich, P., and Hardacre, C. (2010) *J. Chem. Eng. Data 55*, 653–661.
212. Choi, S.U.S., Zhang, Z.G., and Keblinski, P. (1995) Siginer, D.A. and Wang, H.P. (eds), *Developments and Applications of Non-Newtonian Flows*, ASME, New York, FED-Vol. 231/MD-Vol. 66 pp. 99–105.
213. de Dios, A.S. and Díaz-García, M.E. (2010) *Anal. Chim. Acta 666*, 1–22.
214. Vázquez, M. and Paull, B. (2010) *Anal. Chim. Acta 668*, 100.
215. Yang, W.C., Sun, X.H., Pan, T., and Woolley, A.T. (2008) *Electrophoresis 29*, 3429–3435.
216. Cakal, C., Ferrance, J.P., Landers, J.P., and Caglar, P. (2010) *Anal. Bioanal. Chem. 398*, 1909–1917.
217. Sun, X., Weichun, Y., Yanli, G., and Woolley, A.T. (2009) *Lab Chip 9*, 949–953.
218. Shi, Y.N. and Anderson, R.C. (2003) *Electrophoresis 24*, 3371–3377.
219. Righetti, P.G., Gelfi, C., Verzola, B., and Castelletti, L. (2001) *Electrophoresis 22*, 603–611.
220. Dang, F., Zhang, L., Hagiwara, H., Mishina, Y., and Baba, Y. (2003) *Electrophoresis 24*, 714–721.
221. Tanaka, N., Tanigawa, T., Hosoya, K., Kimata, K., Araki, T., and Terabe, S. (1992) *Chem. Lett.* 959–962.
222. Boonsong, K., Caulum, M.M., Dressen, B.M., Chailapakul, O., Cropek, D.M., and Henry, C.S. (2008) *Electrophoresis 29*, 3128–3134.
223. Ballauff, M. and Borisov, O. (2006) *Curr. Opin. Colloid Interface Sci. 11*, 316–323.
224. Rühe, J., Ballauff, M., Biesalski, M., Dziezok, P., Gröhn, F., and Johannsmann, D. (2004) *Adv. Polym. Sci. 165*, 79–150.
225. Rucker, O. and Rühe, J. (1998) *Macromolecules 31*, 592–601.
226. Förster, S., Hermsdorf, N., Böttcher, C., and Lindner, P. (2002) *Macromolecules 35*, 4096–4105.
227. Houbenov, N., Minko, S., and Stamm, M. (2003) *Macromolecules 36*, 5897–5901.
228. Hales, K. and Pochan, D.J. (2006) *Curr. Opin. Colloid Interface Sci. 11*, 330–336.
229. Wang, Y.-C., Choi, M.H., and Han, J. (2004) *Anal. Chem. 76*, 4426–4431.
230. Shah, J.J., Geist, J., Locascio, L.E., Gaitan, M., Rao, M.V., and Vreeland, W.N. (2006) *Electrophoresis 27*, 3788–3796.
231. Woolley, A.T. and Mathies, R.A. (1995) *Anal. Chem. 67*, 3676–3680.
232. Munro, N.J., Huhmer, A.F.R., and Landers, J.P. (2001) *Anal. Chem. 73*, 1784–1794.
233. Lee, J.H., Kopeckova, P., Zhang, J., Kopecek, J., and Andrade, J.D. (1988) *Polym. Mater. Sci. Eng. 59*, 234.
234. Halperin, A. and Leckband, D.E. (2000) *C.R. Acad. Sci. Paris (Ser. IV Phys. Astrophys.) 1*, 1171–1178.

235. Kitagawa, F., Kubota, K., Sueyoshi, K., and Otsuka, K. (2010) *J. Pharm. Biomed. Anal. 53*, 1272–1277.

236. Zhang, Z.W., Feng, X.J., Luo, Q.M., and Liu, B.F. (2009) *Electrophoresis 30*, 3174–3180.

237. Miyaki, K., Zeng, H.-L., Nakagama, T., and Uchiyama, K. (2007) *J. Chromatogr. A 1166*, 201–206.

238. Feng, J.-J., Wang, A.-A., Fan, J., Xu, J.-J., and Chen, H.-Y. (2010) *Anal. Chim. Acta 658*, 75–80.

239. Kolb, H.C., Finn, M.G., and Sharpless, K.B. (2001) *Angew. Chem. Int. Ed. 40*, 2004–2021.

240. Prakash, S., Long, T.M., Selby, J.C., Moore, J.S., and Shannon, M.A. (2007) *Anal. Chem. 79*, 1661–1667.

241. Zhang, Z.W., Feng, X.J., Xu, F., Liu, X., and Liu, B.F. (2010) *Electrophoresis 31*, 3129–3136.

242. Henry, A.C., Tutt, T.J., Galloway, M., Davidson, Y.Y., McWhorter, C.S., Soper, S.A., and McCarley, R.L. (2000) *Anal. Chem. 72*, 5331–5337.

243. Llopis, S.L., Osiri, J., and Soper, S.A. (2007) *Electrophoresis 28*, 984–993.

244. Slentz, B.W., Penner, N.A., and Regnier, F.W. (2002) *J. Chromatogr. A 948*, 225–233.

245. Liu, J., Pan, T., Woolley, A.T., and Lee, M.L. (2004) *Anal. Chem. 76*, 6948–6955.

246. Sun, X., Liu, J., and Lee, M.L. (2008) *Anal. Chem. 80*, 856–863.

247. Xu, L. and Sun, Y. (2008) *Electrophoresis 29*, 880–888.

248. Wu, D., Zhao, B., Dai, Z., Qin, J., and Lin, B. (2006) *Lab Chip 6*, 942–947.

249. Sun, X., Liu, J., and Lee, M.L. (2008) *Electrophoresis 29*, 2760–2767.

250. Pan, T., Fiorini, G.S., Chiu, D.T., and Woolley, A.T. (2007) *Electrophoresis 28*, 2904–2911.

251. Xiao, D., Zhang, H., and Wirth, M. (2002) *Langmuir 18*, 9971–9976.

252. Xiao, D., Le, T.V., and Wirth, M.J. (2004) *Anal. Chem. 76*, 2055–2061.

253. Hagg, M.B. (2000) *J. Membr. Sci. 170*, 173–190.

7

CAPILLARY ELECTROPHORETIC REACTOR AND MICROCHIP CAPILLARY ELECTROPHORETIC REACTOR: DISSOCIATION KINETIC ANALYSIS METHOD FOR "COMPLEXES" USING CAPILLARY ELECTROPHORETIC SEPARATION PROCESS

TORU TAKAHASHI AND NOBUHIKO IKI

7.1 INTRODUCTION

Unlike organic compounds connected with a strong covalent bond, "complexes" formed by bonding or association involving two or more components through weaker bonds or interactions dissociate into a nonequilibrium state. The properties of the complexes in the nonequilibrium state, that is, kinetics, give important information on the "dynamics" of the complexes, which cannot be presumed from the equilibrium information. In many cases, the stability of metal complexes is evaluated from their thermodynamic properties, such as stability constants. Little attention had been paid to their "kinetic" properties. However, in actual chemical systems in which metal complexes participate, it is frequently observed that free ligands and metal ions are steadily separated from the vicinity of metal complexes. Such situations are frequently seen in the case for "functionalized" metal complexes. For example, the kinetic stability, i.e. inertness, is a key factor for the successful application of targeting metal complexes for nuclear magnetic resonance diagnostic imaging. These complexes travel alone in the human body with a fairly long lifetime before excretion. In such a situation, the metal complexes are steadily subjected to exposure to solvolytic dissociation reactions. Evaluation of the dissociation kinetic properties of metal complexes is therefore necessary for the design and development of analytical or diagnostic methods employing them in a "nonequilibrium" environment.

There is no question as to the importance of biomolecular interaction analysis, which is one of the molecular biological approaches to the understanding of the life phenomenon. As for the research trends in biomolecular interaction analysis, structural or thermodynamic approaches are the mainstream. Considering the dynamics of biomolecular complexes is, however, indispensable in understanding their biological meaning because biomolecules are constantly in flux; the temporary interactions between biomolecules or their complexes with noncovalent bonding play important roles in the vital activity in the living cells [1]. In addition, kinetics provides intrinsic information about reaction mechanisms [2].

There have not been many studies on the kinetic analysis of "complexes" in a broad sense, including metal complexes and biomolecular complexes, in spite of its widely recognized importance. This is partly due to the lack of the convenient methodologies for investigating the kinetics of "complexes." Basically, direct measurement of their rate constants is technically challenging because of the extreme difficulty in creating a reaction system, where the reverse reaction to a reaction of interest does not take place, in homogeneous solution. In many cases, therefore, it was necessary to measure them indirectly after repeating many complex operations in the equilibrium state [3,4].

Capillary Electrophoresis and Microchip Capillary Electrophoresis: Principles, Applications, and Limitations, First Edition.
Edited by Carlos D. García, Karin Y. Chumbimuni-Torres, and Emanuel Carrilho.

In contrast to formation kinetics, it is difficult to determine the dissociation rate constant of complex species because the measurement requires a rapid decrease in the concentration of at least one of the components. Such a situation has only been realized in indirect fashions using chemical masking or immobilization of a component. For example, when measuring the dissociation rate of a metal complex, researchers often employ ligand substitution and use a competing ligand to mask the free metal ion. Surface plasmon resonance (SPR) [5] and the quartz crystal microbalance (QCM) [6] are useful in separating a component from a biomolecular complex to observe the dissociation kinetics—a process that sometimes suffers from artifacts caused by mass transfer, nonspecific binding, or the nonnative stability of the complex at the solid–liquid interface [7].

Recently, a review was published pointing out that CE is a promising technique for the investigation of molecular interactions, especially in obtaining kinetic information [8]. We also had exemplified the advantages of CE for kinetic studies in a report in which a unique approach to the dissociation kinetic measurement of metal complexes via a capillary electrophoretic reactor (CER) [9] was described before the review was issued. CER is based on the on-capillary dissociation reaction profile of metal complexes in the precapillary derivatization CE employing an electrophoretic buffer solution without chelating reagent [9,10]. In this CE system, metal complexes, which migrate in their isolated bands along a capillary from those of the free ligand and metal ions, are exposed to an overwhelming force causing dissociation by CE resolution because of the absence of the free ligand in the electrophoretic buffer solution. One can obtain the dissociation rate constant (k_d) of the metal complex by analyzing the first-order decay dissociation degree–time profile directly, readily, and accurately. An approach using CE-based dissociation kinetic analysis methods [9–11] is the only one that permits the direct measurement of the k_d of a metal complex in homogeneous solution because CE is a one-of-a-kind method that can create the driving force for the dissociation reaction of metal complex in homogeneous solution. Another advantage of CER is that the estimation of the dissociation reaction profile of the complex is accurate because the decrease in the peak height of the complex with increase in the migration time is employed for the estimation of the degree of decomposition of the complex.

In this chapter, direct measurements of dissociation rates of "complexes" are described. CER system is successfully exemplified by evaluation of the dissociation rate constants of "complexes" including metal complexes and biomolecular complexes [12]. CER proposed here also serves as an experimental approach for such moderately slow decay kinetics with a half-life of "min." CER-related methods for faster dissociation reactions with a half-life of "sec" and slower dissociation reactions with a half-life of "hour," microchip capillary electrophoretic reactor (μCER) [11] and ligand substitution mode capillary electrophoretic reactor (LS-CER) [13], are also described.

7.2 BASIC CONCEPT OF CER

In measuring the dissociation rate constants of metal complexes, typically chemical methods relying on substitution reactions with protons, metal ions, and other ligands have been employed. However, there is a critical restriction in measuring them by such conventional batch reaction techniques since spectrophotometric or fluorometric detection methods are used. The dissociation reaction of the complexes cannot be monitored without drastic changes in the spectroscopic characteristics involved in the reaction. On the contrary, because the reactant (metal complex) and products (metal ion and free ligand) are mutually separated, the characteristic spectroscopic changes become unnecessary if a highly efficient separation method, capillary electrophoresis (CE), is used for measuring the dissociation rate constants of metal complexes. In addition, the dissociation rate constants of metal complexes are measured indirectly through substitution reactions in the above-mentioned batch reaction method. Therefore, another merit of using CE for the dissociation kinetic analysis of metal complexes is the direct monitoring of the dissociation reactions of the complexes. Thus, CER [9] revolutionized the measurement of the dissociation rate constants of metal complexes since it involves direct measurements. Figure 7.1 shows a schematic illustration of the basic concept of CER.

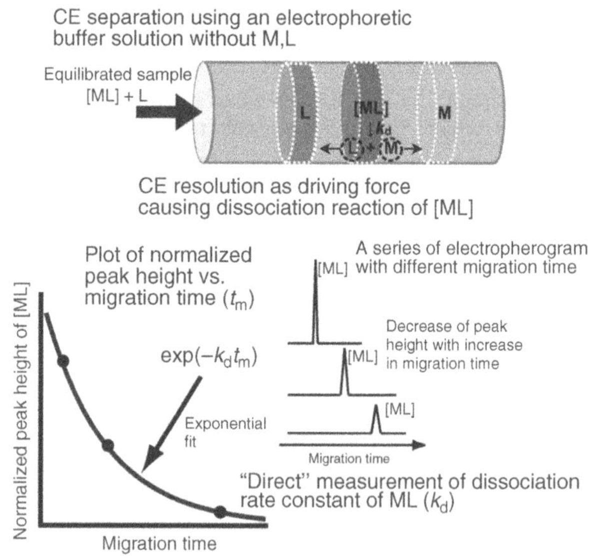

FIGURE 7.1 Schematic illustration of the basic concept of dissociation kinetic analysis using CER. M, metal ion; L, free ligand; ML, metal complex; k_d, dissociation rate constant of ML.

The basic idea of CER is to use the CE separation process as a kind of reactor for the dissociation reaction of metal complexes [10]. A steep reduction in the concentration of free ligand around the metal complex brought about by CE separation serves as the driving force of the dissociation of the metal complex. In CER, an electrophoretic buffer solution with no addition of ligand is employed [14,15]. A preliminary equilibrated sample solution of the metal complex, prepared from metal ion and excess ligand, is introduced into a capillary. Due to the difference in electrophoretic mobilities, the metal complex and excess ligand in the sample solution injected as a single band begin to migrate as two independent bands following the start of CE separation by applying a voltage. In a fairly early stage of the CE separation, the complex does not begin to dissociate because the two bands of the metal complex and the excess ligand are overlapped and the complexation equilibrium is maintained. However, as the separation advances and these two bands become mutually separated, the metal complex is exposed to an overwhelming force to undergo a solvolytic dissociation reaction according to Equation 7.1 since the free ligand is separated from the vicinity of the metal complex:

$$\mathrm{ML} \rightarrow \mathrm{M} + \mathrm{L} \tag{7.1}$$

where solvent molecules are omitted, and ML, M, and L denote the metal complex, metal ion, and free ligand, respectively. In CER, the plug length of the sample solution injected into the capillary must be short enough with respect to the total length of the capillary because it is desirable that the overlapping period of the metal complex band and the excess ligand band is as short as possible. If the difference between the electrophoretic mobility of the metal complex and free ligand is not too small, the period when these bands were overlapped can be considered to be negligibly small compared with the total migration time. Therefore, it can be considered that the migration time of the metal complex is equal to the reaction time of the on-capillary dissociation reaction of the metal complex in CER. Since metal ion and free ligand, which are the products of dissociation reaction, are immediately separated from band of metal complex due to the difference in electrophoretic mobility, backward reaction, that is, the complexation reaction, does not occur in CER. The solvolytic dissociation reaction of the metal complex in the CER follows first-order kinetics according to Equation 7.1. In order to determine the dissociation rate constant of the metal complex, it is necessary to analyze the decrease in the residual ratio of the complex with the increase in reaction time as first-order decay reaction. Hence one can obtain the dissociation rate constant of the metal complex by analyzing the reduction of the residual ratio of the metal complex via the above-described on-capillary dissociation reaction with the increase in migration time as a first-order decay reaction. Its first-order decay profile, the decrease in the absorbance of the peak height of the metal complex with increase in migration time, is obtained by the execution of several CE experiments with a variety of migration times. The variation of migration time was obtained by changing the value of the applied voltage and the capillary length. The first-order decay profile is obtained as the plot of normalized peak height versus migration time. Thus, the dissociation rate constant of the metal complex is obtained by analyzing the plot as a first-order reaction.

7.3 DISSOCIATION KINETIC ANALYSIS OF METAL COMPLEXES USING A CER

In the CE separation process employing an electrophoretic buffer solution with no addition of ligand, the metal complex and its constituents (excess ligand) in a preliminarily equilibrated sample solution begin to move as an independent band according to the difference in their electrophoretic mobilities in the capillary at the same time as the beginning of migration. As a result, the concentration of free ligand in the band of ML in the capillary decreases greatly during the CE separation process when an electrophoretic buffer containing no constituents of the complex was used [14,15]. This separation-induced steep concentration reduction of free ligand around the metal complex is the driving force of the on-capillary dissociation reaction of the complex in the CER. ML and free ligand were present in the injected sample solution. Just after the electrophoresis is started, the zone overlapping between ML and free ligand possibly retards the complex dissociation. In fact, this likelihood is negligible because the zone overlap time is calculated as $<1\%$ of the migration time in all cases. Then, the reaction time of the on-capillary dissociation reaction can be considered to be almost equal to the migration time. The peak height of ML decreases as the migration time increases since the degree of the decrease in ML concentration through the on-capillary dissociation reaction is proportional to the degree of decrease in the peak height of ML.

The dissociation reaction of ML during the CE separation process follows first-order kinetics. The reaction scheme is as follows:

$$\mathrm{ML} \xrightarrow{k_d} \mathrm{M} + \mathrm{L} \tag{7.2}$$

The basic idea of dissociation kinetic analysis using the CER is to measure the residue of the analyte complex with increasing migration time, t_m. In kinetic analysis using a CER, the peak height signals were employed because the peak area data can give unreliable results due to asymmetric peak profiles arising from the on-capillary dissociation reaction. The asymmetrically shaped ML peak is not composed of the complex alone, that is, the leading region of the peak originates in the products, such as free ligand, generated by the above on-capillary dissociation reaction of

Equation 7.2. It is difficult to identify clearly the boundary between the region of the complex and that of the free ligand derived from the on-capillary dissociation reaction. Therefore, the degree of dissociation of the complexes is estimated not from the decrease in the peak area of the complex but from that of the peak height absorbance of the complex as a function of reaction time for accurate estimation. The degree of dissociation–time profiles for the on-capillary dissociation of metal complexes were obtained from several CE experiments with different migration times. The double standardization method [9] is employed for accurate measurement of the dissociation process of ML. Two types of standards employed, internal and external standards, are kinetically inert, that is, their amounts does not change during the CE experiment. The execution of several CE experimental runs is necessary to obtain a series of electropherograms with different migration times. Then an internal standard is added to correct the difference in the injection volumes of the individual experimental runs. On the other hand, an external standard is used for an accurate estimation of the remaining concentration of ML that survived during migration in the capillary because the decrease in the peak height of the complex arises from both the on-capillary dissociation reaction and the zone broadening caused by diffusion. If the decrease in the peak height by zone broadening is not corrected, the decrease in the peak height by the on-capillary dissociation reaction is overestimated. Hence it is necessary to correct the decrease in the peak height by the zone broadening caused by diffusion of the solute. The degree of zone broadening by diffusion depends on the migration time and the molecular size of the solute [9,16]. An external standard that had a migration time and molecular weight close to those of ML was therefore chosen. The electropherograms of the external standard were acquired under the same CE conditions as the electropherograms of the complexes that paired with each of them. The normalized peak height signals of ML with the internal standard and that of external standard in each electropherogram are H_{ML} and H_{ext}, respectively. In this case, the residual ratio of ML can be estimated using H_{ML} and H_{ext}:

$$[ML]/[ML]_0 = AH_{ML}/H_{ext} \tag{7.3}$$

where $[ML]_0$ is the initial concentration of ML and A is the proportionality constant [11]. As the dissociation reaction follows first-order kinetics, the rate law is given by

$$d[ML]/dt = k_d[ML] \tag{7.4}$$

Integrating Equation 7.4 from $t = 0$ to t_m yields Equation 7.5, and then Equation 7.6:

$$\ln([ML]/[ML]_0) = -k_d t_m \tag{7.5}$$

$$[ML]/[ML]_0 = \exp(-k_d t_m) \tag{7.6}$$

Introducing Equation 7.3 into Equation 7.6, we obtain

$$H_{ML}/H_{ext} = \alpha \exp(-k_d t_m) \tag{7.7}$$

where α is $1/A$. We can obtain k_d by fitting Equation 7.7 using the H_{ML}/H_{ext} data at various t_m. Thus, k_d is obtained by analyzing the degree of dissociation–time profile of ML as a plot of the peak height signals of ML normalized with that of external standard (H_{ML}/H_{ext}) against the migration time.

7.3.1 Determination of the Rate Constants of Dissociation of 1:2 Complexes of Al^{3+} and Ga^{3+} with an Azo Dye Ligand 2,2′-Dihydroxyazobenzene-5,5′-Disulfonate in a CER

The solvolytic dissociation rate constants of 1:2 complexes ($[ML_2]^{5-}$) of Al^{3+} ($[AlL_2]^{5-}$) and Ga^{3+} ($[GaL_2]^{5}$) with an azo dye ligand, 2,2′-dihydroxyazobenzene-5,5′-disulfonate (DHABS, H_2L^{2-}), have been evaluated with a CER system [9].

Accurate measurements of the dissociation processes of $[ML_2]^{5-}$ are realized using the double standardization technique. Methyl Orange (MO) is added as an internal standard for correcting for the injection volume, while an external standard, 1:2 complexes of Co^{3+} with DHABS ($[Co^{III}L_2]^{5-}$), is used for the estimation of the concentration of remaining $[ML_2]^{5-}$ that survived during migration in the capillary. In general, a Co^{3+} ion having a low-spin d^6 electron configuration forms kinetically inert complexes [17]. Actually, even in the presence of 10^{-4} mol dm^{-3} EDTA, no evidence was seen for the replacement of the $[Co^{III}L_2]^{5-}$ complex with EDTA at pH 8.0 within 1 h (Iki, unpublished results, 1993). The $[Co^{III}L_2]^{5-}$ complex is suitable as an external standard because it has an electrophoretic mobility value close to those of DHABS complexes of trivalent metal ions studied here, such as Al(III), Ga(III), and In(III).

Typical electropherograms for each complex are shown in Figure 7.2. As can be seen, the t_m values of $[Co^{III}L_2]^{5-}$, $[AlL_2]^{5-}$, and $[GaL_2]^{5-}$ are almost identical. In the case of the Ga system, the leading peak pattern of $[GaL_2]^{5-}$ suggests that the partial dissociation of the $[GaL_2]^{5-}$ complex; the products of the reaction, 1:1 complex, and free ligand, move faster on the electroosmotic flow (EOF) than the $[GaL_2]^{5-}$ ion does. Even at $t_m = 4$ min, the residual ratio of $[GaL_2]^{5-}$ is calculated to be only 36%, which is about 1/e of the initial concentration.

To reduce the degree of dissociation of $[GaL_2]^{5-}$, the reaction time, that is, t_m, should be shortened, driving the sample zone toward the negative side by siphon action before the electrophoresis with a sacrifice of peak sharpness.

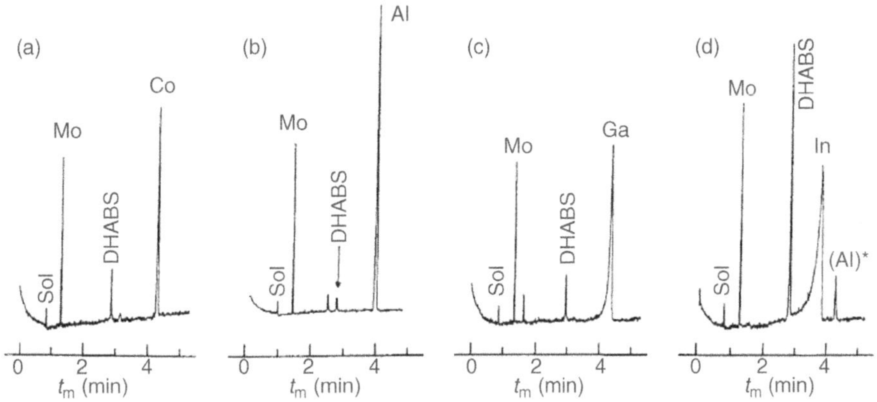

FIGURE 7.2 Typical electropherograms for the $[ML_2]^{5-}$ systems: (a) Co^{3+}; (b) Al^{3+}; (c) Ga^{3+}; (d) In^{3+}. Each solution commonly contains 5.0×10^{-5} mol dm^{-3} MO and 0.02 mol dm^{-3} Tris (pH 8.0). Sol, solvent peak; MO, Methyl Orange; DHABS, free DHABS; In, 1:1 complex of In^{3+} with DHABS and the product species; Al*, $[AlL_2]^{5-}$ from contaminant. Electrophoretic buffer: $[H_3PO_4]_T$ 0.02 mol dm^{-3}, pH 7.0 adjusted with NaOH. Electrophoresis at $T = 303$ K, $i = 20$ μA, $V = 16.2$–16.4 kV, $l = 15$ cm, and $L = 60$ cm. Reprinted with permission from Ref. 9, copyright © 2000 American Chemical Society.

As shown in Figure 7.3, the peaks for MO, $[Co^{III}L_2]^{5-}$, and $[GaL_2]^{5-}$ are slightly broadened; in fact, the leading peak is rather reduced. For the extreme case with the 1:2 complexes of In^{3+} ($[InL_2]^{5-}$) whose dissociation is too fast on the CE time scale, no peaks were found for this complex species even at small t_m values from 1.2 to 2.8 min.

It should be noted that in the $[InL_2]^{5-}$ system the first large peak at 3.6 min is due to the fast-moving products, probably 1:1 complex, and free ligand, and the second one at 4.3 min is due to the $[AlL_2]^{5-}$ species as a contaminant in the In^{3+} solution. In this case, the residence time window in the capillary for the Al and Ga complexes was manipulated by controlling the magnitude of EOF in a capillary by changing the pH of a capillary treatment solution without changing the applied voltage [9]. Some examples are shown in Figure 7.4 and Figure 7.5. The leading peak shape became more significant for $[GaL_2]^{5-}$ at a longer t_m of 6.64 min (Figure 7.4), reflecting an increase in dissociation. At $t_m = 25$ min (Figure 7.5), the $[AlL_2]^{5-}$ peak height is significantly lowered compared with that shown in Figure 7.2a and b, which indicates that ~50% decay of the complex was achieved.

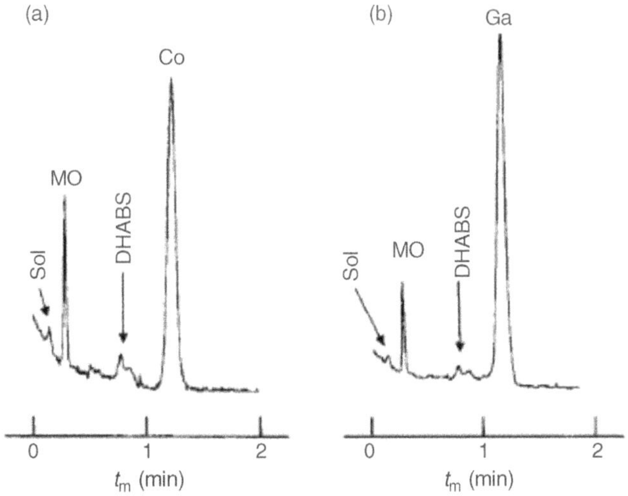

FIGURE 7.3 Typical electropherograms for (a) $[Co^{III}L_2]^{5-}$ and (b) $[GaL_2]^{5-}$ complexes at shorter t_m. The sample zone was driven toward the negative end for 20 min by hydrodynamic action. No conditioning buffer was used in this experiment. Other conditions as in Figure 7.2. Reprinted with permission from Ref. 9, copyright © 2000 American Chemical Society.

In Figure 7.6, where the linear dependence of the logarithmic residual ratio values on t_m is shown, first-order decay kinetics are demonstrated for both Al and Ga.

For the evaluation of the decay rate constants, k_d, Equation 7.8 was derived from Equation 7.7:

$$H_{ML2}/H_{CoL2} = \alpha \exp(-k_d t_m) \tag{7.8}$$

where H_{ML2} and H_{CoL2} are the normalized peak height signals of $[ML_2]^{5-}$ with MO and that of $[Co^{III}L_2]^{5-}$ in each electropherogram, respectively.

Then, k_d values of the complexes were calculated by fitting Equation 7.8 using the data of the plots to be 4.9×10^{-4} s^{-1} for the $[AlL_2]^{5-}$ complex and 3.7×10^{-3} s^{-1} for the $[GaL_2]^{5-}$ complex at 303 K. Just after the electrophoresis is started, the zone overlapping between $[ML_2]^{5-}$ and free DHABS possibly causes retardation of the complex dissociation. In fact, this likelihood is negligible because each zone length is only 1.6 mm and the zone

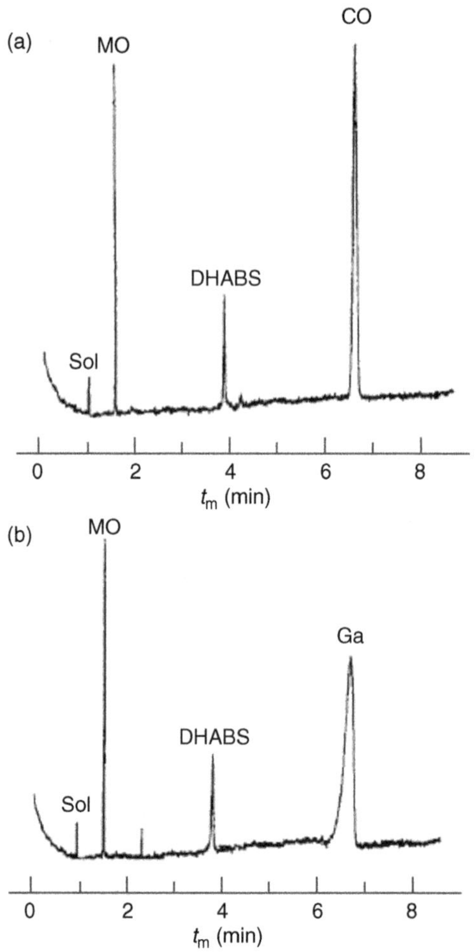

FIGURE 7.4 Typical electropherograms for (a) $[Co^{III}L_2]^{5-}$ and (b) $[GaL_2]^{5-}$ complexes at longer t_m. Other conditions as in Figure 7.2. Reprinted with permission from Ref. 9, copyright © 2000 American Chemical Society.

overlap time is calculated as just 5.6 s for all cases. In this CER system, a decrease in the peak height absorbance of $[ML_2]^{5-}$ is traced and its decay curve is readily obtained. The fact that the high-resolution CE separates the products as soon as they are obtained is very beneficial, in particular when the absorption spectra of the products and the reactant severely overlap. In the DHABS complex systems where the spectrally overlapping species, $[ML_2]^{5-}$, 1:1 complex, and free ligand, are present, the batch reaction system is hardly practical for cross-checking purposes to validate the above k_d values. It should be stressed that the simplicity of CER liberates one from the complicated and tedious task of analyzing time–absorbance data obtained in ordinary solution kinetic experiments.

It has been successfully demonstrated that CE serves as a chemical reactor for measurements of the dissociation kinetics of metal complexes. Also, attractive features of CE such as the plug-flow profile, no interactions with a secondary phase, and the parallel scheme of dissociation and separation allow the simple treatment of data. It is emphasized that

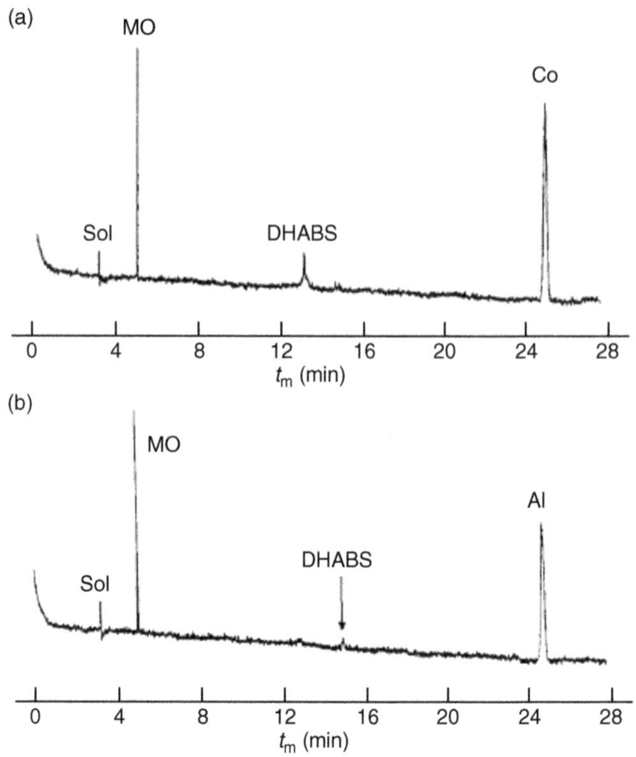

FIGURE 7.5 Typical electropherograms for (a) $[Co^{III}L_2]^{5-}$ and (b) $[AlL_2]^{5-}$ complexes at longer t_m. $l = 45$ cm and $L = 60$ cm. Other conditions as in Figure 7.2. Reprinted with permission from Ref. 9, copyright © 2000 American Chemical Society.

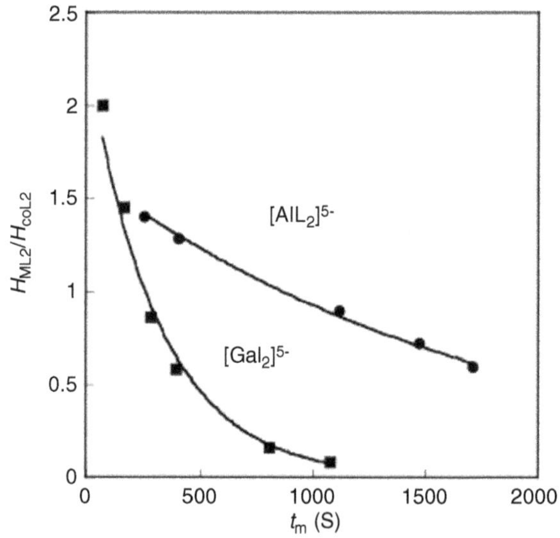

FIGURE 7.6 Reaction profiles obtained for $[AlL_2]^{5-}$ and $[GaL_2]^{5-}$ by CER. Solid lines show the approximation curve for the first-order dissociation reaction profile of each complex by Equation 7.8. Conditions as in Figure 7.5.

suitable t_m reference materials are available; the applications of the CER can actually be extended to a large number of broad-sense "complexes." From a practical viewpoint, the CER will provide useful information for design tactics of kinetically stable complexes for biological targeting use and for HPLC/HPCE methodologies, shedding light on the underlying chemistry of such metal complex functions.

7.4 EXPANDING THE SCOPE OF THE CER TO MEASUREMENTS OF FAST DISSOCIATION KINETICS WITH A HALF-LIFE FROM SECONDS TO DOZENS OF SECONDS: DISSOCIATION KINETIC ANALYSIS OF METAL COMPLEXES USING MICROCHIP CAPILLARY ELECTROPHORETIC REACTOR (μCER)

Recently, miniaturized chemical separation devices have attracted great interest for producing rapid and reliable methods for chemical and biological analyses. Microchip capillary electrophoresis (MCE) [18] has the advantage of reducing the separation time without a reduction in separation efficiency compared with conventional CE systems. From the standpoint of monitoring dissociation kinetic processes on the time scale from several seconds to dozens of seconds, CER is less ideal, since CE separation with general CE apparatus requires minutes or longer to complete. Microchip capillary electrophoretic reactor (μCER) is described that provides a novel method for tracing the dissociation kinetic processes of a metal complex and directly evaluating its dissociation kinetics. The basic idea of μCER is constructed from that of CER, but two attractive features of μCER, which arise from the application of MCE separation process as a chemical reactor for dissociation reactions of metal complexes, are demonstrated. First, the investigation of a relatively fast dissociation kinetic process, which occurs on the time scale from several seconds to dozens of seconds, is performed. This is beyond CER. Second, UV absorption linear imaging detection [19,20] employed with the μCER allows high-throughput data acquisition for dissociation kinetic analysis. In μCER analysis, one can determine the dissociation rate constant of a metal complex with a single MCE analysis. By contrast, repetition of several CE experiments is necessary in CER analysis.

μCER has been successfully demonstrated in the direct evaluation of the solvolytic dissociation rate constant of the complex of Ce^{3+} with a polyaminocarboxylic ligand, 8-amino-2-[(2-amino-5-methylphenoxy)methyl]-6-methoxyquinoline-*N,N,N′,N′*-tetraacetic acid (Quin2) [11]. Figure 7.7 shows the electropherograms extracted at intervals of 12 s from a series of 180 electropherograms obtained successively in a single MCE separation run for a sample containing the 1:1 complex of Ce^{3+} with Quin2 ($[Ce^{III}L]^-$) and 1,3-diaminobenzene (DB). In each electropherogram, two peaks, $[Ce^{III}L]^-$ and DB, are present.

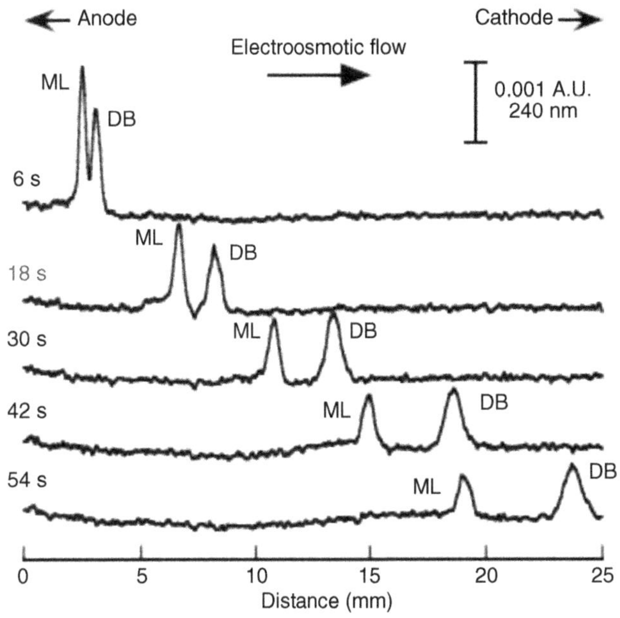

FIGURE 7.7 Typical electropherograms obtained successively in a single MCE separation run for $[Ce^{III}L]^-$. Detection wavelength, 240 nm; applied voltage, 60 V cm^{-1}; separation time, 60 s; electrophoretic buffer, 10 mM borate (pH 9.2); sample, 0.6 mM Quin-2, 0.6 mM Ce^{3+}, 10 mM borate, and 3 mM DB; ML, $[Ce^{III}L]^-$. Reprinted with permission from Ref. 11, copyright © 2005 Royal Society of Chemistry.

There is no independent peak for the free ligand because there is no addition of excess ligand to the sample. The internal standard, DB, is also utilized as a neutral marker. As the migration time is increased, both peaks move from the anode (left) to the cathode (right) and are separated from each other. It is observed that the peak height of $[Ce^{III}L]^-$ decreases as the migration time increases. Of course, the peak height of "inert" DB is slightly reduced owing to the zone broadening due to longitudinal diffusion and other reasons [21]. The extent of the peak height reduction of $[Ce^{III}L]^-$, however, is much greater than that of DB. This shows that the concentration of $[Ce^{III}L]^-$ has been decreasing as it has been exposed to the solvolytic dissociation reaction during the electrophoretic migration separation process: the MCE separation process can work as a reactor for measurement of the dissociation reaction kinetics of a metal complex. The visualization of the dynamic behavior of the on-channel decomposition reaction of the metal complex is achieved with successive acquisitions of UV absorption linear images of the separation channel. The solvolytic dissociation rate constant of $[Ce^{III}L]^-$ is derived from a series of electrophoretic data obtained during a single MCE analysis. The double standardization method [9] is employed for accurate measurement of the dissociation process of $[Ce^{III}L]^-$. The two types of standards employed, anthraquinone-2-sulfonate (AS) and DB, are both

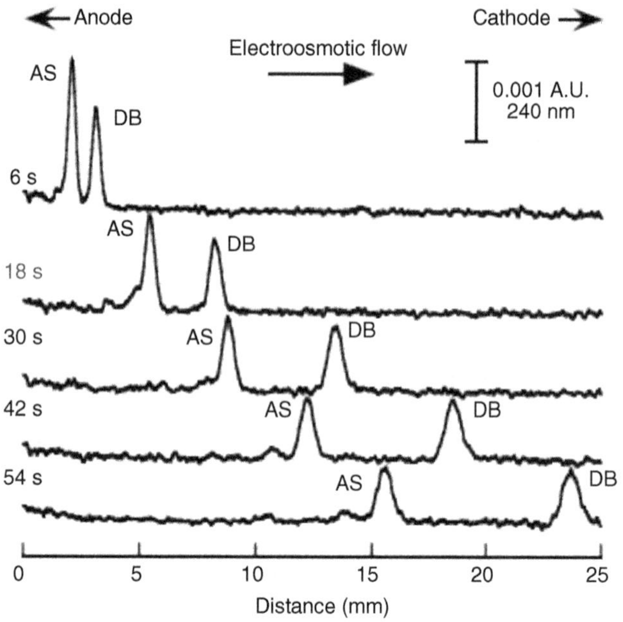

FIGURE 7.8 Typical electropherograms obtained successively in a single MCE separation run for the sample containing AS and DB. Conditions for MCE separation as in Figure 7.7. Sample: 3 mM AS, 10 mM borate, and 3 mM DB. Reprinted with permission from Ref. 11, copyright © 2005 Royal Society of Chemistry.

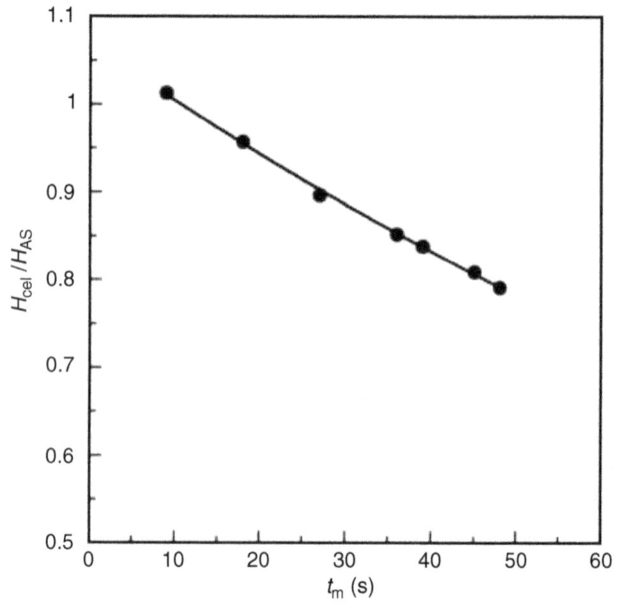

FIGURE 7.9 Reaction profile obtained for $[Ce^{III}L]^-$ by μCER. Solid lines show the approximation curve for the first-order dissociation reaction profile of each complex by Equation 7.9. Reprinted with permission from Ref. 11, copyright © 2005 Royal Society of Chemistry.

kinetically inert. DB is added as an internal standard for correcting the difference in injection volume between two samples, while an external standard, AS, which has an electrophoretic mobility value close to that of $[Ce^{III}L]^-$, is used for estimation of the concentration of remaining $[Ce^{III}L]^-$ that survived during migration in the separation channel. For the kinetic analysis of $[Ce^{III}L]^-$, the peak height signals were employed because the peak area data may give unreliable results caused by asymmetric peak profiles owing to products of the dissociation reaction, such as free ligand [9]. Figure 7.8 shows the extracted electropherograms obtained successively in a single MCE separation run for the sample containing AS and DB. The peak height signals of $[Ce^{III}L]^-$ and AS, normalized to that of DB in each electropherogram, are H_{CeL} and H_{AS}, respectively.

The basic idea with μCER is to measure the residue of the metal complex with increasing migration time, t_m, to obtain a first-order decay curve of dissociation. In this case, the residual ratio of $[Ce^{III}L]^-$ can be estimated using H_{CeL} and H_{AS} similarly to the dissociation kinetic analysis of metal complexes using CER. Then, Equation 7.9 is obtained from Equation 7.7:

$$H_{CeL}/H_{AS} = \alpha \exp(-k_d t_m) \tag{7.9}$$

Then we can obtain k_d by fitting Equation 7.9 using the H_{CeL}/H_{AS} data at various t_m.

Figure 7.9 shows the dependence of H_{CeL}/H_{AS} values on t_m. A simple first-order decay profile for $[Ce^{III}L]^-$ is observed, and the plots are fitted well to Equation 7.9. The dissociation reaction rate constant for $[Ce^{III}L]^-$, k_d, is successfully determined as $(5.97 \pm 0.72) \times 10^{-3}\,s^{-1}$ from five experiments. The k_d value estimated indirectly using the ligand substitution reaction with EDTA in a batch reaction system $(6.16 \pm 0.49) \times 10^{-3}\,s^{-1}$ is in good agreement with that obtained with μCER. This proves the validity of kinetic analysis with μCER.

It is stressed that one can acquire a set of data necessary for kinetic analysis during a single MCE run in this μCER system, whereas one must repeat several CE experiments with a variety of migration times to trace a decrease in the peak height absorbance of $[Ce^{III}L]^-$ and to obtain its decay curve in kinetic analysis with CER. Of course, it is much simpler compared with ordinary kinetic experiments with the requirement for complicated and tedious tasks.

More recently, μCER was successfully applied for the evaluation of the k_d value of an La^{3+} complex with another polyaminocarboxylic ligand with a half-life of seconds [22]. μCER is a new high-throughput methodology for evaluation of the dissociation kinetics rate constant of metal complexes on the time scale of several seconds. Analysis of the determination of the residue of a metal complex within a single MCE separation process as a function of migration time allows the dissociation kinetics of the metal complex to be estimated rapidly.

7.5 EXPANDING THE SCOPE OF CER TO THE MEASUREMENT OF SLOW DISSOCIATION KINETICS WITH A HALF-LIFE OF HOURS

The time range that CER can handle corresponds to the separation time window of CE. In the case of Al(III)– and Ga (III)–DHABS complexes, the half-lives are $t_{1/2} = 24$ and 3 min, respectively; both values are comparable to the time required for a typical CE run. To extend the CER toward the measurement of rapid dissociation with half-lives from seconds to dozens of seconds, MCE employing a miniaturized separation channel was used to realize a μCER. In contrast, the measurement of slow dissociation kinetics with $t_{1/2}$ values of several hours using a CER is not feasible in practice. First, to determine k_d, several CER runs for different t_m values are necessary, meaning that the measurement process would have to be carried out over several days. Second, during such a long CE separation, the peak height of the remaining complex will be affected by fluctuations in the electrophoretic mobility, electroosmotic mobility, and dissociation reaction rate itself, and by adsorption of the complex and/or component species. This would lead to poor reproducibility. Note that conventional techniques such as SPR also require considerable time and have poor reproducibility for similar reasons.

To reduce CER operating time to a reasonable value (several tens of minutes), we developed a new CER mode, the ligand substitution capillary electrophoretic reactor (LS-CER), by introducing a substitution reaction in a capillary to accelerate the slow dissociation. Kinetically stable Ti(IV) complexes with catechin (H_4L, Figure 7.10) and Tiron (H_2R^{2-}) were tested as specimens because one of the isotopes of titanium, ^{45}Ti, is a candidate for a positron emissive nucleus in positron emission tomography (PET) [23,24], for which it is highly desirable to develop a methodology for estimating the dissociation rate constant of the Ti(IV)–ligand complex [13].

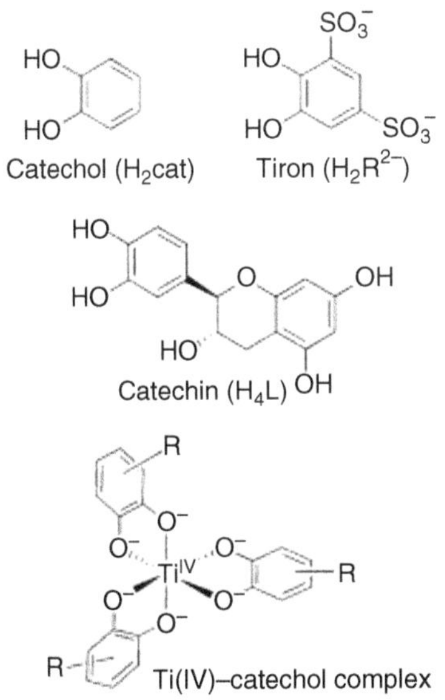

FIGURE 7.10 Structures of catechol ligands and a 1:3 complex with a Ti(IV) ion. R represents functional groups. Reprinted with permission from Ref. 1, copyright © 2009 American Chemical Society.

7.5.1 Principle of LS-CER

For the sake of clarity, we describe the case of the spontaneous dissociation of a 1:1 metal–ligand complex. The discussion should be applicable to the dissociation of molecular and biomolecular complexes with stoichiometries of 1:1, 1:*n*, and *m*:*n* (where *m* and *n* are integers). Given a solution of a 1:1 metal complex (ML) equilibrated with free metal ion (M) and free ligand (L), the complexation equilibrium is

$$ML \underset{k_f}{\overset{k_d}{\rightleftharpoons}} M + L \tag{7.10}$$

where k_d and k_f are dissociation and formation rate constants, respectively. In this system, the rates of formation and dissociation are balanced, and the concentration of ML is determined by the thermodynamic stability constant ($K = k_f/k_d$) of ML and concentrations [M] and [L] as given by

$$[ML] = K[M][L] \tag{7.11}$$

In a conventional batchwise experiment for the estimation of k_d, a competing ligand Y is added to the solution to mask the formation reaction of ML (i.e., the reverse of Equation 7.10). Y then reacts with free M to form the complex MY:

$$M + Y \rightarrow MY \tag{7.12}$$

Here, MY does not necessarily have 1:1 stoichiometry, but the rapid formation and high thermodynamic stability of MY are essential prerequisites to suppress the backward reaction of Equation 7.10. Since the reaction in Equation 7.12 is in competition with the backward reaction of Equation 7.10, the reaction rate of the former should be higher than that of the latter. To fulfill this requirement, a ligand to form MY more stable than ML is used in a large excess relative to L. The use of excess Y over ML is also desirable because of the pseudo-first-order condition. Under such a condition, however, the ligand Y concomitantly attacks ML, and ligand substitution (LS) (i.e., displacement) results in the dissociation of ML:

$$ML + Y \xrightarrow{k_d} MY + L \tag{7.13}$$

where k_Y is the second-order substitution reaction rate constant. Therefore, the LS reaction (Equation 7.13) competes with spontaneous dissociation (Equation 7.10) and the observed rate is

$$-\mathrm{d}[\mathrm{ML}]/\mathrm{d}t = k_{\mathrm{obs}}[\mathrm{ML}] \tag{7.14}$$

where k_{obs} is the apparent dissociation rate constant, given by

$$k_{\mathrm{obs}} = k_\mathrm{d} + k_\mathrm{Y}[\mathrm{Y}] \tag{7.15}$$

Therefore, estimation of a series of k_{obs} values by changing [Y] ought to give both k_d and k_Y values. This is a rather indirect method to obtain k_d, but it is the only method to measure k_d in homogeneous solution without relying on the immobilization of a component. Compared with the LS method, CER is more straightforward because free M and L are directly removed from the vicinity of ML by electrophoresis. Hence, only dissociation (Equation 7.16) occurs in the zone of ML and the reverse reaction does not take place:

$$\mathrm{ML} \xrightarrow{k_\mathrm{d}} \mathrm{M} + \mathrm{L} \tag{7.16}$$

If ML is detected after an elapsed time t_m, the concentration is

$$[\mathrm{ML}]/[\mathrm{ML}]_0 = \exp(-k_\mathrm{d} t_\mathrm{m}) \tag{7.17}$$

In a CE run, the migration time t_m is

$$t_\mathrm{m} = \frac{Ll}{V(\mu_{\mathrm{eo}} + \mu_{\mathrm{ML}})} \tag{7.18}$$

where L and l are the total and effective lengths of the capillary, respectively, V is the applied voltage, μ_{eo} is the electroosmotic mobility, and μ_{ML} is the electrophoretic mobility of ML. For several CE runs with changing conditions such as varying μ_{eo} [9] or V [10,12], the dependence of the residual ratio $[ML]/[ML]_0$ on t_m was obtained to allow estimation of k_d via fitting with Equation 7.17. In addition, whole-column imaging [11] is very useful because a single CER run gives a set of $[ML]/[ML]_0$ versus t_m data and thus k_d. In LS-CER, substitution is introduced to accelerate the apparent dissociation of a very slow reaction with a $t_{1/2}$ of hours to reduce the experimental time substantially to within a usual CE time range (see later). When using an electrophoretic buffer containing a competing ligand Y, a substitution reaction (Equation 7.13) occurs. A set of CER runs with a buffer containing certain concentrations of Y gives k_{obs} from

$$[\mathrm{ML}]/[\mathrm{ML}]_0 = \exp(-k_{\mathrm{obs}} t_\mathrm{m}) \tag{7.19}$$

As discussed in previous reports [9,12], the peak height is used instead of peak area to estimate the residual ratio $[ML]/[ML]_0$ [25–27].[1]

Thus,

$$[\mathrm{ML}]/[\mathrm{ML}]_0 = AH_{\mathrm{ML}}/H_{\mathrm{IS}} \tag{7.20}$$

where H_{ML} and H_{IS} are the peak heights of ML and internal standard (IS), respectively, and A is a proportionality constant [13]. An IS is employed to compensate for the variance of the injection volume of the sample solution and also zone broadening during electromigration. For this purpose, an inert species with size and charge similar to those of ML is used as the IS. Combining Equations 7.19 and 7.20 gives Equation (7.21), which is applied to a set of H_{ML}/H_{IS} versus t_m data to estimate k_{obs}:

$$H_{\mathrm{ML}}/H_{\mathrm{IS}} = A^{-1}\exp(-k_{\mathrm{obs}} t_\mathrm{m}) \tag{7.21}$$

Varying the concentration [Y] in an electrophoretic buffer, a set of k_{obs} versus [Y] data are obtained for the estimation of k_d and k_Y via Equation (7.15). This is the principle of LS-CER.

7.5.2 Application of LS-CER to the Ti(IV)–Catechin Complex

Before applying LS-CER for the measurement of slow dissociation, it was validated using a Ti(IV)–catechin complex for which k_d had been estimated by batchwise LS with spectrometric monitoring [28]. Having a catechol moiety, catechin (H_4L) coordinated to the Ti(IV) ion to form a dianionic 1:3 complex $[Ti(H_2L)_3]^{2-}$ at pH 7.4 (Figure 7.10). Using spectrophotometric titration, the overall stability constant β_3 (Equation 7.22) was determined as $10^{47.95}$ ($I = 0.1$), implying high thermodynamic stability [28].

$$\beta_3 = \frac{[\mathrm{Ti}(\mathrm{H_2L})_3]^{2-}}{[\mathrm{Ti}^{4+}][\mathrm{H_2L_2}]^{3-}} \tag{7.22}$$

In a batchwise LS reaction of $[Ti(H_2L)_3]^{2-}$ with CyDTA as a competing ligand, the dissociation rate constant was estimated to be $k_d = (1.43 \pm 0.04) \times 10^{-4}\,\mathrm{s}^{-1}$ ($t_{1/2} = 1.35$ h) [24], suggesting that measurement of the spontaneous dissociation using CER is time consuming. In LS-CER of $[Ti(H_2L)_3]^{2-}$, EDTA was used as a competing ligand in the electrophoretic buffer. In addition, AS was used as an IS owing to its similar electrophoretic mobility to $[Ti(H_2L)_3]^{2-}$. Figure 7.11 shows typical electropherograms using 6.0 mM EDTA in the electrophoretic buffer. As can be seen, a decrease in the applied voltage increased the migration

[1] We avoided errors derived from uncertainty in the deconvolution of peak area, which is employed in most of the recently reported CE techniques for reaction kinetics [25–27].

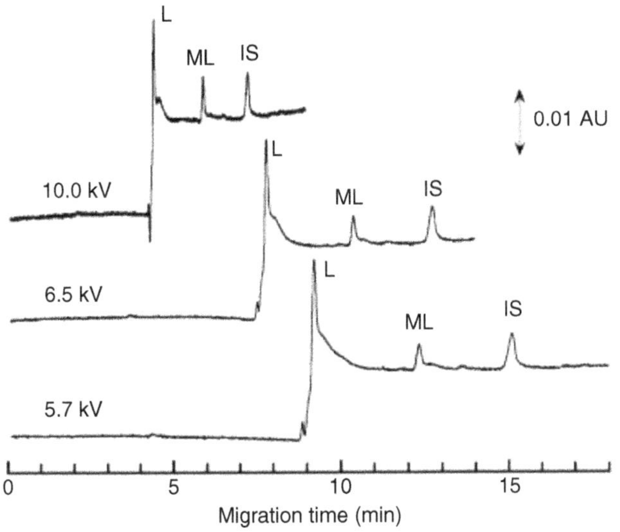

FIGURE 7.11 Typical electropherograms for dissociation of the catechin complex, $[Ti(H_2L)_3]^{2-}$, obtained using LS-CER with EDTA as a competing ligand. L, catechin; ML, $[Ti(H_2L)_3]^{2-}$; IS, AS; sample, $[Ti^{4+}] = 8.0 \times 10^{-5}$ M, $[catechin] = 2.4 \times 10^{-4}$ M, $[AS] = 4.2 \times 10^{-5}$ M, [HEPES] = 20 mM (pH 7.4); electrophoretic buffer, $[NaH_2PO_4] = 20$ mM (pH 7.4), [EDTA] = 6.0 mM; detection wavelength, 210 nm; capillary, FunCap CE/Type S, $L = 43.5$, $l = 31.0$ cm. Applied voltage is shown in the figure. Reprinted with permission from Ref. 13, copyright © 2009 American Chemical Society.

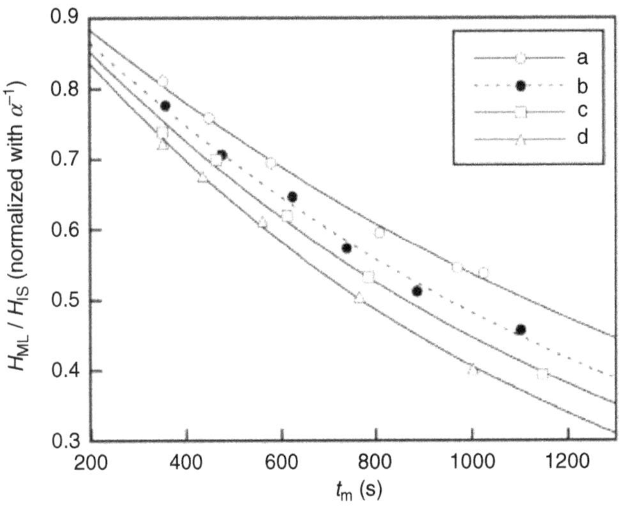

FIGURE 7.12 Dependence of peak height ratio H_{ML}/H_{IS} of $[Ti(H_2L)_3]^{2-}$ on reaction time obtained using LS-CER with EDTA as a competing ligand. Electrophoretic buffer, $[NaH_2PO_4] = 20$ mM (pH 7.4), [EDTA] = (a) 5.0, (b) 6.0, (c) 7.0, and (d) 8.0 mM. Sample composition and detection wavelength as in Figure 7.11. For the sake of clarity, the H_{ML}/H_{IS} ratio was normalized with α^{-1}, and k_d values were obtained by fitting of original H_{ML}/H_{IS} versus t_m data with Equation 7.21. The α^{-1} and $k_d/10^{-4}\,s^{-1}$ values were (a) 1.33, 6.22 ± 0.20, (b) 1.43, 7.29 ± 0.34, (c) 1.37, 8.05 ± 0.28, and (d) 1.19, 9.00 ± 0.21. $R = 0.997.0.999$. Reprinted with permission from Ref. 13, copyright © 2009 American Chemical Society.

time of $[Ti(H_2L)_3]^{2-}$, IS, and free catechin and also broadened each peak. It should be noted that the peak height of $[Ti(H_2L)_3]^{2-}$ became smaller than that of IS as the migration time increased, suggesting that the dissociation of $[Ti(H_2L)_3]^{2-}$ occurred over time. In fact, the dependence of the $H_{[Ti(H_2L)_3]^{2-}}/H_{IS}$ ratio decreased with migration time following a single exponential decay (Figure 7.12b), implying first-order decay according to Equation (7.22). In addition, dissociation of $[Ti(H_2L)_3]^{2-}$ can be monitored within a reasonable operating time (~20 min).

With increasing concentration of EDTA in the electrophoretic buffer, the decay of $H_{[Ti(H_2L)_3]^{2-}}/H_{IS}$ as a function of time became pronounced (Figure 7.12), implying that the dissociation of ML via substitution (Equation 7.23) became more prominent. From the first-order decay curves shown in Figure 7.13, k_{obs} is linearly dependent on [EDTA] (Figure 7.13a), in accordance with Equation (7.16), giving $k_d = (1.64 \pm 0.63) \times 10^{-4}\,s^{-1}$ and $k_Y = (9.17 \pm 0.96) \times 10^{-2}\,M^{-1}\,s^{-1}$. The k_d value estimated with LS-CER agreed with that estimated by batchwise LS, confirming the validity of LS-CER method. In addition, from a technical point of view, it is interesting to compare the time to acquire all necessary data for the estimation of k_d of the Ti(IV)–catechin complex, 16.7 h with a conventional batchwise method [24] and 4.2 h with the present LS-CER. Thus, LS-CER approach is considerably less time consuming than the conventional method.

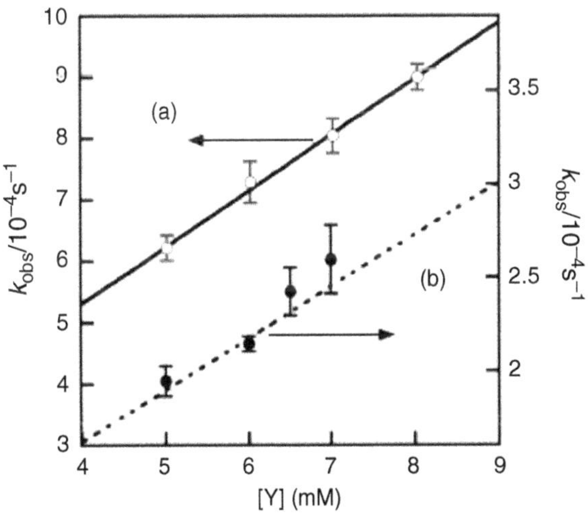

FIGURE 7.13 Relationship between observed dissociation rate constant (k_{obs}) of the Ti(IV) complex and concentration of competing ligand [Y] in electrophoretic buffer. (a) $[Ti(H_2L)_3]^{2-}$, Y = EDTA, $k_d/10^{-4}\,s^{-1} = 1.64 \pm 0.63$, $k_Y/10^{-2}\,M^{-1}\,s^{-1} = 9.17 \pm 0.96$. (b) $[TiR_3]^{8-}$, Y = catechol, $k_d/10^{-4}\,s^{-1} = 0.51 \pm 0.43$, $k_Y/10^{-2}\,M^{-1}\,s^{-1} = 2.78 \pm 0.72$. Fitting with Equation 7.16 was implemented by weighted least-squares regression. Reprinted with permission from Ref. 13, copyright © 2009 American Chemical Society.

7.5.3 Application of LS-CER to the Ti(IV)–Tiron Complex

Tiron (H_2R^{2-}) is a disulfonated derivative of catechol and has been used as a spectrophotometric reagent for Ti(IV) and Fe(III) [29]. Since the thermodynamic stability of a 1:3 complex with Ti(IV), $[TiR_3]^{8-}$ ($\log\beta_3 = 57.6$, $I = 0.138$), is even higher than that of the catechin complex $[Ti(H_2L)_3]^{2-}$, it can be assumed that spontaneous dissociation of $[TiR_3]^{8-}$ is slower than that of $[Ti(H_2L)_3]^{2-}$. This shows the necessity of a method that is more effective than the CER to perform the measurement of k_d in a reasonable experimental time. The reason why the conventional batchwise LS reaction is not applicable to the Tiron complex $[TiR_3]^{8-}$ is the lack of a suitable ligand to replace the coordinating Tiron ligand in $[TiR_3]^{8-}$. For instance, EDTA and CyDTA were unusable because the thermodynamic stabilities of their Ti(IV) complexes are lower than that of $[TiR_3]^{8-}$ [30].[2] In contrast, catechol (H_2cat) forms a thermodynamically more stable complex, $[Ti(cat)_3]^{2-}$, ($\log\beta_3 = 61.6$) [31] than Tiron does, but addition of excess catechol to a solution of Tiron complex $[TiR_3]^{8-}$ resulted in a large absorption band due to catechol that overlaps that of $[TiR_3]^{8-}$ and hinders spectroscopic monitoring of the dissociation process. In the present LS-CER, however, catechol can be employed as the competing ligand because CE allows the detection of $[TiR_3]^{8-}$ as a sharp electrophoretic zone rather than a broad absorption band in the spectral observation. Thiacalix[4]arene tetrasulfonate (TCAS) was used as an IS owing to its anionic charges being as high as 5 in the electrophoretic buffer of pH 7.4 [32]. Typical electropherograms of $[TiR_3]^{8-}$ (Figure 7.14) showed baseline separation from free Tiron and IS. Although the electrophoretic buffer absorbed light at the detection wavelength (210 nm) owing to the dissolved catechol, the peak of $[TiR_3]^{8-}$ was clearly detected. This implies an advantage of the use of CE detection over spectral detection for the substitution reaction.

As the applied voltage decreased, the separation time increased and the peak height of $[TiR_3]^{8-}$ decreased, suggesting that dissociation of $[TiR_3]^{8-}$ occurred. Between the peaks of $[TiR_3]^{8-}$ and IS a small peak was observed, which can be attributed to a 1:2 complex species, $[TiR_2]^{4-}$, remaining in the injected sample owing to incomplete equilibration in the formation of $[TiR_3]^{8-}$. It should be noted that the existence of this different complex species in the sample solution did not interfere with the detection of the dissociation of $[TiR_3]^{8-}$. The dependence of the $H_{[TiR_3]^{8-}}/H_{IS}$ ratio on the reaction time indicated first-order decay (Figure 7.15c), from which k_{obs} was estimated. Although the data were scattered, fitting with a first-order decay was possible and showed more distinct decay at a higher concentration of catechol in the electrophoretic

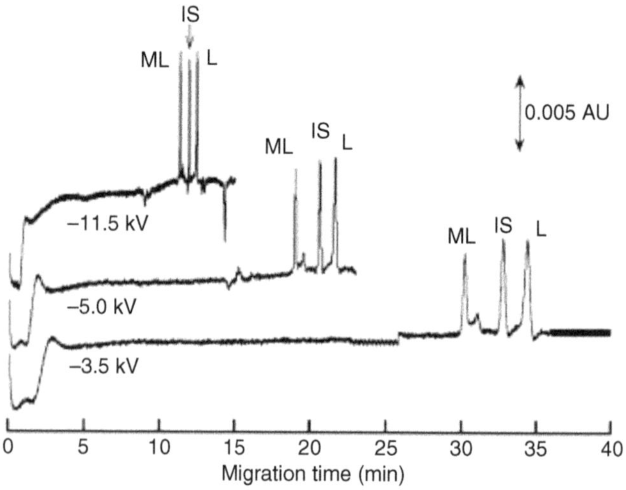

FIGURE 7.14 Typical electropherograms for dissociation of the Tiron complex, $[TiR_3]^{8-}$, obtained using the LS-CER with catechol as a competing ligand. L, Tiron; ML, $[TiR_3]^{8-}$; IS, TCAS; sample, $[Ti^{4+}] = 4.0 \times 10^{-5}$ M, [Tiron] $= 1.2 \times 10^{-4}$ M, [TCAS]) 2.0×10^{-5} M, [HEPES] = 20 mM (pH 7.4); electrophoretic buffer, $[NaH_2PO_4] = 20$ mM (pH 7.4), [catechol] = 6.5 mM; detection wavelength, 210 nm; capillary, polyacrylamide coated, $L = 52.5$, $l = 40.0$ cm. Applied voltage is shown in the figure. Reprinted with permission from Ref. 13, copyright © 2009 American Chemical Society.

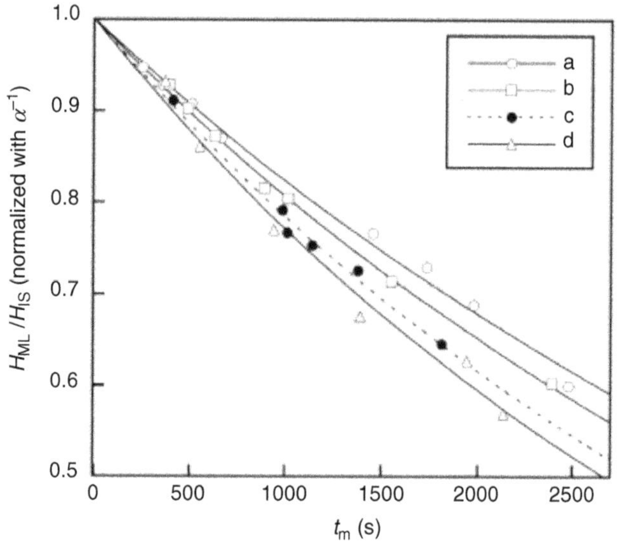

FIGURE 7.15 Dependence of peak height ratio H_{ML}/H_{IS} of $[TiR_3]^{8-}$, normalized with α^{-1}, on reaction time obtained using the LS-CER with catechol as a competing ligand. Electrophoretic buffer, $[NaH_2PO_4] = 20$ mM (pH 7.4), [catechol] = (a) 5.0, (b) 6.0, (c) 6.5, and (d) 7.0 mM. Sample composition and detection wavelength as in Figure 7.14. α^{-1} and $k_d/10^{-4}\,s^{-1}$ values: (a) 1.20, 1.95 ± 0.08, (b) 1.23, 2.42 ± 0.13, (c) 1.17, 2.14 ± 0.05, and (d) 1.19, 2.60 ± 0.18. $R = 0.990$–0.999. Reprinted with permission from Ref. 13, copyright © 2009 American Chemical Society.

[2] $\log\beta$ values (I) 0.1 with $NaClO_4$): $[TiO(edta)]^{2-}$ 17.3; $[TiO(cydta)]^{2-}$ 19.9. Data cited from Ref. 38.

buffer (Figure 7.15). The dependence of k_{obs} on the concentration of catechol further gave dissociation rate constants $k_d = (0.51 \pm 0.43) \times 10^{-4}\,s^{-1}$ and $k_Y = (2.78 \pm 0.72) \times 10^{-2}\,M^{-1}\,s^{-1}$ (Figure 7.13b). From the k_d value, the half-life was estimated to be $t_{1/2} = 3.8$ h, suggesting that normal CER without LS would require an operation time of several hours to obtain an H_{ML}/M_{IS} value for the dissociation of $[TiR_3]^{8-}$. In contrast, LS-CER measurement to give an H_{ML}/M_{IS} ratio was completed within 40 min. Even taking into account the fact that the experiment needs to be run several times (four times in this case) with varying [Y] to obtain the dependence of k_{obs} on [Y] (Equation 7.15), introducing the substitution reaction to CER considerably reduces the experimental time. Thus, the usefulness of the LS-CER in measuring k_d for slow dissociation kinetics has been demonstrated using the Tiron complex $[TiR_3]^{8-}$. In addition to the analytical significance of LS-CER methodology, it is interesting to note the difference in the kinetic stabilities of the two kinds of 1:3 Ti(IV)–catechol complexes; the half-life of the Tiron complex $[TiR_3]^{8-}$ ($t_{1/2} = 3.8$ h) indicates higher kinetic stability than the catechin complex $[Ti(H_2L)_3]^{2-}$ ($t_{1/2} = 1.17$). In spite of two electron-withdrawing groups, $-SO_3^-$, which are attached to the catechol and reduce the basicity of the ligand, the thermodynamic stability of $[TiR_3]^{8-}$ is higher than that of $[Ti(H_2L)_3]^{2-}$, as mentioned before. This can be attributed to the sizes of the substituents; catechin has much bulkier substituents, the thermal motion of which should lead to destabilization of the Ti(IV)–catechol bond. In contrast, the compact and rigid structure of Tiron should lead not only to high thermodynamic stability but also to higher kinetic stability of $[TiR_3]^{8-}$ than $[Ti(H_2L)_3]^{2-}$.

Using an inert Ti(IV)–Tiron complex, it was demonstrated that LS-CER involving LS in an electrophoretic separation field is very useful for measuring the dissociation rate of an extremely slow dissociation within a practically acceptable period. LS-CER methodology was validated by considering k_d for the catechin complex $[Ti(H_2L)_3]^{2-}$, which was estimated using a batchwise LS method. Thus, CER method was successfully extended toward very slow reaction kinetics. The advantages of LS-CER method over CER and/or batchwise substitution are summarized as follows. (i) It can be applied to slow dissociation kinetics with a $t_{1/2}$ of hours to reduce the experimental time considerably. (ii) In principle, it can use a competing ligand Y without MY being more stable than ML because the formation reaction of ML (reverse reaction of Equation 7.1) does not occur owing to electrophoretic separation of the components. (iii) The spectral resolution of Y with ML is not necessary, thus relaxing the prerequisites of the competing ligand. In addition, as a merit of CER over the batchwise method, (iv) the sample solution of ML does not necessarily need to be pure because other components such as complex species of different metal–ligand stoichiometry and metal species are separated and do not interfere with the estimation of remaining [ML]. On the whole, LS-CER method is useful for estimating the slow rate of spontaneous dissociation, which CER and batchwise methods cannot handle. Thus, it is a powerful tool in the rational design of ligands for diagnostic and therapeutic reagents.

7.6 EXPANDING THE SCOPE OF CER TO MEASUREMENT OF THE DISSOCIATION KINETICS OF BIOMOLECULAR COMPLEXES

It is expected that CER theoretically can be applied to the dissociation kinetic analysis of complexes in the broad sense, which are the aggregates of the molecules that are not bound with covalent bonds. In this study, we tried to extend the concept of CER to biomolecular complexes. The dissociation reaction of a protein–DNA complex was chosen as a model system for kinetic analysis by CER. In the replication of DNA, ssDNA-binding protein (SSB) stabilizes single-stranded DNA (ssDNA) through binding with the region consisting of about 30 oligonucleotides on the template ssDNA chain [33]. As SSB repeats binding and dissociation with ssDNA according to the progress of the polymerization of the replicated DNA chain, the dissociation kinetics of the complex of SSB with ssDNA ([SSB–ssDNA]) have biological importance. In this study, it was successfully demonstrated that CE serves as a chemical reactor for measurements of the dissociation kinetics of [SSB–ssDNA] consisting of *Escherichia coli* and ssDNA of 20-mer, 25-mer, and 31-mer [12].

7.6.1 Dissociation Kinetic Analysis of [SSB–ssDNA] Using CER

Figure 7.16 shows a series of electropherograms of [SSB–ssDNA] of 20-mer, 25-mer, and 31-mer ssDNA with different migration times obtained by varying the applied voltage. An electrophoretic buffer containing no constituents of the complex, such as SSB and ssDNA, was used. As shown in Figure 7.16a, the degree of decrease in the peak height of the complex of SSB with s31 ([SSB–s31]) with respect to the increase in migration time is much larger than that of tetraphenylphosphonium (TPP^+), the amount of which does not change during the CE separation period. This indicates that the concentration of [SSB–s31] decreased with increase in the migration time arising from its on-capillary dissociation reaction.

The peak height of [SSB–s31] decreases as the migration time increases since the degree of the decrease in [SSB–s31] through the on-capillary dissociation reaction is proportional to the degree of the decrease in the peak height of [SSB–s31]. As shown in Figures 7.16b and c, a decrease in the peak height of the complexes with increase in migration time

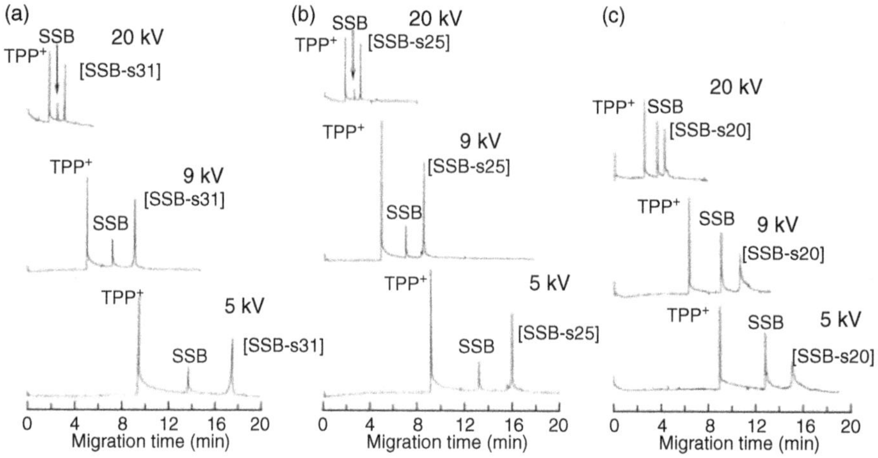

FIGURE 7.16 Series of electropherograms of [SSB–S31], [SSB–S25], and [SSB–S20] with different migration times. (a) [SSB–S31]; (b) [SSB–S25]; and (c) [SSB–S20]. Sample, 0.12 mg mL^{-1} SSB, 4.55 mg mL^{-1} ssDNA (S31, S25, and S20), 10 mM phosphate buffer (pH 7.2), and 0.25 mM TPPCl; capillary, $L = 42.5$ cm, $l = 12.5$ cm; electrophoretic buffer, 15 mM phosphate (pH 7.2). Applied voltage is shown in the figure. Reprinted with permission from Ref. 12, copyright © 2009 Wiley-VCH Verlag GmbH, Weinheim.

according to the on-capillary dissociation reaction of the complexes was also observed in the case of both [SSB–s25] and [SSB–s20], the same as in the case of [SSB–s31]. The degree of the decrease in the peak height of [SSB–s31], [SSB–s25], and [SSB–s20] increased in that order. This indicates that the magnitude of the dissociation rate constant also increases in that order. This agrees with the results of the previous study, in which k_d of [SSB–ssDNA] decreases as the chain length of ssDNA increases [34]. The dissociation reaction of [SSB–ssDNA] during the CE separation process follows first-order kinetics. The reaction scheme is as follows:

$$[\mathrm{SSB-ssDNA}] \xrightarrow{k_d} \mathrm{SSB} + \mathrm{ssDNA} \qquad (7.23)$$

The basic idea of dissociation kinetic analysis using CER is to measure the residue of the analyte complex with increasing migration time, t_m. In the kinetic analysis with the CER, peak height signals were employed because the peak area data can give unreliable results due to asymmetric peak profiles arising from the on-capillary dissociation reaction. The asymmetric-shaped [SSB–ssDNA] peak is not composed of [SSB–ssDNA] alone, that is, the leading region of the peak originates in free SSB generated by the on-capillary dissociation reaction in Equation 7.24. It is difficult to identify clearly the boundary between the region of the complex and that of free SSB derived from the on-capillary dissociation reaction. Therefore, the degree of dissociation of biomolecular complexes is estimated not by the decrease in the peak area of the complex but that of the peak height absorbance of the complex as a function of reaction time for accurate estimation. The degree of dissociation–time profiles for the on-capillary dissociation of three complexes ([SSB–s31], [SSB–s25], and [SSB–s20]) were obtained from several CE experiments with different migration times typically as shown in Figure 7.16. The double standardization method [9] is also employed for accurate measurement of the dissociation process of [SSB–ssDNA]. The two types of standards employed are kinetically inert, that is, their concentration does not change during the CE experiment: transferrin (Tf) for the peak signal standardization and TPP^+ for during the injection volume correction. Typical electropherograms of Tf varying with the applied voltage are shown in Figure 7.17.

The electropherograms of Tf were acquired under the same CE conditions as for the electropherograms of the complexes that paired with each of them. Dissociation kinetic analysis of [SSB–ssDNA] was basically performed in the same manner as that for metal–DHABS complexes. Here we can obtain k_d by fitting Equation 7.24 using the $H_{\mathrm{SSB\text{-}ssDNA}}/H_{\mathrm{Tf}}$ data at various t_m:

$$H_{[\mathrm{SSB-ssDNA}]}/H_{\mathrm{Tf}} = \alpha \exp^{-k_d t_m} \qquad (7.24)$$

where α is a constant. Thus, k_d was obtained by analyzing the degree of dissociation–time profile of [SSB–ssDNA] as the plot of the normalized peak height signals of [SSB–ssDNA] with that of Tf ($H_{\mathrm{SSB\text{-}ssDNA}}/H_{\mathrm{Tf}}$) against the migration time. Simple first-order decay profiles for [SSB–s31], [SSB–s25], and [SSB–s20] are observed, as shown in Figure 7.18, and the plots are fitted well to Equation 7.24. The dissociation rate constants for [SSB–s31], [SSB–s25], and [SSB–s20] are determined as 3.99×10^{-4}, 4.82×10^{-4}, and 1.50×10^{-3} s^{-1}, respectively.

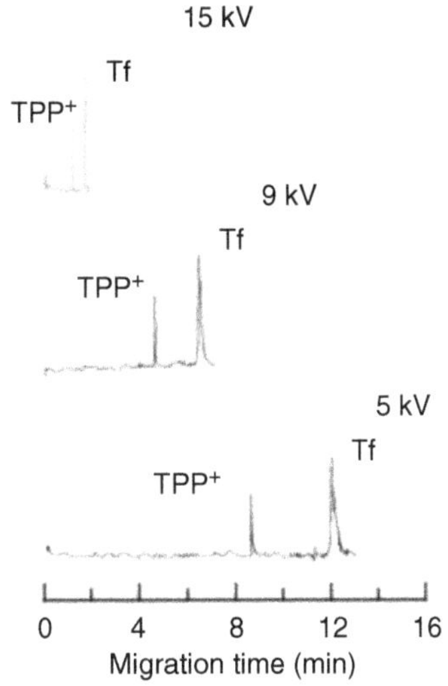

FIGURE 7.17 A series of electrophcrograms of Tf with different migration times. Sample, 0.18 mg mL^{-1} Tf, 10 mM phosphate buffer (pH 7.2), and 0.027 mM TPP^+; capillary, L = 12.5 cm, l = 42.5 cm (5 kV), L = 42.5 cm, and l = 12.5 cm (5 and 9 kV); applied voltage is shown in the figure. Reprinted with permission from Ref. 12, copyright © 2009 Wiley-VCH Verlag GmbH, Weinheim.

The k_d values found and those obtained by various methods are summarized in Table 7.1 [34–36].

In comparison with the k_d values of the longer (70) [36] and shorter (15) [35] chain lengths of their pair ssDNA, the k_d values found were located in the middle of them. The k_d values were found also to decrease with increase in the chain length of ssDNA. This is reasonable because k_d decreases as the chain length of its pair ssDNA increases [34]. Although

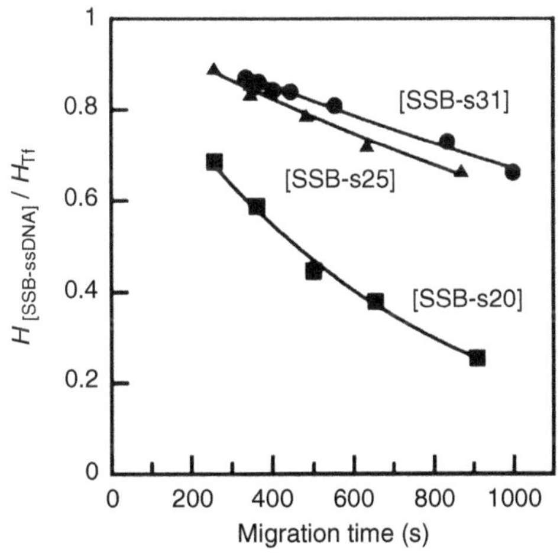

FIGURE 7.18 Reaction profile of [SSB–S31], [SSB–S25], and [SSB–S20] obtained using the CER. Solid lines show the approximation curve for the first-order dissociation reaction profile of each complex by Equation 7.24. Reprinted with permission from Ref. 12, copyright © 2009 Wiley-VCH Verlag GmbH, Weinheim.

TABLE 7.1 Dissociation Rate Constants of [SSB–ssDNA] Obtained by Various Methods

k_d (s^{-1})	Chain Length of ssDNA	Method	Reference
1.3×10^{-5}	70	SPR	35
3.99×10^{-4}	31	CER	12
4.82×10^{-4}	25	CER	12
1.50×10^{-3}	20	CER	12
3.3×10^{-2}	15	CE	34
<1	30–40	Stopped-flow	33
40	16	Temperature jump	33
1700	8	Temperature jump	33

the values of k_d of the complexes of 8, 16, and 35 base ssDNA have also been reported with the stopped-flow method and temperature-jump method [35], they do not agree with those obtained with use of CER, CE, and SPR, as summarized in Table 7.1. "Small" k_d values of the order of $10^{-4}\,s^{-1}$ obtained by CER experiments show that the dissociation reaction of [SSB–ssDNA] is a relatively slow process. This is also supported by the results obtained by other methods [35,36]. In contrasting, the formation reaction of the complex would be a fast process judging from the "large" values of the formation reaction rate constant, k_f [34,36]. In the conventional methods including the stopped-flow method and temperature-jump method, both the formation and dissociation processes occur simultaneously since the experiments for kinetic measurement are performed in the closed batch reaction system. Then, k_f and k_d are obtained simultaneously. Although the simultaneous determination of k_f and k_d is theoretically possible, it is difficult from a practical standpoint to evaluate both k_f and k_d accurately from the experimental data. The error in the evaluation of k_d would be very large since the value of k_d is too small compared with that of k_f. In fact, k_d of the complex with ssDNA of 35 bases was not obtained by the stopped-flow method and temperature-jump method [34]. On the other hand, in the case of the determination of k_d by the proposed method, CE [35,37] and SPR [36], only the dissociation process of the complex is observed and the formation reaction does not occur since these methods are based on separation. The above-mentioned separation-induced concentration jump-like environment, in which the concentration of free SSB or free ssDNA around [SSB–ssDNA] greatly decreases during the separation process, is the driving force of the dissociation reaction of the complex in those systems, and the reverse reaction, that is, the formation reaction, does not occur because the products of the dissociation reaction of the complex (free SSB and free ssDNA) are separated immediately. Compared with conventional "indirect" measurement methods, an accurate measurement of k_d becomes possible by observing the dissociation reaction of the complex directly in these separation-based "direct" measurement methods. The high separation ability of CE gives the clear concentration jump-like property suitable for the direct observation of the dissociation process of the complex to the CE-based kinetic measurement methods, but the sharp concentration jump-like property is not obtained in the SPR-based method [36,38] because diffusion from the solid phase to the mobile phase by the concentration gradient is a driving force of the separation. The clear resolution due to CE separation allows more accurate k_d measurement with CE-based kinetic analysis methods than with the SPR method. In addition, the asymmetric, that is, leading or tailing shaped, target complex peak resulting from the on-capillary dissociation reaction of the complex is observed in CER [9] and other CE-based kinetic analysis methods similar to CER [35,37]. In the estimation of k_d from the ratio of the peak areas [35] or fitting the asymmetric part of the electropherogram [37], the reaction profile might not be accurately described because the boundary between the region of the complex and that of the decomposition products is not clear, and hence the region of the complex and that of the decomposition product in the peak are not clearly distinguished. The kinetic analysis using the peak height absorbance of the complex therefore provides the correct description for the degree of dissociation–time profile of the complexes [9–11], if the mutual separation between the complex and the decomposed products is achieved. CER is a method that can measure k_d most accurately among the existing methods for the measurement of k_d.

In this study, the successful application of CER to the determination of the dissociation kinetics of [SSB–ssDNA] has been demonstrated. The dissociation rate constant of [SSB–ssDNA] in a homogeneous solution can be measured directly and accurately using CER without any troublesome procedures such as immobilization and derivatization. If the proper internal and external standard materials are available, the applications of CER can actually be extended to a wide range of biomolecular complexes.

7.7 CONCLUSIONS

An original CE-based methodology for direct measurement of the dissociation kinetics of metal complexes, CER, is advantageous over other CE-based kinetic analysis methods. CER can estimate the dissociation rate constants of complexes conveniently and accurately by using the separation characteristics of CE and the original standardization method. CER, which is a direct measurement method for dissociation rate constants of metal complexes on the time scale from minutes to dozens of minutes, was successfully extended toward faster reaction kinetics (μCER) and very slow reaction kinetics (LS-CER). In addition, its scope was successfully extended to the dissociation kinetic analysis of biomolecular complexes with half-lives from minutes to dozens of minutes. More recently, the dissociation kinetic analysis of biomolecular complexes with half-lives from seconds to dozens of seconds has been demonstrated using μCER [39]. Now, CER and its related methodologies can treat the reaction kinetics of broad "complexes" on the time scale from seconds to hours. They will be very powerful tools for extracting information on the dynamics of "complexes."

REFERENCES

1. Stryer, L. (1988) In: *Biochemistry*, 3rd ed., W. H. Freeman, New York.
2. Hammes, G.G. (2000) In: *Thermodynamics and Kinetics for the Biological Sciences*, Chapter 4, John Wiley & Sons, Inc., New York.
3. Johnson, K.A., (2003) In: Johnson, K.A. (ed.), *Kinetic Analysis of Macromolecules*, Chapter 1, Oxford University Press, Oxford.
4. Smith, T.W. and Skubitz, K.M. (1975) *Biochemistry 14*, 1496–1502.
5. Hahnefeld, C., Drewianka, S., and Herberg, F.W. (2004) *Molecular Diagnosis of Infectious Diseases*, Methods in Molecular Medicine, 94, 2nd ed., Humana Press, Totowa, NJ.
6. Mori, T. and Okahata, Y. (2005) *Trends Glycosci. Glycotechnol. 17*, 71–83.
7. DiGiacomo, R.A., Xie, L., Cullen, C., and Indelicato, S.R. (2004) *Anal. Biochem. 327*, 165–175.
8. Newman, C.D. and Collins, G.E. (2008) *Electrophoresis 29*, 44–55.
9. Iki, N., Hoshino, H., and Yotsuyanagi, T. (2000) *Anal. Chem. 72*, 4812–4820.
10. Takahashi, T., Takehara, Y., and Hoshino, H. (2007) *Bull. Chem. Soc. Jpn. 80*, 910–915.
11. Takahashi, T., Ohtsuka, K., Iki, N., and Hoshino, H. (2005) *Analyst 130*, 1337–1339.
12. Takahashi, T., Ohtsuka, K., Tomiya, Y., Iki, N., and Hoshino, H. (2009) *Electrophoresis 30*, 3079–3085.
13. Iki, N., Takahashi, M., Takahashi, T., and Hoshino, H. (2009) *Anal. Chem. 81*, 7849–7854.
14. Iki, N., Hoshino H., and Yotsuyanagi, T. (1993) *Chem. Lett.* 701–704.
15. Iki, N., Hoshino H., and Yotsuyanagi, T. (1993) *J. Chromatogr. A 652*, 539–545.
16. Jorgenson, J.W. and Lukacs, K.D. (1981) *Anal. Chem. 53*, 1298–1302.
17. Basolo, F. and Pearson, R.G. (1967) *Mechanisms of Inorganic Reactions*, 2nd ed., Chapter 3, John Wiley & Sons, Inc., New York.
18. Harrison, D.J., Manz, A., Fan, Z., Ludi, H., and Widmer, H.M. (1992) *Anal. Chem. 64*, 1926.
19. Mao, Q. and Pawliszyn, J. (1999) *Analyst 124*, 637.

20. Arai, A., Tanaka, H., Nakane, T., Fujiwake, H., Abe, H., Akechi, M., and Nakanishi, H. (2002) *Shimadzu Rev. 58*, 101.
21. Weinberger, R. (2000) *Practical Capillary Electrophoresis*, 2nd ed., Academic Press, San Diego, CA, p. 64.
22. Ohtsuka K. (2007) Master Thesis, Tohoku University.
23. Vavere, A.L. and Welch, M.J. (2005) *J. Nucl. Med. 46*, 683–690.
24. Vavere, A.L., Laforest, R.M., and Welch, J. (2005) *Nucl. Med. Biol. 32*, 117–122.
25. Newman, C.I.D. and Collins, G.E. (2008) *Electrophoresis 29*, 44–55.
26. Trapp, O. (2008) *Chem. Today 26*, 26–28.
27. Trapp, O. (2006) *Anal. Chem. 78*, 189–198.
28. Takahashi, M. (2007) Master Thesis, Tohoku University.
29. Cheng, K.L., Ueno, K., and Imamura, T. (1982) *CRC Handbook of Organic Analytical Reagents*, CRC Press, Boca Raton, FL.
30. Sillen, L. G., and Martell, A. E. (1964) *Stability Constants of Metal–Ion Complexes*, The Chemical Society, London.
31. Sommer, L. (1963) *Collect. Czech. Chem. Commun. 28*, 210–212.
32. Matsumiya, H., Takahashi, Y., Iki, N., and Miyano, S. (2002) *J. Chem. Soc., Perkin Trans. 2* 1166–1172.
33. Watson, J. D., Hopkins, N.H., Roberts, J.W., Steitz, J.A., and Weiner, A.M. (1987) In: *Molecular Biology of the Gene*, 4th ed., Chapter 10, Addison Wesley Longman, Boston.
34. Krauss, G., Sindermann, H., Schomburg, U., and Maass, G., (1981) *Biochemistry 20*, 5346–5352.
35. Berezovski, M. and Krylov, S.N. (2002) *J. Am. Chem. Soc. 124*, 13674–13675.
36. Fisher, R.J., Fivash, M., Casas-Finet, J., Bladen, S., and McNitt, K. L. (1994) *Methods 6*, 121–133.
37. Krylov, S.N. and Berezovski, M. (2003) *Analyst 124*, 571–575.
38. Karlsson, R., Michaelsson, A., and Mattsson, L. (1991) *J. Immunol. Methods 145*, 229–240.
39. Tomiya, Y. (2011) Master Thesis, Tohoku University.

8

CAPACITIVELY COUPLED CONTACTLESS CONDUCTIVITY DETECTION (C^4D) APPLIED TO CAPILLARY ELECTROPHORESIS (CE) AND MICROCHIP ELECTROPHORESIS (MCE)

José Alberto Fracassi da Silva, Claudimir Lucio do Lago, Dosil Pereira de Jesus, and Wendell Karlos Tomazelli Coltro

8.1 INTRODUCTION

In approximately 12 years, capacitively coupled contactless conductivity detection (C^4D) has emerged from basic academic research to commercially available detectors for capillary electrophoresis (CE), microchip electrophoresis (MCE), and even for capillary high performance liquid chromatography. Although the sensitivity and selectivity are far from some state-of-the-art detection schemes, such as mass spectrometry and laser-induced fluorescence (LIF), C^4D has been applied to several species in different matrices, with the best results being obtained for small high-mobility inorganic ions. Impressive results have been reported for the use of C^4D combined with sample enrichment techniques, such as stacking-mediated concentration, which drops the limits of detection to low parts per billion levels. One great advantage of C^4D is that no derivatization step is required, which simplifies sample preparation before analysis and reduces the total analysis time. Moreover, there is flexibility for the choice of the background electrolyte (BGE). In general, low-conductivity (low-mobility co-ions) BGE can be used for separation of high-mobility analytes, with the clear advantage of low Joule heat generation. The aim of this chapter is to introduce the basics of C^4D for CE and MCE. For this a brief theory section is presented, followed by a review of the literature related to C^4D-CE and C^4D-MCE.

The history of C^4D started in 1928, when Zahn reported that conductance measurements could be carried out without physical contact of the electrodes with the measured solution [1], as reported by Burkhalter [2]. The contactless conductivity technique was popular in the 1960s under the format of "high-frequency titrations" but its use was discontinued in the next decades, mainly due to advances in the instrumentation and methods for chromatography. A novel impulse occurred in the 1980s when "oscillometry" was employed as a detection scheme for flow injection analysis and chromatography [3–5]. A few reports of C^4D applied to isotachophoresis appeared during the 1980s, being the first reports of C^4D for electroseparations [6–8]. In 1998, the use of C^4D was proposed independently by Zemann et al. [9] and by da Silva and do Lago [10]. From this point, the technique underwent fast development, in the words of Martin Pumera "mainly because do Lago published the detailed electronic scheme of the contactless conductivity detector" [11].

8.2 THEORY OF C^4D

8.2.1 Basic Principles of C^4D

As in conventional contact conductivity detection, the technique involves the measurement of a current that is directly proportional to the conductance of the solution. There is no selectivity—the presence of any ions in solution will

Capillary Electrophoresis and Microchip Capillary Electrophoresis: Principles, Applications, and Limitations, First Edition.
Edited by Carlos D. García, Karin Y. Chumbimuni-Torres, and Emanuel Carrilho.

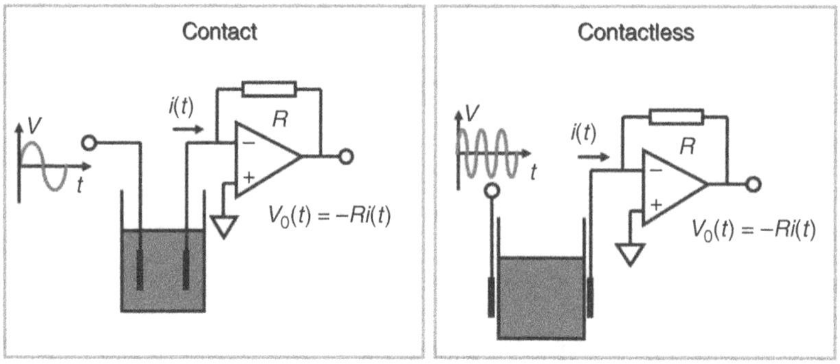

FIGURE 8.1 Comparison between contact and contactless conductivity detection. The current $i(t)$ is proportional to the conductance of the solution, and can be transformed into a voltage $V_0(t)$ by the use of a current-to-voltage converter.

contribute to the conductance of the solution. In other words, conductivity detection is a universal technique that requires a prior separation mechanism to improve selectivity. This is the reason for the success of conductivity-based detection for chromatography and electroseparation techniques.

In direct conductivity, a small-amplitude alternate signal (e.g., a sinusoidal wave) is applied to two inert electrodes (Figure 8.1). As the electrodes are in direct contact with the solution, a low frequency is used, since the capacitance of the electrical double layer (EDL) is relatively high, a few μF per cm^2. If the electrodes were placed outside the reservoir, the capacitance of the wall (C_w) of the reservoir will impose a high resistance to signal transmission or, more accurately, a high capacitive reactance, X_C. X_C is inversely proportional to the operating frequency and the absolute capacitance value. Since C_w is very small, a high-frequency signal must be applied to the electrodes to overcome the capacitive reactance of the reservoir. For example, for CE using a typical capillary (75 μm i.d., 360 μm o.d.), the capacitance of the wall is estimated to be 0.26 pF.

A practical conductivity cell employed for CE can be seen in Figure 8.2a (for microchips, see Section 8.4).

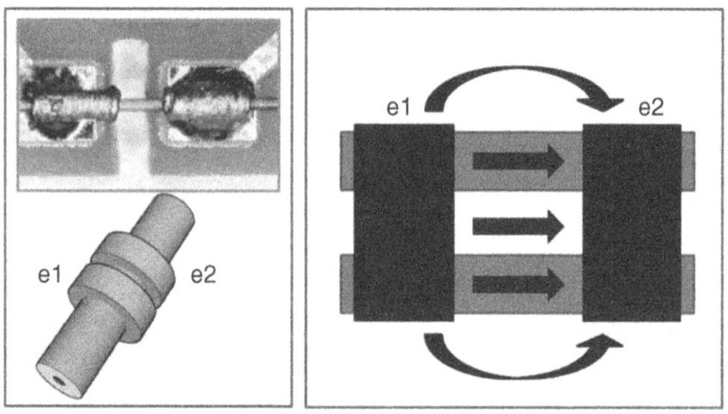

FIGURE 8.2 Conductivity cell for CE. (a) Picture of the mounted tubular electrodes over a circuit board and diagram. (b) Paths to signal transmission from the excitation (e1) to pick-up (e2) electrodes.

Tubular electrodes well fitted to the capillary are needed to minimize the capacitive reactance and avoid noise due to mechanical vibration. There are several paths for signal transmission from the excitation to the pick-up electrodes: (1) directly through the air, (2) directly through the capillary wall, and (3) directly through the solution between the electrodes. For the latter, the signal must be transmitted through the capillary wall twice (Figure 8.2b).

Taking into account these different ways of signal transmission, an equivalent electrical circuit can be proposed [10], as depicted in Figure 8.3. In this simplified model, the external air transmission and the transmission through the capillary can be represented by the capacitance C_0, while the two wall capacitances can be series associated to give a single C_w value. As usual, R_S stands for the resistance of the solution between the electrodes.

8.2.2 Simulation

The set of phenomena inside the capillary separation column during an electrophoresis experiment is quite impressive. Just after sample injection, the chemical composition along the capillary, except for the injection point, is uniform. However, diffusion takes place at the border. This ubiquitous phenomenon changes the concentration of

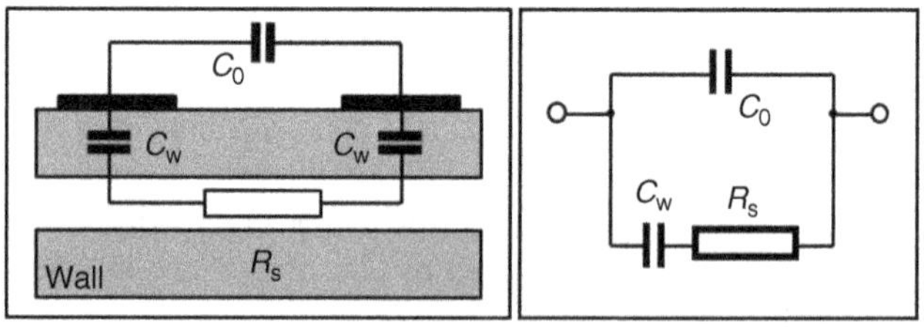

FIGURE 8.3 Equivalent electrical circuit of C^4D. (a) Physical representation of the stray capacitance (C_0), wall capacitance (C_w), and solution resistance (R_S). (b) Simplified equivalent electrical circuit.

all species in the vicinity and shifts chemical equilibriums. The electric field also contributes to move the ionic species and augment even more the complexity of the environment. The electroosmotic flow (EOF) also contributes to the movement of the solution, and sometimes other interfacial phenomena may be present. We may also add two other ingredients: the kinetics of some slow reactions and the omnipresent effect of the temperature over all these phenomena, which becomes more pronounced due to Joule heating. As a consequence, a prediction of what is happening in an electrophoresis experiment is quite difficult.

In this context, computational simulation arises as a fundamental tool to understand, predict, and plan electrophoresis experiments. Although often disregarded, such a tool should be systematically adopted by users—especially those engaged with conductivity detection because the conductivity of a solution virtually depends on the behavior of all ionic and nonionic species. The conductivity (κ) is given by

$$\kappa = F\sum_{i} z_i \mu_i c_i \tag{8.1}$$

where F is Faraday's constant, and z_i, μ_i, and c_i are, respectively, the charge, mobility, and concentration of each ionic species i in the solution. This equation suggests that neutral species ($z_i = 0$) do not contribute to conductivity. However, they may change the mobility of other species by changing their size and the viscosity of the medium.

There are some options of computational programs for CE simulation with different approaches and levels of realism. Similar to other problems, there is an inverse relationship between the realism of the model and the usability of the resulting program. In other words, we should use the simplest model that adequately represents the real world.

A good example of a simulation program is SIMUL 5.0 produced by Gaš and his group [12]. In a few words, the interior of the virtual capillary is segmented, and the physical and chemical processes for each segment are calculated. This kind of simulation is useful to make clear the behavior of the concentration profiles of all ionic species.

Although useful to understand the chemical transformations inside the capillary that define the shape of an electropherogram, programs similar to SIMUL 5.0 demand time and detailed information that impair their daily use. An even simpler model may be advantageously used, in which the shape of the peak is premodeled. This is the case of PeakMaster, also developed by Gaš and his group [13]. After data introduction, the calculation consumes typically <1 s, which allows a constant refining of the proposed experiment.

Despite their great usefulness, one should be aware of an important limitation of this software: the equilibrium positions are reached instaneously. This feature may lead to significant errors, for example, when slow metal complex formation must be considered.

8.2.3 Basic Equation for Sensitivity

Although a complete simulation is enlightening, there is a simpler approach by using an analytical function that describes the sensitivity of a conductivity detector in electrophoresis experiments:

$$\frac{\partial \kappa}{\partial c_A} = \frac{(\mu_A - \mu_S)(\mu_A + \mu_O)}{\mu_A} \cdot F \tag{8.2}$$

where κ is the conductivity of the solution during an electrophoresis experiment, c_A is the concentration of the analyte; μ_A, μ_S, and μ_O are, respectively, the mobilities of the analyte, co-ion, and counter-ion. This equation was obtained by deriving that proposed by Katzmayr et al. [14] and later simplified by Brito-Neto et al. [15]. Only singly charged and fully dissociated species were considered in this formulation, which a priori restricts its use. Even so, this equation is handy for a fast prediction of the behavior of a C⁴D. It is straightforward to show from this equation that sensitivity will be zero if the mobility of the analyte equals that for the co-ion from the buffer. Conversely, one can see that the greater the difference between these mobilities, the higher the sensitivity. Moreover, negative peaks should be obtained for high-mobility co-ions.

Another important issue disregarded in the previous formulations is the noise level, which is not exactly a flaw, because its prediction is an astonishing task. What is important to keep in mind is a general tendency of increasing noise by: (1) increasing buffer concentration, (2) increasing mobilities of co-ion and counter-ion, (3) decreasing pH buffering, and (4) decreasing temperature control. Thus, in some circumstances, there is a tradeoff between the signal intensity and the noise level, and the best compromise would ultimately be obtained by trials.

8.2.4 Equivalent Circuit of a CE-C⁴D System

The representation of the electrophoresis system as an equivalent electrical circuit is an efficient approach to determine the best configuration and/or to identify problems. This is especially true for electrochemical detection because electrochemical detection and the electrophoretic system are coupled. Figure 8.4a shows a convenient equivalent circuit for a CE-C⁴D. In this case, a convenient equivalent circuit means that important features can be understood despite its simplicity. A deeper treatment can be found elsewhere [15,16].

Taking into account that a capacitor acts as a blocking agent for a constant-current (DC) path, one can promptly conclude that a C⁴D does not suffer from interference from

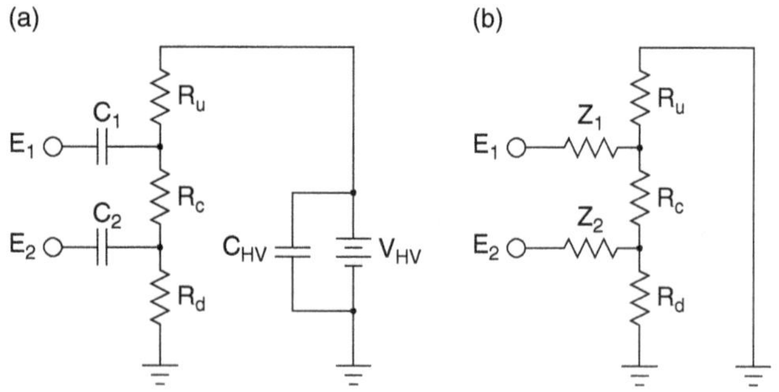

FIGURE 8.4 Equivalent circuit for a CE-C⁴D system (a) and a simplified circuit (b) considering the standpoint of the high-frequency signal of the detector. C_1 and C_2 are the capacitances between the electrodes (E_1 and E_2) and the inner surface of the capillary, which are converted, respectively, to the impedances Z_1 and Z_2 when a fixed frequency is used. R_u, R_c, and R_d stand for, respectively, the portions of solution before, within, and after the detection point. V_{HV} is the high-voltage power supply, which has an inner capacitance C_{HV}.

the electrophoresis system because the high voltages involved in electrophoresis are decoupled. In fact, the decoupling of the high voltage is a great benefit. However, this is an incomplete conclusion because there are other impedances to be taken into account.

A C^4D operates at high constant frequency. Thus, the circuit from Figure 8.4a can be simplified to that shown in Figure 8.4b. The wall capacitance is represented by a generic impedance element (Z_1 and Z_2) to make clear that although the alternating current (AC) flows through the wall, the impedance between the electrode and the inner solution may be as high as 10^6 Ω. Thus, it may be comparable to the impedance R_d if the detector is positioned close to the end of the capillary. In other words, the analytical signal may be affected by the composition of the solution after the detection point, which is highly undesirable. In the worst case, that is, C^4D at the end of the capillary, the analytical signal may be completely suppressed, because of the grounding of the signal.

Although the detector is not affected by the high voltage, the power supply has an inner capacitance that makes it a virtual ground to the AC signal of the detector. This means that the same analysis from the previous paragraph may be applied to the high-voltage end of the capillary. Thus, surprisingly, detector signal will be shunted to the ground when it is positioned close to the high-voltage end of the capillary.

8.2.5 Practical Guidelines

There are some additional considerations that should improve the usage of a CE-C^4D system:

- K.I.S.: *Keep It Simple*: This is a golden rule for the design of a BGE for capillary zone electrophoresis as a whole, and it is particularly true for CE-C^4D. The ideal BGE has only one cation and only one anion. As a tendency, one system eigenzone [17] arises for each additional ion included in the electrolyte composition. For some detection systems, this phenomenon has no apparent consequence because no additional peaks are observed. For conductivity detection, however, virtually always a peak or a transient signal is observed in the electropherogram at the time corresponding to the respective system eigenzone. Consequently, the shape or size of a peak in the vicinity of this system eigenzone is affected, which obviously impairs its identification and/or quantification. However, this feature should not be seen as a disadvantage of C^4D: in fact, it makes clear to the user such an occurrence that sometimes is overlooked in more selective detection schemes.
- Simulation is not only for beginners. Fifteen minutes in front of a computer can save 1 day of hard work.
- The closer the detector is to the end of the capillary, the greater the plate count. This is always true. However, this positioning is not good for a C^4D. For most instruments, there is a natural distance from the end of the capillary that favors detection. However, this is not true for microchip systems. Sometimes, the detector is wrongly positioned at the very end of the channel. The rule is: The closer the detection point to one of the ends of the channel, the lower the wall impedance must be.
- At first glance, there is an inconsistency in selecting species for the BGE. Equation 8.2 shows that, to increase the sensitivity, one should use a co-ion that has mobility as different as possible from that for the analyte. However, exactly the contrary is recommended to minimize electrodispersion. There is, however, a misunderstanding here. Electrodispersion is a severe problem for a concentrated analyte, which is not *a priori* a problem for a low-sensitivity detector. Thus, the analysis of major components may be carried out using a co-ion of similar mobility, which diminishes the electrodispersion and the sensitivity (not a big problem in this case). In this sense, C^4D may be envisioned as a detector with highly flexible sensitivity that conveniently matches the electrodispersion issues.

8.3 C^4D APPLIED TO CAPILLARY ELECTROPHORESIS

Many review papers have been published on this subject in recent years. It is not the goal of this chapter to list all these papers, but some critical reviews offer a good starting point for the discussion of the principles and applications of C^4D in CE and MCE. In 2005, Brito-Neto et al. [15,16] published two tutorial papers where theoretical and practical information were combined with new data. Kuban and Hauser [18] have reviewed 10 years of C^{4}D applied to CE, although these authors have excluded some of the articles published during this period. Other recent review papers also have merit [19–22].

8.3.1 Instrumental Aspects in CE

An advantage of the C^{4}D when compared to spectrophotometric detection is the possibility of use of capillaries made of nontransparent materials, such as poly(ether ether ketone), PEEK [23,24], that provide great flexibility to the CE technique.

Portable equipment has also been developed. Seiman et al. [25] built a portable CE equipment with a cross injector and C^{4}D. A different portable CE equipment containing a miniaturized high-voltage power supply and a C^{4}D detector was proposed by Kuban et al. [26]. The versatility of the constructed equipment was demonstrated through the separation of anions and cations including transition metal ions and arsenic species. Rainelli and Hauser [27] proposed an injection device for short capillaries, used for fast separation of several analytes. The analysis times were comparable to those obtained in microchip separations.

8.3.2 Coupling C^{4}D with UV–Vis Photometric Detectors in CE

The gap between electrodes in C^{4}D can be favorably used to also accommodate a photometric detector, as demonstrated by some authors [28–32]. This is the perfect case of simultaneous detection, since both detectors are positioned at the same point. Sequential detection has also been applied [33–35], including the positioning of the conductivity cell inside commercial CE-UV detection cartridges [36–39]. An on-column photo reactor, followed by UV and C^{4}D detectors, was proposed by Xuan et al. [35] for studies involving iron species in plant metabolism.

A C^{4}D detector positioned at the middle of the capillary has been successfully applied for the simultaneous separation of both positively and negatively charged species [40–43], including the speciation of Cr(III) and Cr(VI) [43]. In this scheme, samples are injected into both ends of the capillary (dual opposite injection DOI). The running conditions must be adjusted to avoid the co-migration of different species.

The coupling of CE with flow injection analysis (FIA) or sequential injection analysis (SIA) is very advantageous, since it makes the system fully automated and brings the possibility of sample treatment before the separation [44–50]. Kuban et al. applied FIA-CE-C^{4}D for the monitoring of inorganic anions in water samples [44], for selenium speciation [45], and for automated dual opposite sample injection [46]. Wang and Fu [47] constructed a poly(dimethylsiloxane), PDMS, body to place together capillary, C^{4}D electrodes and tubing for FIA. Wang et al. [48] proposed a FIA-CE system where negative pressure was applied to the outlet of the capillary with the objective of sample injection. Wuersig et al. [49] used short capillaries in SIA-CE to perform rapid separations. Nine anions or five cations could be separated in 40 or 11 s, respectively.

Another interesting aspect of C^{4}D is the possibility of application in nonaqueous capillary electrophoresis (NACE), mainly in the case of solvents that present strong UV absorption. Muzikar et al. [51] found that the same signal-to-noise ratios are obtained using propylene carbonate, *N,N*-dimethylformamide, and *N,N*-dimethylacetamide as solvents. Propylene carbonate was used as solvent for the separation of aliphatic amines [52]. Porras and Kenndler [53] evaluated the use of formamide as solvent in CE using dual UV and C^{4}D detection and Borissova et al. [54] used ionic liquids in acetonitrile as BGEs in NACE for the separation of organic dyes.

8.3.3 Fundamental Studies in Capillary Electrophoresis Using C^{4}D

Although most papers deal with application and instrumental aspects, interesting studies can be conducted using CE-C^{4}D, including the obtaining of physical constants. Jensen et al. [55] determined the CMC of non-UV absorbing charged surfactants. Henchoz et al. [56] determined the pK_a of 20 amino acids using UV and C^{4}D detectors. Vceláková et al. [57] determined the cationic mobility and the pK_a of 22 amino acids. Gas et al. [58] developed a model for optimization of BGEs for anion separations in CE. It is straightforward to see that system peaks could be easily detected using C^{4}D, since variations in the composition of the BGE reflect directly in variations in solution conductance—the same system peak could not be detected by the use of direct UV detection and, in this case, a strong electrodispersion condition could not be easily detected. The appearance of eigenpeaks was evaluated in univalent weak electrolytes [59] and in methanol [60]. In general, conductivity detectors are useful tools for the monitoring of the EOF. Water plugs generated in the injection step can be promptly detected by C^{4}D. Differently, Muzikar et al. [61] used a pressure-mediated dual-ion technique for monitoring the EOF mobility in CE using organic solvents. An interesting strategy for EOF monitoring involving thermal marks and C^{4}D was introduced by Saito et al. [62]. This strategy was subsequently applied to EOF monitoring in ionic liquid solutions in nonaqueous media [63].

8.3.4 Fundamental Studies on C^{4}D

Many authors have worked on the improvements of C^{4}D instrumentation, which include the electronic circuitry

[64–68] and the geometry of the detection cell [67,69–71]. Johnston et al. [68] reported a 10-fold improvement in the signal-to-noise ratio by the use of a lock-in amplifier, instead of the rectifier/filter stage. A movable detection cell has also been proposed [72,73], but in fact any cell configuration where the electrodes are not fixed to the capillary allows fast positioning of the detection cell. Tuma et al. [74] evaluated the influence of the wall thickness of fused-silica capillaries on the performance of C^4D. Kuban et al. [75] compared different configurations of C^4D, including the TraceDec® commercial C^4D detector (Innovative Sensor Technologies, Austria). Fundamental studies on the parameters that affect the response of C^4D were extensively reported by several authors [15,16,64,67,68,76–78]. In 2006, Johnston et al. [79] proposed a conductivity-based photothermal absorbance detector for capillary separations. In this scheme, a laser beam is focused on the gap of a C^4D detection cell and, if a species absorbs the laser radiation, a local temperature increase is experienced, as well as an increase in the conductivity. Unfortunately, the capabilities of this absorbance/conductivity detection were not demonstrated with a real separation. More recently, Francisco and do Lago [80] proposed an all-in-one version of the C^4D, where the oscillator, detection circuitry, and a high-resolution analog-to-digital converter (ADC) were placed together with the conductivity cell. The absence of moving parts, allied to the low noise of the electronic parts, provides good sensitivity for the detection of alkaline and alkaline earth metal ions. This new version was demonstrated to be effective for most applications in CE-C^4D covering capillaries with internal diameters from 20 to 100 μm and commonly used BGE.

8.3.5 Applications

In more than 10 years of development of C^4D applied to CE, many applications have been developed, as demonstrated in the specialized literature. In general, any charged species can be detected using an appropriate BGE system. As a rule of thumb, high-mobility species are detected using low-conductivity (with low-mobility components) BGEs and the inverse is also true. The main condition for the appearance of a peak in CE-C^4D is the mobility differences between analytes and BGE co-ion (see Equation 8.2). Unfortunately, this intrinsic condition causes the so-called electrodispersion and consequently triangular-shaped peaks. From our experience, the integration of such peaks is not a problem, giving linear responses for at least three orders of magnitude.

Some BGEs cause serious problems of baseline drift or oscillation and, in some cases, the magnitude of the baseline variation exceeds the peak height by many times. In such cases, problems in peak integration and poor reproducibility arise. We have observed that, when the flow of material inside the capillary is very slow, baseline oscillations is prone to appear. These may be a result of poor heat dissipation inside the capillary, but this effect remains without explanation. The conditions of low flow basically occur when the mobility of the EOF is very small (or approaches zero for low pH BGEs) or when the viscosity of the solution inside the capillary is high (as in capillary sieving electrophoresis, CSE). Even so, some authors have reported separations in BGE with low pH (e.g., see Ref. 67 where separation of peptides is achieved in 100 mmol L^{-1} phosphate buffer, pH 2.5). These baseline instabilities may be the reason for so few reports on the use of C^4D in CSE.

Most applications of CE-C^4D have been reported for the separation and detection of inorganic ions, mainly alkaline and alkaline earth metal ions. In fact, a mixture of K^+, Na^+, and Li^+ has been intensively used as the sample model for evaluation of C^4D instrumental setups. Nevertheless, one can find in the literature examples of application of the technique for the separation and detection of inorganic and small organic ions in many samples such as, environmental [81–84], blood serum [85,86], urine [87], electroplating baths [88,89], ethanol fuel [90], coconut water [91], toothpaste [92], explosive residues [93], and fertilizers [94]. Other interesting applications of CE-C^4D involve the separation and detection of transition metal ions [43,95,96] and arsenic species [97]. Amines can be easily detected since cationic species are formed over a wide range of pH values [98]. Nerve agent degradation products were also separated and detected by Xu et al. [99,100] and Goodwin et al. [101]. All these applications were performed using the capillary zone electrophoresis mode or capillary isotachophoresis. Recently, Kuban et al. [102] showed that the deposition of poly(butadiene–maleic acid) copolymer onto the capillary wall can be used for the separation of inorganic cations in the open tubular capillary electrochromatography mode.

The feasibility of detection of short-chain carboxylic acids in different matrices has been demonstrated [103–105]. The analysis of long-chain carboxylic acids, or fatty acids, requires special care for the optimization of the BGE, which includes the addition of organic solvents and surfactants as modifiers to solubilize larger-chain fatty acids and for the resolution of critical pairs [106,107].

The determination of amino acids is of utmost importance in many fields and capillary electrophoresis offers a good alternative for the separation of amino acids in complex matrices. The use of C^4D is very advantageous, since no derivatization reaction is needed. On the other hand, the sensitivity can be a drawback for the analyses of biological samples. Nevertheless, Coufal et al. [108] report the separation of 20 amino acids in several matrices. Additionally, there are reports on separations using CE-C^4D of amino acids in human plasma [109], amniotic fluid [110], urine [111], beer [112], and animal glues after hydrolysis [113]. Reports on the separation and detection of peptides [67,114] and proteins [115,116] can also be found.

Carbohydrates and their derivatives represent classes of compounds with importance in clinical analyses and in the food industry, among others. Carvalho et al. [117] used CE-C4D for the determination of mono- and disaccharides in beverages. Good results were obtained using capillaries of 20 μm of internal diameter. Blanes et al. [118] determined chitin fragments obtained from the reaction with chitinase. The system was able to separate and detect *N*-acetylglucosamide and five *N*-acetyl-chitooligosaccharides (C2—C6) with limits of detection of ca. 3 $\mu mol\,L^{-1}$. Tuma et al. [119] made use of entangled polymers for the separation of the polysaccharide heparin in blood plasma and infusion solutions.

There are a few reports on the detection of synthetic polymers using electromigration techniques and C4D. Oudhoff et al. [120] applied capillary size-exclusion electrokinetic chromatography for the separation of neutral polystyrene standards. Spriestersbach et al. applied capillary zone electrophoresis for the separation of oligomers of the sodium salt of poly(acrylic acid) [121] and of poly (ethylene glycol)–poly(methacrylic acid) copolymers [122].

Although neutral species can be commonly separated using micellar electrokinetic chromatography, as demonstrated for the separation of several aliphatic alcohols [123], the sensitivity is not good enough for application with diluted samples without a preconcentration step. An alternative approach is based on the conversion of the neutral species into a charged one. Schuchert-Shi et al. [124] found that enzymatic reactions can be useful for this purpose, as demonstrated for the monitoring of ethanol, glucose, ethyl acetate, and ethyl butyrate and of urea [125]. Chemical reactions can also be used for the conversion of neutral species, as demonstrated by Rocha et al. [126] for the determination of formaldehyde in indoor and outdoor air through reaction of bisulfite to produce the stable anionic adduct hydroxymethanesulfonate. Nováková et al. [127] used the same reaction for the determination of hydrosulfite and metabisulfite by capillary isotachophoresis. Complexation with caffeic acid or 3,4-dimethoxycinnamic acid was used by Nogueira and do Lago [128] as strategy to create a negative charge on caffeine.

A wide range of applications can be found for pharmaceutical compounds, such as β-blockers and biologically active amines [129], fosfomycin [130], tobramycin [131], chlorhexidine digluconate and polyhexamethylene biguanide [132], salbutamol [133], glucosamine [134], suxamethonium [135], and ethambutol [136]. Quek et al. [137] developed a method for the determination of several pharmaceuticals classified as emerging pollutants. Enantiomeric separations have also been described by many authors [138–143].

Other separations of interest are the determination of vitamin C and preservatives in beverages [144], phenolic acids [145], DNA fragments [146], and flavonoids [147].

8.4 C4D APPLIED TO MICROCHIP CAPILLARY ELECTROPHORESIS

The fundamentals of C4D for conventional CE have also been applied to MCE. Since 2001 and 2002 [148,149], when the first reports were published, the use of C4D has become very popular for detection in microchip electrophoresis. When compared to conventional CE, the most remarkable difference is the electrode format used. In conventional CE, tubular electrodes are placed cylindrically around the outer polyimide coating of the fused-silica capillary. The integration of this geometry on miniaturized devices seemed impractical and, for this reason, planar electrodes are much easier to be integrated into the chip-based devices [11].

Regarding the detection cell arrangement on MCE, basically the electrodes may be embedded in the microchip, may be attached to it (placed either underneath the bottom or on top of the device), or may be external to device [11,20]. In the first option, the electrodes should be covered (i.e., physically separated from contact with the microchannel) with an insulating layer to enable the capacitive coupling. In the second, the bottom plate or top cover thicknesses provide insulation from the microfluidic channel. On the other hand, the use of external electrodes has offered many advantages for measurements in MCE, since they are not part of the microchip itself. In this geometry, electrodes are fabricated in a holder and coupled with a microfluidic device [11,20].

The electrodes for C4D measurements on MCE have been often fabricated by different approaches including photolithographic steps, screen-printing techniques, metal sputtering, or evaporation, as well as using practical tools such as silver paint, aluminum foil, or adhesive metal tapes. The following sections will present a short review about the improvements in detection cell geometry as well as recent applications using C4D on microfluidic devices.

8.4.1 Geometry of the Detection Electrodes

As mentioned before, the electrodes for C4D measurements on MCE can be placed, basically, in three different geometries. Electrodes can be (i) embedded, (ii) attached, or (iii) external to microfluidic device. Figure 8.5 shows examples of each configuration. Due to compatibility with microfabrication techniques, electrodes and channels can be incorporated on the same substrate material. In the pioneering reports, this compatibility was employed to construct microsystems containing microchannels and planar electrodes for electrochemical detection. Regarding C4D measurements, electrodes have been often covered by using thin film deposition technology.

8.4.1.1 Embedded Electrodes The first report about integrating C4D to MCE was published by Guijt et al. [148]. In their work, a four-electrode configuration was fabricated in a

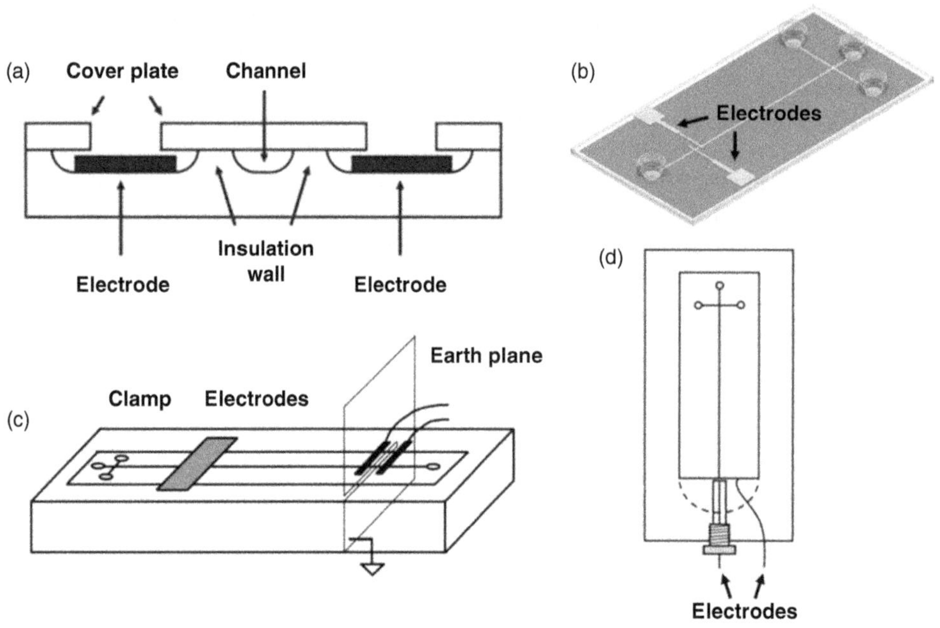

FIGURE 8.5 Detection cell geometry for C^4D on MCE. (a) Representation of in-plane embedded electrodes; (b) design of attached electrodes directly on MCE; (c) scheme of an external holder containing two electrodes for C^4D and their coupling with MCE; and (d) wall-jet configuration for C^4D on chip. (a), (b), (c), and (d) were adapted from Refs. 149, 162, 165, and 169, respectively.

glass substrate and insulated by a 30 nm thick silicon carbide layer. The four-electrode contactless conductivity measurement makes use of two outer electrodes and two inner electrodes. This arrangement enabled the capacitive coupling with the solution inside the microchannel. The wafers containing the electrodes and etched channels were integrated by anodic bonding at 400°C and 100 V for 1 h [148,150]. A very similar approach was reported later by Berthold et al. [151].

Lichtenberg et al. [149] reported a new approach to integrate in-plane electrodes for C^4D to glass channels. Microchannels and electrodes were formed on glass wafers in sequential photolithographic steps. Platinum electrodes were fabricated by first evaporating 20 nm of tantalum as adhesion layer, followed by 150 nm of Pt. A thin glass wall of 10–15 μm (see Figure 8.5a) provided the separation between the detection electrodes and the buffer electrolyte to achieve the electrical insulation necessary for the contactless mode. The wafer containing the etched-glass channels and the electrodes was fused against the other glass plate at 650°C.

Lee et al. [152] reported the construction and the characterization of semicircular electrodes for C^4D on a chip. The off-channel electrodes were deposited around side channels by Au sputtering and patterned using the lift-off process (Figure 8.6). The electrodes were separated from the microfluidic channel at a distance of 40 μm. When compared to planar electrodes, the use of semicircular detection electrodes did provide sensitivity performance in detecting a fluorescent dye as well as inorganic cations in sports drink samples. The difference in sensitivity is attributed to the

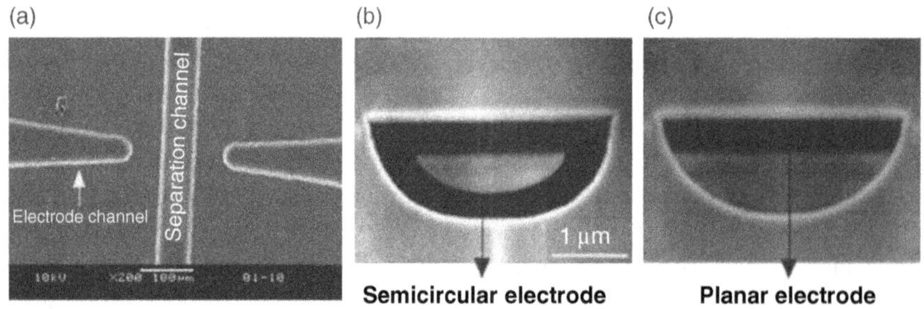

FIGURE 8.6 Scanning electron micrographs showing (a) the detection section of the separation channel and (b) semicircular and (c) planar electrodes in the bonded chips. Reprinted from Ref. 152 with permission.

enlarged area of the semicircular electrodes (see Figure 8.6b and c).

Shang et al. [153] reported the fabrication of a MCE with insulated electrodes for performing C^{4}D measurements. Chromium electrodes were formed using an e-beam evaporator and their electrical insulation was provided by a 100 nm SiO_2 layer. The insulating layer was deposited by plasma-enhanced chemical vapor deposition (PECVD).

A simpler method to fabricate embedded electrodes for performing C^{4}D on MCE was recently demonstrated by Segato et al. [154]. In this simple approach, a very thin PDMS layer (50 μm thick) was explored to provide the electrical insulation of glass channels to sputtered electrodes. This PDMS membrane enabled the capacitive coupling and also allowed the sealing of the glass channels at room temperature [154].

8.4.1.2 Attached Electrodes One of the limitations related to the embedded electrode format is the dependence on sophisticated instrumentation to provide an effective and accurate insulation between the solution channel and the electrode. Noneffective insulation can damage the detector's electronic circuit. For this reason, the search for new and simpler configurations has been the focus of many research groups.

In 2002, Pumera et al. [155] described one of the simplest methods to couple C^{4}D electrodes with MCE. Two rectangular-shaped electrodes, fabricated from aluminum strips, were placed on the outer side of poly(methyl methacrylate) (PMMA) microfluidic chips. In this case, the substrate thickness (125 μm) provided the insulation between the electrodes and the solution. Due to its simplicity and low instrumental requirements, this approach has been adopted by other groups.

Tanyanjiwa and Hauser [156], at the end of 2002, published the use of high-voltage C^{4}D for MCE. In their report, by applying an AC excitation signal of 500 V_{pp} and a frequency of 100 kHz it was possible to carry out detection through a cover plate of 1 mm thickness. For C^{4}D on chip, the authors used two parallel strips, ~0.5 mm wide and 5 mm long, that were formed with silver paint.

As in conventional CE, a movable C^{4}D system was also adapted for MCE by Wang et al. [157]. The rectangular shaped electrodes were fabricated from 10 μm thick aluminum foil and glued around two sides of the 1 mm-thick PMMA plate of the detector. The cover plate of the chip (125 μm thick) was thus mechanically pressed toward the detection electrodes without adhesive. The position of the detection electrodes could be changed by "sliding" the detector holder along the separation channel.

In 2003, do Lago et al. [158] demonstrated the coupling of C^{4}D with polyester-toner (PT) devices. Electrodes were formed using adhesive copper strips that were glued outside the polyester film. The 100 μm thickness allowed capacitive coupling with the solution. This same approach was used for glass-toner devices [159]. Wang et al. [160] demonstrated the use of the screen-printing process for generating highly reproducible sets of silver-ink electrodes for C^{4}D in MCE. Printing of silver electrodes was performed through a patterned stencil on thin PMMA sheets. The PMMA sheets containing the screen-printed electrodes were cut, aligned, and sealed to the channel plate, thus establishing a complete microchip separation device. Ding et al. [161] also showed that tapered rectangular electrodes for C^{4}D can be attached to PDMS chips.

The integration of planar electrodes on PT devices was also demonstrated by Coltro et al. [162]. Aluminum electrodes were sputtered on polyester films using toner masks to define the electrode area. After toner removal, the polyester with incorporated electrodes was laminated against a printed channel at 120°C and 40 cm min^{-1} to create PT devices with integrated electrodes for C^{4}D at a very low cost.

Recently, Mahabadi et al. [163] described a dual top–bottom C^{4}D cell configuration for microchips. This version consists of a pair of top–bottom excitation electrodes and a pair of pickup electrodes disposed onto a very thin plastic microfluidic chip. The electrodes are made of copper strips 1 or 2 mm in width, cut by precision machining. The sensing cell consists of two pairs of copper strips embedded into two polymer blocks and placed inside the housing at the top and bottom. When compared to top–top electrode geometry, the top–bottom geometry exhibited improved detector performance with a detection limit of ca. 0.3 μmol L^{-1} for inorganic cations and ca. 0.15 μmol L^{-1} for anions. The dual electrode geometry in conjunction with thin electrophoresis microchips doubles the total cell capacitance and enables an enhancement effect on the signal coupling into and from the detection volume. Using this strategy, these authors reported the best limits of detection for small inorganic ions found in the literature.

Most recently, an end-to-end differential C^{4}D for MCE was reported by Fercher et al. [164]. This new approach is based on the placement of two distinct detector areas (consisting of two screen-printed silver electrodes) near the ends of the fluid inlet and fluid outlet of the separation channel. Both output signals are subtracted from each other, and the resulting differential signal is amplified and measured. This measurement approach has several advantages over established, single-end detectors: The high baseline level resulting from parasitic stray capacitance and buffer conductivity is reduced, leading to better signal-to-noise ratio and hence higher measurement sensitivity. Furthermore, temperature and, thus, baseline drift effects are diminished owing to the differentiating nature of the system.

8.4.1.3 External Electrodes Tanyanyiwa et al. [165] described the use of external electrodes on the holder to perform high-voltage C^{4}D (HVC^{4}D) on electrophoresis

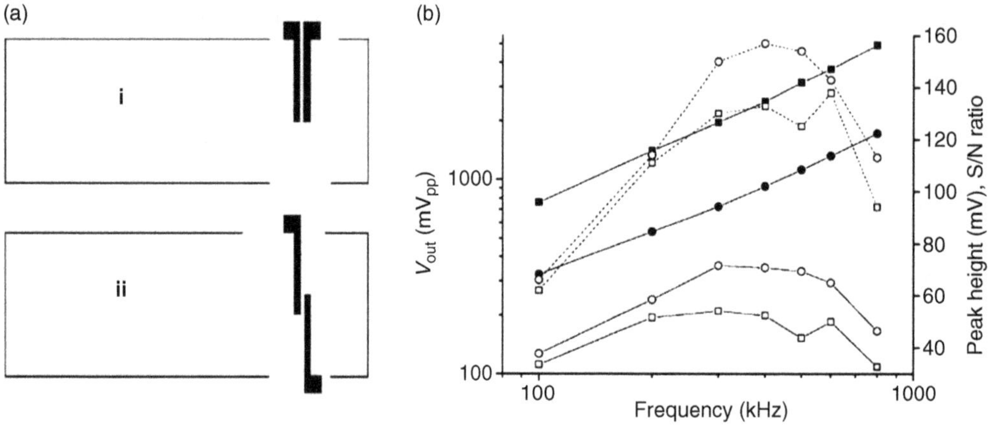

FIGURE 8.7 (a) Schematic drawing of detection electrodes in (i) parallel and (ii) antiparallel orientation. (b) Effect of parallel and antiparallel design of electrodes on output voltage, peak heights and S/N ratios of a lithium peak. V_{out} (parallel) —■—, V_{out} (antiparallel) —●—, peak height (parallel) --□--, peak height (antiparallel) --○--, S/N (parallel) —□—, S/N (antiparallel) —○—. Reprinted from Ref. 166 with permission.

microchips. In this approach, two electrodes, fabricated by using strips of adhesive copper tape, were mounted on a holder constructed from PMMA (see design presented in Figure 8.6c). The holder also features a vertical ground plane, which is employed to minimize direct capacitive coupling between the electrodes. For operation, the chip is placed and then clamped onto its holder. This instrumentation is very simple and allows the use of any kind of substrate material.

Kubán and Hauser [166] investigated the effects of cell geometry as well as operating parameters on the performance of this external system for C^{4}D on MCE. One of the evaluated parameters was the orientation of the electrodes, parallel or antiparallel, as depicted in Figure 8.7a. Using the geometries presented in Figure 8.7a, the cell output voltage, the peak heights, and the signal-to-noise ratio (S/N) were evaluated as functions of the excitation frequency. Due to direct capacitive coupling (stray capacitance) between the electrodes, it was found that output voltage is significantly higher (Figure 8.7b) for the parallel orientation. However, the higher stray capacitance results in reduced peak height and also lower S/N ratio (Figure 8.7b). Due to their better sensitivity, electrodes in antiparallel orientation have been often adopted for C^{4}D measurements on MCE. Hauser's group has significantly contributed to the popularization of C^{4}D on MCE. As will be seen in the following section, this group has presented a lot of examples of applications of chips using this detection mode.

Chen et al. [167] reported the coupling of a glass microchip, bonded with a thin cover glass plate, with an external copper plate containing detection electrodes previously fabricated by wet chemical etching. The authors investigated the effect of sine, square, and triangular waveforms. The sine wave offered the most favorable S/N ratio. This conclusion agrees with the results previously published by Wang et al. [168].

Wang et al. [169] also developed a "hybrid" electrode configuration for performing C^{4}D on MCE. It is based on a "hybrid" arrangement where the receiving electrode is insulated by a thin layer of insulator and placed in the bulk solution of the detection reservoir of the chip, whereas the emitting electrode is in contact with the solution eluted from the channel outlet in a wall-jet configuration (see design depicted in Figure 8.2d). Such an arrangement provided a 10-fold sensitivity enhancement compared to currently used on-channel C^{4}D-MCE as well as low values of noise and easy operation.

8.4.2 Applications

Since 2001, when the first reports about C^{4}D on MCE were published, this technique has led to a large range of applications. Some examples, including bioanalytical assays, on-chip enzymatic reactions, food analysis, and determinations of explosive and chemical warfare agents are listed in the following subheadings.

8.4.2.1 Bioanalytical Applications MCE with C^{4}D has been applied for the determination of a wide range of biochemical species including amino acids, proteins, peptides, DNA, and IgG. Hauser's group showed the first separation and detection of amino acids on MCE using C^{4}D [165,170]. One year later, Abad-Villar et al. [171] reported the detection of a wide range of biochemically relevant species. In their work, 12 essential amino acids were detected in ca. 4 min using a 2.3 mol L^{-1} acetic acid solution (pH 2.1) as BGE. The separation of peptides (leucine enkephalin, bradykinin, angiotensin II, TRH, and oxytocin) was also demonstrated using the same media with addition of 0.05% Tween 20. Electrophoretic separations of the proteins cytochrome c, human serum albumin (HSA), and myoglobin were also shown. For the mentioned examples of applications, separation efficiencies up to 15,000 plates could be obtained and LODs are in the low μmol L^{-1} range, except for immunoglobulin G (IgG) that could be determined down to 0.4 nmol L^{-1}.

Analytical feasibility has been also shown for detecting neurotransmitters (dopamine, ephedrine, and metanephrine), β-blockers (pindolol, atenolol, and acebutolol) and some physiologically active amines (doxylamine, octopamine, noradrenaline, adrenaline, and isoproterenol) using MCE-C^4D [129,170].

Gong and Hauser [140] reported that the baseline separation of the enantiomers of *trans*-cyclohexane-1,2-diamine is less than 2 min. In this case, MCE with C^4D was quite advantageous, since these compounds have neither strong UV-absorbance nor fluorescence.

The determination of 3-methylhistidine (3-MH) and 1-methylhistidine (1-MH) on MCE-C^4D was reported by Tuma et al. [111], in which a silicon microchip with electrodes made of aluminum strips was employed. Both compounds are clinically important.

Kubán and Hauser [172] showed the determination of inorganic ions in clinical samples (serum and urine samples) using MCE-C^4D. Bare electrophoresis chips were used in combination with external electrodes.

8.4.2.2 ***On-Chip Enzymatic Reactions*** Wang et al. [173] described an on-chip enzymatic assay for screening organophosphate (OP) nerve agents, based on a precolumn reaction of organophosphorus hydrolase (OPH), electrophoretic separation of the phosphonic acid products, followed by C^4D measurements on MCE. The complete bioassay was carried in around 3 min including the OPH reaction, separation, and detection of the reaction products.

Abad-Villar et al. [171] performed on-chip enzymatic digestions of HSA using pepsin. As the enzymatic digestion proceeds, the amount of different fragments varies, resulting in changes in the peak heights and shapes. The products of the enzymatic reaction were successfully monitored with MCE-C^4D. Schuchert-Shi et al. [124] successfully showed the conversion of glucose to gluconate using glucose oxidase and the enzymatic hydrolysis of ethyl acetate generating acetate ions. These same authors reported 1 year later the monitoring of the production of ammonium from urea in serum samples [125].

Vázquez et al. [174] have demonstrated the ability of C^4D to monitor the detection of peroxynitrite degradation products. Peroxynitrite ($ONOO^-$) is a strong oxidizing agent that is produced in cells from the reaction of superoxide (O_2^-) and nitric oxide (NO). It degrades rapidly into its main metabolites, that is, nitrate (NO_3^-) and nitrite (NO_2^-). Both metabolites were successfully detected on MCE by C^4D and amperometric detection coupled in-line.

8.4.2.3 ***Food Analysis*** A quantitative analysis of inorganic and organic ions in alcoholic (beer and wine) and nonalcoholic (potable water, fruit juices, and milk) beverages has been performed by MCE with C^4D [175]. Law et al. [176] reported the application of MCE with C^4D to determine food additives in soft drinks and vitamin C tablets. For the vitamin C tablet, besides the ascorbate (that is the major component) small amounts of chloride, citrate/maleate, and acetate were also found. For the soft drinks, sorbate and benzoate are the major additives present, accompanied by very small amounts of acetate and lactate.

In addition, Lee et al. [152] reported the determination of major inorganic cations in commercial sports drinks, mineral waters, and a red wine. Chen et al. [177] used a movable C^4D to detect sodium and chloride ions in seawater samples. Lu et al. [178] reported that the rapid and quantitative determination of fluoroacetate in diluted fruit juices and tap water was performed by MCE-C^4D.

8.4.2.4 ***Explosives and Chemical Warfare Agents*** MCE with C^4D has appeared a very promising tool to detect major explosive-related ions (ammonium, methylammonium, potassium, sodium, perchlorate, chloride, and nitrate) as well as their residues at the sites of terrorist attacks. Wang et al. [179] described the use of MCE-C^4D for the rapid separation and detection of explosive-related cations and anions using the same microchannel and running buffer. Using a movable C^4D [180,169] and a dual-opposite injection strategy, the separation/detection of these explosive-related ions was also demonstrated.

Nerve agents derive their name from their adverse effect on the nervous system. In response to recent terrorist activity and to the ratification of the Chemical Warfare Convention, there are urgent demands for rapid and reliable methods for the determination of chemical warfare agents (CWA) and their degradation products [179]. Wang et al. [181] demonstrated the application of MCE with C^4D for the screening of organophosphonate nerve agent degradation products. The miniaturized system relies on an efficient chip-based separation of alkyl methylphosphonic acids (breakdown products of Sarin, Soman, and VX nerve agents). Applicability to analysis of natural (river) water samples was demonstrated. Wang et al. [157] demonstrated the same application using a movable C^4D. Nerve agents were also used as model analytes to evaluate a MCE with dual-opposite injection strategy [180]. Ding et al. [161] have also shown the applicability of a PDMS device with C^4D for the determination of organophosphonate nerve agent degradation products. Applicability of this method for natural (lake and tap) water samples was also demonstrated.

Ding and Rogers [182] reported the use of a MCE with C^4D for the determination of nitrogen mustard gas degradation products. Three alkyl ethanolamines: *N*-methyldiethanolamine (MDEA), *N*-ethyldiethanolamine

(EDEA), and triethanolamine (TEA), (degradation/ precursor products of HN-1, HN-2, and HN-3 blister agents) were analyzed by MCE-C^{4}D. The original PDMS channel was coated with poly(ethyleneimine) (PEI) to improve the separation of the three ethanolamines. Experimental conditions for the separation and detection processes have been optimized to yield a well-defined separation and good sensitivity.

8.4.2.5 Other Applications Tanyanyiwa et al. [170] reported the separation of sucrose and fructose using a 10 mmol L^{-1} sodium hydroxide electrolyte or a 100 mmol L^{-1} triethylamine buffer at pH 12. Better sensitivity was obtained with the sodium hydroxide electrolyte. Mühlberger et al. [183] also reported an analytical method to separate carbohydrates on MCE-C^{4}D. Some carbohydrates such as glucose, fructose, and sucrose were determined in fruit nectar, fruit juice, and cola samples using PEEK chips [183]. In addition, detection of organic acids and pharmaceutical substances [170], as well heavy metals [165], has also been described by other groups using MCE with C^{4}D.

Walsh et al. [184] reported the use of scanning C^{4}D (sC^{4}D) for the evaluation of the structural homogeneity and density of both packed and monolithic stationary phases in microfluidic chips. The chip was placed on top of the substrate containing the copper electrodes and, by sliding the chip from mark 0 to 11, the packed columns were sequentially scanned along their whole length, from beginning to end (Figure 8.8a). To continue scanning another channel, the chip was simply displaced accordingly to the left or right.

The resulting sC^{4}D profiles are shown in Figure 8.8b. As can be seen, the dramatic increase in the C^{4}D response observed in the region between 6 and 8 mm for the column packed in channel 1 (C #1), and between 6 and 7 mm for the column packed in channel 2 (C #2), allows a straightforward determination of the end of the packed columns. However, the increase in the C^{4}D response for C #1 is more gradual than the one observed for C #2. This indicates that C #1 was not as densely packed in the last ca. 2 mm as the rest of the column.

8.5 CONCLUDING REMARKS

C^{4}D can be considered a breakthrough in CE detection because of its instrumental simplicity and flexibility. As UV–Vis transparent window detection is not required in C^{4}D, capillaries made with materials other than fused silica can be used. The limits of detection generally attained by CE-C^{4}D for species that lack a strong UV–Vis-absorbing moiety are considerably better than those obtained by indirect photometric detection. On the order hand, the strong influence of the temperature on the C^{4}D signal can be pointed out as the major drawback. A well-established theory about the physical and chemical principles involved in C^{4}D is found in the literature (see, e.g., [15,16]) and allows one to understand how some parameters affect the response of the detector. The optimization of the BGE composition is an important step to achieve a good compromise between sensitivity and separation efficiency.

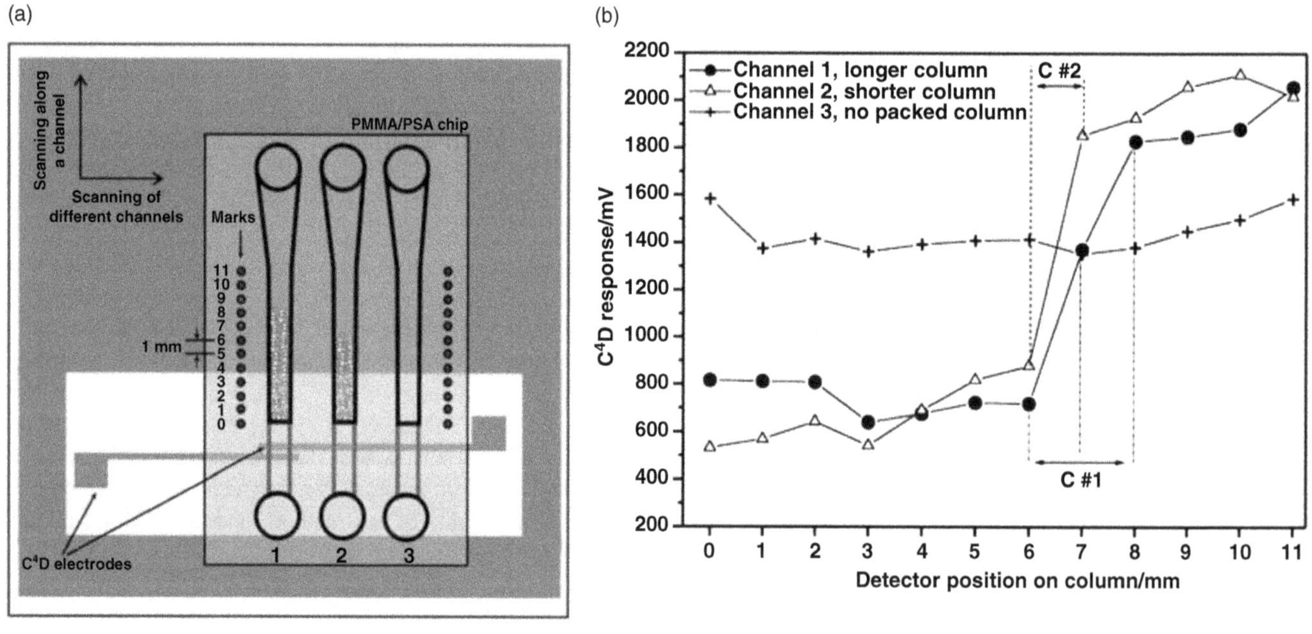

FIGURE 8.8 Representation of the (a) measurement setup for sC^{4}D and (b) typical scan results for packed microfluidic channels. Reprinted from Ref. 184 with permission.

The implementation of C^4D on MCE fabricated with different substrate materials with conventional and alternative microfabrication techniques was described. As the electrodes used in the detection are not in contact with the solutions, inexpensive materials (e.g., aluminum, carbon, or copper) can be used to manufacture the electrodes. So, low-cost and disposable MCEs with C^4D can be produced and used for fast biochemical assays, diagnostics, homeland security, food safety, and many other applications.

The wide range of successful applications described in the literature for CE-C^4D and MCE-C^4D demonstrate the versatility of this detection method, and the fact that commercial detectors are now available suggests that there will be much more in the coming years.

ACKNOWLEDGMENTS

Professor Carol H. Collins is thanked for manuscript revision and language assistance.

REFERENCES

1. Zahn, H. (1928) *Z. Physik 51*, 350–354.
2. Burkhalter, T.S. (1964) High frequency conductometric (impedimetric) titrations, In: Wilson, C.L. and Wilson, D. W. (eds) *Comprehensive Analytical Chemistry, IIA, Electrical Methods, Chapter V*, Elsevier, Amsterdam.
3. Pungor, E., Pal, F., and Toth, K. (1983) *Anal. Chem. 55*, 1728–1731.
4. Alder, J.F., Fielden, P.R., and Clark, A.J. (1984) *Anal. Chem. 56*, 985–988.
5. Pal, F., Pungor, E., and Kovats, E. (1988) *Anal. Chem. 60*, 2254–2258.
6. Gas, B. and Vacik, J. (1980) *Chem. Listy 74*, 652–658.
7. Gas, B., Demjanenko, M., and Vacik, J. (1980) *J. Chromatogr. 192*, 253–257.
8. Vacik, J., Zuska, J., and Muselasova, I. (1985) *J. Chromatogr. 320*, 233–240.
9. Zemann, A.J., Schnell, E., Volgger, D., and Bonn, G.K. (1998) *Anal. Chem. 70*, 563–567.
10. da Silva, J.A.F. and do Lago, C.L. (1998) *Anal. Chem. 70*, 4339–4343.
11. Pumera, M. (2007) *Talanta 74*, 358–364.
12. Hruška, V., Jaroš, M., and Gaš, B. (2006) *Electrophoresis 27*, 984–991.
13. Jaroš, M., Soga, T., van de Goor, T., and Gaš, B. (2005) *Electrophoresis 26*, 1948–1953.
14. Katzmayr, M.U., Klampfl, C.W., and Buchberger, W. (1999) *J. Chromatogr. A 850*, 355–362.
15. Brito-Neto, J.G.A., da Silva, J.A.F., Blanes, L., and do Lago, C.L. (2005) *Electroanalysis 17*, 1198–1206.
16. Brito-Neto, J.G.A., da Silva, J.A.F., Blanes, L., and do Lago, C.L. (2005) *Electroanalysis 17*, 1207–1214.
17. Gaš, B., Hruška, V., Dittmann, M., Bek, F., and Witt, K. (2007) *J. Sep. Sci. 30*, 1435–1445.
18. Kuban, P. and Hauser, P.C. (2009) *Electrophoresis 30*, 176–188.
19. Matysik, F.M. (2008) *Microchim. Acta 160*, 1–14.
20. Kuban, P. and Hauser, P.C. (2008) *Anal. Chim. Acta 607*, 15–29.
21. Sonlinova, V. and Kasicka, V. (2006) *J. Sep. Sci. 29*, 1743–1762.
22. Opekar, F. and Stulik, K. (2006) *Electroanalysis 18*, 1282–1288.
23. Knjazeva, T., Kulp, M., and Kaljurand, M. (2009) *Electrophoresis 30*, 424–430.
24. Tanyanyiwa, J., Leuthardt, S., and Hauser, P.C. (2002) *J. Chromatogr. A 978*, 205–211.
25. Seiman, A., Jaanus, M., Vaher, M., and Kaljurand, M. (2009) *Electrophoresis 30*, 507–514.
26. Kuban, P., Nguyen, H.T.A., Macka, M., Haddad, P.R., and Hauser, P.C. (2007) *Electroanalysis 19*, 2059–2065.
27. Rainelli, A. and Hauser, P.C. (2005) *Anal. Bioanal. Chem. 382*, 789–794.
28. Chvojka, T., Jelínek, I., Opekar, F., and Stulik, K. (2001) *Anal. Chim. Acta 433*, 13–21.
29. Pumera, M., Jelínek, I., and Jindrich, J. (2001) *Fresenius J. Anal. Chem. 369*, 666–669.
30. Novotny, M., Opekar, F., Jelínek, I., and Stulik, K. (2004) *Anal. Chim. Acta 525*, 17–21.
31. Tan, F., Yang, B., and Guan, Y. (2005) *Anal. Sci. 21*, 583–585.
32. Zikmundová, J., Tuma, P., and Opekar, F. (2008) *J. Sep. Sci. 31*, 353–355.
33. Petr, J., Maier, V., Horáková, J., and Sevcik, J. (2006) *Electrophoresis 27*, 4735–4745.
34. Urban, P.L., Goodall, D.M., Carvalho, A.Z., Bergström, E.T., Van Schepdael, A., Neil C., and Bruce, N.C. (2008) *J. Chromatogr. A 1206*, 52–63.
35. Xuan, Y., Weber, G., and Manz, A. (2006) *J. Chromatogr. A 1130*, 212–218.
36. Vourinen, P.S., Jussila, M., Sirén, H., Palonen, S., and Riekkola, M.-L. (2003) *J. Chromatogr. A 990*, 45–52.
37. Lopez-Avila, V., van de Goor, T., Gas, B., and Coufal, P. (2003) *J. Chromatogr. A 993*, 143–152.
38. Jensen, H., Østergaard, J., Thomsen, A.E., and Hansen, S.H. (2007) *Electrophoresis 28*, 322–327.
39. Tuma, P., Opekar, O., Samcova, E., and Stulik, K. (2008) *Electroanalysis 20*, 477–484.
40. Kuban, P., Kuban, P., and Kuban, V. (2002) *Electrophoresis 23*, 3725–3734.
41. Kuban, P., Karlberg, B., Kuban, P., and Kuban, V. (2002) *J. Chromatogr. A 964*, 227–241.
42. Kuban, P., Reinhardt, M., Müller, B., and Hauser, P.C. (2004) *J. Environ. Monit. 6*, 169–174.

43. Kuban, P., Kuban, P., and Kuban, V. (2003) *Electrophoresis 24*, 1397–1403.
44. Kuban, P., Kuban, P., and Kuban, V. (2003) *Electrophoresis 24*, 1935–1943.
45. Kuban, P., Kuban, P., and Kuban, V. (2004) *Anal. Bioanal. Chem. 378*, 378–382.
46. Kuban, P., Kuban, P., Hauser, P.C., and Kuban, V. (2004) *Electrophoresis 25*, 35–42.
47. Wang, L.X. and Fu, C.G. (2004) *Instrum. Sci. Technol. 32*, 303–309.
48. Wang, J., Cai, P., Mo, J. and Chen, Z. (2005) *Anal. Lett. 38*, 857–867.
49. Wuersig, A., Kuban, P., Khaloo, S.S., and Hauser, P.C. (2006) *Analyst 131*, 944–949.
50. Sprung, C., Sirén, H., Rovio, S., and Työppönen, T. (2008) *Sep. Sci. Technol. 15*, 3856–3872.
51. Muzikar, J., van de Goor, T., Gas, B., and Kenndler, E. (2001) *J. Chromatogr. A 924*, 147–154.
52. Muzikar, J., van de Goor, T., Gas, B., and Kenndler, E. (2002) *Anal. Chem. 74*, 428–433.
53. Porras, S.P. and Kenndler, E. (2004) *Electrophoresis 25*, 2946–2958.
54. Borissova, M., Gorbatsova, J., Ebber, A., Kaljurand, M., Koel, M., and Vaher, M. (2007) *Electrophoresis 28*, 3600–3605.
55. Jensen; H., Østergaard, J., and Hansen, S.T. (2007) *Electrophoresis 28*, 2975–2980.
56. Henchoz, Y., Schappler, J., Geiser, L., Prat, J., Carrupt, P.-A., and Veuthey, J.-L. (2007) *Anal. Bioanal. Chem. 389*, 1869–1878.
57. Vceláková, K., Zusková, I., Kenndler, E., and Gas, B. (2004) *Electrophoresis 25*, 309–317.
58. Gas, B., Coufal, P., Jaros, M., Muzikar, J., and Jelinek, I. (2001) *J. Chromatogr. A 905*, 269–279.
59. Stedry, M., Jaros, M., Vcelakova, K., and Gas, B. (2003) *Electrophoresis 24*, 536–547.
60. Vcelakova, K., Zuskova, I., Porras, S.P., Gas, B., and Kenndler, E. (2005) *Electrophoresis 26*, 463–472.
61. Muzikar, J., van de Goor, T., Gas, B., and Kenndler, E. (2002) *J. Chromatogr. A 960*, 199–208.
62. Saito, R.M., Neves, C.A., Lopes, F.S., Blanes, L., Brito-Neto, J.G.A., and do Lago, C.L. (2007) *Anal. Chem. 79*, 215–223.
63. Seiman, A., Vaher, M., and Kaljurand, M. (2008) *J. Chromatogr. A 1189*, 266–273.
64. da Silva, J.A.F., Guzman, N., and do Lago, C.L. (2002) *J. Chromatogr. A 942*, 249–258.
65. Zhang, S., Wang, L., Dang, Z., Li, T., Liu, X., Huang, X., Deng, X., and Tang, X. (2007) *Instrum. Sci. Technol. 35*, 275–294.
66. Tanyanyiwa, J., Galliker, B., Schwarz, M.A., and Hauser, P. (2002) *Analyst 127*, 214–218.
67. Baltussen, E., Guijt, R.M., van der Steen, G., Laugere, F., Baltussen, S., and van Dedem, G.W.K. (2002) *Electrophoresis 23*, 2888–2893.
68. Johnston, S.E., Fadgen, K.E., Tolley, L.T., and Jorgenson, J. W. (2005) *J. Chromatogr. A 1094*, 148–157.
69. Tuma, P., Opekar, F., and Jelinek, I. (2001) *Electroanalysis, 13*, 989–992.
70. Tuma, P., Opekar, F., and Stulik, K. (2002) *Electrophoresis 23*, 3718–3724.
71. Novotny, M., Opekar, F., and Stulik, K. (2005) *Electroanalysis 17*, 1181–1186.
72. Unterholzner, V., Macka, M., Haddad, P.R., and Zemann, A. (2002) *Analyst 127*, 715–718.
73. Macka, M., Hutchinson, J., Zemann, A., Shusheng, Z., and Haddad, P.R. (2003) *Electrophoresis 24*, 2144–2149.
74. Tuma, P., Samcova, E., and Stulik, K. (2009) *Electroanalysis, 21*, 590–594.
75. Kuban, P., Evenhuis, C.J., Macka, M., Haddad, P. R., and Hauser, P.C. (2006) *Electroanalysis 18*, 1289–1296.
76. Gas, B., Zuska, J., Coufal, P., and van de Goor, T. (2002) *Electrophoresis 23*, 3520–3527.
77. Kuban, P. and Hauser, P.C. (2004) *Electrophoresis 25*, 3387–3397.
78. Kuban, P. and Hauser, P.C. (2004) *Electrophoresis 25*, 3398–3405.
79. Johnston, S.E., Fadgen, K.E., and Jorgenson, J.W. (2006) *Anal. Chem. 78*, 5309–5315.
80. Francisco, K.J.M. and do Lago, C.L. (2009) *Electrophoresis 30*, 3458–3464.
81. Rocha, F.R., da Silva, J.A.F., do Lago, C.L., Fornaro, A., and Gutz, I.G.R. (2003) *Atmos. Environ. 37*, 105–115.
82. Fornaro, A. and Gutz, I.G.R. (2003) *Atmos. Environ. 37*, 117–128.
83. Richter, E.M., Fornaro, A., do Lago, C.L., and Angnes, L. (2007) *Quim. Nova 30*, 1147–1152.
84. Xu, Y., Wang, W., and Li, S.F.Y. (2007) *Electrophoresis 28*, 1530–1539.
85. da Silva, J.A.F., Ricelli, N.L., Carvalho, A.Z., and do Lago, C. L. (2003) *J. Braz. Chem. Soc. 14*, 265–268.
86. Wan, Q.J., Kuban, P., Tanyanyiwa, J., Rainelli, A., and Hauser, P.C. (2004) *Anal. Chim. Acta 525*, 11–16.
87. Tuma, P., Samcová, E., and Duska, F. (2008) *J. Sep. Sci. 31*, 2260–2264.
88. Zhang, L., Khaloo, S.S., Kuban, P., and Hauser, P.C. (2006) *Meas. Sci. Technol. 17*, 3317–3322.
89. Wang, L.-S., Zhang, S.-F., Dang, Z., Liu, X.-X., Huang, X.-J., Xiao, M.-W., and Chen, Z.-W. (2007) *Talanta 72*, 1342–1347.
90. Muñoz, R.A.A., Richter, E.M., de Jesus, D.P., do Lago, C.L., and Angnes, L. (2004) *J. Braz. Chem. Soc. 15*, 523–526.
91. Richter, E.M., de Jesus, D.P., Muñoz, R.A.A., do Lago, C.L., and Angnes, L. (2005) *J. Braz. Chem. Soc. 16*, 1134–1139.
92. Guimarães, I.C., Rezende, C.C., da Silva, J.A., and de Jesus, D.P. (2009) *Talanta 78*, 1436–1439.
93. Hutchinson, J.P., Johns, C., Breadmore, M.C., Hilder, E.F., Guijt, R.M., Lennard, C., Dicinoski, G., and Haddad, P.R. (2008) *Electrophoresis 29*, 4593–4602.

94. Nováková, M., Krivánková, L., Skalicka, I., Bartos, M., and Vytras, K. (2008) *Chromatographia 67*, S63–S68.

95. Tanyanyiwa, J. and Hauser, P.C. (2002) *Electrophoresis 23*, 3781–3786.

96. Tan, F., Yang, B., and Guan, Y. (2005) *Anal. Sci. 21*, 955–958.

97. Nguyen, H.T.A., Kuban, P., Pham, V.H., and Hauser, P.C. (2007) *Electrophoresis 28*, 3500–3506.

98. Gong, X.Y. and Hauser, P.C. (2006) *Electrophoresis 27*, 468–473.

99. Xu, L., Gong, X.Y., Lee, H.K., and Hauser, P.C. (2008) *J. Chromatogr. A 1205*, 158–162.

100. Xu, L., Hauser, P.C., and Lee, H.K. (2008) *J. Chromatogr. A 1214*, 17–22.

101. Goodwin, L., Hanna, M., Startin, J.R., Keely, B.J., and Goodall, D.M. (2002) *Analyst 127*, 204–206.

102. Kuban, P., Kuban, P., Kuban, V., Hauser, P. C., and Bocek, P. (2008) *J. Chromatogr. A 1190*, 377–382.

103. Law, W.S., Zhao, J.H., Hauser, P.C., and Li, S.F.Y. (2007) *J. Sep. Sci. 30*, 3247–3254.

104. Belin, G.K., Krähenbühl, S., and Hauser, P.C. (2007) *J. Chromatogr. B 847*, 205–209.

105. Gong, X.Y., Kuban, P., Scholer, A., and Hauser, P.C. (2008) *J. Chromatogr. A 1213*, 100–104.

106. Oliveira, M.A.L., do Lago, C.L., Tavares, M.F.M., and da Silva, J.A.F. (2003) *Quim. Nova 26*, 821–824.

107. Surowiec, I., Kaml, I., and Kenndler, E. (2004) *J. Chromatogr. A 1024*, 245–254.

108. Coufal, P., Zuska, J., van de Goor, T., Smith, V., and Gas, B. (2003) *Electrophoresis 24*, 671–677.

109. Samková, E. and Tuma, P. (2006) *Electroanalysis 18*, 152–157.

110. Tuma, P., Samková, E., and Andelová, K. (2006) *J. Chromatogr. B 839*, 12–18.

111. Tuma, P., Samková, E., Opekar, F., Jurka, V., and Stulik, K. (2007) *Electrophoresis 28*, 2174–2180.

112. Tanyanyiwa, J., Schweizer, K., and Hauser, P.C. (2003) *Electrophoresis 24*, 2119–2124.

113. Harrison, S.M., Kaml, I., Prokoratova, V., Mazanek, M., and Kenndler, E. (2005) *Anal. Bioanal. Chem. 382*, 1520–1526.

114. Sázelová, P., Kasicka, V., Solínová, V., and Koval, D. (2006) *J. Chromatogr. B 841*, 145–151.

115. Abad-Villar, E.M., Tanyanyiwa, J., Fernandez-Abedul, M.T., Costa-Garcia, A., and Hauser, P.C. (2004) *Anal. Chem. 76*, 1282–1288.

116. Abad-Villar, E.M., Kuban, P., and Hauser, P.C. (2006) *J. Sep. Sci. 29*, 1031–1037.

117. Carvalho, A.Z., da Silva, J.A.F., and do Lago, C.L. (2003) *Electrophoresis 24*, 2138–2143.

118. Blanes, L., Saito, R.M., Genta, F.A., Donegá, J., Terra, W.R., Ferreira, C., and do Lago, C.L. (2008) *Anal. Biochem. 373*, 99–103.

119. Tuma, P., Samcová, E., Opekar, F., and Stulik, K. (2008) *Collect. Czech. Chem. Commun. 73*, 187–200.

120. Oudhoff, K.A., Macka, M., Haddad, P.R., Schoenmakers, P.J., and Kok, W.T. (2005) *J. Chromatogr. A 1068*, 183–187.

121. Spriestersbach, K.-H., Rittig, F., and Pasch, H. (2008) *Electrophoresis 29*, 4407–4411.

122. Spriestersbach, K.-H., Rittig, F., and Pasch, H. (2009) *Int. J. Polym. Anal. Charact. 14*, 196–209.

123. da Silva, J.A.F. and do Lago, C.L. (2000) *Electrophoresis 21*, 1405–1408.

124. Schuchert-Shi, A., Kuban, P., and Hauser, P.C. (2007) *Electrophoresis 28*, 4690–4696.

125. Schuchert-Shi, A. and Hauser, P.C. (2008) *Anal. Biochem. 376*, 262–267.

126. Rocha, F.R., Coelho, L.H.G., Lopes, M.L.A., Carvalho, L.R. F., da Silva, J.A.F., do Lago, C.L., and Gutz, I.G.R. (2008) *Talanta 76*, 271–275.

127. Nováková, M., Krivánková, Bartos, M., Urbanová, V., and Vytras, K. (2007) *Talanta 74*, 183–189.

128. Nogueira, T. and do Lago, C.L. (2007) *Electrophoresis 28*, 3570–3574.

129. Tanyanyiwa, J. and Hauser, P.C. (2004) *Electrophoresis 25*, 3010–3016.

130. Petsch, M., Mayer-Helm, B.X., Sauermann, R., Joukhadar, C., and Kenndler, E. (2004) *Electrophoresis 25*, 2292–2298.

131. Law, W.S., Kuban, P., Yuan, L.L., Zhao, J.H., Li, S.F.Y., and Hauser, P.C. (2006) *Electrophoresis 27*, 1932–1938.

132. Abad-Villar, E.M., Etter, S.F., Thiel, M.A., and Hauser, P.C. (2006) *Anal. Chim. Acta 561*, 133–137.

133. Felix, F.S., Quintino, M.S.M., Carvalho, A.Z., Coelho, L.H. G., do Lago, C.L., and Angnes, L. (2006) *J. Pharm. Biomed. Anal. 40*, 1288–1292.

134. Jac, P., Los, P., Spácil, Z., Pospísilová, M., and Polásek, M. (2008) *Electrophoresis 29*, 3511–3518.

135. Nussbaumer, S., Fleury-Souverain, S., Rudaz, S., Bonnabry, P., and Veuthey, J.-L. (2009) *J. Pharm. Biomed. Anal. 49*, 333–337.

136. da Silva, J.A.F., Castro, N.V., de Jesus, D.P., Faria, A.F., de Souza, M.V.N., and de Oliveira, M.A.L. (2010) *Electrophoresis 31*, 570–574.

137. Quek, N.M., Law, W.S., Lau, H.F., Zhao, J.H., Hauser, P.C., and Li, S.F.Y. (2008) *Electrophoresis 29*, 3701–3709.

138. Gong, X.Y., Kuban, P., Tanyanyiwa, J., and Hauser, P.C. (2005) *J. Chromatogr. A 1082*, 230–234.

139. Gong, X.Y. and Hauser, P.C. (2005) *J. Chromatogr. A 1094*, 196–199.

140. Gong, X.Y. and Hauser, P.C. (2006) *Electrophoresis 27*, 4375–4382.

141. Maier, V., Petr, J., Knob, R., Horáková, J., and Sevcík, J. (2007) *Electrophoresis 28*, 1815–1822.

142. Gong, X.Y., Dobrunz, D., Kümin, M., Wiesner, M., Revell, J. D., Wennemers, H., and Hauser, P.C. (2008) *J. Sep. Sci. 31*, 565–573.

143. Lecoeur-Lorin, M., Delépée, R., and Morin, P. (2009) *Electrophoresis 30*, 487–498.

144. Law, W.S., Kuban, P., Zhao, J.H., Li, S.F.Y., and Hauser, P.C. (2005) *Electrophoresis 26*, 4648–4655.
145. Kuban, P., Sterbová, D., and Kuban, V. (2006) *Electrophoresis 27*, 1368–1375.
146. Xu, Y. and Li, S.F.Y. (2006) *Electrophoresis 27*, 4025–4028.
147. Bachmann, S., Huck, C.W., Bakry, R., and Bonn, G.K. (2007) *Electrophoresis 28*, 799–805.
148. Guijt, R.M., Baltussen, E., Van Der Steen, G., Frank, J., Billiet, H.A.H., Schalkhammer, T., Laugere, F., Vellekoop, M.J., Berthold, A., Sarro, P.M., and Van Dedem, G.W.K. (2001) *Electrophoresis 22*, 2537–2541.
149. Lichtenberg, J., Rooij, N.F., and Verpoorte, E. (2002) *Electrophoresis 23*, 3769–3780.
150. Laugere, F., Guijt, R.M., Bastemeijer, J., Van Der Steen, G., Berthold, A., Baltussen, E., Sarro, P., Van Dedem, G.W.K., Vellekoop, M.J., and Bossche, A. (2003) *Anal. Chem. 75*, 306–312.
151. Berthold, A., Laugere, F., Schellevis, H., de Boer, C.R., Laros, M., Guijt, R.M., Sarro, P.M., and Vellekoop, M.J. (2001) *Electrophoresis 22*, 3511–3519.
152. Lee, C.Y., Chen, C.M., Chang, G.L., Lin, C.H, and Fu, L.M. (2006) *Electrophoresis 27*, 5043–5050.
153. Shang, T., Teng, E., Woolley, A.T., Mazzeo, B.A., Schultz, S. M., and Hawkins, A.R. (2010) *Electrophoresis 31*, 2596–2601.
154. Segato, T.P., Coltro, W.K.T., Almeida, A.L.J., Piazetta, M.H. O., Gobbi, A.L., Mazo, L.H., and Carrilho, E. (2010) *Electrophoresis 31*, 2526–2533.
155. Pumera, M., Wang, J., Opekar, F., Jelínek, I., Feldman, J., Löwe, H., and Hardt, S. (2002) *Anal. Chem. 74*, 1968–1971.
156. Tanyanyiwa, J. and Hauser, P.C. (2002) *Anal. Chem. 74*, 6378–6385.
157. Wang, J., Chen, G., and Muck, A. (2003) *Anal. Chem. 75*, 4475–4479.
158. do Lago, C.L., Da Silva, H.D.T., Neves, C.A., Brito-Neto, J. G.A., and Da Silva, J.A.F. (2003) *A. Anal. Chem. 75*, 3853–3858.
159. do Lago, C.L., Neves, C.A., Jesus, D.P., da Silva, H.D.T., Brito-Neto, J.G.A., and da Silva, J.A.F. (2004) *Electrophoresis 25*, 3825–3831.
160. Wang, J., Chen, G., Chatrathi, M.P., Wang, M., Rinehart, R., and Muck, A. (2008) *Electroanalysis 20*, 2416–2421.
161. Ding, Y., Garcia, C.D., and Rogers, K.R. (2008) *Anal. Lett. 41*, 335–350.
162. Coltro, W.K.T., Da Silva, J.A.F., and Carrilho, E. (2008) *Electrophoresis 29*, 2260–2265.
163. Mahabadi, K.A., Rodriguez, I., Lim, C.Y., Maurya, D.K., Hauser, P.C., and de Rooij, N.F. (2010) *Electrophoresis 31*, 3511–3519.
164. Fercher, G., Haller, A., Smetana, W., and Vellekoop, M.J. (2010) *Anal. Chem. 82*, 3270–3275.
165. Tanyanyiwa, J., Abad-Villar, E.M., Fernandéz-Abedul, T., Costa-García, A., Hoffmamn, W., Guber, A.E., Herrmann, D., Gerlach, A., Gottschlich, N., and Hauser, P.C. (2003) *Analyst 128*, 1019–1023.
166. Kubáñ, P. and Hauser, P.C. (2005) *Lab Chip 5*, 407–415.
167. Chen, Z., Li, Q., Li, O., Zhou, X., Wei, Y., and Mo, J. (2007) *Talanta 71*, 1944–1950.
168. Wang, J., Pumera, M., Collins, G., and Mulchandani, A. (2002) *Anal. Chem. 74*, 621–624.
169. Wang, J., Chen, G., and Muck, A. (2009) *Talanta 78*, 207–211.
170. Tanyanyiwa, J., Abad-Villar, E.M., and Hauser, P.C. (2004) *Electrophoresis 25*, 903–908.
171. Abad-Villar, E.M., Kubáñ, P., and Hauser, P.C. (2005) *Electrophoresis 26*, 3609–3614.
172. Kubán, P. and Hauser, P.C. (2008) *Lab Chip 8*, 1929–1936.
173. Wang, J., Chen, G., Muck Jr., A., Chatrathi, M.P., Mulchandani, A., and Chen, W. (2004) *Anal. Chim. Acta 505*, 183–187.
174. Vázquez, M., Frankenfeld C., Coltro, W.K.T., Carrilho E., Diamond D., and Lunte S.M. (2010) *Analyst 135*, 96–103.
175. Kubán, P. and Hauser, P.C. (2005) *Electrophoresis 26*, 3169–3178.
176. Law, W.S., Kubán, P., Zhao, J.H., Li, S.F.Y., and Hauser, P.C. (2005) *Electrophoresis 26*, 4648–4655.
177. Chen, Y., Yang, P., Li, J., Chen, D., and Chen, G. (2006) *Anal. Bioanal. Chem. 384*, 683–691.
178. Lu, Q., Wu, P., and Collins, G.E. (2007) *Electrophoresis 28*, 3485–3491.
179. Wang, J., Pumera, M., Collins, G., Opekar, F., and Jelínek, I. (2002) *Analyst 127*, 719–723.
180. Wang, J., Chen, G., Muck Jr., A., and Collins, G. (2003) *Electrophoresis 24*, 3728–3734.
181. Wang, J., Pumera, M., Collins, G., and Mulchandani, A. (2002) *Anal. Chem. 74*, 6121–6125.
182. Ding, Y. and Rogers, K.R. (2008) *Electroanalysis 20*, 2192–2198.
183. Muhlberger, H., Wonhee H., Guber, A.E., Saile, V., and Hoffmann, W. (2008) *IEEE 8*, 572–579.
184. Walsh, Z., Vazquez, M., Benito-Lopez, F., Paull, B., Macka, M., Svecd, F., and Diamond, D. (2010) *Lab Chip 10*, 1777–1780.

9

CAPILLARY ELECTROPHORESIS WITH ELECTROCHEMICAL DETECTION

Blánaid White

9.1 PRINCIPLES OF ELECTROCHEMICAL DETECTION

Three types of electrochemical detection have been demonstrated for CE, measuring current (amperometric detection), voltage (potentiometric detection), or conductivity (conductimetric detection) [1].

9.1.1 Amperometric Detection

Amperometric detection (AD) is based on the oxidation or reduction of an analyte on a working electrode (WE). As a consequence, only electroactive species can be detected. However, very low limits of detection (LOD) can be achieved. Analyte compounds undergo oxidation or reduction reactions through the loss or gain, respectively, of electrons at the electrode surface. Oxidation is most commonly employed, due to interference from dissolved oxygen in reductive mode. The electrical current arising from the electrons passed to or from the electrode is recorded and is proportional to the concentration of analyte present. The current (i) that is measured in amperometry is related to A the electrode area, n the number of electrons exchanged, F the Faraday constant, D the diffusion coefficient, δ_N the thickness of the diffusion layer c, and the concentration of the species as illustrated in Equation 9.1.

$$i = -AnFD\frac{c}{\delta_N} \tag{9.1}$$

The applied potential must be positive or negative enough to induce oxidation or reduction of the analyte species, respectively. In bulk amperometric analysis, the concentration of the analyte that is oxidized or reduced is sufficiently small as to not affect the concentration of the analyte in the bulk solution. Given the small cell volumes used in CE, however, this may no longer be true. The fraction of the analyte that is consumed at the detection electrode is often expressed as coulometric efficiency, as methods where complete bulk electrolysis occurs are generally known as coulometry. A high coulometric efficiency is desirable as this relates to the sensitivity of the detection.

AD, also called chronoamperometry, typically measures current with respect to time, and utilizes a three-electrode setup (working electrode (WE), reference electrode (RE), counter electrode) controlled by a potentiostat [2]. This instrument controls the voltage across the working electrode-counter electrode pair, and adjusts this voltage by maintaining the potential difference between the working and the reference electrodes. Practically, this is achieved by adjusting the potential of the solution (as measured by the RE) up or down by adding or removing charge at the counter electrode. To create this current, a driving voltage is required to overcome the resistance of the ions in solution to move toward the appropriate electrode, and this potential required is known as iR drop or ohmic potential drop, which leads to an error in the working potential of this system [3]. In CE–EC, any residual voltage gradient due to the electrophoretic separation voltage will add to this bias, and as it is usually in the order of several hundred millivolts, it is not negligible

Capillary Electrophoresis and Microchip Capillary Electrophoresis: Principles, Applications, and Limitations, First Edition.
Edited by Carlos D. García, Karin Y. Chumbimuni-Torres, and Emanuel Carrilho.

and has to be compensated for by adjusting the working potential setting accordingly. To find the correct value, hydrodynamic voltammograms with standards must be acquired. Practically, the iR drop is usually minimized by placing the RE as close to the WE as possible, but is not completely eliminated. (In CE–EC, the WE is also referred to as the detector electrode.)

In addition to minimizing the potential gradient between the WE and the RE to minimize iR drop, it is also necessary to minimize the potential gradient in the solution across the width of the WE itself. If not, at part of the electrode surface, the potential gradient between the solution and the electrode surface may lead to the reduction of oxidation of water, and thus an unacceptable background current, given that the majority of systems are aqueous. If this gradient exceeds 1.2 V, water oxidation and reduction might even occur concurrently at opposite ends of the same electrode.

The oxidation or reduction of the analyte species happens on the WE surface, and therefore the current generated is dependent on the size of this electrode. The electrochemical current generated by the oxidation or reduction of the analyte species at the detector electrode can increase six or more orders of magnitude less than the current produced in the capillary upon application of the separation current [4]. Electrically isolating the separation voltage from the detector is therefore a challenge when interfacing AD to CE. A number of methodologies have emerged to overcome this challenge.

Pulsed amperometric detection (PAD) is a method commonly utilized to overcome electrode fouling that rapidly accumulates with successive analyses. Here, electrodes are frequently cleaned by first applying a high positive potential in which any reaction products that have accumulated on the surface, and the electrode surface itself, are oxidized. This is followed by a negative potential that reduces the electrode surface back to pure metal, and then the actual working potential. The cleaning and reactivation potentials can be identified by performing cyclic voltammograms in the appropriate background electrolyte and identifying the electrode oxidation and reduction potentials. The electrode must be capable of undergoing repeated oxidation and reduction cycles without significant loss of electrode material, and for this reason, gold (Au) and platinum (Pt) are extensively utilized. PAD is especially useful for analytes such as carbohydrates and amines that tend to foul the electrode surface.

9.1.2 Potentiometric Detection

Potentiometric detection (PD) is based on the measurement of the Nernst potential either across an ion selective barrier or at an electrode surface with respect to a reference electrode. It is the simplest of the EC detection methods, as it is not necessary to apply a potential, as with AD, but merely to measure a voltage between the working and reference electrodes. PD usually utilizes miniaturized ion-selective electrodes (ISEs) as selective sensors for specific ions. Herein lies the biggest limitation of PD, namely that ISEs only exist for a limited range of analytes. In practice, however, no ISE is perfectly selective, and as they all respond to a range of different ions, they can be used as detectors in CE for a broader range of analytes.

The voltage that is measured from an ISE is generated by a charge formation at the WE due to the migration of an ion of interest through a permselective membrane. This electrode potential can be described by the Nernst equation in Equation 9.2:

$$E = E^0 + s \cdot \log(a_X + K \cdot a_Y) \qquad (9.2)$$

where E is the electrode potential, E^0 a constant dependent on the type and construction of the electrode, s is the slope (58 mV for a single charged ion), a_X is the activity of X (the ion to be determined), K is the selectivity coefficient for Y, and a_Y is the activity of Y, where Y is any other ion present in solution. Care must be taken to choose a buffer for which the electrode shows little selectivity, to prevent buffer ions dominating the electrode response. Electrode size and sample volume have no direct bearing on the signal, so PD is ideally suited for miniaturization. The voltage at the WE is measured relative to a second electrode, the reference electrode (RE), which must also be included in the detector cell setup. It is also possible to use the electrophoretic ground electrode as a pseudo-RE, as water reduction or oxidation occurring at the electrophoretic electrode leads to a stable potential. Regardless of whether a separate RE or a pseudo-RE is used, the measured voltage also includes a constant contribution from the residual electrophoretic field between the WE and the RE. To prevent the flow of electrophoretic separation current through the WE, the ISE membranes usually have a high resistance, and the measuring electronics have a high-input impedance, which blocks current flow.

9.1.3 Conductivity Detection

Conductivity detection (CD) is a universal form of detection, commonly used in ion chromatography. The most commonly utilized form of CD, capacitively coupled contactless conductivity detection (C^4D), is discussed elsewhere in this book, and so only a brief discussion of CD is given at this point. CD is based on the measurement of the conductivity or resistance of a solution between two indicator electrodes when a small current is applied to them. To suppress the influence of any redox reaction taking place at the electrodes, conductometric measurements are usually carried out using AC. voltages at a frequency of typically 1 kHz. The conductance of a solution (L) is defined in Equation 9.3:

$$L = \frac{A}{l}\sum \lambda_i c_i \qquad (9.3)$$

where A is the electrode area, l is the electrode distance, c is the concentration of the charge carriers, and λ their mobility in the electric field. As the conductance is directly proportional to the electrode area and distance, the detector cell volume affects the measured signal, so that in contrast to AD and PD, CD is a bulk detector. The mobility of the ions, as with conventional CE, is a function of the radius of the hydrated ion, and the number of charges. This has the significant advantage of ensuring that a response is generated from all ions, in contrast to other forms of EC detection, such as AD, where only electroactive species are detected, and as such, CD is a far more universal detector. Unfortunately, this also means that the ions present in the CE background electrolyte (BGE) are detected by the CD, which can have implications on the detection limits obtainable. Decreasing the ionic strength of the BGE while maintaining the analyte ion concentrations can therefore result in an increase in detector sensitivity.

9.2 INTERFACING AMPEROMETRIC DETECTION TO CAPILLARY ELECTROPHORESIS

Amperometric detection (AD) is the most widely used for the three EC techniques within CE–EC. Often the term EC in the context of chromatography and electrophoresis has the exclusive connotation of AD [5], which is discussed for the remainder of this chapter. Capacitively coupled contactless conductivity detection (C^4D) is the other rapidly increasing form of EC detection being coupled to CE and MCE, but as discussed elsewhere in this book, it will not be discussed further here.

For EC detection, it is usually necessary to place at least the detection electrode into, or very close to the end of, the separation capillary to avoid extracolumn bandbroadening. This implies that this electrode is placed ahead of the electrophoretic ground electrode, and thus the electrophoretic field is present at the detector cell. This has two consequences. First, a voltage bias occurs in the solution between the different electrodes belonging to the detector, and second, a voltage gradient is present in the solution across the width of the detector electrode [6]. Two main types of AD currently exist for CE: end-column and off-column. In end-column detection, the WE is placed at the end of the separation capillary, which usually does not require any decoupling of the electrophoretic separation current and the detection current. In off-column detection, a conductive junction between the separation and the detection components is formed, which involves grounding the separation voltage before it reaches the detector, usually necessitating a decoupler.

Given that microelectrodes are compatible in size with the internal diameters typically utilized in capillary electrophoresis, it is not surprising they are frequently employed in CEEC systems [7]. Using microelectrodes, the detection cell can easily be miniaturized, usually without a loss of sensitivity, as the noise at a microelectrode decreases more rapidly than the signal when the size of the electrode is decreased. The advantages, disadvantages, and applications of multiple types of microelectrode in EC detection have been extensively reviewed [8].

9.2.1 Off-Column Detection

AD typically records currents of electroactive species that are in the order of nA–pA, which are many orders of magnitude smaller than the currents generated from the applied separation potential. To prevent interferences from the separation current, a number of decouplers have been developed to allow the EC detector to be interfaced to the capillary itself.

The use of a decoupler to interface AD with CE was first demonstrated in 1987 by Wallingford and Ewing [9]. The authors carefully constructed a fracture at the cathodic end of the separation capillary and then joined both parts of the capillary with porous glass capillary to create an electrically conductive porous glass joint that decouples the electrochemical detector from the electrophoretic separation current. The electrophoretic separation potential dropped across the capillary prior to the porous joint and the resulting electroosmotic flow acted as a "pump," pushing both buffer and solute bands through the short section of capillary after the joint to the EC detector, as illustrated in Figure 9.1.

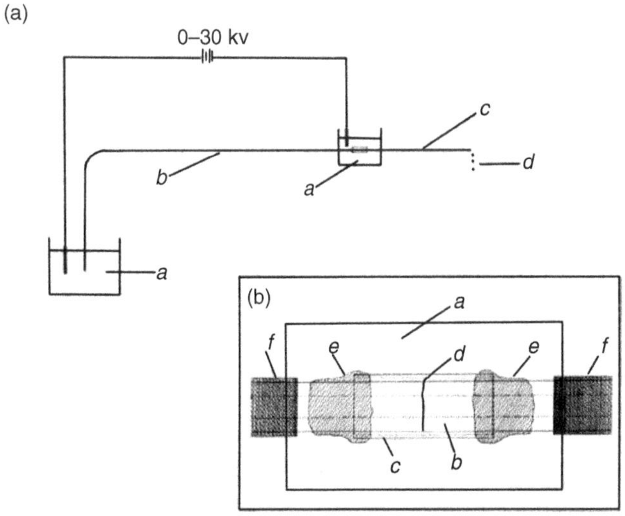

FIGURE 9.1 (a) Schematic of CE system coupled to off-column AD: (*a*) buffer reservoir; (*b*) separation capillary; (*c*) detection capillary; (*d*) eluent. (b) Detailed schematic of porous joint: (*a*) microscope slide; (*b*) fused silica capillary; (*c*) porous glass capillary; (*d*) joint; (*e*) epoxy; (*f*) polymer coating.

This section of the capillary had ground potential applied so that there was very little interference from the high potential field on the electrodes used in the detection cell. This AD detector consisted of a carbon fiber WE of 10 μm diameter and 0.1–1 mm length, with a saturated calomel RE and stainless steel CE. Careful construction of the glass joint and improvements in design reduced band broadening between the decoupler and the detector to the point where it was considered negligible [10–12]. In terms of careful construction, close alignment of the separation and detection capillaries at the joint was required, to avoid dead volume and a resulting decrease in efficiency. Additionally, the porous glass capillary utilized in this decoupler design was not readily available. This challenge was overcome by using nafion tubing to create an on column joint to connect the separation and detection capillaries, which resulted in a 100% success rate for larger capillaries [13,14]. This system was successfully applied for the analysis of amino acids in microdialysate samples [15]. Similarly, heat shrunk Teflon was also utilized to decouple the electrophoretic separation and detection currents [16].

Nafion was also used to fabricate the next-generation decouplers [17]. This design removed the issues of alignment of separation and detection capillaries around a fracture. Instead, tungsten wire was threaded into the end of the separation capillary and through the detection capillary. The distance between capillaries was chosen based on requirements, and then Nafion was cast over the separation capillary-tungsten wire-detection capillary junction. Once the Nafion membrane had set, the tungsten wire was simply pulled out, leaving continuous aligned channel between the separation and the detection capillaries. This design resulted in lower dead volumes at the decoupler and improved resolution of the separation and detection currents. Mechanical stability was added by gluing the capillaries to an epoxy board. Similar decoupler designs using cellulose acetate and palladium (Pd) have also been developed [8,18,19]. In 2003, a decoupler was designed utilizing a single continuous capillary, which was constructed by etching a series of holes through the side of the separation capillary with a CO_2 laser and then coating the holes with cellulose acetate [20]. This improved decoupler exhibited excellent stability and effective decoupling, as well as a high mechanical stability and ease of manufacture.

9.2.2 End-Column Detection

Alternatively, to avoid the use of a decoupler, end-column EC detection can be utilized, where the WE is placed at the end of the separation capillary. In either scenario, the distance between the electrophoretic ground (EG) of the separation current and the working electrode of the detection cell (WE) is critical. The key is to reduce this distance so that broadening of the sample bands is minimized, but not

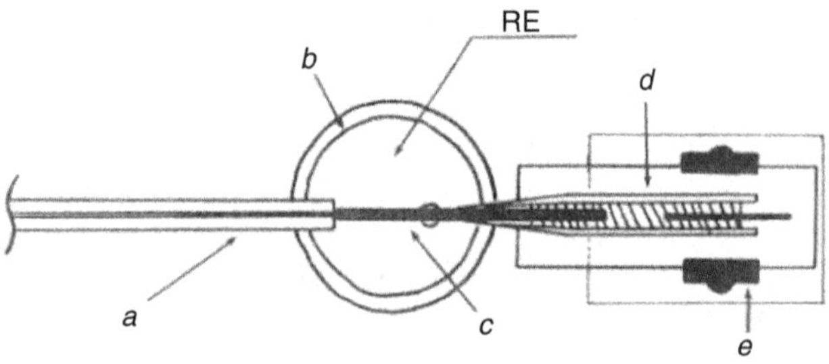

FIGURE 9.2 Schematic drawing of CZE with end-column AD: (*a*) capillary; (*b*) cathodic buffer reservoir and electrochemical cell; (*c*) carbon fiber electrode; (*d*) electrode assembly; (*e*) micromanipulator; (RE) reference electrode.

bringing the WE so close that a high electrophoretic current interference is observed. Therefore, the optimal distance is also a function of the separation field strength. Moreover, it was demonstrated that electrophoretic separation voltage had a greater effect on noise when capillaries with large, rather than small, internal diameters were used [16].

End-column detection, which does not require a decoupler, has been predominantly used in CE with AD. One of the first end column detectors, which did not require a decoupler, based on the judicious choice of separation capillary diameter, which was utilized for both CD and AD was introduced by Huang et al. [21] For CD, the WE consisted of a 50 μm Pt wire housed in a fused silica capillary, while for AD, the WE consisted of 10 μm carbon fibers, as shown in Figure 9.2.

The need for a decoupler for AD was removed by utilizing a separation capillary of just 5 μm diameter, where the electrophoretic current through the capillary was very small (1–15 nA) and electrical interference was minimized. Because the diameter of the WE was larger than the diameter of the capillary itself, good oxidation efficiency was observed. Because CD did not appear to be as sensitive to electrophoretic currents, larger separation capillaries were successfully interfaced to this detector without the requirement of a decoupler. While there was an increase in band broadening and therefore a decrease in separation efficiency, this was offset against the ease of interfacing the detector to any CE system. The reproducible alignment of the WE was subsequently addressed, by etching a conical shaped bore at the detector end of the electrophoresis capillary, which prevented movement of the WE once it was aligned [22]. However, this work was also carried out on narrow separation capillaries, in this case 2 μm. This is consistent with research where it was shown that the background current in a capillary electrophoresis experiment scales as the square of the capillary inner diameter, as illustrated in Figure 9.3 [10,23].

It was subsequently discovered, using carbon microdisc electrodes, that capillaries up to 50 μm could be used without a joint for CE–EC [24]. Unfortunately, the LODs

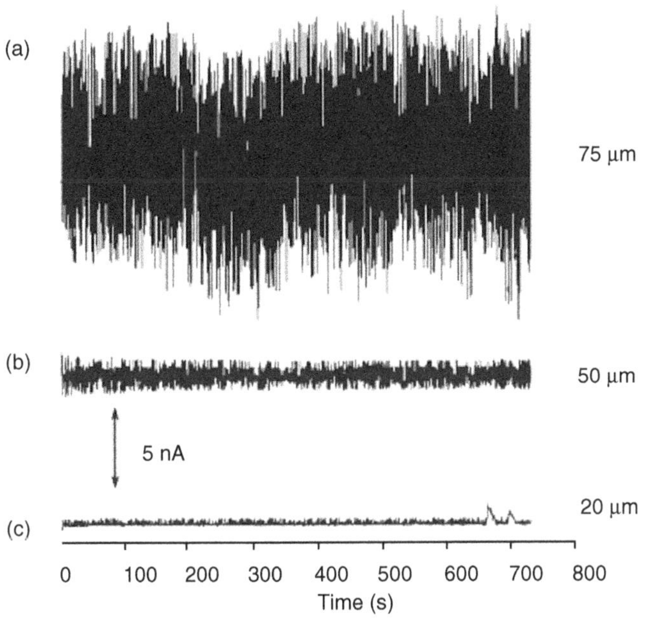

FIGURE 9.3 Capillary electropherograms of solutions containing 500 μM 2,3-dihydroxybenzoic acid and 500 μM 3,4-dihydroxyphenylacetic acid with 75 μm (a), 50 μm (b), and 25 μm (c) inner diameter capillaries, UV detection, and a 15 kV separation voltage.

were higher in these instances [25]. Improvements in separation efficiencies and LOD were achieved by etching both the carbon fiber WE and the inside of the narrow capillary, as this further improved the WE alignment [26]. The capillary was etched using HF, while a flame was used for the WE. Applying a square wave electrochemical signal, as opposed to a constant potential, during AD detection was also shown to discriminate against background current, and this may be useful for larger capillaries to lower the LOD achievable [23].

An alternative to align the WE precisely with the end of the detection capillary in the instrumentation setups above is to utilize larger electrodes in a wall-jet setup, as was demonstrated utilizing Cu electrodes up to 300 μm by Ye and Baldwin [27]. However, the increase in WE area was also accompanied by an increase in band broadening due to the larger electrode surface area, and an increasing possibility of variability of the potential across the electrode surface. Nonetheless, the advantage of disc electrodes of not having to precisely align them with the separation capillary resulted in their use increasing sharply once they were introduced. Additionally, they are more mechanically robust and less fragile than carbon fiber electrodes, and unlike carbon fiber electrodes, the entire surface of disc electrodes are in contact with the eluting solute zones. Ewing and co-authors noted in 1994 that there was still a 1–2 order of magnitude loss of sensitivity with end-column detection compared to off-column detection [28]. Further advantages of end-column detection were exploited with the development of designed cells, so that capillaries and detector electrodes could be aligned without the use of micromanipulators and microscopes [29–31]. In 2000, improvement in wall-jet detection configurations had resulted in systems comparable to decoupled off-column detection arrangements [16,24].

When a decoupler is not used, a number of practical optimizations should be carried out prior to utilize the CE–EC for analysis. Care should be given to the grounding of the CE apparatus and the potentiostat utilized in the detector setup, as ground loops may interfere with the measurements obtained. It has been shown that the separation voltage (and not the resulting separation current, or a change in the reference potential) can cause a shift in the half wave potential of the analytes being detected [32]. Therefore, an isolating transformer, to ensure galvanic separation between the high-voltage power supply and the potentiostat, is recommended [33]. The different WE configurations commonly utilized during end-column detection were analyzed by Matysik, as illustrated in Figure 9.4 [34].

The iR drop in the detection zone caused by the electrophoretic current should be determined by cyclic voltammograms in the presence and the absence of the separation high voltage, and accounted for by adjusting the potential at the WE (via the potentiostat) [24,33]. An alternative solution to eliminate the influence of the separation electric field is to position the working and reference electrodes in such a way that both electrodes are exposed to very similar local potentials due to the presence of the CE voltage [35]. This was achieved by positioning the WE and the RE very close to each other on a chip that was positioned close to the end of the capillary. One challenge for this concept remains to implement a suitable miniaturized reference electrode exhibiting good long-term stability of the potential and independence of the redox potential of the solution [36].

Additionally, because the electroosmotic flow rate is proportional to the high voltage applied for separation, the zone volume of the solvent increases with the square of the capillary diameter [34]. For this reason, capillary to electrode distances of greater than 25 μm can be utilized for capillaries of internal diameter of greater than 50 μm.

Indirect AD has also been achieved, by adding a cationic electrophore to the CE BGE [37]. However, while it increased with application range of AD, it was less sensitive than direct detection, and has not been pursued extensively. Indirect AD was also explored for nonaqueous CE, with two ions, $[FcMTMA]^+$ and $[Co(phen)_3]^{3+}$, found to be suitable as electroactive buffer additives [38].

9.2.3 Use of Multiple Detection Electrodes

Another key advantage of EC detection is that multiple electrodes can be used in series, increasing the range of species which can be detected almost simultaneously. An

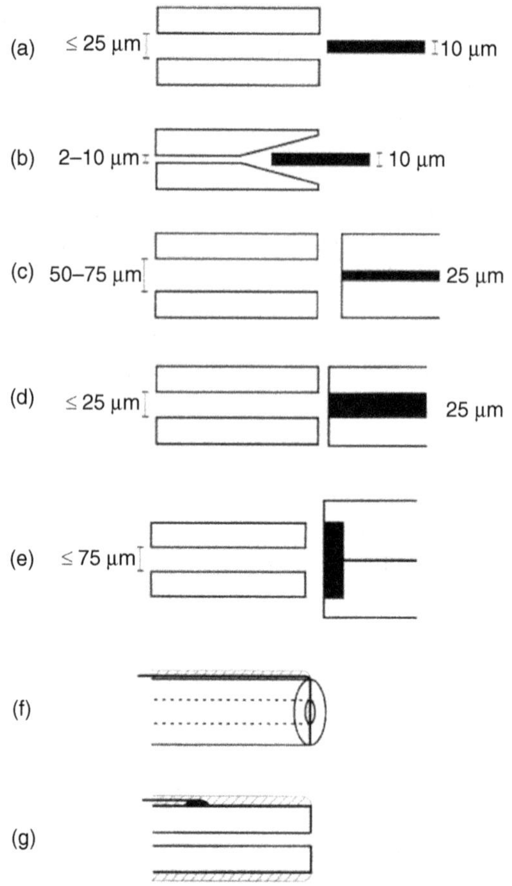

FIGURE 9.4 Typical capillary-to-electrode arrangements used for CE–ED without decoupling: (a) carbon or metal fiber electrodes positioned close to the capillary outlet, (b) fiber electrode placed in a conically etched capillary end, (c–e) disc electrodes with dimensions smaller, equal or larger than the capillary outlet, (f) on-capillary arrangement with a metal wire mounted onto the capillary end, (g) on-capillary arrangement based on a metal layer sputtered onto the exit of the capillary.

early example of this is the simultaneous detection of cysteine and cystine using two Au/mercury amalgam electrodes placed in series and set at −1.00 V and at +0.20 V versus Ag/AgCl, respectively [39]. Additionally, if the electrodes are set in a parallel configuration, they can also be employed for voltammetric characterization of the analyte as has been demonstrated using both ring disc and parallel disc configurations [40]. However, these electrodes may be more difficult to construct and align than those employed for single electrode detection [7]. Microfabricated arrays of electrodes have also been successfully applied to EC detection, where not only they have been reported to enhance selectivity, but through redox cycling, improvements in sensitivity were also reported [41]. The total electrode area needs to be kept small (relative to the capillary diameter) to minimize band broadening.

9.2.4 Pulsed Amperometric Detection

In 1993, pulsed amperometric detection (PAD) was successfully demonstrated for carbohydrate analysis, utilizing off-column detection with a nafion decoupler and a Au wire microelectrode for detection [42]. Unfortunately, separation efficiencies were somewhat low, and LODs in the order of 10^{-7} M were obtained. A similar analysis with 10 μm Au disc electrodes reported soon after obtained improved separation efficiencies and LODs [43]. LODs and separation efficiencies better than UV and comparable to or even better than laser-induced fluorescence were reported for CE–PAD for the analysis of biogenic amines [44].

Integrated pulsed amperometric detection (iPAD) is a modification of PAD that uses an integrated potential scan for the detection step [45]. A triangular wave allows the oxide formation produced during the positive scan to be compensated for with the oxide dissolution during the negative scan. In iPAD, a much lower baseline stabilization time can be achieved and the signal obtained is mainly produced by the irreversible oxidation of the analyte during the positive scan. iPAD was found to be superior to PAD as a universal detector for carbohydrates, amino acids, and peptides, as it generated a more stable baseline with fewer oxide-induced artifacts [46]. PAD was also demonstrated as a mode of detection in conjunction with micellar electrokinetic chromatography, although sensitivity was low, due to the need to dilute the sample after separation and before detection [47].

Surfactants added to the CE background electrolyte can interact with the surface of the WE and may facilitate the electron transfer process [48]. This phenomenon is most apparent for PAD, possibly because the formation of surface aggregates, which facilitate electrode surface–analyte interactions, is greater at higher electrode potentials commonly used in PAD. Surfactants may influence the redox potential as well as the stability of the electrogenerated products [10].

9.2.5 Nonaqueous EC Detection

AD has also been successfully demonstrated in nonaqueous solutions, which opens up the possibility to detect compounds that are not water soluble, or not detectable within the potential window (of 1.2 V) of aqueous buffers [33,49]. AD in nonaqueous systems has also the advantage that capillaries of internal diameter of up to 75 μm can be configured within the end-column wall-jet configuration, and were shown to have lower LODs than UV across all the capillary diameters [50].

9.2.6 Electrode Material

A key parameter for EC detection is the choice of WE material, as it will determine the signal-to-background characteristics as well as the reproducibility of the response obtained. Judicious choice of electrode material (and applied potential) can influence the selectivity of the electrode. The most recent developments in WE materials are discussed below. Carbon is the most commonly utilized WE material, employed for various analysis. Nitroaromatic compounds have been analyzed utilizing mesoporous carbon electrodes (2010) [51]. This end-column detection utilized in-house prepared carbon electrodes, simply connecting pencil lead to Cu wire. Pencil lead was also used for end-column AD of flavonoids and ascorbic acid, where the optimum distance between the end of the separation channel and the WE was determined experimentally as approximately 40 μm [52]. Ascorbic acid was also measured from lysed cells using CE–EC with a carbon fiber microdisc bundle electrode [53], which consisted of approximately 40 carbon fibers [54], in a refinement of an earlier carbon fiber bundle electrode [55,56]. This carbon fiber AD cell has also successfully been used for detection of DNA damage biomarkers [57,58]. Carbon nanotubes (CNTs) have also been incorporated into WE with significant advantages obtained, including a significant reduction in electrode overpotential and a reduction in surface fouling [59]. Various CNT-based electrodes have been fabricated, including CNT-modified screen-printed carbon electrodes [60], CNT/nano-Fe_3O_4 microelectrodes [61], CNT/Teflon composite electrodes [62], and CNT screen-printed electrodes [63]. The primary method for preparing these WEs is surface modification, that is, dispersing the CNTs in solutions that are then cast onto the WE surface [64]. These CNT-modified WEs have been utilized for various applications [65].

By nanostructuring the surface modification of the WE, a higher sensitivity can be obtained because of the greater surface area of the electrode [66]. Higher sensitivity is not directly linked to the lower detection limits because, in some cases, the background noise levels increase with the same or greater magnitude than does the signal. In cases where the signal increases in greater magnitude than does the noise, the LOD is improved.

Chemically modified carbon electrodes have also been utilized in EC detection, with various compounds, for example, Vitamin B_{12} for the analysis of antioxidants [67]. Boron-doped diamond has also been utilized as an effective WE material. These electrodes display attractive properties such as a wide potential window, low and stable background currents, negligible adsorption of organic compounds, and low noise [59]. Boron-doped diamond WEs with an electrode surface area of $300 \times 50\,\mu m^2$ displayed lower LODs than carbon fiber WEs, and did not require any polishing between analyses [68]. Alternatively, microwave-assisted chemical vapor deposition was utilized to deposit thin films of polycrystalline diamond on electrochemically sharpened Pt wires [69]. These microelectrodes, without any activation pretreatment, exhibited low background current (~100 pA) and noise (e.g., 3 pA).

The ability to replace the WEs is important for analyses where electrode fouling routinely occurs, and so has been explored by a number of groups. Screen-printed electrodes (SPEs) are cost-effective, reproducible, and easy to manufacture, and so are ideal disposable electrodes. Carbon ink SPEs were successfully utilized in a wall-jet end-column configuration, positioned 25 μm from the end of the separation capillary for the analysis of uric acid [70]. A custom detection cell, where the WE slotted in perpendicular to the capillary outlet without requiring micropositioning, was also developed, and excellent reproducibility between SPEs was reported. Similar disposable electrodes of carbon fiber, Pt,and Au wires fabricated on polyester/polyethylene laminating foils have also successfully been utilized as WEs in AD detection configurations for nonaqueous CE [71].

To overcome the repeated polishing required with carbon-based WE, a number of other materials have also been explored. Three-hundred-micrometer diameter Cu disc electrodes have been successfully utilized for a number of analyses, including glycosides and sugars [72] and diethylene glycol [73]. Cu wires were used in an end-column configuration for the detection of creatinine and uric acid [74]. A heated Cu microdisc electrode has been successfully applied to the end-column detection of polyhydroxy compounds for both CE and MCE [75]. Heating the electrodes prevents the fouling that usually occurs as oxidation products accumulate on the Cu surface. Au and Pt have also been used as WE substrates. An Au microelectrode WE of approximately 60 μm, housed within a custom-designed AD cell that aligned the WE approximately 20 μm from the end of the capillary, was successfully utilized for mercury determination [76]. Before every run, however, the microelectrode had to be held at −500 mV for 5 min to ensure that the impurities were reduced and stripped from the electrode surface. Au electrodes have also been enzymatically modified to enhance their detection selectivity; for example, horseradish peroxidase cross-linked in an osmium-based redox polymer hydrogel (HRP-Os) was physically adsorbed on Au microelectrodes for neurotransmitter detection [77]. Pt microelectrodes have also been extensively employed, especially for NACE–EC [78,79], and CE immunoassays with EC detection [80,81].

9.2.7 Dual Conductivity and Amperometric Detection

Capacitively coupled contactless conductivity detection (C^4D), which can be viewed as a universal detector, and amperometric detection (AD), which detects electroactive species, have also been coupled in series, for example, for the detection of urinary phenylalanine metabolites [82]. The end-column AD WE consisted of a carbon disc electrode. In this setup, AD was used to detect the electroactive species, whereas the C^4D was used to monitor the sample matrix.

9.3 INTERFACING ELECTROCHEMICAL DETECTION TO MICROFLUIDIC CAPILLARY ELECTROPHORESIS

Microfluidic capillary electrophoresis (MCE) has several demonstrated advantages over standard capillary electrophoresis, as explored in other chapters within this book. A potential problem with micrototal analytical systems is to miniaturize the detection components, but electrochemical detection is particularly amenable to such miniaturization, as it is independent of optical path length. Moreover, it is compatible with micromachining technologies. As a result, MCE with integrated EC is particularly advantageous. A number of challenges exist when integrating EC detection with MCE, similar to those of conventional CE, primarily isolating the separation and detection currents and aligning the WE with the end of the capillary.

9.3.1 End-Column Detection

As with conventional CE analysis, in the case of MCE, both decoupled and end-column (in microchips, this is commonly referred to as end-channel) EC detection cells have been developed. As with CE, the most common method is end-channel, where the detection cell is placed just outside the separation channel [83]. For MCE–EC with end-channel detection, two alignment modes have emerged: off-chip and on-chip. Off-chip end-channel detection allows for convenient replacing and easy cleaning of the WEs, while on-chip end-channel detection integrates the WEs on the microchip.

Gavin and Ewing [84,85] originally demonstrated EC detection for separations in open-channel electrophoresis chips, using detection cells consisting of Pt microelectrode arrays and an Ag/AgCl pellet RE for continuous monitoring of analytes that had been sampled by a capillary. In 1998, Woolley et al. [86] published the first MCE with integrated EC detection. Capillaries were microfabricated using photolithography, wet chemical etching, and thermal bonding methods, and the detection cell consisted of plasma sputtered Pt WE and CE, with an Ag/AgCl RE placed as close as possible to the integrated electrodes. In this flow-by design, the surface of the band Pt electrode was parallel to the flow direction. In 1999, Wang et al. developed screen-printed "detection strips" for EC detection in MCE, enabling raid facile switching of the detector as required [87]. The detection strip, placed at the right angles to the end of the separation channel, comprised a screen-printed carbon WE, Ag/AgCl wire RE, and a Pt wire CE, held in place with a plastic screw. Catecholamines were successfully detected, with LOD's comparable to conventional CE systems obtained. These electrodes could also be chemically modified to exploit electrocatalysis [88]. This design, which allowed for easy removal of the WE, was also utilized with glassy carbon electrodes, for the detection of antioxidants [89]. Replaceable carbon fiber microdisc electrodes for MCE–EC were also reported by Zeng et al. [90] Disc electrodes such as these are typically aligned with the separation channel in a wall-jet configuration. Therefore, the detector operates in coulombic mode, as the detector volume is typically significantly greater than the injection volume [91]. Under these conditions, an improvement in separation efficiency was obtained when the detector potential was increased, with the rationale that at high potentials the analytes were immediately oxidized, while at lower potentials, this oxidation was slower, resulting in broader analyte peaks. An alternative to utilizing disc electrodes aligned at the exit of the separation channel is to place the electrode on the end of the separation channel itself. It was demonstrated by using electroless deposition to deposit an Au film WE (containing Au aggregates of 100–150 nm) directly on the end of the separation channel [92]. This design obviated the need for photolithographic electrode fabrication or careful channel/electrode alignment. Electroless deposition also formed the fabrication basis for Au nanoelectrode integration on a MCE–EC platform [93]. This resulted in a lower LOD for ammonia sensing than conventional bulk Au WEs.

In 2000, a dual-electrode AD configuration with Au electrodes on a glass microchip reversibly bonded to a PDMS microchip containing the electrophoretic injection and separation channels was reported [94]. Practical advantages of this design included the elimination of the high-temperature bonding step, which was potentially damaging to the electrodes. Having integrated WEs close to the separation capillary but macro REs in the detection cell can result in excessive detection potential shifts [95]. To minimize this, REs should be fabricated close to the WEs. In 2002, a fully integrated MCE–EC device, which required no external electrodes, was demonstrated [96]. All Pt electrodes were formed by a photolithographic liftoff process, with the RE coated electrochemically with an Ag/AgCl layer. Pyrolysed photoresist films have also been used as substrates for carbon WEs for dual-electrode AD [97]. While separation efficiencies were similar to those reported for carbon fiber electrodes, they were considerably higher than those for dual Au electrodes.

A feature of MCE–EC with end-channel detection is often high background currents and decreased separation efficiency, due to diffusion of the analyte in the region between the end of the separation channel and the WE. To overcome this problem, a sheath-flow supported MCE–EC system that allows efficient detection of analytes at a distance of up to 250 μm from the separation channel exit was developed [98]. An alternative method to overcome these limitations was illustrated by Martin et al. [99] In this instance, the WE was placed in the end of the separation channel, but the use of a decoupler was avoided by utilizing an electrically isolated "floating" potentiostat. Using this configuration, separation efficiencies comparable to laser-induced fluorescence were obtained. This type of configuration was termed "*in-channel*." The shift in the half wave potential of the catecholamine being detected observed during in channel AD was compensated by using a large detection potential. In-channel AD exhibited a 4.6-fold decrease in plate height while lowering the peak skew by a factor of 1.3 compared to end-channel detection [100]. Additionally, for end-channel AD, the channel cross-sectional area strongly affects the high voltage interference with the AD [101]. It was shown that the iR drop generated at the detection cell decreased significantly when the channel dimensions were reduced from width 49 μm and depth 14 μm to width of 39 μm and depth of 9 μm, which enhanced both the separation efficiency and the LODs obtained.

In CE–EC, as discussed previously, the separation efficiency can be maintained when the distance between the WE tip and the capillary outlet is less than the diameter of the capillary. This important rule has also been proven by experimental results with silica microchannels [87,96,102]. For a PDMS microchannel with a width of 50 μm, the optimal electrode-outlet distance was found to be 10 μm, due to the relatively slow speed of EOF in the PDMS microchannel [103].

Just as for CE–EC, a two-electrode system, where the electrophoretic ground electrode acts both as a pseudoreference and as the counter electrode is plausible and has been demonstrated for MCE–EC [104].

9.3.2 Pulsed Amperometric Detection

As discussed previously, pulsed amperometric detection (PAD) exploits the electrocatalytic activity of noble metal (e.g., Au and Pt) electrode surfaces to oxidize various polar functional groups while maintaining uniform and reproducible electrode activity [46]. PAD was successfully demonstrated on MCE–EC in 2002 [105]. A hybrid poly-(dimethylsiloxane)/glass device was coupled with a Pt WE for detection of glucose, maltose, and xylose. However, the separation efficiency for this analysis was quite low. PAD was subsequently performed with an Au wire electrode with a new chip design, where the electrode was aligned at the end of the separation channel [106]. The alignment of the electrode was facilitated by simply placing the wire in a premolded PDMS channel. The effects of solution pH, buffer concentration, and SDS on carbohydrate separation and detector response were studied, with results indicating that both iPAD and PAD were affected in a similar manner by changes in solution conditions [107]. PAD was also carried out using MCE–EC incorporating a Pd decoupler [108]. A Pd wire was used as the cathode for decoupling, and a second Au or Pt wire was used as the WE for either EC or PAD. Using PAD, sub-μM detection limits for several carbohydrates and thiols were achieved.

Indirect AD was also demonstrated on MCE–EC, using dissolved oxygen as the electroactive indicator [109]. The potential difference induced by the CE separation electric field between the WE and the RE resulted in a change of the reduction potential of oxygen, which was used to determine nonelectroactive analytes.

9.3.3 Off-Channel Detection

To avoid the band broadening often observed in end-channel AD, endeavors are ongoing to incorporate the WE in the end of the separation channel. In 2000, a completely polymeric device, decoupling the separation and detection currents (by means of a microhole drilled in the microchannel cover), and utilizing a carbon WE, with an Ag/AgCl RE wire electrode was reported [110]. The authors noted that they had yet to optimize the device, and the relatively high limits of detection could be caused by either incomplete decoupling or analyte leakage. Also in 2000, a Pd decoupler was published [111]. In this case, Pd was evaporated onto a polymeric microchip for both the decoupler and the WE. While low separation efficiencies were generated, possibly due to the fabrication procedure, the decoupler did not create dead volumes or sample leakage. Pd decouplers have the advantage of absorbing hydrogen evolved on the capillary electrophoretic ground electrode, and so problems due to the formation of hydrogen gas bubbles can be minimized [95]. Improved fabrication procedures were utilized to fabricate an integrated MCE–EC chip utilizing a Pd decoupler [112]. While efforts were made to optimize the decoupler length, it was concluded that an enhanced performance would be best achieved by a more judicious choice of WE material. Subsequently, dual Au WEs demonstrated unpredicted signal enhancements due to redox cycling of the analytes between them [113]. Dual carbon ink electrodes were also utilized in an MCE–EC chip with a Pd decoupler [114].

In 2003, an MCE–EC where the electrodes and the decoupler were fully integrated onto a glass chip was demonstrated [115]. The electrodes were constructed by a 450 nm-thick Au layer deposited on a seed layer of 50 nm-thickness Ti, with one decoupling-ground electrode as the cathode of CE electric field positioned in front of the three-electrode electrochemical system. Pt nanoparticles were electrically deposited on the ground electrode and the reference electrode to be the decoupler and the pseudoreference electrode, respectively. A novel approach for off-channel AD, where the electrode surface could be renewed by pulling the WE strip 3 mm so that fresh surface area was exposed within the channel, resulted in highly reproducible detection with a substantial increase in the lifetime of the microchip [116].

The laser etched decoupler developed by Osbourn and Lunte for standard CE–EC [20] was also adapted for use with MCE–EC [117]. This decoupler was constructed by aligning a series of 20–30 μm holes through the coverplate of the microchip with the separation channel and casting a thin film of cellulose acetate within the holes. This achieved a limit of detection of 25 nM for dopamine at a carbon fiber electrode, the lowest reported for a microchip CEEC device.

Recently, researchers have begun to pursue multiple separation channel separations in conjunction with AD [118]. It can be a system where in the same configuration of detecting electrodes, one capillary with the flow of BGE only, serves as a reference channel [119]. The configuration of a dual-channel microchip with an end-capillary detection based on a metal wire WE aligned at the outlet of the separation channel can be also employed for carrying out two simultaneous CE separations [120].

9.3.4 Electrode Material

Judicious choice of electrode material (and applied potential) can influence the selectivity of the electrode. WEs utilized in have been fabricated from a number of materials, including carbon ink [87,110,121,122], carbon paste [123], carbon fiber [124], glassy carbon [125], Au [104,126,127], Pt [84–86], Pd [111], Cu [104], and diamond [128,129] and have been extensively reviewed [83,100]. As with conventional CE–AD, carbon is the most widely utilized material due to its minimal fouling, low overpotential, and wider EC range for organic compounds than metals [83]. Screen-printed electrodes (SPEs) have become known as "think-film electrodes," while sputter coated electrodes are "thin-film electrodes" [130]. Graphite-epoxy composite electrodes resulted in lower limits of detection than SPEs [131]. Carbon nanotubes (CNTs) were used to modify carbon SPEs, and the resulting chemically modified electrodes were demonstrated to result in significantly lower operating potentials and yield substantially enhanced signal-to-noise characteristics, reflecting the electrocatalytic activity and resistance to surface fouling of CNT materials [60]. CNT-modified WEs have been extensively investigated for MCE–EC, with numerous novel methods of preparation reported [59]. The electrocatalytic activity of different carbon materials (single wall CNT (SWCNT), multiple wall CNT (MWCN, and carbon powder) cast on different substrate electrode materials (glassy carbon (GC), Au, and Pt) has also been investigated, and it was found that in all cases, enhanced resolution between catecholamines was detected [132]. A lower detection potential could be employed using the modified WEs, which allowed for faster electron transfer and a resulting reduction in peak tailing, which improved the resolution.

For catechol-estrogen-derived DNA adduct analysis, photolithography and chemical wet etching were used to pattern Au microelectrodes, which were made of a 10 nm layer of titanium as an adhesion layer below a 200 nm layer of Au on a PDMS/glass hybrid microchip [133]. This microchip was previously developed for catecholamine detection (2006) [134]. Here a separation voltage of 800–1500 V yielded ~1 nA of background current. The hybrid microchip consisted of a glass plate, on which the electrodes are fabricated, bonded to a PDMS layer that had the injection and separation channels micromolded into it. The two pieces were sealed with RF-plasma treatment. A Pd decoupler was subsequently added to this design, which reduced the background to a stable current of approximately 5 pA [135]. A distance of 600 μm was deemed optimal for a 10 mM borate buffer and a separation voltage of 150–300 V cm^{-1}.

Usually, AD involves anodic oxidation of the electroactive species, but anodic reduction of some species, particularly peroxides, can also be exploited. For anodic reduction, however, all oxygen must first be removed from the eluent. While reductive AD of peroxides with an Au WE, an Ag/AgCl RE, and Pt wire CE coupled with standard CE was demonstrated in 2000 [136], it was subsequently demonstrated for MCE with a 3 mm Au electrode positioned 50 μm in line with the end of the separation channel for the detection of organic peroxides [137]. An Ag/AgCl RE and Pt wire CE were placed underneath the WE in the waste reservoir of the chip holder.

Relatively straightforward coupling of EC detector cells to microchip capillary electrophoresis (MCE) is now routinely demonstrated in the literature. As an example, the EC cell used by Dossi et al. [138] consists of a 3 mm glassy carbon WE positioned 20 μm from the end of the separation channel, along with an Ag/AgCl RE and a Pt wire as a CE, controlled using a Palmsens electrochemical analyzer. EC detection technology is sufficiently mature to be used as an analysis tool for complex analyses, without recourse to further refinement of optimization. Kovachev et al. [139] use a similar EC system to Dossi et al. for complex antioxidant determination.

9.3.5 Portable CE and MCE Systems

Portable systems offer unrivalled potential for *in situ* analysis. Microchip CE systems with integrated EC detection offer a realistic possibility for truly portable analysis systems. In 2001, Kappes and co-workers demonstrated the feasibility of a field portable capillary electrophoresis instrument with EC detection for determination of inorganic ions [140]. Utilizing a 25 μm capillary with a wall-jet detection configuration (no decoupler), amperometric, potentiometric, and conductometric detection were demonstrated. In 2003, Jackson et al. [141] developed both a portable battery-powered dual-source high-voltage power supply that allowed independent voltage control of both the CE injection and the separation channels on the microchip and a battery-powered potentiostat for potential control and current monitoring for AD. In 2008, an MCE–EC system with PAD was integrated into a mobile platform, to enable remote wireless analysis [142]. This proof-of-concept system was used to demonstrate a successful separation of phenol and ferulic acid; sensitivity and LOD were not investigated. Fernández-la-Villa presented a portable MCE system with end channel Au electrodes, which can operate in single or dual amperometric mode in 2010 [143]. Compared to alternative EC detection potentiostats, LODs for catecholamines similar to previously reported MEC systems were reported. Alternative MCE systems were not evaluated for comparison.

9.3.6 Applications of CE–MCE with AD

CE and MCE are particularly suited for the analysis of very small volumes of sample, as typically a new nanoliter of sample is all that is required. For this reason, the analysis of biological samples has been extensively explored using CE. Moreover, online preconcentration techniques, such as pH-mediated sample stacking [145,145] and field-amplified sample stacking [15,146,147], and its facile coupling to sampling techniques such as microdialysis [130a,132], have resulted in the successful analysis of samples in complex biological media such as urine [66,72,133], plasma [134], and serum [46,113]. Additionally, CE generates very small volumes of waste, and so it is ideal for the analysis of hazardous samples, such as explosives [51,128,151]. Use of MCE for these analyses is also advantageous, as the entire analysis chip can be disposed of appropriately, if required [71,152–154].

The incorporation of AD with CE analysis further enhances the advantages of CE. Applications of both CE–EC and MCE–EC are vast, spanning a range of analytes, from organic and inorganic ions to nucleic acids, proteins, cellular components, and intact cells, and have been reviewed by multiple authors [100,118,155–117]. The following discussion is not meant as a comprehensive review of the multiple application areas. Instead examples of CE–EC applications that illustrate the broad applicability of this analysis technique are explored.

A huge number of biological analytes of interest, including neurotransmitters, amino acids, ascorbic acid, and DNA biomarkers such as 8-hydroxy-2-deoxy-guanosine (8OHdG), are electroactive and therefore ideal for EC detection. Amino acids have been detected both using carbon paste electrodes for CE–EC and both Cu [158] electrodes and carbon paste electrodes modified with multiwall carbon nanotubes and Cu oxide [159] for MCE–EC. Amino acids were derivatized with naphthalene dicarboxaldehyde to allow for detection using bare carbon paste electrodes [123]. 8OHdG is a widely used biomarker of oxidative damage to DNA. In 2004, elevated levels of this biomarker in urine were correlated to cigarette smoking, using CE–AD with end-column detection (LOD 4.3 nM) [57]. A year later, both the base and the deoxy-sugar form of this biomarker were detected in rat cerebral cortex microdialysate samples using off-column AD with a laser etched decoupler, with a reduced LOD of 0.5 nM [160]. The basal level of the sugar, 8-oxoguanine, was determined to be 5.5 ± 1.3 nM, but a basal concentration of 8OHdG was not detected. This analysis also demonstrated how pH-mediated sample stacking is ideal for on-capillary preconcentration of analyte in a biological matrix, which is typically of high ionic content. A microfluidic device, also utilizing a decoupler, of electroplated Pd, obtained an LOD of 100 nM for 8OHdG, significantly higher than with CE [135]. Elevated levels of urinary 8OHdG have also been correlated with cancer patients using end-column CE–EC [161]. While elevated levels of taurines have been correlated with numerous diseases, their analysis is quite limited due to its low concentration in biological fluids. CE–EC is ideal, however, and a method has been developed for their analysis, using PAD with Au electrodes, which resulted in an LOD of 4×10^{-2} μM, which was comparable to laser-induced fluorescence detection [150]. Trace analysis of DNA has also been demonstrated on MCE–EC [162]. The WE was modified with conducting polymer (poly(5,2′:5,2″-terthiophene-3′-carboxylic acid) (pTTCA)) and Au nanoparticles, to enhance the polymer's electrocatalytic activity for the direct detection of DNA, and in conjunction with field-amplified sample stacking and injection (FASS and FASI), a sensitivity of 0.20 nA $(\text{fg}/\mu\text{L})^{-1}$ and a detection limit of 5.7 amol in a 50 μL sample were achieved.

A range of analytes of interest in foods and as food additives, such as antioxidants [67], flavonoids, and herbicides, have been detected using CE–AD. Antioxidant concentrations have been measured in food [139] such as tomatoes [163], sweet potato [164], and pears [89]. The LOD for the herbicide maleic hydrazide analyzed in potato samples utilizing CE–EC was 1.3 $\mu\text{g g}^{-1}$; this value is lower than 50 $\mu\text{g g}^{-1}$, which is the maximum residue level (MRL) permitted for this plant growth regulator in potato samples [165]. The LOD for the pesticide amitrole in apples was 0.4 $\mu\text{g g}^{-1}$, which was unfortunately above the MRL of 0.01 $\mu\text{g g}^{-1}$ [166]. The authors reported, however, that it was one of the lowest LODs reported for amitrole analysis. MCE–EC was utilized for the analysis of food dyes, with an LODs of 3.8–15.1 μM obtained [167]. CE–EC has also been used to investigate for the presence of both legal and illegal antibiotics in animal tissues. Using carbon disc electrodes and end-column CE–AD, β-agonists, illegal growth promoters, were determined in pig feed, liver, and urine [168]. A similar analysis was used to determine sulfonamide antibiotics in animal tissue [169]. The MRL permitted by the

E.U. for total sulfonamides is $100\,\text{ng g}^{-1}$, and using CE–EC, individual LODs of $66–157\,\text{ng g}^{-1}$ were obtained.

Additionally, electroactive ingredients in pharmaceutical drugs [170,171] and Chinese medicines [172–176] have been extensively detected using M/CE with AD, as have nitroaromatic explosives. Using a 300 μm carbon disc electrode in end-column configuration for CE–AD, a number of active ingredients in Chinese medicines, including isoflavins [177], phenols [172], phenolic acids [171,173], and their derivatives [178], have been extensively explored. The commonly used NSAID's salicylic acid, acetaminophen, diflunisal, and diclofenac were detected using MCE–EC with an Au wire electrode, with LODs of approximately $0.26\,\mu\text{g mL}^{-1}$ [179]. Detection limits of between 3.0 and $4.7\,\mu\text{g L}^{-1}$ were observed for 2,4,6-trinitrotoluene (TNT), 1,3,5-trinitrobenzene (TNB), 2,4-dinitrotoluene (DNT), and 1,3-dinitrobenzene (DNB) using CE–AD with a 500 μm carbon disc electrode modified with mesoporous carbon material [51]. These LODs are an order of magnitude lower than those observed by Wang et al. [128] 7 years earlier for dinitrobenzene and 2,4-dinitrotoluene using a chemically vapor-deposited boron-doped diamond (BDD) film band (0.3×6.0 mm) electrode.

MCE–CE with AD is surprisingly broad in its range of applicability for a range of industrial analytes of environmental interest including mercury levels in coal samples (utilizing Au electrodes) [76] and textile dyes in water samples (using Pt electrodes and NACE) [78]. Using Cu electrodes [75], or nonaqueous CE (NACE), carbohydrates have also been extensively investigated using CE with AD.

CE with EC detection has also been utilized for the analysis of cells and cellular components, including cells [54] and synaptic vesicles [180], as a significant number of biologically important analytes, such as antioxidants and neurotransmitters, are naturally electrochemically active. Individual rat hepatocytes were lysed in the fused silica capillaries, and the released ascorbic acid was detected using carbon fiber microdisc bundle electrodes [54]. The concentration of ascorbic acid was measured to be 28–63 fmol. For detection of neurotransmitters in synaptic vesicles, a hybrid device was developed, where vesicles were separated in a conventional capillary that was then interfaced to a microfluidic chip for lysis of the vesicles and subsequent AD of dopamine loaded lipsosomes [180]. The neurotransmitter dopamine was also detected (after stimulated release) from a confluent layer of rat PC12 cells using a microchip with a Pd decoupler that also featured integrated microdialysis sampling. Using at a perfusate flow rate of $0.3\,\mu\text{L min}^{-1}$ and obtaining a 77% recovery, a 9 μM LOD was achieved.

CE-based immunoassays that employ EC detection have also been developed. In one example, based on the *o*-aminophenol (OAP)-H_2O_2-horseradish peroxidase (HRP) system, after the enzyme immunoreaction, the HRP-labeled antibody or HRP-labeled antigen catalyzed the enzyme substrate OAP and H_2O_2 [81]. The product of the enzymatic catalysis reaction 2-aminophenoxazine-3-one (AP) was determined using electrochemical detection on a Pt electrode, and an LOD of $0.48\,\text{ng mL}^{-1}$ for alpha-fetoprotein and $1.0\,\text{ng mL}^{-1}$ for thyroxine was obtained. This biological system was also manipulated for the detection of three tumor markers, carcinoembryonic antigen (CEA), prostate-specific antigen (PSA), and human chorionic gonadotropin (HCG), and achieved LODs of 0.22, 0.17, and $0.30\,\text{ng mL}^{-1}$, respectively [80]. These results were comparable in terms of linear proportionality to the standard spectrophotometric ELISA method, although LODs were not compared. Similar analyses have also been developed on microchip. Alpha-fetoprotein was also detected using a microchip enzyme immunoassay with EC detection, with an LOD of $31.8\,\text{ng mL}^{-1}$, one order of magnitude higher than

TABLE 9.1 Special Setup and Characteristics of Three Modes of Isolating Detection

Detection Mode		Special Setup	Characteristic
End-column–end-channel	Off-chip	Setup with precise electrode positioning is needed.	The working electrode can be conveniently replaced and easily cleaned. The channel–electrode distance can be adjusted.
	On-chip		It has the limitations that the detection electrode is not exchangeable and, when poisoned, it cannot be replaced easily. This mode is suitable for single-use, disposable microchips.
Off-column–off-channel		An additional decoupler is needed.	Some decouplers can absorb hydrogen evolved on the capillary electrophoretic ground electrode. Problems due to the formation of hydrogen-gas bubbles can be minimized.
In-channel		An electrically isolated potentiostat is needed.	High separation efficiencies comparable to those obtained with laser-induced fluorescence detection can be obtained. A large detection potential should be used.

conventional CE–EC, recorded [181]. Another microchip assay analyzed the concentration of histamine released in whole blood samples, as a measure of allergic reactions [125]. Free Fc-conjugated IgG antibody (Fc-IgG) was separated from antibody–antigen complexes on the basis of differences in isoelectric point (*p*I) using a multichanneled matrix column coated with cation-exchange resin. Again, a good correlation was found between the response of the electrochemical immunoassay, described in this chapter, and the conventional radioimmunoassay.

9.3.7 Future Directions for CE–MCE with EC Detection

AD is a proven detection method, for both CE and MCE, due to its high sensitivity and ability to be miniaturized without loss of performance. However, care must be taken during the design of the interface between CE–MCE and AD to minimize or adjust for interferences due to the presence of the CE separation voltage. To isolate the separation electric field, three strategies have been used for AD: end-column (CE)–end-channel (MCE); off-column (CE)–off-channel (MCE); and in-channel with an electrically isolated potentiostat, and are summarized in Table 9.1 [102].

While commercial detectors are available for C^4D detection, as yet, there are no commercial detectors for AD detection. However, attempts have been made to produce detectors that have this potential. One such example is the plexiglas detection cell designed to hold both SPE and glassy carbon disc working electrodes, and to be interfaced to commercial CE instruments [182]. It is anticipated that additional commercializable AD cells will be published in the near future. Nanomaterial-based electrodes in both conventional CE and microfluidics have been demonstrated to be very useful for the analysis of a wide variety of samples. Their advantages in improving the sensitivity, limits of detection, and peak capacity should continue to be exploited in the near future. A major obstacle to the realization of real-life applications of such devices is that only electroactive analytes can be detected directly. Other analytes need derivatization before detection. Nonetheless, the challenge of integrating EC detectors with CE and MCE separations has been sufficiently met to allow these analysis tools to be used for more complex and sophisticated analysis. Recent publications in this area are beginning to hint at the future possibilities for such devices. Incorporating sampling techniques such as microdialysis and monolithic SPE, on-capillary preconcentration techniques such as pH-mediated sample stacking (for high conductivity sample matrices) or field-amplified sample stacking (for low conductivity sample matrices) with intelligent detection electrode material and design will continue to increase its sensitivity, selectivity, and ultimately its application for complex analysis.

REFERENCES

1. Kappes, T. and Hauser, P.C. (1999) *J. Chromatogr. A. 834*, 89–101.
2. Bard, A.J. and Faulkner, L.R. (2001) *Electrochemical Methods: Fundamentals and Applications*, 2nd ed., Wiley, New York.
3. Kuban, P. and Hauser, P.C. (2009) *Electrophoresis 30*. 3305–3314.
4. Yik, Y.F. and Li, S.F.Y. (1992) *Trends Anal. Chem. 11*, 325–332.
5. Kappes, T. and Hauser, P.C. (2000) *Electroanalysis 12*, 165–170.
6. Kubáň, P. and Hauser, P.C. (2009) *Electrophoresis 30*, 3305–3314.
7. Holland, L.A. and Lunte, S.M. (1998) *Anal. Commun. 35*, 1H–4H.
8. You, T., Yang, X., and Wang, E. (1999) *Electroanalysis 11*, 459–464.
9. Wallingford, R.A. and Ewing, A.G. (1987) *Anal. Chem. 59*, 1762–1766.
10. Wallingford, R.A. and Ewing, A.G. (1988) *Anal. Chem. 60*, 258–263.
11. Wallingford, R.A. and Ewing, A.G. (1988) *Anal. Chem. 60*, 1972–1975.
12. Wallingford, R.A. and Ewing, A.G. (1989) *Anal. Chem. 61*, 98–100.
13. O'Shea, T.J., Greenhagen, R.D., Lunte, S.M., Lunte, C.E., Smyth, M.R., Radzik, D.M., and Watanabe, N. (1992) *J. Chromatogr. 593*, 305–312.
14. Park, S. and Lunte, C.E. (1995) *Anal. Chem. 67*, 4366–4370.
15. O'Shea, T.J., Weber, P.L., Bammel, B.P., Lunte, C.E., Lunte, S.M., and Smyth, M.R. (1992) *J. Chromatogr 608*, 189–195.
16. Lu, W. and Cassidy, R.M. (1994) *Anal. Chem. 66*, 200–204.
17. Park, S., Lunte, S.M., and Lunte, C.E. (1995) *Anal. Chem. 67*, 911–918.
18. Huang, X. and Kok, W.T. (1995) *J. Chromatogr. A 716*, 347–353.
19. Kok, W.T. and Sahin, Y. (1993) *Anal. Chem. 65*, 2497–2501.
20. Osbourn, D.M. and Lunte, C.E. (2001) *Anal. Chem. 73*, 5961–5964.
21. Huang, X., Zare, R.N., Sloss, S., and Ewing, A.G. (1991) *Anal. Chem. 63*, 189–192.
22. Sloss, S. and Ewing, A.G. (1993) *Anal. Chem. 65*, 577–581.
23. Hooper, S.E., Roach, D., and Anderson, M.R. (2008) *Electroanalysis 20*, 102–106.
24. Matysik, F.-M. (1996) *J. Chromatogr. A 742*, 229–234.
25. Matysik, F.M. and Backofen, U. (1996) *Fresen. J. Anal. Chem. 356*, 169–172.
26. Powell, P.R., Woods, L.A., and Ewing, A.G. (2005) *J. Sep. Sci. 28*, 2540–2545.
27. Ye, J. and Baldwin, R.P. (1993) *Anal. Chem. 65*, 3525–3527.

28. Ewing, A.G., Mesaros, J.M., and Gavin, P.F. (1994) *Anal. Chem. 66*, 527A–537A.
29. Fermier, A.M., Gostkowski, M.L., and Colón, L.A. (1996) *Anal. Chem. 68*, 1661–1664.
30. Chen, M.-C. and Huang, H.-J. (1995) *Anal. Chem. 67*, 4010–4014.
31. Lunte, S.M. and Zhong, M. (1998) *Anal. Commun. 35*, 209–212.
32. Wallenborg, S.R., Nyholm, L., and Lunte, C.E. (1998) *Anal. Chem. 71*, 544–549.
33. Matysik, F.-M. (1998) *J. Chromatogr. A 802*, 349–354.
34. Matysik, F.-M. (2000) *Electroanalysis 12*, 1349–1355.
35. Klett, O., Björefors, F., and Nyholm, L. (2001) *Anal. Chem. 73*, 1909–1915.
36. Matysik, F.-M. (2008) *Microchim. Acta 160*, 1–14.
37. Olefirowicz, T.M. and Ewing, A.G. (1990) *Anal. Chem. 62*, 1872–1876.
38. Matysik, F.-M., Marggraf, D., Gläser, P., and Broekaert, J.A. C. (2002) *Electrophoresis 23*, 3711–3717.
39. Lin, B.L., Colon, L.A., and Zare, R.N. (1994) *J. Chromatogr. A 680*, 263–270.
40. Zhong, M., Zhou, J., Lunte, S.M., Zhao, G., Giolando, D.M., and Kirchhoff, J.R. (1996) *Anal. Chem. 68*, 203–207.
41. Matysik, F.-M., Björefors, F., and Nyholm, L. (1999) *Anal. Chim. Acta 385*, 409–415.
42. Oshea, T.J., Lunte, S.M., and Lacourse, W.R. (1993) *Anal. Chem. 65*, 948–951.
43. Lu, W.Z. and Cassidy, R.M. (1993) *Anal. Chem. 65*, 2878–2881.
44. Sun, X.H., Yang, X.R., and Wang, E.K. (2003) *J. Chromatogr. A 1005*, 189–195.
45. García, C.D. and Henry, C.S. (2005) *Electroanalysis 17*, 1125–1131.
46. LaCourse, W.R. and Modi, S.J. (2005) *Electroanalysis 17*, 1141–1152.
47. Hübner, J., Nguyen, A., Turcu, F., Melchior, D., Kling, H.-W., Gäb, S., and Schmitz, O. (2006) *Anal. Bioanal. Chem. 384*, 259–264.
48. Mora, M.F., Felhofer, J., Ayon, A., and Garcia, C.D. (2008) *Anal. Lett. 41*, 312–334.
49. Matysik, F.-M. (1998) *Electrochim. Acta 43*, 3475–3482.
50. Blasco, S., Kortz, L., and Matysik, F.-M. (2009) *Electrophoresis 30*, 3355–3360.
51. Nie, D., Li, P., Zhang, D., Zhou, T., Liang, Y., and Shi, G. (2010) *Electrophoresis 31*, 2981–2988.
52. Qian, X., Zhang, Q., Zhang, Y., and Tu, Y. (2010) *Anal. Sci. 26*, 557–560.
53. He, P., Niu, Y., Mei, Z.-H., Bao, J.-F., and Sun, X.-M. (2010) *J. Chromatogr. B 878*, 1093–1097.
54. Sun, X., Niu, Y., Bi, S., and Zhang, S. (2008) *J. Chromatogr. B 870*, 46–50.
55. Jin, W., Yu, D., Dong, Q., and Ye, X. (2000) *Electrophoresis 21*, 925–929.
56. Jin, W., Weng, Q., and Wu, J. (1997) *Anal. Chim. Acta 342*, 67–74.
57. Yao, Q.-H., Mei, S.-R., Weng, Q.-F., Zhang, P.-D., Yang, Q., Wu, C.-Y., and Xu, G.-W. (2004) *Talanta 63*, 617–623.
58. Zhang, S.W., Zou, C.J., Luo, N., Weng, Q.F., Cai, L.S., Wu, C. Y., and Xing, J. (2010) *Chinese Chem. Lett. 21*, 85–88.
59. Chen, G. (2007) *Talanta 74*, 326–332.
60. Wang, J., Chen, G., Chatrathi, M.P., and Musameh, M. (2003) *Anal. Chem. 76*, 298–302.
61. Qu, S., Wang, J., Kong, J., Yang, P., and Chen, G. (2007) *Talanta 71*, 1096–1102.
62. Wang, J. and Musameh, M. (2003) *Anal. Chem. 75*, 2075–2079.
63. Wang, J., Nair, U.S., and Christopher, S.A. (2004) *J. Geophys. Res. D Atmos. 109*, 1–11.
64. Zhang, M., Smith, A., and Gorski, W. (2004) *Anal. Chem. 76*, 5045–5050.
65. Agüí, L., Yáñez-Sedeño, P., and Pingarrón, J.M. (2008) *Anal. Chim. Acta 622*, 11–47.
66. Pumera, M. and Escarpa, A. (2009) *Electrophoresis 30*, 3315–3323.
67. Dong, S., Chi, L., He, P., Wang, Q., and Fang, Y. (2009) *Talanta 80*, 809–814.
68. Shin, D., Sarada, B.V., Tryk, D.A., Fujishima, A., and Wang, J. (2002) *Anal. Chem. 75*, 530–534.
69. Cvačka, J., Quaiserová, V., Park, J., Show, Y., Muck, A., and Swain, G.M. (2003) *Anal. Chem. 75*, 2678–2687.
70. Tsai, D.-M., Tai, H.-W., Shih, P.-R., Chang, R.-L., Wu, H., Conte, E.D., and Zen, J.-M. (2005) *Electroanalysis 17*, 1991–1994.
71. Blasco, S., Gläser, P., and Matysik, F.-M. (2005) *Electroanalysis 17*, 1187–1191.
72. Chen, G., Zhang, L., and Zhu, Y. (2006) *J. Pharm. Biomed. Anal. 41*, 129–134.
73. Xing, X., Shi, X., Zhang, H., Wang, W., and Ye, J. (2009) *Microchim. Acta 167*, 297–302.
74. Xing, X., Shi, X., Zhang, M., Jin, W., and Ye, J. (2008) *Chromatographia 67*, 985–988.
75. Chen, Q.-Z., Fang, Y.-M., Wei, H., Huang, Z.-X., Chen, G.-N., and Sun, J.-J. (2010) *Analyst (Cambridge, U.K.) 135*, 1124–1130.
76. Martin, L.G., Jongwana, L.T., and Crouch, A.M. (2010) *Electrochim. Acta 55*, 4303–4308.
77. Mukherjee, J. and Kirchhoff, J.R. (2009) *Anal. Chem. 81*, 6996–7002.
78. Peláez-Cid, A.-A., Blasco-Sancho, S., and Matysik, F.-M. (2008) *Talanta 75*, 1362–1368.
79. Psurek, A., Matysik, F.-M., and Scriba, G.K.E. (2006) *Electrophoresis 27*, 1199–1208.
80. Zhang, S., Li, X., and Zhang, F. (2007) *Electrophoresis 28*, 4427–4434.
81. Li, X.-M., Zhang, F., and Zhang, S.-S. (2008) *J. Sep. Sci. 31*, 336–340.

82. Zhang, D.-L., Li, W.-L., Zhang, J.-B., Tang, W.-R., Chen, X.-F., Cao, K.-W., Chu, Q.-C., and Ye, J.-N. (2010) *Electrophoresis 31*, 2989–2996.

83. Lacher, N.A., Garrison, K.E., Martin, R.S., and Lunte, S.M. (2001) *Electrophoresis 22*, 2526–2536.

84. Gavin, P.F. and Ewing, A.G. (1996) *J. Am. Chem. Soc. 118*8932–8936.

85. Gavin, P.F. and Ewing, A.G. (1997) *Anal. Chem. 69*, 3838–3845.

86. Woolley, A.T., Lao, K., Glazer, A.N., and Mathies, R.A. (1998) *Anal. Chem. 70*, 684–688.

87. Wang, J., Tian, B., and Sahlin, E. (1999) *Anal. Chem. 71*, 5436–5440.

88. Wang, J., Chatrathi, M.P., Tian, B., and Polsky, R. (2000) *Electroanalysis 12*, 691–694.

89. Blasco, A.J., Barrigas, I., González, M.C., and Escarpa, A. (2005) *Electrophoresis 26*, 4664–4673.

90. Zeng, Y., Chen, H., Pang, D.-W., Wang, Z.-L., and Cheng, J.-K. (2002) *Anal. Chem. 74*, 2441–2445.

91. Pumera, M., Merkoçi, A., and Alegret, S. (2006) *Electrophoresis 27*, 5068–5072.

92. Hilmi, A. and Luong, J.H.T. (2000) *Anal. Chem. 72*, 4677–4682.

93. Liao, K.-T., Chen, C.-M., Huang, H.-J., and Lin, C.-H. (2007) *J. Chromatogr. A 1165*, 213–218.

94. Martin, R.S., Gawron, A.J., Lunte, S.M., and Henry, C.S. (2000) *Anal. Chem. 72*, 3196–3202.

95. Nyholm, L. (2005) *Analyst 130*, 599–605.

96. Baldwin, R.P., Roussel, T.J., Crain, M.M., Bathlagunda, V., Jackson, D.J., Gullapalli, J., Conklin, J.A., Pai, R., Naber, J.F., Walsh, K.M., and Keynton, R.S. (2002) *Anal. Chem. 74*, 3690–3697.

97. Fischer, D.J., Vandaveer, W.R., Grigsby, R.J., and Lunte, S.M. (2005) *Electroanalysis 17*, 1153–1159.

98. Ertl, P., Emrich, C.A., Singhal, P., and Mathies, R.A. (2004) *Anal. Chem. 76*, 3749–3755.

99. Martin, R.S., Ratzlaff, K.L., Huynh, B.H., and Lunte, S.M. (2002) *Anal. Chem. 74*, 1136.

100. Vandaveer, W.R., Pasas-Farmer, S.A., Fischer, D.J., Frankenfeld, C.N., and Lunte, S.M. (2004) *Electrophoresis 25*, 3528–3549.

101. Wang, Y. and Chen, H. (2005) *J. Chromatogr. A 1080*, 192–198.

102. Xu, J.-J., Wang, A.-J., and Chen, H.-Y. (2007) *Trends Anal. Chem. 26*, 125–132.

103. Bao, N., Xu, J.J., Dou, Y.H., Cai, Y., Chen, H.Y., and Xia, X. H. (2004) *J. Chromatogr. A 1041*, 245–248.

104. Schwarz, M.A., Galliker, B., Fluri, K., Kappes, T., and Hauser, P.C. (2001) *Analyst 126*, 147–151.

105. Fanguy, J.C. and Henry, C.S. (2002) *Analyst 127*, 1021–1023.

106. Garcia, C.D. and Henry, C.S. (2003) *Anal. Chem. 75*, 4778–4783.

107. García, C.D. and Henry, C.S. (2005) *Electroanalysis 17*, 223–230.

108. Vickers, J.A. and Henry, C.S. (2005) *Electrophoresis 26*, 4641–4647.

109. Xu, J.-J., Bao, N., Xia, X.-H., Peng, Y., and Chen, H.-Y. (2004) *Anal. Chem. 76*, 6902–6907.

110. Rossier, J.S., Ferrigno, R., and Girault, H.H. (2000) *J. Electroanal. Chem. 492*, 15–22.

111. Chen, D.-C., Hsu, F.-L., Zhan, D.-Z., and Chen, C.-H. (2001) *Anal. Chem. 73*, 758–762.

112. Lacher, N.A., Lunte, S.M., and Martin, R.S. (2004) *Anal. Chem. 76*, 2482–2491.

113. Lai, C.C.J., Chen, C.H., and Ko, F.H. (2004) *J. Chromatogr. A 1023*, 143–150.

114. Mecker, L.C. and Martin, R.S. (2006) *Electrophoresis 27*, 5032–5042.

115. Wu, C.-C., Wu, R.-G., Huang, J.-G., Lin, Y.-C., and Chang, H.-C. (2003) *Anal. Chem. 75*947–952.

116. Lin, K.-W., Huang, Y.-K., Su, H.-L., and Hsieh, Y.-Z. (2008) *Anal. Chim. Acta 619*, 115–121.

117. Osbourn, D.M. and Lunte, C.E. (2003) *Anal. Chem. 75*, 2710–2714.

118. Trojanowicz, M. (2009) *Anal. Chim. Acta 653*, 36–58.

119. Chen, C. and Jong, H.H. (2007) *Anal. Chem. 79*, 7182–7186.

120. Pozo-Ayuso, D.F., Castaño-Álvarez, M., Fernández-la-Villa, A., García-Granda, M., Fernández-Abedul, M.T., Costa-García, A., and Rodríguez-García, J. (2008) *J. Chromatogr. A 1180*, 193–202.

121. Wang, J., Chatrathi, M.P., and Tian, B. (2000) *Anal. Chim. Acta 416*, 9–14.

122. Wang, J., Pumera, M., Prakash Chatrathi, M., Rodriguez, A., Spillman, S., Martin, R.S., and Lunte, S.M. (2002) *Electroanalysis 14*, 1251–1255.

123. Martin, R.S., Gawron, A.J., Fogarty, B.A., Regan, F.B., Dempsey, E., and Lunte, S.M. (2001) *Analyst 126*, 277–280.

124. Gawron, A.J., Martin, R.S., and Lunte, S.M. (2001) *Electrophoresis 22*, 242–248.

125. Lim, T.K., Ohta, H., and Matsunaga, T. (2003) *Anal. Chem. 75*, 3316–3321.

126. Wang, J., Tian, B., and Sahlin, E. (1999) *Anal. Chem. 71*, 3901–3904.

127. Martin, R.S., Gawron, A.J., and Lunte, S.M. (2000) *Anal. Chem. 72*, 3196–3202.

128. Wang, J., Chen, G., Chatrathi, M.P., Fujishima, A., Tryk, D.A., and Shin, D. (2003) *Anal. Chem. 75*, 935–939.

129. Wang, J., Chen, G., Muck, A., Shin, D.C., and Fujishima, A. (2004) *J. Chromatogr. A 1022*. 207–212.

130. Castaño-Álvarez, M., Fernández-Abedul, M.T., and Costa-García, A. (2006) *J. Chromatogr. A 1109*, 291–299.

131. Pumera, M., Merkoçi, A., and Alegret, S. (2006) *Electroanalysis 18*, 207–210.

132. Pumera, M., Merkoçi, A., and Alegret, S. (2007) *Electrophoresis 28*, 1274–1280.

133. Bani-Yaseen, A.D., Kawaguchi, T., Price, A.K., Culbertson, C.T., and Jankowiak, R. (2011) *Anal. Bioanal. Chem. 399*, 519–524.

134. Dawoud, A.A., Kawaguchi, T., Markushin, Y., Porter, M.D., and Jankowiak, R. (2006) *Sens. Actuators B Chem. 120*, 42–50.
135. Dawoud, A.A., Kawaguchi, T., and Jankowiak, R. (2007) *Anal. Bioanal. Chem. 388*, 245–252.
136. Rüttinger, H.-H. and Radschuweit, A. (2000) *J. Chromatogr. A 868*, 127–134.
137. Wang, J., Escarpa, A., Pumera, M., and Feldman, J. (2002) *J. Chromatogr. A 952*, 249–254.
138. Dossi, N., Toniolo, R., Susmel, S., Pizzariello, A., and Bontempelli, G. (2010) *Electrophoresis 31*, 2541–2547.
139. Kovachev, N., Canals, A., and Escarpa, A. (2010) *Anal. Chem. 82*, 2925–2931.
140. Kappes, T., Galliker, B., Schwarz, M.A., and Hauser, P.C. (2001) *Trends Anal. Chem. 20*, 133–139.
141. Jackson, D.J., Naber, J.F., Roussel, T.J., Crain, M.M., Walsh, K.M., Keynton, R.S., and Baldwin, R.P. (2003) *Anal. Chem. 75*, 3643–3649.
142. Berg, C., Valdez, D.C., Bergeron, P., Mora, M.F., Garcia, C.D., and Ayon, A. (2008) *Electrophoresis 29*, 4914–4921.
143. Fernandez-la-Villa, A., Pozo-Ayuso, D.F., and Castano-Alvarez, M. (2010) *Electrophoresis 31*, 2641–2649.
144. Hoque, M.E., Arnett, S.D., and Lunte, C.E. (2005) *J. Chromatogr. B 827*, 51–57.
145. Jacobson, S.C. and Ramsey, J.M. (1995) *Electrophoresis 16*, 481–486.
146. Weng, Q., Xu, G., Yuan, K., and Tang, P. (2006) *J. Chromatogr. B 835*, 55–61.
147. Guan, Q. and Henry, C.S. (2009) *Electrophoresis 30*, 3339–3346.
148. Lunte, S.M., Malone, M.A., and Zuo, H. (1994) *Curr. Sep. 13*, 75–79.
149. Wang, W., Zhang, Y., Wang, J., Shi, X., and Ye, J. (2010) *Meat Sci. 85*, 302–305.
150. da Silva, D., Mrestani, Y., Rüttinger, H.-H., Wohlrab, J., and Neubert, R. (2008) *Chromatographia 67*, 813–817.
151. Wang, J. and Pumera, M. (2002) *Anal. Chem. 74*, 5919–5923.
152. Wang, J., Pumera, M., Chatrathi, M.P., Escarpa, A., Konrad, R., Griebel, A., Dörner, W., and Löwe, H. (2002) *Electrophoresis 23*, 596–601.
153. Rossier, J.S., Schwarz, A., Reymond, F., Ferrigno, R., Bianchi, F., and Girault, H.H. (1999) *Electrophoresis 20*, 727–731.
154. Kim, J.-H., Kang, C.J., and Kim, Y.-S. (2005) *Biosens. Bioelectron. 20*, 2314–2317.
155. Baldwin, R.P. (2000) *Electrophoresis 21*, 4017–4028.
156. Holland, L.A. and Leigh, A.M. (2002) *Electrophoresis 23*, 3649–3658.
157. Rashid, M., Dou, Y.-H., Auger, V., and Ali, Z. (2010) *Micro Nanosyst. 2*, 108–136.
158. Schwarz, M.A., Galliker, B., Fluri, K., Kappes, T., and Hauser, P.C. (2001) *Analyst 126*, 147–151.
159. Dong, S., Zhang, S., Chi, L., He, P., Wang, Q., and Fang, Y. (2008) *Anal. Biochem. 381*, 199–204.
160. Arnett, S.D., Osbourn, D.M., Moore, K.D., Vandaveer, S.S., and Lunte, C.E. (2005) *J. Chromatogr. B 827*, 16–25.
161. Zhang, S.-W., Xing, J., Cai, L.-S., and Wu, C.-Y. (2009) *Anal. Bioanal. Chem. 395*, 479–487.
162. Shiddiky, M.J.A. and Shim, Y.-B. (2007) *Anal. Chem. 79*, 3724–3733.
163. Peng, Y., Zhang, Y., and Ye, J. (2008) *J. Agric. Food Chem. 56*, 1838–1844.
164. Guan, Y., Wu, T., Lin, M., and Ye, J. (2005) *J. Agric. Food Chem. 54*, 24–28.
165. Chicharro, M., Bermejo, E., Ongay, S., and Zapardiel, A. (2008) *Electroanalysis 20*, 534–541.
166. Moreno, M., Bermejo, E., Sánchez, A., Chicharro, M., and Zapardiel, A. (2008) *Anal. Bioanal. Chem. 391*, 867–872.
167. Dossi, N., Piccin, E., Bontempelli, G., Carrilho, E., and Wang, J. (2007) *Electrophoresis 28*, 4240–4246.
168. Wang, W., Zhang, Y., Wang, J., Shi, X., and Ye, J. (2010) *Meat Sci. 85*, 302–305.
169. Chu, Q., Zhang, D., Wang, J., and Ye, J. (2009) *J. Sci. Food Agric. 89*, 2498–2504.
170. Zhang, Z.-X., Zhang, X.-W., and Zhang, S.-S. (2009) *Anal. Biochem. 387*, 171–177.
171. Zhang, L., Dong, S., Yang, Z., Wang, Q., He, P., and Fang, Y. (2008) *J. Pharm. Biomed. Anal. 48*, 198–200.
172. Chu, Q., Jiang, L., and Ye, J. (2010) *J. Anal. Chem. 65*, 103–108.
173. Jin, S., Guo, X., Zhang, Y., Li, P., He, P., Wang, Q., and Fang, Y. (2009) *Chinese J. Chem. 27*, 773–776.
174. Gong, W., Cao, Y., and Wang, Y. (2007) *Chromatographia 66*, 767–771.
175. Chen, G., Xu, X., Zhu, Y., Zhang, L., and Yang, P. (2006) *J. Pharm. Biomed. Anal. 41*, 1479–1484.
176. Tang, Z., Zhou, Y., Zeng, Y., Zang, S., He, P., and Fang, Y. (2006) *J. Pharm. Biomed. Anal. 40*, 484–489.
177. Peng, Y.Y. and Ye, J.N. (2006) *Fitoterapia 77*, 171–178.
178. Tang, Z.-X., Zeng, Y.-K., Zhou, Y., Zang, S.-L., He, P.-G., and Fang, Y.-Z. (2006) *Chinese J. Chem. 24*, 1378–1383.
179. Ding, Y. and Garcia, C.D. (2006) *Electroanalysis 18*, 2202–2209.
180. Omiatek, D.M., Santillo, M.F., Heien, M.L., and Ewing, A.G. (2009) *Anal. Chem. 81*, 2294–2302.
181. Zhang, S., Cao, W., Li, J., and Su, M. (2009) *Electrophoresis 30*, 3427–3435.
182. Arribas, A.S., Moreno, M., Bermejo, E., Ángeles Lorenzo, M., Zapardiel, A., and Chicharro, M. (2009) *Electrophoresis 30*, 3480–3488.

10

OVERCOMING CHALLENGES IN USING MICROCHIP ELECTROPHORESIS FOR EXTENDED MONITORING APPLICATIONS

SCOTT D. NOBLITT AND CHARLES S. HENRY

10.1 INTRODUCTION

Analytical instrumentation for routine online monitoring of the chemical composition of dynamic systems has exhibited a continual increase in demand for a variety of applications. One specific area of particular importance is clinical biology, where studies of transient biological processes are desired. Another area of interest is in environmental monitoring, where both atmospheric and water quality monitoring are in high demand. Online measurement systems are needed in these areas because these applications often exhibit rapid compositional changes as well as short-term analyte stability. Traditional offline measurement techniques may not be sufficient because their poor temporal resolution and inherent delay between sample collection and compositional analysis can result in measurement artifacts and/or the loss of information on important, short-lived events. This loss or distortion of chemical information can affect the evaluation of reaction mechanisms or potential impact on human and environmental health. Often, higher-frequency offline sampling can only partly address these issues. Thus, replacing offline measurements with online sampling systems can increase both data quality and quantity. Because online instruments operate at nearly steady-state conditions, they are less subject to systematic errors originating from transient operation. Additionally, online instrumentation, once installed, often requires less manual operation and intervention than offline analyses and is consequently less prone to human error.

Spectroscopic instrumentation is commonly employed for online monitoring systems because these instruments have few, if any, moving parts and offline instruments can often be easily adapted to online operation. However, for many applications involving complex matrices, the matrix complexity forbids spectroscopic analysis because multiple component signals overlap. Consequently, a separation step is required to measure relevant analytes. Commonly used separation techniques for online monitoring include gas chromatography (GC) and liquid chromatography (LC). These technologies are mature and inherently use pressure-driven flow as a driving force. Thus, they can be integrated into online instrumentation using existing methods. However, both of these separation methods suffer from limitations that prevent their use in some online applications. GC requires volatile analytes or analytes that can be derivatized to be made volatile, and most derivatization reactions are difficult or impossible to perform online. Additionally, GC requires compressed gas and heat sources for operation and thus typically possesses a large footprint that is inconvenient for field applications. LC (and its subset, ion chromatography, IC) requires a sample size that can be too large for some specialized applications, although their typical injection volumes of 50–100 μL are not an issue in most analyses. Many LC methods also require significant quantities of mobile phase. As an example, a mobile phase flow rate of 1 mL min^{-1} requires 10 L of eluent for 1 week of operation. For operation in the field, this amount of liquid waste is at the least inconvenient. Additionally, many LC

Capillary Electrophoresis and Microchip Capillary Electrophoresis: Principles, Applications, and Limitations, First Edition.
Edited by Carlos D. García, Karin Y. Chumbimuni-Torres, and Emanuel Carrilho.

and GC methods are unable to perform at the desired sampling rate or must undergo periodic stationary phase regeneration that interferes with continuous analyses. An alternative, faster separation technique with no stationary phase would therefore prove advantageous for some online monitoring applications.

Capillary electrophoresis (CE) presents a promising instrumental technique for online monitoring because of its simple instrumentation, lack of a solid stationary phase, small energy footprint, small sample and reagent consumption, and short separation times [1–3]. However, CE presents some challenges to interfacing to online analyses. First, traditional LC injection methods either are incompatible with or have to be modified to work with CE because of the small sample injections (0.1–10 nL) involved. Second, online integration requires decoupling of the high separation potential from the sampling interface. Furthermore, hydrodynamic flow in CE typically degrades the separation and is unwanted. If the online system utilizes a continuously flowing sample, either the CE method has to be engineered to compensate for this flow or the flow has to be isolated from the separation capillary. Additionally, even though CE possesses no solid stationary phase, some components in the sample can adsorb to the capillary surface and interfere with the reproducibility of the separation. Finally, CE analyses alter the composition of the background electrolyte (BGE) over time through ion depletion and electrolytic reactions. If ignored, these effects can lower the reproducibility of the separation and quantitation. Several research groups have worked to interface CE instrumentation to online monitoring systems. Much of their work is discussed in review articles on the subject, and we direct interested readers to those reviews [4–9]. It should be noted that none of this work has focused on increasing the longevity of the BGE. One reason for this is that traditional CE does not suffer from BGE degradation as severely as microchip electrophoresis because of the larger buffer volumes. Another explanation for this lack of discussion is that many of these methods do not operate unattended for several days and instead were only tested for a few hours. Also, some methods avoid the issue of BGE degradation by employing a continuous flow that constantly replenishes the BGE solution. However, several independent reports did focus specifically on instrumental modifications to increase separation longevity. Macka et al. demonstrated the importance of positioning the electrophoresis electrode at an adequate distance from the capillary entrance [10]. This distance is important because it allows sufficient diffusional distance for the buffer to consume compositional changes generated at the electrodes, thereby increasing the usefulness of using large BGE reservoir volumes. The authors monitored changes in the buffer pH using indicator dyes and concluded that a capillary tip to electrode distance of 1 mm was sufficient to avoid deleterious effects. The necessary distance likely depends on a variety of factors, including the EOF magnitude and direction, total separation current, buffering capacity of the BGE, and diffusion constants of the BGE constituents. Independent reports by Desiderio et al. and de Jesus et al. illustrated the utility of employing BGE reservoirs with two compartments [11,12]. The separation capillary is positioned in one reservoir compartment, while the electrode is present in the other. The two reservoirs are connected, but via a medium that inhibits mass transport of electrolysis products from the electrodes. Desiderio utilized a cotton plug to inhibit mass transport, while de Jesus employed a salt bridge. Unfortunately, to the authors' knowledge, no comprehensive reviews on the subject of method longevity have been written. The closest publication to this type of review was an editorial by Mayer [13]. Even though a plethora of application-oriented CE reviews have been authored, they neglect to include discussion on long-term monitoring. However, one review, also by Mayer, does discuss some of the aspects of method longevity as subtopics [14].

Microchip electrophoresis (MCE) provides an enticing alternative to traditional CE for interfacing to online monitoring systems. MCE has several advantages in addition to those of CE, including small size and portability, shorter analysis times, lower sample and energy consumption, ease of multiplexing several functions into a single device, facile incorporation of electrochemical detection, and potentially simpler interfacing to online flows due to MCE's native ability to interconnect multiple microfluidic channels into a single network [15–22]. MCE does possess several drawbacks that hinder online monitoring however, including lower BGE volumes (less than 100 μL in MCE vs. several mL for traditional CE) that yield shorter operational times between BGE replacement, difficulty or inability to automate hydrodynamic rinsing of the separation channel between injections, less ideal capillary surface chemistry, and difficulty in reproducible quantitation due to sample ion depletion, small injection volumes, and changes in the surface or BGE during extended operation [23]. As with traditional CE, multiple groups have developed interfaces between flowing sample streams and MCE, permitting online operation [24–33]. Most of these instruments were not characterized with respect to long-term operation, although Büttgenbach and Wilke found that their instrument was limited to about 20 injections (about 24 min), which they attributed to analyte adsorption to the capillary surface [25]. Fang et al. also tested their system for a longer time interval and observed a migration time relative standard deviation (RSD%) of 4.9% for 166 injections performed during a 4 h run [27]. Independent of the development of these online systems, several research groups have made progress in overcoming one or more of the obstacles to extended monitoring. One highlight is the work by Oki et al. where the authors nearly eliminated pH changes in the BGE reservoirs through the use of salt bridges or active neutralization via EOF pumping [34]. Several papers by Kennedy's group document and illustrate the importance of long-term operation. The group focused on analyzing insulin production from individual islets of Langerhans using MCE. Initial efforts yielded a method that could operate for 30 min [35], and a later system improved this value to 2 h [36]. Continued development eventually led to an instrumental method capable of 24-h continuous operation via the continuous perfusion of all reagents [37], which is a major breakthrough in the field of microfluidics and has relevance in this application because the insulin production exhibits cycles of various lengths. Unfortunately, the overall approach taken by Kennedy's group is the exception rather than the typical development path. Instead, the majority of reported MCE methods do not discuss operational longevity, and no follow-up efforts are made to increase the operation time. The review by Revermann and coworkers thoroughly covers many of the limitations of MCE and some solutions to these issues [23]. The review has the additional benefit of discussing some of the early traditional CE literature that investigates the fundamental phenomena behind these limitations.

In this chapter, we discuss the challenges in designing MCE instrumentation and zone electrophoresis separations for extended monitoring applications. Particular focus is given to ensuring the ability to perform rapid sequential injections in real samples where the analytes of interest are high-mobility ions (absolute infinite dilution ionic mobilities greater than $4 \times 10^{-4}\,cm^2\,V^{-1}\,s^{-1}$). Specific topics covered include the following:

1. Choosing a buffer system and microchip design that allow long-term operation.
2. Achieving rapid sequential injection performance by selecting appropriate injection procedures and avoiding system peaks from the BGE.
3. Ensuring robust quantitation by protecting the surface of the capillary from contamination, accounting for and avoiding sample ion depletion effects, and choosing an appropriate internal standard to help compensate for fluctuations in conditions.

The discussion will focus on operation with sample solutions that are spatially static, but much of this subject matter can be directly applied to systems that monitor flowing sample streams.

10.2 BACKGROUND ELECTROLYTE (BGE) LONGEVITY

The composition of the BGE is perhaps the most crucial component in electrophoretic separations. Several review articles discuss the important role of the BGE and rules of thumb in their preparation for CE [38–40]. The basics of buffering covered in these reviews should be understood before attempting BGE design for electrophoresis. However, the long-term stability of the BGE is a critical issue in online MCE that is often ignored in discussions on buffering. Although depleted BGE can be replaced with fresh solution via either manual replacement or an automated drain/fill procedure, this process temporarily terminates the near-steady-state separation process, resulting in data gaps and an increase in the potential for instrumental and systematic errors during each BGE replacement. Even in instruments employing automated refilling procedures, higher-longevity BGEs will allow these instruments to operate for longer periods between the replenishment processes, reducing the number of data gaps. For systems using a continuous refilling procedure, replacing a normal BGE with a long-lasting one will lower the required perfusion rate, decreasing sample consumption and the amount of hydrodynamic interference. Consequently, increased buffer longevity is desirable for all online MCE systems. This section will discuss why certain BGEs exhibit poor longevity and reproducibility as well as ways to lengthen the useful BGE operational time by proper manipulation of buffer composition and microchip physical parameters.

To understand the importance of BGE stability in electrophoresis, the relationship of BGE composition with analyte migration time must be considered. Normally, capillary zone electrophoresis (CZE) assumes a uniform composition throughout the capillary. During an electrophoretic run, however, electric current passing through the capillary changes the BGE composition, a process termed "buffer depletion" [41]. When this occurs, intra- or inter-run migration times are affected. In addition to complicating qualitative identification in MCE, migration time shifts also affect peak areas. This phenomenon is due to the fact that peak areas are proportional to migration time for many CE detection methods (methods where this is not true are those where analyte is consumed during detection, including mass spectrometry and amperometry) [14]. Another complication arises from irreproducible migration times when quantifying analytes from electrokinetically biased injections. This is due to the injected volume of analyte being proportional to the migration velocity of the analyte; thus, the injected sample quantity exhibits similar variance as the migration time. Thus, this mechanism uniformly affects all separations regardless of the detector type.

Potentially, the most important factor in buffer depletion is pH, which is the primary reason that BGEs must be well buffered. During CE, electrolysis occurs at the separation electrodes, resulting in the anodic reservoir solution increasing in acidity and the cathodic solution increasing in basicity. These electrolytic reactions can cause an overall change in the pH of the system as well as the establishment of a pH gradient across the separation capillary. The change in pH directly affects the ionic mobility (μ, $m^2 V^{-1} s^{-1}$) of weak acids and bases. For a particular ionizable group, the average ionic charge of an acid (z_A) is given by Equation 10.1.

$$z_A = \frac{10^{pH-pK_a}}{1+10^{pH-pK_a}} \tag{10.1}$$

Similarly, the average ionic charge of a basis moiety (z_B) is given by Equation 10.2. (Note: In this discussion, all references to the pK_a of a base actually refer to the acid dissociation constant for the protonated conjugate acid of that base.)

$$z_B = \frac{1}{1+10^{pH-pK_a}} \tag{10.2}$$

Because the ionic mobility is proportional to the charge of the ion, some separations, particularly those operating near the pK_a of an analyte(s), are very sensitive to BGE pH changes. As an example, a cation with one ionizable group will exhibit an ionic mobility that is 4.7% faster at 0.02 pH units below its pK_a than at 0.02 units above it (assuming no change in hydrodynamic radius with protonation percentage). pH changes can also affect migration times for strong electrolytes by altering the electroosmotic flow (EOF). EOF is proportional to the zeta potential of the capillary surface, which in turn is related to the surface charge. Therefore, changes in the charge state of ionizable surface groups will affect all components of the separation. Strong electrolytes can also be affected by pH changes when they are complexed by a BGE component that is pH sensitive because the fraction of the analyte undergoing complexation changes with pH.

In addition to pH changes, individual buffer species concentrations in the BGE can also change during operation. Just as with pH, this can alter separations where complexation reactions are utilized. More importantly, nearly all separations will be affected by concentration and pH changes via ionic strength effects. This is particularly important when simultaneously analyzing both monovalent and polyvalent species. More highly charged species are more highly affected by ionic strength than less highly charged ones, and unintended ionic strength changes can cause previously resolved species of different valence to comigrate. The mobility dependence on ionic strength is classically estimated by assuming a reduction in mobility that is directly proportional to the magnitude of the analyte charge and proportional to the square root of the BGE ionic strength. This approximation is not strictly valid due to the finite size of real ions. Li and coworkers performed a thorough investigation of this phenomenon and found that infinite-dilution mobilities can be more accurately adjusted for finite ionic strengths using Equation 10.3 [42]. Here, μ_0 is the infinite dilution mobility, i is the ionic strength, and j is a constant that should be determined empirically but can be estimated using the Pitts equation [43].

$$\mu \approx \mu_0 - \frac{jz\sqrt{i}}{1+2.4\sqrt{i}} \tag{10.3}$$

The important points of this equation are that changes in mobility are more pronounced for polyvalent species and that ionic strength changes have a larger relative effect on BGEs with a lower nominal ionic strength.

With some of the mechanisms of BGE failure under extended electrophoresis elucidated in the preceding text, discussion can focus on engineering BGEs to minimize these effects. Several articles have been written on the empirical consequences of buffer depletion, as well as the theory behind it [41,44–46]. The important factors affecting the percentage of BGE compositional change with time are total BGE volume, separation current, and buffer capacity. Of these, volume of the BGE is the easiest to discuss and understand. Put simply, BGE volume impacts none of the depletion processes on a mass or molar basis. Therefore, all fractional or percentile changes to the BGE, the relevant measure when observing empirical phenomena, are inversely proportional to volume. Consequently, the acceptable BGE longevity should be exactly proportional to volume. This has a dramatic importance in microfluidic devices. For instance, an MCE system with a 2 mm diameter, 1 mm tall fluid reservoir can only hold about 3.1 μL of BGE. Changing the reservoir size to a 10 mm diameter and the fluid height to 2 mm increases the volume to 157 μL, a 50-fold increase. If the small reservoir could support a 5 min analysis, the larger one would last over 4 h, even without any other changes to the system. Additional benefits of larger microchip reservoirs included reduced Laplace (meniscus) pressures and a lower surface area-to-volume ratio in the reservoir [47]. Both of these factors decrease the reproducibility of the separation as they increase in magnitude; thus, reducing them improves the integrity of the separation.

The electrical current induced by the separation potential (s) is a primary driver of buffer depletion and should be minimized. This is due to the electrical current being directly proportional to the electrolysis occurring at the electrodes, and thus proportional to the hydronium and hydroxide being produced at the anode and cathode, respectively. The separation current, I (A), is given by Equation 10.4, where R_e is the electrical resistance of the channel (Ω), U is the applied potential (V), ρ is the resistivity of the BGE (Ω m), A is the capillary cross-sectional area (m^2), L is the total capillary length (m), E is the applied electric field ($V m^{-1}$), and κ is the solution conductivity ($S m^{-1}$).

$$I = \frac{U}{R_e} = \frac{UA}{\rho L} = \kappa EA \tag{10.4}$$

While this equation is based on electrical current and the resulting electrolysis, similar results can be achieved by starting from the ionic mobility of the buffer constituents and modeling the system from a charge balance perspective (with hydronium and hydroxide production needed to maintain charge balance from the exiting ions). Due to the intrinsic dependence of electrical current on the mobility of the electrolyte components, these are actually equivalent models that are just mathematically derived from different approaches. Consequently, depletion effects from individual components electrophoretically migrating out of the reservoir should correlate well with current as long as all the ionic BGE components are similar in both mobility and concentration. Thus, all further discussion will use separation current as a proxy for all buffer depletion effects. It is important to note that Equation 10.4 contains both voltage and capillary length, and thus it is not the absolute magnitude of the applied voltage that decides the depletion rate, but instead it is the applied electric field. Consequently, post-detection column length does not change the depletion rate despite increasing the required separation potential. However, it should be realized that for a given required maximum separation time, shorter capillaries yield reduced depletion effects because they utilize lower field strengths (and also have the cost of decreased resolving power). Another important factor is that most MCE systems employ multiple BGE reservoirs and electrophoretic channels, so proper design of the microfluidic network is needed to ensure that none of the other channels have a considerably higher applied field than does the separation channel, as that high-field channel and reservoir may then become the limiting factor in operational longevity. This is especially true when utilizing gated injection, as this method employs continuous flows in all four capillary segments [48]. Discussion of proper-gated injection design has been provided in several publications [49–51], and these approaches can be applied to designing microchips for long-term analysis. One effective way to decrease separation current and therefore increase operation time is to employ narrower capillaries, a change that is easily accomplished with many MCE fabrication protocols. Decreasing cross-sectional area has the additional benefits of decreasing any Joule heating effects, increasing hydrodynamic resistance, and decreasing volumetric EOF pumping that can lead to hydrodynamic artifacts over time (and will be discussed later). Smaller channel cross-sections do have several potential drawbacks, including reduced detection sensitivity, increased analyte adsorption to the capillary surface, and an increase in relative channel dimension uncertainty. The choice of a final channel size is therefore dependent on an appropriate compromise of the aforementioned effects.

The final term in Equation 10.4 is the only chemistry-specific term in the equation—solution conductivity. The conductivity of the BGE is inherently tied to its buffer capacity and so the two will be discussed together. Intuitively, it is easy to assume that higher buffer capacity will lead to increased resistance to buffer depletion, which has been previously reported [41]. However, this is not always effective, as was shown in Bello's work for a buffer operating at its pK_a [44]. Intuitively, this is rationalized by noting that depletion effects are proportional to the solution conductivity, and conductivity is roughly proportional to the concentration of the BGE (with some deviations—as ionic strength increases, the molar conductivity decreases, resulting in small longevity benefits). Therefore, the ratio of buffer capacity to solution conductivity is the relevant figure of merit, as was discussed by Reijenga et al. [52]. This is shown graphically in Figure 10.1, where PeakMaster 5.2 simulation results are shown for the conductivity, buffer capacity, and buffer capacity-to-conductivity ratio as a function of the concentration of a BGE composed of an equimolar mixture of 2-(*N*-morpholino)ethanesulfonate (MES) and histidine (HIS) [53]. The ratio of conductivity and buffer capacity is representative of buffer longevity, and this plot shows it to only be weakly dependent on buffer concentration, agreeing with the results of Bello [44]. It should be stated, however, that this result is only valid when all the components added to the BGE are buffering in nature (either a buffering ion or a titrating species). When nonbuffering salts (i.e., NaCl) are added to the BGE to alter ionic strength or perform complexation chemistry, then an improved buffer capacity-to-conductivity ratio will be achieved when increasing the

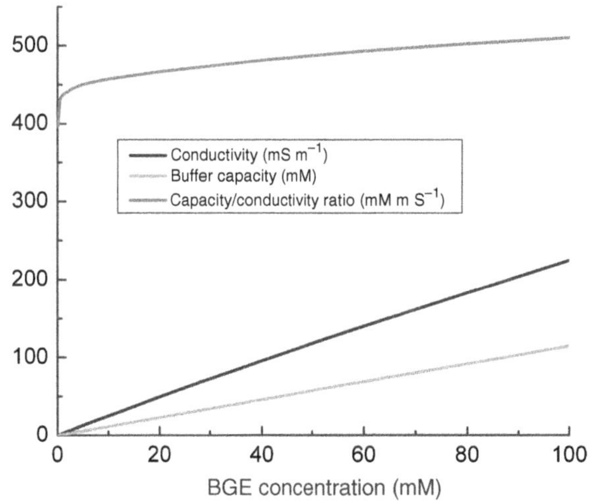

FIGURE 10.1 PeakMaster 5.2 simulation results for an equimolar mixture of MES and HIS (*x*-axis is concentration of each component, not total concentration). Buffer capacity increases proportionally with concentration, whereas solution conductivity increases with a nearly linear response and only a slight decrease in molar conductivity with increasing concentration. The ratio of buffer capacity to conductivity is effectively proportional to BGE longevity (with respect to pH change) and shows only weak dependence on BGE concentration. The *y*-axis units for each trace are given in the legend.

concentration of the buffer as long as the nonbuffering ion concentration does not increase. Therefore, if a predetermined ionic strength is required for desired selectivity or EOF, then it is best to achieve higher ionic strength by increasing the concentration of buffering ions instead of adding nonbuffering salts. Although the weak dependence of buffering capacity-to-conductivity ratio on the concentration of the BGE indicates that an arbitrary concentration can be selected, the effects of sample matrix on the BGE are not concentration independent, and therefore the BGE concentrations need to be decided with the sample composition in mind.

As stated in the preceding text, buffer concentration and conductivity are inherently related. Therefore, the conductivity of the BGE must be decreased via a method other than total concentration to increase buffer longevity. The solution to this is intentional and well-planned buffer choices. First and most importantly, the buffer should be composed of both a weak acid and a weak base at a pH in both buffering regions. This permits every ion in solution to contribute to buffer capacity instead of only half of the ions, essentially doubling buffer capacity. The next step in preparing a low-conductivity BGE is to choose buffering compounds that have low molar conductivities. Foremost in this approach is to ensure that each component has a net charge of one or less. Additional charge typically does not increase buffering capacity (because multiple acidity constants typically do not have overlapping buffering regions), but it often greatly increases solution conductivity because molar conductivity is nearly proportional to the square of the ionic charge (due to the molar conductivity being proportional to both the charge and the mobility, with the mobility itself also being proportional to charge). Polyvalent ions are also known to lead to other problems in the BGE, particularly the formation of additional system zones [54,55]. The next step in lowering molar conductivity is to utilize large buffering molecules. Ionic mobility is inversely proportional to the hydrodynamic radius, so larger molecules are less conductive. A good rule of thumb is to target compounds with mobilities below $3\times10^{-8}\,m^2\,V^{-1}\,s^{-1}$, as significant further improvements are difficult to achieve because very few common buffers have mobilities below $2\times10^{-8}\,m^2\,V^{-1}\,s^{-1}$. It is common to find claims that zwitterionic buffers should be employed because their net charge is nearly zero. This is misleading, as the real reason that zwitterions often exhibit low molar conductivities is that charged moieties require more solvating molecules than neutral structures; thus, they considerably increase the hydrodynamic volume in zwitterionic species. However, this low net charge claim is related to the final step in lowering conductivity, which is rarely discussed, and that is to employ a buffering acid with a pK_a above the BGE pH and a base with a pK_a below the operating pH [56]. Intuitively, the benefit to this approach is apparent by realizing that when the acid is below its pK_a, it is mostly uncharged and vice versa for bases. To show this graphically, PeakMaster 5.2 was used to simulate hypothetical buffers composed of equimolar acid and base components, each having mobilities of $2.5\times10^{-8}\,m^2\,V^{-1}$ and pK_a values averaging 6.0 [53]. Buffer capacity was kept constant at 15 mM, and the buffer concentration to attain this capacity and the resulting conductivity are plotted as a function of buffer pK_a in Figure 10.2. The x-axis in this plot is the acid pK_a minus pH, or equivalently, pH minus the base pK_a. It is clear that the aforementioned rule of thumb of keeping the acid pK_a above the pH and the base pK_a below the pH is a powerful way to reduce buffer conductivity while maintaining constant buffer capacity. Figure 10.2 also shows that the common perception of operating at a buffer's pK_a to yield best results is not always true. For instance, utilizing an acid with a pK_a one unit above the BGE pH and a base with a pK_a one unit below the pH results in a conductivity that is 45% lower than when pH $= pK_a$, increasing buffer longevity by 83%. Additionally, reported buffer capacity is derived from a differential computation (infinitesimally small addition of acid or base), and for larger, finite additions of perturbing acid or base, the lower conductivity BGE should actually undergo even less pH change than in the BGE where pK_a = pH. Figure 10.2 also illustrates that the buffering approach recommended here for long-term monitoring is also beneficial for most conductivity detection applications in electrophoresis, where a low background signal is desired. As stated in the preceding text, the conductivity background drops by 45% when using a pH–pK_a differential of 1. The

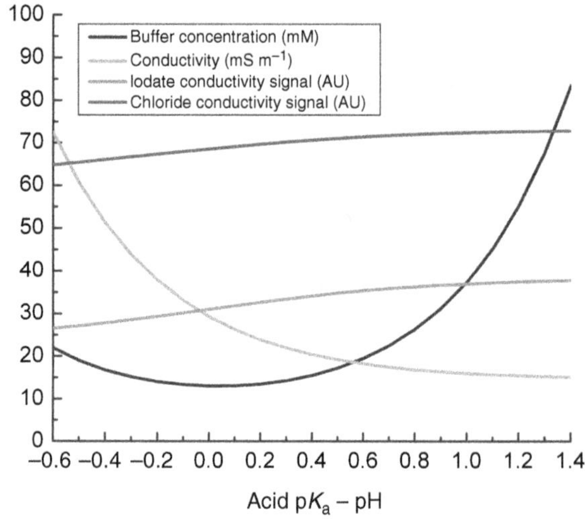

FIGURE 10.2 PeakMaster 5.2 simulation results for hypothetical pH 6.0 BGEs consisting of equimolar acid and base having pK_a values at equal but opposite values from 6.0 and electrophoretic mobilities of $2.5\times10^{-4}\,cm^2\,V^{-1}\,s^{-1}$. The x-axis in this plot is pK_a, acid minus pH, or equivalently, pH minus base pK_a, base. Buffering capacity was kept constant and the necessary buffer concentration to achieve these conditions and the resulting conductivity are shown. The calculated conductivity detection signals for chloride and iodate are also shown.

PeakMaster simulation shows that the same changes in conditions result in a 6% increase in signal for chloride due to changes in the mobility and displacement of the BGE co-anion [53]. Because the mobility of iodate is closer to the mobility of the BGE co-anion, iodate shows an even larger percent increase in signal, 19%. Assuming the background noise is proportional to the background signal, these improvements combine to yield a signal-to-noise ratio increase of 93% and 118% for chloride and iodate, respectively. The similarities in BGE conditions for conductivity detection and BGE longevity make conductivity detection an appealing option for extended monitoring applications.

The plot in Figure 10.2 indicates that indefinitely increasing the pK_a difference from the operating pH would result in the best performance. However, in practice, a variety of complications arise that limit this approach. One of these is solubility, which imparts a physical limitation on the BGE composition. Also, changes in solution viscosity or unexpected, significant complexation phenomena may present problems at high buffer concentrations. Another issue is ionic strength. Although not shown in Figure 10.2, the ionic strength follows the same behavior as the conductivity, dropping from 17.97 mM at −0.6 pH units to a value of 3.40 mM at +1.4 pH units. This drop in ionic strength will need to be accounted for when designing separation conditions because polyvalent species will migrate faster relative to monovalent ions at the lower ionic strength. The reduced ionic strength can also cause an increased EOF magnitude. The lower ionic strength and conductivity also lead to increased peak broadening during sample overloading, limiting resolution and/or useful linear range. Perhaps the most problematic issue with increasing buffer concentration is trace buffer contamination. As an example, if a HEPES buffer (MW = 238.3) has 200 ppm sulfate impurity, this equates to 50 μM sulfate in a BGE containing 100 mM HEPES. While not enough to significantly increase the background conductivity of the solution, this concentration is high enough to interfere with sensitive conductivity or indirect UV detection (particularly at low ionic strength) through competitive displacement, high blank values, or the introduction of unwanted system zones. Consequently, we recommend employing buffers that have pK_a values displaced 0.5–1.0 units from the pH. Relative to the case where $pH = pK_a$, this increases buffer longevity by 52–83% and only requires a buffer concentration increase of 32–102%.

Although the above-mentioned discussion on buffer depletion does not appear to be affected by EOF, investigations have shown long-term electrophoresis stability to be affected by the EOF [41]. Confounding this issue is that not all the studies tied the concentration dependence of the BGE stability to the EOF even though that was a core factor. For instance, we mentioned earlier that some research found the BGE longevity significantly depended on buffer concentration even though Bello's work indicated this to be untrue [44]. The reported concentration dependence is in fact an EOF effect, as increasing BGE concentration (and thus ionic strength) decreases the EOF magnitude. Several mechanisms contribute to the reproducibility and longevity being dependent on EOF magnitude. First, the EOF pumps solution from one reservoir to another, lowering the volume in one reservoir while increasing the volume in the other. The change in volume affects both head height and Laplace pressures, inducing unwanted hydrodynamic flow [47,57]. This pressure-driven flow changes migration times and peak areas, and it can also increase band broadening. Because these changes lower reproducibility and induce systematic changes with time, they give the appearance of buffer depletion even if buffer composition remains constant. A second mechanism to the reduction in reproducibility with an increased EOF is that migration time uncertainty is inherently tied to the EOF uncertainty. This becomes clear from examining the equation for migration time in electrophoresis, Equation 10.5, where L_{eff} (m) is the effective separation distance, t_{mig} (s) is the migration time, and μ is mobility ($m^2\,V^{-1}\,s^{-1}$) for both the ion and EOF.

$$t_{\text{mig}} = \frac{L_{\text{eff}}}{E(\mu + \mu_{\text{EOF}})} \tag{10.5}$$

Typically, EOF is the primary factor in migration time uncertainty because other factors, including ionic mobility (dependent on temperature and viscosity), electric field (dependent on total length, applied potential, and a uniform BGE conductivity through the capillary), and effective separation length, are nominally constant. Therefore, with otherwise ideal conditions, the uncertainty (σ) in the migration time is given by Equation 10.6.

$$\sigma_{t_{\text{mig}}} = \frac{\sigma_{\mu_{\text{EOF}}} L_{\text{eff}}}{E(\mu + \mu_{\text{EOF}})^2} \tag{10.6}$$

Relative standard deviation (RSD) is often a better indicator of separation reproducibility than absolute uncertainties, and combining Equations 10.5 and 10.6 and rewriting them in terms of RSD values yields Equation 10.7.

$$\text{RSD}_{t_{\text{mig}}} = \frac{\text{RSD}_{\text{EOF}}|\mu_{\text{EOF}}|}{|\mu + \mu_{\text{EOF}}|} \tag{10.7}$$

The results of this equation are graphically illustrated in Figure 10.3 where each trace shows the RSD of the migration time induced by a 3% RSD in the electroosmotic mobility for a range of typical ionic mobility values. Separations with analytes in regions with large EOF values will show higher sensitivity to BGE degradation because small changes in BGE composition induce changes in the EOF, and these separations are more sensitive to EOF fluctuations. The traces in Figure 10.3 indicate that the most reproducible

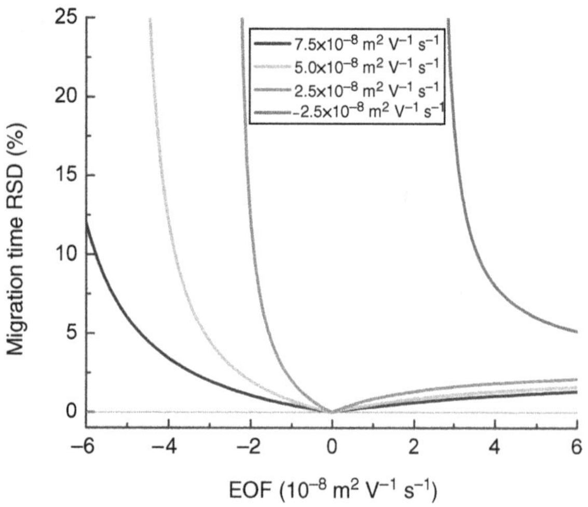

FIGURE 10.3 Expected migration time uncertainty (RSD) as a function of EOF using Equation 10.6 and assuming $RSD_{EOF} = 3\%$. Note that equivalent plots are obtained with oppositely charged ions and a mirrored x-axis.

results will come from separations where the EOF is near zero. For higher EOF magnitudes, using a co-EOF approach yields superior migration-time reproducibilities when compared to counter-EOF methods. It can also be concluded from the poor reproducibilities of the low-mobility anions in Figure 10.3 that these species would be better analyzed by reversing the electrophoresis polarity and analyzing with a slow co- or counter-EOF than by using a high EOF to "drag" anions to the cathode with normal polarity.

Note that, for Figure 10.3, equivalent results are obtained when plotting oppositely charged ions using a mirrored x-axis. A third and final mechanism for EOF effects to reduce the BGE longevity and/or reproducibility is one where the EOF actually changes the rate of BGE compositional change. This occurs when the volume reduction in one reservoir due to EOF pumping is large enough that it significantly increases the rate of compositional change in the BGE caused by electrolysis. The volumetric flow from EOF (Q_{EOF}, $m^3\,s^{-1}$) is easily calculated using Equation 10.8.

$$Q_{EOF} = E\mu_{EOF}A \tag{10.8}$$

Typically, these values are small, but extended monitoring can make these values significant in extreme cases, such as a combination of high EOF magnitude, small BGE reservoirs, and large capillary cross-sectional areas. For instance, at 300 V cm^{-1} and an EOF of $6\times10^{-8}\,m^2\,V^{-1}\,s^{-1}$ in a 50 μm × 50 μm channel, 1 h of operation will transfer about 16 μL of solution. That volume is significant in most MCE applications, and the volume loss may result in depletion rates of up to 50% higher than expected, neglecting EOF when the BGE volume in one reservoir is 30–40 μL, as is common in MCE. Overall, the easiest way to minimize EOF contributions to separation degradation is to suppress the EOF using surfactants, polymers, and/or covalent modification. In addition to reducing the volumetric transfer rate from one reservoir to another (inducing hydrodynamic flow through the channel and increased depletion in the reservoir the EOF exits from), lowering EOF should also improve reproducibility when small fluctuations in EOF occur, as shown in Figure 10.3.

Although BGE longevity and reproducibility in MCE are typically evaluated with respect to operation time, any other factors that affect composition, pH, or ionic strength can also limit BGE lifetime even if electrophoresis is not occurring at all times. Some of these include temperature, carbonate dissolution, and component volatility, reactivity, and stability. Temperature not only directly affects ionic mobility and EOF, but also changes the pK_a of buffers and therefore the pH. Typically, higher temperatures lower the pK_a of a molecule, and amine moieties tend to exhibit higher temperature dependencies than carboxylates [39,58]. Additionally, higher temperatures yield higher electrophoretic currents, so any change in temperature should be taken into consideration when determining BGE operational times. One often overlooked factor concerns changes in the ambient temperature throughout the day. Obviously, operation outdoors is highly susceptible to this phenomenon. In some outdoor environments, normal day–night changes can be 20 K or larger. This is particularly troublesome when ambient temperatures go below the freezing point of water. Therefore, rigorous outdoor MCE monitoring must utilize a container with temperature control. Such a container is often needed indoors also, as indoor temperatures can also show large variance (5 K is not unreasonable). Changes indoors can be due to external forcing effects, energy-saving thermostating systems, changes in the quantity of personnel in the building throughout the day, and seasonal differences. Because of all the unknown factors contributing to the ambient temperature, temperature monitoring alongside the MCE system during operation may be needed for rigorous performance evaluation and troubleshooting.

Carbonate dissolution from atmospheric carbon dioxide is a well-understood process, although it is often ignored, and it causes serious reproducibility problems in MCE at high pH. An excellent discussion on atmospheric carbonate contamination in CE is given in a buffering review by Persat [39], so only the most pertinent points will be summarized here. Below pH 5.5, contamination from carbon dioxide is negligible except for the most rigorous applications. Above this, bicarbonate concentration becomes significant, and by pH 8, the divalent carbonate concentration reaches significant levels. In addition to pH and ionic strength changes, carbonate can complex with both buffer and sample cations and even form insoluble precipitates. Carbonate contamination also makes sensitive conductivity detection of anions above pH 8.5 nearly impossible due to carbonate's high molar conductivity that increases background signal and induces relatively high mobility system zones. To overcome carbonate dissolution effects, three options are readily available. The first and easiest is operation at low pH. This solution works well for strong inorganic acids as well as most inorganic cations (where the low pH will also help avoid interference from hydroxide). Clearly, this solution will not work as well when pH selectivity requirements demand operation near the pK_a of protonated amines (8–12) or when detection modes require high pH (pulsed amperometric detection, for instance). A second option is to engineer and operate the BGE for use in fully equilibrated carbon dioxide environments. This approach requires additional calculations in the BGE design as well as an additional wait time after the BGE solution is prepared to ensure equilibration with the atmosphere. This method is only recommended for intermediate pH values where the BGE will not be dominated by carbonate and will not work in cases where insoluble carbonates will form. A third possibility to preventing carbonate interference is to prepare and operate the BGE in a carbon dioxide–free environment such as a glovebox or glovebag. Note that the source water will need to be properly degassed prior to use with this method. Although a variety of difficulties may be introduced from this approach, if the continuous operation already requires rigorous environmental control, then the additional requirement of a carbon dioxide–free environment may not be too severe.

In addition to carbonate and temperature effects, the chosen BGE constituents themselves can also lead to unintended compositional changes over time. Component volatility is a leading culprit here. Even if none of the specific BGE components are volatile, all aqueous solutions will still undergo evaporation of water with time, leading to concentration and ionic strength increases as well as possible changes in surfactant behavior. Operating in an environment with nearly 100% relative humidity can minimize water evaporation, although care must be taken to avoid arcing between the high-voltage connections used for electrophoresis when applying potentials in a high-humidity environment. Another option is to employ BGE reservoir covers that minimize the mass transfer cross-sectional area, reducing evaporation effects approximately proportionally to this reduction in cross-section (or even reduce them beyond the proportional size if turbulent mixing effects are eliminated with this step). Noblitt and coworkers successfully utilized this approach in an online monitoring instrument employing MCE for aerosol analysis [59]. In that instrument, the covering lids were airtight and interconnected to permit isobaric operation. However, in the majority of cases, airtight seals are not desired because they can hydrodynamically close the system, leading to unwanted pressure build-up. In addition to water evaporation, other volatile components can be problematic. This is especially true when organic co-solvents are utilized. Traditional organic modifiers such as acetonitrile, methanol, ethanol, and acetone should not be used in extended online monitoring applications unless their volatilization can be limited or, preferably, eliminated. Less volatile organics such as dimethylsulfoxide (DMSO) can likely still be used. Although typically less prone to volatilization problems, even buffering ions and complexation additives can suffer from volatilization over extended times. Acetic acid and 15-crown-5, respectively, are examples of commonly used constituents that are volatile. In addition to volatilization effects, BGEs can also succumb to component degradation or reactions. Clearly, components that react with oxygen and carbon dioxide should be avoided. Compounds that hydrolyze on the timescales of interest should not be employed. Primary amines (unless completely protonated) and reduced sulfur species are often relatively reactive and should be avoided in BGEs. Large macromolecules can behave unpredictably and irreproducibly, especially with regards to channel surface chemistry. Heavy metal ions should be avoided because they can change surface chemistries with time, and some heavy metals undergo slow ligand exchange, changing the BGE composition over hours or days. Perhaps the most pernicious, least reproducible, and hardest to address problem in extended monitoring comes from microbial action on BGE solutions. This issue is ubiquitous, and although BGE filtering protocols and the inclusion of biocides in the BGE can limit microbial action, in some cases, the problem is unavoidable and can only be worked around by empirically determining operation time. This measurement should be performed in an environment as closely resembling the working environment as possible, and systems that exhibit stability in clean indoor air may be quickly consumed in unfiltered outdoor air. In summary, great care must go into design of the BGE for extended online monitoring with MCE even beyond buffer depletion considerations. If the separation functions properly initially but fails after longer operation, each component needs to be evaluated with regards to the above-mentioned issues and problematic components replaced with acceptable alternatives.

Overall, the preceding paragraphs discuss the pertinent points in assembling a BGE that is robust with respect to both electrophoretic and nonelectrophoretic degradation. The theoretical approach developed by Bello provides a way to calculate buffer longevity by determining the time needed to change the BGE pH by a predetermined amount [44]. However, we recommend measuring buffer longevity empirically because properly estimating the acceptable amount of pH change can be difficult and other effects such as EOF alterations, ionic strength variance, and changes in the amount of complexation can also result from BGE degradation. To test BGE longevity, sequential analysis of a standard solution while monitoring migration times and peak areas works well. (Note that sample depletion can also affect peak areas and relative peak areas, so take measures to ensure this is not occurring during the longevity measurement through a sample replacement scheme.) Because the final MCE device and BGE may be targeting several days of operation, it may be better to utilize a modified MCE device operating under less-ideal longevity conditions to perform the test. The maximum operational time for the final device (O_2) can then be calculated using the proportionality in Equation 10.9 and the maximum operational time for the testing device (O_1). In this equation, V (m^3) represents volume of the BGE reservoirs.

$$O_2 = O_1 \frac{V_2 E_1 A_1}{V_1 E_2 A_2} \tag{10.9}$$

Utilizing this approach comes with some limitations because performance may unexpectedly change with the different conditions. For example, changing the buffer volume may cause differences in Laplace pressures or evaporation rates. Operating at higher electric fields can alter behavior through Joule heating. Changing the channel cross-sectional area is particularly risky, as this may affect both the EOF and the surface adsorption effects. Other factors simply cannot be accounted for except with a true extended run, including BGE component degradation or reactions, capillary surface fouling, and evaporation effects. Consequently, Equation 10.9 should only be considered approximate. Only a true extended analysis at the exact operating conditions can ensure long-term performance integrity. As a final example of the impact that the methods discussed in this section can accomplish, the combined benefits from the above-mentioned discussion will be applied for a typical MCE system. The starting system in this case is a BGE of potassium phosphate at pH 7.0, a separation channel of 50 μm × 50 μm, and a 3 mm diameter buffer reservoir 2 mm in height. Changing the buffer system to a HEPES/Bis-Tris system reduces the background conductivity by 95.3% (calculation done with PeakMaster 5.2 for buffers of equal 15 mM buffering capacity) [53]. Decreasing the channel size to 25 μm × 25 μm lowers the buffer consumption by 75%. Increasing the BGE reservoir size to 12 mm diameter by 3 mm tall changes the volume from 14 μL to 339 μL. The cumulative effect of these three changes results in an expected BGE longevity increase of over 2000-fold. If the original method was acceptable for 5 min, the adjusted method will operate for about 7 days. Additional changes such as suppressing the EOF, minimizing evaporation, and lowering the separation field strength will only further improve the system.

10.3 ACHIEVING RAPID SEQUENTIAL INJECTIONS

For many online monitoring instruments, short sampling intervals are required. Even when it is not required, high frequency collection is often still beneficial because it permits increased data averaging and/or improved confidence intervals. In separations, the frequency of data collection is inversely proportional to the minimum acceptable time between sequential injections *with a real sample*. Knowledge of the sample matrix is important in this endeavor because the appearance of peaks late in the separation, even if they are not important in the analysis, can interfere with analytes of interest if another injection has already been performed. In chromatography, frequently, the only solution is to increase the interval between injections and allow the late-eluting peak to exit the column. Traditional CE, however, can avoid this limitation by hydrodynamically rinsing the capillary between analyses to remove any late-migrating ions or system zones. Often, several capillary volumes of solution can be flushed in less than 1 min. Initially, one might expect MCE, due to its fast separation times, to be an ideal technique for performing rapid sequential injections during online monitoring. However, many MCE systems are incapable of automated hydrodynamic flushing. Consequently, they suffer from the same problems as chromatography systems with respect to late-migrating peaks. Arguably, MCE systems may be even more limited in this aspect than chromatography, because many electrophoresis separations exhibit late-migrating system zones. This section discusses approaches to minimizing or even eliminating unwanted, late-migrating ion peaks and system zones. Specifically, discussion will focus on design of the BGE, choice of the EOF direction, injection mode, and optimizing injection timing for maximum sampling frequency. Because the detection technique has ramifications on the sensitivity to system peaks and unwanted sample peaks, this will also be covered.

The first step in achieving rapid subsequent injections is to reduce the number of system zones (also known as system peaks, eigenzones, or eigenpeaks) due to BGE composition. The occurrence and prediction of system zones in CE have been covered by a variety of authors and publications [38,60–69]. Therefore, only the pertinent conclusions and rules of thumb will be discussed here. System zones are created by the interruption of the otherwise continuous BGE during sample introduction. This BGE perturbation then propagates through the capillary in the same manner as a true ion peak. The detector can then record the resulting system zones as peaks and/or dips, thus interfering with quantitation of analytes of interest. System zones can also induce peak broadening in nearby analyte peaks. Consequently, even separations employing detectors that do not directly detect the system zones (fluorescence, amperometry, and direct UV absorbance, for instance) can still be negatively impacted by their presence. When performing rapid sequential injections without hydrodynamically rinsing the system between injections, any system zones migrating significantly slower than the analytes of interest that were injected with a previous injection can interfere with the behavior of the faster analytes injected in later runs. The best way to minimize system zones is to construct the BGE using only one base (cation) and one acid (anion). As long as both species are monovalent, this approach should yield two system zones that effectively migrate with or near the EOF, termed the injection zone. It should be noted that operation in either highly basic or highly acidic conditions could shift these two system zones away from the injection zone due to the interactions with hydroxide and hydronium, respectively. If the required selectivity is such that a BGE with only one acid and base cannot be utilized, then the next best choice is to utilize a BGE with one co-ion and two (or more) counter-ions. Employing this BGE will result in no interfering system zones in the separation near the ions with the polarity of interest, and $n-1$ system zones (apart for the injection zone) of the opposite polarity will be present, where n is the number of counter-ions. The opposite-polarity system zones can then be avoided by proper selection of the EOF direction and magnitude (discussed later). As a last resort, BGEs containing two or more co-ions can be used. These BGEs will exhibit system zones (following the $n-1$ rule) with the same migration direction as the polarity of interest. Note that if the low-conductivity BGE suggestions from the last section are followed, the resulting system zones should also be of a low mobility because system zones will appear between the mobilities of the two BGE co-ions. We recommend employing simulation software such as Peak-Master to predict the appearance of system zones when designing the BGE [53,70]. Software simulators will predict the location of the system zones with relative accuracy, and they will also quantitatively indicate when the operating pH is extreme enough to generate substantial system zones from either hydronium or hydroxide ions. Note that simulation software may fail to predict system zones correctly if the BGE components are sufficiently impure so as to have a significant quantity of additional ions present. Therefore, only highly pure BGE components should be used. Choosing only the purest BGE components also helps increase laboratory-to-laboratory reproducibility. As a final recommendation, if BGE additives or modifiers are being used, we suggest choosing nonionic compounds if possible, as these species will not add additional system zones to the separation.

Truly eliminating all system zones is an impossible task, as even the simplest BGE will always exhibit one effective system zone, the injection zone (or "water peak") that migrates with the EOF. As mentioned in the preceding text, system peaks can affect separations even when they are invisible to the detector, and operating in a sequential injection mode complicates the issue further. Even if all the system zones could be eliminated, early injections might still interfere with later injections if unimportant, late-migrating ions are present in the sample. However, an easy way to eliminate both issues is through proper selection of EOF direction and magnitude. For instance, even a very slow counter-EOF flow will remove the injection zone, ions of opposite polarity, and any system zones resulting from multiple counter-ions. If there are slow, interfering ions with the same polarity as the analytes or system zones from multiple co-ions, then the EOF magnitude will need to be sufficiently high to remove these as well. Control of the EOF direction and magnitude is possible through dynamic modification with small molecules, surfactant micelles, and/or polymer additives. It can also be performed with static surface changes, including covalent modification, neutral polymers, and polyelectrolyte multilayers. These surface modification procedures are covered thoroughly in the literature and therefore will not be discussed here [71–77]. Despite its ability to eliminate system peaks, counter-EOF approaches are not commonly used. Possibly, the main reasoning behind this is the increase in analysis time caused by a counter-EOF flow. This argument, however, is a bit misleading because resolution increases with increasing counter-EOF magnitude, and the improved resolving power indicates that shorter capillaries can be employed that will somewhat offset the slower net analyte mobility [1,2,78,79]. Consequently, a quantitative understanding of the impact of EOF on resolution is needed to design highly optimized separation systems. Because selection and control of EOF magnitude is very important in long-term monitoring applications and also directly affects resolution, Equation 10.10 we show how to calculate the theoretical resolution in MCE as a function of EOF. As an example, the resolution of three closely migrating ion pairs (cesium/potassium, barium/1,5-diaminopentane, and sodium/trimethylamine) is calculated for a range of EOF values and a 5 cm MCE channel with an effective potential of 1500 V, 0.5 mm injection length (1% of the capillary length), and 0.5 mm detection zone (also 1% of the capillary length). To perform this calculation, the following derivation and assumptions were made. Separation efficiency, N, is given by Equation 10.10, where $w_{0.5}$ (s) is the peak width at half height, w (s) is the baseline width of the peak, and the square of σ_{total} is the total variance of the peak in the separation.

$$N = 5.54\left(\frac{t_{\text{mig}}}{w_{0.5}}\right)^2 = 16\left(\frac{t_{\text{mig}}}{w}\right)^2 = \frac{L_{\text{eff}}^2}{\sigma_{\text{total}}^2} \tag{10.10}$$

Resolution between peaks 1 and 2, R_{sep}, is given by Equation 10.11.

$$R_{\text{sep}} = \frac{2\left(t_{\text{mig},2} - t_{\text{mig},1}\right)}{w_1 + w_2} \tag{10.11}$$

Combining Equations 10.9 and 10.10 results in Equation 10.12.

$$R_{sep} = \frac{t_{mig,2} - t_{mig,1}}{2}\left(\frac{\sqrt{N_1N_2}}{t_{mig,1}\sqrt{N_2} + t_{mig,2}\sqrt{N_1}}\right) \quad (10.12)$$

Migration time can be written similarly to Equation 10.5, but with the electric field separated into effective length and effective voltage (V_{eff}), resulting in Equation 10.13.

$$t_{mig} = \frac{L_{eff}^2}{V_{eff}(\mu + \mu_{EOF})} \quad (10.13)$$

Substituting Equation 10.13 into Equation 10.12 gives Equation 10.14.

$$R = \frac{(\mu_1 - \mu_2)\sqrt{N_1N_2}}{2(\sqrt{N_2}(\mu_1 + \mu_{EOF}) + \sqrt{N_1}(\mu_2 + \mu_{EOF}))} \quad (10.14)$$

Substituting the right-hand side of Equation 10.10 into Equation 10.14 results in Equation 10.15.

$$R = \frac{L_{eff}(\mu_1 - \mu_2)}{2(\sigma_{total,2}(\mu_1 + \mu_{EOF}) + \sigma_{total,1}(\mu_2 + \mu_{EOF}))} \quad (10.15)$$

Ideally, the only significant contributors to variance in an electrophoretic separation are the injection length, diffusional broadening, and the detection zone length. Thus, the total variance is given by Equation 10.16. (Note that other variance sources, including analyte–surface interactions and detection electronics, can also exist but are ignored here.)

$$\sigma_{total}^2 = \sigma_{inj}^2 + \sigma_{diff}^2 + \sigma_{det}^2 \quad (10.16)$$

The diffusional variance is given by the Einstein equation in Equation 10.17. The injection and detection contributions are similar for rectangular profiles and are given in Equations 10.18 and 10.19. In these three equations, D is the diffusion constant, L_{inj} (m) is the physical length of the injected sample plug, and L_{det} (m) is the physical length of the detection window [80].

$$\sigma_{diff}^2 = 2Dt_{mig} \quad (10.17)$$

$$\sigma_{inj}^2 = \frac{L_{inj}^2}{12} \quad (10.18)$$

$$\sigma_{det}^2 = \frac{L_{det}^2}{12} \quad (10.19)$$

The diffusion constant is related to the mobility through the Nernst–Einstein relationship, allowing Equation 10.17 to be rewritten as Equation 10.20, where F is the Faraday constant (98,485 A s mol^{-1}), R_{gas} is the gas constant (8.314 J K^{-1} mol^{-1}), and T is temperature (K).

$$\sigma_{diff}^2 = \frac{2\mu R_{gas}Tt_{mig}}{zF} \quad (10.20)$$

Inserting Equations 10.16, 10.18, 10.19, and 10.20 into Equation 10.15 and simplifying gives Equation 10.21.

$$R_{sep} = \frac{L_{eff}(\mu_1 - \mu_2)}{2\left(\sqrt{\frac{2\mu_2R_{gas}TL_{eff}^2}{V_{eff}z_2F(\mu_2 + \mu_{EOF})} + \frac{L_{inj,2}^2}{12} + \frac{L_{det}^2}{12}}(\mu_1 + \mu_{EOF}) + \sqrt{\frac{2\mu_1R_{gas}TL_{eff}^2}{V_{eff}z_1F(\mu_1 + \mu_{EOF})} + \frac{L_{inj,1}^2}{12} + \frac{L_{det}^2}{12}}(\mu_2 + \mu_{EOF})\right)} \quad (10.21)$$

Equation 10.21 gives the expected resolution for two compounds using the easily measured terms effective length, effective voltage, ionic mobilities, electroosmotic mobility, injection length, detection length, and ionic charge. The equation is somewhat cumbersome; however, with some assumptions, it can be simplified. Assuming that the two separation efficiencies are similar (N_{avg}), the ionic mobility plus electroosmotic mobility quantities are similar ($\mu_{avg} + \mu_{EOF}$), and the injection lengths ($L_{inj,avg}$) are equal (not true for electrokinetically biased injections) results in Equation 10.22, which is easier to compute than Equation 10.21. Note that the assumption of average separation efficiency is usually accurate when the two species being compared have equivalent charges and similar mobilities. The assumption of equal ionic mobility plus electroosmotic mobility quantities is relatively accurate unless a counter-EOF approach with an EOF magnitude near that of the ionic mobilities is employed.

$$R_{sep} \approx L_{eff}\sqrt{\frac{12V_{eff}(\mu_{avg} + \mu_{EOF})}{(24\mu_{avg}R_{gas}TL_{eff}^2/z_{avg}F) + V_{eff}(\mu_{avg} + \mu_{EOF})(L_{inj,avg}^2 + L_{det}^2)}}\frac{(\mu_1 - \mu_2)}{4(\mu_{avg} + \mu_{EOF})} \quad (10.22)$$

Equation 10.21 was used to calculate the expected resolution between the three pairs of closely migrating cations mentioned in the preceding text for a range of EOF values.

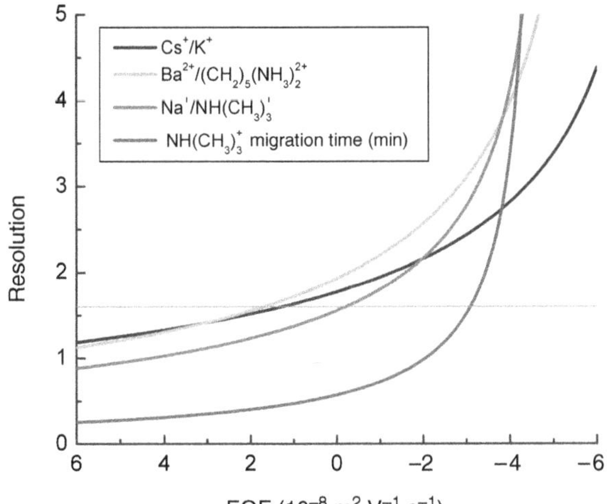

FIGURE 10.4 Calculated resolution between three closely migrating cation pairs as a function of EOF. The BGE was 250 mM acetic acid. Mobilities were calculated by PeakMaster 5.2 as cesium = 7.8524, potassium = 7.4645, barium = 6.2508, 1,5-diaminopentane = 5.9764, sodium = 5.0450, and trimethylamine = 4.8258 (all units $10^{-8}\,m^2\,V^{-1}\,s^{-1}$) [53]. The migration time for the slowest analyte, trimethylamine, is also shown to illustrate the expected increase in analysis time. The gray horizontal line at a resolution of 1.6 represents complete baseline resolution. Calculations were carried out using Equation 10.21 for $L_{eff} = 5$ cm, $V_{eff} = 1500$ V, $L_{inj} = 0.5$ mm, $L_{det} = 0.5$ mm, and $T = 25°C$.

PeakMaster 5.2 was used to calculate the mobilities for these analytes for a chosen BGE of 250 mM acetic acid (commonly used for indirect detection of cations; pH 2.68, $I = 2.2$ mM) [53,81]. The resolution between each cation pair and the migration time of the slowest analyte (trimethylamine) are shown in Figure 10.4. At an electroosmotic mobility of $6\times10^{-8}\,m^2\,V^{-1}\,s^{-1}$ (a common value for an unsuppressed EOF on many materials), none of the analyte pairs exhibit baseline resolution. By $2\times10^{-8}\,m^2\,V^{-1}\,s^{-1}$, a common value for suppressed EOF, both cesium/potassium and barium/1,5-diaminopentane show a resolution above 1.5, nearly baseline resolution. At a slightly reversed EOF of $-3\times10^{-9}\,m^2\,V^{-1}\,s^{-1}$, all of the species are baseline resolved and show a resolution improvement of 54% (cesium/potassium) to 83% (sodium/trimethylamine), relative to an EOF of $6\times10^{-8}\,m^2\,V^{-1}\,s^{-1}$. The migration time of the slowest species, trimethylamine, also increases by 88%. However, the injection zone (the only system zone in this BGE) will not be observed, so subsequent injections can be performed immediately after the final analyte has migrated through the detection zone.

As mentioned in the preceding text, in addition to system zones interfering with analyses, unwanted peaks from ions that are not of interest can also cause additional problems in sequential injection analyses. Prediction of the appearance of these interfering peaks can only be made when there is robust knowledge of the sample composition. In complex matrices such as biological samples, this is not always possible. However, some general rules can be applied to minimize interfering peaks by reducing their mobilities so that they are removed by a counter-EOF approach. In effect, the goal is to make the separation as chemically selective possible. One type of analyte system where this is easily accomplished is in the analysis of strong electrolytes. In this case, the majority of interferences can be removed by operating at a pH that will reduce or eliminate the charge on most weak electrolytes of similar polarity. For instance, if inorganic anions are the target analytes, a low pH BGE should be selected because most inorganic anions have pK_a values below 2. By operating at pH 4 or below, most organic acids will be nearly fully protonated. Using acetic acid as an example, at 10 mM ionic strength and pH 6.0, its mobility is $-3.68\times10^{-8}\,m^2\,V^{-1}\,s^{-1}$. At pH 4.5, it has reduced in mobility to a value of $-1.47\times10^{-8}\,m^2\,V^{-1}\,s^{-1}$. At pH 4.0, it has dropped to $-0.63\times10^{-8}\,m^2\,V^{-1}\,s^{-1}$, and by pH 3.5, its mobility is a very low $-0.22\times10^{-8}\,m^2\,V^{-1}\,s^{-1}$, slow enough to not interfere in practically any counter-EOF separation (all calculations performed in PeakMaster for a BGE composed of benzoic acid and pyridine) [53]. Note that the general approach of slowing down unwanted ions to reverse their net migration is not limited to pH methods. Complexation via small molecule binding or with high-affinity micelles can also be performed. However, whether by complexation or pH, this approach should be undertaken with the knowledge that an incomplete reduction in mobility of the unwanted ions might cause them to appear as very slow and broad peaks migrating at excessively long migration times, potentially interfering with subsequent injections.

Because selective separation chemistry often cannot eliminate all low-mobility interfering compounds, extra selectivity via the injection method is desirable. Additional consideration should be given to the injection method because some approaches require long injection times and/or can interfere with the pseudo-steady-state MCE operation during extended monitoring. Three injection methods are common in MCE and will be discussed here. First, there are several variants of hydrodynamic sample introduction [24–26,31,82,83]. These injections ideally yield an unbiased sample plug (identical injection lengths/volumes for all species). This injection mode is excellent for separations with analytes having a wide range of mobilities. Hydrodynamic injections are typically relatively short, which is a desirable quality for rapid sequential injection systems. However, hydrodynamic injections offer no selectivity gain and can often increase the complexity of the microfluidic instrument. Gated injection offers an alternative that is nominally the fastest of the injection methods (injection times shorter than 1% of the total separation time are common) and is operated with a simple, temporary change of applied potentials [48]. Theoretically, quantitation from separations using gated injection should suffer less from migration time drift (from EOF or pH changes) because while peak areas for most detection methods in CE are proportional to migration time, gated injection volumes are inversely proportional to migration time and the two effects should negate each other [84]. Because an analyte's injected volume is proportional to its apparent mobility, any unwanted, late-migrating peaks will only have a small fraction of the injected volume of early peaks. Thus, gated injection is likely the best option for performing rapid sequential injections in extended monitoring applications. However, the inherent electrokinetic biasing can also be undesirable when relevant analytes cover a large range of migration times. Additionally, gated injection has the drawback of continuously driving sample from the sample reservoir into a waste solution. For spatially static samples, this means that the sample is subjected to the same depletion phenomena that affect the BGE. To illustrate the differences between unbiased and biased injections, Figure 10.5 compares the relative signal when using electrokinetically biased injections and unbiased injections as a function of ionic mobility for monoanions in an example separation.

The assumed conditions were a BGE of 30 mM Bis-Tris/15 mM mandelic acid with an EOF of $1.5\times10^{-8}\,m^2\,V^{-1}\,s^{-1}$. Both the response from a generic detection method (equivalent sensitivity to all analytes) and the response with conductivity detection (which is inherently tied to the analyte mobility) are shown. Also shown are relative migration time and relative peak width. This figure illustrates why selectivity in the injection and/or detection steps is required. Without additional selectivity, an ion with a mobility of $-3\times10^{-8}\,m^2\,V^{-1}\,s^{-1}$ will give a peak that is 91.7% of the height of an equimolar ion with a mobility of $-7\times10^{-8}\,m^2\,V^{-1}\,s^{-1}$ while taking 3.67 times longer to reach the detection zone and exhibiting a peak that is wider by a factor of 3.03. With gated injection biasing, the relative injection volume for the slower ion is reduced by 72.7%, and when combined with conductivity detection biasing, the final peak height for an equimolar sample would be only 3.7% as large as that for the higher-mobility ion. The "background" signal from such a short, broad peak could be subtracted from the baseline if a faster ion from a subsequent peak co-migrated with it. Other detection methods, particularly fluorescence and electrochemical detection, can offer increased selectivity over conductivity detection, although their selectivities are structure-dependent instead of mobility-dependent. Consequently, it can be surmised that the often unwanted biasing from gated injection, conductivity detection, or other detection methods can be advantageous if properly utilized when performing a rapid sequential injection analysis.

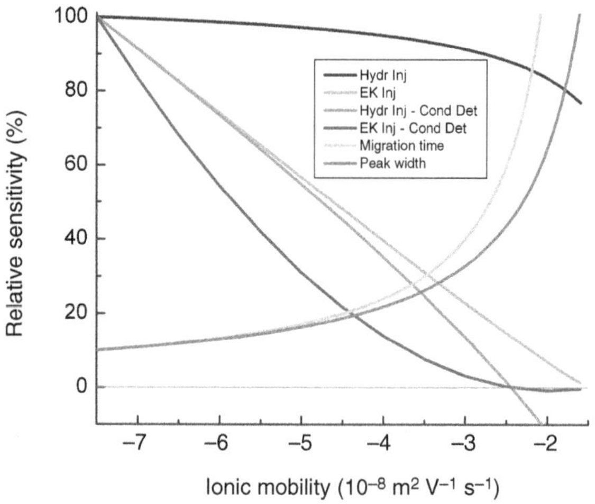

FIGURE 10.5 Relative sensitivity (peak heights) for two injection and detection modes and relative migration times and peak widths as a function of ionic mobility for a diffusion-limited separation. A BGE of 30 mM Bis-Tris and 15 mM mandelic acid are assumed. EOF for plot is $1.5\times10^{-8}\,m^2\,V^{-1}\,s^{-1}$. The generic detector results assume constant molar detection sensitivity. Relative conductivity detection sensitivity was obtained from the "conductivity signal" calculated using simulation results from PeakMaster 5.2. Signal was assumed to be proportional to the injection volume and inversely proportional to the physical peak width, which increases with time due to diffusional broadening.

The third injection mode discussed here is pinched injection [85]. This approach to injection uses electric fields to drive the injection, but it attempts to remove electrokinetic biasing by defining an injection segment with a set length and then operating the injection for a long enough time to ensure that the injection volume is representative of actual sample concentrations. Consequently, under standard operation, the pinched injector is possibly the worst for rapid sequential injections because it has the longest injection process, interferes with the steady-state operation of the MCE separation, and does not provide any selectivity in the injection process. However, theoretically, a pinched injection should be able to provide a relatively sharp mobility cutoff (thus, additional selectivity) if the injection time is set to permit only certain analytes to reach the injector, although the authors are unaware of this phenomenon previously being intentionally exploited. Essentially, this approach transforms the system into a pseudo two-dimensional separation system, with the first dimension being the injection process, which acts like a high-pass filter with respect to the velocity of the ions exiting the sample channel. Assuming the sample matrix is similar to the separation BGE, the injection time can be estimated using Equation 10.5, where L_{eff} represents the sample-to-injector distance and E is the electric field during the injection process for the channel connecting the sample reservoir and double-T injector. The potential gain in selectivity using this approach was evaluated theoretically for a microchip possessing a 1-cm injection channel, 0.5-mm injector length (typically defined by a double-T offset), injection field of $-500\,V\,cm^{-1}$, EOF of $1.5\times10^{-8}\,m^2\,V^{-1}\,s^{-1}$, and BGE of 15 mM MES and 15 mM histidine. Chloride ($-7.435\times10^{-8}\,m^2\,V^{-1}\,s^{-1}$) reaches the start of the injector in 3.37 s and fills the injector at 3.54 s. Nitrate (-6.995×10^{-8} m^2 V^{-1} s^{-1}), however, takes 3.67 s to reach the injector and fills it completely at 3.85 s. Because of diffusional broadening and experimental uncertainties, these two species likely could not be separated by selective injection. However, both analytes should be separable from acetate (-3.737×10^{-8} m^2 V^{-1} s^{-1}), which takes 8.94 s to reach the injector and fills it completely at 9.39 s. Note that these calculations assume no diffusional broadening, which would need to be accounted for in real applications. Nevertheless, the predicted behavior illustrates that this method should be useful for some applications and needs to be further evaluated with real tests. The goal of such a test would be to exclude late-migrating peaks from the separation so that rapid sequential injections could be performed without interference. Presumably, longer injection channels and well-buffered sample solutions will need to be utilized to reproducibly succeed with this approach.

A final way to increase the sequential injection sampling frequency is to employ an injection interval that is shorter than the total separation time and utilizing a method termed the peak overlap technique [86]. For some separations, the peak overlap technique is impossible because of sample complexity. However, when analyzing relatively simple sample matrices with only a few closely spaced peaks, or with several sets of closely spaced peaks separated with large areas of flat baseline, sampling frequency can be increased considerably by proper injection interval selection. Kuban and coworkers showcased this technique using a traditional CE system in continuous flow mode [86]. Specifically, they analyzed chloride, sulfate, nitrate, and carbonate in tap water using a thiosulfate internal standard. Carbonate migrated near 3.8 min, and the other ions appeared in closely spaced peaks totaling about 15 s wide and centered at a migration time of about 2.6 min. They demonstrated the ability to perform three rapid sequential injections spaced by 20 s that resulted in an electropherogram with three resolved groups of faster ions followed by three slow, broad carbonate peaks. Thus, they were able to obtain three separations in about 4.9 min, whereas a single injection would require about 4.2 min without the peak overlap approach. At steady state, this method allowed three injections to performed every 2.5 min, for an average sampling rate of 72 h^{-1} even though sampling rate without peak overlap would be about 14 h^{-1}, an improvement of roughly a factor of 5. An illustrative example of the peak overlap technique can be seen in Figure 10.6. The top electropherogram in this figure shows the separation of four anions plus an internal standard compiled by Noblitt et al. specifically for use in online monitoring systems for aerosol composition 56. The first peak in this separation begins around 26 s and the final peak terminates before 45 s. Because the window from first peak to last peak is shorter than the window from injection to the first peak, a second injection can be started

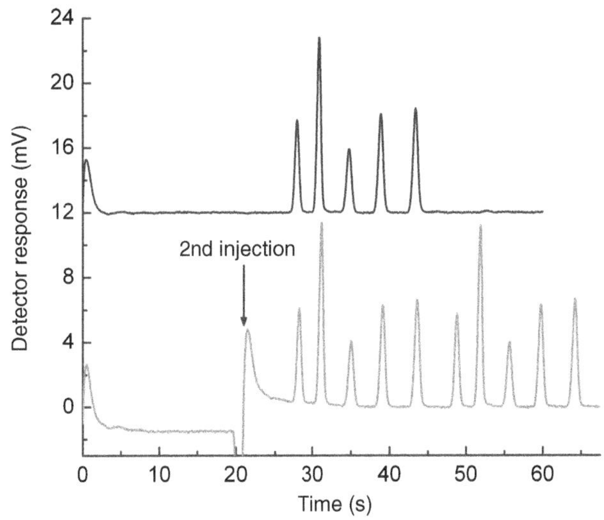

FIGURE 10.6 Electropherograms showing the advantage of optimizing injection interval timing to maximize sampling frequency where all analytes are detected after the second injection. The top trace shows a single injection. The bottom trace shows two injections with the second injection occurring at 21 s, prior to any of the peaks reaching the detection zone. Peak order is chloride, sulfate, nitrate, oxalate, and 1,3-propanedisulfonate (then repeated). BGE conditions and microchip design are those given by Noblitt et al. except for the following conditions: 5 μM analytes; 1.2 s gated injection; 20 μm platinum detection electrodes, and a 20 Hz effective collection rate. Note that the baseline perturbation and step change occur from capacitive changes during injection when using contact conductivity detection.

shortly before the first peak from the first injection reaches the detection zone. This approach is shown in the second electropherogram in Figure 10.6, where an injection scheme with alternating intervals of 21 s and 48 s is utilized. Without the peak overlap technique, a sampling rate of 75 h^{-1} is possible. Utilizing the peak overlap technique permits a throughput of 104 h^{-1}, a 39% improvement. Most sampling applications cannot utilize such ambitious use of the peak overlap technique, particularly when unwanted peaks are present in the separation. The top electropherogram in Figure 10.7 illustrates this by showing the previous separation with the addition of an undesirable formate peak (which can appear as a gas phase impurity in aerosol sampling). The formate peak prevents the use of the sampling interval used in Figure 10.6, and waiting until the formate peak exits the capillary requires a sampling interval of at least 58 s. However, the original 48 s interval can still be applied here, a sampling frequency increase of 21% over the 58 s interval imposed by the addition of formate. Note that this approach can be used even if formate quantitation is desired, as the integrity of this peak is maintained. Strictly speaking, the examples given in Figures 10.6 and 10.7 are not utilizing the peak overlap technique because the faster ions in the second injection do not overtake the slower ions in the first

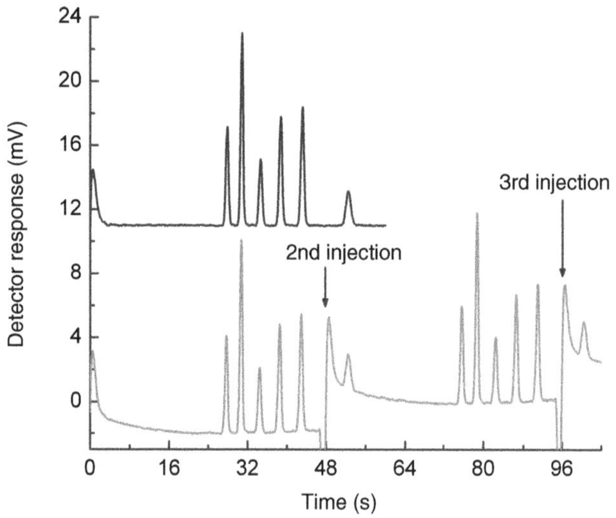

FIGURE 10.7 Electropherograms showing the advantage of optimizing injection interval timing to maximize sampling frequency where the final peak is detected after the subsequent injection. The top trace shows a single injection. The bottom trace shows three injections using a 48 s injection intervals. All conditions are the same as in Figure 10.6, with the additional sixth peak being formate.

injection (although such an example is easily possible with this separation, for instance if malate was present in the sample, it migrates at a time of ~85 s and would appear after the faster ions in a second injection). However, the approach and results are the same as a true peak overlap method, and both Figures 10.6 and 10.7 illustrate the general concept. As a final cautionary note, when applying the peak overlap method, one should confirm that quantitation is not significantly affected by either slow, broad peaks from the first injection or the injection procedure itself. Thus, shorter injection methods such as gated injection are preferred over the longer and more perturbing methods such as pinched injection. To evaluate the effect of faster ions passing through zones of slower ions, only empirical testing can confirm quantitative accuracy. For proper evaluation, analyte concentrations during the test-runs should be similar to those expected in real samples.

To conclude for this section, higher injection sampling frequencies result in higher-quality data from either higher temporal resolution or increased precision via averaging. Achieving rapid sequential injections requires proper design of the BGE and robust knowledge of the sample matrix to avoid system zones and minimize undesirable, late-migrating ion peaks. Addition of specific complexing agents or strict pH selection can dramatically reduce the number of late-migrating species in some separations, making the separation highly selective. Proper selection of the EOF through either static or dynamic surface modification is needed to ensure that unwanted peaks migrate away from the detection zone. A weak counter-EOF separation is recommended and has the benefit of increasing resolution between peaks with only a modest increase in migration time. The available injection modes were discussed with respect to their advantages and disadvantages. Gated injection is recommended because of its short injection times and biasing that reduces the peak heights of unwanted, late migrating interferences. However, gated injection has the drawbacks of continuously depleting the sample solution and also may not be acceptable when working with samples having a large range of migration times because the injection biasing may prevent acceptable sensitivity for slower species. Additional selectivity against slower species can be gained from certain detection methods. Conductivity detection was shown as an example, and its sensitivity toward ions is roughly proportional to the difference between the analyte's mobility and the BGE co-ion's mobility. Finally, the peak overlap technique was exhibited and showed how subsequent injections can be performed prior to all analytes being detected, increasing the injection rate. The combination of two or more of the aforementioned approaches to removing unwanted, late-migrating peaks will likely be needed to achieve desired results, and for more complex samples with a wide range of desired analytes, rapid sequential injections may not be possible and only a long sampling interval will be possible without a hydrodynamic flushing system.

10.4 ROBUST QUANTITATION

One of the most frequently cited limitations of CE and MCE is its poor reproducibility in quantitative analysis [14]. This is especially true during extended monitoring because accumulated changes to the BGE composition, alterations of the capillary surface conditions, depletion or matrix changes in the sample, and/or ambient temperature changes can lead to shifts in migration times. Because peak areas are proportional to the migration time for many detection methods in electrophoresis, variance in migration time will lower quantitative reproducibility. Additionally, injection volumes may drift in MCE due to changes in viscosity or fouling of the capillary surface. Several techniques for improving the quantitative reproducibility of MCE, particularly in the case of extended online monitoring, will be discussed in this section. Much of the discussion will assume a static sample solution, but some of the topics also pertain to sampling from a continuous sample flow.

The most important step in attaining long-term quantitative reproducibility is to ensure that both the BGE and sample solutions are robustly buffered. The reasoning and methods for buffering the BGE were described earlier and will not be rehashed here. Buffering of the sample is critical because the sample composition affects injection quantities for many injection procedures. pH is a typical example of how sample matrix affects quantitation. Equations 10.1 and 10.2 describe how pH changes an analytes protonation state, and ionic mobility is proportional to the charge of an ion. In electrokinetic injections, the injected volume is proportional to the effective mobility and is therefore very sensitive to pH when operating near the pK_a of an analyte. Thus, as concluded for the BGE pH, sample pH should be kept away from the analytes' pK_a values if possible. Additionally, many samples are not naturally buffered or at a consistent pH, so mixing of the sample with a buffering agent is often required. Another issue with sample composition is the bulk solution conductivity. Differences in the solution conductivity between the sample and BGE can induce stacking or destacking [87]. Although a low-conductivity matrix may be tempting to use because of enhanced sensitivity through stacking, its use in extended monitoring applications is risky because any compositional changes induced by the sample will be relatively larger and the sample ions will also deplete more quickly because of the stacking. Instead, a sample matrix closely matching the BGE or even one that is of a higher concentration than the BGE will yield more reproducible results. In general, the amount of sample buffering is application-dependent, and only general knowledge of the sample composition and empirical testing will confirm the validity and reproducibility of a method.

Another important factor in quantitative reproducibility is the stability of the capillary surface chemistry. Clearly, this will affect the EOF, and the effects of EOF on quantitation were already shown in Equation 10.7 and Figure 10.3. However, even aside from the EOF effects, surface changes can alter quantitation. The mechanism behind this is a change in the specific interactions occurring between particularly analytes and the capillary surface. One prime example is the adsorption of heavy metal ions onto negatively charged silanol groups on glass, fused silica, and poly (dimethylsiloxane) (PDMS) surfaces. If these interactions are reversible and occur on relatively fast timescales, then any surface changes may result in small shifts in migration times or a change in the amount of tailing present in the electropherogram. If the interactions are irreversible, then the apparent detection sensitivity likely will increase with time as the surface becomes saturated and the number of potential binding sites dwindles. Overall, the best way to avoid quantitation problems caused by slowly changing surface conditions is to try to eliminate all unintentional specific (and even nonspecific) interactions between the sample and BGE species and the capillary surface. One easy way to minimize these interactions with small molecules is to modify the capillary surface so that its charge is of the same polarity as the analytes of interest. This approach has the additional benefit of inducing the counter-EOF flow that was recommended earlier for eliminating extraneous, late-migrating peaks. To stabilize both the surface conditions and EOF, either surfactant dynamic coatings or polymeric static coatings are often sufficient. Effectively, these species out-compete other species for surface interactions and thus yield a more consistent surface status. However, care must be taken when using static coatings, as most static coatings slowly de-adsorb with time. Addition of small amounts of the static coating agent into the BGE can solve this issue. For separations of macromolecules, protecting the surface is more difficult and often requires special procedures or coatings. Much work has gone into protecting capillary surfaces from nonspecific macromolecular adsorption, particularly by Lucy's group, and we direct interested readers to those publications [71–73,77,88–99]. One interesting surface-stability problem that has not been explored thoroughly is the contamination of the surface from trace impurities unintentionally present in the BGE. These impurities can bind to the surface and then interact with analyte species during the separation, thus causing nominally noninteracting ions to show wall adsorption behavior. For instance, Gassner et al. showed that trace amounts of iron (III) in the BGE adsorb to the capillary surface and subsequently bind some anions [100]. Potentially, microbial growth caused by not replacing the BGE can also generate BGE contamination that alters surface conditions, sometimes irreversibly. This issue can be difficult to diagnose and often does not appear in early laboratory tests. Problems from trace contaminants are not limited to the BGE; trace species in the sample can also complex sample ions and interfere with proper injection of those ions. With metal ions, acidic rinses would likely maintain the integrity of the capillary surface with respect to contaminants in the BGE, but this approach will not solve the issue of unwanted complexation in the sample solution and also interferes with steady-state operation during extended monitoring applications. An alternative to this approach that solves both the capillary surface and sample matrix issues is to utilize a BGE containing a species that competes with the analytes for binding to the contaminant species. The BGE binding species does not have to possess a higher binding constant to the contaminant than do the analytes (although this helps) as long as the BGE species is in sufficient excess to bind all significant amounts of the contaminant. Noblitt et al. utilized this approach for the analysis of anions in an MCE separation specifically designed for extended monitoring [56]. Picolinic acid, a common ligand for heavy metals not only served as the buffering acid but was also used to bind any metal ions present in the sample or on the capillary surface. Results showed increased oxalate sensitivity when using picolinic acid compared to its noncomplexing isomer nicotinic acid, indicating that the picolinic acid was protecting oxalate from slow or irreversible ligand exchanges with surface-bound metal contaminants.

Although the above-mentioned considerations can help minimize irreproducibility in quantitative measurements, even very small, uncontrollable changes in sample composition, instrument temperature, or surface conditions will cause unacceptably large variance in response. To account for these changes, an internal standard is required. The concentration of an unknown can then be calculated from the peak area ratio relative to the internal standard, as given by Equation 10.23 where C (μM) is the solution-phase concentration of the analyte (i) and internal standard (IS), a is the peak area, and m_i is the slope of the calibration curve (relative response of the analyte with respect to the internal standard).

$$C_i = \frac{C_{IS}}{m_i}\frac{a_i}{a_{IS}} \tag{10.23}$$

Several excellent studies have shown the importance of using internal standards [101–104], so only the overall conclusions will be discussed here. The propagated uncertainty in the determined concentration from using Equation 10.23 is given by Equation 10.24. The impacts of the various parameters in this equation are not immediately apparent and are more easily comprehended when represented as RSD values, as shown in Equation 10.25.

$$\sigma_{C_i} = \frac{\sqrt{\sigma^2_{C_{IS}} a_i^2 + \sigma^2_{a_i} C^2_{IS} + (\sigma^2_m C^2_{IS} a_i^2/m_i^2) + (\sigma^2_{a_{IS}} C^2_{IS} a_i^2/a^2_{IS})}}{m_i a_{IS}} \tag{10.24}$$

$$\mathrm{RSD}_{C_i} = \sqrt{\mathrm{RSD}^2_{C_{IS}} + \mathrm{RSD}^2_{a_i} + \mathrm{RSD}^2_{m_i} + \mathrm{RSD}^2_{a_{IS}}} \tag{10.25}$$

Note that rigorous sample preparation procedures can reduce the RSD of the internal standard concentration to nearly zero, and although evaporation effects will affect the internal standard concentration, they affect the analyte concentration equally, so this term can often be ignored. Sample depletion does affect the internal standard concentration and therefore the measured concentration of the analyte, but this behavior will be discussed in detail later. Equation 10.25 also shows that reducing the relative uncertainty of the peak integration (thus, reducing uncertainty in a) will improve results. The precision in this measurement is dependent on the integration procedure and the signal-to-noise ratio of the peak. For the analyte peak, the signal-to-noise is dependent on the concentration of the analyte, of which the user has limited control. However, the uncertainty in the measured sample concentration is also dependent on the reproducibility of the internal standard peak integration. The user does have control over this uncertainty contribution, as larger internal standard peaks give more reproducible integrations because of the smaller role of the noise in the quantitation. Therefore, we recommend using the highest possible internal standard concentration before significant overloading effects, such as band broadening, fronting, or tailing, occur. Often the largest source of irreproducibility is the uncertainty in the slope of the calibration curve, m_i. Although rigorous calibration procedures can give the appearance of high certainty in m_i, differences in the sample matrix or operating conditions between the calibration standards and actual samples can change the relative response. Without an internal standard, injection volume is commonly the largest contributor to measurement uncertainty (by effectively changing m_i), as it can vary with temperature, viscosity, solution conductivity, injection time, and in the presence of unwanted hydrodynamic flow. However, application of an internal standard compensates for nearly all injection volume variance (the primary exception being for biased injections where the biasing factor changes by different amounts for the analyte and standard). Changes in the applied potential or from bulk solution evaporation are also fully corrected with an internal standard. Fluctuations in EOF, matrix composition, capillary surface interactions, hydrodynamic flow, or BGE conditions may not be entirely resolved with internal standard correction. To best account for variance in these aspects of the method, an internal standard that is as similar as possible to the analytes is best. Consequently, one should target an internal standard that is similar to the analytes in pK_a, valence, and mobility. One successful approach for improving reproducibility is to normalize the peak area by dividing it by the migration time [14,105–109]. The theory behind this approach is the cancellation of the dependence of peak area on migration time, so it may be less successful for detection techniques that consume the sample, including mass spectrometry and some electrochemical methods. Another study found that transforming the x-axis of the electropherogram into the mobility domain resulted in higher qualitative and quantitative precision [110]. Presumably, these improvements were realized because this transformation accounted for some of the changes in peak width that are induced by changes in EOF or hydrodynamic flow. This conclusion further supports the recommendation of a low-magnitude EOF. With a decreased EOF, there is lower uncertainty in migration time and also a reduced chance for the intrusion of unwanted hydrodynamic flow induced by fluid head heights or Laplace pressures. Additionally, some studies have shown that utilizing multiple internal standards is necessary to achieve acceptable reproducibility levels, particularly when using electrokinetic injection protocols [111,112]. However, these studies were performed using traditional CE instrumentation, which incorporates the physical movement of either the capillary or the buffer vials, a process that decreases reproducibility. From the authors' experience, MCE gated and pinched injections do not suffer from this additional physical step and therefore acceptable reprodubilities can be obtained with one internal standard except for the most rigorous applications. Finally, it should be realized that changes in some specific molecular interactions such as adsorption to capillary surfaces and complexation reactions could occur to certain analytes and not affect the internal standard. Consequently, those changes will affect the value of m_i and thus analyte quantitation while not altering the observed peak from the standard. The end user should be cognizant of these possibilities and take proper precautions to protect the surface and ensure a stable concentration of complexation agent in both the sample and BGE.

As alluded to earlier, one quantitation issue that is mostly limited to extended monitoring studies from static solutions and therefore has been mostly ignored in the literature is the phenomenon of sample ion depletion. The general mechanisms behind sample depletion are the same as those of buffer depletion, described in detail earlier. However, sample depletion directly affects the concentrations of specific ions in the sample solution, and the resulting quantified peaks will be representative of the depleted concentrations instead of the original ones. Although an internal standard will also undergo depletion, each species depletes at a unique rate and therefore internal standard correction is only approximate. Consequently, the apparent relative response factor shown in Equation 10.23 will systematically deviate with time, although the physical mechanism is unwanted changes in both internal standard and analyte concentrations and not a change in the actual detection. This phenomenon is unavoidable when using static solutions. For injection methods utilizing transient sample consumption, such as pinched injection, the effect is typically negligible. For injection techniques that continually deplete the sample, including gated injection and some types of hydrodynamic injection, sample depletion is a major issue. Therefore, the user should be aware of the effect, take measures to minimize it, and calculate the longest acceptable operating time from a single sample solution. This calculation was performed by Noblitt et al. for an extended online monitoring system exhibiting both sample depletion and accumulation [59], and the derivation for depletion only is given here. The velocity of an ion leaving a static solution is calculated using Equation 10.26.

$$v = E(\mu_i + \mu_{\mathrm{EOF}}) \tag{10.26}$$

The change in moles of analyte i in the sample reservoir with time ($\mathrm{d}n_i/\mathrm{d}t$) equals that exiting velocity multiplied by the channel cross-sectional area and the analyte concentration (C, $\mathrm{mol\,L^{-1}}$), as given in Equation 10.27. Note that the value of C in this equation is the concentration actually exiting the reservoir and therefore in systems exhibiting stacking or destacking will not be equal to the bulk solution concentration.

$$\frac{\mathrm{d}n_i}{\mathrm{d}t} = -1000 v_i A C_i \tag{10.27}$$

The change in concentration is inversely proportional to sample volume, which also changes with time. The volume change depends on the EOF. Ignoring any changes due to evaporation, condensation, or density alteration, the volume only depends on the volumetric flow rate of the EOF, which is given by Equation 10.8. Combining Equations 10.8, 10.26, and 10.27, results in Equation 10.28, which is the change in concentration of a sample species due to depletion with respect to time, where V_0 is the sample volume at $t = 0$. It should be realized that the electric field in this equation is that for the channel connected to the sample reservoir and is often not equal to the separation field.

$$\frac{\mathrm{d}C_i}{\mathrm{d}t} = \frac{-EAC_i(\mu_i + \mu_{\mathrm{EOF}})}{V_0 - EA\mu_{\mathrm{EOF}}t} \tag{10.28}$$

This equation is only approximate in real applications, as any changes to the BGE or sample composition will affect the stacking factor, electric field, and/or mobilities. However, in the absence of significant stacking effects and in the presence of a relatively simple sample matrix, Equation 10.28 should be sufficient for quantitative depletion estimates. Note that the fractional change in concentration is directly proportional to the electric field, channel cross-section, and effective mobility. It is inversely proportional to sample volume. Because the sample ions and internal standard all possess different effective mobilities, they will deplete at different rates. Consequently, Equation 10.23 only yields the correct result at $t = 0$ and after that will show systematic deviations for all ions having a mobility different from the internal standard. To illustrate this quantitatively, Figure 10.8 shows the determined concentration using Equation 10.23 for a theoretical depletion scenario at a variety of ionic mobilities and an internal standard mobility of $5.0 \times 10^{-8}\,\mathrm{m^2\,V^{-1}\,s^{-1}}$.

As expected, ions faster than the internal standard deplete more rapidly and show negative systematic deviations, whereas the slower ions show positive deviations. Specifically, after 1 h an ion with a mobility of $7 \times 10^{-8}\,\mathrm{m^2\,V^{-1}\,s^{-1}}$ would have depleted by 35.5% but be measured as 15.4% lower than the starting concentration using the internal standard. Conversely, an ion with a mobility of $3 \times 10^{-8}\,\mathrm{m^2\,V^{-1}\,s^{-1}}$ would have decreased in concentration by 9.9%, but according to the internal standard, the concentration would have increased by 18.2%. For an ion with the same mobility as the internal standard, no deviations are observed, clearly illustrating one benefit of choosing a standard species that is similar to the analytes in mobility. While the simulation in Figure 10.8 is instructive, its practical applicability is low because no compositional changes are occurring other than those from electrophoretic depletion. In reality, static sample solutions can be used to monitor reaction rates or collected samples when the sample collection process has a negligible

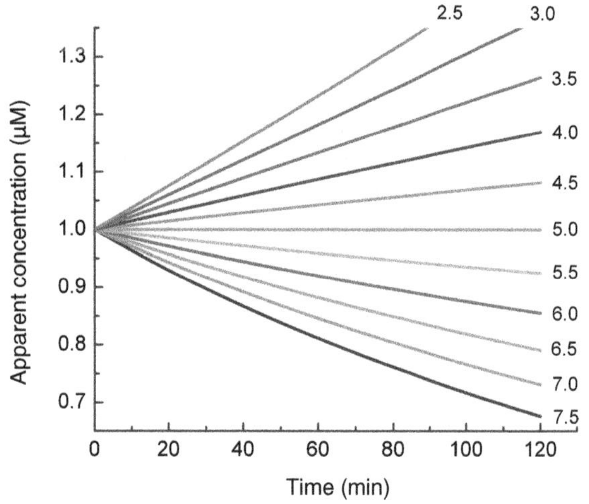

FIGURE 10.8 Effect of sample depletion on the quantitation of a static solution. The apparent concentration as calculated using Equation 10.23 is shown assuming an internal standard mobility of $5.0\times10^{-8}\,m^2\,V^{-1}\,s^{-1}$. Each trace represents an ion with the labeled mobility (in $10^{-8}\,m^2\,V^{-1}\,s^{-1}$). Simulation conditions: initial concentration = 1 μM, initial volume = 30 μL, channel cross-section = $2.5\times10^{-9}\,m^2$, electric field = $300\,V\,cm^{-1}$, EOF = $-1.75\times10^{-8}\,m^2\,V^{-1}\,s^{-1}$. Equations were solved using a first-order iterative approximation (Euler's Method) with a step size of 2 s.

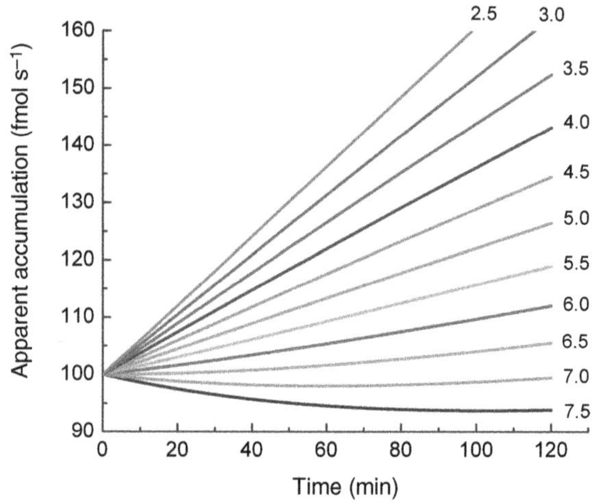

FIGURE 10.9 Effect of sample depletion on quantitation of a static solution accumulating sample ions when quantitation is performed using a differential method. Actual accumulation rate for sample ions = $100\,fmol\,s^{-1}$. Initial sample concentrations = 0. An internal standard mobility of $5.0\times10^{-8}\,m^2\,V^{-1}\,s^{-1}$ and initial concentration of 1 μM are used. Each trace represents an ion with the labeled mobility (in $10^{-8}\,m^2\,V^{-1}\,s^{-1}$). All simulation conditions are the same as in Figure 10.8.

impact on the total solution volume. To achieve high-time resolution in these applications, quantitation via a differential method (differences in point-to-point concentration measurements) is needed. The effect of sample depletion on differential quantitation is somewhat different than on a solution of otherwise constant sample concentrations, and simulation results for a differential method are shown in Figure 10.9. For this simulation, Equation 10.28 was used but with the addition of an accumulation term equal to $100\,fmol\,s^{-1}$ in the numerator. With this approach, deviations are systematically more positive than shown in Figure 10.8. This result is due to the analytes starting at a concentration of zero, whereas the standard began at a concentration of 1 μM (arbitrary and has no effect on the results unless internal standard accumulates also). Consequently, the standard begins with a high depletion rate, while the effect of the early depletion on the analyte concentrations is negligible. This is most apparent when considering an analyte with the same mobility as the internal standard ($5\times10^{-8}\,m^2\,V^{-1}\,s^{-1}$). With finite starting concentrations and no accumulation, an ion with this mobility exhibits no systematic quantitative deviations. However, with zero initial concentration and constant accumulation, the differential measurement yields results that are 6.7, 13.4, and 19.4% high at analysis times of 30, 60, and 90 min, respectively. The difference in this case is the initial concentration of the analytes versus the internal standard. The internal standard starts at a set concentration, depletes at its highest rate initially, and does not accumulate any new material during the analysis. Slower ions show even higher positive deviations than ions with the same mobility as the standard. Faster ions exhibit less positive deviations, and the fastest ions even exhibit negative deviations early in the analysis. With increasing analysis time, however, even the fastest ions will eventually begin to show positive deviations. This phenomenon is due to all sample ions approaching a steady-state concentration where accumulation rate equals depletion rate. The internal standard, however, does not have an accumulation source and therefore all ions will eventually show an increasing concentration relative to the internal standard when undergoing constant accumulation. The most prominent conclusion from Figure 10.9 is that the traditional rule-of-thumb of choosing a standard with a mobility as close as possible to the analytes is not necessarily valid when performing extended analyses with a differential quantitation method. Instead, an internal standard that is slower than the analytes is desirable, and calculating the actual depletion results similarly to what is done in Figure 10.9 is recommended. However, the best way to limit effects of sample depletion in injection methods that continually deplete the sample is to take the same measures used to limit buffer depletion. Larger sample volumes, smaller sample channels, and lower electric fields all reduce sample depletion. This approach requires a compromise, as MCE is often praised for its small sample volumes and short analysis times. However, for most systems that continuously deplete the sample during operation, the electric field in the sample channel can be decreased without sacrificing the electric field strength in the separation channel, and it is the separation field that determines the analysis time and separation efficiency.

Overall, achieving robust quantitation in MCE is difficult even without the additional complication of extended online operation. Stability of the BGE composition and surface chemistry are required to avoid analyte-specific effects that cannot be accounted for with an internal standard as well as changes to the injection volume or EOF that can only be partially accounted for using an internal standard. Employing an internal standard greatly increases the robustness of quantitation. For best results with an internal standard, the chosen standard should be similar to the analytes in pK_a, valence, and mobility. The internal standard concentration should be the highest concentration that does not result in a significantly overloaded (broad or triangular-shaped) peak because larger peaks give more reproducible integration values. A problem that is specific to long-term monitoring from a static sample solution is the issue of sample depletion. This phenomenon can be minimized using the same approach as that used to reduce buffer depletion. When accumulating sample ions into a static solution, a differential analysis method is often employed, and sample depletion results in some unique systematic deviations when using internal standards. The majority of ions show positive systematic deviations with increasing analysis time, and only ions considerably faster than the internal standard show negative deviations. Even these ions will eventually show positive deviations as the sample solution approaches a steady-state concentration of analyte while the internal standard concentration continues to diminish. When attempting a differential analysis, it is therefore recommended that the chosen internal standard be slower than the analytes.

10.5 CONCLUSIONS

Extended monitoring applications requiring high time resolution represent an enticing field that may be addressable by MCE because the instrumentation has short analysis times with a small physical footprint and high portability. However, MCE is notorious for its inability to maintain robust performance for extended monitoring times. This shortcoming is due to a variety of issues, particularly small BGE volumes, high electric fields, susceptibility to hydrodynamic flow interference, and unstable capillary surface conditions. Although most MCE analyses appear to be very rapid, many are often plagued by late-migrating system zones or unimportant sample ion peaks, limiting the method's ability to perform subsequent analyses at a high sampling frequency. This chapter discussed how to lengthen the longevity of the BGE integrity, improve the robustness of the sample quantitation, and increase the sampling frequency by avoiding system zones and late-migrating interfering peaks. Each of these issues can be reduced or even eliminated with proper design of the MCE instrumentation and the separation BGE. To increase operation time, larger solution volumes, smaller capillary cross-sections, and a high buffer capacity to background conductivity ratio are important. The recommended BGE components are a low-mobility weak acid and base pair with the acid having its pK_a above the operating pH and the base having a pK_a below it. Other factors, including evaporation and component reactivity, also need to be minimized or eliminated. BGE longevity depends on a variety of factors, including the specific buffer and sample components and desired tolerance to fluctuations in migration time and quantitation. Therefore, we recommend empirically evaluating the final separation conditions and then scaling as appropriate using the proportionality given in Equation 10.9. Additionally, the EOF should be suppressed to avoid any degrading effects caused by large discrepancies in reservoir head heights or meniscus pressures. These same BGE conditions naturally lead to separations with a minimal number of system zones, as the system zones will migrate with or near the injection zone. To remove this system zone entirely, a weak counter-EOF flow is suggested. Counter-EOF separations typically require longer migration times and result in broader peaks, but higher resolution between species is observed. Additionally, unwanted late-migrating sample ion peaks can be minimized by appropriate choice of operating pH, which will lead to many weak electrolytes migrating near the EOF, so they will also never be detected. Employing high-affinity surfactant micelles or selective small molecule binding agents to complex unwanted species is another method for removing unwanted peaks. To further increase selectivity against late-migrating species, biased injection and detection methods can be utilized. Knowledge of the sample matrix is helpful, as it permits the development of these highly selective separations and allows determination of the applicability of the peak overlap technique. To achieve maximum quantitative reproducibility, it is essential to utilize an internal standard to account for changes in applied field, injection volume, and EOF. Typically, the chosen standard should be similar to the analytes in mobility, valence, and pK_a. When sampling static solutions in extended monitoring applications, sample ion depletion is an important issue. Sample depletion can be reduced using the same approaches used to minimize buffer depletion. However, sample depletion always occurs in finite amounts and results in systematic deviations in quantitation. An internal standard only partly accounts for this, and when a differential analysis method is used, the internal standard should be a slower ion than the analytes of interest. For a more complete prediction of the effects of depletion, we recommend solving Equation 10.28 (including any necessary accumulation terms) and processing the results with Equation 10.23 to evaluate the systematic deviations in quantitation induced by sample depletion.

It should be noted that every extended monitoring application is unique and the direction of the development depends heavily on the analytes of interest, the detection method, and the sample composition and conductivity. Thus, we recommend against employing traditional BGEs and microchip designs simply because they have been shown to work in short-term analyses. Typically, these systems will have little or no optimization for long-term monitoring. Often, only a few of the suggestions given in this chapter will be appropriate in a given system. In particular, complex sample matrices present severe issues with capillary surface stability, achieving rapid sequential injections, and employing the peak overlap technique. For difficult matrices, reproducible, steady-state operation with rapid sequential injections may be practically impossible. For these systems, it may be better to employ multiple microchips that alternate sampling duties while the other chips undergo hydrodynamic flushes and/or surface treatment. Overall, the area of extended monitoring with MCE is still in its infancy, and the authors expect the number of successful applications in this field to increase. As experience in this field increases, new solutions to some of the hurdles specific to MCE extended monitoring will likely not only be realized, but also addressed.

REFERENCES

1. Jorgenson, J.W. and Lukacs, K.D. (1981) *Anal. Chem. 53*, 1298–1302.
2. Jorgenson, J.W. and Lukacs, K.D. (1983) *Science 222*, 266–272.
3. Mikkers, F.E.P., Everaerts, F.M., and Verheggen, T. (1979) *J. Chromatogr. 169*, 11–20.
4. Chen, X.G., Fan, L.Y., and Hu, Z. (2004) *Electrophoresis 25*, 3962–3969.
5. Fang, Z.L. (1999) *Anal. Chim. Acta 400*, 233–247.
6. Fang, Z.L., Chen, H.W., Fang, Q., and Pu, Q.S. (2000) *Anal. Sci. 16*, 197–203.
7. Kuban, P. and Karlberg, B. (1998) *Trends Anal. Chem. 17*, 34–41.
8. Kuban, P. and Karlberg, B. (2009) *Anal. Chim. Acta 648*, 129–145.
9. Santos, B., Simonet, B.M., Rios, A., and Valcarcel, M. (2006) *Trends Anal. Chem. 25*, 968–976.
10. Macka, M., Andersson, P., and Haddad, P.R. (1998) *Anal. Chem. 70*, 743–749.
11. Desiderio, C., Fanali, S., and Bocek, P. (1999) *Electrophoresis 20*, 525–528.
12. de Jesus, D.P., Brito-Neto, J.G.A., Richter, E.M., Angnes, L., Gutz, I.G.R., and do Lago, C.L. (2005) *Anal. Chem. 77*, 607–614.
13. Mayer, B.X. and Muller, M. (2000) *LC GC N. Am. 18*, 694-+.
14. Mayer, B.X. (2001) *J. Chromatogr. A 907*, 21–37.
15. Willauer, H.D. and Collins, G.E. (2003) *Electrophoresis 24*, 2193–2207.
16. Dolnik, V., Liu, S.R., and Jovanovich, S. (2000) *Electrophoresis 21*, 41–54.
17. Lacher, N.A., Garrison, K.E., Martin, R.S., and Lunte, S.M. (2001) *Electrophoresis 22*, 2526–2536.
18. Manz, A., Graber, N., and Widmer, H.M. (1990) *Sensor. Actuators B Chem. 1*, 244–248.
19. Li, S.F.Y. and Kricka, L.J. (2006) *Clin. Chem. 52*, 37–45.
20. Dolnik, V. and Liu, S.R. (2005) *J. Sep. Sci. 28*, 1994–2009.
21. Felhofer, J.L., Blanes, L., and Garcia, C.D. (2010) *Electrophoresis 31*, 2469–2486.
22. Faure, K., Pravda, M., and Glennon, J.D. (2006) *Anal. Lett. 39*, 435–449.
23. Revermann, T., Gotz, S., Kunnemeyer, J., and Karst, U. (2008) *Analyst 133*, 167–174.
24. Attiya, S., Jemere, A.B., Tang, T., Fitzpatrick, G., Seiler, K., Chiem, N., and Harrison, D.J. (2001) *Electrophoresis 22*, 318–327.
25. Buttgenbach, S. and Wilke, R. (2005) *Anal. Bioanal. Chem. 383*, 733–737.
26. Chen, S.H., Lin, Y.H., Wang, L.Y., Lin, C.C., and Lee, G.B. (2002) *Anal. Chem. 74*, 5146–5153.
27. Fang, Q., Xu, G.M., and Fang, Z.L. (2002) *Anal. Chem. 74*, 1223–1231.
28. Fang, Z.L. and Fang, Q. (2001) *Fresen. J. Anal. Chem. 370*, 978–983.
29. Li, M.W., Huynh, B.H., Hulvey, M.K., Lunte, S.M., and Martin, R.S. (2006) *Anal. Chem. 78*, 1042–1051.
30. Lin, Y.H., Lee, G.B., Li, C.W., Huang, G.R., and Chen, S.H. (2001) *J. Chromatogr. A 937*, 115–125.
31. Mecker, L.C. and Martin, R.S. (2008) *Anal. Chem. 80*, 9257–9264.
32. Reschke, B.R., Schiffbauer, J., Edwards, B.F., and Timperman, A.T. (2010) *Analyst 135*, 1351–1359.
33. Wang, M., Roman, G.T., Perry, M.L., and Kennedy, R.T. (2009) *Anal. Chem. 81*, 9072–9078.
34. Oki, A., Takamura, Y., Ito, Y., and Horiike, Y. (2002) *Electrophoresis 23*, 2860–2864.
35. Roper, M.G., Shackman, J.G., Dahlgren, G.M., and Kennedy, R.T. (2003) *Anal. Chem. 75*, 4711–4717.
36. Shackman, J.G., Dahlgren, G.M., Peters, J.L., and Kennedy, R.T. (2005) *Lab Chip 5*, 56–63.
37. Reid, K.R. and Kennedy, R.T. (2009) *Anal. Chem. 81*, 6837–6842.
38. Beckers, J.L. and Bocek, P. (2003) *Electrophoresis 24*, 518–535.
39. Persat, A., Chambers, R.D., and Santiago, J.G. (2009) *Lab Chip 9*, 2437–2453.
40. Persat, A., Suss, M.E., and Santiago, J.G. (2009) *Lab Chip 9*, 2454–2469.
41. Kelly, M.A., Altria, K.D., and Clark, B.J. (1997) *J. Chromatogr. A 768*, 73–80.

42. Li, D.M., Fu, S.L., and Lucy, C.A. (1999) *Anal. Chem. 71*, 687–699.
43. Pitts, E. (1953) *Proc. R. Soc. Lond. A Mat. 217*, 43–70.
44. Bello, M.S. (1996) *J. Chromatogr. A 744*, 81–91.
45. Hows, M.E.P. and Perrett, D. (1998) *Chromatographia 48*, 355–359.
46. Stoyanov, A.V. and Pawliszyn, J. (2004) *Analyst 129*, 979–982.
47. Crabtree, H.J., Cheong, E.C.S., Tilroe, D.A., and Backhouse, C.J. (2001) *Anal. Chem. 73*, 4079–4086.
48. Jacobson, S.C., Koutny, L.B., Hergenroder, R., Moore, A.W., and Ramsey, J.M. (1994) *Anal. Chem. 66*, 3472–3476.
49. Ermakov, S.V., Jacobson, S.C., and Ramsey, J.M. (2000) *Anal. Chem. 72*, 3512–3517.
50. Zhang, G.S., Du, W., Liu, B.F., Hisamoto, H., and Terabe, S. (2007) *Anal. Chim. Acta 584*, 129–135.
51. Jacobson, S.C., Ermakov, S.V., and Ramsey, J.M. (1999) *Anal. Chem. 71*, 3273–3276.
52. Reijenga, J.C., Verheggen, T., Martens, J., and Everaerts, F.M. (1996) *J. Chromatogr. A 744*, 147–153.
53. Jaros, M., Hruska, V., Stedry, M., Zuskova, I., and Gas, B. (2004) *Electrophoresis 25*, 3080–3085.
54. Beckers, J.L., Gebauer, P., and Bocek, P. (2001) *J. Chromatogr. A 916*, 41–49.
55. Beckers, J.L. and Bocek, P. (2002) *Electrophoresis 23*, 1942–1946.
56. Noblitt, S.D., Schwandner, F.M., Hering, S.V., Collett, J.L., and Henry, C.S. (2009) *J. Chromatogr. A 1216*, 1503–1510.
57. Yan, D.G., Yang, C., and Huang, X.Y. (2007) *Microfluid. Nanofluid. 3*, 333–340.
58. Reijenga, J.C., Gagliardi, L.G., and Kenndler, E. (2007) *J. Chromatogr. A 1155*, 142–145.
59. Noblitt, S.D., Lewis, G.S., Liu, Y., Hering, S.V., Collett, J.L., and Henry, C.S. (2009) *Anal. Chem. 81*, 10029–10037.
60. Gas, B. and Kenndler, E. (2004) *Electrophoresis 25*, 3901–3912.
61. Hruska, V., Stedry, M., Vcelakova, K., Lokajova, J., Tesarova, E., Jaros, M., and Gas, B. (2006) *Electrophoresis 27*, 4610–4617.
62. Beckers, J.L. (1994) *J. Chromatogr. A 662*, 153–166.
63. Beckers, J.L. and Everaerts, F.M. (1997) *J. Chromatogr. A 787*, 235–242.
64. Beckers, J.L. and Bocek, P. (2005) *Electrophoresis 26*, 446–452.
65. Beckers, J.L., Gebauer, P., and Bocek, P. (2001) *Electrophoresis 22*, 3648–3658.
66. Beckers, J.L., Urbanek, M., and Bocek, P. (2005) *Electrophoresis 26*, 1869–1873.
67. Gebauer, P., Beckers, J.L., and Bocek, P. (2002) *Electrophoresis 23*, 1779–1785.
68. Poppe, H. (1999) *J. Chromatogr. A 831*, 105–121.
69. Sellmeyer, H. and Poppe, H. (2002) *J. Chromatogr. A 960*, 175–185.
70. Hruska, V., Jaros, M., and Gas, B. (2006) *Electrophoresis 27*, 984–991.
71. Dolnik, V. (2004) *Electrophoresis 25*, 3589–3601.
72. Horvath, J. and Dolnik, V. (2001) *Electrophoresis 22*, 644–655.
73. Lucy, C.A., MacDonald, A.M. and Gulcev, M.D. (2008) *J. Chromatogr. A 1184*, 81–105.
74. Zhou, J.W., Ellis, A.V., and Voelcker, N.H. (2010) *Electrophoresis 31*, 2–16.
75. Znaleziona, J., Petr, J., Knob, R., Maier, V., and Sevcik, J. (2008) *Chromatographia 67*, S5–S12.
76. Melanson, J.E., Baryla, N.E., and Lucy, C.A. (2001) *Trends Anal. Chem. 20*, 365–374.
77. Liu, J.K. and Lee, M.L. (2006) *Electrophoresis 27*, 3533–3546.
78. Lucy, C.A. and McDonald, T.L. (1995) *Anal. Chem. 67*, 1074–1078.
79. Lucy, C.A., Yeung, K.K.C., Fu, S.L., Li, D.M., Henselwood, T.L., and Underhill, R.S. (1999) *Can. J. Chem. 77*, 281–290.
80. Jones, H.K., Nguyen, N.T., and Smith, R.D. (1990) *J. Chromatogr. 504*, 1–19.
81. Gong, X.Y. and Hauser, P.C. (2006) *Electrophoresis 27*, 468–473.
82. Backofen, U., Matysik, F.M., and Lunte, C.E. (2002) *Anal. Chem. 74*, 4054–4059.
83. Price, A.K. and Culbertson, C.T. (2009) *Anal. Chem. 81*, 8942–8948.
84. vanderMoolen, J.N., Boelens, H.F.M., Poppe, H., and Smit, H.C. (1996) *J. Chromatogr. A 744*, 103–113.
85. Jacobson, S.C., Hergenroder, R., Koutny, L.B., Warmack, R.J., and Ramsey, J.M. (1994) *Anal. Chem. 66*, 1107–1113.
86. Kuban, P., Engstrom, A., Olsson, J.C., Thorsen, G., Tryzell, R., and Karlberg, B. (1997) *Anal. Chim. Acta 337*, 117–124.
87. Beckers, J.L. and Bocek, P. (2000) *Electrophoresis 21*, 2747–2767.
88. Badal, M.Y., Wong, M., Chiem, N., Salimi-Moosavi, H., and Harrison, D.J. (2002) *J. Chromatogr. A 947*, 277–286.
89. Baryla, N.E. and Lucy, C.A. (2000) *Anal. Chem. 72*, 2280–2284.
90. Cunliffe, J.M., Baryla, N.E., and Lucy, C.A. (2002) *Anal. Chem. 74*, 776–783.
91. Huang, B., Kim, S., Wu, H., and Zare, R.N. (2007) *Anal. Chem. 79*, 9145–9149
92. Kamande, M.W., Fletcher, K.A., Lowry, M., and Warner, I.M. (2005) *J. Sep. Sci. 28*, 710–718.
93. Liu, Q., Li, Y.Q., Tang, F., Ding, L., and Yao, S.Z. (2007) *Electrophoresis 28*, 2275–2282.
94. Liu, Q., Yang, Y.M., and Yao, S.Z. (2008) *J. Chromatogr. A 1187*, 260–266.
95. Luo, Y.Q., Huang, B., Wu, H., and Zare, R.N. (2006) *Anal. Chem. 78*, 4588–4592.
96. Melanson, J.E., Baryla, N.E., and Lucy, C.A. (2000) *Anal. Chem. 72*, 4110–4114.

97. Wang, C.Z. and Lucy, C.A. (2004) *Electrophoresis 25*, 825–832.
98. Xiao, Y., Yu, X.D., Xu, J.J., and Chen, H.Y. (2007) *Electrophoresis 28*, 3302–3307.
99. Yeung, K.K.C. and Lucy, C.A. (1997) *Anal. Chem. 69*, 3435–3441.
100. Gassner, B., Friedl, W., and Kenndler, E. (1994) *J. Chromatogr. A 680*, 25–31.
101. Altria, K.D. (2002) *LC GC Europe 15*, 588–594.
102. Haber, C., VanSaun, R.J., and Jones, W.R. (1998) *Anal. Chem. 70*, 2261–2267.
103. Leube, J. and Roeckel, O. (1994) *Anal. Chem. 66*, 1090–1096.
104. Altria, K.D. and Fabre, H. (1995) *Chromatographia 40*, 313–320.
105. Nielen, M.W.F. (1991) *J. Chromatogr. 588*, 321–326.
106. Altria, K.D. (1993) *Chromatographia 35*, 177–182.
107. Hjerten, S., Elenbring, K., Kilar, F., Liao, J.L., Chen, A.J.C., Siebert, C.J., and Zhu, M.D. (1987) *J. Chromatogr. 403*, 47–61.
108. Huang, X.H., Coleman, W.F., and Zare, R.N. (1989) *J. Chromatogr. 480*, 95–110.
109. Watzig, H. and Dette, C. (1993) *J. Chromatogr. 636*, 31–38.
110. Schmitt-Kopplin, P., Garmash, A.V., Kudryavtsev, A.V., Menzinger, F., Perminova, I.V., Hertkorn, N., Freitag, D., Petrosyan, V.S., and Kettrup, A. (2001) *Electrophoresis 22*, 77–87.
111. Dose, E.V. and Guiochon, G.A. (1991) *Anal. Chem. 63*, 1154–1158.
112. Dang, Q.X., Yan, L.X., Sun, Z.P., and Ling, D.K. (1993) *J. Chromatogr. 630*, 363–369.

11

DISTINCTION OF COEXISTING PROTEIN CONFORMATIONS BY CAPILLARY ELECTROPHORESIS

Hanno Stutz

11.1 INTRODUCTION

Three decades ago, capillary electrophoresis (CE) started with the promising and enthusiastic perspectives of so many novice techniques in analytical chemistry. Although it has met, and in major parts even exceeded, initial expectations, other aspects have remained mediocre or still suboptimal. Since most analytes are accessible by at least one of the available CE-modes, CE separations are actually expected to have an even more widespread application. The numerous parameters provided by versatile separation modes are both blessing and curse. For example, they allow for selectivity tuning of matchless quality and specific separation tailoring even for highly diverging analyte combinations. Versatility of CE may turn into the Achilles heel, due to the fact that the extensive number of accessible optimization parameters makes separation optimization less straightforward than for chromatographic techniques. The complex interrelation of manifold separation parameters excludes a true single factor testing during the optimization process. Therefore, additionally to a rich practical experience operators should also possess comprehensive knowledge of fundamentals referring not only to CE but also to biomolecules/proteins as well as to surface and interface phenomena related to protein adsorption effects in order to entirely exploit the inherent potential of CE. The poor side of CE emerges if separations degrade or run out of control with versatility then turning into a boomerang since elucidation of the responsible factors can degenerate into Sisyphean task.

However, if optimized and well-controlled separation systems are available, the unrivaled potential of CE in separating closely related proteins becomes accessible. This is particularly required if protein conformations have to be distinguished and quantified. In this context, biotechnological products, diseases involving protein aggregation and fibril formation, and biopharmaceuticals are concerned. In case of recombinant products, integrity of protein structure and conformity to their natural counterparts have to be assured proving absence of folding derivates and aggregates. An elucidation of fundamentals in (un)folding processes, of intermediate products and the formation of protein seeds, which induce further aggregation, is essential. Generally, this would support understanding of the evolvement of—mainly neuropathologic—conformational disorders. Consequently, pharmaceuticals that could stabilize native (or early intermediate) folding states would offer an option at least to retard the progress of neurodegenerative diseases. This approach would prevent the aberrant aggregation, abandon it at early stages, or even re-dissolve protein aggregates and convert released monomers again into native conformation.

This chapter intends to discuss the current position of CE in the distinction of different protein conformations addressing strengths and flaws by numerous examples and compare it to HPLC. Therefore, theoretical aspects of *in vivo* protein folding, unfolding, and misfolding will be discussed including kinetic and thermodynamic issues. Special attention is given to conformations contained in recombinant products since this technology is operating at the intersection of *in vivo* (i.e., protein expression in host cell) and *in vitro* (i.e., protein extraction, purification, and possible refolding) techniques. A brief survey of (mainly) spectroscopic

Capillary Electrophoresis and Microchip Capillary Electrophoresis: Principles, Applications, and Limitations, First Edition.
Edited by Carlos D. García, Karin Y. Chumbimuni-Torres, and Emanuel Carrilho.

techniques frequently applied in the characterization of protein structures will be followed by a discussion of HPLC applications in the investigation of conformations both on the intact protein as well as on the peptide level in combination with various conformation-related detection systems. The major part comprises a survey of general aspects and pitfalls of CE in conformational separations and also addresses fundamentals of thermodynamic and kinetic aspects. Subsequent parts refer to CE investigations of various denaturant agents and their effects on proteins, relating data to molecular information and corresponding spectroscopic results. The final section compares CE with HPLC and addresses their individual features in the separation of protein conformations.

11.1.1 Theoretical Aspects of *in vivo* Protein Folding

The correct physiological function of proteins is frequently related to their selective interaction with natural target ligands and thus relies on a defined three-dimensional structure, that is, their conformation. In this context, the biological activity of proteins has to combine conformational stability with intrinsic structural flexibility of functional domains and groups that interact with these ligands [1]. For instance, protease inhibitors, such as α1-antitrypsin and antithrombin, naturally circulate in a metastable conformation. By interaction with the target protease a movement of reactive loops is induced, changing the inhibitor into a relaxed conformation, thus blocking the target enzyme for further interactions. Mutations in the shutter domain can evoke a premature conformational shift [2]. The propensity for a (partial) release of the integral protein structure has somewhat simplistically been categorized by the terms "hard" and "soft" protein and is based on changes in the Gibbs free energy (which accompanies unfolding (ΔG_U) of (quasi) native structures. According to the adhesive–cohesive model of protein compressibility, competition between intraprotein interactions (i.e., cohesive forces) and protein–solvent interactions (i.e., adhesive forces) via surface-exposed amino acids governs the structural stability. In addition, enthalpic contributions from charged atoms buried in interior parts are also relevant. In case both forces are well balanced, compressibility is low and thus structural integrity high [3]. This delicate balance between stabilizing and destabilizing forces can only be realized by distinct primary sequences, which have been selected by molecular evolution [1]. In general, the self-organization process of proteins is driven by their attempt to achieve a minimum of free energy (G) whereby the unique topology of the free energy landscape, the accessible funnels that are passed during folding events, and the native conformation are already encoded in the primary sequence via the side chains of amino acids (Figure 11.1) [4–8].

In fact, protein conformation can be considered the secret and unraveled part of the genetic code [1]. Prediction of the final protein conformation hold after translation merely on

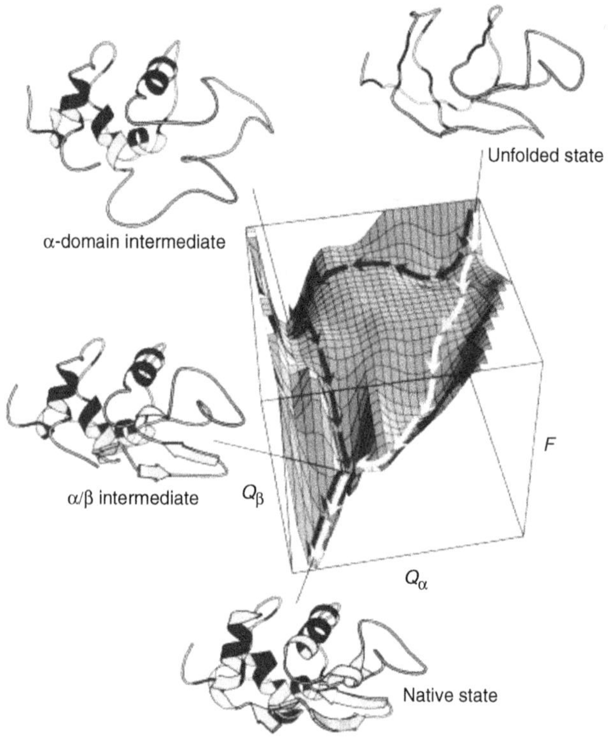

FIGURE 11.1 Diagram representing a schematic free energy landscape for the folding process of hen lysozyme, which consists of two domains annotated as α and β. Qα and Qβ is the number of native contacts in the α- and β-domain, respectively. Arrows denote two alternative folding pathways. (i) The bright track is a fast-folding pathway with concurrent formation of α- and β-domain, whereby the α/β intermediate is populated only transiently. The dark track gives an alternative slow-folding track, where first a trapped long-living α-domain intermediate is formed. The further transformation toward the α/β intermediate requires either transition of an energy barrier or partial unfolding to get access to the fast-foding track. Native-like regions (α-helix, β-sheet) are given as ribbon representation. Further details are given elsewhere. Reprinted from Ref. 7, with permission from Elsevier.

the basis of the primary sequence is still one of the most intriguing challenges of structural biology [6,8]. Several sequence motifs, however, have been related to particular secondary structures [5]. In this context, mini-motifs of <40 amino acids, representing simple mimics of proteins, have been designed and characterized revealing basic rules for formation of secondary and tertiary structures [9].

In proteins, theoretically an astronomic number of possible conformations can be passed on the protein's stochastic journey from random coil to native state [4]. The discrepancy between the time required for a systematic screening of all possible conformations (10^{11} years for a protein of 100 residues) and the actual observed fast folding events (ms-s) has dominated discussion as the *Levinthal paradoxon* over three decades. In fact, the initially unfolded polypeptide chain collapses rapidly to an ensemble of semicompact conformations that represent the starting point in the subsequent random search for native contacts. Thereby, the number of searched transitions is drastically reduced, which reveals the aforementioned paradox [7,8]. The unordered protein structure possesses high states of entropy (S) and free energy (G). During the folding process, enthalpy and entropy get progressively reduced, with the latter delivering an important contribution. The balance between enthalpy and entropy decrease will determine the free energy surface for protein folding [5,8,10]. Thermodynamically, the native conformation represents the most stable state that can be held under physiological or prevailing conditions with a minimum in free energy [10,11]. Multiple moves of the polypeptide chain and residue–residue interactions performed successively will result in a folding trajectory on the protein's free-energy surface with each trajectory comprising series of microscopic minima and barriers in G [7].

The critical folding events can be restricted by several checkpoints within the energetic funnel, that is, energetic barriers. These barriers are bottlenecks occurring when entropy decreases faster than the average enthalpy and are represented by a small number of distinct core residues, namely arrays of polar and hydrophobic amino acids, which stochastically have to establish specific intraprotein contacts when constituting a so-called folding nucleus [4,5,7,10,12]. Intrachain diffusion, that is, the rate of intrachain contact formation between two residues of the polypeptide chain, is crucial for early stages in protein folding events and thus for energy barriers. This diffusion has been probed by triplet–triplet energy transfer between two reporter groups attached to specific sites. In this case, these groups are linked to the N- and C-termini of polypeptides, and end-to-end diffusion can be addressed. Time constants of initial contact formation for the generation of loops, α-helices, and β-hairpins are between 5 and 30 ns [13]. Local secondary structure elements, such as α-helices and β-turns, are formed between <100 ns up to 1 μs, whereas β-sheet formation is more time-consuming [4]. This might be related to the fact that contrary to α-helices fairly distant amino acids have to be brought in contact in β-sheets, explaining also diverging folding kinetics for proteins that vary in their secondary structure elements [7,9]. Consistently, folding is considered a modular process with independent formation of secondary structure elements and their progressive assembly to local (sub-) domains, subsequently [1,4]. The interaction of preformed modular domains, most likely by diffusional motion as proposed by the "diffusion–collision" model, eventually establishes the correct global conformation in an ultimate cooperative step via "core contacts." The final native structure may form by establishing missing "native contacts" with some retardation in comparison to the formation of individual domains [4–7]. The interconversion between compact states is also slower than between extended. Small proteins with a single folding nucleus mostly show a two-state transition, whereas larger proteins comprising several nucleation sites hold multistep unfolding via intermediates [7]. However, the complex interplay of folding microevents, including the comprehensive energetic aspects during the actual folding process, has not completely been revealed [1]. Transitional states, so-called folding intermediates, possess already a highly congruent topology with the native state, although their flexibility is higher [4,5,12,14]. Consistently, their stability is only slightly lower than for native states [1]. Consequently, transitional state ensembles, that is, a limited number of folding intermediates, are populated at the top of energetic barriers, which separate folded and unfolded states [4,12]. By changing frame conditions, for example, the temperature, accessible folding pathways as well as their number will be changed [7]. For proteins, multiple pathways toward the native state might be accessible, which differ also in their kinetics due to local maxima in G (see Figure 11.1). Remarkably, alternative folding pathways have been shown to cross identical energy barriers [12].

Depending on the protein, *in vivo* folding can occur cotranslational, in the cytoplasm or in specific compartments [5]. The *in vivo* self-assembly of proteins is accelerated enzymatically, for example, by peptidyl prolyl *cis–trans* isomerases, which promote *cis–trans* isomerization of peptide bonds, and disulfide isomerases for increased disulfide bond formation [1,4–6,15]. Disulfide bonds can even form before protein translation is complete [6]. In addition, chaperones, which mainly belong to three conserved protein families (i.e., hsp 70, hsp 90, and charperonin), prevent misfolding by providing molecular cavities, where proteins can finalize their folding in a protected microenvironment within a crowded intracellular pool of macromolecules [5,6].

11.2 PROTEIN MISFOLDING AND INDUCTION OF UNFOLDING

Alternatively to the described regular folding process, proteins might also develop non-native interactions, thus entering a misfolding trap. The terms "folding" and "misfolding" strictly have to be distinguished. According to Dobson, "folding" refers to structure formation from completely or partially unfolded state to ultimate native topology, whereas "misfolding" describes the generation of non-native interactions in a sufficient high number to stabilize diverted conformations and affect the function [5]. Changes in the cellular microenvironment or mutations—point mutations as well as deletions—can modify the topology of the free energy landscape of protein folding [7,12], but in some instances, the gross topology of the energy landscape can also remain unchanged as exemplified by computer simulations of thermal unfolding [16].

Clearance of proteins that are misfolded or fail to hold their correct topology is governed by a cell internal quality control system involving molecular chaperones, proteases, and the final ubiquitin-proteasome degradation system (UPS) [5,17]. Chaperones recognize misfolded/partial (un)folded proteins and decide about their further doom—whether they are secreted, retained in the endoplasmatic reticulum, or degraded. Binding to chaperones might prevent secretion of ill-structured variants [18]. Equally, chaperones can act as rescue molecules and solubilizers for misfolded and aggregated proteins [5]. Indeed, 15–30% of translated cellular proteins are nonproperly folded, that is, either unassembled, unfolded, or misfolded, fail to pass the quality check, and have to be selectively destructed to peptides by means of the UPS [5,8,17].

11.3 CONFORMATIONAL PATHOLOGIES

Destabilization of the native state with (partial) unfolding and the induction of misfolded variants results in a concomitant loss of protein function and the development of diseases [12]. Several pathologies are related to defectively folded proteins and are therefore described as so-called conformational diseases [11]. Proteins with a pronounced structural lability are initially released in natural conformation with structural changes occurring subsequently, for example, as late-phase or episodic onset [2]. As (partially) unfolded proteins are thermodynamically unfavorable, they possess an enhanced propensity for aggregation (Figure 11.2), especially if they expose hydrophobic protein domains, carry a low net charge, or possess a β-sheet predisposition [11]. β-Sheets are of particular relevance in this context since they implement both hydrophilic and hydrophobic residues, the latter promoting the aggregation tendency [9].

Conformational diseases comprise various forms of amyloidosis, where conformational aberrants organize in characteristic fibrils and plaques. This is particularly relevant in cardiomyopathies, neurodegenerative diseases—for example, Alzheimer's (AD), Parkinson's (PD), and Huntington's disease (HD), transmissible spongioform encephalopathies (TSE) including bovine spongioform encephalopathy (BSE), Scrapie, and Creutzfeldt–Jakob Disease (CJD), amyotrophic lateral sclerosis (ALS), late-onset diabetes type II, and hemoglobinopathies (sickle-cell disease and inclusion-body anemias) [2,5,10,11,19]. In physiological dysfunctions, a rapid degradation of misfolded proteins is prevented since highly abundant aberrant conformations can obstruct the quality control system or induce molecule malfunctions when implemented in oligomeric proteinaceous complexes [11]. Instead of misfolding clearance a continuous intra- and extracellular self-association and accumulation of protein aggregates occur [2]. Aggregation of prion-like proteins requires (partial) unfolding and the presentation of previously hidden amyloidogenic sequences [5,20]. A β-sheet

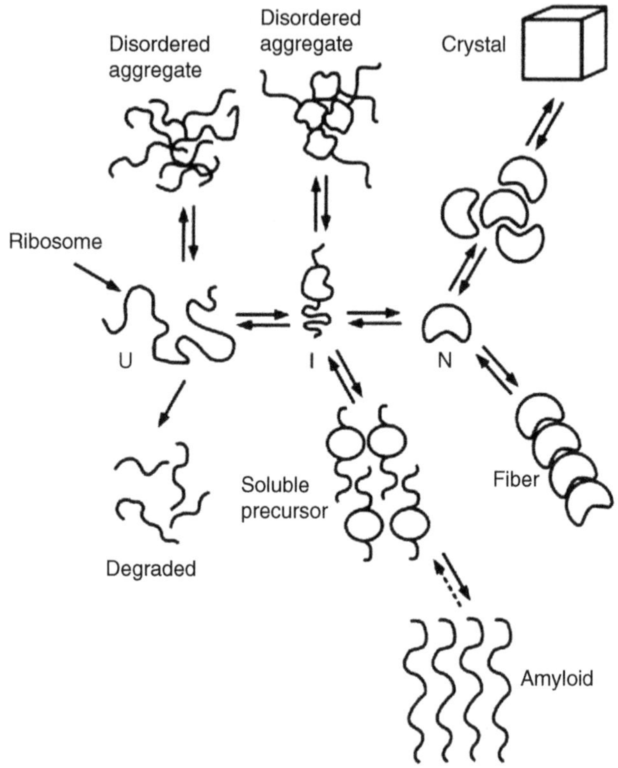

FIGURE 11.2 Scheme representing conformational states that are accessible for polypeptide chains in the course of biosynthesis. The monomeric polypeptide can fold from its disordered and unfolded state (U), via a partially structure intermediate (I) toward the native conformation (N). The native state can form aggregates including even highly ordered crystals and fibres, which, in either case, retain protein structure. I and U can generate highly disordered aggregates. In addition, amyloid fibrils may form by nucleation and elongation, which most likely occur from I. Reprinted from Ref. 8, with permission from Elsevier.

linkage with the strand of one donor protein becoming incorporated among strands of a neighbor protein can promote aggregation, and the exchange is fostered by particular amino acid sequences [2,21]. This so-called three-dimensional domain swapping might represent a mechanism for amyloid formation since it has been found in crystallized human prion, and it is also suggested for β_2-microglobulin (β_2m), and superoxide dismutase (SOD) [20,21]. The formation of β-sheet dominated structures, including also hydrophobic amino acids, enhances the inclination for oligomerization. Amyloid fibril formation starts with initial small bead-like amorphous aggregates and proceeds via protofilaments/protofibrils, which finally associate laterally (Figure 11.2). The core structures of these fibrils are β-sheets [11,20]. Recent results indicate prion-like characteristics, such as accumulation of fibrillar aggregates, template-related self-propagation, and transmission of misfolded conformations, for amyloid-β, α-synuclein, and tau- and polyglutamine proteins. Transcellular propagation of protein misfolding occurs (i) via the release of protein aggregates subsequent to cell death, (ii) by exocytosis, or (iii) via transmission over synaptic gaps and uptake by adjacent cells [22]. All amyloid fibrils possess the tendency to intercalate the dye Congo red (CR) and thioflavin derivatives [5,11,23,24]. (Neuro)toxic effects are associated with early aggregates (so-called prefibrillar structures) that might act as prokaryotic toxins and disrupt intracellular ion homeostasis by creating membrane pores [5].

Causative sources for the induction or enhancement of a conformational crisis resulting in misfolding are manifold and encompass mutations; temperature increases by 2–3 °C, as in the case of fever; glycation or oxidation, which affects mainly apolipoproteins in the case of artheriosclerosis; changes in metallation status; pH extremes; high ionic strength; and contact with air–liquid interfaces or negatively charged solid surfaces [2,10,25,26]. These deviations from normal physiological conditions induce (partial) unfolding and may shift the relative thermodynamic stability of various folding states [5,10]. Unfolded/misfolded states might therefore possess a higher G than the native state and represent a thermodynamically unfavorable state. Aggregation and fibril formation can reduce G again [10] (Figure 11.2).

Actually, the early induced misfolding event determines the final amyloid conformation as well as the polymerization rate. Similar effects have been described for Alzheimer's peptide Aβ(1–40) whereby the misfolded conformation could even be induced by seeds [19]. Furthermore, substantially increased concentrations in plasma can induce amyloid formation for immunoglobulins or β_2m [2]. Even nondisease associated proteins, for example, myoglobin or lysozyme, can form fibrils at increased concentrations and this form of aggregation can be generic to polypeptides [5]. Owing to the particular emphasis on selected patho-physiological relevant biomolecules in subsequent CE sections, namely β_2m, prion protein, and SOD, molecular fundamentals of these biomolecules have to be discussed in brief to improve intelligibility of the analytical sections.

β_2m, a 12 kDa protein, is folded in a β-sandwich structure with seven β-strands and is implemented as the light chain in the major histocompatibility complex I (MHC I) at cell surfaces and stabilizes the heavy chain. Disulfide bonds are established between Cys25 and Cys80. Patients with chronic kidney malfunctions and frequent hemodialysis encounter 20- to 60-fold increased β_2m serum concentrations in comparison to a healthy control group. Under these conditions, β_2m forms aggregates that are deposited mainly in the musculoskeletal system. These generated aggregates cause dialysis-related amyloidosis (DRA) syndrome with joint pain, destructive arthropathies, pathological bone fractures, and nerve compression, for example, carpal tunnel syndrome. By lowered pH, a condition prevailing during (acute) inflammation in synovial fluid, β_2m and especially coexisting folding intermediates of reduced stability slowly precipitate and form mixed fibrillar deposits. These deposits are stabilized by collagen fibers. Similar effects are induced by increased temperature and/or binding of Cu^{2+}. The (un) folding intermediate (I_2) of β_2m was considered to possess a high extent of tertiary structure, but to reconvert slowly (within 10–15 min at pH 7.4) to the native conformer. The slow refolding has been attributed to a *cis–trans* isomerization at Pro32, since the refolding was abolished when Pro32 was substituted [21,27–29]. Besides wildtype (WT) β_2m different variants have been constructed, all possessing unique features and stability. Since they will be addressed in subsequent CE sections, individual variants will be discussed all together for reasons of completeness. The variant ΔK58-β_2m is thought to be generated *in vivo* by enzymatic cleavage since it was detected at low concentrations in hemodialysis patients. This variant was most likely acting as a seed for fibril formation if pH was $\leq$6.8. ΔK58-β_2m was cleaved at K58 and lost this amino acid. The resulting peptides 1–57 and 59–99 were linked by a disulfide bridge. cK58-β_2m represents a cleaved β_2m variant with Lys58 still attached [29].

Most prominently, prion proteins (PrP), which seem to be involved in copper homeostasis, are known to exist in a normal cellular form (PrP^C) and a deviating disease eliciting conformation (PrP^{Sc}), which acts as a conformational template to propagate the structural aberration [26,30]. The formation of PrP^{Sc} involves profound conformational changes in the N-terminal part entailing a reduction of the α-helix content and an increase in β-sheets. Both prion conformations differ in their solubility and heat resistance [26]. Post-translational *in vitro* modifications of PrP^C mimicking oxidative stress, that is, nitration, resulted in changes in the secondary structure inducing properties corresponding with PrP^{Sc} [31].

In case of SOD, the metallation state is involved in the stabilization of the native state, and at increased concentrations, non-native dimerization occurs by so-called intermonomer domain swapping similar to β_2m [5,20]. Metal loss from SOD mutants associated with familial ALS can contribute to misfolding even at physiological temperatures [25]. The misfolded structures either allow for a further elongation in the aggregation process or get closed and generate dead-end products [20].

11.4 DISTINCTION BETWEEN CONFORMATIONS

Unraveling principles and mechanisms of conformational changes, formation of folding intermediates, misfolded structures, and aggregation, including fibril and plaque formation, are crucial issues from manifold viewpoints. Relevant aspects cover issues of fundamental research, such as triggers (e.g., additives, concentrations, environmental conditions) for

(i) unfolding/refolding and (ii) aggregation/disassembly, of (iii) disease-related research (e.g., for prion-pathologies), (iv) aspects of diagnostics including detection of disease relevant conformational aberrations, (v) quality assurance for biopharmaceutical products, as well as (vi) assistance in the establishment of appropriate production processes and storage conditions to avoid both misfolded products and aggregates, for example, by in process analysis.

11.5 RELEVANCE OF CONFORMATIONS FOR BIOTECHNOLOGICAL PRODUCTS

Protein unfolding and aggregation represent crucial problems in biotechnology and can occur during purification, refolding, or storage of recombinant proteins [32,33]. Both effects can decrease the efficiency of biopharmaceuticals and particularly aggregates might cause adverse effects when administered. Whereas native proteins can be affected by aggregation in the course of freeze drying, aggregation of denatured proteins is more frequent especially when proteins are re-solubilized from inclusion bodies (IBs) by denaturants [33,34]. The assurance of native-like protein conformations and the identification of additional coexisting non-native structures is an essential issue in biotechnology, in which proteins are expressed in alien-host cells (e.g., *Escherichia coli*, *Pichia pastoris*, *Saccharomyces cerevisiae*, *Caenorhabditis elegans*). The artificial frame conditions prevailing during recombinant expression are primarily optimized for high protein yield, which might not necessarily foster correct folding. High yield production of recombinant proteins frequently entails the formation of IBs in bacteria and aggresomes in eukaryotic cells [5,35], that is, insoluble intracellular protein deposits in the cytosol. This was originally intended as an evolutionary defense mechanism [11], in which highly enriched proteins are protected from proteolytic activity [35]. To remedy this situation, applicable disaggregation procedures would be helpful, for example, conditions that stabilize the native conformation, and could in turn also provide preliminary concepts for treating amyloidic diseases by clearance of protein deposits [32]. However, refolding of proteins gained from IB is cumbersome since the inner hydrophobic core of proteins is exposed in IBs and increases the susceptibility for aggregation during the refolding process. Aggregation is favored especially under high protein concentrations [36]. Protein recovery from these aggregates comprises washing steps and re-solubilization by application of chaotropic reagents, for example, urea or guanidinium chloride (GdmCl), detergents, for example, SDS, *N*-cetyltrimethylammonium chloride (CTAB), Triton X-100, and/or the employment of extreme pH values [35]. In addition, aggregation inhibitors, such as arginine (or urea), and molecular chaperons have been applied [36]. Consequently, subsequent protein renaturation to regain biological active protein conformation requires the removal of added agents and/or an re-adjustment to neutral pH [35]. Denaturants have to be removed gradually to avoid loss of protein activity. If dilutions are performed too fast, protein aggregation is induced [36]. Final products have to be characterized including their assessment of conformity to wild-type (WT)-like structures ensuring appropriate function in particular of enzymes, therapeutic, and diagnostic proteins. In general, comparison of folding dynamics, stability, and activity of recombinant proteins with their WT-counterparts is mandatory, revealing also bound (competitive) ligands remaining from previous affinity purification [37,38]. This is of particular relevance if proteins are intended to become applied a bioscavengers requiring an empty binding cavity. In addition, protein–ligand complexes might elicit immunogenicity [38]. Furthermore, the design of proteins/enzymes of enhanced conformational stability either under physiological conditions or under pH and temperature extremes is another industrial endeavor [39]. In this context, time- and cost-efficient methods for a distinction between coexisting conformations, for the detection of dimers, oligomers, and aggregates, as well as for a determination of the thermodynamic stability of proteins are required.

11.6 CONFORMATIONAL ELUCIDATION—AN OVERVIEW OF ALTERNATIVE METHODS TO CE

Recently, Righetti and Verzola have surveyed methods applied to illuminate (un)folding/refolding events of proteins discussing their strengths and drawbacks in comparison with capillary zone electrophoresis (CZE) [40]. Frequently applied methods comprise (i) far and near UV–circular dichroism (UV–CD) spectroscopy [41,42], (ii) Fourier transform infrared spectroscopy (FT-IR) [31,43], (iii) NMR spectroscopy [44,45], (iv) small-angle X-ray scattering [46], (v) fluorescence spectroscopy [32,47], (vi) differential scanning calorimetry (DSC) [38,41], (vii) H/D exchange mass spectrometry [28,48,49], (viii) size exclusion chromatography (SEC) [36,50], (ix) slab gel electrophoresis in denaturant gels [51], (x) atomic force microscopy (AFM) [32], and (xi) immunochemical techniques [52–54].

A comprehensive discussion of these techniques is beyond the scope of this chapter. Instead, particular attention will be given to CE and HPLC applications in the separation of protein conformations gained during unfolding/(re)folding processes. Mediocre concentration detection limits and repeatability of migration times (t_m), which is less straightforward achievable for proteins, have generally hampered the application of CE as an alternative or even substitute of HPLC. Therefore, both techniques will be surveyed in the light of their applicability but also of their individual limitations in resolving conformational mixtures. Final conclusions provide a resumé defending CE in terms of favorable and unique aspects but equally addressing pitfalls, and restrictions, finally giving an outlook for desirable innovations.

11.7 HPLC IN CONFORMATIONAL DISTINCTION

11.7.1 Intact Proteins

11.7.1.1 Reversed-Phase (RP)–HPLC Due to the mass transfer between mobile and stationary phase a preservation of protein conformations originally hold in the injection sample cannot *a priori* be presumed during the chromatographic separation process [55]. Especially hydrophobic phases exert a catalytic effect on protein unfolding [56]. Events proposed to take place during adsorption comprise an increase in helical content and a reduction of tertiary contacts [57]. In general, extended contact times of proteins with the stationary phase will foster unfolding, particularly of labile species [56,58]. In addition, column temperature and solvents applied in the reversed-phase (RP) mode can destabilize protein structures and affect both melting temperature (T_m) and elution order [55,56,59]. The group of Karger provided one of the first discussions of conformation-based separations on moderately hydrophobic C1–C4 columns [56,58]. Bovine pancreatic ribonuclease A (RNase A) resulted in a broad signal followed by a sharp peak that represented native and unfolded states, respectively. The height of the unfolded conformation increased with faster flow rates under conditions keeping the capacity factor k constant. Further (un)folding intermediates were expected, but due to their short lifetime in the ms regime, they were not detected. Depending on the separation temperature RNase A adhered onto the stationary phase either exclusively in the unfolded state (>20 °C) or at low temperature (<20 °C) concomitantly also in folded structure. Changes in the gradient and k induced alterations in the rate constant of refolding [58]. Conversely, the native conformer of α-lactalbumin resulted in a narrow well-defined peak, whereas a broad peak at increased retention time (t_r) comprised a population of various (partially) unfolded conformations. Spectral differences to the native state corroborated the prevalence of these unfolded variants. Application of lower temperatures, for example, down to 5 °C, proved reversibility of unfolding. The Ca^{2+}-depleted apo-form showed a higher temperature susceptibility, whereby T_m was decreased by ~27 °C. However, when Ca^{2+} was added to the mobile phase a less hydrophobic conformation of higher stability was re-induced. This effect was absent in case of Mg^{2+} addition [56]. On the contrary, an irreversibly denatured variant occurred for papain, eluting more than 10 min after the native conformation. Enhanced retention was due to a pronounced adsorption of the more stretched unfolded variant and the induction of a slow denaturation at the surface of the stationary C4-phase. The ratio of both conformations changed in the tested range of pH 2.2–3.0 and the temperature interval between 5 and 30 °C, favoring the native variant at lower temperature and higher pH [60].

Hydrophobic proteins are known to unfold in RP–HPLC entailing denaturation and aggregation with final (quasi) irreversible binding onto the stationary phase. Frequently, ghost peaks are detected in subsequent blank runs due to disaggregation and successive release of monomers from adhered aggregates. Structural rearrangements induced by hydrophobic surfaces and the organic solvent composition generate multiple conformations, all showing different interaction with alkyl groups of the stationary phase and thus distinct t_r. Propan-2-ol (IPA) and methanol both induced conformational changes in concanavalin A methacylamide (Con A-MA), even at 10% (v/v) IPA. Acetonitril (ACN) was less disruptive. In case of interferon (IFN)-α2b 60% (v/v) ACN induced loss of α-helix but not of β-sheet. Trifluoroethanol (TFE) shifted native β-sheet of Con A-MA in α-helical conformation. Systematic denaturation studies revealed deviations from physiological pH, increase of temperature, prolonged contact time with stationary phase, as well as type and concentration of applied electrolytes to exert an influence on the integrity of the tested proteins, that is, bovine insulin and Con A-MA. Altogether, these parameters induced populations of intermediates and unfolded species reflected in heterogeneous separation profiles [55]. Molecular dynamics transition has been performed for a 46-bead β-barrel model protein consisting of hydrophobic, hydrophilic, and neutral amino acids, which was probed in a confined pore thus mimicking a porous adsorbent. Process kinetics, that is, conformational changes during adsorption/desorption in hydrophobic interaction chromatography (HIC) and RP–HPLC, have been modeled for a protein concentration of 3.4 mmol L^{-1} at 300 K. Assuming a "collapse-rearrangement" mechanism native, unfolded, and intermediate conformations were formed in solution as well as protein aggregates. Simulations provided also time-dependent snapshots of the evolution of individual states over 10 ns. During adsorption, hydrophobic interactions drive proteins toward the stationary model surface with subsequent adsorption evoking full conformational extension, which results in well-dispersed unfolded states inhibiting aggregation in the adhered state. During elution, protein desorption with instant refolding and parallel formation of aggregates from intermediate and unfolded states occurred, with aggregates growing over time and becoming dominant. For the optimization of the elution process in terms of high yield of native protein by desorption, gradient elution in RP–HPLC and HIC reduced aggregate formation [61].

Walcher et al. described an apparent conformational change of an intact protein in RP–HPLC with poly(styrene-divinylbenzene) (PS-DVB) monolithic columns. Trypsin inhibitor showed a remarkable change in the elution order, abruptly shifting its position between 40.0 and 50.0 °C. This has been attributed to a pronounced conformational transition [62]. Results are in line with previous data for trypsin inhibitor, showing a cooperative unfolding with a sharp thermal transition. Whereas the native protein contains a mixture of left- and right-handed helices and some random coil content, helices are converted to random coil at elevated temperatures. T_m was highly depending on pH and ionic strength [63].

11.7.1.2 Size Exclusion (SEC)–HPLC SEC–HPLC was applied to investigate GdmCl-induced multiphase unfolding of bovine carbonic anhydrase II (Ca II) proceeding via native-like molten globule (MG) and a more extended pre-MG, both intermediates lacking biological activity. A reduction of the column temperature to 4 °C allowed for the distinction between an early eluting asymmetric peak, comprising two or more rapidly inter-converting unfolded variants, and the native conformation giving a narrow symmetric peak. Data were in agreement with results derived from near and far UV–CD. [50]. SEC–HPLC has also been utilized to follow structural transition of the HIV protein gp41. Conformational alteration proceeds via an intermediate finally forming a so-called fusion state, where N- and C-terminal ectodomains generate a six-helix bundle, which represents the fusogenic core conformation and initiates the fusion between viral and host cell membrane. Model peptides derived from the relevant gp41 domains were applied in the formation of the six-helix bundle structure by three heterodimers and separated from monomers [64]. Gu et al. [36] described refolding of denatured lysozyme by SEC with a linear urea gradient. At high protein concentrations aggregates formed during refolding and were attached onto the stationary phase before re-dissolved by the urea front resulting in additional peaks after the native monomer. Shallow urea gradients and low elution flow-rates were essential for an efficient refolding [36]. Meanwhile, so-called protein folding liquid chromatography has gained increased importance in order to recover bioactivity of recombinant proteins as outlined elsewhere comprehensively [65].

11.7.1.3 Ion-Exchange–HPLC Herbold et al. [57] investigated unfolding of cytochrome *c* on an anion exchange column of sulfo propyl type. Two peaks were detected and attributed to reduced and oxidized variants, with varying ratio in dependence of the temperature. Higher temperatures increased t_r due to a reduction in tertiary contacts, a reversible expansion, and thus an increase in accessible adsorption sites on the protein surface. At temperatures >70 °C and residence times >5 min on the stationary phase, irreversible protein losses were observed. Adsorbed molecules possessed a considerably lower T_m in comparison to bulk proteins [57].

11.7.2 HPLC with Detectors Sensitive for Conformations and Aggregates

To combine both separations of different conformations (of monomers) and small aggregates with structure-related information, Luykx et al. [66] coupled SEC or RP–HPLC with a dual detector setup, which arranged UV—in series either with scanning fluorescence—or CD-detector. The latter detectors were intended to indicate changes in tertiary and secondary structure, respectively. Whereas in SEC native separation conditions are applied but allow mainly for a distinction between (native) monomers and small aggregates or multimers, RP–HPLC reveals differences in hydrophobicity, but might also induce structural changes *per se* [66,67]. Besides recombinant IFN-α2b and *N*-methionyl human growth hormone, anti-tetanus immunoglobulin was investigated. Structural changes were judged on the basis of the signal ratio of UV at 214 nm with CD at 220 nm or with fluorescence emission maximum (F_{max}) after excitation at 295 nm, respectively. Fluorescence emission addressed Trp residues. ACN contents <20% in the mobile phase hardly changed protein structure, whereas above 30% ACN alterations in tertiary structure and loosening in secondary structure occurred. Contrary to other proteins, for example, cytochrome *c*, in this particular case 0.1% trifluoroacetic acid (TFA) stabilized the protein structures and increased the β-sheet content [66]. SEC in combination with multiangle laser light scattering (MALLS) allowed for a distinction between monomers and aggregates of whey protein induced by heating. This approach allowed for an absolute mass determination of formed aggregates and revealed changes in the size of the latter especially in the M_r region of 10^5–10^6. Native and heat-induced metastable monomers, however, could not be resolved [67].

11.7.3 Peptides as Model Compounds for Hydrophobic Stationary Phases in HPLC

As discussed in Sections 11.7.1.1 and 11.7.2, conformational distinction on intact protein level without interfering artifacts originating from mass-transfer effects is cumbersome in HPLC. Instead, changes in protein structures related to mutations have been investigated indirectly on the basis of the primary sequence using reporter peptides containing the relevant mutation. Accordingly, two PrP^{TSE} allotypes have been distinguished from PrP^{C} by tryptic peptides applying RP–HPLC–MS [68]. Khew et al. [59] investigated collagen-like structures, which excel in the unique triple helix configuration bridged by disulfide bonds. The analysis of spontaneous misalignments in the collagen assembly by CD and NMR was cumbersome. Therefore, RP–HPLC was applied to track different conformations of synthetic collagen-like peptides—(Pro–Hyp–Gly)$_{10}$ and (Pro–Pro–Gly)$_{10}$—in unassembled and triple helical structures (THPs). Model peptides possessed different hydrophobicities in single-strand state (SSP). In addition, an unordered control peptide was included in the study. Thermal titration gave evidence for a progressive shift of heterogeneous THP peaks toward SSP peak, which was also reflected in cooperative melting curves. THP heterogeneity was due to misalignments in the course of triple-helix self-assembly. Furthermore, THPs of the above stated two 10-block peptides differed because of their divergent conformational stability. THPs and SSP were in equilibrium and the elimination of either form entailed refolding or unfolding re-settling the balance. In freshly prepared standard solutions, THP peaks were underrepresented and increased over several days by slow folding. Curing misfolded THPs required complete disintegration to individual SSPs. The retention mechanism and thus elution order were governed by hydrophobicity of THPs and SSPs. Moreover, additional factors, such as differences in volume and surface of varying conformations, resulting in steric effects and alterations in adsorption/desorption kinetics are involved in RP–HPLC [59].

Antimicrobial and cytotoxic peptides offer another attractive substance class, since they alter their secondary structure in hydrophobic (e.g., membrane lipid) environment. Depending on their preferred conformation subsequent to their integration in membrane systems, α-helical and β-sheet peptides can be distinguished [69]. Organic solvents applied to mimic the hydrophobic membrane core (i.e., methanol, ethanol, ACN, TFE, and 1,1,1,3,3,3-hexafluoroisopropanol) usually stabilize this preferred secondary structure of peptides [69,70]. Interaction of these peptides with C18 modified surfaces was able to induce α-helical and β-sheet structures. In case of adhered β-sheet peptides, a competition between an intermolecular interaction with dissolved peptides and the stationary phase occurred, which lowered the binding affinity to C18. This competition did not occur for induced α-helix conformations. Deviations between experimental and theoretically derived t_r were related to structural shifts of peptides induced by their adsorption (compare Section 11.7.2). Owing to the aforementioned differences in binding affinities, β-sheet peptides will elute earlier, whereas α-helix peptides will be retarded in comparison to theoretically predicted values. Results are reviewed in detail elsewhere [69]. Chen et al. [70] described effects of separation temperatures on model peptides in RP–HPLC. Selected peptides comprised random coil decapeptides and amphipathic ~100% α-helical octadecapeptides substituted at a defined position by various D- and L-amino acids. D-Amino acids were selected due to their α-helix disrupting properties. Retention of random coil peptides is governed by side-chain hydrophobicity. Retention of all model peptides was reduced with increased temperature due to increased solubility and mass transfer and decreased solvent viscosity. Nevertheless, temperature effect was more pronounced for α-helical peptides due to induced conformational changes [70]. Conformations of peptides with 21–31 residues have also been investigated in dependence of the applied parenteral solutions with RP–HPLC off-line coupled to FT-IR spectrometry and ESI–MS [71].

11.8 CAPILLARY ELECTROPHORESIS (CE) IN CONFORMATIONAL SEPARATIONS

11.8.1 Fundamental Aspects and Survey of Pitfalls

Most techniques addressing the elucidation of conformational changes measure a cross-section over coexisting folding species and consistently provide information about structural properties averaged over the entire population of coexisting conformations. Chromatographic separation techniques, such as RP–HPLC, SEC–HPLC, and HIC, theoretically have the potential to separate various folding species. However, the integral equilibrium constant for analyte mass transfer between mobile and stationary phase is prone to falsify conformational distributions as shown in Section 11.7. This can shift the conformation equilibrium originally prevailing in the sample solution or even induce folding artifacts. Therefore, corrective measures for the mass transfer would be desirable. Furthermore, the stationary phase might be influenced by parameters applied to induce unfolding (e.g., temperature and denaturating agents) [72].

The separation principle of CE, referring mainly to CZE, should overcome these limitations *per se* since neither a mass transfer term nor organic solvents is principally involved in CE and thus a true image of coexisting structural variants can theoretically be achieved. However, one has to be aware that at least in the fast-time regime CE also reflects conformations preferably hold under the separation conditions [73]. Contrary to most other applied techniques, CE would allow for both a calculation of kinetics related to structural conversion based on encountered peak shapes and a quantitation of structural variants populated under the prevailing separation conditions. In addition, the outstanding efficiency and selectivity is predestined for a distinction of closely related protein variants and conformations. Even accelerated separations are accessible by either a reduction in the capillary length or an increase in the separation voltage. Therefore, folding intermediates of low lifetime which are thus unavailable in HPLC could become detectable. However, possible drawbacks of CE approaches have to be considered as well. High field strength has to be excluded as a source that could affect the (un)folding equilibrium [74]. Nevertheless, the inherent tendency of proteins for adhesion onto the inner wall of fused silica capillaries could fraudulently alter the encountered separation profiles by spurious generation of peak broadening, shifts in the apparent electrophoretic mobility, or even occurrence of novel peaks [74–76], which might altogether be misinterpreted as novel conformations [76,77]. Additionally, wall adsorption/desorption cycles can also cause unfolding/refolding processes [72]. A causative distinction of peak tailing due to protein adhesion or (un)folding in the fast-time regime is sometimes cumbersome [72,78]. Therefore, measures have to be taken to avoid these misleading artifacts and provide a true image of conformational ensembles. Reduction in protein adsorption has been realized by operation at pH extremes or high-ionic-strength conditions [79], although pH extremes have been shown to trigger unfolding events resulting in slightly relaxed (quasi) native structures at best [47,80]. In addition, this might also result in disrupture of quaternary structure with subsequent release of subunits and induction of novel peaks [76,81,82]. Ideally, the confirmation of structural integrity of quasi native structures under the selected separation conditions has to be performed by orthogonal methods. Prevention of protein adsorption is frequently achieved by beneficial electrolyte systems, for example, so-called Good buffers, electrolyte additives, which probably have no effect on protein structure but shield anchor sites on the surface, and wall passivation either by physical adhesion of neutral or charged polymers or by covalent wall modifications [79,83–86]. However, for the latter case an entanglement of individual protein molecules to a different extent in coating polymers with subsequent changes in mobility has been discussed [87].

11.8.2 Electrophoretic Mobility of Proteins

The electrophoretic mobility μ of a spherical ion i in CE can be described by means of the Stoke's equation:

$$\mu_{i,\text{o}} = \frac{ze}{6\pi r_{\text{H}}\eta} \tag{11.1}$$

where z is the charge number, e the elementary charge, r_{H} the Stoke's radius, and η the viscosity of the buffer. However, this simplistic expression assumes zero ionic strength (annotated as $\mu_{i,\text{o}}$) and neglects thus ionic effects, for example, analyte–solvent interactions, on mobility [72,88]. Implementation of the Debey–Hückel theory accounts for these ionic effects but requires knowledge of the hydrodynamic radius (r_{H}) and the actual valence Z_{a} of the protein, which are both highly influenced by electrolyte ions and the separation conditions, for example, dielectric constant, ionic strength, and pH. Therefore, improved equations for proteins have implemented M_{r} or the number of amino acids. Altogether, these modified models considered parameters that are not affected by structural alterations and therefore succeeded only in describing μ for peptides with random coil structure or proteins that are approximate entirely unfolded [72,89]. Unfortunately, models failed when trying to predict μ of folded or MG protein structures. A more advanced model of Compton [89] considered several aspects relevant in protein folding including electrostatic surface charge suppression, for example, by shielding polar and hydrophobic patches on protein surfaces, and the alteration of pK_{a}-values of amino acid side chains in dependence of the proteinaceous microenvironment in comparison to individual amino acids [72,73,89]. The charge suppression model assumes—in a somewhat simplistic approach—equal suppression for all surface charges. However, due to nonuniform charge distribution, knowledge of actual spatial charge arrangement is required. Consequently, theoretical calculation of μ is a complex matter due to the zwitterionic character and secondary structures held by proteins. According to the Compton model, the transformation of the calculated valence Z_{c} into the actual valence Z_{a} requires the application of a proportionality constant F_{Z}, which depends, among other factors, on Z_{a}, M_{r} and the ionic strength I. Alternatively, F_{Z} can be determined from the ratio of the theoretically calculated mobility (derived from equations) to the mobility measured in CE. Deviations between Z_{c} and Z_{a} have been attributed to charge suppression effects and calculations for eight model proteins revealed substantial differences between both parameters [89]. Actually, the practical predictability of protein mobilities for native and MG states is limited by this discrepancy [72]. With the derived model equation, protein valence, frictional ratio, ionic strength, partial specific protein volume, and Henry's function are considered, which means that μ for distinct conformations become calculable if experimentally derived values for valence (Z_{a}), size, hydration state, and shape are accessible [89]. These theoretical considerations have to be associated with effects taking place on the molecular level not only during unfolding/refolding but also in the course of CE separations. Actually, unfolding events can increase the accessibility of natively buried amino acid residues, thus changing their pK_{a} values and subsequently also the protein's pI. In addition, pK_{a} values will also be controlled by the surrounding solvent and separation temperature [72]. Compton and O'Grady described the interdependence by means of the following equation:

$$\mu = \frac{Z_{\text{C}}}{C' M_{\text{r}}^{m}} \tag{11.2}$$

where C' represents the amino acid content and m represents a constant varying between 1/3 and 2/3 depending on the ionic strength I and M_{r} [89]. Furthermore, interaction of charged protein sites with countercharged electrolyte ions, as described by several authors [90,91], can evolve deviations and even reversals of experimental mobilities from theoretically predicted. Therefore, specific ionic interaction between analyte and electrolyte ions can influence the resolution (R_s) between (conformational) protein species [89].

The various modes of CE can provide information about unfolding processes by either shifts in r_H, the p*I*, the mobility, the shape of peaks, and/or altered interactions with specific monoclonal antibodies in affinity strategies. All these parameters can be affected by conformational changes and/or protein aggregation. Mobility decreases during the unfolding process are related to a reduction of the protein net charge and/or a molecular expansion, thus increasing r_H. Increases in mobility can be related to an increase in protein net charge either by exposing charged residues or a reduced charge shielding [72].

Point mutations in proteins, as described for *Staphylococcus aureus* nuclease, where basic or neutral amino acids were replaced by (other) neutral amino acids impressively proved the selectivity of CE in case appropriate preventive means against protein adsorption were taken. Even variants with identical net charges were separated, although their resolution was naturally lower than for variants with charge differences. Kálmán et al. [92] have also calculated the r_Hs for native and MG structures as well as for random coil variants, clearly demonstrating an increase for the latter. Subtle changes in the tertiary fold, for example, by losses of intramolecular H-bonds or changes in *trans–cis* configuration of peptide bonds, were capable to shift mobility despite similar sizes and net charges. Effective protein size and shape inherently also consider the solvent corona associated with the analyte (e.g., the hydration shell). Resolvable mobility differences of $2\text{–}3 \times 10^{-10}\,\text{m}^2\,\text{V}^{-1}\,\text{s}^{-1}$ have been related to shifts of 26–38 pm in r_H, respectively, which is <2% of the examined protein radius [92]. Most changes in mobility, however, are not evoked by changes in the molecular size or the net charge exclusively, but by a combination of both effects. Information about individual contributions of changes in charge and size are missing and not accessible in most cases by sole CE measurements. Even the absence of changes in mobility is technically no proof for a lack of unfolding events, since shifts in protein size and changes in p*I* could compensate each other.

11.8.3 Peak Profiles and Derivable Thermodynamic Aspects of Protein Re-/Unfolding

Peak shapes and profiles encountered for reversible folding/unfolding transition in CE are highly dependent on the kinetics of interconversion. Computer simulations and their comparisons with experimental electropherograms have augmented the understanding how thermodynamic parameters and separations profiles are interrelated (Figure 11.3).

Derived models relate the rate of interconversion between different folding structures to the total analysis time required by CE [72]. Interconversion between protein conformers is related to rotational isomerization, that is, reversible rotation of peptide bonds between the nitrogen atom of Pro and the carbonyl of a preceding amino acid. These rotations entail *cis–trans* transitions [93]. On the basis of the ratio between interconversion rate and CE analysis time, kinetics of isomerization can be classified as (i) slow-time, (ii) intermediate-time, and (iii) fast-time regimes resulting in distinct separation profiles, that is, (i) isolated peaks, (ii) distinct peaks with an interconnecting plateau, and (iii) single peak with shifting mobility and bandwidth depending on the prevailing ratio of coexisting conformations (Figure 11.3). In case of (i), CE profiles reflect the conformational equilibrium established in the sample prior to its injection in CE, since no interconversion occurs during the separation and species prevail quasi frozen for their quantification. For (ii), some percentage of protein molecules will undergo isomerization (giving the interconnecting plateau), whereas others remain unaffected and quantification is impossible in this case. Peak shapes and areas depend on the initial distribution of isomer fractions and the sum of ($k_1 + k_2 = k_3$) (Equation 11.3). Fast isomerization (iii) ensures repetitive un-/refolding processes of all individual molecules during CE separation with the final time-averaged peak in the electropherogram covering all conformations. Direct measurement of conformational populations present at the detection time, calculation of thermodynamic parameters, and the relative contribution of respective folding states are merely possible for (i) slow-time and (iii) fast-time interconversions [72]. Hilser and Freire [72] have simulated CE profiles for (1) a reversible two-state transition for all time regimes (Equation 11.3) and (2) a reversible three-state transition model, for slow interconversion kinetics where unfolding proceeds via a folding intermediate I (Equation 11.4):

$$\mathrm{N} \underset{k_2}{\overset{k_1}{\rightleftharpoons}} \mathrm{U} \tag{11.3}$$

$$\mathrm{N} \underset{k_2}{\overset{k_1}{\rightleftharpoons}} \mathrm{I} \underset{k_{2'}}{\overset{k_{1'}}{\rightleftharpoons}} \mathrm{U} \tag{11.4}$$

Thereby, k_1 (and $k_{1'}$) and k_2 (and $k_{2'}$) represent unfolding and refolding kinetic rates, respectively, and their sum ($k_1 + k_2$) is addressed as $k_{3'}$. Their quotient gives the unfolding equilibrium constant K_U:

$$K_U = \frac{k_1}{k_2} = \frac{P_U\,\text{area}_U}{P_N\,\text{area}_F} \tag{11.5}$$

With P_U and P_N representing fractions of unfolded and native protein conformations, area_U and area_F giving peak areas of unfolded and native conformations [72,94]. The interconversion models selected by Hilser and Freire represent situations where the energetic landscape solely allows for the population of two or three folding states [72]. However, folding pathways according to the Lumry–Eyring

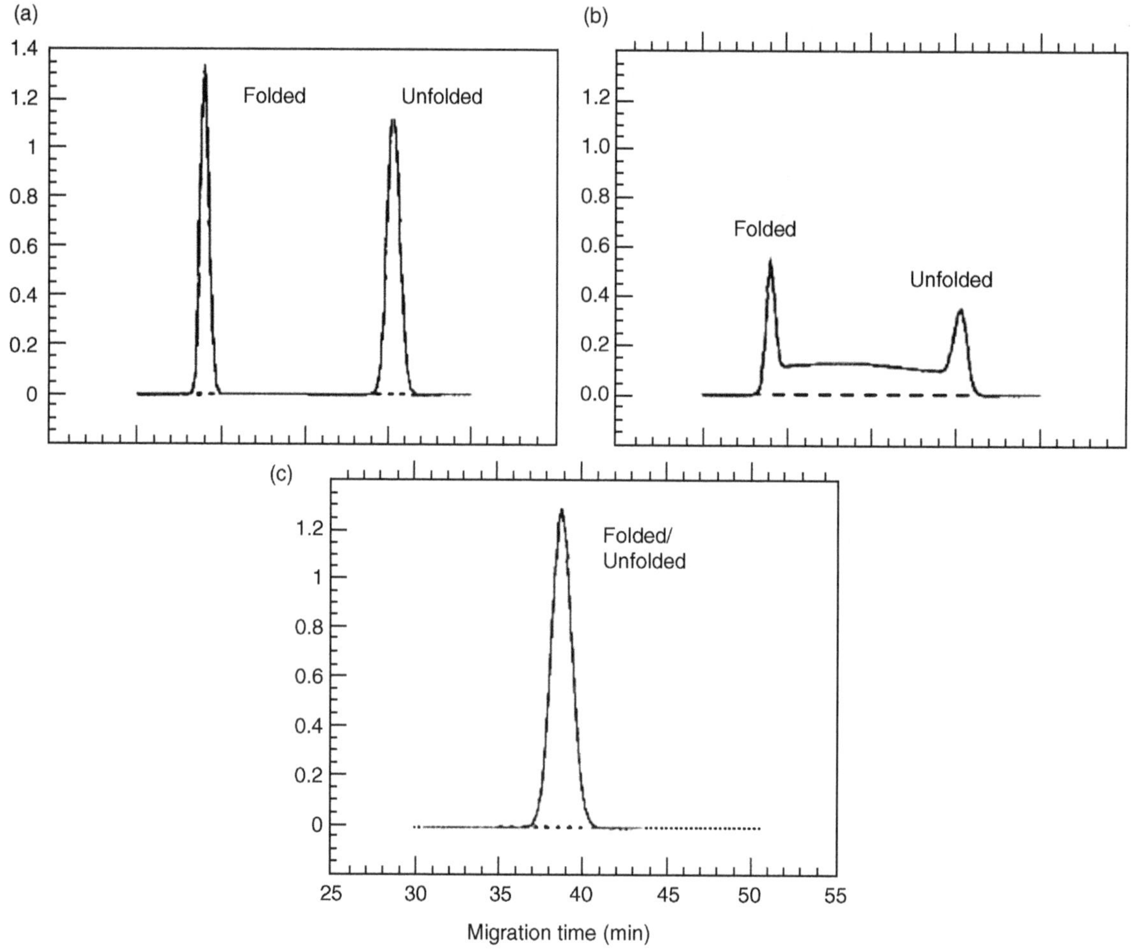

FIGURE 11.3 Simulated electropherograms of macromolecules, for example, proteins, subject to reversible isomerization with different time regimes. Conformations are present in equivalent amounts at the start of the separation. The equilibrium constant $k_1/k_2 = 1$. (a) Slow-time regime, (b) intermediate-time regime, and (c) fast-time regime. Dashed lines give extrapolated baselines. Reprinted from Ref. 72, with permission from Elsevier.

model with an initial reversible unfolding (U) followed by a final irreversible denaturation (D)

$$\mathrm{N} \underset{k_2}{\overset{k_1}{\rightleftharpoons}} \mathrm{U} \xrightarrow{k_{1'}} \mathrm{D} \tag{11.6}$$

might occur for some proteins but have not been considered in these models [95].

The population of individual conformational states can be derived from the integrated peak areas in slow-time systems. However, as these distributions are recorded as electropherograms at a fixed migration distance (i.e., effective capillary length to detector) but with various migration times (t_m), peak areas fail to image the true situation. Peaks of lower velocity will reside longer in the detection region and thus exhibit increased areas. Therefore, peak areas have to be divided by their respective t_m to compensate for this effect. Additionally, slowly migrating peaks will possess increased width and a reduced height due to prolonged longitudinal diffusion [72,96,97]. Based on time-corrected peak areas ($\mathrm{area}_{t_m\,\mathrm{corr}}$), true quantities of conformational populations are directly measurable and can be applied for the calculation of K_U [72]:

$$K_U = \frac{\mathrm{area}_{U,t_m\,\mathrm{corr}}}{\mathrm{area}_{F,t_m\,\mathrm{corr}}} = e^{\frac{-\Delta G_i^\circ}{RT}} \tag{11.7}$$

where ΔG_i° is the change in Gibbs free energy in the course of un-/(re)folding events. In case unfolding is induced thermally, van't Hoff plots for thermal protein denaturation, that is, plotting of $\ln K_U$ versus $1/T$, obey a linear course within the thermal transition region. This allows for a calculation of the related enthalpy change of unfolding ΔH_m at T_m—derived from the slope—and the transition temperature T_m [39,95,98] (Figure 11.4).

T_m can also be calculated by temperature induced changes in $\Delta\mu$ (acquired, e.g., by the correction with a peptide marker). Linearity of mobility would differ in the pre- and

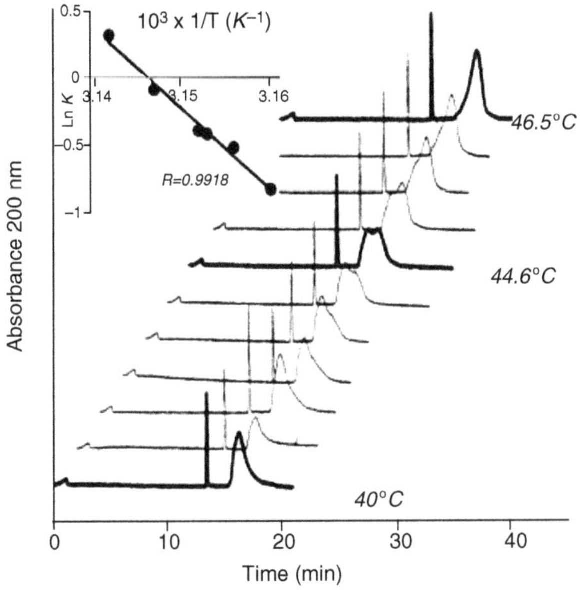

FIGURE 11.4 Thermal unfolding of acetylcholine esterase (AChE) of a snake between 40.0 and 46.5 °C. Separation was performed in 100 mmol L^{-1} phosphate buffer at pH 7.3 with 75 μA constant current and 200 nm. Inset depicts van't Hoff plot (ln K_U against $1/T$) yielding $T_m = 44.6$ °C and $\Delta H = 511$ kJ/mol. T_m can also be derived visually due to equivalent states of N and U. Reproduced with permission from Ref. 94. Copyright Wiley-VCH Verlag GmbH & Co KGaA.

post-transition region. Alternatively, equivalent peak areas for folded and unfolded protein conformations would equally indicate T_m (Figure 11.4) [74]. However, detection has to be performed around 200 nm since absorption at this wavelength is independent of the conformation [75]. Identical t_m-corrected peak areas would therefore indicate equivalent fractions of both conformers. Entropy change (ΔS_m) at the unfolding midpoint T_m can be calculated due to the fact that $\Delta G = 0$ at T_m [39,95,98]:

$$\Delta S_m = \frac{\Delta H_m}{T_m} \tag{11.8}$$

$\Delta G_{(T)}$ for unfolding events, experimentally determined as a function of T, can be fitted to the form of the Gibbs–Helmholtz equation:

$$\Delta G_{(T)} = \Delta H_m(1 - T/T_m) - \Delta C_p[(T - T_m) + T\ln(T/T_m)] \tag{11.9}$$

with ΔC_p representing the change in the heat capacity between native and denatured conformation. Heat capacity change for the unfolding reaction can experimentally be derived by plotting ΔH_m against T_m at different pH values via the resulting slope [39,95] according to

$$\Delta C_p = d(\Delta H)/d(T) \tag{11.10}$$

In a slightly modified approach (compare with Equation 11.7) thermodynamic parameters for conformational shifts induced by denaturant reagents are designed:

$$K_U = \frac{\text{area}(c)_{U,t_m\,\text{corr}}}{\text{area}(c)_{F,t_m\,\text{corr}}} = e^{\frac{-\Delta G_i^\circ - m_i.c}{R.T}} \tag{11.11}$$

with c annotating the concentration of the denaturating reagent and m_i giving a temperature-derived factor that is related to the solution free energy of the respective conformation [72]. In general, any number of transitional states is applicable in this model for slow-time folding.

In fast-time regime reversible unfolding processes only one observable is provided, that is, mobility. The observed single peak represents a time-averaged image of coexisting fast-interconverting conformational ensembles and fails to provide direct access to the partition function K_U as it is offered by the slow-time processes. Therefore, CE strategies in fast-time regime have naturally to be confined to two-state transitions. The observed mobility is averaged from mobility contributions of all coexisting folding states. With proceeding unfolding transition the peak will continuously change not only its mobility but also the shape. During the unfolding progress the peak broadens until at a 50% equilibrium between both states the peak width reaches its maximum and narrows again when the equilibrium shifts further to the novel conformation. On the basis of the observed mobility changes, van't Hoff analysis can be performed by plotting mobility against temperature in the transition region to derive thermodynamic unfolding data. However, peak asymmetries due to unfavorable k_3 will falsify calculated thermodynamic data. Instead of relating t_m to the peak apex, t_m where 50% of all conformations have passed the detection window has to be used for mobility calculation. K_U can indirectly be calculated from individual mobilities of the folded and unfolded state, which have to be derived by linear extrapolation [72].

11.8.4 Dipeptides as a Case Study for Isomerization

Newman and McGuffin [93] applied the theoretical plate model to calculate rate constants and equilibrium constants for isomerization of two model dipeptides (Ala–Pro and Phe–Pro) based on plate heights and velocity, respectively. Zone profiles were generated by Monte Carlo simulations varying the rate constant and the effective capillary length. Separations were performed at two time regimes: (i) fast time analysis allowed for a distinction between individual isomers and their respective mobilities that were utilized to calculate equilibrium constants. (ii) At extended analysis times, where steady state has been reached, both variants have merged and migrated as a single zone. This approach allowed for the calculation of rate constants. Changes in the molar enthalpy (ΔH) and molar entropy (ΔS) as well as the forward and reverse rate constants (k_f, k_r) of isomerization were calculated by these approaches. Moreover, via the Arrhenius equation the molar activation energy (E_A) for either *cis*- or *trans*-isomerization became accessible by measuring k_f and k_r at different temperatures:

$$\ln k = \ln A - \frac{\Delta E_A}{RT} \tag{11.12}$$

By plotting $\ln k_f$ (or $\ln k_r$) against $1/T$ the slope results in respective E_A whereas the factor A in Equation 11.12, which is assumed to be temperature independent, is derived from the intercept [93].

11.8.5 Denaturation Factors and Strategies Applied in CE

Denaturant factors tested in CE so far comprise sample composition, chaotropic reagents, pH, organic solvents, separation temperature, ionic strength, electrical field or current, detergents, and ligands [73] (Figure 11.5). Notably, changing individual separation parameters, in this particular context denaturing agents, will concomitantly affect a much higher number of interrelated separation parameters than in HPLC, with their interconnection frequently not evident at first glance. Temperature increases in the course of thermal denaturation would affect viscosity, pH, and buffer capacity of the applied separation electrolyte, as well as pK_a-values of amino acid side chains and the electroosmotic flow (EOF). Therefore, changes in protein mobility due to these factors have to be corrected, for example, by appropriate marker compounds, to reveal conformational changes [73]. Beside, accurate temperature adjustment and control are mandatory [75]. If denaturation by urea or organic solvent addition is addressed, changes in pH, slight shifts in ionic strength, and—most pronounced—in viscosity become relevant (Figure 11.6) [37,73,99] as shown for the urea denaturation of human serum albumin (HSA) [33,100]. In either case, measures of corrections for these shifts are required for deriving accurate thermodynamic data [73].

Unfolding/refolding events can be performed by equilibrium or dynamic strategies [101]. In equilibrium approaches proteins are unfolded by incubation prior to their injection with the denaturant agent present in equivalent concentrations in the injection solution as well as in the separation buffer. Protein unfolding/(re)folding by dynamic concepts occurs during the CE separation process. In this case, denaturants are only applied to the sample or the buffer [101]. General aspects of protein folding/unfolding in CE have been reviewed by several authors [72,94,102].

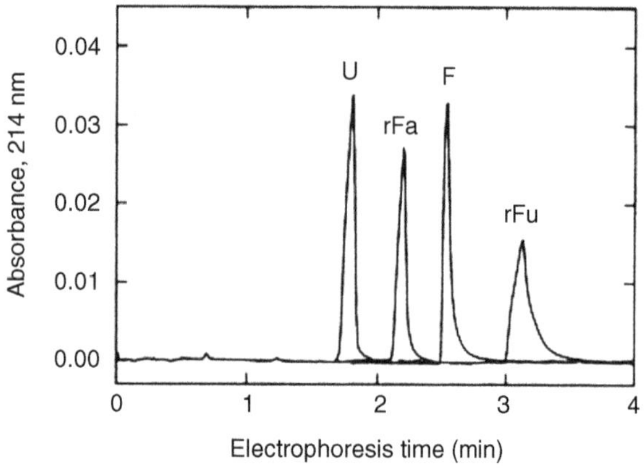

FIGURE 11.5 Superposition of individual electropherograms recorded for horse heart cytochrome c in 40 mmol L^{-1} isoelectric aspartic acid buffer (Asp). Peaks represent various conformations gained under different unfolding/(re)folding conditions. Annotations: (F) peak of folded protein variant gained at 15 °C; (U) unfolded conformation induced by capillary thermostatization to 65 °C. rFa and rFu depict peaks induced by dynamic refolding events after injecting the unfolded protein in the aforementioned separation buffer at 25 °C. (rFa) refolded conformation after acid denaturation in 10 mmol L^{-1} HCl; (rFu) refolded state gained subsequent to unfolding in 20 mmol L^{-1} Asp and 6 mol L^{-1} urea. Reprinted from Ref. 101, with permission from Elsevier.

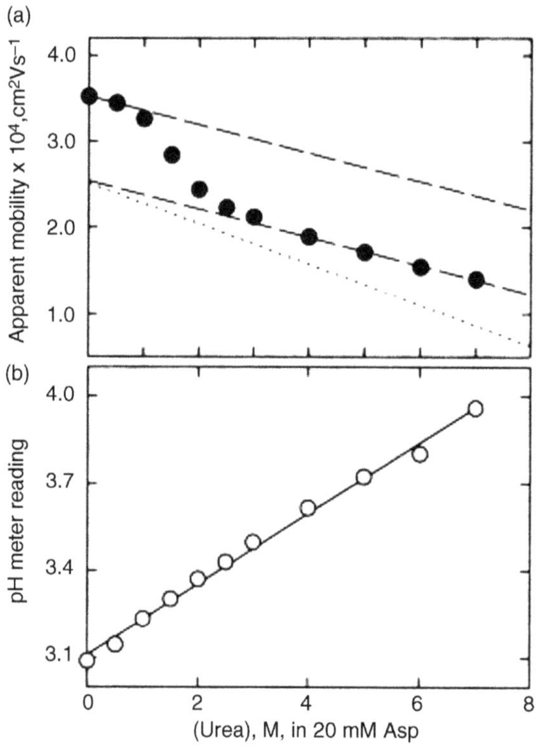

FIGURE 11.6 Effect of variable urea concentrations on the unfolding of cytochrome c and the pH. Measurements were done in 20 mmol L^{-1} Asp. (a) Changes in the apparent protein mobility under the influence of added urea with upper and lower dashed line giving folded and unfolded state, respectively. Dotted line considers the effect of viscosity changes by urea addition for U. (b) Changes in pH induced by urea addition. Reprinted from Ref. 101, with permission from Elsevier.

11.8.5.1 Separation Electrolyte, Injection Solution, and Sample Storage The selected separation buffer exerts an essential influence on (re)folding/unfolding processes in CE, since it contributes to the stability of proteins and the onset of conformational changes. Phosphate buffers and phosphate-buffered saline (PBS) were shown to stabilize the native protein conformation in dissolved state and to reduce protein self-association [103,104]. In addition, phosphate provides low ionization heat, that is, minor changes in pH with temperature increase, and high UV transparency at wavelengths around 200 m, which are favorable for protein detection [39,75,105]. Amphoteric isoelectric buffers, as used by Stellwagen et al. [101], reduced ionic strength and Joule heating.

When freeze-dried samples of placental alkaline phosphatase (PLAP), a dimeric enzyme of 58 kDa, were stored at room temperature, the enzymatic activity was merely slightly mitigated, but clear changes in separation profiles became evident, most likely due to aggregation [34]. In the case of β_2m, a reduction of ionic strength in the sample solution shifted the equilibrium from native conformation toward folding intermediates. However, this was related to the field-amplified stacking effect and the applied fast current ramp time, which increased the temperature in the sample up to 50 °C and promoted unfolding of two β_2m variants that both had a lower thermal stability than the WT. The temperature gradient mostly occurred in the initial centimeters of the capillary before entering the cartridge [28].

11.8.5.2 Denaturation by Urea, Dithiothreitol, and GdmCl Induction of protein unfolding by urea or GdmCl is one of the standard methods in revealing mechanisms and kinetics of protein (un)folding reactions and determination of structural stability. The critical urea concentration for triggering complete unfolding is protein specific. However, onset of unfolding sometimes occurred at lower urea concentrations than theoretically predicted, which is probably related to additional changes in the protein tertiary structure [33]. The observed linear correlation between $\Delta G_U{}^\circ$ and the denaturant concentration $[D]$ is expressed by m_{eq}

$$m_{eq} = \frac{\Delta G_U{}^\circ}{[D]} \tag{11.13}$$

which reflects the strengths of the denaturant. Similar interrelations have been revealed for free activation energies for folding and unfolding with related kinetic m-values [106]. Urea and GdmCl act similarly by (i) a weak binding to the polypeptide backbone and thus an increase in solubility of most protein parts and by (ii) an hyperbolic increase in the solvent viscosity [13,107]. Denaturant interaction with proteins mitigates intramolecular interactions (H-bonds and van der Waals forces) leading to an extended polypeptide chain of lower rigidity. This is in contradiction with the linear decrease in intra-chain diffusion with increasing denaturant concentration. However, the reduced chain contact formation is due to the increased viscosity and the increased end-to-end distance of the stretched polypeptide [13]. Some authors have also proposed urea binding to carbonyl groups disrupting α-helices [33]. The different effects of GdmCl and urea on the reduction of the rate constants of polypeptide chain end-to-end-diffusion are due to their different binding affinities [107]. The number of binding sites is equal for urea and GdmCl, but they differ in their affinity for denaturants. During unfolding progressively more binding sites are exposed, although individual affinities are assumed to remain identical. The residence time of both denaturants on the polypeptide chain is only a few nanoseconds [107]. As for proteins, chaotropic reagents also disrupt water structures, thus increasing the hydrophobicity of water [33]. A thermodynamic discussion of encountered effects is given elsewhere [106].

Kilár and Hjerten [108] monitored unfolding of different glyco- and metalloforms of human transferrin (Tf) in the presence of urea. A series of different iron-free glycoforms, that is, 2-sialo to 6-sialo Tf, were distinguished, with the occurrence of novel peaks in the 3–6 mol L^{-1} urea domain. This was due to the coexistence of each glycoform in two further conformations, namely intermediate and unfolded state. Apparently, Tf transition from folded, over intermediate to unfolded state occurred within this urea concentration region and different states coexisted clearly resolved due to their low unfolding kinetics. A selective unfolding of the C-terminal Tf-lobe has been proposed in this context. Iron-saturated metallo-forms were shown to be more resistant against urea unfolding than apo-forms due to their initially more compact structure and no MGs were detected. Encountered differences in t_m were due to changes in net charge density [108]. Urea-induced conformational changes from folded to unfolded states were also induced in bovine serum albumin (BSA) and reduced the accessibility of negative charges exposed on the surface [109]. Conformational transition of cytochrome c in isoelectric aspartic acid (Asp) buffer (Figure 11.5) was tracked by a sigmoidal change in the mobility with a reduction in the mobility for the unfolded variant due to a larger friction (Figure 11.6a). Equal conformational populations were achieved at 1.5 mol L^{-1} urea [101]. In all described cases an equilibrium unfolding approach was selected, with identical urea concentrations in sample and separation buffer [101,108,109].

Urea unfolding of muscle acylphosphatase in isoelectric glutamic acid buffer (pH 3.2) including 1 mmol L^{-1} tetraethylene pentamine (TEPA) revealed a rate constant for unfolding of 3.0 ± 0.06 s and a concentration of 4.56 mol L^{-1} urea for midpoint transition. Both values were in excellent agreement with independent spectrofluorometric studies. Mobilities were corrected for viscosity and pH increments due to urea addition [110]. In case of human IgG, which was assumed to comprise a mixture of IgG 1–4, changes in peak shape and relative intensity occurred, but not in terms of peak number when urea was increased (up to 8 mol L^{-1}) in the sample and the buffer as well. An investigation of the samples by dynamic light scattering proved the presence of IgG monomers, dimers, and larger aggregates in the absence of urea, the latter two accounting for less than 1% of total protein content. With 8 mol L^{-1} urea the size of monomers increased from 10.9 to 13.9 nm and that of smaller aggregates from 33.2 to 54.5 nm due to unfolding events in either case. In parallel, IgG large aggregates disappeared, since they were broken apart. The size of (dimer) aggregates remained relatively constant with increasing urea due to their formation from denatured monomers. The aggregate fraction(s) could not unambiguously be assigned to peaks in CE. Complementary aqueous two-phase partitioning revealed a change in the IgG charge at 4 mol L^{-1} urea [33]. Skelsey and Bushey [111] compared unfolding/refolding of β-lactoglobulin B after different exposure times to either 8 mol L^{-1} urea, 0.1 mol L^{-1}dithiothreitol (DTT) or to a combination of both reagents and subsequent dialysis. In case of urea denaturation, mobility was decreased due to an increase in r_H. Unfolding by a disruption of disulfide bonds via DDT increased the negative mobility, since the pI was lowered from 5.38 to 5.22. The combined urea/DTT application resembled the results gained by sole urea addition [111].

The effect of ligand binding on protein stability in the presence of urea was investigated via a recombinant regulatory subunit of cyclic AMP-proteinkinase (RIα). Under discontinuous conditions (i.e., dynamic approach) rapid dissociation of cAMP from the receptor occurred in less than 20 ms during electromigration in ligand-deficient buffer. Under equilibrium conditions with excess of cAMP added to the separation electrolyte deviating peak profiles were encountered. This was accompanied by an increase of C_m (i.e., the urea concentration inducing unfolding in 50% of the protein population) and $\Delta G_U{}^{\circ}$. The onset of unfolding occurred at higher urea concentrations when the ligand (cAMP) was present in the separation buffer. The CZE approach allowed for a distinction between folded cAMP-saturated, folded cAMP-stripped, and unfolded RIα. In addition to the thermodynamic stability, the dissociation constant of the complex was determined by means of an integrated CE approach [37].

11.8.5.3 Effects of pH and Organic Solvents When PLAP was exposed to extreme acidic (pH 1–4) or alkaline (pH 12) conditions, the major peak of native PLAP disappeared, either with no novel peaks occurring in parallel, which indicates precipitation, or with emergence of a novel peak at reduced t_m due to oligomerization, respectively [34]. β_2m variants (WT and ΔK58-β_2m) possessed specific susceptibility against acidic shifts in terms of their aggregation propensity. An acidification to pH 5 caused both conformations (native and intermediate) of ΔK58-β_2m to disappear nearly completely within 2 h, since the induced intermediates (I_2) aggregated, then precipitated irreversibly, and thus were continuously removed from solution. Oppositely, WT and its intermediate remained unaffected. An aggregation tendency for the β_2m variant was also observed at pH 6.8 and 37 °C, which was attenuated when temperature and protein concentration were reduced. The results were confirmed by parallel turbidity measurements [29]. Bovine SOD was subject of an pronounced peak broadening with two small peaks interconnected by a plateau when separated by CZE in propionic- and acetic acid both at pH 2.3. The observed pattern presumably indicates unfolding of intermediate-time regime. Comparable effects were absent in CZE separations with formic- and orthophosphoric acid at pH < 2.0, where apparently only the intermediate or unfolded conformation is present [81]. An increased peak width for the same protein was also observed in phosphate buffers at pH 2.50 and 20 °C, with considerable reduction in width and increase in peak height when the capillary temperature was increased from 20 to 35 °C [76].

Organic solvents, such as ACN, TFE, and ethanol, evoked conformational changes in β_2m, which became evident from the induction of a novel slow migrating peak (so-called s-peak, which corresponds to I_2) occurring in addition to the native fast peak (so-called f-peak). The area of this novel peak was positively correlated with the percentage of organic solvent included in the sample [112]. Due to evaporation of the solvent over consecutive runs refolding was observed, whereby the reversion to the native peak was also influenced by the separation temperature [112,113]. Reversibility of unfolding depended on the type of the organic solvent, its percentage, and the applied electric field. Lower fields, which are synonymous with increased t_m, promoted the refolding by a removal of denaturant from the protein environment in the course of migration. With 33% TFE an irreversible unfolding occurred independent of t_m (Figure 11.7) [112].

11.8.5.4 Temperature The transition temperature T_m, where half of the investigated protein species exists in unfolded state, is for most proteins not accessible by commercial CE instruments due to their upper temperature limit at ~60 °C. Therefore, either mutant proteins of reduced T_m, modified heating devices, or separation conditions that reduce T_m (e.g., low pH) have to be selected to access thermal unfolding in CE [39,75,114]. For RNase A, thermal unfolding was investigated at pH 2.3, 2.7, and 3.1 applying a marker peptide to compensate for temperature-related mobility increases. Mobility differences between RNase A and the marker peptide increased during the unfolding transition, which was due to an increased presentation of

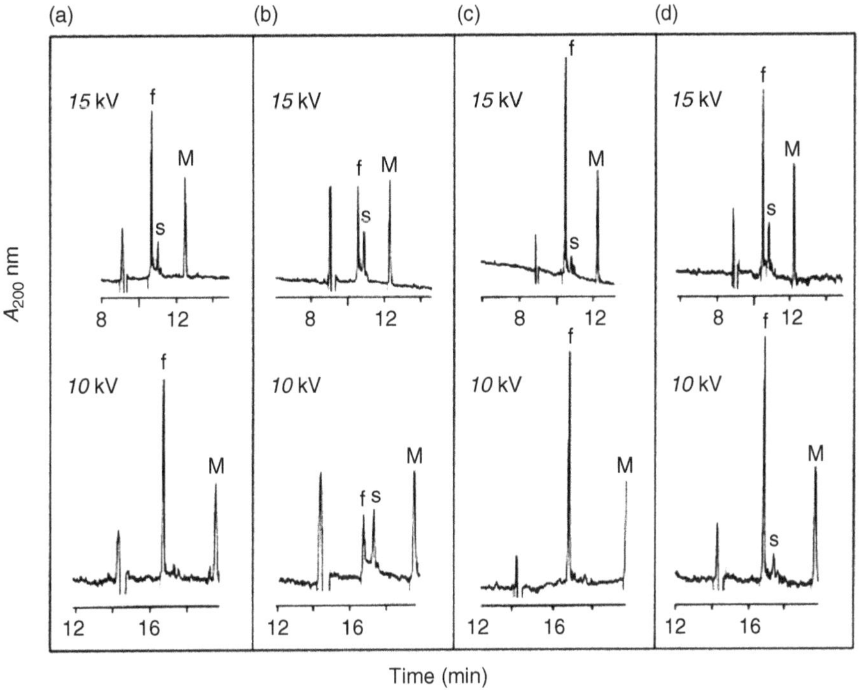

FIGURE 11.7 Investigation of reversibility of β_2m-unfolding in dependence of organic solvent content and analysis time. (a) 13% (v/v) trifluoroethanol (TFE); (b) 33% (v/v) TFE; (c) 13% (v/v) ACN; (d) 33% (v/v) ACN. Analysis times were varied by reduction of separation voltage from 15 to 10 kV. Peaks: (f) native β_2m conformation; (s) β_2m unfolding intermediate; (M) marker peptide. Elevated baseline between (f) and (s) indicates intermediate-time regime. Reprinted from Ref. 112, with permission from Elsevier.

positive charges after unfolding. A pH increase from 2.3 to 3.1 entailed an T_m increase from 35.5 ± 0.3 °C to 48.6 ± 0.7 °C, which was determined by plotting mobility differences against temperature. This resulted in the typical sigmoid-shaped transitions. T_m values determined with CD were 2–3 °C higher. An addition of 0.55 mol L^{-1} sorbitol or sucrose to the separation buffer as well as to the sample increased T_m by 9.4 °C and acted as stabilizers. Polyethylene glycol (PEG) and 2-methyl-2,4-pentanediol (MPD), which are classified as destabilizing excipients, bind specifically to denatured proteins. However, in the current case T_m was not affected, since PEG and MPD were repelled from the charged surface of the native protein [114]. Rochu et al. [39,75] investigated β-lactoglobulin isoforms A and B in terms of their thermal stability. An external thermo-device with a thermocouple microprobe situated in the CE cartridge allowed for a thermostatization of the entire capillary between 20 and 95 °C with high precision and accuracy. Depending on the prevailing temperature, the pH, and the protein concentration the lactoglobulin isoforms exist in different mono- and oligomeric states, with monomers and dimers prevailing at pH 2.0 and between pH 5 and 7, respectively, the latter dissociating at ~40 °C. At pH 2.0 MGs existed, which progressively formed soluble aggregates when temperature was increased to 80 °C. This was related to partial unfolding and the formation of intermolecular H-bonds and hydrophobic interactions, which were disrupted at elevated temperature. The occurrence of an additional peak has been attributed to wall adherence phenomena. Experiments performed around the neutral pH duplicated results. In either case, narrow unfolding transition regions expanding only over 2–4 °C were encountered. In addition, unfolding was accompanied by a blue shift of the absorbance maximum from 278 to 265 nm. ΔH_m, ΔS_m, ΔC_p, and ΔG were calculated from CE data according to aforementioned approaches. Stability curves were established by plotting ΔG against the respective temperature, which allowed for a prediction of the cold denaturation temperature (T'_m) and the temperature of maximum protein stability (T_{max}) [39,75].

Thermal stability of acetylcholine esterase (AChE) from *Bungarus fasciatus* alone or complexed with various inhibitors has been compared over a range of 25–70 °C by CZE and DSC comparing the ligand-free and ligand-bound enzyme. With the addition of inhibitors heat stability of AChE was increased as evident from increased T_m. This is due to a change in conformational plasticity subsequent to ligand binding. The pursued approach allowed also for a ranking of the investigated ligands in terms of their stabilization potential. However, divergent T_m were observed for

DSC and CZE in case ionic ligands were attached reversibly. This was due to the fact that contrary to DSC, ligands were released in CZE under the influence of the electrical field. In case that neutral ligands were attached reversibly or inhibitors were bound covalently, enzyme–inhibitor complexes and conjugates could be analyzed by CZE [38]. Ishihama et al. [115] compared T_m derived from CZE and far-UV—CD spectroscopy measurements for eight proteins. Disturbing Joule heating effects were compensated by dividing mobility by the separation current averaged over the analysis time. For proteins with a valid two-state transition model, that is, lysozyme and cytochrome *c*, thermodynamic parameters of both methods corroborated. Deviations occurred either in case of neglected three-state transitions or if investigated proteins contained impurities that erroneously affected CD—but not CZE measurements [115]. For PLAP up to 65 °C neither loss in activity nor changes in CZE profile were observed. An exposure to 67.5 °C, however, induced already the formation of deactivated products, either by chemical degradation or by conformational changes [34]. Temperature variations between 5 and 27 °C clearly demonstrated a shift from native toward intermediate conformation for two variants of β_2m. These results were corroborated by H/D exchange MS. At temperatures >30 °C native and intermediate state started to merge changing also the interconnecting plateau between both peaks. Temperatures above 40 °C caused irreversible aggregation of both conformers with a concomitant decrease in peaks. Furthermore, computer simulations of separation profiles were given for various temperatures allowing also for a calculation of the activation enthalpies for unfolding and refolding. The results differed from H/D exchange MS data due to diverging experimental conditions and the temperature uncertainty in CE [28].

11.8.5.5 Electrical Field Proteins with anisotropic charge distribution may show conformational susceptibility to an externally applied electrical field. AChE from *Bungarus fasciatus* possesses a rather low T_m around 40 °C. Increasing temperatures led to rapid irreversible aggregation of the enzyme according to the Lumry–Eyring model. The exceptional high dipole moment of ~1000 Debye prompted this protein to field-induced unfolding when increasing the voltage from 12 to 30 kV at 20 °C. Field-induced Joule heating was experimentally excluded as a possible trigger. For AChE, α-helices and β-sheets were shown to be oriented along the dipole axis. Most likely, destructive field-stimulated transitions entailing displacements of charged side chains induced a different orientation of secondary structures and protein stretching. The reduction in mobility was related to an associated electrostatic shielding (i.e., charge suppression). Depending on the selected buffer concentration, the applied voltage, and temperature additional peaks were induced most likely due to capillary wall interactions [74]. Similar effects with apparent transition between two conformations have been encountered for transferrin, when separated in 10–50 mmol L^{-1} $(NH_4)_2HPO_4$. Equilibrium between putative conformations could be shifted by voltage variation between 15 and 20 kV in combination with different temperatures. Cycling back to the initial conditions re-established the original peak pattern, proving the reversibility of the process. The encountered peak pattern indicated an intermediate unfolding regime [81]. A reduction of the separation current can introduce a shift from intermediate to native states as documented for β_2m when the protein was dissolved in water. Apparently, this is an indirect effect and related to the increased t_m, which enhance the contact time with high ionic strength buffer, this way promoting refolding [116].

11.8.5.6 Detergents Conformational relaxation of proteins under the influence of (ionic) detergents in CE is primarily focused on the effect of SDS. This is related to the widespread application of SDS, which is routinely applied in SDS-PAGE and two-dimensional electrophoresis in order to unfold proteins. Due to an approximate constant mass ratio, that is, 1.0–1.4 g between protein and attached SDS [117], negative charges are added proportional to M_r and provide a mass-based separation. However, since identical mass-to-charge ratios are achieved in solutions of high SDS concentrations, separation of protein–SDS complexes requires gels as sieving matrices and was thought to fail in free solution [118]. Furthermore, SDS is also employed as a protein solubilizing agent in CE and biotechnology, for example, to disrupt IBs [35,119,120]. Whereas 1–8 mol L^{-1} of chaotropic reagents, most prominently urea or GdmCl, are frequently required to unfold proteins, SDS concentrations in the low mmol L^{-1} region are only necessary to unfold most proteins [119]. Binding isotherms reveal adhesion of individual SDS molecules onto specific high-energy sites of proteins, initially exerting no unfolding effects but instead even stabilizing the native conformation at low concentration. However, at slightly increased concentrations a minor expansion and increase in r_H occur. Exceeding the so-called critical aggregation concentration induces cooperative SDS binding with gross protein denaturation and statistical arrangement of SDS clusters in pearl-necklace structure along the protein backbone generating random coil structures. When attaining the so-called polymer saturation point (psp), SDS-micelles also occur in the bulk solution [46,121–123].

One of the first descriptions of protein–SDS interactions in free solution was reported by Kenndler and Schmidt-Beiwl [124] when adding SDS only to protein samples running a discontinuous system. However, merely an increase in t_m was encountered, which has been attributed to isotachophoretic (ITP)-like effects [124]. Meanwhile, Kenndler and coworkers [125] have exposed (i) recombinant concatemers of a very low density lipoprotein (VLDL)-receptor which was fused to a maltose-binding protein (MBP) as well as (ii) a viral capsid protein of human rhinovirus to 100 mmol L^{-1} borate with varying concentrations of SDS and dodecyl-PEG-ether and diluted the sample with the respective separation buffer. In the first case (i) mobility and t_m increased until a plateau was gained when both the detergent and the amino acid concentration, which was calculated from the primary sequence, were equimolar. In the presence of SDS two minor peaks at extended t_m were induced in MBP, most likely representing MGs or aggregates. When SDS was added to the sample this effect was even more pronounced and changes in profiles developed over a sample incubation interval of 90 min due to gradual conformational changes and a progressive exposure of SDS binding sites. With SDS concentrations exceeding the critical micelle concentration (cmc) exceptional narrowing of peaks with plate numbers up to 1×10^7 occurred at the expense of R_s, due to a merging of all of protein peaks [125]. Similar effects have also been described by our group, most impressive for plasma amine oxidase [96]. As mentioned earlier, Kremser et al. [125] attributed this phenomenon to a discontinuous electrical field caused by the higher mobility of SDS micelles in comparison to borate ions, creating a lower electrical field in the sample zone than in the buffer. Consequently, capillary isotachophoresis (CITP)-like conditions developed with micelles acting as leading- and borate as terminating ion, entailing massive enrichment of protein–SDS complexes with artificial plate numbers. In the second case (ii) SDS induced a disintegration of viral capsid proteins, but contrary to the fused receptor module proteins, preserved mobility differences even at high SDS concentrations. Protein-specific interactions with SDS have been proposed as an explanation for this divergent behavior [125]. Sasa and Takeda [126] described conformational changes of BSA in the presence of SDS revealing four distinct mobilities in the course of increased SDS concentrations, which were related to distinct BSA–SDS complexes. Tadey and Purdy [120] applied detergents, namely, SDS and sodium deoxycholate, to improve solubility and R_s of hydrophobic plasma apolipoproteins in uncoated capillaries at pH 9. Both detergents increased mobilities of the proteins, changing r_H in an equivalent way, but resulted in different elution orders. In the case SDS concentrations remained below a critical value, broad series of peaks appeared. In the light of nowadays knowledge, peak series most likely represent some folding intermediates with different detergent binding stochiometries. In continuous systems, similar separation results were achieved, indicating a strong SDS–protein interaction [120]. Complexes between SDS and BSA have also been applied as pseudo-stationary phases in electrokinetic capillary chromatography. Depending on the selected BSA–SDS ratio different selectivities in the separation of phenols were induced, most likely due to the different conformations hold by BSA [127]. The application of a discontinuous approach with SDS merely added to the protein containing injection solution but not to the separation buffer induced the formation of novel peaks at increased t_m for selected metal-binding proteins. Evidently, novel peaks represented MGs that were created by cooperative unfolding in the sample solution. However, as different separation profiles were encountered for continuous and discontinuous separation systems, SDS detachment from proteins took place during the migration in SDS deficient buffer, thus triggering (initial) refolding events. Nine different metal-binding proteins were tested for their SDS-induced unfolding susceptibility (Table 11.1). At increased SDS concentrations peak sharpening and reduction of previously existing mobility differences occurred [96], corroborating results of Kenndler's group [124,125]. At the selected pH and ionic strength apparently intermediate refolding kinetics is favoured for myoglobin. Conformational changes derived by changing CZE profiles, which were gained for myoglobin under the influence of SDS, have been confirmed by far UV–CD spectroscopy. These CD results proved a reduction in α-helix as well as an increase in β-sheet and in unordered structures with elevating SDS concentrations (Figure 11.8) [96] which corroborated previous CD results [128]. Currently, it is not clear whether the increased β-sheet portion would also enhance the aggregation propensity. Whereas MGs of myoglobin mostly lacked the prosthetic heme group, the MG cluster were constantly flanked by peaks that possessed a side maximum >400 nm corresponding to holo-myoglobin forms. The side maximum of these holo-forms showed some blue shift of the Soret band due to a relaxation of the heme-pocket. In addition, free heme was detected, proving its release during SDS incubation (Figure 11.8) [96].

Both, plotting the fraction of non-native myoglobin or of free heme against added SDS concentration provided sigmoid transition and allowed for deriving the SDS concentration that induced unfolding in 50% of the initial protein population. Since derived graphs represented combined unfolding/refolding curves, calculation of thermodynamic data was not considered. SDS susceptibility was highly protein specific (Table 11.1). Apparently, the number of primary SDS binding sites, their accessibility, and also the quaternary structure are key factors [96].

The relation between SDS concentration and mobility, shape, and number of novel peaks, respectively, could be related with previously established binding isotherms [96]. Approaches with a sequential-injection CE system automatically mixing myoglobin and SDS has duplicated and confirmed the results of our group. In addition, infrared spectroscopy proved a shift of the amide I bound and also a reduction in α-helix content [129]. Subsequently, a complementary discontinuous approach was tested with SDS included at various concentrations in a Tris–glycine buffer (pH 8.4) but with injected protein samples deficient of

TABLE 11.1 Peak Duplication of Various Metal-Binding Proteins Under the Influence of Various SDS Concentrations in the Injection Solution

MBP	SDS-Free Solvent	Percentage of 2nd Protein Peak in 2 mmol L^{-1} SDS (±95% C.I.)	Percentage of 2nd Protein Peak in 10 mmol L^{-1} SDS (±95% C.I.)
Albumin	H_2O	93.9 ± 0.5	98.5 ± 0.4
Carbonic anhydrase	10% v/v EG[a]	42.6 ± 2.2	92.1 ± 0.2
Conalbumin (=ovotransfer)	10% v/v EG	31.5 ± 1.5	95.4 ± 4.5
Ferritin type I	H_2O	6.9 ± 0.3	21.9 ± 1.3
Hemoglobin	10% v/v EG	77.1 ± 1.6	100[b]
Myoglobin	H_2O	88.5 ± 1.1	97.3 ± 0.5
Plasma amine oxidase	10% v/v EG	79.5 ± 0.6	91.2 ± 1.8
Superoxide dismutase	10% v/v EG	3.7 ± 0.1	9.6 ± 0.9
Transferrin	H_2O	57.0 ± 1.2	96.9 ± 0.4

The 2nd peak was introduced by protein interaction with SDS. Area is given in percent of the total peak area corresponding to the area of all peaks encountered for one protein species. All areas were corrected by migration times. 95% C.I. = 95% confidence interval.

[a]EG Ethylene glycol.

[b]The peak of quasi native hemoglobin was not detected at 10 mmol L^{-1} SDS.

Source: Reproduced with permission from Ref. 96. Copyright Wiley-VCH Verlag GmbH & Co. KGaA.

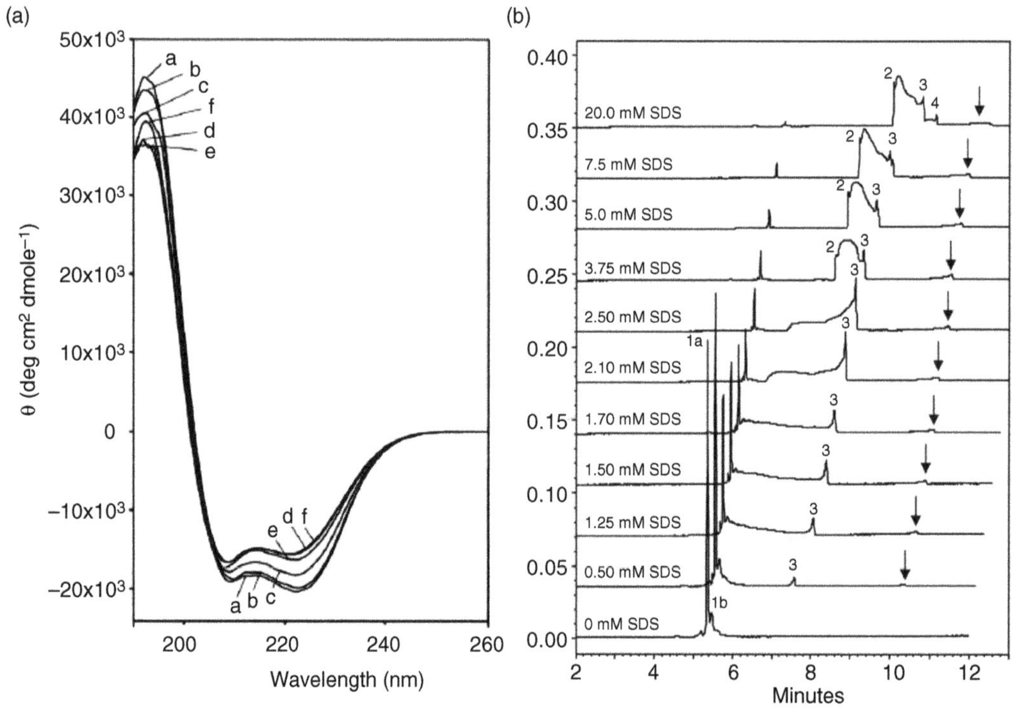

FIGURE 11.8 Induction of myoglobin un-/refolding conformations under the influence of various SDS concentrations. (a) Far-UV–CD spectra of Mb dissolved in various aqueous SDS concentrations. With (*a*) 0, (*b*) 0.10, (*c*) 0.30, (*d*) 0.42, (*e*) 0.50, and (*f*) 1.50 mmol L^{-1} SDS. At 0.50 mmol L^{-1} SDS (referring to 2.50 mmol L^{-1} SDS in CZE due to a fivefold reduction in concentration of both SDS and myoglobin) α-helix content became reduced from 68 to 56% concomitantly inducing β-sheet and increasing unordered structural elements from 23 to 35%. (b) Myoglobin peak pattern in CZE induced by a discontinuous system where SDS was only added to the sample solution. Dilution in the course of migration in detergent-deficient buffer induced refolding events. Peaks: (1*a*) quasi native state with (1*b*) minor myoglobin fraction; (2–4) SDS induced refolding intermediates all containing heme. Plateaus before peak 2 and between peak 2 and 3 miss the prosthetic group and thus also the Soret band. Arrows indicate free heme released during the unfolding process. Reproduced with permission from Ref. 96. Copyright Wiley-VCH Verlag GmbH & Co. KGaA.

the detergent. Encountered mobility shifts were assigned to protein–SDS complexes formed at various stochiometries [130], clearly confirming previously reported protein specific SDS susceptibilities [96,131]. Protein unfolding in the presence of SDS occurred faster than refolding [96,130].

CTLA4Ig, an immunglobulin fusion protein, consisting of the CTLA4 receptor and the constant CH_2/CH_3 domains of human IgG 1 was characterized by MEKC employing 14 mmol L^{-1} borate, 17 mmol L^{-1} SDS, and 3.5% ACN at pH 9.4. Two major species of equivalent mass were postulated as conformational variants. Heat denaturation of the fusion protein was performed between 25 and 85 °C in absence of SDS and with the detergent added. Increasing temperature promoted conversion toward the second peak additionally revealing a contribution of heating time. In the presence of SDS, the onset of conformational transition was considerably shifted to lower T_m. Far UV–CD measurements confirmed the induction of conformational changes at the temperature where changes in the MEKC profile were observed [118]. Andersen et al. [123] investigated myoglobin–SDS interactions by multiple techniques including isothermal titration calorimetry, fluorescence spectroscopy, and pyrene interactions. Several transition points were revealed in enthalpograms with increasing SDS concentrations. This is an indication for a complex unfolding process with multiple unfolding regimes, involving different myoglobin–SDS stochiometries and parallel unfolding pathways. With 0.15 mmol L^{-1} SDS, changes in the coordination of heme and moderate conformational changes occurred. SDS increases to 0.4–1.2, 1.2–2.0, and 2.0 mmol L^{-1}—cmc induced binding of 16, 21, and 61 SDS molecules to myoglobin, respectively. SDS binding was accompanied by changes in absorbance at 410 nm and fluorescence of Trp residues [123]. SDS-induced effects and derived binding stoichiometries with the detergent perfectly corroborate the results and conclusions of our group published previously [96]. A recent publication proposed an SDS-related unfolding pathway for ubiquitin (UBI) by combining CZE, CD, and H/D-exchange data. In CZE, six groups of populated UBI-SDS complexes were distinguished. On the basis of the protein charge ladders, which were obtained by acylation of ε-lysine residues, a "charge ruler" allowed for deriving the number of bound SDS molecules and a correlation with mobility. A pronounced change in the secondary structure occurred at a critical SDS concentration, but contrary to myoglobin with a substantial conversion from β-sheet to α-helix [132].

11.8.5.7 Ligands and Ions—Case Studies on Potential Amyloidogenic β_2m Several authors have described changes—mostly a stabilization—in the conformational stability and in unfolding kinetics for proteins when interacting with ligands. These ligand interactions are of paramount interest, since the desired stabilization effect could prevent or at least mitigate protein unfolding tendencies and consequently also amyloidosis. Thus, appropriate ligands might represent potent therapeutic means in tackling amyloid diseases. De Lorenzi and co-workers [24] have investigated the therapeutic effects of three sulfonated candidate compounds, namely, CR, suramin, and a so-called compound 573, on the stability of native and intermediate conformation (I_2) of β_2m by affinity CE (ACE). Binding affinities were cross checked by surface plasmon resonance. Ligands were added to the separation buffer and their binding to β_2m caused a shift in t_m that could also be used as a measure for interaction. Whereas suramin bound preferentially to the intermediate (I_2), compound 573 showed a stronger interaction with the native structure. In the course of pH unfolding/refolding experiments (by acidic/neutral pH cycling), ~5% of I_2 persisted after refolding. With suramin, I_2 was retained to an even higher extent in the electropherogram, whereas with compound 573 refolding was promoted (retaining only 3% I_2). The effect of CR depended on the applied concentration, for example, with CR concentrations equimolar to the protein the refolding rate was accelerated twofold [24,133]. Heegard et al. [113] tested the intercalation tendency of anionic CR (i.e., congophilicity) in native and partially unfolded human β_2m. Therefore, the dye was added to the separation buffer, clearly proving an interaction with native β_2m by anionic shifts in the mobility of the protein–dye complex as well as by peak broadening and height reduction. Mobility shifts were related to a noninteracting marker peptide. The ACN-induced folding intermediate possessed a considerably augmented affinity for CR with mobility shifts induced at already 10-fold lower CR concentrations where native β_2m was not yet affected. The encountered congophilicity would provide an additional indication for the amyloid formation propensity of β_2m folding intermediates. When $\Delta(1/t_m)$ were plotted against the applied CR concentrations, the dissociation constant K_d could be derived after linear transformation [113]. Okun et al. [134] investigated the thermal denaturation of the capsid of human rhinoviruses. Denaturation occurred within a narrow temperature window of 3–4 °C creating novel peaks. Conversion depended also on the applied incubation time. With the addition of antiviral compounds (e.g., hydrophobic heterocycles) the decay could be retarded 8- to 30-fold, depending on the tested drug. A simultaneous application of both SDS and the various drugs abolished this effect, which is most likely due to a competitive binding of SDS [134].

Similarly, inorganic ions can influence the conformational stability. In this context, Cu^{2+} binding was shown to destabilize the conformation and increase the amyloidogenicity of β-amyloid peptides, but also of immunoglobulin light chain, prion protein, and β_2m (Figure 11.9). A comparison of WT, N-terminal truncated, and the cleaved variant $\Delta K58\text{-}\beta_2m$ revealed their specific structural susceptibility against Cu^{2+}. Both β_2m variants were less stable than the

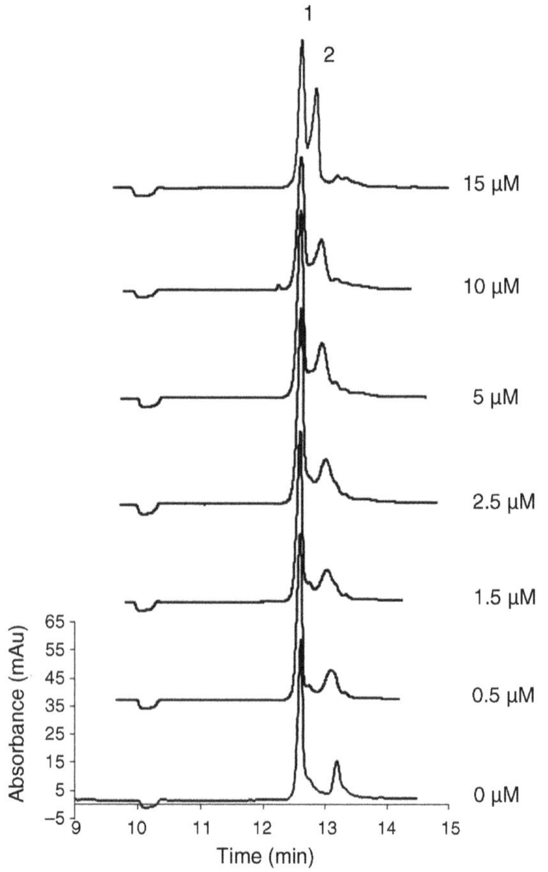

FIGURE 11.9 Selective copper binding of WT β_2m conformations in affinity CE. $CuCl_2$ was added in different concentrations directly to the separation buffer (Cu^{2+} concentrations are given in the figure). Migration order of peaks: (1) WT-β_2m; (2) unfolding intermediate of β_2m. (1) is unaffected by Cu^{2+}. The intermediate (2) encountered a cationic mobility shift accompanied by a transient peak broadening prior to a final peak sharpening due to binding of Cu^{2+}. Reproduced with permission from Ref. 133. Copyright Wiley-VCH Verlag GmbH & Co. KGaA.

WT [27]. Cu^{2+}-induced effects on WT-β_2m resembled the influence of organic solvents both resulting in an increase of the s- and a parallel decrease of the f-peak (compare Section 11.8.5.3). The effect was accompanied by a reduction in total peak area, which indicated a general loss in solubility [27]. These effects were not observed by addition of Zn^{2+}, Ca^{2+}, and Na^+ and seem to be ion specific [27,133]. In parallel, the aggregation tendency of WT-β_2m was increased due to a 4.5-fold decrease in the refolding kinetics of the intermediate resulting in an enhanced stabilization of this conformation at around 15–20% of the WT peak area. ΔK58-β_2m showed *per se* a slower refolding kinetics, which was not affected by Cu^{2+} [27]. Besides WT- and ΔK58-β_2m, truncated variants of β_2m lacking the first three (ΔN3-β_2m) and the first six N-terminal amino acids (ΔN6-β_2m) were investigated together with a point mutant with a His residue exchanged against Tyr at position 31 (His31Tyr-β_2m) [133]. Especially ΔN6-β_2m showed a pronounced amyloidogenic propensity, with 50% precipitation after 15 h. Structural analysis of this variant revealed the conversion of some β-strands in disordered segments with one strand remaining intra-molecularly unpaired and thus acting as a propagator for intermolecular assembly with adjacent molecules [116]. WT-β_2m, ΔN3-β_2m, and His31Tyr-β_2m all showed the aforementioned I_2 intermediate in CZE, whereas only a single peak was detected in the case of ΔN6-β_2m, since this variant formed a trimer of I_2-like conformations, whereas the native form was absent (Figure 11.10) [133].

GdmCl denaturation in combination with far UV–CD revealed a distinct hierarchy for the tested variants in their resistance against unfolding: ΔN6-β_2m < ΔN3-β_2m < WT-β_2m < His31Tyr-β_2m, which was also evident from the relative height of the intermediate peak encountered for the tested β_2m variants, respectively, under identical CZE conditions (Figure 11.10). Contrary to WT-β_2m, a promoted unfolding in the presence of Cu^{2+}, which would entail enhanced fibril formation, was not observed for His31-Tyr-β_2m. This stresses the importance of His31 for Cu^{2+} binding. In addition, refolding from intermediate to native state occurred 10-times faster for this mutant than for WT-β_2m [133]. The results have been surveyed in a further publication including additional β_2m variants with point mutations at position 3 and 31 [116].

11.8.6 β-Amyloid Peptides

β-Amyloid peptides (Aβ), particularly peptides 1–40 and 1–42 originating from proteolytic cleavage of amyloid precursor protein, play a crucial role in the progress of Alzheimer's disease. Subsequent to their dissolution both peptides were freeze-dried, reconstituted in phosphate buffer (pH 7.4), and centrifuged with the supernatant either injected directly into CE or immediately after filtration by means of Microcon filters with different M_r cut-offs. CZE separations of Aβ 1–42 in 80 mmol L^{-1} phosphate pH 7.4 allowed for the distinction between several fast-migrating peaks representing mono- to undecamers with $M_r < 50{,}000$ and a low cytotoxicity, and a slow-migrating broad peak comprising dodecamers and bigger aggregates ($M_r > 50{,}000$) of high cell toxicity. This fraction increased over 200 min at the expense of the smaller oligomers attaining a constant ratio after 6 days with peaks vanishing afterwards most likely due to precipitation via fibril formation. In case of Aβ 1–40 one major peak (mass between 10 and 30 kDa) was detected, which continuously disappeared indicating an aggregation process distinctive from Aβ 1–42. Operation voltage had no disruptive effect on the oligomers. Therefore, CE was applicable in the detection of intact aggregates, which serve as nucleation seeds as well and thus might be of diagnostic relevance [135].

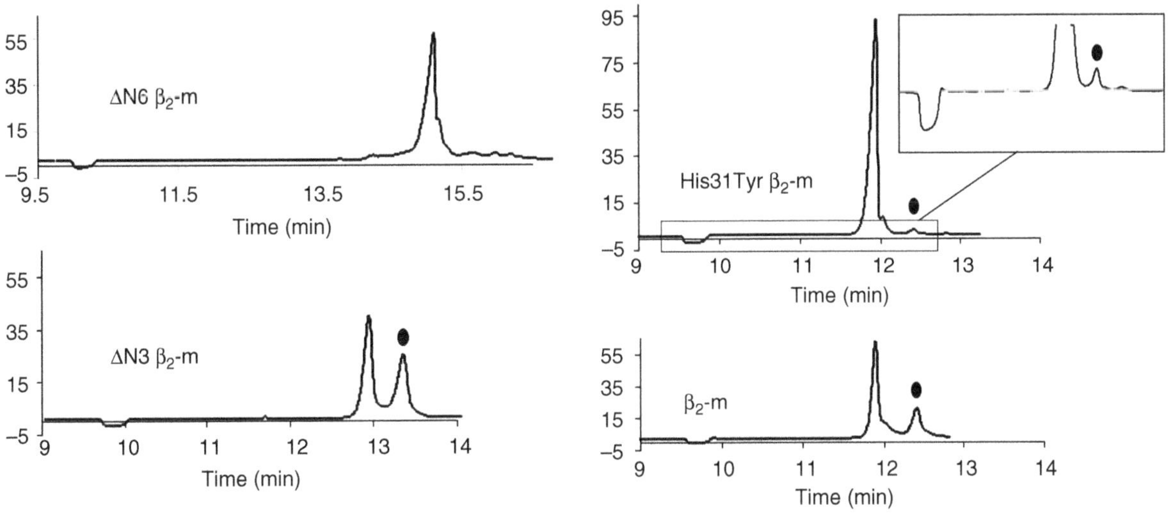

FIGURE 11.10 CZE analysis of various β_2m variants. Relative stability of the tested variants can be derived from the relative height of the respective unfolding intermediate indicated by the filled circle (•). In case of the ΔN6 β_2m mutant the single peak represents a trimer. Reproduced with permission from Ref. 133. Copyright Wiley-VCH Verlag GmbH & Co. KGaA.

11.8.6.1 Prions CIEF has been used to profile prion proteins extracted from healthy (PrP^C) and Scrapie infected sheep and hamster brains (PrP^{Sc}). Proteins were solubilized in 1% *n*-octylglucoside before submitted to CIEF in order to prevent precipitation. Samples of normal brains showed minor signals between p*I* 5.90 and p*I* 9.03. In infected brains, both hamster and sheep showed increased protein content subsequent to extraction with prominent peaks in the acidic region of p*I* ~ 4 and p*I* ~ 3, respectively [136]. SDS–CGE also revealed differences for healthy controls and Scrapy infected sheep brains. A peak representing PrP^{Sc} with ~19.2 kDa was only detected in prominent amounts in infected brain material, whereas in control groups this peak was hardly detectable or absent at all [52]. In both presented cases a differential proof of different conformations was performed via the absence PrP^C. This is mainly related to the different susceptibility of both conformations against proteinase K digest and heat application during the sample preparation [52,136].

11.9 COMPARISON BETWEEN CE AND HPLC

Direct comparisons of CE and HPLC in terms of their selectivity for protein conformations have scarcely been reported. Heegard et al. [112] compared the performance of CE, HPLC, and HPLC–MS in separating variants of β_2m. Whereas a less hydrophobic Met99-oxidized β_2m variant could be separated from its nonoxidized parent compound by RP–HPLC on a C18 column, CE failed in this context. However, in case the respective HPLC-fractions of this protein were reinjected into CE, each peak could be further separated in a fast-migrating native and a slower-migrating unfolded conformation. Thus, distinction was impossible in LC and LC–MS (Figure 11.11) [112].

11.10 CONCLUSIVE DISCUSSION AND METHOD EVALUATION

11.10.1 General Aspects

Differentiation between conformational variants of proteins is getting progressively in the focus of research. This would address not only changes in the folding state *per se* but equivalently distinction, identification, and—ideally—quantification of different protein conformations, including also di- and multimers up to amyloid fibrils [4,5,75]. Spectroscopic techniques in stand-alone operation are merely capable of addressing structural changes averaged over the entire protein population, that is, all coexisting conformations. Therefore, separation methods are required to resolve conformations and make them individually addressable. In this context, HPLC and CE with their various modes of separation represent primary candidates in tackling this task. A performance evaluation of both techniques is practical, because of their widespread application, natural rivalry, and their orthogonal separation principles. A direct comparison of the current performance standard for CE and HPLC would, strictly spoken, require the administration of aliquots stemming from the same sample to both separation systems revealing individual strengths and drawbacks in a direct concurrency situation. Since numerous modes of separation are available in CE and HPLC, an even more differentiated approach should envisage an inter-system comparison of

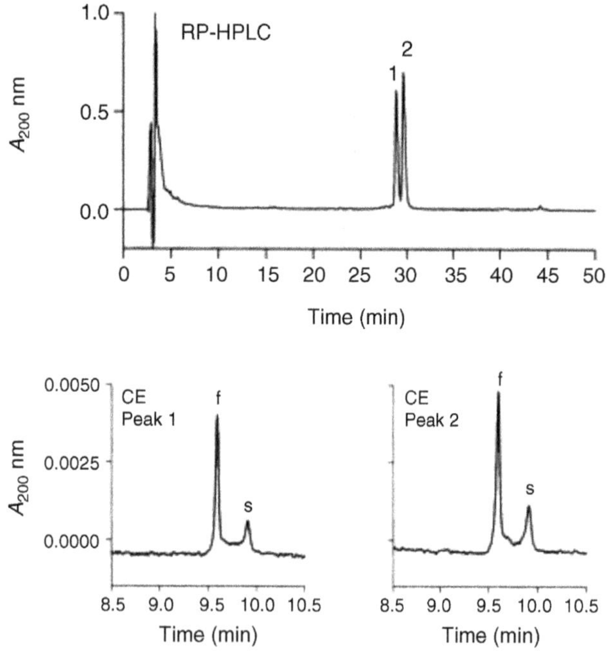

FIGURE 11.11 RP–HPLC separation of β_2m on a C18 column. Peaks 1 and 2 represent β_2m and a related variant oxidized at Met 99, respectively. Both compounds were collected separately and subsequent to an appropriate sample preparation injected in CE. CZE separations revealed both a native conformation (f) and a coexisting unfolding intermediate (s) in either LC-fraction. These conformations could not be distinguished in RP–HPLC. Reprinted from Ref. 112, with permission from Elsevier.

either related modes or separation systems specifically optimized for a particular mixture of conformations. However, since these proposed strategies have scarcely been realized in the context of conformational separations, general features of both CE and HPLC have to be stressed.

11.10.2 HPLC

In HPLC, RP–HLPC [55,56,59] and SEC–HPLC [36,50] are most frequently applied in separating protein conformations. In general, t_r in HPLC are more extended than those in CE and HPLC might additionally require lower separation temperatures to reduce kinetics of interconversion or even preventing it completely [56], probably shifting kinetics in the slow-time regime. However, in RP–HPLC an artificial induction of conformational changes by protein interaction with the stationary phase and/or the organic solvent(s) represents an inherent pitfall [55] and is considerably more likely than in CE. An increase in contact time with the stationary phase, as expected at reduced temperatures, will promote unfolding of adsorbed proteins. Subsequent desorption from stationary phase and induction of refolding might induce a variety of folding intermediates, increasing the propensity of forming conformation-specific aggregates [61,66]. Although protein adsorption onto silica surfaces cannot be excluded *a priori* in CE, attachment to hydrophobic surfaces is always stronger and endangers a higher risk of unfolding [83,84]. Furthermore, the appropriate solvent/-gradient selection in RP–HPLC is of paramount importance (i) due to protein-specific susceptibility to organic solvents and (ii) to promote refolding of desorbed proteins [55]. If both aspects are not under control, they might dismantle or even increase the complexity of conformational protein ensembles originally present in injected samples. Theoretically, the separation principle for conformers in RP–HPLC addresses conformation-specific differences in hydrophobicity as well as in protein volume and/or surface and thus steric effects. Both aspects would influence adsorption/desorption kinetics and thus resolution. SEC–HPLC provides separation conditions promoting the preservation of (quasi) native protein structures by omitting unfolding stressors as in RP–HPLC. However, SEC–HPLC mainly targets separation of (native) monomers, dimers, and oligomers/aggregates, but might fail in addressing monomers with subtle changes in r_H as well as in secondary and tertiary structures [67].

If conformational problems could be solved *pars pro toto* on peptide level, HPLC provides an appropriate application tool, since several investigations have already addressed theoretical aspects of peptide conformations. Most straightforward, reporter peptides can be tackled serving as indicators for mutations that are known to affect conformation on intact protein level [68]. An undeniable strength of HPLC approaches is the numerous detection systems (e.g., CD, fluorescence [66], FT-IR [71], multiple laser light scattering (MALLS) [67]) that can be coupled on-line to HPLC either alone or in combination, providing access to changes in secondary and tertiary structures.

11.10.3 CE

Absence of mass transfer, minute sample consumption, fast migration times, and exceptional high efficiency predestine CE to resolve closely related proteins including conformations. In CZE, which is preferably applied, proteins are resolved based on their different r_H (which is related to protein size, shape, and solvent interaction) and the p*I*-related net charge. Both parameters are highly sensitive even to subtle conformational changes [72,92]. Beside monomeric conformations, dimers, multimers, and aggregates can be separated. Indirectly, also the generation of bigger aggregates and fibrils can be derived in cases of precipitation simply by a continuous vanishing and final disappearance of peaks [34,116,135]. Contrary to other approaches, the calculation of thermodynamic parameters is directly accessible from electropherograms for (i) two-state slow-time and fast-time regime and (ii) three-state slow-time regime of interconversion, even allowing a cross-check by application of complementary methods [72]. The transition temperature T_m during thermal unfolding can, for instance, be determined by changes in t_m-corrected peak areas. Alternatively, changes in mobility with subsequent application of van't Hoff plots or the adjustment of the temperature to achieve equal peak areas for native (N) and unfolded state (U) can be utilized to derive T_m. However, approaches are error afflicted in case mobility changes in the transition region are steep or if poor resolution between N and U is gained, respectively. Moreover, intermediate or unfolded conformations might possess an increased tendency for capillary wall adhesion. This can result in aberrant peak profiles and an erroneous misinterpretation of the underlying (un)folding time regime [74,94,95,98]. Protein adsorption as well as contained impurities might also result in mobility shifts, peak tailing, or even peak duplication. Protein wall adhesion is an inherent problem in CE separation of proteins causing limitations in achieving repeatable t_m [76,84]. Furthermore, the complex interrelation of separation parameters, as exemplified in previous sections, frequently prevents single factor testing. This is due to a concomitant influence of a variety of other separation parameters that have to be identified either to consider them or to take measures of correction. Differentiation between signals related to conformational transition states (MG, U) and dimers/oligomers/aggregates or assignation of profile changes due to (chemical) degradation products, for example, peptides, deamidated, or truncated proteins, merely by CE with UV detection is intriguing. Comprehensive background information about the investigated protein or orthogonal analytical methods are mostly required [33,34]. However, due to a general lack of appropriate detection systems only limited information about the nature of conformational changes is accessible in common CE–UV strategies. Until now, complementary spectroscopic methods have only been applied off-line to provide complementary information on changes in secondary and tertiary structures [38,96,114,132]. In general, on-line hyphenation with detectors designed for distinguishing conformations has, contrary to HPLC, rarely been applied. LIF and MS would be commercially available addressing either changes in fluorescence and thus in the accessibility of Trp and Tyr residues [66] or shifts in multiple charged envelopes in ESI–MS, showing, for instance, bi- or trimodal distributions that are principally indicative for coexisting conformational states [137,138]. CE–FT-IR has been tested as a laboratory-made prototype [139]. Particularly, a CE-compatible CD detector, providing a wavelength scanning option fast enough to cope with narrow CE peaks, would represent a milestone in further establishing CE as a conformation-resolving technique. This would allow not only for resolution and quantification of coexisting individual conformations, but also for addressing changes in secondary and tertiary structures of individual structural variants.

REFERENCES

1. Jaenicke, R. (1991) *Biochemistry 30*(13), 3147–3161.
2. Carell, R.W. and Lomas, D.A. (1997) *Lancet 350*(9071), 134–138.
3. Dadarlat, V.M. and Post, C.B. (2003) *Proc. Natl. Acad. Sci. U.S.A. 100*(25), 14778–14783.
4. Dobson, C.M. (2003) *Nature 426*(6968), 884–890.
5. Dobson, C.M. (2004) *Semin. Cell Dev. Biol. 15*(1), 3–16.
6. Gething, M.-J. and Sambrook, J. (1992) *Nature 355*(6355), 33–45.
7. Dinner, A.R., Sali, A., Smith, L.J., Dobson, C.M., and Karplus, M. (2000) *Trends Biochem. Sci. 25*(7), 331–339.
8. Dobson, C.M. and Karplus, M. (1999) *Curr. Opin. Struct. Biol. 9*(1), 92–101.
9. Imperiali, B. and Ottesen, J. (1999) *J. Pept. Res. 54*(3), 177–184.
10. Herczenik, E. and Gebbink, M.F.B.G. (2008) *FASEB J. 22*(7), 2115–2133.
11. Gregersen, N., Bolund, L., and Bross, P. (2005) *Mol. Biotechnol. 31*(2), 141–150.
12. Dokholyan, N.V. (2006) *Curr. Opin. Struct. Biol. 16*(1), 79–85.
13. Krieger, F., Fierz, B., Bieri, O., Drewello, M., and Kiefhaber, T. (2003) *J. Mol. Biol. 332*(1), 265–274.
14. McCoy, L.F., Rowe, E.S., and Wong, K.-P. (1980) *Biochemistry 19*(21), 4738–4743.
15. Hinderaker, M.P. and Raines, R.T. (2003) *Protein Sci. 12*(6), 1188–1194.
16. Day, R., Bennion, B.J., Ham, S., and Daggett, V. (2002) *J. Mol. Biol. 322*(1), 189–203.
17. Tanaka, K., Suzuki, T., Hattori, N., and Mizuno, Y. (2004) *Biochim. Biophys. Acta (BBA)—Mol. Cell Res. 1695*(1–3), 235–247.
18. Sörgjerd, K., Ghafouri, B., Jonsson, B.-H., Kelly, J.W., Blond, S.Y., and Hammarström, P. (2006) *J. Mol. Biol. 356*(2), 469–482.
19. Saupe, S.J. and Supattapone, S. (2006) *EMBO Rep. 7*(3), 254–258.
20. Khare, S.D., Wilcox, K.C., Gong, P., and Dokholyan, N.V. (2005) *Proteins Struct. Funct. Bioinf. 61*(3), 617–632.
21. Esposito, G., Corazza, A., Viglino, P., Verdone, G., Pettirossi, F., Fogolari, F., Makek, A., Giorgetti, S., Mangione, P., Stoppini, M., and Bellotti, V. (2005) *Biochim. Biophys. Acta (BBA)—Proteins Proteomics 1753*(1), 76–84.
22. Frost, B. and Diamond, M.I. (2010) *Nat. Rev. Neurosci. 11*(3), 155–159.
23. McLaughlin, R.W., De Stigter, J.K., Sikkink, L.A., Baden, E. M., and Ramirez-Alvarado, M. (2006) *Protein Sci. 15*(7), 1710–1722.
24. Carazzone, C., Colombo, R., Quaglia, M., Mangione, P., Raimondi, S., Giorgetti, S., Caccialanza, G., Bellotti, V., and De Lorenzi, E. (2008) *Electrophoresis 29*(7), 1502–1510.
25. Rodriguez, J.A., Valentine, J.S., Eggers, D.K., Roe, J.A., Tiwari, A., Brown, R.H., and Hayward, L.J. (2002) *J. Biol. Chem. 277*(18), 15932–15937.

26. Prusiner, S.B. (1997) *Science 278*(5336), 245–251.
27. De Lorenzi, E., Colombo, R., Sabella, S., Corlin, D.B., and Heegaard, N.H.H. (2008) *Electrophoresis 29*(8), 1734–1740.
28. Heegaard, N.H.H., Jørgensen, T.J.D., Cheng, L., Schou, C., Nissen, M.H., and Trapp, O. (2006) *Anal. Chem. 78*(11), 3667–3673.
29. Corlin, D.B., Johnsen, C.K., Nissen, M.H., and Heegaard, N.H.H. (2009) *Biochem. Biophys. Res. Commun. 381*(2), 187–191.
30. Perrett, S. and Jones, G.W. (2008) *Curr. Opin. Struct. Biol. 18* (1), 52–59.
31. Dear, D.V., Young, D.S., Kazlauskaite, J., Meersman, F., Oxley, D., Webster, J., Pinheiro, T.J.T., Gill, A.C., Bronstein, I., and Lowe, C.R. (2007) *Biochim. Biophys. Acta (BBA)—Proteins Proteomics 1774*(7), 792–802.
32. Calamai, M., Canale, C., Relini, A., Stefani, M., Chiti, F., and Dobson, C.M. (2005) *J. Mol. Biol. 346*(2), 603–616.
33. Bermudez, O. and Forciniti, D. (2004) *J. Chromatogr. B 807* (1), 17–24.
34. Eriksson, H.J.C., Wijngaard, M., Hinrichs, W.L.J., Frijlink, H.W., Somsen, G.W., and de Jong, G.J. (2003) *J. Pharm. Biomed. Anal. 31*(2), 351–357.
35. Clark, E.D.B. (2001) *Curr. Opin. Biotechnol. 12*(2), 202–207.
36. Gu, Z., Su, Z., and Janson, J.-C. (2001) *J. Chromatogr. A 918* (2), 311–318.
37. Gavina, J.M.A., Das, R., and Britz-McKibbin, P. (2006) *Electrophoresis 27*(21), 4196–4204.
38. Rochu, D., Cléry-Barraud, C., Renault, F., Chevalier, A., Bon, C., and Masson, P. (2006) *Electrophoresis 27*(2), 442–451.
39. Rochu, D., Ducret, G., and Masson, P. (1999) *J. Chromatogr. A 838*(1–2), 157–165.
40. Righetti, P.G. and Verzola, B. (2001) *Electrophoresis 22*(12), 2359–2374.
41. Vermeer, A.W.P. and Norde, W. (2000) *Biophys. J. 78*(1), 394–404.
42. Borén, K., Andersson, P., Larsson, M., and Carlsson, U. (1999) *Biochim. Biophys. Acta (BBA)—Protein Struct. Mol. Enzymol. 1430*(1), 111–118.
43. Bertrand, A., Brito, R.M., Alix, A.J.P., Lancelin, J.M., Carvalho, R.A., Geraldes, C.F.G.C., and Lakhdar-Ghazal, F. (2006) *Spectrochim. Acta A Mol. Biomol. Spectrosc. 65*(3–4), 711–718.
44. Jonasson, P., Kjellsson, A., Sethson, I., and Jonsson, B.-H. (1999) *FEBS Lett. 445*(2–3), 361–365.
45. Dyson, H.J., Wright, P.E., *Nat. Struct. Mol. Biol. 5*(7), 499–503.
46. Santos, S.F., Zanette, D., Fischer, H., and Itri, R. (2003) *J. Colloid Interface Sci. 262*(2), 400–408.
47. Jamin, M. and Baldwin, R.L. (1998) *J. Mol. Biol. 276*(2), 491–504.
48. Engen, J.R. and Smith, D.L. (2001) *Anal. Chem. 73*(9), 256 A–265 A
49. Deng, Y. and Smith, D.L. (1999) *Anal. Biochem. 276*(2), 150–160.
50. Uversky, V.N. and Ptitsyn, O.B. (1996) *J. Mol. Biol. 255*(1), 215–228.
51. Goldenberg, D.P. and Creighton, T.E. (1984) *Anal. Biochem. 138*(1), 1–18.
52. Schmerr, J.M., Jenny, A., and Cutlip, R.C. (1997) *J. Chromatogr. B Biomed. Sci. Appl. 697*(1–2), 223–229.
53. Dullnig, V., Weiss, R., Amon, S., Rizzi, A., and Stutz, H. (2009) *Electrophoresis 30*(13), 2337–2346.
54. Thalhamer, T., Dobias, H., Stepanoska, T., Pröll, M., Stutz, H., Dissertori, O., Lackner, P., Ferreira, F., Wallner, M., Thalhamer, J., and Hartl, A. (2010) *J. Allergy Clin. Immunol. 125*(4), 926–934. e10.
55. Sawicka, K., Sahota, T., Taylor, M.J., and Tanna, S. (2006) *J. Chromatogr. A 1132*(1–2), 117–123.
56. Wu, S.-L., Figueroa, A., and Karger, B.L. (1986) *J. Chromatogr. A 371*, 3–27.
57. Herbold, C.W., Miller, J.H., and Goheen, S.C. (1999) *J. Chromatogr. A 863*(2), 137–146.
58. Lu, X.M., Benedek, K., and Karger, B.L. (1986) *J. Chromatogr. A 359*, 19–29.
59. Khew, S.T. and Tong, Y.W. (2007) *J. Chromatogr. B 858*(1–2), 79–90.
60. Cohen, S.A., Benedek, K.P., Dong, S., Tapuhi, Y., and Karger, B.L. (1984) *Anal. Chem. 56*(2), 217–221.
61. Zhang, L., Lu, D., and Liu, Z. (2009) *J. Chromatogr. A 1216* (12), 2483–2490.
62. Walcher, W., Toll, H., Ingendoh, A., and Huber, C.G. (2004) *J. Chromatogr. A 1053*(1–2), 107–117.
63. Wu, Y.V. and Scheraga, H.A. (1962) *Biochemistry 1*(5), 905–911.
64. Liu, S., Zhao, Q., and Jiang, S. (2003). *Peptides 24*(9), 1303–1313.
65. Geng, X., and Wang, C. (2007) *J. Chromatogr. B 849*(1–2), 69–80.
66. Luykx, D.M.A.M., Goerdayal, S.S., Dingemanse, P.J., Jiskoot, W., and Jongen, P.M.J.M. (2005) *J. Chromatogr. B 821*(1), 45–52.
67. Flett, K.L. and Corredig, M. (2009) *Food Chem. 115*(4), 1479–1485.
68. Cartoni, C., Schininà, M.E., Maras, B., Nonno, R., Vaccari, G., Di Bari, M., Conte, M., De Pascalis, A., Principe, S., Cardone, F., Pocchiari, M., and Agrimi, U. (2007) *J. Chromatogr. B 849*(1–2), 302–306.
69. Blondelle, S.E., Lohner, K., and Aguilar, M.-I. (1999) *Biochim. Biophys. Acta (BBA)—Biomembr. 1462*(1–2), 89–108.
70. Chen, Y., Mant, C.T., and Hodges, R.S. (2003) *J. Chromatogr. A 1010*(1), 45–61.
71. Venkateshwaran, T.G., Stewart, J.T., de Haseth, J.A., and Bartlett, M.G. (1999) *J. Pharm. Biomed. Anal. 19*(5), 709–723.
72. Hilser, V.J. and Freire, E. (1995) *Anal. Biochem. 224*(2), 465–485.
73. Piaggio, M.V., Peirotti, M.B., and Deiber, J.A. (2007) *Electrophoresis 28*(13), 2223–2234.

74. Rochu, D., Pernet, T., Renault, F., Bon, C., and Masson, P. (2001) *J. Chromatogr. A 910*(2), 347–357.
75. Rochu, D., Ducret, G., Ribes, F., Vanin, S., and Masson, P. (1999) *Electrophoresis 20*(7), 1586–1594.
76. Stutz, H., Bordin, G., and Rodriguez, A.R. (2003) *Anal. Chim. Acta 477*(1), 1–19.
77. Fang, N., Zhang, H., Li, J., Li, H.-W., and Yeung, E.S. (2007) *Anal. Chem. 79*(16), 6047–6054.
78. Ermakov, S.V., Zhukov, M.Y., Capelli, L., and Righetti, P.G. (1995) *J. Chromatogr. A 699*(1–2), 297–313.
79. Rodriguez, I. and Li, S.F.Y. (1999) *Anal. Chim. Acta 383*(1–2), 1–26.
80. Gilmanshin, R., Dyer, R.B., and Callender, R.H. (1997) *Protein Sci. 6*(10), 2134–2142.
81. Stutz, H., Bordin, G., and Rodriguez, A.R. (2004) *Electrophoresis 25*(7–8), 1071–1089.
82. Lillard, S.J., Yeung, E.S., Lautamo, R.M.A., and Mao, D.T. (1995) *J. Chromatogr. A 718*(2), 397–404.
83. Doherty, E.A.S., Meagher, R.J., Albarghouthi, M.N., and Barron, A.E. (2003) *Electrophoresis 24*(1–2), 34–54.
84. Stutz, H. (2009) *Electrophoresis 30*(12), 2032–2061.
85. Lucy, C.A., MacDonald, A.M., and Gulcev, M.D. (2008) *J. Chromatogr. A 1184*(1–2), 81–105.
86. Kronsteiner, B., Malissa, H., and Stutz, H. (2007) *Electrophoresis 28*(13), 2241–2251.
87. Nichols, E.R. and Craig, D.B. (2008) *Electrophoresis 29*(20), 4257–4269.
88. Jouyban, A. and Kenndler, E. (2006) *Electrophoresis 27*(5–6), 992–1005.
89. Compton, B.J. and O'Grady, E.A. (1991) *Anal. Chem. 63*(22), 2597–2602.
90. Menon, M.K. and Zydney, A.L. (1998) *Anal. Chem. 70*(8), 1581–1584.
91. Rabiller-Baudry, M. and Chaufer, B. (2003) *J. Chromatogr. B 797*(1–2), 331–345.
92. Kálmán, F., Ma, S., Fox, R.O., and Horváth, C. (1995) *J. Chromatogr. A 705*(1), 135–154.
93. Newman, C.I.D. and McGuffin, V.L. (2006) *Electrophoresis 27*(3), 542–552.
94. Rochu, D. and Masson, P. *Electrophoresis* (2002) *23*(2), 189–202.
95. Rochu, D., Georges, C., Répiton, J., Viguié, N., Saliou, B., Bon, C., and Masson, P. (2001) *Biochim. Biophys. Acta (BBA)—Protein Struct. Mol. Enzymol. 1545*(1–2), 216–226.
96. Stutz, H., Wallner, M., Malissa, H., Bordin, G., and Rodriguez, A.R. (2005) *Electrophoresis 26*(6), 1089–1105.
97. Schmitt-Kopplin, P., Garmash, A.V., Kudryavtsev, A.V., Menzinger, F., Perminova, I.V., Hertkorn, N., Freitag, D., Petrosyan, V.S., and Kettrup, A. (2001) *Electrophoresis 22* (1), 77–87.
98. Hilser, V.J., Worosila, G.D., and Freire, E. (1993) *Anal. Biochem. 208*(1), 125–131.
99. Verzola, B., Perduca, M., Mezô, G., Hudecz, F., and Righetti, P.G. (2003) *Electrophoresis 24*(5), 794–800.
100. Candiano, G., Ghiggeri, G.M., Delfino, G., Queirolo, C., and Vecchio, G. (1986) *J. Chromatogr. B Biomed. Sci. Appl. 374*, 279–288.
101. Stellwagen, E., Gelfi, C., and Righetti, P.G. (1999) *J. Chromatogr. A 838*(1–2), 131–138.
102. Gavina, J.M.A. and Britz-McKibbin, P. (2007) *Curr. Anal. Chem. 3*, 17–31.
103. Kurrat, R., Prenosil, J.E., and Ramsden, J.J. (1997) *J. Colloid Interface Sci 185*(1), 1–8.
104. Docoslis, A., Rusinski, L.A., Giese, R.F., and van Oss, C.J. (2001) *Colloids Surf. B: Biointerfaces 22*(4), 267–283.
105. Wehr, T., Rodriguez-Díaz, R., and Zhu, M. (1999) *Capillary Electrophoresis of Proteins*. Vol. *80*, Marcel Dekker, New York, p. 286.
106. Myers, J.K., Nick Pace, C., and Martin Scholtz, J. (1995) *Protein Sci. 4*(10), 2138–2148.
107. Möglich, A., Krieger, F., and Kiefhaber, T. (2005) *J. Mol. Biol. 345*(1), 153–162.
108. Kilár, F. and Hjertén, S. (1993) *J. Chromatogr. A 638*(2), 269–276.
109. Strege, M.A. and Lagu, A.L. (1994) *Am. Lab.* February, 48C–48D.
110. Verzola, B., Chiti, F., Manao, G., and Righetti, P.G. (2000) *Anal. Biochem. 282*(2), 239–244.
111. Skelsey, K.R. and Bushey, M.M. (1996) *J. Chromatogr. Sci. 34*(2), 85–91.
112. Heegaard, N.H.H., Rovatti, L., Nissen, M.H., and Hamdan, M. (2003) *J. Chromatogr. A 1004*(1–2), 51–59.
113. Heegaard, N.H.H., Sen, J.W., and Nissen, M.H. (2000) *J. Chromatogr. A 894*(1–2), 319–327.
114. McIntosh, K.A., Charman, W.N., and Charman, S.A. (1998) *J. Pharm. Biomed. Anal. 16*(6), 1097–1105.
115. Ishihama, Y., Oda, Y., Asakawa, N., and Iwakura, M. (1997) *Anal. Sci. 13*(6), 931–938.
116. Heegaard, N.H.H. and Lorenzi, E.D. (2005) *Biochim. Biophys. Acta (BBA)—Proteins Proteomics 1753*(1), 131–140.
117. Reynolds, J.A. and Tanford, C. (1970) *Proc. Natl. Acad. Sci. U.S.A. 66*(3), 1002–1007.
118. Greve, K.F., Hughes, D.E., Richberg, P., Kats, M., and Karger, B.L. (1996) *J. Chromatogr. A 723*(2), 273–284.
119. Jones, M.N. (1992) *Chem. Soc. Rev. 21*(2), 127–136.
120. Tadey, T. and Purdy, W.C. (1993) *J. Chromatogr. A 652*(1), 131–138.
121. Kelley, D. and McClements, D.J. (2003) *Food Hydrocoll. 17*(1), 73–85.
122. Turro, N.J., Lei, X.-G., Ananthapadmanabhan, K.P., and Aronson, M. (1995) *Langmuir 11*(7), 2525–2533.
123. Andersen, K.K., Westh, P., and Otzen, D.E. (2007) *Langmuir 24*(2), 399–407.
124. Kenndler, E. and Schmidt-Beiwl, K. (1991) *J. Chromatogr. A 545*(2), 397–402.
125. Kremser, L., Bilek, G., and Kenndler, E. (2007) *Electrophoresis 28*(20), 3684–3690.

126. Sasa, K. and Takeda, K. (1993) *J. Colloid Interface Sci.* *157*(2), 516–517.

127. Gaillon, L., Cozette, S., Lelievre, J., and Gaboriaud, R. (2000) *J. Chromatogr. A* *876*(1–2), 169–182.

128. Takeda, K., Wada, A., Yamamoto, K., Hachiya, K., and Batra, P.P. (1988) *J. Colloid Interface Sci.* *125*(1), 307–313.

129. Kulka, S., Quintas, G., and Lendl, B. (2006) *Analyst* *131*(6), 739–744.

130. Gudiksen, K.L., Gitlin, I., and Whitesides, G.M. (2006) *Proc. Natl. Acad. Sci.* *103*(21), 7968–7972.

131. Steinhardt, J., Scott, J.R., and Birdi, K.S. (1977) *Biochemistry* *16*(4), 718–725.

132. Schneider, G.F., Shaw, B.F., Lee, A., Carillho, E., and Whitesides, G.M. (2008) *J. Am. Chem. Soc.* *130*(51), 17384–17393.

133. Lorenzi, E.D., Grossi, S., Massolini, G., Giorgetti, S., Mangione, P., Andreola, A., Chiti, F., Bellotti, V., and Caccialanza, G. (2002) *Electrophoresis* *23*(6), 918–925.

134. Okun, V.M., Nizet, S., Blaas, D., and Kenndler, E. (2002) *Electrophoresis* *23*(6), 896–902.

135. Sabella, S., Quaglia, M., Lanni, C., Racchi, M., Govoni, S., Caccialanza, G., Calligaro, A., Bellotti, V., and De Lorenzi, E. (2004) *Electrophoresis* *25*(18–19), 3186–3194.

136. Schmerr, M.J., Cutlip, R.C., and Jenny, A. (1998) *J. Chromatogr. A* *802*(1), 135–141.

137. Dobo, A. and Kaltashov, I.A. *Anal. Chem.* (2001) *73*(20), 4763–4773.

138. Mohimen, A., Dobo, A., Hoerner, J.K., and Kaltashov, I.A. (2003) *Anal. Chem.* *75*(16), 4139–4147.

139. Kulka, S., Quintás, G., and Lendl, B. (2006) *Vib. Spectrosc.* *42*(2), 392–396.

12

CAPILLARY ELECTROMIGRATION TECHNIQUES FOR THE ANALYSIS OF DRUGS AND METABOLITES IN BIOLOGICAL MATRICES: A CRITICAL APPRAISAL

Cristiane Masetto de Gaitani, Anderson Rodrigo Moraes de Oliveira, and Pierina Sueli Bonato

12.1 INTRODUCTION

Bioanalytical methods play a predominant role in several areas including drug discovery and development, pharmacokinetic and metabolism studies of prescription medicines, therapeutic drug monitoring, clinical and forensic toxicology, and so on. The discovery and development of new drugs require bioanalytical methods to support all the stages of the process. In the discovery step, generic, simple, and high-throughput methods are required to screen a large number of drug candidates. In preclinical and clinical stages, the methods are more specific and complete validation is required to obtain unequivocal analytical data [1]. Bioanalytical methods for prescription medicines are also required because pharmacokinetic and metabolism studies are continuously carried out to obtain a better understanding of drug action, drug interactions, influence of disease, and so on [2]. Furthermore, therapeutic drug monitoring, based on the measurements of the drug concentration in blood or other matrices, can be used to improve the therapeutic response, to evaluate patient compliance, and to minimize toxicity [3,4]. Reliable qualitative and quantitative bioanalytical methods for toxicological analyses are also the basis of clinical and forensic toxicology. The screening and confirmation of drugs of abuse in body fluids can be used to identify potential users of illegal drugs and to monitor drug addicts following withdrawal therapy [5].

Over the past decades, there were important advancements in the area of separation sciences, allowing obtaining highly reliable methods to support the application studies. Among the separation techniques, high-performance liquid chromatography (HPLC) is the most matured, universal, and robust, and therefore it has been the most used technique for the determination of drugs, metabolites, and other small molecules in complex biological matrices for several decades. Liquid chromatography coupled to tandem mass spectrometry (LC–MS/MS) in particular has played an important role in pharmacokinetic and metabolism studies due to its sensitivity and selectivity. These characteristics allow to obtain high-throughput bioassays, necessary to deal with the great number of samples derived from various stages of new chemical entities discovery as well as from studies carried out with prescription medicines [6].

However, capillary electromigration techniques (capillary electrophoresis (CE) in its different modes of operation, capillary electrochromatography (CEC) and capillary electrophoresis on microchips) can show real advantages over chromatographic methods for the analysis of drugs and metabolites in biological matrices.

Among the advantages of capillary electromigration techniques, the high efficiency obtained could be considered the main one. This feature permits the easy separation of similar compounds, avoiding the interference of metabolites, other coadministered drugs, and endogenous compounds. The separation of enantiomers, isomers with the same physical and chemical properties, could be obtained simply by adding a chiral selector to the running buffer. This high efficiency also contributes to the detectability of the methods by increasing the signal/noise ratio [7].

Capillary Electrophoresis and Microchip Capillary Electrophoresis: Principles, Applications, and Limitations, First Edition.
Edited by Carlos D. García, Karin Y. Chumbimuni-Torres, and Emanuel Carrilho.

CE is a very versatile technique allowing the analysis of ionic, ionizable, neutral, small molecules or macromolecules without any need for special instrumental modification. The desired separation could be obtained simply by changing the electrolyte solution inside the capillary. In addition, method development could be rapidly and automatically performed due to the simplicity of the technique. Due to this high versatility, CE can be used for the analysis of low-molecular-weight drugs, polar metabolites including phase II biotransformation products, biopharmaceuticals, peptides, and so on. Although CEC is not as versatile as CE, it has been used for the analysis of several classes of drugs and metabolites, including neutral ones.

Although CE methods could be developed with nonaqueous electrolyte solutions [8], the use of aqueous solutions is much more frequent. In this case, the good compatibility between electrolyte solution and aqueous samples is normally observed, allowing the direct injection of the sample when detectability difficulties are not a concern. CEC is frequently performed in reversed-phase conditions and aqueous samples could also be directly analyzed. The use of small amounts or no organic solvents has an additional advantage: the low production of organic waste. This feature is still better for microchip [9]. The low use of organic solvents is a very important advantage of capillary electromigration techniques when compared to HPLC, characterized by the use of large amounts of organic solvents.

In routine use, bioanalytical methods are normally developed to be applied to a large number of samples. The short analysis time, frequently found in capillary electromigration techniques, is an important feature of these techniques. CE with microchip is still more attractive due to the better heat dissipation, which allows the use of high electric fields, yielding faster separations. In addition, multiple steps such as sample collection, pretreatment, injection, and separation can be performed on the same chip at any scheduled time and location [7].

Finally, CE is considered a low-cost technique. Even when high-cost reagents are necessary for a specific analysis, such as chiral selectors for enantiomer separations, the consumption of these reagents does not prevent the use of the technique for routine analysis [10]. CEC requires the use of a capillary column and due to this, the cost of this technique is higher. In spite of this, the use of CEC is still advantageous due to the small amount of solvents.

One of the main drawbacks of the capillary electromigration methods for bioanalysis applications is related to the lack of detectability, caused by the low amount of sample injected into the capillary and short pathway of the light in UV–Vis detection systems, the most frequently used detection system. The enhancement of sensitivity could be obtained by using off-line or online sample treatment techniques, online preconcentration techniques, and/or more sensitive detection systems such as laser-induced fluorescence (LIF) [7]. However in some cases, especially when the drug/metabolite is present in the biological matrix at sub ng mL^{-1} levels, these approaches could be not enough to obtain acceptable methods.

Another concern in using electromigration methods for routine applications is the low repeatability of migration times, resolution of the peaks, and areas. However, in these cases, the use of controlled procedures, such as capillary conditioning, is normally enough to obtain the desired repeatability.

In this chapter, strategies to obtain reliable bioanalytical methods for the analysis of drugs and metabolites in biological matrices will be critically discussed, using recent examples and comparison to other analytical techniques.

12.2 STRATEGIES TO OBTAIN RELIABLE CAPILLARY ELECTROMIGRATION METHODS FOR THE BIOANALYSIS OF DRUGS AND METABOLITES

The analysis of drugs and metabolites in biological matrices is a very complicated task due to the minimal amount of analytes available to be extracted, limitations in the sample quantity available for the analysis and the presence of several biological interferences. Therefore, during method development, some method parameters such as detectability (limit of quantification and limit of detection), selectivity, and repeatability should be carefully evaluated and controlled to obtain high-quality methods.

12.2.1 Selectivity and Detectability

These two parameters will be discussed simultaneously because their evaluations are very close during method development. Selectivity is the ability of an analytical method to differentiate and to determine the analyte in the presence of other constituents in the sample and refers directly to the ability of the method to produce a response for a single analyte [11]. Capillary electrophoresis is a very selective technique mainly due to its high efficiency, resulting in enough resolution of the target analyte from the other compounds present in the matrices. In addition, last migrating compounds do not interfere in the analysis because they are usually removed from the capillary during the washing procedure performed before each analysis. CE is able to analyze compounds with different physico-chemical properties from small ions to macromolecules just by changing the running buffer. In HPLC analyses, the versatility of the technique is only possible by changing the column. Although the development of new stationary phases such as those designed for the analysis of basic and hydrophilic analytes can bring additional selectivity to HPLC methods

[12], CE is still better under this aspect. In addition, the conventional uncoated silica capillary employed in CE is much cheaper than the columns used in HPLC.

The evaluation of method selectivity for the analysis of drugs and metabolites usually involves the analyses of blank samples of the appropriated biological matrix, and, according to official guidelines, samples from six different sources must be evaluated to ensure the quality of this parameter [11,13]. Figure 12.1 compares an electropherogram and a chromatogram for drug free human plasma samples obtained by a CE method and by an HPLC method employing the same wavelength for UV-detection and the same sample preparation technique [14]. As it can be seen, the baseline obtained by CE analysis is much cleaner than that obtained by HPLC.

Several strategies can be used to increase the selectivity in capillary electromigration methods designed for drug analysis in biological samples. The increase in selectivity could be obtained by using more selective detectors such as LIF and mass spectrometry (MS) detectors, other separation modes, or a more selective sample preparation procedure. The use of more selective detection systems and procedures more selective for sample preparation may also affect the method detectability and will be discussed further in this chapter.

Drug analysis in biological samples is normally carried out by using capillary zone electrophoresis (CZE) or micellar electrokinetic chromatography (MEKC). CZE could also be performed under nonaqueous conditions (nonaqueous capillary electrophoresis—NACE). Vuorensola et al. compared aqueous CE–MS and NACE–MS for the separation of catecholamines. The authors observed that the electrophoretic mobilities of the catecholamines and the electroosmotic flow decreased from water to 1-propanol and correlated with the ratio of dielectric constant to viscosity of the solvent. In addition, superior separation efficiency (plate number six times higher) was achieved in alcoholic background electrolytes [15]. New approaches such as the use of isoelectric buffers and new additives such as ionic liquids have also been used to increase separation resolution and consequently method selectivity [16].

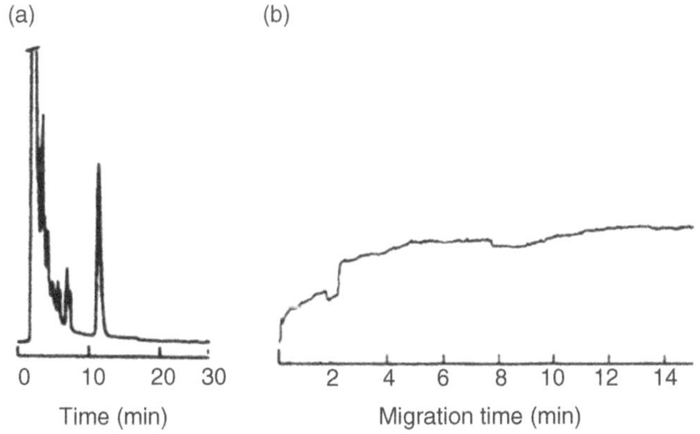

FIGURE 12.1 Blank plasma for the analyses of hydroxymebendazole enantiomers in human plasma after liquid–liquid extraction with diisopropyl ether:ethyl acetate (8:2, v/v) and analysis by (a) HPLC and (b) CE. Wavelength set at 290 nm. Retention and migration times for hydroxymebendazol enantiomers: enantiomer 1 = 22.4 and 10.4 min; enantiomer 2 = 25.9 and 10.7 min for HPLC analysis and CE analysis, respectively. Reproduced and adapted with permission from Ref. 14. Wiley-VCH Verlag GmbH & Co. KGaA.

The selectivity in electromigration techniques can be raised employing capillary electrochromatography (CEC), which combines the advantages of HPLC and CE, that is, stationary-phase versatility and great separation efficiency. Many research studies on CEC have been published in which different types of pharmaceutical compounds are used as model analytes to study column performance. However, in only few papers, CEC is used for the analysis of drugs in body fluids, and attention is given to analytical performance parameters such as the repeatability of retention times and quantitative results [17]. In recent years, monolithic capillary columns have gained great attention for use in CEC. The main benefits of using the monolithic columns are the elimination of problems associated with packed columns, such as difficulty in packing and the bubble formation in the running due to the supporting frits of bed [18]. Molecularly imprinted polymers (MIP) have also been used as stationary phase in CEC, giving to this technique a unique selectivity [19].

Another important application of capillary electromigration methods is in enantioselective analysis. Usually chiral analysis is very complex to be performed in gas chromatography (GC) or HPLC. In HPLC, the high cost of the chiral columns and the high amount of organic solvents used make this technique very expensive. In addition, better chiral separation is accomplished in normal phase or polar organic phase modes characterized by the use of mobile phases constituted of only organic solvents. This fact results in the production of high amounts of residue leading to waste harms. The most important drawbacks faced by chiral GC analysis are the limited number of chiral columns, the limited maximum temperature allowed, and the lack of chiral stability of some enantiomers at high temperatures yielding racemization [20]. Enantioselective analysis is easier to be performed in CE than in HPLC or GC. It is performed just by adding a few micrograms of a chiral selector in the running buffer [21]. Nowadays there are several chiral selectors commercially available. This increases the selectivity of the separation and makes this technique suitable for several kinds of analytes. In addition, CEC can also be employed to separate enantiomers. The chiral separation is frequently accomplished by the addition of a chiral mobile phase additive in connection with achiral supports such as commercially available reversed-phase capillary column. Alternately, CEC could be employed by using a chiral stationary phase [22].

Due to the numerous advantages of microdevices for CE (decrease of sample, electrolyte and reagent consumption, lower analysis time, etc), microchip capillary electrophoresis has been extensively used in the last decade. Many papers have highlighted the possibility to increase resolution (and consequently selectivity) by using these microdevices, since microchip fabrication permits to adapt the chip design according to the requirements. Microdevices are also very attractive for two-dimensional separations [16].

It was mentioned that selectivity and detectability are much related to each other. The cleaner the baseline, the bigger signal/noise (S/N) ratio is, therefore improving the method detectability.

Limit of quantification (LOQ) and limit of detection (LOD) are both used for the evaluation of method detectability. For bioanalytical methods, LOQ is defined as the lowest amount of analyte in a sample that can be quantified with suitable precision (20%) and accuracy (80–120%). It can be determined based on the signal/noise ratio (S/N ratio = 10) and also employing the standard deviation of the response (σ) and the slope of the analytical curve (S), according to the following expression: [11,23]

$$\mathrm{LOQ} = \frac{10\sigma}{S}$$

In contrast to LOQ, the LOD should not be precise neither accurate. It is also determined using the signal/noise ratio, which should be at least 3. As described for the LOQ, LOD can be estimated based on the standard deviation of the response and the slope of the analytical curve, according to the following expression:

$$\mathrm{LOD} = \frac{3.3\sigma}{S}$$

Considering the definitions described above, several factors can influence the detectability of a bioanalytical method, by influencing the baseline signal/noise ratio. Between those factors, (i) efficiency, (ii) sample preparation technique and (iii) detection methods are the most important ones.

12.2.1.1 Efficiency Usually CE with UV detection is considered a technique with limited detectability. This is mainly due to the small diameter of the capillary resulting in short detection pathway (25–100 μm) and also the small injection volume (around nanoliters). However, CE shows higher efficiency for theoretical plate numbers than HPLC [24]. That high efficiency yields thinner peaks improving the S/N ratio and allowing to obtain LOQ or LOD almost similar to those obtained by HPLC. This effect can be explained by the absence of some van Deemter factors that usually are responsible for the peak broadening in HPLC. Equation 12.1 shows the van Deemter factors and its meaning [25].

$$h = A + B/\mu_x + C\mu_x$$

where h, height equivalent to a theoretical plate; A, Eddy diffusion (multipath term); B, longitudinal diffusion; C, mass transfer to and from stationary phase or/and mobile phase; μ_x, flow rate.

In conventional CE, there is no stationary phase; therefore, the factors that contribute to the peak enlargement are minimal; only B factor (diffusion) can contribute to the peak broadening in conventional CE. However, other sources of dispersion, including joule heating, capillary wall binding, injection, detection, and electromigration dispersion, lead to fewer theoretical plates than the simple theory predicts [26].

Wang et al. analyzed primary and tertiary amines by gas chromatography–mass spectrometry (GC–MS), high-performance liquid chromatography–ultraviolet detection (HPLC–UV), and capillary electrophoresis–micellar electrokinetic chromatography (CE–MECK–UV). In this work, the authors compared these techniques by evaluating their plate numbers. The authors noticed that GC–MS gave the best plate numbers ($1.2 \times 10^6 - 1.5 \times 10^6$ for primary amines), followed by CE–MECK–UV ($4.2 \times 10^5 - 4.4 \times 10^5$ for primary amines), and HPLC–UV ($2.5 \times 10^5 - 2.8 \times 10^5$ for primary amines) [27]. Despite the better plate number obtained with GC–MS, this technique is only suitable for volatile or semivolatile compounds. After derivatization, it is possible to analyze nonvolatile compounds; however, this procedure is time-consuming and, frequently, an individual derivatization procedure for each compound in the sample is necessary. The majority of compounds analyzed in biological matrices are nonvolatile compounds. Consequently, CE and HPLC are better suitable for these ones. The lower plate number observed in HPLC when compared to CE could be bypassed by using stationary phases with particle size sub-2 μm (UHPLC, ultra-high-performance liquid chromatography) [28].

The efficiency and also the detectability can be improved to a great extent in CE using some online preconcentration technique. The main online preconcentration techniques in CE are stacking, transient isotachophoresis, dynamic pH junction, and sweeping [29]. Each of these techniques is based on a particular concentration mechanism, which depends on differences in the properties of the sample solution and the running buffer. An interesting application of one of these techniques was achieved in the work of Grard et al. In this work, the authors analyzed a new chiral adrenoreceptor antagonist in plasma using a field-amplified sample injection (FASI–CE–UV) procedure. The detectability was improved 180 times compared with normal hydrodynamic injection (Figure 12.2). Using this technique, it was possible to obtain quantification limits of 2 ng mL^{-1} for each enantiomer [30].

Another interesting preconcentration method was accomplished by Shihabi to analyze mycophenolic acid, an immunosuppressant drug, in human serum. Shihabi compared the mycophenolic acid analyses by HPLC and CE. In this work, the detectability and the efficiency in CE were greatly

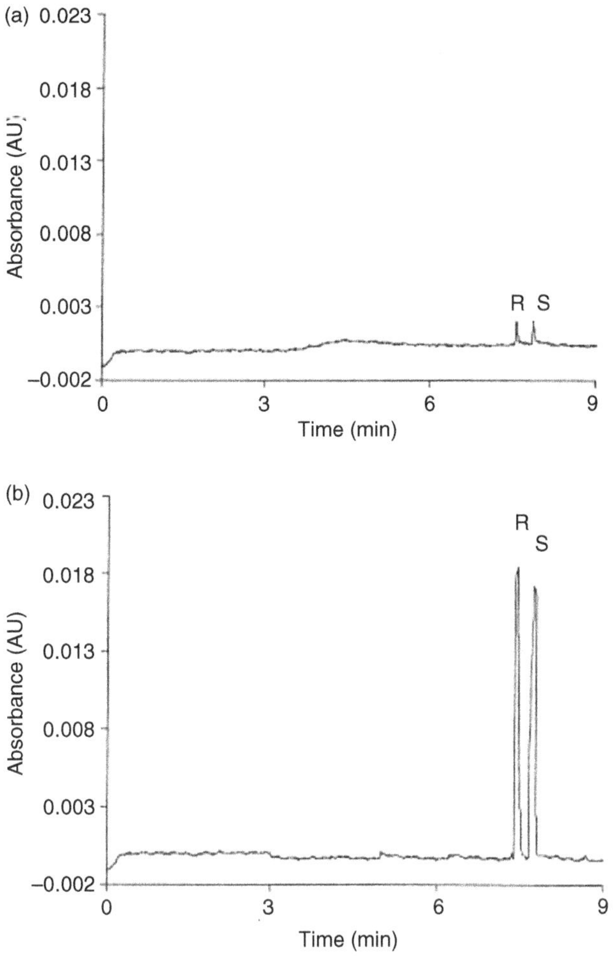

FIGURE 12.2 Comparison of normalized electropherograms of racemic drug obtained by hydrodynamic injection and FASI. Capillary column, 57 cm × 50 μm i.d.; applied voltage, +20 kV; temperature 25 °C; buffer, formic acid-ammonia (pH 4, ionic strength 50 mmol L^{-1}) and 3.5 mmol L^{-1} HP-β-CD; and drug concentration, 1 μg mL^{-1}. Injection, (a) hydrodynamic 5 s, 0.5 psi; (b) FASI, hydrodynamic injection of the preinjection plug (water) for 5 s then electrokinetic injection of sample dissolved in water with +10 kV for 5 s. Reproduced with permission from Ref. 30. Wiley-VCH Verlag GmbH & Co. KGaA.

improved by using two strategies: (i) a transient pseudo ITP by adding acetonitrile in the sample and (ii) a wider capillary (75 μm i.d.). The combination of these two strategies offered a greater sensitivity, which matches that of the HPLC (the minimum detection level was 95 μg L^{-1} for CE versus 125 μg L^{-1} for HPLC) [31].

Another application of an interesting online sample preconcentration procedure was described by Fang et al. In this work the authors described an online centrifuge microextraction back-extraction field-amplified sample injection capillary electrophoresis system (CME–OLBE–FASI–CE) for determining trace ephedrine derivatives in urine and serum using UV–Vis detection. CME and OLBE–FASI were two separate concentration units. The method was validated with detection limits between 0.15 and 0.25 ng mL^{-1} [32].

12.2.1.2 Sample Preparation Sample preparation has been recognized as one of the most important steps of a bioanalytical method, especially when trace analysis is required. The increasing demand for the analysis of very low concentrations of analytes in biological matrices requires the elimination of a high amount of interferences and the reduction of final volumes of extracts to obtain a higher concentration of analytes and then facilitate sensitive detection, decreasing the LOQ or LOD. Since the amount of analyte that reaches the detector in electromigration techniques is very low, the sample preparation employed should be able to recover as high as possible the analyte present in the sample. Classical techniques such as liquid–liquid extraction (LLE) and solid-phase extraction (SPE) have been the main used extraction techniques in bioanalytical capillary electromigration methods [21,33]. These techniques are very used due to their simplicity and high recoveries. More details about sample preparation techniques can be found in references [33,34]. In spite of using the same sample preparation technique, the amount of analytes injected in HPLC is much more due to the higher injection volume. At first, it seems an advantage for HPLC methods, since lower LOD can be obtained easily even if the recovery value is not high. In contrast, the high amount of extracted interferences that reaches the detector can disrupt the analysis, even increasing the LOQ/LOD, or can interfere in the analyte signal detection. Therefore, during a bioanalytical method development, the sample preparation technique has to be carefully chosen.

Nowadays, the environmental pressure demands for more sustainable sample preparation techniques, in which the volume of organic solvents should be reduced. In this context, some microextraction techniques have been developed. Among them, hollow-fiber liquid-phase microextraction (HF-LPME) and solid-phase microextraction (SPME) have stood out.

SPME was introduced by Arthur and Pawliszyn [35] in the early 1990s and became commercially available in 1993. SPME is based on the partitioning of the analytes between a coated fiber and a sample matrix. The coated fiber is made of a small fused-silica rod coated with a thin layer of a sorbing material. SPME can be performed exposing the fiber to a vapor phase (headspace) or by direct immersion of the fiber in the matrix. The extraction is governed by the sorption process and, depending on the type of the fiber, absorption, adsorption, or a mixture of these sorption processes can take place [36]. After a predetermined extraction time, the analytes should be desorbed from the fiber to be analyzed. The desorption process can be performed thermally by exposing the fiber in the gas chromatography injector or by immersing the fiber in an appropriated solvent and further injection of the extract into the chromatography–electrophoretic system. SPME has several advantages over other sample preparation techniques, but the most important ones are the low amount of organic solvent consumed and its simplicity. In many cases, such as when the analysis is performed by GC, the amount of organic solvent consumed is zero. Actually, the need for organic solvent is in the desorption step, when it is performed off-line with an organic solvent.

There is a great interest in the miniaturization of liquid–liquid extractions. The major idea behind this is to make easy the automation, to speed up extractions, and to reduce the consumption of organic solvents. Miniaturized liquid–liquid extraction was first introduced in 1996, and was based on a droplet of organic solvent hanging at the end of a micro syringe needle, called single-drop microextraction (SDME) [37]. After suspending, the droplet is immersed in the aqueous sample and the extraction is performed. After a well-controlled time, the drop is withdrawn into the needle and the extract is injected directly into a GC system or the drop is evaporated to dryness and the residue is dissolved in an appropriated solvent. As expected, this extraction technique is not suitable for systems in which the extraction is performed under agitation because the drops can be lost during the extraction, leading to bias. To improve this device, Pedersen-Bjergaard and Rasmussen developed a liquid-phase microextraction technique in which the organic solvent was protected by a polypropylene hollow fiber (HF–LPME, hollow-fiber liquid phase microextraction) [38]. In this case, the organic solvent was immobilized in the wall pores of the hollow fiber, providing a supported liquid membrane, and an aqueous acceptor solution was held within its lumen. Analytes were extracted into the intermediary organic phase and then subsequently into the aqueous phase. This is thus considered three-phase HF–LPME. Another mode of HF–LPME is based on a two-phase system in which the same organic solvent is used to fill both the wall pores and the hollow fiber lumen. HF–LPME is a technique compatible with the main analysis instruments. If the acceptor phase is an aqueous solution (three-phase mode), it is completely compatible with CE and, in some cases, with MS [39] and HPLC [40]. On the other hand, if the acceptor phase is an organic solvent (two-phase mode), it becomes compatible with GC [41]. There are several possible formats to perform HF–LPME [42]. In the conventional configuration, the needles of two microsyringes are inserted through a septum and the two needles are connected with a piece of a hollow fiber. The microsyringes could also be replaced by two gel-loading round tips [43]. In another format, called as "rodlike," one side of a piece of a hollow fiber is connected to one microsyringe that is used for collection and injection of the acceptor phase whereas the other fiber side is closed [44].

HF–LPME and SPME are both attractive sample preparation techniques; however, the recovery values obtained in SPME are usually very low so its use for sample preparation in capillary electromigration methods is restricted to the analyses of compounds present in high concentrations in the samples ($\mu g\ mL^{-1}$). De Oliveira et al. developed an enantioselective method to analyze hydroxychloroquine (HCQ) and its metabolites in urine by HPLC using SPME for sample preparation. The recovery values ranged from 9.0 to 15%, and the LOQ was $42\ ng\ mL^{-1}$ for the metabolites and $50\ ng\ mL^{-1}$ for HCQ enantiomers [45]. The same authors developed another enantioselective method for the analysis of HCQ and metabolites, but now, employing CE and HF–LPME [46]. Employing this latter microextraction technique, it was possible to obtain recovery values close to 90%, yielding LOQ of $10\ ng\ mL^{-1}$ for HCQ and $21\ ng\ mL^{-1}$ for the metabolites. Despite the low detectability power, the capillary electrophoresis LOQ was lower than that obtained by HPLC due to the use of a more efficient extraction technique.

Usually, the analysis of drugs and metabolites in biological matrices is performed in systems almost completely automated. The handling is often in the sample preparation step. In addition, in the majority of applications, there are several samples to be analyzed. So, high-throughput systems are desired where the manual handling is minimal. Except LLE, the other main sample preparation techniques can be coupled with CE, for example, SPME, SPE, and HF–LPME [47–49]. The first SPME–CE interface was developed by Wang and Pawliszyn [50]. This system is essentially a SPME fiber assembly, an interface made of a Teflon block and a CE system. Another system for coupling SPME and CE was developed by Santos et al [51]. This interface was made of a methacrylate material and was actually a modification of a vertical split-flow interface with an additional channel closed with a septum that permits the introduction of the fiber in the interface [47]. Both cases show that it is possible to couple SPME and CE; however, all extraction steps are still performed off-line that is not so attractive when there are several samples to be sampled. SPE coupled with CE has been shown to be interesting and several modes have been described, for example, off-line, at-line, in-line, and online modes. However, the complete high throughput with minimal manual sample handling is accomplished only with in-line and online modes. In in-line systems, the sorbent material is introduced in the CE capillary; in online systems, the SPE column is separated from the CE capillary and therefore requires an interface that allows the introduction of compatible volumes of desorption solvent in the CE [49]. Finally, HF–LPME can also be coupled with CE using the same modes: at-line, online, and in-line [48]. In at-line mode, the acceptor solution is transferred to a CE vial using a mechanical interface. In online mode, the acceptor solution is driven by a stream to a split-flow interface. In this case, only a fraction of the acceptor solution is introduced into the capillary. In the last mode, in-line mode, there is a direct contact of the acceptor solution with the capillary, that is, the

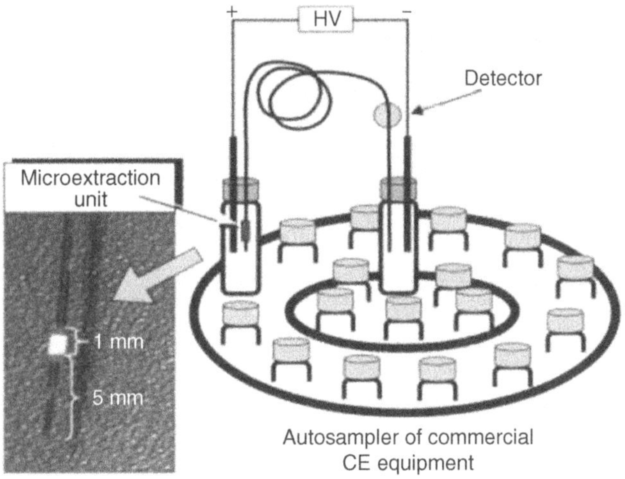

FIGURE 12.3 In-line LPME–CE arrangement for the determination of NSAIDs in urine. Reproduced with permission from Ref. 52. Wiley-VCH Verlag GmbH & Co. KGaA.

capillary inlet is directly inserted into the acceptor solution or the extraction unit and acceptor solution is introduced jointly into the CE capillary [48].

Nozal et al. developed an in-line extraction HF–LPME–CE system (Figure 12.3) for the analysis of nonsteroidal anti-inflammatory drugs in human urine. The developed system permitted analyte extraction, preconcentration, and electrophoretic separation to be automatically performed in commercial CE equipment and resulted in LODs lower than 2 ng mL^{-1} [52].

Another feature of CE is the feasibility of direct biological sample injections. Besides its low use due to mainly the low preconcentration power, it has been done by some researchers. Wei et al. analyzed metformin in human plasma by CE–UV and compared the validation results with a developed method by HPLC–UV for which a LLE procedure was used for sample preparation. In this work, the human plasma (180 μL) was mixed with 20 μL of an aqueous solution containing metformin. After that, an aliquot of 30 μL was transferred to a CE vial and analyzed. The comparison of validation parameters showed no differences between both methods with linearity ranging from 0.1 to 2.0 μg mL^{-1} [53].

12.2.1.3 Detectors Another factor that can influence the method detectability is the detector employed. A variety of detectors are available for CE. However, the most employed one for bioassays is the UV–Vis detector. LIF, electrochemical, and mass spectroscopy (MS) detectors have also been used.

UV–Vis Detector. Various features such as simplicity, commercial availability, versatility, and low cost make this detector very attractive. UV–Vis detector yields LODs in capillary electromigration methods corresponding to a few femtomoles of analyte (at subpicomole levels), that is, the detector is characterized by a high mass sensitivity [54]. There is a misunderstanding that UV–Vis detector is not sensible enough and therefore CE lacks sensitivity. Clohs and McErlane, in a paper reporting a chiral method to analyze carvedilol enantiomers in serum by HPLC–fluorescence and CE–UV, obtained LOD of 50 ng mL^{-1} using CE–UV and 1 ng mL^{-1} by HPLC–FL. Clearly, method detectability by HPLC was better. However, the volume injected in the HPLC system was 10 μL, whereas the injection volume for CE was around 18 nL. The amount of drug at the detector was approximately 0.003 and 0.04 ng of each enantiomer for CE and HPLC, respectively. Thus, it can be seen that the sensitivity at the detector was higher for CE than for HPLC (better mass sensitivity) [55].

Several strategies can be used to increase the absorbance signal to overcome poor concentration sensitivity of UV–Vis–CE, such as: (a) increasing the path length by using a wider capillary or some special detection cell (bubble cells, high-sensitivity cells, and Z-cell); (b) selecting lower UV wavelengths (many organic molecules can be detected at 195–210 nm with high absorption); and (c) indirect absorbance detection for analytes with low values of molar absorbance [56,57]. In addition, the use of some online preconcentration procedure (discussed before) can enable the application of UV–Vis detector in sensitive bioanalysis [58–60].

Laser-Induced Fluorescence (LIF). LIF is the most sensitive detector developed to date. The LODs of the best CE–laser-based systems are well below 10^{-13} mmol L^{-1}, producing mass detection of less than 10 molecules [57]. CE fluorescence detection can be performed by direct on-column detection, by post-column detection, and by indirect fluorescent detection. Analysis of nonfluorescent analytes or with fluorophores that cannot be excited usually requires a chemical derivatization step; the kinetics of derivatization reaction often contributes to detection limitations [61]. In addition, interferences that have the same migration times as those of the target analytes may also be labeled and provide fluorescence properties comparable to those of the analytes. In terms of detectability, the analysis of drugs and metabolites in biological fluids by CE–LIF has been shown to be equivalent to or better than those obtained by HPLC. Musenga et al. developed a method to analyze sertraline and its metabolite *N*-desmethylsertraline in human plasma by CE–LIF. This method showed excellent LOQ with values of 3.0 and 4.0 ng mL^{-1}, respectively [62]. Captopril, an antihypertensive drug, was analyzed by CE–LIF in pharmaceutical preparations and human urine after derivatization with the fluorescent label 5-iodoacetamidofluorescein. The method showed excellent linearity range of 3.5–6000 ng

mL^{-1} [63]. Duloxetine, an antidepressant drug, was analyzed in human plasma by CE–LIF with LOQ of 2.5 ng mL^{-1} [64].

All these data indicate that CE–LIF can be used in cases where the amount to be quantified is very low or cannot be detected by a conventional CE–UV–Vis detection. LIF detection is also an interesting approach for CE in microchips [7]. The bottleneck faced in CE–LIF drug bioanalysis is the time required for derivatization of analytes, when this reaction has to be performed.

Electrochemical Detectors. Electrochemical detection is a good detection alternative for capillary electromigration techniques due to its sensitivity, versatility (detection in universal or selective modes) [10], nondestructivity, and greatly reduced background noise if the electrodes of the detector are not in contact with the running buffer (contactless conductivity detection) [7]. However, the complicated fabrication and handling procedures, the need of accurate spatial alignment of the electrodes within the capillary, the limited sensor lifetime and requirement of recalibration, and the possibility of strong adsorption of the intermediate reaction products of the analyte to the electrode surface are limiting factors for the use of these detectors. The amperometric mode is used for electroactive analytes, with LODs in the range of 10^{-10}–10^{-11} mol L^{-1}. Conductometric mode is the most frequently used electrochemical detection for CE; however, the LODs are in the range of 10^{-6}–10^{-7} mol L^{-1} [10].

Zheng et al. described a highly sensitive method for the analysis of doping substances (methylephedrine (ME), celiprolol, sotalol, and indapamide) in urine. By using amperometric detection and transient isotachophoresis as preconcentration technique, they obtained detection limits in the range of 10^{-13}–10^{-14} mol L^{-1} [65].

Mass Spectrometer (MS) Detectors. MS detector is a common device employed coupled to GC and HPLC and also to CE. The most important applications of CE–MS are in the area of proteome and disease biomarkers. CE–MS has been shown to be useful in drug bioanalysis mainly for qualitative purpose, such as in forensic area [66–69]. In pharmacokinetics or therapeutic drug monitoring, the use of CE–MS is still modest.

Electrospray ionization (ESI) is the most commonly used ionization interface in CE. The three configurations commonly used for coupling CE and ESI–MS are coaxial sheath liquid, liquid junction, and sheathless or nanospray. All settings involve the closure of the high-voltage circuit in the outlet end of the capillary separation. These settings differ from each other in the position of the electrodes and the nature of ionization (the presence of sheath liquid or not) [70]. Although MS is likely more sensitive than UV–Vis detector, the CE–MS interfaces employing make-up liquids dilute the CE efflux, resulting in lower concentration sensitivity when compared to LC–MS or GC–MS [56].

Boatto et al. described the development of a method to screen for and to quantify four 2,5-methylenedioxy-derivatives of 4-thio-phenylethylamine in human plasma. These compounds are phenethylamine designer drugs that are being introduced on the illicit drug market because these compounds are not covered by existing legislation. The use of capillary electrophoresis coupled with electrospray ionization–mass spectrometry (CE–ESI–MS) provided data that permitted the unambiguous confirmation of these drugs in human plasma. LOQs were in the range of 27.3–43.0 ng mL^{-1} [71].

12.2.2 Repeatability

As commented before, capillary electromigration techniques have been considered complementary and alternative techniques to HPLC and offer high separation efficiency, versatility, simplicity of instrumentation, fast and economic analysis, and use of minimum amounts of samples. However, one of the main negative points of CE and CEC claimed in the past was the poor precision of both migration times [72] and peak areas ($\sim$5% relative standard deviation) [73] which is mainly caused by a sum of factors such as variance of electroosmotic flow (EOF) and injection volume, buffer depletion, change in the analyte mobility, capillary wall interaction, temperature, overloading, and peak integration [74]. However, by performing the experiments under controlled conditions, enough precision could be obtained to validate the methods following the official guidelines [75,76].

In routine analyses, it is essential to keep migration times constant for peak identification. In addition, peak area is proportional to both sample concentration and migration time, so changes in migration times will result in variations in the areas and failure in the quantitative analysis. These problems can be solved by using an internal standard or peak area normalization [73], which will be discussed later.

The electrolyte solution changes during the analysis due to the extensive application of voltage and this may result in a loss of repeatability [76]. This phenomenon is named "buffer depletion" and is responsible for decreased analysis precision. In aqueous solutions, electrolysis of water produces soluble protons at the anode and hydroxide ions at the cathode. In consequence of this, the solution in the anode reservoir will become increasingly acidic, whereas the solution in the cathode reservoir becomes increasingly alkaline. The result of this buffer depletion is a pH gradient formed into the capillary that will have direct impact on the ionization of the species and on the magnitude of EOF, affecting peak areas, migration times, resolution, and peak efficiency [74,77]. The electrolysis depends on the buffering capacity of the buffer system, so it is important to choose an electrolyte with strong buffering capacity and appropriate pH value. The electrolysis also depends on other factors that must be checked for each method developed. It has been known that high ionic strength reduces the effects of buffer depletion and improves resolution and detectability. However, the use of both high voltage and ionic strengths beyond long run times increases the pH oxidation-reduction at the electrodes [73,77]. In addition, when the electrolyte solution is very concentrated, heating problems can occur due to the Joule effect, producing unacceptable levels of current and temperature gradient into the capillary and thus damaging the analysis. An alternative is the use of zwitterionic buffers that allows the increase in buffer concentration without the generation of excessive current and heating [77]. The choice of buffer concentration should be made taking into account other important aspects such as size of the capillary, applied voltage, running time, and conductivity of the buffer system [78]. To avoid depletion effects, a limited number of runs using the same run buffer vials can be established [76]. Figure 12.4 shows the effect of buffer depletion on the analysis of albendazole metabolites. The resolution between the internal standard (phenacetin) and the first eluted enantiomer of albendazole sulfoxide (ASOX) was seriously affected after six injections without changing the running buffer.

Another approach to improve the separation efficiency and run-to-run repeatability is to establish a suitable procedure for capillary preconditioning. This step makes the system returns to the initial conditions of the analysis, minimizing the variation of migration times by reducing the influence of the EOF variation [73,74,78–80].

The preconditioning procedure is more important when a biological sample such as plasma, serum, or urine is directly injected. In bioanalytical methods, a rinsing procedure of the capillary between-run with sodium hydroxide solution is usually performed to remove adsorbed material from the capillary wall. The washing step with sodium hydroxide is followed by washing with water and then the capillary is equilibrated with the electrolyte solution. This washing procedure is very important and should be carefully tested to find the optimum conditions. Some methods require a long time to wash and to re-equilibrate the charge of the capillary because incomplete reaction of siloxane to silanol groups can occur. In addition, strong conditions are sometimes necessary to reduce

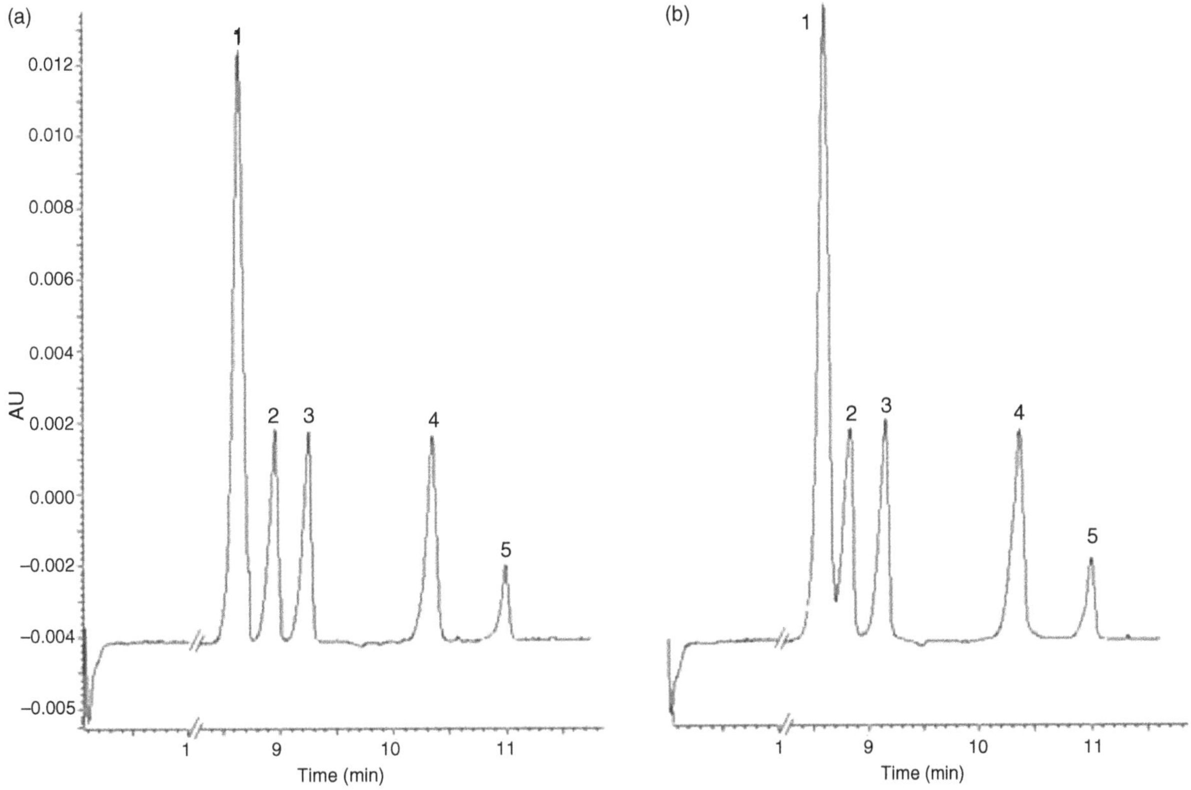

FIGURE 12.4 The effect of buffer depletion in albendazole metabolites analysis by CE. (a) first injection; (b) sixth injection. 1-Phenacetin (IS); 2-ASOX E1; 3-ASOX E2; 4-ABZsulfone; 5-ABZaminesulfone. Experimental conditions: running buffer: tris(hydroxyaminomethane) 50 mmol L^{-1} pH = 9.3 plus sulfated-β-cyclodextrin (2.0%, w/v). Applied voltage and temperature of analysis: +15 kV and 20 °C, respectively. Analyte concentrations: 0.2 mg mL^{-1} (unpublished results).

EOF variations and to obtain reproducible migration times and peak areas [81].

Instrumental aspects related with the injection, temperature control, and detection are also important for precision. In capillary electromigration techniques, unlike HPLC, there is not a fixed loop injector, so it is difficult to introduce a small sample volume (nL) with precision. The injection is automatic and the two modes most used of injection are hydrodynamic by pressure and electrokinetic. In general, the hydrodynamic injection is more precise and robust than electrokinetic injection [74,76,82].

In the hydrodynamic injection the sample is injected by pressure and the volume injected depends on the viscosity and the temperature. On the other hand, in the electrokinetic injection, the sample is injected by voltage application and the volume injected depends on EOF, electrophoretic mobility of the analytes in the sample, and the temperature. Many factors can affect the EOF and, therefore, the amount of analyte injected. The conductivity of biological samples such as plasma, serum, urine, and cerebrospinal fluid can vary between samples due to variations in endogenous component concentrations and ionic strength. In addition, changes in the analyte charge and problems in the current generation also make the quantification difficult [74,82]. However, a positive point of the electrokinetic injection is the selectivity. As, in this kind of injection, the amount injected depends on the electrophoretic mobility of the analyte, it is possible to manipulate the analyte charge by optimizing the pH and discriminate the ionic species of interest that will be introduced into the capillary. Detailed discussion about electrokinetic and hydrodynamic injection can be found in the paper of Breadmore [83].

During the hydrodynamic injection, a pressure across the capillary is generated for a set period of time. Due to the difficulty in precisely controlling the pressure for short time periods (sometimes 1 or 2 s), the injection volume is often affected. So, to obtain injection repeatability, longer injection times are advised. However, an increase in the injection time will increase the sample amount injected and can compromise drastically the separation efficiency [73]. Of course, the analyst needs to consider both sample concentration and volume. In biological analyses, it is necessary to quantify low sample concentrations with high precision, and then special techniques for injection such as sample stacking should be used [29].

Schappler et al. discuss about the combination of sample preparation procedure and the injection mode. The researchers used the technique of protein precipitation before hydrodynamic injection and LLE followed by an electrokinetic injection and concluded that for more sensitive analysis, electrokinetic injection is recommended because it allows on-capillary sample preparation. In addition, the electrokinetic injection combined with LLE increased both the sensitivity and the selectivity of the method [84,85].

Staňová et al. evaluated two injection modes for the analysis of buserelin in urine using CE–ESI–MS and observed that electrokinetic injection was better than hydrodynamic injection. Relative standard deviations of migration times were between 1 and 6.1% for hydrodynamic injection and 1% for electrokinetic injection. The same results were observed for peak areas (7.2–12% for hydrodynamic injection and 5.4–8.1% for electrokinetic injection) [86].

The temperature also plays an important role in many separations due to its relation with the electrophoretic mobility and the EOF. The electrophoretic mobility and EOF increase with the increase in temperature due to the decrease in viscosity, causing a reduction in the migration times and also reducing the peak area. In addition, the viscosity decreases with increasing temperature and more amounts of solution enter the capillary for the same injection time. So, it is very important to keep constant the capillary, electrolyte reservoir, and sample vial temperature [73,74].

The use of internal standard (IS) to correct injection volume fluctuations is highly recommended in electromigration techniques. In addition, the IS also corrects dilution errors and errors during the extraction procedure, improving the precision of peak area. The IS should have some characteristics such as similar chemical structure to the analyte, known purity, and stability and, mainly, it should not comigrate with compounds of interest or interferent compounds. An ideal IS should also have similar mobility to the analytes to compensate for EOF changes during the run [87]. Altria et al. showed that the use of internal standard improved the injection precision from 2.8% down to 0.2% RSD [73].

In capillary electromigration techniques, differently from chromatography, the speed of the molecules that pass through the detector is not constant, so molecules with higher migration times spend more time in the detection cell and therefore give a detector response for a longer time [74,88]. This phenomenon is best viewed during the analysis of enantiomers. Theoretically, a racemic mixture consists of equimolar amounts of two enantiomers and then it is expected to obtain equal peak areas. However, we observe that the second enantiomer area is slightly larger. To compensate this effect, it is a common practice to use the normalized peak area, which is calculated by dividing the peak area by its corresponding migration time.

12.3 SELECTED APPLICATIONS OF CAPILLARY ELECTROMIGRATION TECHNIQUES IN BIOANALYSIS

12.3.1 Pharmacokinetics and Metabolism Studies

Pharmacokinetics and *in vivo* metabolism studies are carried out by analyzing serial samples of plasma/serum, and less frequently urine, collected from volunteers or patients after administration of single dose or multiple doses of the drug. By statistical analysis of the results, it is possible to obtain information about the absorption, distribution, biotransformation, and excretion of the drug. Pharmacokinetic studies require sensitive assays to quantify the drug and metabolites in low concentrations frequently found in the samples, especially for drugs with high distribution volume and that are administered in low doses. Concentrations as low as sub ng mL^{-1} are frequently observed. Selectivity is another requirement for the methods to avoid the interference of metabolites, endogenous compounds, and coadministered drugs when the study is performed in patients under polytherapy. Due to its selectivity, detectability, and high throughput characteristics, LC–MS/MS is the main analytical technique used for this kind of study in the present [6].

In spite of the several methods appearing in the literature showing the development and application of CE or CEC for pharmacokinetic and metabolism studies, these techniques are not yet used as a routine tool for this kind of application. The major drawback of these techniques that impairs their use in this case is their low detectability when UV–Vis detection is used. Although more sensitive detection systems such as LIF or electrochemiluminescence could be used, they are suitable only for specific analytes. Liu et al. developed a simple, rapid, and sensitive method for the determination of the methylephedrine (ME) and pseudoephedrine (PSE) in human urine using CE with electrochemiluminescence detection. The method was applied to the determination of ME and PSE in human urine samples and for the pharmacokinetic study of PSE [89]. In another paper, a CE method based on electrochemiluminescence detection was applied with success for the analysis of proline (detection limit of 0.02 μg mL^{-1}) and fleroxacin (detection limit of 0.3 ng mL^{-1}) in human urine of patients with renal insufficiency and chronic uremia [90].

MS detection is also an alternative but the improvement in detectability is not always enough to obtain the detectability required for the method. As a consequence, CE and CEC could only be used as an alternative technique for drugs whose sample concentrations are in the range of ng to μg mL^{-1}.

The determination of nine drugs of abuse (amphetamine, methamphetamine, 3,4-methylenedioxyamphetamine, 3,4-methylenedioxy-methamphetamine, 3,4-methylenedioxyethylamphetamine, cocaine, codeine, heroin, and morphine) was performed using CEC–MS with a capillary column packed with CN modified silica particles. Figure 12.5 shows the electrochromatograms of urine samples spiked with the

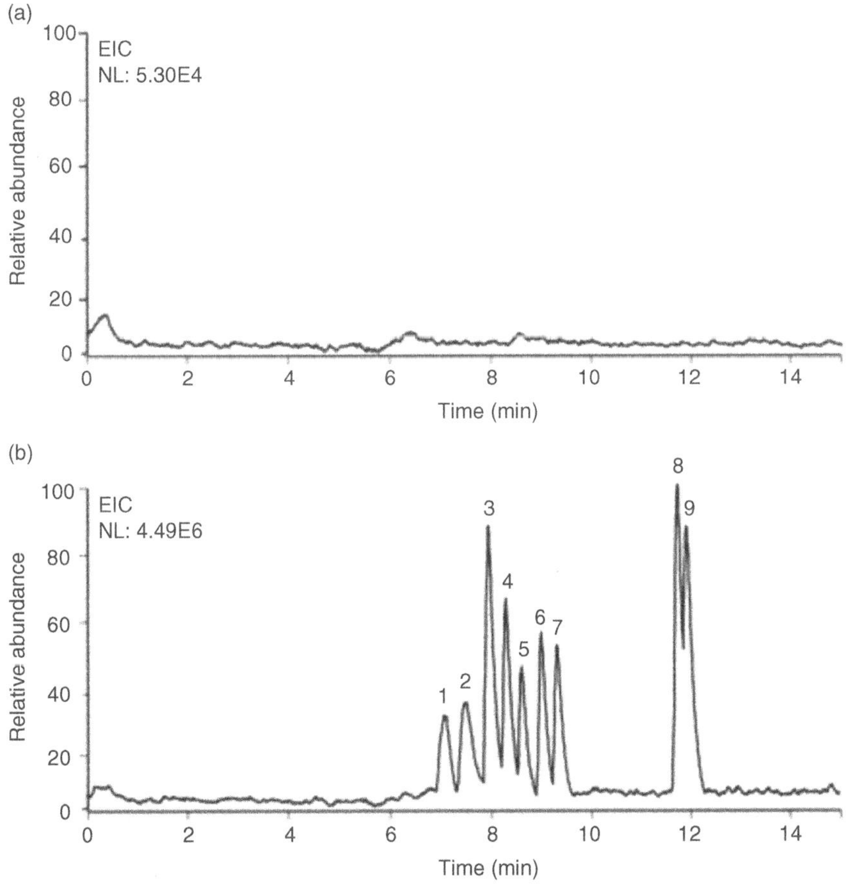

FIGURE 12.5 Extracted ion electrochromatograms of (a) blank urine sample and (b) blank urine sample spiked with the drugs of abuse mixture, after SPE procedure. Samples were electrokinetically injected (12 kV, 10 s). Reproduced with permission from Ref. 91. Wiley-VCH Verlag GmbH & Co. KGaA.

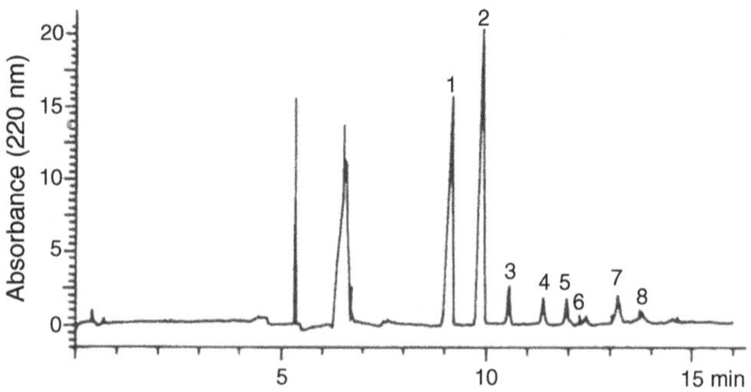

FIGURE 12.6 Electropherogram for microsomes incubated with rac-HCQ: (1) (−)-(*R*)-HCQ, (2) (+)-(*S*)-HCQ, (3) (−)-(*R*)-DHCQ, (4) (+)-(*S*)-DHCQ, (5) (+)-(*S*)-DCQ, (6) (−)-(*R*)-BDCQ, (7) (−)-(*R*)-DCQ, and (8) (+)-(*S*)-BDCQ. Reproduced with permission from Ref. 92. Wiley-VCH Verlag GmbH & Co. KGaA.

analytes. The quantification limit obtained was between 2.5 and 12.5 ng mL^{-1} for the studied drugs and the method was applied to the analysis of drugs of abuse in spiked urine samples [91].

Capillary electromigration techniques seem to be very attractive for *in vitro* metabolism studies. In this case, the concentrations of the drugs and metabolites in the samples are normally high and detectability is not the major concern. For this kind of application, the analyst could take benefit of CE and CEC efficiency, which allows the resolution of several metabolites normally formed. Cardoso and Bonato used CE to evaluate the *in vitro* chiral metabolism of hydroxychloroquine (HCQ). The drug and metabolites enantiomers were resolved using hydroxypropyl-β-cyclodextrin as chiral selector. Figure 12.6 shows the electropherogram of the HCQ enantiomers and its metabolites desethylchloroquine (DCQ), desethylhydroxychloroquine (DHCQ), and bisdesethylchloroquine (BDCQ) after incubation with microsomes obtained from rats liver homogenates. The quantification limits were 125 ng mL^{-1} for all enantiomers [92].

12.3.2 Enantioselective Analysis of Drugs and Metabolites

Stereochemistry has become a critical issue in drug development because, in many cases, one enantiomer has therapeutic activities whereas the other one is inactive or frequently responsible for the side effects. As a result, the separation, identification, and quantification of enantiomers are now very important analytical harms. Among the analytical techniques based on separation processes suitable for chiral analysis, capillary electromigration techniques have attracted great attention in the last decade. Enantiomer separation by CE is carried out by adding a suitable chiral selector to the running buffer. In spite of the great number of chiral selectors available for enantiomer separation by CE, cyclodextrins (CD) and their derivatives are the preferred ones. In CEC, the resolution is obtained by using capillaries packed with chiral stationary phases, frequently the same that are used in HPLC. The great success of CEC and mainly CE for chiral separation are the high resolution power (extremely high separation efficiency and selectivity of the several available chiral selectors), straightforward optimization of separation conditions, fast separations, and low cost due to the small amount of solutions required for the analysis [10].

The analysis of enantiomers in biological matrices faces the same challenges regarding detectability. In this case, the situation is still worse because the concentration of the enantiomers of a chiral drug is at least half the concentration of the racemate. Several innovative aspects of electromigration techniques such as MS hyphenation, microchip technology, online sample preparation techniques, and so on have been explored for the chiral analysis of drugs [10], but the application of these innovations to the analysis in biological matrices is still modest.

The enantiomers of seven amphetamines were analyzed using CE–MS and low concentration of highly sulfated gamma-cyclodextrin (HS-γ-CD) as chiral selector in less than 6 min. LLE was used to clean up and concentrate the sample. The detection limits obtained were in the range of 0.1–0.4 μg mL^{-1} and the method was suitable for amphetamines analysis in clinical and forensic samples [93].

Tang et al. developed a method for the analysis of thalidomide enantiomers in human urine by CEC using online preconcentration on an octyl monolithic column and enantiomer separation on a norvancomycin-bonded monolithic column. To improve detectability, solvent gradient effect was also studied. Figure 12.7a shows the electrochromatogram obtained using an norvancomycin-bonded monolith column for 500 mg L^{-1} racemic thalidomide (dissolved in TEAA buffer, electrokinetic injection at 3 kV for 3 s and 0.2 mm sample plug), and Figure 12.7b shows the performance of the hyphenated monoliths for 2.0 mg L^{-1} thalidomide enantiomers (dissolved in the TEAA buffer, electrokinetic injection at 3 kV for 7 min and 34.8 mm sample plug). Using this condition, 278- and 298-fold enrichment factors and detection limits of 90 and 94 μg L^{-1} for the two thalidomide enantiomers were obtained [94].

In the review published by Erny and Cifuentes, the authors discuss LC, CE, and CEC for the analysis of chiral drugs and their metabolites in biological fluids and detection by MS. The authors clearly show the advantages of CE and CEC techniques coupled with MS and the promissory use of the microchip-MS coupling [95].

12.3.3 Biopharmaceuticals or Biotechnology-Derived Pharmaceuticals

Due to the advances in protein and peptide production through recombinant techniques, the number of biotechnology-derived pharmaceuticals or biopharmaceuticals has increased.

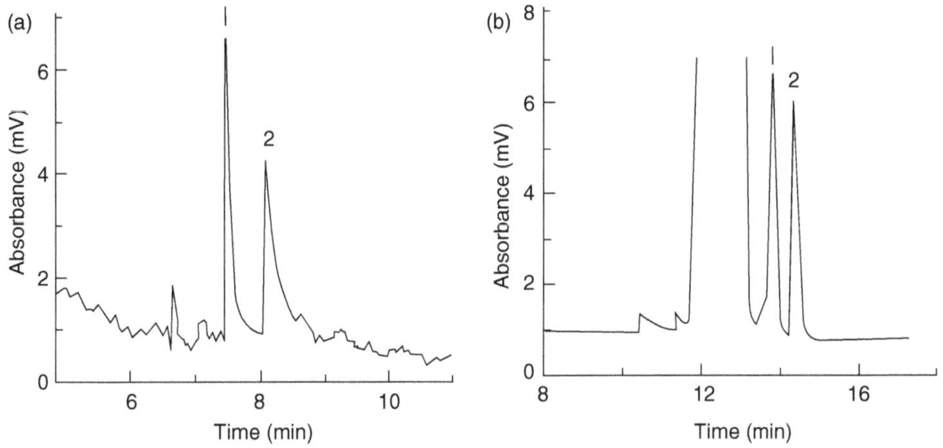

FIGURE 12.7 Comparison of thalidomide enantiomers with (a) a conventional sample plug length (0.2 mm) and (b) the proposed CEC method for a long sample plug length (34.8 mm) with self-concentration and solvent gradient effects. Conditions: sample matrix, 0.1% TEAA buffer (pH 4.8); sample concentration, (a) 500 mg L^{-1}, (b) 2 mg L^{-1}. Reproduced with permission from Ref. 94. Wiley-VCH Verlag GmbH & Co. KGaA.

The assessment of product quality for biopharmaceuticals is a great challenge due to the presence of variants or isoforms of the bulk biopharmaceuticals in addition to other possible impurities.

At the moment, the major application of CE for biopharmaceuticals analysis is for the quality control to evaluate the presence of impurities, degradation products, and to control batch-to-batch reproducibility. Wielgos et al. analyzed heparin samples obtained from porcine intestine and separated an impurity, the oversulfated chondroitin sulfate (OSCS). A high speed and high resolution was achieved using 600 mmol L^{-1} lithium phosphate, pH 2.8. Under these conditions, the detection limit for OSCS was below 0.1%. In addition, a linear relationship between peak area and official assay for biological activity (USP) was obtained, showing that CE method can be an alternative to the bioassay [96].

The analysis of biopharmaceuticals in biological matrices is still more difficult to carry out and no example could be found in the consulted literature.

12.3.4 Therapeutic Drug Monitoring

In therapeutic drug monitoring, drug concentrations in whole blood, plasma, or serum, and sometimes in other matrices such as saliva, are measured at specific times. The results obtained can be used for dose adjustment, to evaluate patient compliance, drug interactions, and so on. Therapeutic drug monitoring is required for drugs with narrow therapeutic ranges, such as some antibiotics, immunosuppressants, antiepileptic drugs, digoxin, and so on [3]. In these cases, the drug quantity required to be effective is very close to the quantity related with significant side effects and/or signs of toxicity.

The bioanalytical method used for this application should be sensitive enough to measure trough concentrations (sample collected immediately before the next dose), required for some drugs. In addition, method selectivity is also very important because the patients are normally under polytherapy. In this case, other drugs and metabolites could interfere with the determination of the target analyte. The high efficiency of capillary electromigration techniques is a strength allied for the analyst to obtain the required selectivity.

Therapeutic drug monitoring of methotrexate and its eight metabolites was realized by Cheng et al. for the evaluation of response and adverse effects of the drug in the therapy of acute lymphoblastic leukemia. The method was developed using solid-phase extraction and online CE stacking, combining three procedures: large-volume sample stacking (LVSS), sweeping, and dynamic pH junction (DypH). Using this stacking procedure, the detectability of the method could be improved by 40-fold (detection limit was 0.1 μmol L^{-1} for methotrexate) in relation to CZE mode in biological matrix [97].

Lin et al. used MECK for the determination of free levels and total levels of phenytoin in human plasma of epileptic patients. The method was optimized and parameters such as buffer pH, ionic buffer concentration, surfactant (SDS) percentage, and injection were evaluated. In addition, an HPLC method was also developed and both methods were compared, with good correlation between them (Figure 12.8). The CE method was applied for therapeutic drug monitoring of phenytoin to evaluate patient's compliances, helping dose adjustment, efficacy, and safety in the therapy [98].

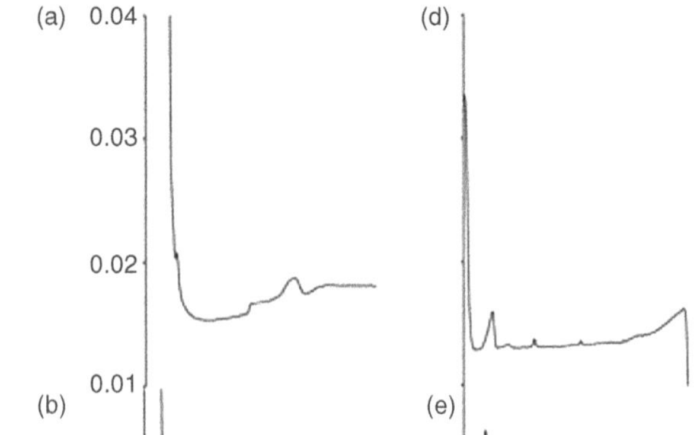

FIGURE 12.8 Electropherograms of total-level (a–c) and free-level (d–f) phenytoin determinations. (a) Plasma blank; (b) spiked with phenytoin 20.0 μg mL^{-1} and ketorolac 25.0 μg mL^{-1} (IS); (c) plasma sample from patient who receive phenytoin therapy; (d) plasma blank; (e) spiked with phenytoin 5.0 μg mL^{-1} and chloramine T (IS) in human plasma ultrafiltrate for free-level determination; (f) plasma sample from patient who receive phenytoin therapy for free-level determination. Peaks: 1, phenytoin; 2, chloramine T; 4, ketorolac. Reproduced with permission from Ref. 98. Wiley-VCH Verlag GmbH & Co. KGaA.

12.3.5 Clinical and Forensic Toxicology

Clinical toxicology is the branch of toxicology that is concerned with human poisoning, due to drug overdoses (pharmaceuticals or drugs of abuse) and toxic environmental, occupational, or accidental exposures. Forensic toxicology involves the use of toxicology and other disciplines to aid medical or legal investigation of death, poisoning, and drug use. In clinical toxicology, the specimen identity is presumed and a screening result is sufficient for medical decision. On the other hand, in forensic toxicology, as the results are used for legal actions, the identity of specimen must be proved and only confirmed results can be considered positive.

Capillary electrophoresis is a mature technique and classified as almost a standard tool for forensic analysis of DNA for which automated multicapillary configurations are available. In other areas of clinical and forensic toxicology, the use of capillary electromigration methods is still insipid. However, the advancement in coupling of CE to MS will greatly influence the acceptance of the technique, particularly for the analysis of thermally labile, highly polar, or charged drugs or metabolites [99].

The analysis of cocaine (COC) and its five metabolites in human urine was done using CE coupled to MS via electrospray ionization (Figure 12.9). The method was validated for the parameters detectability, linearity, precision, accuracy, recovery, and quantification limit (250–500 ng mL^{-1}) and showed clearly the application of CE–MS in real samples and its potentialities in clinical and forensic toxicology [100].

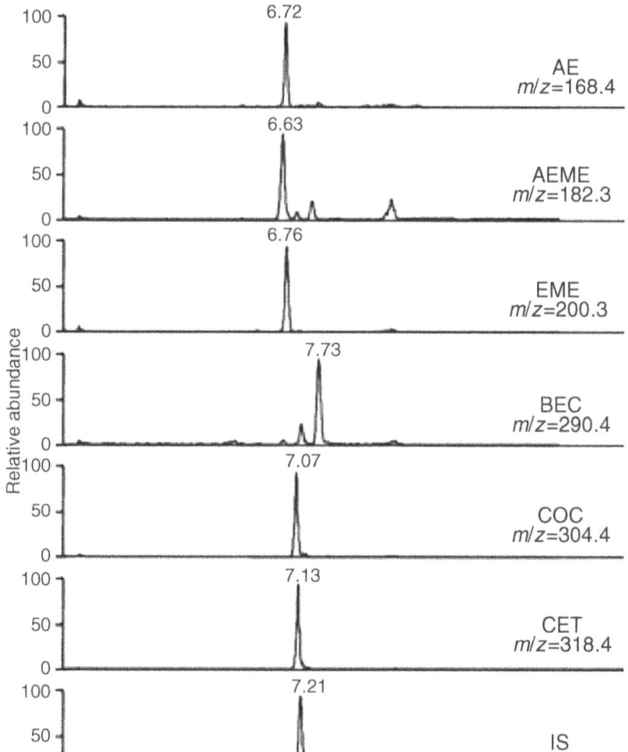

FIGURE 12.9 Extracted ion electropherogram of a urine blank sample spiked with COC and metabolites at 5000 ng mL^{-1}. Analytical conditions: background electrolyte composition, 1 mol L^{-1} formic acid (pH 1.9); t_{inj} = 10 s at 4 psi; V = 30 kV; L_t = 80 cm; i.d. = 50 μm; sheath liquid composition and flow = 0.05 mol L^{-1} formic acid in 1:1 methanol:water, 60% methanol and 5 μL min^{-1}, respectively; sheath gas: nitrogen at 15 arb. units; VESI = +3 kV; T_{cap} = 200 °C. Reproduced with permission from Ref. 100. Wiley-VCH Verlag GmbH & Co. KGaA.

12.4 CONCLUDING REMARKS

Capillary electromigration techniques have proved to be very attractive for the analysis of drugs and metabolites in biological matrices. The lower concentration sensitivity of the capillary electromigration techniques when UV detection is employed can be partly compensated by using several strategies of sample concentration or more sensitive detection system as discussed earlier. In addition, the improvements in the equipments and the use of controlled conditions for capillary conditioning and during the analytical procedure make capillary electromigration methods as reproducible as GC or HPLC methods.

In spite of this, the use of CE and CEC as techniques for routine analysis in bioanalytical laboratories is still insipid. The great majority of the papers reported in literature show only the application of the developed methods in the analysis of some selected samples, with the purpose to demonstrate the applicability of the methods. The widespread use of these developed methods is still the great aspiration of the researchers who believe in the great potential of capillary electromigration techniques. In our opinion, two main reasons are responsible for this situation. The first one is the distrust of application analysts in discarding conventional and well-known techniques and to adhere to a new one. Changes always require new investments in equipment and mainly in the formation and training of the involved staff in the analysis. It is worth to remember that capillary electromigration techniques are based on different separation principles than that occurring in the chromatographic techniques normally used for bioanalytical applications, bringing additional difficulties for new users. The other reason is the great success obtained in the development of LC–MS technique that nowadays can be considered the best technique for the analysis of drugs and metabolites in biological matrices.

In this aspect, the recent developments in coupling CE and CEC to MS can be the great attractiveness for the introduction of these techniques as routine tools for the analysis of drugs and metabolites in biological matrices. Other innovations such as in-line coupling of sample preparation techniques and the use of microchips are still rarely used for the analysis of drugs, but they could be primordial resources to obtain faster and totally automatized methods.

REFERENCES

1. Srinivas, N.R. (2008) *Biomed. Chromatogr.* *22*, 235–243.
2. Liu, X.D. and Jia, L. (2007) *Curr. Drug Met.* *8*, 815–821.
3. Fernandez, E.L., Pares, L., Ajuria, I., Bandres, F., Castanyer, B., Campos, F., Farre, C., Pou, L., Queralto, J.M., and To-Figueras, J. (2010) *J. Clin. Chem. Lab. Med.* *48*, 437–446.
4. Musenga, A., Saracino, M.A., Sani, G., and Raggi, M.A. (2009) *Curr. Med. Chem.* *16*, 1463–1481.
5. Smith, M.L., Vorce, S.P., Holler, J.M., Shimomura, E., Magluilo, J., Jacobs, A.J., and Huestis, M.A.J. (2007) *Anal. Toxicol.* *31*, 237–253.
6. Xu, R.N., Fan, L., Rieser, M.J., and El-Shourbagy, T.A. (2007) *J. Pharm. Biom. Anal.* *44*, 342–355.
7. Sánchez-Hernández, L., García-Ruiz, C., Marina, M.L., and Crego, A.L. (2010) *Electrophoresis* *31*, 28–43.
8. Geiser, L. and Veuthey, J-L. (2009) *Electrophoresis* *30*, 36–49.
9. Guihen, E. and O'Connor, W.T. (2010) *Electrophoresis* *31*, 55–64.
10. Mikus, P. and Maráková, K. (2009) *Electrophoresis* *30*, 2773–2802.
11. U. S. Department of Health and Human Services, Food and Drug Administration, Guidance for industry, bioanalytical method validation, 2001. http://www.fda.gov/cder/guidance/index.htm
12. Ramautar, R., Somsen, G.W., and de Jong, G.J. (2009) *Electrophoresis* *30*, 276–291.
13. European Medicines Agency, Guideline on validation of bioanalytical methods, 2009. http://www.ema.europa.eu/docs/en_GB/document_library/Scientific_guideline/2009/12/WC500018062.pdf
14. Paias, F.O. and Bonato, P.S. (2002) *J. Sep. Sci.* *25*, 1063–1069.
15. Vuorensola, K., Sirén, H., Kostiainen, R., and Kotiaho, T. (2002) *J. Chromatogr. A* *979*, 179–189.
16. Varenne, A. and Descroix, S. (2008) *Anal. Chim. Acta* *628*, 9–23.
17. Huo, Y. and Kok, W.T. (2008) *Electrophoresis* *29*, 80–93.
18. Dong, X., Wu, R., Dong, J., Wu, M., Zhu, Y., and Zou, H. (2009) *Electrophoresis* *30*, 141–154.
19. Hunag, Y-P., Liu, Z-S., Zheng, C., and Gao, R-Y. (2009) *Electrophoresis* *30*, 155–162.
20. Schurig, V. (2001) *J. Chromatogr. A* *906*, 275–299.
21. Bonato, P.S. (2003) *Electrophoresis* *24*, 4078–4094.
22. Preinerstorfer, B., Lammerhofer, M., and Lindner, W. (2009) *Electrophoresis* *30*, 100–132.
23. Holzgrabe, U., Borst, C., Büttner, C., and Bitar, Y. (2010) Quantitative analysis in pharmaceutical analysis, Chapter 11, In: Van Eeckhaut, Ann and Michote, Yvette (eds), *Chiral Separations by Capillary Electrophoresis*, CRC Press, Boca Raton, pp. 313–340.
24. Kitagishi, K. (1997) Comparison with other analytical methods, In: Shintani, H. and Polonshy, J. (eds), *Handbook of Capillary Electrophoresis Applications*, Blackie Academic & Professional Publishers, New York, pp. 137–145.
25. Carr, P.W., Wang, X., and Stoll, D.R. (2009) *Anal. Chem.* *81*, 5342–5353.
26. Weinberger, R. (2000) *Practical Capillary Electrophoresis.* 2nd ed., Academic Press, San Diego.
27. Wang, M.J., Liu, J.T., Chen, H.M., Lin, J.J., and Lin, C.H. (2008) *J. Chromatogr. A* *1181*, 131–136.
28. Maldaner, L. and Jardim, I.C.S.F. (2009) *Quím. Nova* *32*, 214–222.

29. Kartsova, L.A. and Bessonova, E.A. (2009) *J. Anal. Chem. 64*, 326–337.
30. Grard, S., Morin, P., and Ribet, J.P. (2002) *Electrophoresis 23*, 2399–2407.
31. Shihabi, Z.K. (2009) *Electrophoresis 30*, 1516–1521.
32. Fang, H., Zeng, Z., and Liu, L. (2006) *Anal. Chem. 78*, 6043–6049.
33. Hernández-Borges, J., Miquel-Borges, T.M., Delgado-Rodríguez, M.A., and Cifuentes, A. (2007) *J. Chromatogr. A 1153*, 214–226.
34. Nerín, C., Salafranca, J., Aznar, M., and Batlle, R. (2009) *Anal. Bioanal. Chem. 393*, 809–833.
35. Arthur, C.L. and Pawliszyn. J. (1990) *Anal. Chem. 62*, 2145–2148.
36. Alpendurada, M.F. (2000) *J. Chromatogr. A 889*, 3–14.
37. Jeannot, M.A. and Cantwell, F.F. (1996) *Anal. Chem. 68*, 2236–2240.
38. Pedersen-Bjergaard, S. and Rasmussen, K.E. (1999) *Anal. Chem. 71*, 2650–2656.
39. Halvorsen, T.G., Pedersen-Bjergaard, S., Reubsaet, J.L.E., and Rasmussen, K.E. (2001) *J. Sep. Sci. 24*, 615–622.
40. Esrafili, A., Yamini, Y., and Shariati, S. (2007) *Anal. Chim. Acta, 604*, 127–133.
41. Chiang, J.S. and Huang, S.D. (2008) *J. Chromatogr. A 1185*, 19–22.
42. Rasmussen, K.E. and Pedersen-Biergaard, S. (2004) *Trends Anal. Chem. 23*, 1–11.
43. Magalhães, I.R.S. and Bonato, P.S. (2009) *Anal. Bioanal. Chem. 393*, 1805–1813.
44. Müller, S., Möder, M., Schrader, S., and Popp, P. (2003) *J. Chromatogr. A 985*, 99–106.
45. De Oliveira, A.R.M. and Bonato, P.S. (2007) *J. Sep. Sci. 30*, 2351–2359.
46. De Oliveira, A.R.M., Cardoso, C.D., and Bonato, P.S. (2007) *Electrophoresis. 28*, 1081–1091.
47. Kumar, A. and Malik, A.K. (2009) *Critical Rev. Anal. Chem. 39*, 81–88.
48. Arce, L., Noza, L., Simonet, B.M., Ríos, A., and Valcárcel, M. (2009) *Trends Anal. Chem. 28*, 842–853.
49. Ramautar, R., Somsen, G.W., and de Jong, G. (2010) *J. Electrophoresis 31*, 44–54.
50. Wang, C.W. and Pawliszyn, J. (1998) *Anal. Commun. 35*, 353–356.
51. Santos, B., Simonet, B.M., Rios, A., and Valcárcel, M. (2007) *Electrophoresis 28*, 1312–1318.
52. Nozal, L., Arce, L., Simonet, B.M., Rios, A., and Valcárcel, M. (2007) *Electrophoresis 28*, 3284–3289.
53. Wei, S-Y., Yeh, H-H., Liao, F-F., and Chen, S-H. (2009) *J. Sep. Sci. 32*, 413–421.
54. Sánchez-Hernández, L., Grego, A.L., Marina, M.L., and Ruiz-Garcia, C. (2008) *Electrophoresis 29*, 237–251.
55. Clohs, L. and McErlane, K.M. (2003) *J. Pharm. Biom. Anal. 31*, 407–412.
56. Hempel, G. (2000) *Electrophoresis 21*, 691–698.
57. Swinney, K. and Bornhop, D.J. (2000) *Electrophoresis 21*, 1239–1250.
58. Mikus, P., Kubacak, P., Valaskova, I., and Havranek, E. (2006) *Talanta 70*, 840–846.
59. Li, H., Wang, P.H., Li, C., Wang, H., and Zhang, H.S. (2008) *Microchem. J. 89*, 34–41.
60. Marak, J., Mikus, P., Marakova, K., Kaniansky, D., Valaskova, I., and Havranek, E. (2007) *Electrophoresis 28*, 2738–2747.
61. Waterval, J.C., Lingeman, H., Bult, A., and Underberg, W.J. M. (2000) *Electrophoresis 21*, 4029–4045.
62. Musenga, A., Kenndler, E., Mercolini, L., Fanali, S., and Raggi, M.A. (2007) *Electrophoresis 28*, 1823–1831.
63. Ruiz-Pérez, T., Lozano-Martínez, C., and Galera, R. (2006) *Electrophoresis 27*, 2310–2316.
64. Musenga, A., Amore, M., Mandrioli, R., Kenndler, E., de Martino, L., and Raggi, M.A. (2009) *J. Chromatogr. B. 877*, 1126–1132.
65. Zheng, L.H., Zhang, L., Tong, P., Zheng, X.Y., Chi, Y.W., and Chen, G.A. (2010) *Talanta 81*, 1288–1294.
66. Schappler, J., Guillarme, D., Prat, J., Veuthey, J-L., and Rudaz, S. (2008) *Electrophoresis 29*, 2193–2202.
67. Gottardo, R., Bortolotti, F., De Paoli, G., Pascali, J.P., Miksik, I., and Tagliaro, F. (2007) *J. Chromatogr. A 1159*, 185–189.
68. Gottardo, R., Fanigliulo, A., Bortolotti, F., De Paoli, G., Pascali, J.P., and Tagliaro, F. (2007) *J. Chromatogr. A 1159*, 190–197.
69. Wey, A.B. and Thormann, W. (2001) *J. Chromatogr. A 916*, 225–238.
70. Assunção, N.A., Bechara, E.J.H., Simionato, A.V.C., Tavares, M.F.M., and Carrilho, E. (2008) *Quím. Nova 31*, 2124–2133.
71. Boatto, G., Nieddu, M., Dessi, G., Manconi, P., and Cerri, R. (2007) *J. Chromatogr. A 1159*, 198–202.
72. Blanco-Heras, G.A., Turnes-Carou, M.A., López-Mahía, P., Muniategui-Lorenzo, S., Prada-Rodríguez, P., and Fernández-Fernández, E. (2007) *J. Chromatogr. A 1144*, 275–278.
73. Altria, K.D. (1996) *Capillary Electrophoresis Guidebook: Principles, Operation and Applications*. Humana Press, Totowa, NJ.
74. Mayer, B.X. (2001) *J Chromatogr. A 907*, 21–37.
75. Holzgrabe, U., Brinz, D., Kopec, S., Weber, C., and Bitar, Y. (2006) *Electrophoresis 27*, 2283–2292.
76. Ali, I., Aboul-Enein, Hassan, Y., and Gupta, V.K. (2006) *Anal. Lett. 39*, 2345–2357.
77. Kelly, M.A., Altria, K.D., and Clark, B.J. (1997) *J. Chromatogr. A 768*, 73–80.
78. Jianping, Z., Changqin, H., and Wenying, L. (2006) *Chin. J. Chromatogr. 24*, 396–401.
79. Faller, T. and Engelhardt, H. (1999) *J. Chromatogr. A 853*, 83–94.
80. Kaupp, S., Bubert, H., Baurd, L., Nelson, G., and Wätzig, H. (2000) *J. Chromatogr. A 894*, 73–77.
81. Mayer, B.X., Hollenstein, U.M., Brunner, H.G., and Eichler, Müller, M. (2000) *Electrophoresis 21*, 1558–1564.
82. Schaeper, J. and Sepaniak, M.J. (2000) *Electrophoresis 21*, 1421–1429.

83. Breadmore, M. (2009) *Bioanalysis 1*, 889–894.

84. Schappler, J., Guillarme, D., Prat, J., Veuthey, J-L., and Rudaz, S. (2006) *Electrophoresis 27*, 1537–1546.

85. Schappler, J., Guillarme, D., Prat, J., Veuthey, J-L., and Rudaz, S. (2008) *Electrophoresis 29*, 2193–2202.

86. Staňová, A., Marák, J., Maier, V., Ranc, V., Znaleziona, J., Ševčík, J., and Kaniansky, D. (2010) *Electrophoresis 31*, 1234–1240.

87. Hettiarachchi, K. and Cheung, A.P. (1995) *J. Chromatogr. A 717*, 191–202.

88. Watzig, H., Degenhardt, M., and Kunkel, A. (1998) *Electrophoresis 19*, 2695–2752.

89. Liu, Y.M., Tian, W., Jia, Y.X., and Yue, H.Y. (2009) *Biom. Chromatogr. 23*, 1138–1144.

90. Sun, H.W., Li, L.Q., and Wu, Y.Y. (2009) *Drug Test. Analysis 1*, 87–92.

91. Aturki, Z., D'Orazio, G., Rocco, A., Bortolotti, F., Gottardo, R., Tagliaro, F., and Fanali, S. (2010) *Electrophoresis 31*, 1256–1263.

92. Cardoso, C.D., Jabor, V.A.P., and Bonato, P.S. (2006) *Electrophoresis 27*, 1248–1254.

93. Rudaz, S., Geiser, L., Souverain, S., Prat, J., and Veuthey, J-L. (2005) *Electrophoresis 26*, 3910–3920.

94. Tang, A.N., Wang, X.N., Ding, G.S., and Yan, X.P. (2009) *Electrophoresis 30*, 682–688.

95. Erny, G.L. and Cifuentes, A. (2006) *J. Pharm. Biom. Anal. 40*, 509–515.

96. Wielgos, T., Havel, K., Ivanova, N., and Weinberger, R. (2009) *J. Pharm. Biom. Anal. 49*, 319–326.

97. Cheng, H.L., Liao, Y.M., Chiou, S.S., and Wu, S.M. (2008) *Electrophoresis 29*, 3665–3673.

98. Lin, P.C., Hsieh, Y.H., Liao, F.F., and Chen, S.H. (2010) *Electrophoresis 31*, 1572–1582.

99. Tagliaro, F., Pascali, J., Fanigliulo, A., and Bortolotti, F. (2010) *Electrophoresis 31*, 251–259.

100. da Costa, J.L., Tonin, F.G., Zanolli, L.A., Chasin, A.A.M., and Tavares, M.F.M. (2009) *Electrophoresis 30*, 2238–2244.

13

CAPILLARY ELECTROPHORESIS AND MULTICOLOR FLUORESCENT DNA ANALYSIS IN AN OPTOFLUIDIC CHIP

Chaitanya Dongre, Hugo J.W.M. Hoekstra, and Markus Pollnau

13.1 INTRODUCTION

A lab on a chip (LOC) [1–4] squeezes the functionalities of a biological/chemical laboratory onto a single substrate through a network of microfluidic (MF) channels, reservoirs, valves, pumps and sensors. Its advantages are speed of analysis, low sample consumption, and measurement automation and standardization. This concept promises dramatic advances both in basic research and in clinical applications, enabling, for example, low-cost diagnostic tools.

Capillary electrophoresis (CE) is a powerful method for biomolecule separation and analysis. The sorting and sizing of DNA molecules within the human genome project [5] has been enabled largely by CE separation and analysis [6]. The human genome project has also led to the genetic mapping of various human illnesses [7]. Thanks to the high suitability of CE for miniaturization and LOC integration [8–10] and the understanding of DNA separation by such microchip CE (MCE), on-chip integration of DNA sequencing [11–14] as well as genetic diagnostics [15,16] has become feasible. Nevertheless, a number of challenges remain to be solved.

The traditional techniques, for example, using a bulk glass capillary filled with agarose gel [17], provide a high separation resolution. However, they typically require longer analysis times and bulky instrumentation. This issue has been addressed by the development of MCE. Laser-induced fluorescence (LIF) is the most popular MCE monitoring technique, allowing a low limit of detection (LOD) of, for example, 250 pM [18] or 500 pM [19], using visible or even near-infrared [20] fluorescent dye labels. Conventional instruments use confocal setups [21,22] to focus the excitation light and to collect the resulting fluorescence. Such schemes can provide high sensitivity, with the vision of going down to interrogating at the single-molecule level [23,24]; however, they require accurate mechanical alignment of the optics to the MF channels and are sensitive to mechanical vibrations and drifts. The need to use MCE in combination with a massive benchtop instrument, such as an optical microscope, frustrates many of the MCE advantages; in particular, it strongly limits device portability and prevents in-field or point-of-care applications. Much of the commercial success of MCE will critically depend on the ability to successfully integrate optical detection schemes, efforts toward which have started almost a decade ago [25,26].

Several attempts have been made to integrate micro-optical components in MF LOCs to perform on-chip optical detection [27–39]. In addition to the integration of optical waveguides (WGs), other techniques have been explored for optofluidic integration in LOCs, for example, the use of the MF channel itself as a liquid-core optofluidic WG [37,38,40,41] or even a liquid-core optofluidic laser [42,43], silicon-based optoelectronic transducer elements [44], optical fibers for fluorescence excitation [45], integration of lasers for fluorescence excitation [46], optofluidic lenses for fluorescence collection [47], photodetectors for fluorescence detection [48], and so on. Contrary to some authors who use the term "optofluidics" for optical propagation controlled by fluidics [49], many other researchers and also we use this term in a more general way for research combining fluidics and optics. Integrated optical WGs allow

Capillary Electrophoresis and Microchip Capillary Electrophoresis: Principles, Applications, and Limitations, First Edition.
Edited by Carlos D. García, Karin Y. Chumbimuni-Torres, and Emanuel Carrilho.

to confine and transport light in the chip, directing it to a small volume of the MF channel and collecting the emitted/transmitted light, as has recently been applied to monitor on-chip DNA sequencing using zero-mode WG sensors [12,50] in a now commercialized DNA sequencer [14]. The integration of optical WGs or other photonic components with MF channels typically needs additional processing steps. Depending on the substrate of choice, different WG fabrication processes are possible. Several such approaches are reported in the literature and include SiON [51], silica on silicon [18,26,52–54], ion exchange in soda lime glasses [55,56], photolithography in polymers [57–60], liquid core WGs [19,37,38,54,61,62], and femtosecond (fs) laser written WGs in glass [63–71].

The focus of this chapter will be essentially to demonstrate the value addition of integrated optics to MCE-based DNA analysis in LOCs, for example, by showing potential enhancement of the separation resolution, enhancement of the sensitivity—enabling us to detect low concentrations of permanently, exclusively end-labeled DNA molecules, and so on, and the challenges encountered as a result of the optofluidic integration. The sensitive optofluidic functionality finally enabled the implementation of multicolor fluorescent analysis of DNA probe samples from two different genes, corresponding to two different illnesses, analyzed unambiguously during a single MCE experiment. This is achieved by the combination of two existing technologies, namely the clean-room-based fabrication of MF chips [72], and the writing of optical WGs with high-intensity fs-laser pulses [69–71], providing a platform for MCE with integrated optical sensing functionalities. This is a first step toward addressing the more complex problem of integrating optical detection on an electrophoretic microchip toward a truly portable analysis system.

13.2 OPTOFLUIDIC INTEGRATION IN AN ELECTROPHORETIC MICROCHIP

13.2.1 Sample Fabrication

The layout of our optofluidic chip is presented in Figure 13.1a. Chips were fabricated in a two-step procedure. First, the MF channel network and MF reservoirs were patterned photolithographically and wet-etched in fused silica glass and then sealed off by bonding another piece of fused silica glass on top. The chip has dimensions of 55 mm × 5.5 mm × 1 mm, and the MF channels have a cross-section of ~110 μm width and ~50 μm depth [72].

In a second step, the optical WG was inscribed into the bulk of such a fused silica chip by fs-laser writing [74] at a translation speed of 20 $\mu m\,s^{-1}$, using a Ti:Sapphire laser operating at 800 nm wavelength with 150 fs, 4 μJ pulses at a repetition rate of 1 kHz [71]. Employing tunable, astigmatic beam shaping [75], an elliptical cross-section of the written WG was obtained, with a major diameter of ~50 μm in the vertical direction, to excite the maximum possible volume of the MF channel, while the minor diameter in the horizontal direction is ~12 μm to retain a high spatial resolution along the direction of DNA migration and separation.

The WG displays a graded refractive index profile and a maximum refractive index increase of 2×10^{-3}. Fiber-to-chip coupling of laser light with coupling losses of <2 dB was achieved and the WG propagation losses were measured to be in the range of 0.5–0.9 dB cm^{-1} at the wavelength of 543 nm. The excitation WG intersects the CE separation channel orthogonally at a distance of ~3.6 cm from the MF crossing junction at which the separation commences, toward the end of the MF CE separation channel, close to reservoir 4, where the plug separation is highest. The fluorescence signal was detected through a small detection window to reduce the background noise. Constant, inherent, mutual alignment of the excitation WG with its detection window has the potential to render our system more compact, faster to operate, and more reproducible than conventional LIF approaches [58,76].

13.2.2 Optofluidic Characterization

As a proof of principle for the optofluidic integration in a dynamic MCE flow and separation experiment, a sample containing two different fluorescent dyes—Rhodamine-6G and Rhodamine-B (absorption maxima at 530 and 540 nm, respectively)—was used to demonstrate MCE separation and its fluorescence monitoring [73]. These analyte molecules were introduced into reservoir 1 of the CE microchip (see Figure 13.1a).

The MF channels were filled with a buffer (20 mM MES/20 mM His, pH 6.2). The CE sample loading and separation protocol was based on optimized actuation voltages of up to 1.5 kV, delivered by Pt electrodes integrated into the MF reservoirs. The entire on-chip migration was controlled with a LabVIEW (National Instruments, Inc.) script steering an MF control system (Capilix BV) [77]. Application of a high voltage forced the analyte molecules to migrate into the CE injection channel from sample reservoir 1 to waste reservoir 3 (see Figure 13.1). By switching the voltages at all four reservoirs simultaneously, a well-confined plug containing the mixture of analyte molecules—with a volume of ~605 pL at the crossing junction of the two MF channels—was injected into the CE separation channel, from the MF crossing junction toward waste reservoir 4, and the analyte molecules contained in the plug volume were separated according to their size. The 543 nm line from a green continuous-wave (CW) He–Ne laser was coupled into the on-chip integrated WG. Distinct fluorescent segments appeared and faded away as the two plugs of separated dye molecules migrated across the excitation WG, as shown in

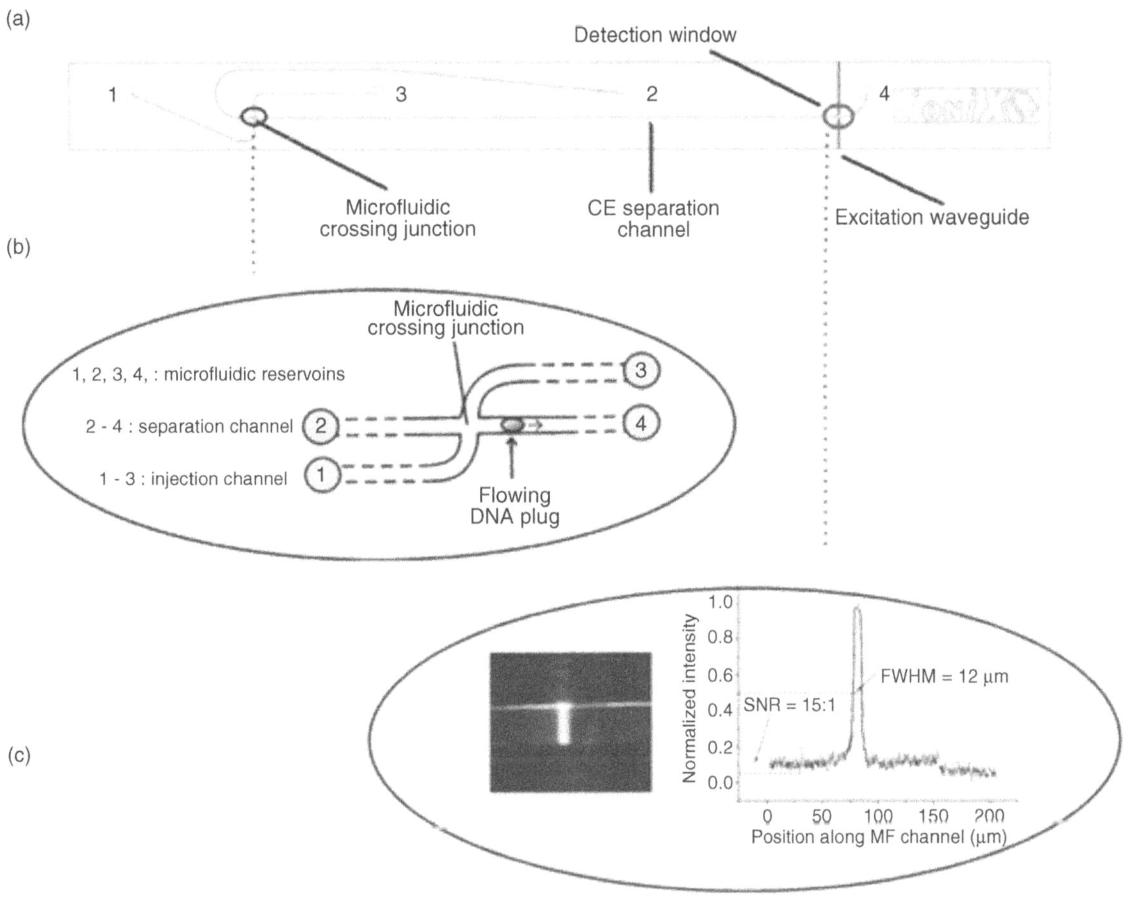

FIGURE 13.1 (a) Schematic of the optofluidic chip showing reservoirs 1–4, sample injection channel (reservoir 1 → reservoir 3) and CE separation channel (reservoir 2 → reservoir 4), as well as the integrated optical WG and detection window. (b) Injection from the MF crossing junction and migration of a DNA molecule plug in the MCE separation channel. (c) WG excitation of Fluorescein dye filling the MF channel, and the emitted fluorescence intensity as a function of the position along the MF channel (indicating the 12 μm spatial resolution). Figure modified from Ref. 73.

Figure 13.2a. On plotting the fluorescence intensity at the WG–MF-channel intersection as a function of time, we obtained the electropherogram in Figure 13.2b, where the two sharp peaks correspond to the two separated species.

13.3 FLUORESCENCE MONITORING OF ON-CHIP DNA SEPARATION

13.3.1 Experimental Materials and Methods

Prior to the experimental runs the inner walls of the MF channel network were coated with an epoxy-poly(dimethylacrylamide) (EPDMA)-based polymer [78,79] to suppress the electroosmotic flow and to minimize adsorption of DNA molecules on the MF channel wall. Subsequently, the channels were filled with a sieving gel matrix consisting of hydroxyethyl cellulose (HEC) [80] (2% wt/vol), dissolved in 20 mM MES/20 mM His buffer (pH 6.2) to maximize the resolution of the DNA CE separation. The intercalating dye SYBR Green I (Molecular Probes Inc.) was added according to the manufacturer's protocol. The reagents were sterilized, filtered (0.22 μm pore size), and stored at 269 K. The same CE sample loading and separation protocol as described in the preceding text was then applied.

Double-stranded DNA molecules were injected as a single mixture plug into the MF CE separation channel. For the movie snapshots of Figure 13.3, Hg arc lamp background illumination was applied to visualize migration of the plugs, and the fluorescence was recorded by a CCD camera. Excitation through the optical WG was provided by the 488 nm line from a CW Ar-ion laser. For ultrasensitive fluorescence detection, a photon-counting photomultiplier tube (PMT) (H7421-40, Hamamatsu Photonics K.K.) cooled down to 203 K was built onto the output port of an inverted microscope (DM5000, Leica Microsystems GmbH) and

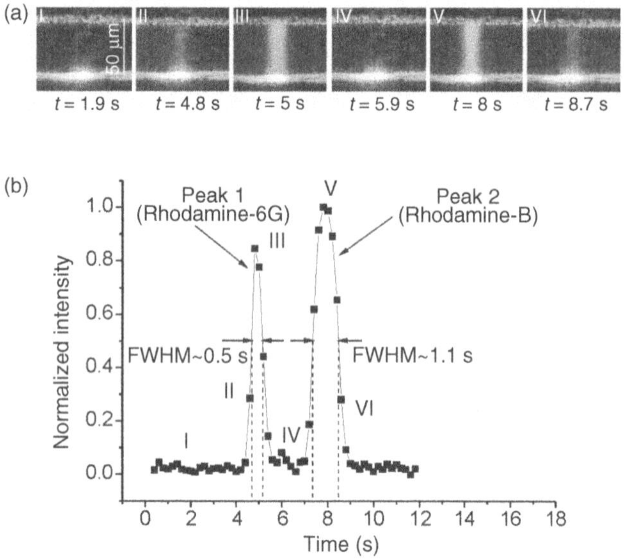

FIGURE 13.2 (a) Passage in a MF channel and across an excitation WG of fluorescent dyes Rhodamine-6G and Rhodamine-B. (b) The corresponding time-dependent fluorescence intensity (the electropherogram) at the WG—MF channel intersection after injection into the separation channel. Figure modified from Ref. 73.

aligned to collect the fluorescence signal through the detection window. Combination of excitation/emission band filters (K3, Leica Microsystems GmbH, and XF57, Omega Optical, Inc.) in the fluorescence collection path ensured that only the fluorescence signal emitted by the fluorescently labeled DNA molecules reached the PMT, while all other sources of background signal, including the scattered laser excitation signals, were rejected.

13.3.2 Experimental Results and Analysis

A commercial DNA ladder consisting of molecules with 17 different sizes (in the range of 50–3000 bp) was injected and separated, and the individual plugs were detected [81]. A movie (for snapshots, see Figure 13.3) depicts the migration of the separated DNA plugs across the excitation WG under additional background Hg arc lamp illumination.

The four snapshots in Figure 13.3 correspond to the migration of the first well-separated DNA plug across the excitation WG. At this point in time, the remaining well-separated 16 plugs were not yet in the field of view of the CCD camera. On CW laser excitation of such a plug through the WG, a sharp fluorescent segment $\sim 12\,\mu m$ in width (matching well with the minor diameter of the WG cross-section) can be observed along the WG–MF-channel intersection. Plotting the fluorescence intensity at the WG–MF-channel intersection as a function of time during the MCE separation leads to an electropherogram, which is further numerically postprocessed (as described in Section 13.4.2) to obtain the electropherogram with a high signal-to-noise ratio (SNR) in Figure 13.4a.

The LOD of DNA analysis in the integrated optofluidic chip was explored under integrated-WG laser excitation. Replacing the CCD camera by an ultrasensitive PMT allowed us to record the fluorescence intensity versus migration time, as depicted in the electropherogram of Figure 13.4a. We estimate the LOD by considering that an injected DNA concentration of 10 nM diluted by, first, the spatial separation of 17 different molecule sizes and, second, plug broadening from 110 μm to at least 248 μm provided a fluorescence signal that is 125 times the fourfold standard deviation of the detected background noise. Following the numerical postprocessing (described in Section 13.4.2) a further 10-fold enhancement of the SNR was obtained. Using the above numbers, our conservative estimate of the LOD is $10\,nM/(17 \times 248\,\mu m/110\,\mu m \times 125 \times 10) \approx 220\,fM$, which surpasses previously reported LOD values in laboratory setups involving on-chip-integrated fluorescence monitoring [19,71].

After fitting the fluorescence intensity peaks in Figure 13.4a with Gaussian profiles, the peak widths obtained from Figure 13.4b were plotted versus the square root of the DNA molecule sizes (see Figure 13.5a). The square-root dependence, confirmed by the linear fit, indicates that the migration of DNA molecules follows Poissonian statistics, the dispersion being governed by the number of collisions a molecule typically undergoes as it traverses through the sieving gel matrix [83]. The Gaussian fit is nevertheless justified, as for high expectation values the Poissonian approaches a Gaussian shape.

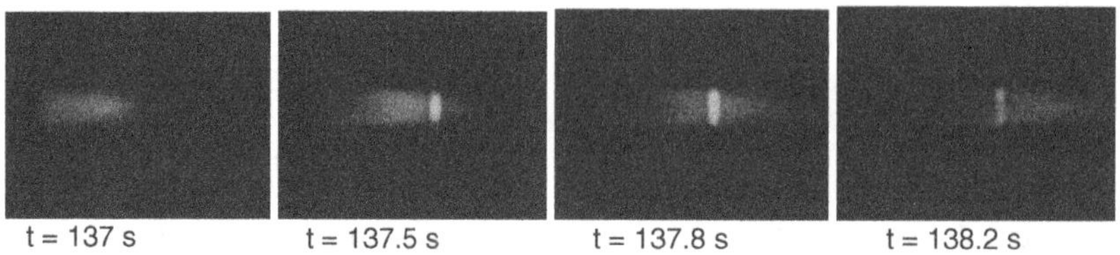

FIGURE 13.3 Snapshots of a movie recorded with a CCD camera showing transient fluorescence from several molecule plugs formed by CE separation of a DNA ladder consisting of molecules with 17 different sizes (50–3000 base pairs), fluorescence-labeled with an intercalating dye, as these plugs pass by the point of integrated-WG laser excitation at 488 nm. Figure taken from Ref. 81.

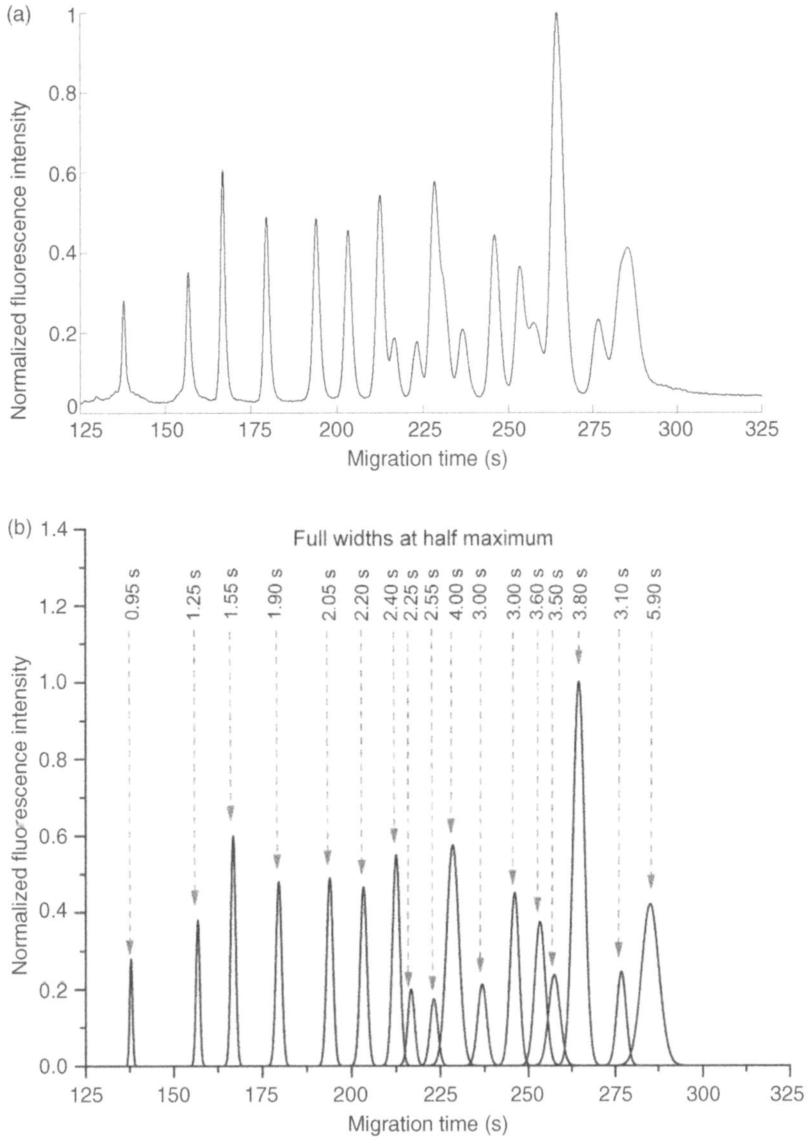

FIGURE 13.4 (a) Electropherogram depicting the normalized fluorescence intensity versus migration time with integrated-WG laser excitation. (b) Gaussian fits to the electropherogram peaks; the peak widths are expressed as full-width-at-half-maximum. Figure modified from Refs. 81, 82.

Optimization of the coating and sieving gel matrix led to excellent MCE separation of the DNA molecules under integrated-WG laser excitation (see Figure 13.4a). When plotting the temporal occurrence of the fluorescence peaks of the 17 molecules in Figure 13.4a versus their *a priori* known base-pair sizes, the two smallest (insufficient interaction with the sieving matrix) as well as the two largest DNA molecules (tendency to fold) exhibit clear deviations (and have not been shown in Figure 13.5), while the peaks of all DNA molecules with sizes of 150–1000 base pairs, in the diagnostically relevant region, are resolved with a sizing accuracy higher than 99% (Figure 13.5b, left-hand ordinate). The high sizing accuracy is comparable to state-of-the-art performance in microchip CE-based DNA sequencing without integrated-WG laser excitation [9,84].

Each measured fluorescence peak in the electropherogram is a convolution—along the migration axis—of the Gaussian WG mode-field cross-section Δx_{WG} and the diffusion-induced physical width Δx_{MF} of the MF plug. The resulting spatial plug widths Δx are related to the temporal electropherogram peak widths Δt (provided in Figure 13.5a as full width at half maximum) as

$$\Delta x = \frac{L}{t}\Delta t = \sqrt{\Delta x_{WG}^2 + \Delta x_{MF}^2}$$

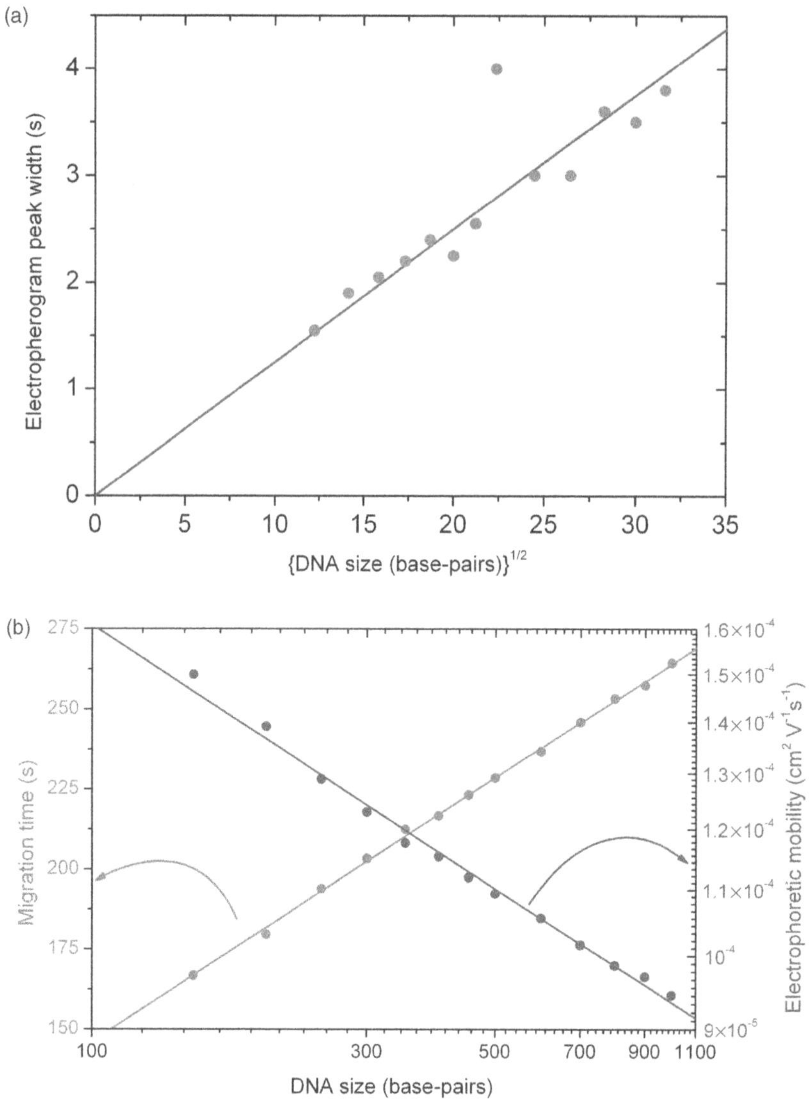

FIGURE 13.5 (a) Peak widths obtained from the Gaussian fit versus square root of the base-pair size for the experimentally measured electropherogram. (b) Migration time (left-hand ordinate) and electrophoretic mobility (right-hand ordinate) versus the logarithm of the base-pair size for the measured electropherogram. Figure modified from Ref. 81.

where $L = 3.6$ cm and t (data on the abscissa of Figure 13.4a) are the migration length and time, respectively. In our experiment, the minimum electropherogram peak width was found to be $\Delta x = 248\ \mu m$ (the very first peak in Figure 13.4a). With $\Delta x_{WG} = 12\ \mu m$, we obtain $\Delta x_{MF} = 247.7\ \mu m \gg \Delta x_{WG}$, that is, the current MCE separation protocols impose limitations on the achievable spatial resolution due to the broadening of the MF plug width during the MCE separation. The integrated-WG mode profile would in principle allow for a 20-fold better resolution. MCE technological advances, for example, by improving the MF channel-wall coatings and sieving gel matrices, will lead to smaller MF plug widths, thereby fully exploiting the MCE monitoring resolution achieved with integrated optical WGs. Clearly, once the challenges of MF plug broadening have been addressed by the MCE community, integrated optical instrumentation promises to be well-suited to handle the resolution demands of future MF-chip generations.

From the measured migration time t versus DNA size (Figure 13.5b, left-hand ordinate), we derive the electrophoretic mobility μ according to the equation

$$\mu = \frac{L}{Et}$$

where L is the distance between the MF crossing junction and the detection window and E is the electric field strength in the MCE separation channel. The observed linear decrease (on double logarithmic scale) of the electrophoretic mobility with increasing DNA size (Figure 13.5b, right hand ordinate) matches well with observations in conventional bulk capillary-based DNA gel electrophoresis setups [85].

When repeating these experiments, a number of times in the same and then in a different microchip, using different DNA samples, it was observed that the relative peak heights as well as their relative positions in time varied slightly from experiment to experiment owing to slight changes in the condition of the sieving gel matrix, ambient temperature, and so on. Also the slope of the straight-line fit (on logarithmic scale) for DNA migration time versus DNA size was found to vary from experiment to experiment [81]. This variation of power dependence points toward different flow mechanisms in these experiments, which again potentially depend on ambient temperature and/or condition of the sieving gel matrix, a result that questions whether a general, simple model of electrophoretic migration in microcapillaries can be established and universally applied.

13.4 TOWARD ULTRASENSITIVE FLUORESCENCE DETECTION

13.4.1 Optimization of the Experimental Setup

In this section, we describe the various approaches applied to increase the sensitivity of the setup. These include the various steps taken to prepare the chip for the experiment, carefully choosing the wavelength filters to pass only the desired fluorescence signals to the detector, and so on. Before starting to work with the MF chip, its back-surface (the surface on the side of the MF reservoirs, that is, the surface opposite to the surface facing the detector) was painted black with a permanent ink. By sealing the MF reservoirs before this step we made sure that no ink particles or droplets could enter the MF channels through the reservoirs, thereby clogging the chip or contaminating the analyte or the buffer solutions. Almost all the stray light that may be originating from, for example, autofluorescence of the polymeric MF cartridge was absorbed by the black ink, thereby reducing the background noise otherwise added to the fluorescence emitted by the analyte molecules. A multiband filter set (XF57 from Omega Optical, Inc.) was used to pass only the wavelength bands corresponding to the emission spectra of the fluorescent dyes in question. The well-enclosed optical path within the inverted microscope ensures once more that no stray signal interferes with the fluorescence signal from the analyte. The PMT was fitted directly via a C-mount adapter to the eyepiece of the microscope to capture the fluorescence photons released from the microscope eyepiece. The housing of the setup ensured an optimum, completely closed measurement system, with the measurements taking place in complete darkness (with all the possible sources of light, including LED displays on power meters, PC-monitors, and so on, in the laboratory being either turned off or covered with black paper).

13.4.2 All-Numerical Postprocessed Noise Filtering

If an excitation beam is modulated at a specific frequency, its temporal signature leads to a similar temporal excitation of and emission by the fluorescent molecule that absorbs light at this wavelength, as long as the excited-state lifetime of the fluorescent molecule is short compared to the inverse of the modulation frequency. The Fourier spectrum of the detected fluorescence signal contains a peak, corresponding to the specific modulation frequency of the excitation beam. The decoding of the detected signal consists of filtering out this peak in the Fourier domain corresponding to the modulation frequency, to extract only the signal corresponding to the specific excitation wavelength. By applying inverse Fourier transformation, the corresponding individual time-domain signal is regenerated, minus the broadband background noise. This results in an enhancement of the SNR, and thereby the sensitivity of the setup. We implement this entire procedure numerically by means of a postprocessing program that can be applied to any measured, especially nonmodulated fluorescence signal. We therefore present a numerical analog of the so-called phase-sensitive detection or lock-in amplification [86–89], a technique that has found its way also in numerous biochemical detection strategies, including those involving sensitive fluorescence detection [90,91]. We have applied this technique in the context of fluorescence monitoring during MCE separations to enable the detection of the electropherogram peaks buried under the background noise.

This procedure consists of the following steps. First, we multiply the measured signal with a square-wave function, let us call it SQ1 (to simulate the real-life situation of a mechanical chopper cutting off and turning on the laser beam at regular time intervals during the measurement, with one as the maximum value, and zero as the minimum), having a period equal to a multiple of the sampling period of the signal. The resulting signal is modulated *a posteriori*, but it is as if it were modulated *a priori*. Second, the signal is processed in the Fourier domain. Discrete Fourier transform of the modulated signal leads to its frequency spectrum, which is essentially a frequency-shifted (due to the modulation) version of the original spectrum of the signal. Third, we filter out the single peak around the modulation frequency and perform an inverse Fourier transform on the filtered signal to return to the time-domain signal with a lower noise. Furthermore, we multiply the original signal by a square wave SQ2, which is simply a temporally shifted, out-of-phase version of the square wave SQ1, and apply the same processing steps as described earlier. As a final step, we add the absolute values of the two signals obtained by such

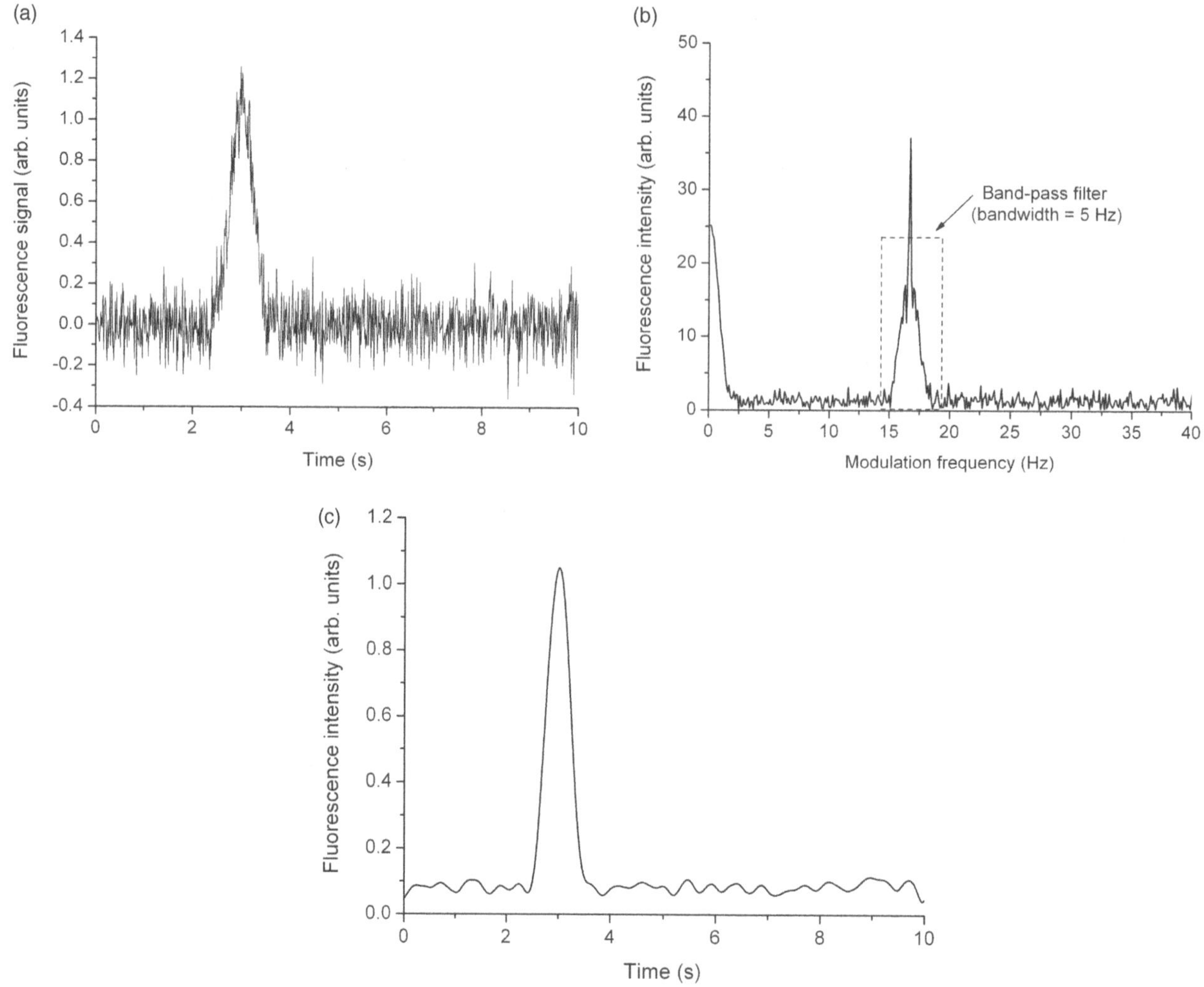

FIGURE 13.6 Illustrations to the presented noise filtering procedure. (a) A computer-generated Gaussian signal with added random noise. (b) A part of the Fourier spectrum of the fluorescence signal after the *a posteriori* modulation by multiplication with a square wave; the dotted line shows the rectangular band-pass filter function. (c) Fluorescence signal generated after the all-numerical lock-in amplification procedure. Figure modified from Ref. 82.

postprocessing with the two out-of-phase square waves. By applying only one square wave, we obtain a sevenfold SNR enhancement. The addition of the result from processing with the second, phase-shifted square wave delivers another factor of $\sqrt{2}$, thereby increasing the total enhancement to 10-fold, that is, almost by an order of magnitude.

As an illustration, we apply the procedure described earlier to a computer-generated Gaussian signal with an added random noise (see Figure 13.6a). The standard deviation of the noise around its average in Figure 13.6a is ~0.095, corresponding to a SNR of 12. We multiply this signal by a square wave, in this case having a period of 60 ms, equaling a frequency of 16.67 Hz. The spectrum of the resulting modulated signal is shown in Figure 13.6b. The spectrum of the signal around the modulation frequency of 16.67 Hz is extracted by means of a rectangular band-pass filter (bandwidth 5 Hz) centered at 16.67 Hz. The time-domain signal resulting from an inverse Fourier transform is shown in Figure 13.6c. The standard deviation of the noise around its average in the processed signal is ~0.0149, corresponding to a SNR of 74. In comparison to the signal in Figure 13.6a, we observe a sevenfold improvement in the SNR. A further improvement by a factor of $\sqrt{2}$ is obtained by applying the same processing steps with an out-of-phase modulating square wave and adding the two signals, leading to a 10-fold enhancement.

Then we apply this technique to the measured electropherogram of Figure 13.7a of the 17-molecule DNA ladder, which was already presented in Section 13.3.2. A 10-fold SNR enhancement was realized without inducing significant broadening even of the narrowest peaks. This procedure leads to a LOD of 220 fM, which corresponds to the

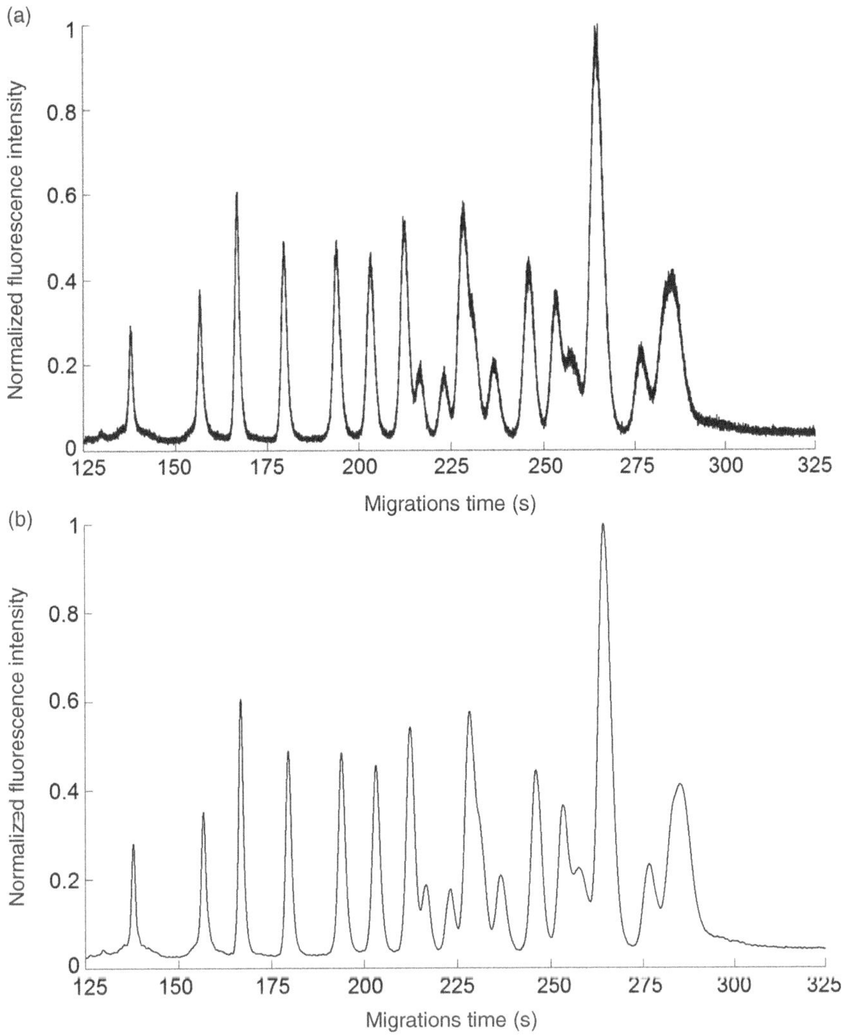

FIGURE 13.7 Application of Fourier transformation to SNR enhancement. (a) Originally measured fluorescence signal. (b) Fluorescence signal generated after applying the all-numerical lock-in amplification procedure to the measured electropherogram for a 17-molecule DNA ladder. Figure modified from Ref. 82.

detection of 8–9 DNA molecules in the excitation volume [82]. This example demonstrates the capability of this approach to visualize signals that would otherwise have been lost in the background noise. It further paves the way toward the high sensitivities that we require for the detection of covalently fluorescent-end-labeled, diagnostically relevant DNA molecules during their MCE separation and analysis, as will be described in Section 13.5.3. As this technique does not imply any modification in an existing fluorescence detection setup, but only Fourier analysis of already measured fluorescence data, it is an attractive and cost-effective approach to achieve SNR enhancement. Besides, it does not require the use of bulky equipment, such as a mechanical chopper, to modulate the excitation laser beam or lock-in amplifier, helping to miniaturize fluorescence detection and portability of a commercial optofluidic LOC device based on this measurement principle.

13.5 MULTICOLOR FLUORESCENT DNA ANALYSIS

To enhance analysis capabilities, multiwavelength fluorescence sensing has been implemented, mostly in bulk capillaries [92], for example, by use of a broadband light source such as a Xe lamp and extracting the multiple wavelengths with monochromators and complex optical schemes to achieve an unambiguous separation of different wavelengths. Other reports include the use of multiple photodiodes [93], external wavelength-selective gratings [94], color-CCD arrays [95], or external spectrum analyzers

[96]. In this section, we demonstrate two novel, simple and cost-effective approaches that allow dual- or multicolor fluorescence monitoring in a miniaturized LOC.

13.5.1 Dual-Point, Dual-Wavelength Fluorescence Monitoring

In several fundamentally important diagnostic applications, only two independent samples, each consisting of a number of differently sized DNA molecules, have to be monitored simultaneously; hence, only two excitation beams and detection of fluorescence signals at merely two wavelengths are required. These applications include, firstly, internal calibration of the setup during a MCE experiment by adding to the sample under investigation a reference sample with several DNA molecules of known base-pair sizes, thereby making the experiment insensitive to environmental influences such as operating temperature, condition of the sieving gel matrix and inner wall coating of the MF channels, or changes of the applied electric field. Secondly, an unknown, potentially malign sample exhibiting single base-pair insertions or deletions can be separated together with its healthy counterpart, thereby providing unprecedented sizing accuracy to the MCE experiment.

Here, we propose a simple method for spatial and wavelength duplexing, which applies dual-point, dual-wavelength detection from either one or two detection windows (DWs) [97]. It does not require any external apparatus to separate the wavelengths, thereby allowing for fluorescence detection with a single, ultrasensitive PMT and enabling an ultra-low LOD.

The layout of our optofluidic chip is presented in Figure 13.8. Each of the two DWs consists of a region in which two WGs carrying two different excitation wavelengths intersect the MCE separation channel in plane at two nearby locations. From the MF crossing junction, at which the separation commences, the four WGs are distanced by 2.0 cm and 2.1 cm (WG1 and WG2 within DW1, respectively) as well as 3.5 cm and 3.6 cm (WG3 and WG4 within DW2, respectively). Depending on the application, only a single DW or both DWs can be used. In the latter case, the order of the two excitation wavelengths is swapped between the two DWs to ensure unambiguous analysis. The same experimental protocol and detection optics as in Section 13.3.1 were implemented.

Dual-point, dual-wavelength fluorescence monitoring is valuable in diagnostic applications to compare an unknown sample, for example, from a gene region under investigation for base-pair insertion/deletion, with a reference sample, for example, from the same gene region corresponding to a healthy person. The base-pair difference between the corresponding analyte DNA molecules would be small (1–2 bp), if any, thus challenging the resolution capabilities of standard MCE systems. Here we determine these differences by exclusively labeling the two samples, investigating both samples simultaneously in a single migration experiment, and monitoring their separation with dual-point, dual-wavelength fluorescence excitation and detection through a single DW (in this case DW2 in Figure 13.8). If no insertion/deletion is present, the two species migrate with the same speed, but will nevertheless be optically separated due to the spatial separation between the two excitation WGs in the DW.

While diagnostically relevant genetic probes are usually double-stranded DNA molecules sized >100 bp, clinically produced by a polymerase chain reaction, for this proof of principle we used commercially produced (Molecular Probes, Inc.) single-stranded DNA primer molecules sized 19 and 20 nucleotides (nt), end-labeled with either Alexa Fluor 647 (AF647) or Cyanine 3 (Cy3). These two fluorophores can be selectively excited in the red (He–Ne laser at 633 nm, coupled through WG3) and in the green (He–Ne laser at 543 nm, coupled through WG4), respectively.

Initially, only a single species (19-nt-AF647) was injected in the separation channel, with either the red or green He–Ne laser turned on. In the former case, a strong signal peak was observed when the fluorescently labeled DNA plug crossed the excitation WG, while in the latter

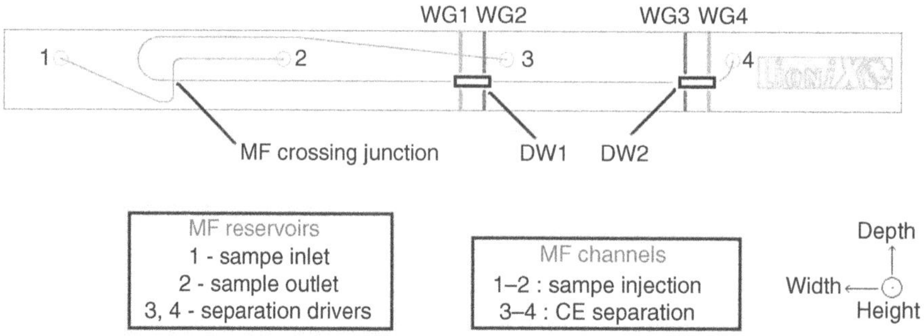

FIGURE 13.8 Layout of the optofluidic chip indicating the MF reservoirs, MF channels, and the two DWs, each comprising two WGs crossing the MCE separation channel perpendicularly in plane. Figure taken from Ref. 97.

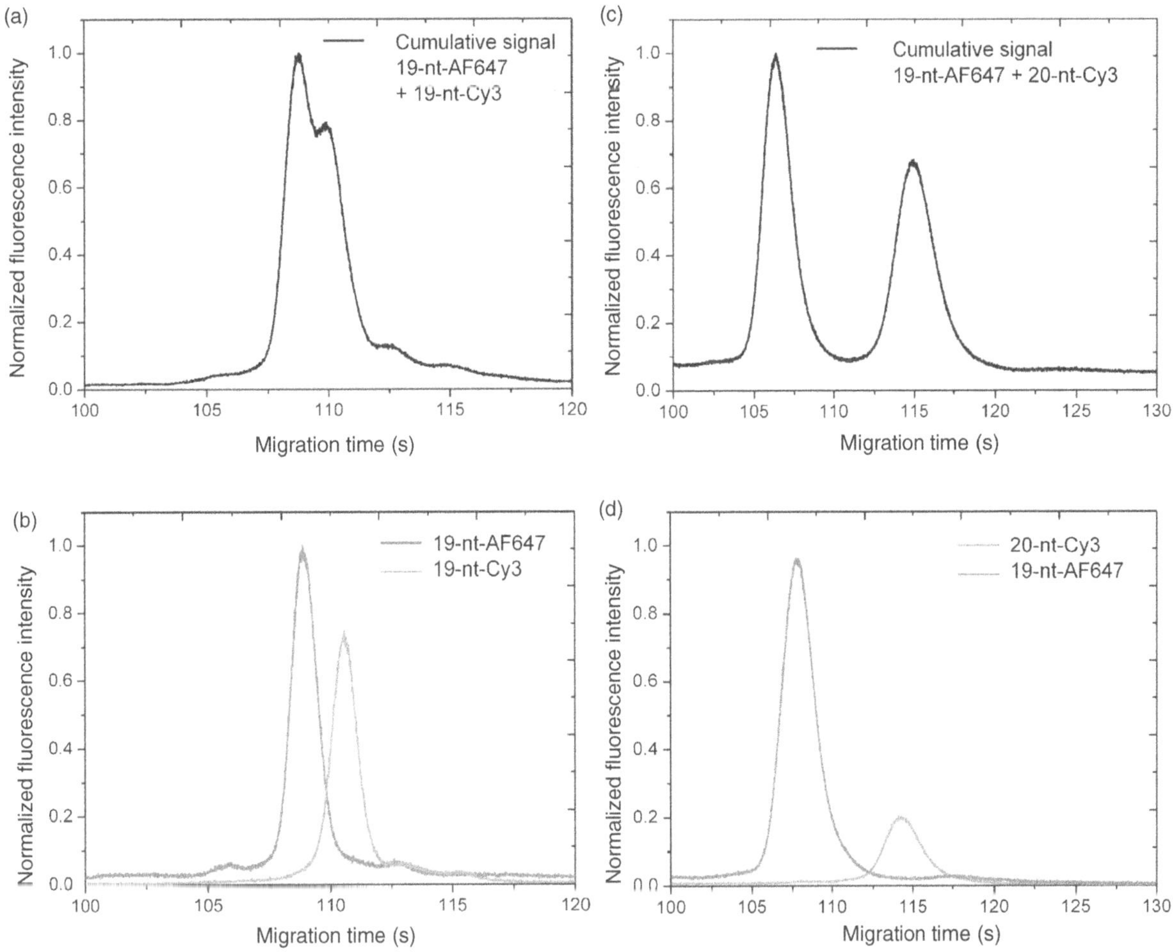

FIGURE 13.9 Electropherograms depicting the MCE separation of two fluorescently labeled DNA molecules: (a) cumulative signal during simultaneous dual-wavelength excitation of migrating 19-nt-AF647 and 19-nt-Cy3 molecules. (b) Individual signals detected during single-wavelength excitation with only one of the two lasers switched on, temporally superimposed on each other. (c and d) The same for 19-nt-AF647 and 20-nt-Cy3 molecules. Figure taken from Ref. 97.

case, the measured signal equaled the PMT baseline signal. This procedure was repeated for the same molecule labeled with the other dye (19-nt-Cy3), leading to the complementary result (i.e., strong signal peak with the green He–Ne laser and no signal with the red He–Ne laser). This confirmed that cross-excitation between the two fluorophores was, for all practical purposes, absent.

The two labeled DNA molecules were then mixed and the resulting polychromatic mixture was resolved into the individual monochromatic DNA components by dual-point, dual-wavelength excitation in the MCE separation channel. Figure 13.9a depicts the corresponding electropherogram. Since cross-excitation is absent, the two peaks appearing after a migration time of ~108–110 s and separated by ~1.8 s correspond to the two species, each excited by the corresponding laser wavelength of 633 nm or 543 nm coupled into the adjacent WGs. This is further confirmed by repeating the MCE separation twice, alternately turning off one of the lasers. The corresponding electropherograms displayed in Figure 13.9b show a good match with the corresponding peak heights and positions in Figure 13.9a. To visualize the case of single base-pair insertion/deletion, we repeat the experiment with a new sample mixture (19-nt-AF647 + 20-nt-Cy3). The corresponding electropherograms are shown in Figure 13.9c,d, with an evidently larger peak separation of ~6.2 s.

A unique advantage of integrated dual-point, dual-wavelength excitation is its capability of achieving MCE separation of equally sized, but differently labeled species, which naturally migrate at the same speed, by the inherent spatial separation of the two WGs carrying the corresponding unique excitation wavelengths as well as highly accurate detection of single-nucleotide insertion/deletion, as is relevant in genetic diagnostics—detection of anomalies in genetic samples obtained from a patient with respect to their healthy genetic counterparts. The lack of perfect match in the peak positions and intensities between Figure 13.9a,b and Figure 13.9c,d is attributed to slight, unintentional differences in the experimental flow parameters. Such effects justify the need for an internal calibration/referencing.

MCE separation and, consequently, sizing accuracy depend critically on multiple flow parameters, such as fluid and chip temperature, condition of the sieving gel matrix and MF wall coating, and applied electric field. Hence, for achieving single-base-pair accuracy of DNA analysis—for molecules in the diagnostically relevant range of 150–1000 bp—in an optofluidic chip, reliable intrinsic calibration of the CE separation setup is required. In our approach, a set of sample molecules under investigation and a set of well-known reference molecules (for calibration) can be exclusively labeled with two different fluorophores and separated simultaneously in a single MCE experiment. The reference molecules provide an intrinsic "ruler," allowing one to accurately sort the sample molecules according to their size independently of the fluctuating experimental conditions. Building on the experimental results presented in Section 13.3, we discuss our approach for intrinsic calibration by means of a numerical simulation [81].

Monitoring through two well-separated DWs, each comprising a WG pair carrying two different excitation wavelengths whose spatial order is swapped in DW2 (Figure 13.8), provides the following advantages. (i) An unfortunate size difference between two differently labeled DNA species may occur, such that the difference in migration speed exactly compensates for the distance between the two excitation WGs within one DW. As a result, the two fluorescence peaks are simultaneously detected by the color-blind PMT in this DW. However, the peaks do not coincide in the other DW because the swapping of the excitation wavelength would indeed enhance the effect of different migration speeds instead of creating a degeneracy. (ii) From the double detection, additional information can be deduced about migration speed, plug broadening, and so on. (iii) Low-concentrated species that disappear under the background noise owing to plug broadening while migrating to DW2 will still give rise to weak fluorescence peaks at DW1, thus providing extra information that would otherwise be lost.

In principle, such a setup can be operated with a single PMT, for example, by collecting the fluorescence from the two DWs by two fibers glued to the top of the chip [71] or two 3D-integrated WGs [98], and transporting the signals to the same PMT, thereby simplifying the whole setup and reducing its size. In this case, to achieve an unambiguous detection, it is a critical prerequisite to choose a sufficiently large distance between the two DWs, such that the first peak in DW2 appears only after the last peak in DW1 has disappeared. Since our experimental setup requires detection by a PMT through a microscope, only one of the two DWs can be monitored at a time. Hence, it is currently not possible in our setup to monitor both DWs simultaneously.

Therefore, to demonstrate the potential of dual-point, dual-wavelength monitoring for internal size calibration, we present a simulation of a MCE experiment based on a relationship between DNA size, migration speed, and plug broadening that was established in an earlier experiment [81]. We consider the separation of a DNA sample mixture consisting of green-labeled molecules S1 (250 bp), S2 (300 bp), S3 (450 bp), and S4 (700 bp), migrated together with a reference DNA mixture consisting of red-labeled molecules R1 (150 bp), R2 (355 bp), and R3 (1000 bp). Figure 13.10 shows simulated electropherograms that would be detected from the two DWs with mutually swapped excitation wavelengths. The sizes of the reference molecules are chosen such that the smallest and largest base-pair sizes belong to reference molecules, and the third, medium-sized reference molecule provides an indication of the experimental deviation from the expected migration due to parameter fluctuations. The distance between the two DWs is chosen such that the R3 peak in DW1 occurs before the R1 peak in DW2. While the R2 and S3 peaks incidentally overlap in DW1 owing to their unfortunate size difference, they are well resolved in DW2 as a result of swapping the excitation wavelengths. In such a setting, it is straightforward to identify each peak in both parts of the diagram, enabling unambiguous analysis of two sets of

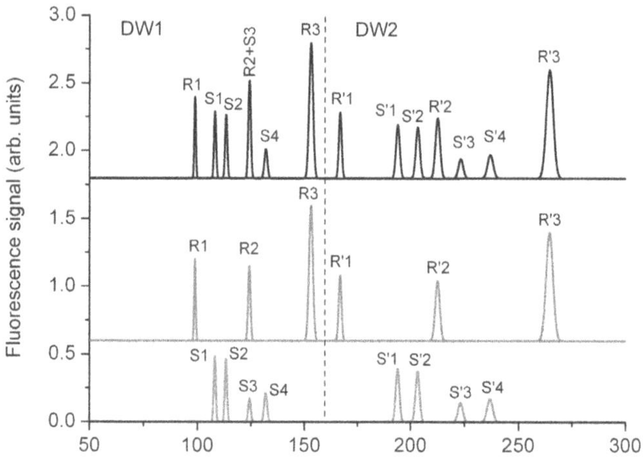

FIGURE 13.10 Simulated electropherograms as would be detected from the two DWs with swapped excitation wavelengths during an experiment with internal calibration using a green-labeled DNA sample mixture ("S" shown in the lower trace) consisting of four different molecule sizes (250 bp, 300 bp, 450 bp, and 700 bp) and a red-labeled reference mixture ("R" shown in the middle trace) consisting of three different molecule sizes (150 bp, 355 bp, and 1000 bp). Figure taken from Ref. 97.

exclusively color-labeled DNA molecules of different origin by detection with a single color-blind PMT.

Such a separation experiment in a setup that uses two DWs for internal calibration/referencing of DNA sizes can be a highly effective approach for intrinsically eliminating the undesirable effects of several flow parameters on calibration of a MCE separation system by migrating a well-known reference sample simultaneously together with the unknown sample.

13.5.2 Modulation-Frequency Encoded Multiwavelength Fluorescence Sensing

If an excitation beam is modulated at a specific frequency, its temporal signature leads to a similar temporal excitation of, and emission by, the fluorescent molecule that absorbs light at this wavelength, as long as the excited-state lifetime of the fluorescent molecule is short compared to the inverse of the modulation frequency. In other words, the fluorescence emitted by the molecule is encoded with the same frequency as the exciting beam. In the case of simultaneous presence of multiple analytes with distinct absorption bands, each analyte can be excited by a corresponding beam with a distinct modulation frequency, as shown in the schematics of Figure 13.11. If the different modulation frequencies are harmonically uncorrelated with one another, all the resulting fluorescence signals can be unambiguously addressed to the presence of the corresponding analyte molecule, even if all these fluorescences happen to occur at the same wavelength or if the fluorescences occur at different wavelengths but a wavelength-insensitive detection system, for example, an ultrasensitive PMT without monochromator, is used. As a result, the Fourier spectrum of the detected cumulative fluorescence will be composed of individual, nonoverlapping peaks, each corresponding to the specific modulation frequency of its exciting beam. This generic method of modulation-frequency encoded multiwavelength sensing [99,100] can be applied not only to fluorescence monitoring, but also to other optical measurement principles, for example, absorption and interferometry.

Decoding of the detected signal, therefore, consists of filtering out the peaks in the Fourier domain corresponding to the different modulation frequencies, to unambiguously extract the signal corresponding to a specific excitation wavelength. By applying inverse Fourier transform, the corresponding individual time-domain signals are regenerated.

It must be noted, though, that such numerical analysis will typically involve a discrete Fourier transform, resulting in a well-defined, bounded range of frequencies (defined, for example, by the number of temporal data points). All the higher harmonics and all the possible linear combinations thereof will essentially fall in this range, or will be folded back around the right-hand extremity of the range, to

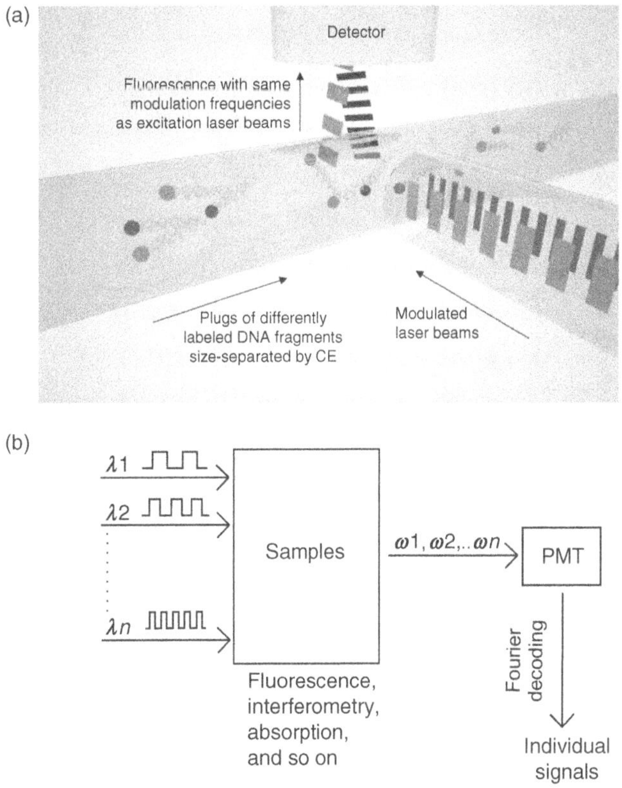

FIGURE 13.11 (a) Schematic showing plugs of exclusively fluorescence-labeled molecules migrating through the MF separation channel, intersecting the excitation WG which guides laser light of different wavelengths and modulation frequencies, and a plug containing DNA molecules with two different labels emitting fluorescence light at the two modulation frequencies while crossing the excitation WG. (b) A block diagram depicting the generic principle of the method. Figure modified from Refs. 99, 100.

nevertheless appear to be in the range. Such aliasing effects must also be taken into account, while choosing the correct frequency peaks in the Fourier domain for applying the inverse Fourier transform.

As a proof of principle, we monitored the migration of sets of differently sized biomolecules through an excitation/detection area. Here two sets of fluorescence-labeled biomolecules were used, each set consisting of two single-stranded DNA molecules with sizes of 19 and 23 nucleotides, which migrate with different speeds. One set was red-labeled, the other blue-labeled. Two laser beams at the absorption wavelengths of the labels, intensity-modulated at frequencies f_1 and f_2, respectively, were launched into the excitation/detection area. For fluorescence detection, a PMT (H7421-40, Hamamatsu Photonics K.K.) cooled down to 203 K was aligned to collect the fluorescence signal from the excitation/detection area.

The modulation of each laser induced an accordingly time-varying fluorescence signal emitted by those dye molecules absorbing this specific laser wavelength, and

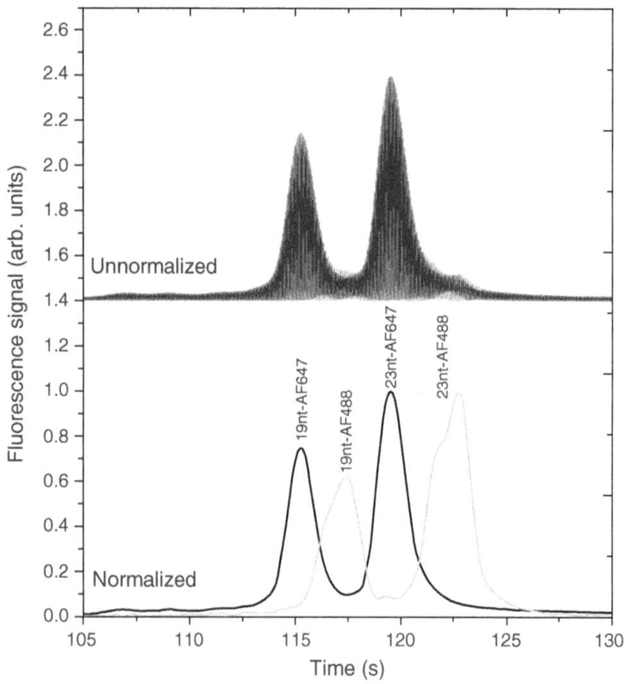

FIGURE 13.12 In a proof-of-principle experiment, fluorescence intensity versus time as measured by a color-blind PMT during the migration of two sets of blue- (shown in light gray) and red-labeled (shown in black) biomolecules through the excitation/detection area under modulated excitation: cumulative signal from all four molecules (upper part) and corresponding individual signals resolved by Fourier decoding (lower part). Figure modified from Refs. 99, 100.

the resulting time-modulated fluorescence signal at both wavelengths was simultaneously recorded by the ultrasensitive, albeit color-blind PMT, as shown in the upper parts of Figure 13.12. Fourier analysis was implemented in Matlab (MathWorks, Inc.) for back-end processing of the recorded total fluorescence signal, resulting in a frequency spectrum that consists of signals at the different modulation frequencies.

Only a narrow band of frequencies ($\Delta f = 2\,\text{Hz}$) around each modulation frequency ($f_1 = 17\,\text{Hz}$, $f_2 = 31\,\text{Hz}$) was selected by a flat-top transfer function and extracted back into the time domain by an inverse Fourier transformation, leading to the blue- and red-colored graphs in the lower parts of the images in Figure 13.12. During this analysis, the intensities at the higher harmonics of the fundamental frequencies were neglected during the inverse Fourier transformation, as this procedure ensured the best SNR. The selected band of $\Delta f = 2\,\text{Hz}$ is wide enough to cover the frequencies corresponding to the different response peaks, which have a typical width of 2–3 s in the time domain. On the other hand, this selected bandwidth is small enough to reject most of the noise.

From this example, we learn that dual-wavelength fluorescence monitoring of MCE separation is feasible without cross-excitation and, equally importantly, no physically unrealistic features are introduced to the time-domain signal as a result of the Fourier decoding. This easily applicable optical detection scheme enables exciting possibilities in many fields, whenever multiplexing can add value to a measurement system, for example, when it provides an advantage or is even imperative to investigate different sets of biomolecules in a single sample volume and to identify their origin. The described example shows that the simultaneous investigation of several independent and exclusively labeled biomolecule sets is feasible in a single experiment.

13.5.3 Application to Multiplex Ligation-Dependent Probe Amplification

Here we present a more advanced experimental example of this fundamentally distinct separation scheme for multiplex diagnostic applications, potentially leading to a paradigm shift in DNA analysis. The sets of individual DNA fragments originate from different diagnostically relevant regions of one or several genes and are analyzed in a single experiment, based on the exclusive wavelength labeling of the different genetic origins and encoding of their fluorescence by different modulation frequencies.

While in the DNA analysis presented in Section 13.3.2, we employed intercalating dye molecules, which bind to DNA molecules in multiple copies, thereby producing a large fluorescence signal, thanks to the improved sensitivity, as described in Section 13.4, we are now able to detect even singly or doubly end-labeled DNA molecules, which owing to the decrease in the total number of fluorescence emitters per analyte molecule produce a far lower fluorescence intensity. Contrary to intercalating dyes, which usually unspecifically bind and unbind at short time intervals to any DNA molecules in their vicinity [101–104] (exceptions can be found in [105]), end labels are attached covalently and permanently during a polymerase chain reaction (PCR). This allowed us to detect exclusively end-labeled double-stranded DNA molecules in the diagnostically relevant size range and unambiguously identify their genetic origin by exciting solely a specific set of molecules at the absorption wavelength of its permanent end label. In our approach, we launch simultaneously several laser wavelengths through a single WG, encode the fluorescence emitted by each set of end-labeled DNA molecules by exclusively time-modulating its excitation laser beam, and subsequently decode the emitted and detected fluorescence by Fourier analysis. In this manner, we separate differently encoded fluorescence signals from each other, thus in principle enabling simultaneous multicolor interrogation of multiple sets of DNA fragments of different genetic origin in a single small detection volume with a single ultrasensitive, albeit color-blind PMT.

We separated in a single experiment two diagnostically relevant samples obtained by multiplex ligation-dependent probe amplification (MLPA)—a set of 12 (origin: ERBB2 gene, 17th human chromosome, associated with breast cancer) and a set of 23 (origin: RPS19 gene, 19th human chromosome, associated with Diamond–Blackfan anemia) exclusively end-labeled DNA molecules sized 130–480 base pairs [106]. The results of this diagnostic experiment will be described in the following paragraphs.

Of the well-known PCR, MLPA [107] is a recent high-resolution variation. It consists of optimized thermal cycles during which multiple, correlated primer probes parallely target multiple DNA sites, ligate in pairs, and are amplified, resulting in multiple amplicon sets from different regions of a gene, each with its unique fluorescent label and comprising multiple DNA fragments having different lengths. The pattern of peaks witnessed in the subject (patient-gene) electropherogram on size-dependent MCE separation of these amplicons must be compared to a reference (healthy gene) electropherogram to draw accurate conclusions concerning the presence/absence of specific genetic mutations or chromosome aberrations in the region of the gene that was being targeted by the probes. In our experiments, we worked with commercially available MLPA samples [106] that were further amplified in a conventional PCR (Zebra Bioscience BV) [108] to reach a typical concentration of 10 nM to detect the fluorescence emitted from end-labeled DNA molecules with our optofluidic MCE setup. Ligated MLPA probes were amplified with PCR readymix (P4600, Sigma-Aldrich). Custom oligonucleotide SALSA primers (synthesized by Invitrogen) were modified at their two (5′ and 3′) ends with Alexa fluor 488 or 647. The PCR reaction mix used for amplification contained 400 nM of labeled forward and reverse SALSA primers. In this specific example, 12 DNA molecules were MLPA-extracted from the ERBB2 gene (associated with breast cancer) on the 17th human chromosome, and 23 DNA molecules were MLPA-extracted from the RPS19 gene (associated with Diamond–Blackfan anemia) on the 19th human chromosome, as a sample mixture. Each probe set was exclusively fluorescently labeled during PCR amplification, the former with Alexa Fluor 488 and the latter with Alexa Fluor 647. A three-step PCR protocol was employed for synthesis utilizing an iCycler (Bio-Rad). Final amplicons contained fluorescent species at both 3′ and 5′ ends of the DNA. In principle, MLPA–PCR multiplexing and exclusive fluorescent labeling can also be simultaneously applied to probes from different regions of the same gene in a single reaction [109].

The same experimental protocol as described in Section 13.3.1 was implemented, except that in this case no intercalating dye was used. The exclusively end-labeled MLPA-probe DNA molecules were injected as a single mixture plug into the MCE separation channel to be mutually separated and wavelength-selectively excited and detected. Two laser beams, the 488 nm line from an Ar-ion laser (mechanically amplitude-modulated at 17 Hz by means of a chopper) and the 635 nm line from a red laser diode (electrooptically amplitude-modulated at 31 Hz), were combined by an external fiber combiner and simultaneously launched into the excitation WG. For fluorescence detection, an ultrasensitive, but color-blind PMT was used.

The cumulative fluorescence signal as measured by the PMT is shown in Figure 13.13a. With the two excitation lasers modulated with different, harmonically uncorrelated frequencies, Fourier analysis of the detected signal results in the Fourier spectrum shown in Figure 13.13b, which resolves the spatio-temporally overlapping contributions at the two different wavelengths, thus separating similarly or even equally sized DNA molecules by identifying their origins via exciting their permanent end labels with unique temporal signatures, as shown in Figure 13.13c,d. The 12 blue-labeled breast-cancer probes and the 23 red-labeled anemia probes are unambiguously identified [99,100].

In addition to the earlier example, the modulation-encoding and Fourier-decoding scheme described earlier enables exciting possibilities, as described already in Section 13.5.1. We can intentionally add a standard reference to a diagnostic DNA sample, thereby providing intrinsic molecule-size calibration to the setup, making it independent of experimental parameters such as applied voltage, gel-matrix condition, and ambient temperature. Besides, we can also migrate an unknown, potentially malign sample including DNA molecules of different sizes from a diagnostically relevant genomic region together with their healthy counterparts of otherwise identical molecules, all exclusively color-labeled, thus providing unprecedented resolution to the experiment.

In combination with MLPA, this identification scheme will potentially become a powerful tool. Multiple MLPA samples from different genomic regions, associated with predispositions to different diseases, each exclusively color-end-labeled, as envisioned in Figure 13.14, can, in principle, be prepared in a single processing step [109]. Simultaneous investigation of several diagnostically relevant regions of the same or different genes becomes feasible in a single few-minute experiment, as principally illustrated in this section, ultimately leading to fast sampling of the complete genome for various defects. The results in Figure 13.13 are an example of this principle (although in this specific experiment, the two sets of DNA molecules were prepared in two separate MLPA experiments and then mixed).

This elegant approach introduces the concept of parallel processing to optical detection in LOCs. The experiment involves only two spectrally well-separated wavelengths to avoid cross-excitation of overlapping absorption bands of two dyes by a single excitation wavelength. However, ongoing research toward the development and commercial mass-production of tunable, narrow-excitation-band [110]

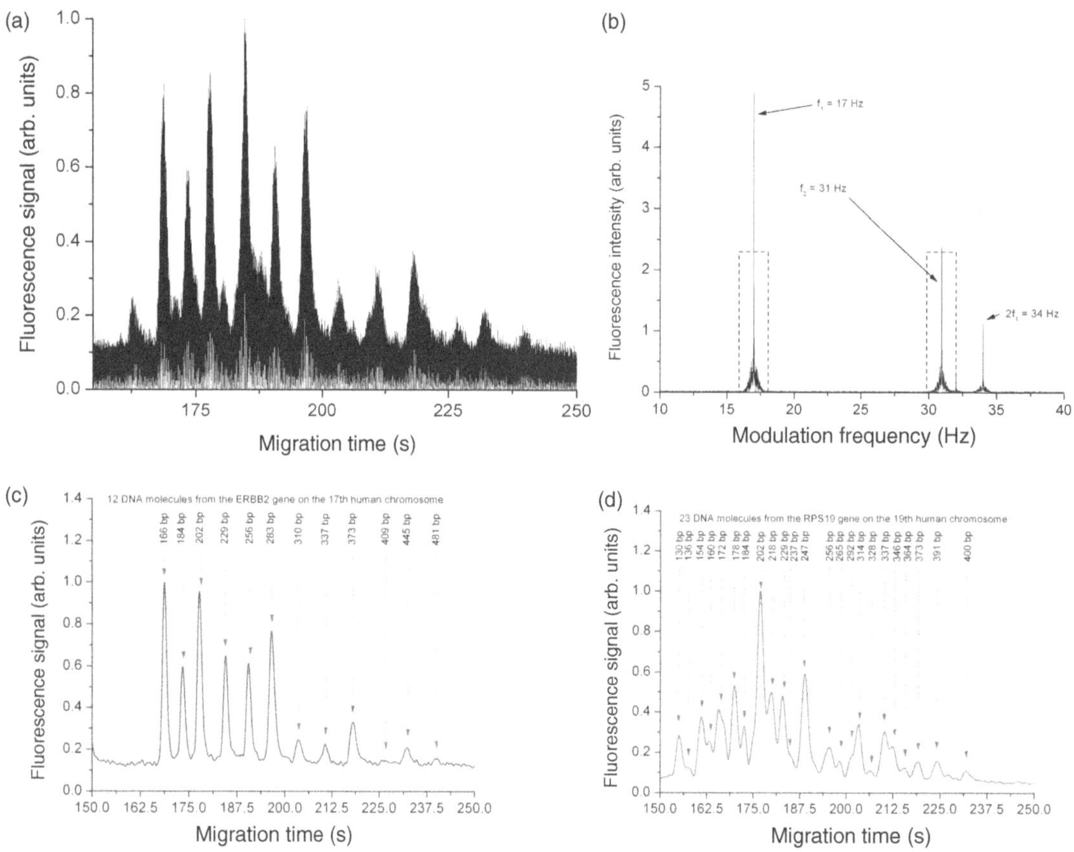

FIGURE 13.13 Illustration of the decoding of signals originating from two differently labeled, migrating samples with encoded excitation. (a) PMT-detected fluorescence signal from 35 end-labeled DNA molecules (consisting of 12 and 23 MLPA-PCR amplified DNA molecules from two chromosome regions) versus migration time. (b) Fourier spectrum and applied transfer function (indicated by the dashed line); individual signals separated by Fourier analysis. (c) 12 DNA molecules from a breast-cancer gene (dark gray). (d) 23 DNA molecules from a Diamond–Blackfan anemia gene (light gray). Several fluorescence peaks are below the noise level in part a, but are resolved in parts c and d by Fourier analysis. Figure modified from Refs. 99, 100.

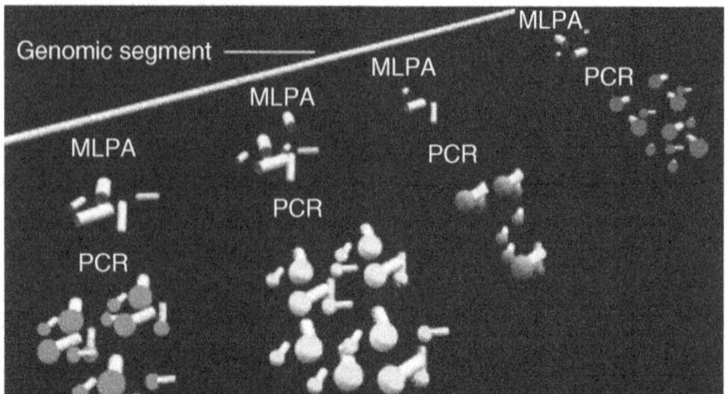

FIGURE 13.14 MLPA extraction and exclusive permanent end-labeling during PCR of sets of DNA fragments from different gene regions. Figure taken from Ref. 100.

II–VI semiconductor quantum-dot labels [111] can ultimately give rise to much higher degrees of spectral multiplexing, thereby fully exploiting this extra dimension in on-chip biomolecule detection. The envisioned compact and portable DNA super-analyzers may address global public health challenges [112] cost-effectively, for example, in cancer research by early identification of genomic deletions/insertions and in prenatal chromosome-aberration diagnosis, but also whenever time or sample-consumption efficiency is at stake; for example, imagine an automated sampling robot collecting and analyzing samples on distant planets as pointers to the existence of life, a paleontologist decoding the DNA signature from a Jurassic fossil, or a virologist simultaneously analyzing multiple samples faster than the viruses mutate.

13.6 CONCLUSIONS AND OUTLOOK

Large-scale CE holds several advantages when compared to other molecular separation techniques, such as liquid-phase chromatography [113]. These advantages include a high reproducibility even of highly complex samples. MCE is particularly beneficial in a clinical environment due to its high speed (a few-minutes duration) and the small volumes required for analysis [114]. Furthermore, accurate laser excitation and optical read-out of fluorescence from labeled molecules migrating in the MF chips can be provided by integrated optical waveguides.

The fast (few minutes per chip) optical postprocessing of CE microchips by femtosecond-laser inscription of optical channel waveguides into the microchip fully exploits the existing microfabrication technologies for the production of MF chips. Optofluidic characterization of the chips confirmed their excellent functionality and showed the feasibility of our setup to implement diagnostically relevant DNA separation/analyses. The functionality thereby obtained in an optofluidic CE chip enabled integrated-WG LIF excitation during MCE-based DNA analysis, although the chip-to-chip and experiment-to-experiment variations of electropherogram peak positions and relative peak heights—in other words, experimental reproducibility—is still an issue that needs to be addressed. High (>99%) DNA sizing accuracy was achieved. The potential of integrated optics may allow for a further 20-fold enhancement in the separation resolution in future CE microchips if the corresponding challenges of minimizing the MF plug broadening are overcome, thereby increasing the spatial MCE separation resolution. The low intrinsic LOD of 2.1 pM provided a sensitivity higher than the values reported in the literature for other integrated optofluidic approaches for monitoring dynamic electrophoretic migration experiments.

While promising results of electrophoretic separation with high sensitivity have been described, they all share a single limitation, namely the use of intercalating fluorescent dyes. These attach and detach at numerous regular positions along an analyte DNA molecule, leading to a relatively high fluorescence signal as compared to end-labeled molecules. However, due to the transitory nature of the attachment, the unique color-coding by exclusive fluorescent labeling of specific sets of DNA probes would be rendered impossible. This problem would have made impossible the construction of a setup that could perform multiplex DNA analyses in a single electrophoretic experiment, with a unique color-coding of the different sets of exclusively end-labeled DNA probes. Because end-labeled DNA molecules provide a smaller fluorescence signal compared to those intercalated with multiple dye molecules, it was therefore critical to enable the detection of ultra-low concentrations of fluorescently end-labeled DNA molecules. The steps toward such higher sensitivities were described in Section 13.4.2. Specifically, the all-numerically implemented lock-in amplification delivered a 10-fold SNR enhancement, leading to an LOD of 220 fM. This improvement set the stage for the detection of end-labeled molecules, and made it realistically possible to perform the envisioned multicolor fluorescent DNA analysis.

Our principle of parallel optical processing was based on color encoding by the unique modulation of multiple excitation laser beams to excite accordingly modulated fluorescence in the individual DNA probe sets, and the subsequent Fourier analytical decoding of the signal measured by a color-blind, but ultrasensitive PMT. The multicolor signals corresponding to the individual DNA probe-sets, which in our example correspond to probes from anemia and breast-cancer genes, respectively, were thereby simultaneously, nevertheless unambiguously unraveled. This demonstrates multicolor fluorescent DNA analysis in an optofluidic chip based on color-coding of the fluorescence signals emitted by the analyte molecules, thanks to modulation of the excitation laser beams. This approach introduces the concept of parallel processing for optical detection in MCE, and can also be applied elsewhere. Our proof of principle involves only two spectrally well-separated wavelengths to avoid cross-excitation of overlapping absorption bands of two dyes by a single excitation wavelength. However, applying tunable narrow-band labels could give rise to much higher degrees of spectral multiplexing, thereby fully exploiting this extra dimension in on-chip biomolecule detection. The envisioned compact and portable DNA analyzers may address global public health challenges cost-effectively, for example, in cancer research, by early identification of genomic deletions/insertions and in prenatal chromosome-aberration diagnosis, but also whenever time or sample-consumption efficiency is at stake within possibly wide-ranging application examples in space research, paleontology, virology, and so on.

Two aspects where this work could be extended in the future are (i) integration of optical detection on the chip, for example, as a hybrid flip-chip—containing a microlens or even an integrated WG for collecting and focusing the fluorescence signal emitted from the MF channels with a single electron-multiplying CCD or an avalanche photodiode for the sensitive detection of the fluorescence—that can be attached on top of the optofluidic chip; (ii) integration of optical excitation on the chip, once again, for example, as a hybrid flip-chip—containing solid-state light sources, such as LEDs, diode-lasers, or integrated optical WG lasers, for example, based on rare-earth-ion (e.g., Pr^{3+}) doping—that can be attached to the side-facet of the optofluidic chip as to directly couple to the WGs.

These developments will represent a significant advance not only in scientific terms, but also from a commercial and a broader societal point of view, whereby highly functional, point-of-care optofluidic devices would be deployed in the field, as enlightened biochemical labs-on-chips of the future.

ACKNOWLEDGMENTS

The authors gratefully acknowledge financial support through the EU project HIBISCUS, fruitful collaborations and discussions with J. van Weerd, R. van Weeghel, G.A.J. Besselink, M.J. Gilde, R. Dekker, R.G. Heideman, H.H. van den Vlekkert, N. Bellini, R. Martinez Vazquez, D. Nolli, R. Osellame, G. Cerullo, R. Ramponi, M. Cretich, and M. Chiari, as well as technical support from A.J.F. Hollink, M. Dijkstra, and H.A.G.M. van Wolferen.

REFERENCES

1. Manz, A., Graber, N., and Widmer, H.M. (1990) *Sens. Actuators B 1*, 244–248.
2. Reyes, R.D., Iossifidis, D., Auroux, P.A., and Manz, A. (2002) *Anal. Chem. 74*, 2623–2636.
3. Auroux, P.A., Reyes, D.R., Iossifidis, D., and Manz, A. (2002) *Anal. Chem. 74*, 2637–2652.
4. Jakeway, S.C., de Mello, A.J., and Russell, E.L. (2000) *Fresen. J. Anal. Chem. 366*, 525–539.
5. Lander, E.S. et al. (2001) *Nature 409*, 860–921.
6. Slater, G.W., Kenward, M., McCormick, L.C., and Gauthier, M.G. (2003) *Curr. Opin. Biotechnol. 14*, 58–64.
7. Altshuler, D., Daly, M.J., and Lander, E.S. (2008) *Science 322*, 881–888.
8. Harrison, D.J., Fluri, K., Seiler, K., Fan, Z.H., Effenhauser, C.S., and Manz, A. (1993) *Science 261*, 895–897.
9. Landers, J.P. (2003) *Anal. Chem. 75*, 2919–2926
10. Bruin, G.J.M. (2000) *Electrophoresis 21*, 3931–3951.
11. Fredlake, C.P., Hert, D.G., Kan, C.W., Chiesl, T.N., Root, B.E., Forster, R.E., and Barron, A.E. (2008) *Proc. Natl. Acad. Sci. USA 105*, 476–481.
12. Eid, J. et al. (2009) *Science 323*, 133–138.
13. Pile, D. (2009) *Nat. Photon 3*, 79–80.
14. http://www.pacificbiosciences.com. Accessed on Nov. 12, 2012.
15. Lagally, E.T. and Mathies, R.A. (2004) *J. Phys. D Appl. Phys. 37*, 245–261.
16. Easley, C.J., Karlinsey, J.M., Bienvenue, J.M., Legendre, L.G., Roper, M.G., Feldman, S.H., Hughes, M.A., Hewlett, E.L., Merkel, T.J., Ferrance, J.P., and Landers, J.P. (2006) *Proc. Natl. Acad. Sci. USA 103*, 19272–19277.
17. Brody, J.R. and Kern, S.E. (2004) *Anal. Biochem. 333*, 1–13.
18. Hubner, J., Mogensen, K.B., Jorgensen, A.M., Friis, P., Telleman, P., and Kutter, J.P. (2001) *Rev. Sci. Instrum. 72*, 229–234.
19. Bliss, C.L., McMullin, J.N., and Backhouse, C.J. (2007) *Lab Chip 7*, 1280–1287.
20. Soper, S.A., Flanagan, J.H., Legendre, B.L., Williams, D.C., and Hammer, D.N. (1996) *J. Select. Topics Quantum Electron. 2*, 1129–1140.
21. Dittrich, P.S. and Schwille, P. (2003) *Anal. Chem. 75*, 5767–5774.
22. Campbell, L.C., Wilkinson, M.J., Manz, A., Camilleri, P., and Humphreys, C.J. (2004) *Lab Chip 4*, 225–229.
23. Craighead, H.G. (2006) *Nature 442*, 387–393.
24. Dittrich, P.S. and Manz, A. (2005) *Anal. Bioanal. Chem. 382*, 1771–1782.
25. Ruano, J.M., Benoit, V., Aitchison, J.S., and Cooper, J.M. (2000) *Anal. Chem. 72*, 1093–1097
26. Friis, P., Hoppe, K., Leistiko, O., Mogensen, K.B., Hübner, J., and Kutter, J.P. (2001) *Appl. Opt. 40*, 6246–6251.
27. Verpoorte, E. (2003) *Lab Chip 3*, 42–52.
28. Goetz, S. and Karst, U. (2007) *Anal. Bioanal. Chem. 387*, 183–192.
29. Hunt, H.C. and Wilkinson, J.S. (2008) *Microfluid. Nanofluid. 4*, 53–79.
30. Myers, F.B. and Lee, L.P. (2008) *Lab Chip 8*, 2015–2031.
31. Lien, V., Berdichevsky, Y., and Lo, Y.H. (2004) *IEEE Photon. Technol. Lett. 16*, 1525–1527.
32. Lien, V., Zhao, K., Berdichevsky, Y., and Lo, Y.H. (2005) *IEEE J. Select. Topics Quantum Electron. 11*, 827–834.
33. Leeds, A.R., van Keuren, E.T., Durst, M.E., Schneider, T.W., Currie, J.F., and Paranjape, M. (2004) *Sens. Actuators A 115*, 571–580.
34. Cleary, A., Garcia-Blanco, S., Glidle, A., Aitchison, J.S., Laybourn, P., and Cooper, J.M. (2005) *IEEE Sens. J. 5*, 1315–1320.
35. Psaltis, D., Quake, S.R., and Yang, C. (2006) *Nature 442*, 381–386.
36. Lambeck, P.V. (2006) *Meas. Sci. Technol. 17*, 93–102.
37. Yin, D., Deamer, D.W., Schmidt, H., Barber, J.P., and Hawkins, A.R. (2006) *Opt. Lett. 31*, 2136–2138.
38. Yin, D., Lunt, E.J., Rudenko, M.I., Deamer, D.W., Hawkins, A.R., and Schmidt, H. (2007) *Lab Chip 7*, 1171–1175.
39. Applegate, R.W., Jr. Squier, J., Vestad, T., Oakey, J., Marr, D. W.M., Bado, P., Dugan, M.A., and Said, A.A. (2006) *Lab Chip 6*, 422–426.
40. Mach, P., Dolinski, M., Baldwin, K.W., Rogers, J.A., Kerbage, C., Windeler, R.S., and Eggleton, B.J. (2002) *Appl. Phys. Lett. 80*, 4294–4296.
41. Wolfe, D.B., Vezenov, D.V., Mayers, B.T., Whitesides, G.M., Conroy, R.S., and Prentiss, M.G. (2005) *Appl. Phys. Lett. 87*, 181105.
42. Li, Z., Zhang, Z., Emery, T., Scherer, A., and Psaltis, D. (2006) *Opt. Express 14*, 696–701.
43. Vezenov, D.V., Mayers, B.T., Conroy, R.S., Whitesides, G.M., Snee, P.T., Chan, Y., Nocera, D.G., and Bawendi, M.G. (2005) *J. Am. Chem. Soc. 127*, 8952–8953.
44. Misiakos, K., Kakabakos, S.E., Petrou, P.S., and Ruf, H. (2004) *Anal. Chem. 76*, 1366–1373.

45. Lin, C.H., Lee, G.B., Fu, L.M., and Chen, S.H. (2004) *Biosens. Bioelectron. 20*, 83–90.
46. Cran-McGreehin, S.J., Dholakia, K. and Krauss, T.F. (2006) *Opt. Express 14*, 7723–7730.
47. Song, C., Nguyen, N.T., Tan, S.H., and Asundi, A.K. (2009) *Lab Chip 9*, 1178–1184.
48. Namasivayam, V., Lin, R., Johnson, B., Brahmasandra, S., Razzacki, Z., Burke, D.T., and Burns, M.A. (2004) *J. Micromech. Microeng. 14*, 81–90.
49. Tang, S.K., Stan, C.A., and Whitesides, G.M. (2008) *Lab Chip 8*, 395–401.
50. Levene, M.J., Korlach, J., Turner, S.W., Foquet, M., Craighead, H.G., and Webb, W.W. (2003) *Science 299*, 682–687.
51. Mogensen, K.B., Friis, P., Hübner, J., Petersen, N., Jorgensen, A.M., Telleman, P., and Kutter, J.P. (2001) *Opt. Lett. 26*, 716–718.
52. Mogensen, K.B., Klank, H., and Kutter, J.P. (2004) *Electrophoresis 25*, 3498–3512.
53. Dumais, P., Callender, C.L., Noad, J.P., and Ledderhof, C.J. (2005) *Photon. Technol. Lett. 17*, 441–443.
54. Dumais, P., Callender, C.L., Ledderhof, C.J., and Noad, J.P. (2006) *Appl. Opt. 45*, 9182–9190.
55. Vieillard, J., Mazurczyk, R., Morin, C., Hannes, B., Chevelot, Y., Desbene, P.L., and Krawczyk, S. (2007) *J. Chromatogr. B 845*, 218–225.
56. Mazurczyk, R., Vieillard, J., Bouchard, A., Hannes, B., and Krawczyk, S. (2006) *Sens. Actuators B 118*, 11–19.
57. Mogensen, K.B., El-Ali, J., Wolff, A., and Kutter, J.P. (2003) *Appl. Opt. 42*, 4072–4079.
58. Kuswandi, B., Nuriman H.J., and Verboom, W. (2007) *Anal. Chim. Acta 601*, 141.
59. Lee, G.B., Lin, C.H., and Guan, G.L. (2003) *Sens. Actuators A 103*, 165–170.
60. Wang, Z., Sekulovic, A., Kutter, J.P., Bang, D.B., and Wolff, A. (2006) *Electrophoresis 27*, 5051–5058.
61. Olivares, J.A., Stark, P.C., and Jackson, P. (2002) *Anal. Chem. 74*, 2008–2013.
62. Bliss, C.L., McMullin, J.N., and Backhouse, C.J. (2008) *Lab Chip 8*, 143–151.
63. Davis, K.M., Miura, K., Sugimoto, N., and Hirao, K. (1996) *Opt. Lett. 21*, 1729–1731.
64. Miura, K., Qiu, J., Inouye, H., Mitsuyu, T., and Hirao, K. (1997) *Appl. Phys. Lett. 71*, 3329–3331.
65. Liu, X., Du, D., and Mourou, G. (1997) *IEEE J. Quantum Electron. 33*, 1706–1716.
66. Bellouard, Y., Said, A., Dugan, M., and Bado, P. (2004) *Mater. Res. Soc. Symp. Proc. 782*, A3.2.1–A3.2.6.
67. Itoh, K., Watanabe, W., Nolte, S., and Schaffer, C. (2006) *MRS Bull. 31*, 620–625.
68. Eaton, S.M., Chen, S.W., Zhang, L., Zhang, H., Iyer, R., Aitchison, J.S., and Herman, P.R. (2006) *IEEE Photon. Technol. Lett. 18*, 2174–2176.
69. Osellame, R., Maselli, V., Martinez Vazquez, R., Ramponi, R., and Cerullo, G. (2007) *Appl. Phys. Lett. 90*, 231118.
70. Gattass, R.R. and Mazur, E. (2008) *Nat. Photon. 2*, 219.
71. Martinez Vazquez, R., Osellame, R., Nolli, D., Dongre, C., van den Vlekkert, H.H., Ramponi, R., Pollnau, M., and Cerullo, G. (2009) *Lab Chip 9*, 91–96.
72. http://www.lionixbv.nl. Accessed on Nov. 12, 2012.
73. Dongre, C., Dekker, R., Hoekstra, H.J.W.M., Pollnau, M., Martinez Vazquez, R., Osellame, R.C., Erullo, G., Ramponi, R., van Weeghel, R., Besselink, G.A.J., and van den Vlekkert, H.H. (2008) *Opt. Lett. 33*, 2503–2505.
74. Osellame, R., Hoekstra, H.J.W.M., Cerullo, G., and Pollnau, M. (2011) *Laser Photon. Rev. 5*, 442–463.
75. Osellame, R., Taccheo, S., Marangoni, M., Ramponi, R., Laporta, P., Polli, D., de Silvestri, S., and Cerullo, G. (2003) *J. Opt. Soc. Am. B 20*, 1559–1567.
76. Johnson, M.E. and Landers, J.P. (2004) *Electrophoresis 25*, 3513–3527.
77. http://www.capilix.com. Accessed on Nov. 12, 2012.
78. Cretich, M., Chiari, M., Pirri, G., and Crippa, A. (2005) *Electrophoresis 26*, 1913–1919.
79. Martinez Vazquez, R., Osellame, R., Cretich, M., Chiari, M., Dongre, C., Hoekstra, H.J.W.M., Pollnau, M., van den Vlekkert, H.H., Ramponi, R., and Cerullo, G. (2009) *Anal. Bioanal. Chem. 393*, 1209–1216.
80. Tian, H. and Landers, J.P. (2002) *Anal. Biochem. 309*, 212–223.
81. Dongre, C., van Weerd, J., Besselink, G.A.J., van Weeghel, R., Martinez Vazquez, R., Osellame, R., Cerullo, G., Cretich, M., Chiari, M., Hoekstra, H.J.W.M., and Pollnau, M. (2010) *Electrophoresis 31*, 2584–2588.
82. Dongre, C., Pollnau, M., and Hoekstra, H.J.W.M. (2011) *Analyst 136*, 1248–1251.
83. Krawczyk, M.J., Dulak, J., and Kulakowski, K. (2002) *Electrophoresis 23*, 182–185.
84. Paegel, B.M., Emrich, C.A., Wedemeyer, G.J., Scherer, J.R., and Mathies, R.A. (2002) *Proc. Natl. Acad. Sci. USA 99*, 574–579.
85. Dolnik, V. and Gurske, W.A. (1999) *Electrophoresis 20*, 3373–3380.
86. Horlick, G. (1975) *Anal. Chem. 47*, 352–354.
87. Blair, D.P. and Sydenham, P.H. (1975) *J. Phys. E 8*, 621–627.
88. Temple, P.A. (1975) *Am. J. Phys. 43*, 801–807.
89. Wolfson, R. (1991) *Am. J. Phys. 59*, 569–572.
90. Novak, L., Neuzil, P., Pipper, J., Zhang, Y., and Lee, S. (2007) *Lab Chip 7*, 27–29.
91. Rivera, L., Puyol, M., Villuendas, F., and Alonso, J. (2008) *Sens. Actuators B 134*, 863–868.
92. Iten, R. (2007) Eur. Patent 1962084.
93. Zhu, L., Stryjewski, W.J., and Soper, S.A. (2004) *Anal. Biochem. 330*, 206–218.
94. Götz, S. and Karst, U. (2007) *Sens. Actuators B 123*, 622–627.
95. Karger, A.E., Harris, J.M., and Gesteland, R.F. (1991) *Nucleic Acids Res. 19*, 4955–4962.
96. Sugino, H., Ozaki, K., Shirasaki, Y., Arakawa, T., Shoji, S., and Funatsu, T. (2009) *Lab Chip 9*, 1254–1260.

97. Dongre, C., van Weerd, J., Bellini, N., Osellame, R., Cerullo, G., van Weeghel, R., Hoekstra, H.J.W.M., and Pollnau, M. (2010) *Biomed. Opt. Express 1*, 729–735.

98. Crespi, A., Gu, Y., Ngamsom, B., Hoekstra, H.J.W.M., Dongre, C., Pollnau, M., Ramponi, R., van den Vlekkert, H.H., Watts, P., Cerullo, G., and Osellame, R. (2010) *Lab Chip 10*, 1167–1173.

99. Dongre, C., van Weerd, J., Besselink, G.A.J., Martinez Vazquez, R., Osellame, R., Cerullo, G., van Weeghel, R., van den Vlekkert, H.H., Hoekstra, H.J.W.M., and Pollnau, M. (2010) High-resolution, multi-wavelength fluorescent DNA analysis in an optofluidic chip, in *Conference on Lasers and Electro-Optics, San José, California, 2010, Technical Digest*, Optical Society of America, Washington, DC, paper CTuN7.

100. Dongre, C., van Weerd, J., Besselink, G.A.J., Martinez Vazquez, R., Osellame, R., Cerullo, G., van Weeghel, R., van den Vlekkert, H.H., Hoekstra, H.J.W.M., and Pollnau, M. (2011) *Lab Chip 11*, 679–683.

101. Lei, H.J. and Crothers, D.M. (1969) *J. Mol. Biol. 39*, 461–477.

102. Zipper, H., Brunner, H., Bernhagen, J., and Vitzthum, F. (2004) *Nucleic Acids Res. 32*, e103.

103. Olson, N.A., Khandurina, J., and Guttman, A. (2004) *J. Chromatogr. A 1051*, 155–160.

104. Biver, T., de Biasi, A., Secco, F., Venturini, M., and Yarmoluk, S. (2005) *Biophys. J. 89*, 374–383.

105. Sulzbacher Caruso, C., Lanças, F.M., and Carrilho, E. (2003) *Electrophoresis 24*, 78–85.

106. http://www.mrc-holland.com. Accessed on Nov. 12, 2012.

107. Schouten, J.P., McElgunn, C.J., Waaijer, R., Zwijnenburg, D., Diepvens, F., and Pals, G. (2002) *Nucleic Acids Res. 30*, 57–69.

108. http://www.zebrabioscience.nl. Accessed on Nov. 12, 2012.

109. Bunyan, D.J., Skinner, A.C., Ashton, E.J., Sillibourne, J., Brown, T., Collins, A.L., Cross, N.C., Harvey, J.F., and Robinson, D.O. (2007) *Mol. Biotechnol. 35*, 135–140.

110. Walker, B.J., Nair, G.P., Marshall, L.F., Bulovic, V., and Bawendi, M.G. (2009) *J. Am. Chem. Soc. 131*, 9624–9625.

111. Resch-Genger, U., Grabolle, M., Cavaliere-Jaricot, S., Nitschke, R., and Nann, T. (2008) *Nat. Methods 5*, 763–775.

112. Yager, P., Edwards, T., Fu, E., Helton, K., Nelson, K., Tam, M. R., and Weigl, B.H. (2006) *Nature 442*, 381–393.

113. Dakna, M., He, Z.Y., Yu, W.C., Mischak, H., and Kolch, W. (2009) *Chromatogr. B 877*, 1250–1258.

114. Kohlheyer, D., Eijkel, J.C.T., van den Berg, A., and Schasfoort, R.B.M. (2008) *Electrophoresis 29*, 977–993.

14

CAPILLARY ELECTROPHORESIS OF INTACT UNFRACTIONATED HEPARIN AND RELATED IMPURITIES

ROBERT WEINBERGER

14.1 INTRODUCTION

Heparin is one of the oldest drugs still in use [1–2]. This naturally occurring anticoagulant is administered intravenously to acutely treat deep vein and pulmonary thrombosis, unstable angina, and during coronary bypass surgery. In addition, the surface coating properties of heparin are used to prevent coagulation within blood collection tubes and kidney dialysis machines. A countless number of doses (millions) have been administered to patients over the years with relative safety. In any given month, 1–4 deaths were reported in patients treated with heparin. This was considered the baseline mortality of this already ill population subset; however, something went terribly wrong. The number of deaths began to increase in October 2007 when seven were reported. The peak mortality was January–February 2008 when 31 and 28 deaths were reported to Food and Drug Administration (FDA), respectively. Perhaps 10 times more patients were sickened as well. Massive recalls were undertaken on a worldwide basis.

All of the affected raw heparin had been imported from China. According to the *New York Times*, the *FDA* had earlier admitted that it violated its own policy by failing to inspect the Changzhou plant before allowing it, in 2004, to make and export heparin's active ingredient to Baxter in the United States, where the finished product was made [3].

Heparin is a natural product that is extracted from animal tissues. Years ago, beef intestine was the tissue of choice. Fear of mad cow disease shifted production from cattle to swine. Since 50% of the world's pigs are in China, the country became the world's largest producer of heparin. The production of heparin is a cottage industry in China where there are hundreds if not thousands of small shops that extract the heparin-containing pulp from pig intestines. Consolidators collect the pulp and deliver material to factories for extraction and purification of raw heparin. Because of the unique way heparin is gathered and produced, even a thorough inspection might not have averted the crisis.

Heparin is a naturally occurring polydisperse glycosoaminoglycan (GAG) with the highest negative charge density of any known biological molecule. This and other GAGs are found in many animal tissues and some are potential contaminants in heparin active pharmaceutical ingredient (API). Structural differences between GAGs are often determined by measuring the disaccharide or oligosaccharide content of each material. Despite their polydispersity, it is still desirable to perform separations on intact GAGs. The broad peak found for heparin was not a wall effect but reflected the reality of a very complex mixture of similar compounds. Figure 14.1 shows the major repeating units of each of these molecules. Chondroitin sulfate normally has a sulfate group in either the 4 position (chondroitin A) as shown or the 6 position (chondroitin C). If one or more of the R groups are sulfated, the chondroitin sulfate is considered oversulfated or oversulfated chondroitin sulfate (OSCS). If an OSCS contains sulfate groups at all of the R positions, it is

Capillary Electrophoresis and Microchip Capillary Electrophoresis: Principles, Applications, and Limitations, First Edition.
Edited by Carlos D. García, Karin Y. Chumbimuni-Torres, and Emanuel Carrilho.

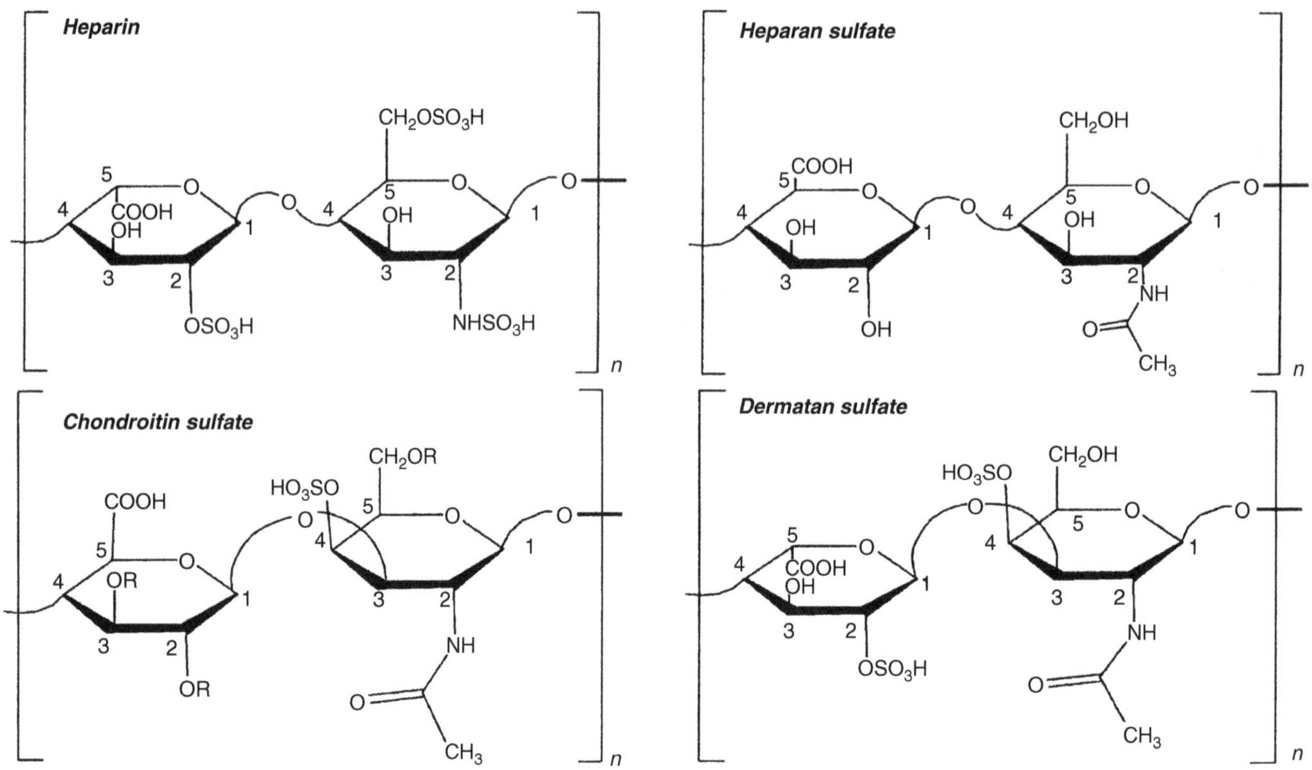

FIGURE 14.1 Chemical structures of heparin, heparan sulfate, chondroitin sulfate A, and DS major repeating units. OSCS will have at least one but possibly up to all three (fully sulfated) R groups as sulfate groups. Reprinted with permission from Ref. 4.

considered fully sulfated or fully sulfated chondroitin sulfate (FSCS) [4]. Dermatan sulfate (DS) was at one time referred to as chondroitin sulfate B.

An intense effort in industry, academia, and government was undertaken to determine the cause of the increase in mortality and morbidity because of intravenous administration of certain lots of heparin. Starting from a blank canvas, over 60 different analytical techniques were employed in an attempt to distinguish "good" heparin from "bad." Searches for heavy metals, biological contamination, proteins, nucleic acids, small molecules, and solvents were undertaken. An unknown substance in "bad" heparin inhibited enzyme digestion using either heparinase or chondroitinase. By March 2008, only two techniques, capillary electrophoresis (CE) and nuclear magnetic resonance (NMR) spectroscopy, were able to identify "bad" heparin. At that time, the offending impurity was as yet unidentified. "Good" or "bad" heparin was determined entirely on whether a particular lot of heparin elicited an adverse effect. In April 2008, the impurity was identified as OSCS [5], a related GAG.

The CE technique developed in early 2008 at Baxter Healthcare was an emergency method, badly in need of optimization. Figure 14.2 shows the separation of the unidentified impurity—now known as OSCS—and heparin. "Bad" lots of heparin gave the unknown peak while the peak was absent in "good" heparin. This method, which appeared on the FDA website on March 6, 2008, could be used to identify high levels of OSCS contamination in heparin samples.

Once the emergency methods appeared on the FDA website and screening of all heparin was accomplished, the number of adverse reactions quickly returned to baseline. The FDA's CE and NMR methods were validated and became official USP methodology on June 18, 2008. Both CE and NMR served as an identity test for OSCS with a specification of none detectable.

I was asked by numerous stakeholders in March 2008 first to run the FDA method and then to determine whether the separation between heparin and OSCS could be improved. I later discovered that others had tried to improve the method without success. It turned out that their focus was on sample concentration and field strength, neither of which was critical to the separation. Therein became an important consideration for the general acceptance of CE for heparin analysis; the user base was not sufficiently expert despite 23 years of commercial availability of instrumentation.

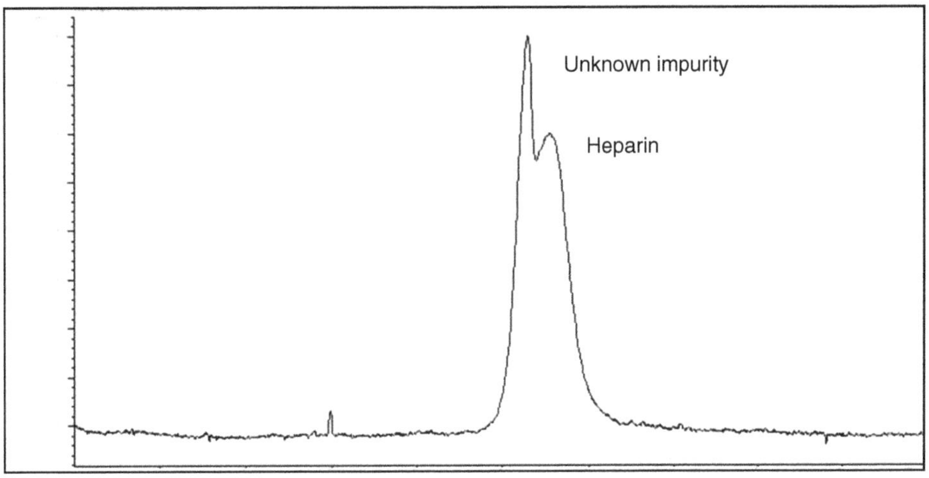

FIGURE 14.2 CE separation of heparin and impurity. Buffer: 36 mM sodium phosphate, pH 3.5 capillary: 50 μm, 64.5 cm total, 56 cm effective length. Voltage: −30 kV. Injection: 500 mbar s (mbs). Wavelength: 200 nm. Heparin concentration: 10 mg mL^{-1}. Heparin migrates toward the positive electrode.

14.2 CAPILLARY ELECTROPHORESIS AND HEPARIN

From 1991 to 2010, there were close to 200 papers on CE of heparin and related compounds. Some of these papers dealt with binding of heparin to proteins or the use of heparin in chiral recognition, neither of which was pertinent to heparin analysis or the determination of impurities in heparin. The first use of CE for heparin analysis was reported in 1991 when following depolymerization, the disaccharide composition of dermatan, chondroitin, heparin, and heparan sulfates was determined [6,7]. Disaccharide analysis can be helpful in elucidation of the structure of these complex molecules. For example, a combination of galactosamine and glucosamine was found in "bad" heparin, whereas "good" heparin contained only glucosamine. Most of the papers of this era employed depolymerization and disaccharide and/or oligosaccharide separations. Depolymerization is a degradation technique; however, such techniques allowed in part that contaminated heparin to enter the world's supplies. When determining impurities in the API, it is often best to determine the intact molecule if a sensitive and selective technique is available because degradation techniques can be deceived by unscrupulous individuals at any point in the supply chain.

The first report of an intact GAG separation was found somewhat hidden in a 1994 publication [8]. What was remarkable about one of the reported separations was partial resolution of the microheterogeneous structure of heparin. Nevertheless, separations of intact heparin with sufficient resolving power to separate the toxic impurities of 2008 and others remained elusive for some time. In 1996, Malsch et al. [9] reported separations of low molecular mass heparins using 20 mM phosphate buffer, pH 3.5. In 2008, Patel et al. used 50 mM phosphate buffer, pH 3.5 to compare separations of stressed and unstressed low-molecular-weight heparin [10]. These latter two separations employed countermigration CE at low pH with reversed polarity. The electroosmotic flow (EOF) was low enough such that the negatively charged GAGs would migrate toward and reach the positive electrode. However simple, these conditions served as a starting point for method optimization.

14.3 METHOD DEVELOPMENT IN CAPILLARY ELECTROPHORESIS

Methodology in CE is often quite simple. The idea is to take a minimalist approach and do just what is necessary avoiding getting overly complicated. This advice is often unheeded. The important experimental criteria are described in more detail in Table 14.1 along with the optimal conditions for separation of intact heparin from many of its impurities. Depending on the application, certain parameters become critical for the separation and these become apparent during scouting runs. The buffer pH is often the most important parameter but this was not the case for heparin. For heparin, high-concentration buffers were critical, while for other applications, lower-concentration buffers may work best. In any event, one has to stay disciplined and work methodically regardless the challenge. For heparin, feasibility studies were complete in a day or so with the optimization quickly completed. That being said, method development in CE breaks down into four main areas: (i) capillary selection, (ii) buffer development, (iii) instrumental parameters, and (iv) sample preparation.

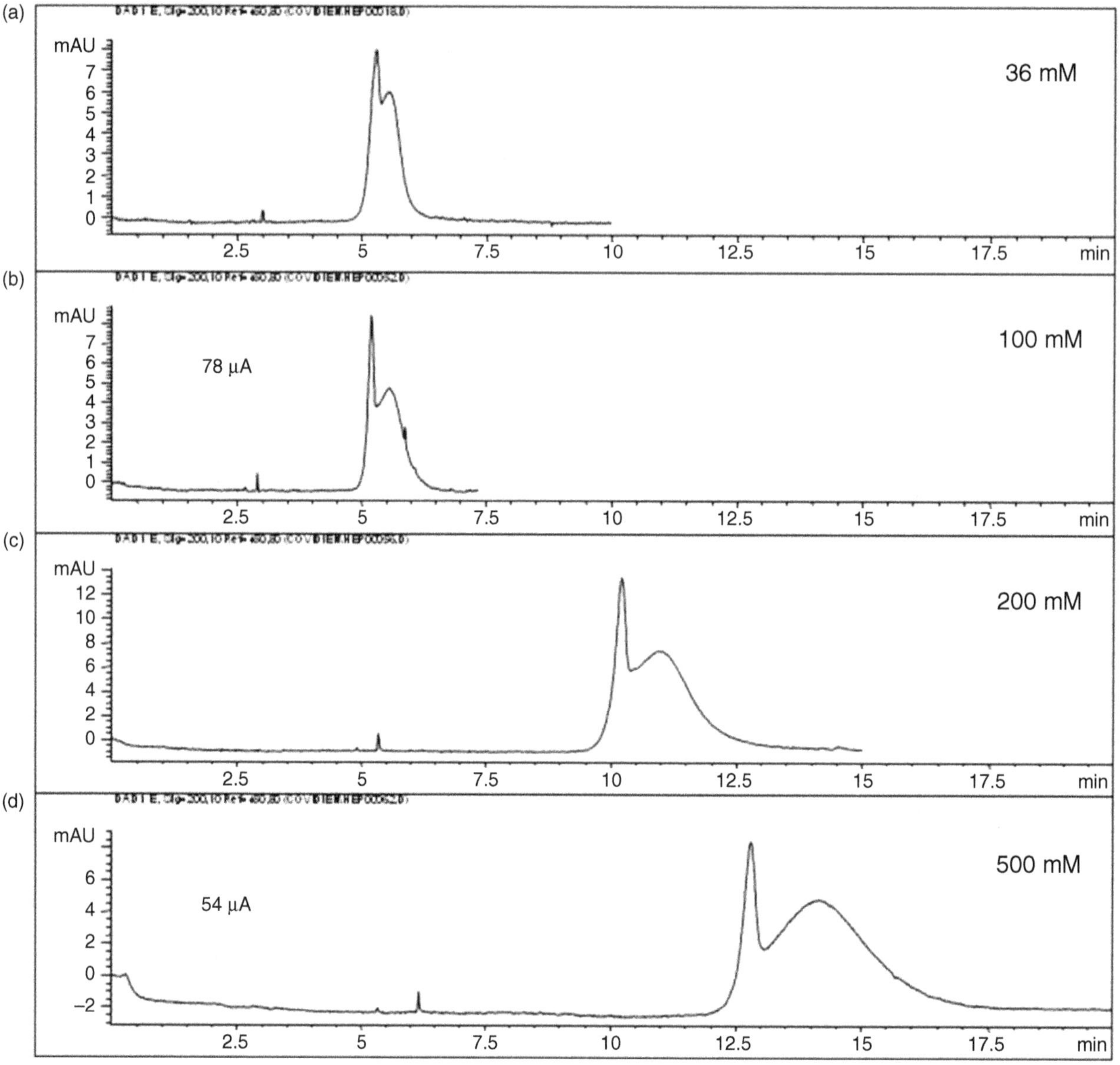

FIGURE 14.3 Impact of buffer concentration on the migration time and resolution of OSCS and heparin. Buffer pH: 3.5 and temperature: 25°C. (a and b) Capillary: 50 μm, 64.5 cm total length, 56 cm effective length; voltage: −30 kV; injection: 500 mbs. (c and d) Capillary: 25 μm, 80.5 cm total length, 72 cm effective length; voltage: −30 kV; injection: 2000 mbs. Reprinted with permission from Ref. 4.

Heparin was polyanionic as were all of the related impurities. Since the compound was anionic, a bare silica capillary was selected, since the charge on the capillary wall was slightly negative at low pH. The initial inner diameter was 50 μm, but 25 μm was quickly adapted as the current rose when high-concentration buffers were employed. At pH 3.5, a buffer concentration study was performed and these results are shown in Figure 14.3 [4]. As the phosphate buffer concentration was increased from 36 to 100 mM, the migration times remained constant but the OSCS peak sharpened. The sharpening was due to sample stacking, which was mediated by the increase in buffer concentration. The migration times did not change because by coincidence, both the solute mobility and EOF were decreased by the same amount. As the buffer concentration was further increased, the migration times for heparin and OSCS now increased, since the EOF approached zero but the solute mobility continued to decrease. The added time on the capillary resulted in an improvement in resolution. For this application, the high-concentration buffer was the

TABLE 14.1 Analytical Conditions for Heparin Method

Capillary selection	
a. Surface	Bare silica
b. Diameter	25 μm i.d.
c. Length	8.5 or 24.5 cm (33 cm total length)
d. Reconditioning	1 min 1 N phosphoric acid 1 min 0.1 N sodium hydroxide
Buffer development	
a. Co-ion	Phosphate
b. pH	2.5
c. Concentration	0.6–1 M
d. Counterion	Lithium
d. Buffer additives	None required
Instrumental parameters	
a. Voltage	−14 kV
b. Injection size	600 mbs
c. Temperature	20°C
d. Detection	200 nm
Sample	
a. Concentration	50 mg mL^{-1}
b. Matrix	Water
c. Sample preparation	Filter, 0.21 μm cellulose acetate

critical parameter. Because such a high buffer concentration was required, a 25 μm i.d. capillary was used to reduce the current to acceptable levels. A pH of 2.5 was later found to be optimal.

The next important parameter was the buffer counterion. Since pH 2.5 was used, phosphate was the logical choice for co-ion, since its buffer capacity was good and the solution was almost UV transparent at 200 nm. Normally, sodium is chosen as the counterion but it was ideal to maximize the field strength. By selecting the lower-mobility ion, lithium as counterion, the field strength could be increased. Another group pursuing the same application chose tris as the counterion [11].

Another unusual feature of the method was the high concentration of heparin in the injection solution, 30–50 mg mL^{-1}. Such a high solute concentration was possible only because of the high buffer concentration. The high heparin concentration contributed to the high sensitivity of the method toward OSCS and other impurities.

To speed the separation, the capillary length should be reduced and the buffer optimized to the shortened length. In this case, the capillary length can be the short-end on the Agilent CE which is 8.5 cm [12]. The optimal buffer concentration for the short capillary was 1 M lithium phosphate, pH 2.5. Higher-concentration buffers became too conductive and the currents ran high despite the 25 μm i.d. capillary. Alternatively, a 24 cm capillary can be used with 0.6 M lithium phosphate or higher. Here, the resolution was somewhat higher albeit with longer separation times.

Since the EOF approached zero, the reproducibility of the separation was excellent. A calibration run is shown in Figure 14.4. Thiosulfate could serve as an internal standard but this was found to be unnecessary. The limit of detection for OSCS was well below 0.1%. The optimized CE method on an 8.5 cm capillary is shown in Figure 14.5 as an overlay of five separate runs. Needless to say, the reproducibility is excellent. The largest source of error is peak integration. The run time was limited to 5 min, the heparin concentration was 50 mg mL^{-1}, OSCS was 0.4%, and DS was 2.0% of the heparin sample. The amount of OSCS was 125 times

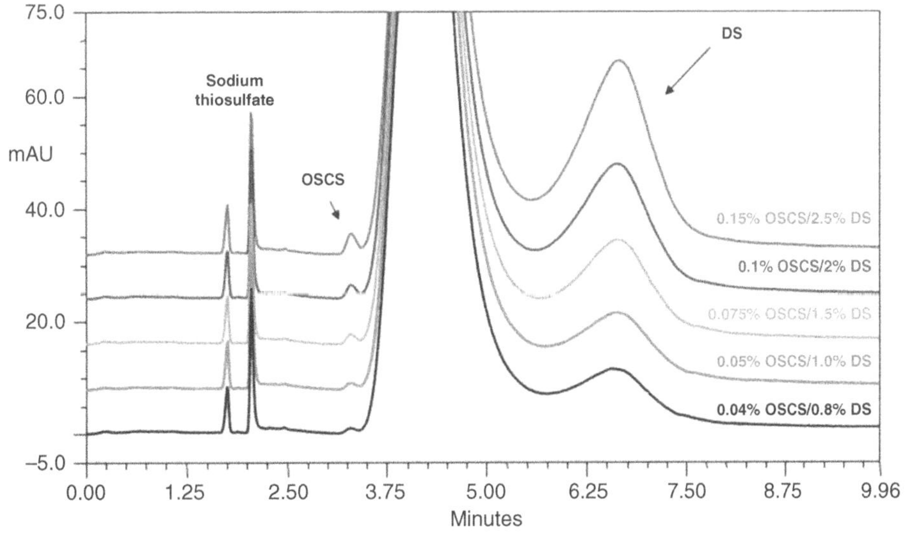

FIGURE 14.4 Buffer: 600 mM Li phosphate pH 2.5; capillary: 25 μm i.d. × 21.5 cm effective × 30 cm total length; temperature: 20°C; injection: 1000 mbs; voltage: −14 kV. Solute concentrations: OSCS: 0.1%, DS: 2.0%. Sodium thiosulfate can serve as an internal standard if needed.

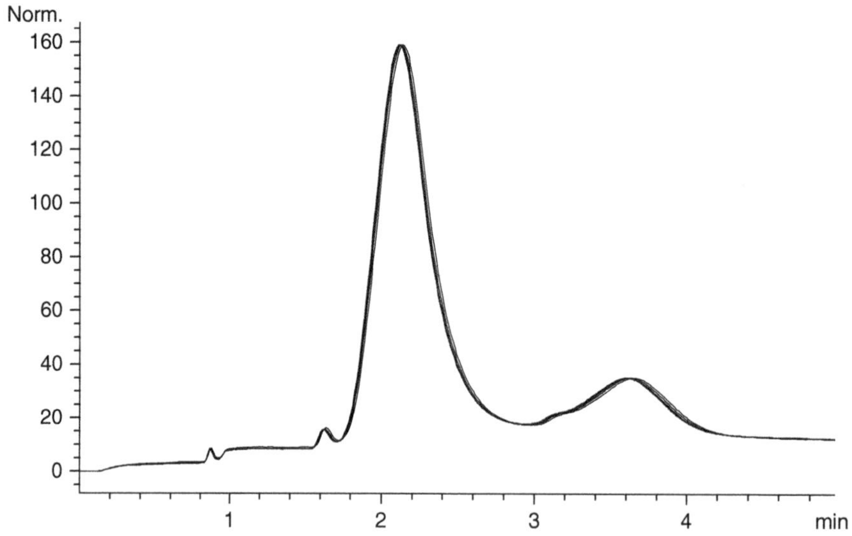

FIGURE 14.5 Overlay of five runs on the short end. Buffer: 600 mM Li phosphate pH 2.5; capillary: 25 μm i.d. × 24.5 cm effective × 33 cm total length, temperature: 20°C, injection: 600 mbs; voltage: −14 kV. The total run time is 5 min.

lower than the heparin concentration. This simulates the type of samples commonly found in the field with the impurities near their limit of detection (LOD) or specification limit.

A comparison of quantitation of OSCS by both CE and NMR methods was performed [12]. There was no significant difference between the two methods for 30 tested samples ranging in concentration from 0% to 32% OSCS. The CE method was at least 10× more sensitive compared to the NMR method.

14.4 COMMON IMPURITIES FOUND IN HEPARIN

While OSCS was the toxic impurity found in contaminated heparin, other GAGs were often found, particularly in crude heparin. These include chondroitin A, DS, and heparan sulfate as well as their oversulfated versions. A number of these structures are shown in Figure 14.1. The USP has a zero limit test for OSCS. Otherwise, only DS has a specification limit in the USP. The separation of some of these impurities from a crude heparin is shown in Figure 14.6.

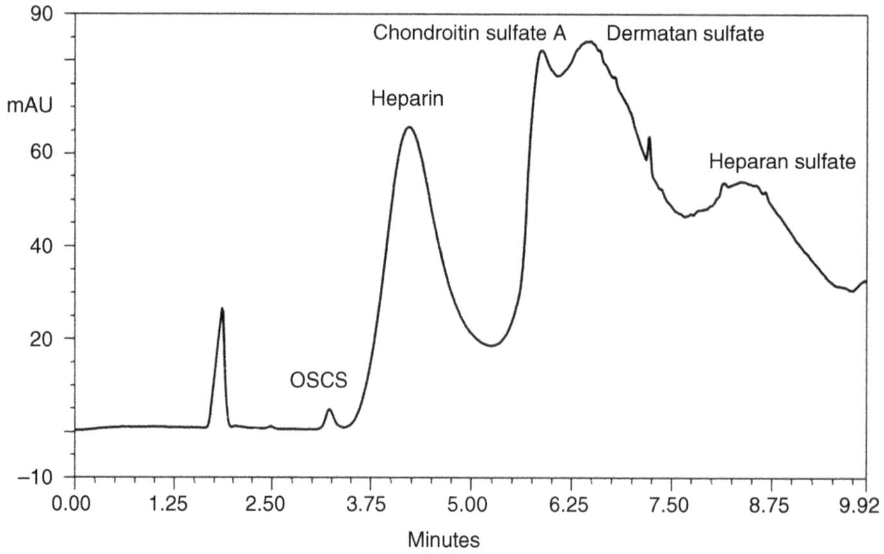

FIGURE 14.6 Separation of impurities in crude heparin. Buffer: 600 mM Li phosphate pH 2.5; capillary: 25 μm i.d. × 24.5 cm effective × 33 cm total length, temperature: 20°C, injection: 1000 mbs; voltage: −14 kV.

It was probably best to dissolve heparin samples in 10 mM SDS instead of water to enhance solubility but this was not studied in detail. Proteins and other insoluble material can lower the recoveries of impurities such as OSCS. Oversulfated DS and oversulfated heparin coeluted with each other but were separable from OSCS and heparin, respectively [12].

14.5 THE UNITED STATES PHARMACOEPIA AND CE OF HEPARIN

The USP specifications for heparin were woefully inadequate prior to the crisis. The monograph was lacking both chromatographic and spectroscopic methodologies. Originally issued in the 1950s prior to the instrumentation era, the monograph relied on a bioassay for blood coagulation time. From this bioassay, the potency of each individual batch of heparin could be determined and expressed as international units (IU). There are approximately 200 IU per mg of heparin but this varies depending on the potency of the drug. Degradation tests such as Kjeldahl total nitrogen were also part of that specification. In general, degradation tests are more easily deceived if that is the aim of the perpetrator. Indeed, that was the case for contaminated heparin. The specifications and tests for heparin API at the time of the crisis are shown in Table 14.2. While the specifications and tests were thorough for inorganic impurities, bacterial contamination, and residual proteins, they were deficient to test for related organic impurities. While this may have been difficult to anticipate, the lack of separation and spectroscopic techniques for a widely prescribed drug substance was highly unusual at this day and age.

Shortly after inclusion of CE as the official method, the USP opted to replace the CE method with strong anion-exchange (SAX) chromatography [13]. The use of CE as an identity test method was criticized because of migration time drift, variability between systems, variability of heparin from different sources, buffer incompatibility for use with heparin calcium, and resolution and sensitivity limitations for some impurities [14]. It did not appear that the regulators made a diligent attempt to overcome these issues. The new non-CE methodologies became official on October 1, 2009.

Indeed, the bar for acceptance of CE is very high. The speed, resolution, and sensitivity of SAX were poor compared to optimized CE. Ion-exchange columns are expensive as well and are prone to failure. A separation of heparin, DS, and OSCS using the revised USP SAX method is shown in Figure 14.7. The sloping baseline is from changes in the background UV absorption from gradient elution. The run time is over an hour. The peaks are very broad, especially OSCS.

The FDA was very active in promoting the chromatographic separations. Keire et al. [15] concluded that CE was not amenable to the separation of impurities in crude heparin but only employed the obsolete emergency CE method even though more advanced and appropriate methods were at their disposal at that time. They spent their time optimizing SAX including sample preparation.

A limit test for DS was also included in the latest USP specification. Here, the heparin is acid degraded and galactosamine/glucosoamine ratio is determined by a second HPLC method, in this case using a pulsed ampeometric detection. A limit of 1% DS was established but the test can be easily deceived by the addition of galactosamine to the untested API, thereby lowering the galactosamine/glucosoamine ratio. The CE method illustrated in Figures 14.3 and 14.4 also determines DS simultaneously with OSCS. There is neither a degradation step required nor an additional test protocol, thus making a strong point to CE methods.

TABLE 14.2 USP Tests at the Time of the Crisis

Test	Purpose
Identification	Tests for the counterion (sodium, calcium, etc.)
Nitrogen content	Kjeldahl nitrogen for total nitrogen content
pH	Determine pH of aqueous solution
Sterility	Culture for microbial organisms
Loss on drying	Determine volatile organic content
Residue on ignition	Determine inorganic impurities
Protein	Determine residual protein
Heavy metals	Measure metallic impurities
Bacterial endotoxins	Screen for sugars from bacterial cell walls that can trigger anaphylactic reactions
Anti-factor Xa assay	Tests for binding to antithrombin III to inhibit factor Xa
Blood plasma assay	Determine anticoagulant activity in sheep blood

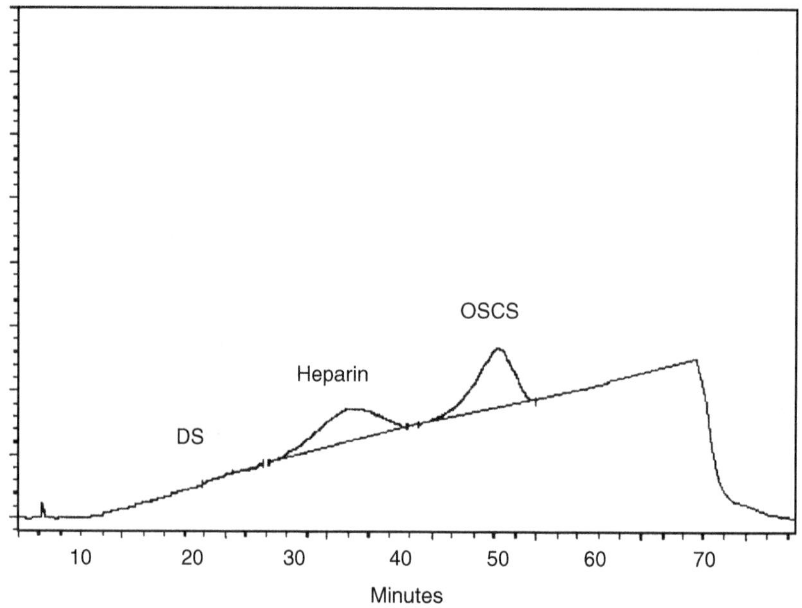

FIGURE 14.7 SAX chromatogram of dermatan sulfate, heparin, and OSCS using the USP method for impurity identity.

14.6 INTERLABORATORY COLLABORATIVE STUDY

Eight laboratories from industrial, academic, and regulatory settings took part in an interlaboratory collaborative study. The purpose of this study was to show a skeptical world that CE methods could be transferred and reproduced in laboratories around the world. To reduce experimental error, all of the laboratories performed separations of a set of uniformly prepared samples. Six sets of samples were employed. Each laboratory performed five injections of each sample on three separate days. Although the USP designated OSCS as none detectable, it was decided by the group to test the quantitative aspects of the method.

Analysis of variance (ANOVA) was performed on the sample whose electropherogram is shown in Figure 14.8. All of the organizations had excellent results for migration time variation. The percentage relative standard deviations (%RSDs) were less than 0.5% for seven of eight laboratories (0.1–0.8%), the outlier being 0.8%. For DS, the migration time %RSDs were less than 1.3% (0.1–2.7%) for seven of eight laboratories. An outlier from the same laboratory as the previous outlier was 2.7%. The %RSDs were less than 1% for six of eight laboratories.

The peak areas showed more variation as expected. The concentration of OSCS was only 0.4%, so this was essentially trace analysis. Instruments from Agilent and Beckman were used and peak integration was left to the discretion of the user. The %RSD ranged from 1% to 21%. Four of the laboratories ranged from 1% to 7% while the others were higher. In my own laboratory (RSD = 3%), the error of integration, obtained by manually integrating

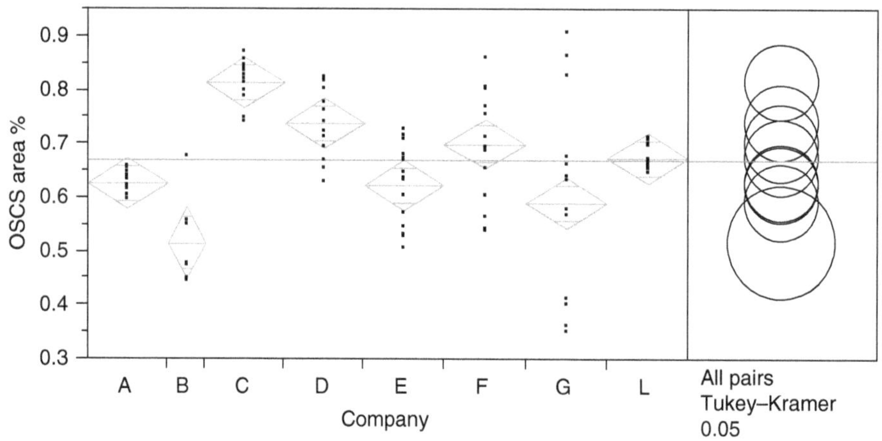

FIGURE 14.8 One-way ANOVA of OSCS area%. Refer to footnote 1.

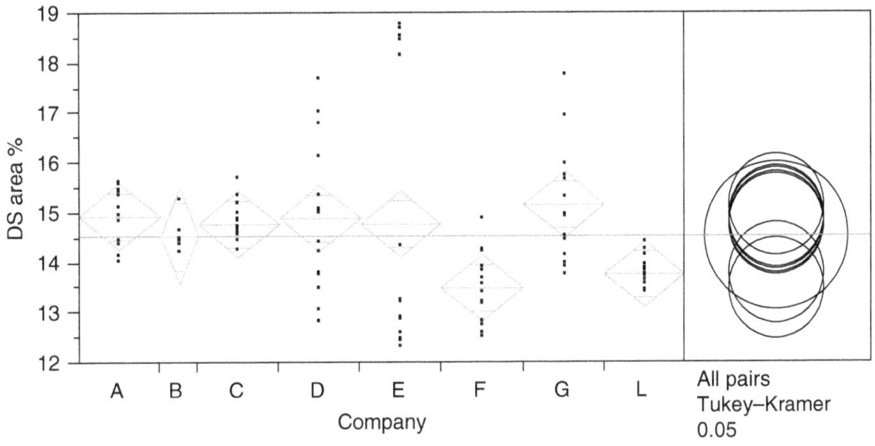

FIGURE 14.9 One-way ANOVA of OSCS area%. Refer to footnote 1.

the same electropherogram five times, was 1.5%. Figure 14.9 shows ANOVA of the OSCS peak area results. It is seen that laboratories E, F, and G had more variation than the others. Despite that variance, the Tukey–Kramer analysis shows that the results had substantial overlap. This means the results were indicative of a single population.

The DS peak area %RSD was less than 10% for seven of eight laboratories. Despite this scatter, the average peak areas were all very similar. The Tukey–Kramer analysis showed even more overlap compared to OSCS, indicating the samples were all from the same population. These results, and those for OSCS, were typical for interlab studies, even width compare HPLC.[1]

14.7 CONCLUSIONS

CE is appropriate as a method to determine impurities in heparin API. This was once again reiterated in a recent review article [16]. While CE may be optimal in speed and resolution, there are often other factors that drive the selection of analytical methods. These include familiarity and availability of appropriate instrumentation. Some of the regulatory scientists in the United States and around the world plus some of the heparin manufacturers were distinctly uncomfortable with CE. Others advocated CE for heparin separations. In addition, there was a need to harmonize methodology around the world for heparin separations. These in part were the driving forces to replace CE with HPLC for heparin impurity identification and impurity limit tests, despite the clear superiority of CE methodology.

[1] "A collaboration between various laboratory organizations and regulatory authorities concerning the analysis of Heparin using capillary electrophoresis." Unpublished data.

REFERENCES

1. http://www.healthheritageresearch.com/Heparin-Conntact9608.html. Originally published in CONNTACT August 1996 (Vol. 9 No. 4) (Employee News of Connaught Laboratories Limited).
2. Wardrop, D. and Keeling, D. (2008) *Br. J. Haematol.* *141*, 757–763.
3. Bogdanich, W. (2008) "Tainted Blood Drug to Face Import Tests", *New York Times*, March 15.
4. Wielgos, T., Havel, K. et al. (2009) *J. Pharm. Biomed. Anal.* *49* (2), 319–326.
5. Guerrini, M., Beccati, D., Shriver, Z., Naggi, A., Viswanathan, K., Bisio, A., Capila, I., Lansing, J.C., Guglieri, S., Fraser, B., et al. (2008) *Nat. Biotechnol.* *26*, 669–675.
6. al-Hakim, A. and Linhardt, R.J. (1991) *Anal. Biochem.* *195*(1), 68–73.
7. Ampofo, S.A., Wang, H.M., et al. (1991) *Anal. Biochem.* *199* (2), 249–55.
8. Stefansson, M. and Novotny, M. (1994) *Anal. Chem.* *66*, 3466–3471.
9. Malsch, R., Harenberg, J. et al. (1996) *J. Chromatogr. B Biomed. Appl* *685*, 223–231.
10. Patel, R.P., Narkowicz, C. et al. (2008) *J. Pharm. Biomed. Anal.* *46*(1), 30–35.
11. Somsen, G.W., Tak, Y.H. et al. (2009) *J. Chromatogr. A* *1216* (18), 4107–4112.
12. Wielgos, T., Szabo, C., Ray, J., Mampreian, D., Weinberger, I., and Weinberger, R. (2010) *J. Liq. Chromatogr. Related Tech.* *33*(18), 1649–1665.
13. Weinberger, R. (2009) *Am. Lab. On-Line Ed.* June.
14. Trehy, M.L., Reepmeyer, J.C., Kolinski, R.E., Westenberger, B.J., and Buhse, L.F. (2009) *J. Pharm. Biomed. Anal.* *49*(3), 670–673.
15. Keire, D.A., Trehy, M.L., Reepmeyer, J.C., Kolinski, R.E., Ye, W., Dunn, J., Westenberger, B.J., and Buhse, L.F. (2010) *J. Pharm. Biomed. Anal.* *51*(4), 921–926.
16. Beni, S., Limtiaco, J.F., and Larive, C.K. (2011) *Anal. Bioanal. Chem.* *399*(2), 527–539.

15

MICROCHIP CAPILLARY ELECTROPHORESIS FOR *IN SITU* PLANETARY EXPLORATION

Peter A. Willis and Amanda M. Stockton

15.1 INTRODUCTION

Among the highest-level science goals described in the NASA Science Mission Directorate's *2007–2016 Science Plan* are the following objectives for space exploration [1][1]:

(1) Advance scientific knowledge of the potential for life elsewhere.
(2) Search for life elsewhere.
(3) Understand the hazards and resources available as humans explore space.

Microchip capillary electrophoresis (μCE) is uniquely suited to address these science goals because of its ultra high sensitivity when coupled with laser-induced fluorescence detection and its inherent low mass, volume, power, and reagent consumption requirements.

Meeting the goals listed above requires *in situ* chemical analyses of complex mixtures. Examples of such analyses could involve studies of distributions of complex "prebiotic" organics in lakes on Titan (i.e., Goal 1), analyses of the chiral ratios of amino acids present in the sedimentary deposits of an ancient Martian riverbed that may have once harbored microbial colonies (i.e., Goal 2), or the determination of whether it is safe for an astronaut to melt and drink an icy portion of the surface of a comet (i.e., Goal 3) during a manned mission to a primitive solar system body. All of these analyses require portable, wet-chemistry instrumentation capable of surviving exposure to harsh environmental conditions during spaceflight operations. Propulsion and life support technologies currently prohibit human operators further from Earth than low-Earth orbit or the moon. This implies that present-day *in situ* science investigations conducted on sites further away must be executed by an autonomous probe designed for a one-way trip to a distant body. Furthermore, the instrumentation aboard a robotic explorer (rover, lander, or possibly boat, submarine, or balloon) must be capable of completely autonomous function in order to acquire the data of interest for transmission back to investigators on Earth.

The process of *in situ* chemical exploration begins once a spacecraft has safely landed on the surface of an extraterrestrial body (such as a planet or moon) and identified a specimen of interest for analysis. Broadly speaking, this process comprises the following five steps, although in some instances, certain steps may be omitted:

Step 1: Sample Acquisition. Acquisition and handling of material from local environment (i.e., drilling, scooping, and sipping).

Step 2: Sample Preprocessing. Temperature/pressure treatment of material (i.e., melting and extraction of organic material).

Step 3: Sample Processing. Additional processing of desired fraction of sample (i.e., desalting, concentrating, diluting, fluorescent derivatization).

Step 4: Separation. Chromatographic separation of analytes.

Step 5: Detection. Detection/identification of analytes.

[1] The reader interested in learning in greater detail of NASA's astrobiology goals (two of which are given at a very high level as points 1 and 2 here) is referred to NASA's Astrobiology Roadmap (2007).

Capillary Electrophoresis and Microchip Capillary Electrophoresis: Principles, Applications, and Limitations, First Edition.
Edited by Carlos D. García, Karin Y. Chumbimuni-Torres, and Emanuel Carrilho.

TABLE 15.1 The Steps in an End-to-End Chemical Analysis

		Instrumentation Scenarios for *In Situ* Chemical Exploration				Individual Circumstances of Potential Extraterrestrial Targets of *In Situ* Chemical Exploration		
		Low Complexity	Medium Complexity	State-of-the-Art	Future State-of-the-Art	Mars	Titan	Europa
		Example instrument systems				Overarching science questions addressed by wet chemistry		
		VCAM [2], Rosina [3,4] (simple fly-by MS)	MOMA [5] (laser-desorption–MS)	SAM [6], Viking [7,8] (pyrolysis–GC–MS)	Urey-like [9], MOA [14,15] (wet-extraction–derivatization–μCE–LIF)	Organics present? Terrestrial-like? Oxidized?	Tholin structure? Dune composition? Dissolved organics? Organic oxidation?	Organics present? Organic correlation with sulfates? Radiolytic "gardening" products?
Step	Instrument Subsystem	How analytical steps are addressed				Unique challenges of extraterrestrial target		
1.	Sample acquisition	None, open shutter	None	Drilling–scraping ice–rock	Drilling–scraping ice–rock	Drilling regolith challenging	Environment, landscape not well characterized	Drilling challenging, possibly ionized material
2.	Sample preprocessing	None	Laser desorption	Pyrolysis	Crush–melt–extract	Low-level organics requires efficient extraction	Native organics will volatilize in warm space craft	Melting required
3.	Sample processing	None	None	None	Dilute–chemical derivatization–spiking	Low-level organics require low-loss processes	Native organics will volatilize in warm space craft	Must be performed in shielded warm box
4.	Separation	None	None	Gas chromatography	Microfluidic liquid chromatography	Done in warm box to keep wet chemistry liquid	Done in warm box to keep wet chemistry liquid	Done in warm box to keep wet chemistry liquid
5.	Detection	Electron impact (EI) mass spectrometry (MS)	EI or LASER ionization–MS	VIS–IR spectroscopy; EI–MS	LASER-induced fluorescence, conductivity; electrospray–MS	MS requires a pump	MS requires a pump	Extreme radiation environment make shielding a requirement

Various instrumentation scenarios are shown for *in situ* chemical exploration, and some of the unique challenges associated with each, based on the extraterrestrial target of interest. In our discussion, we focus on Steps 3–5 of the proposed "future state-of-the-art" instrument suite. The bulk of the past work in this area has been focused on future implementation on Mars, although the authors are currently developing analogous instrumentation for Titan as well.

Instruments with varying levels of complexity implement each of these steps differently (or, in some cases, omit them entirely). Examples of portable, automated instruments that meet the mass/power/volume constraints imposed upon spaceflight payloads, with how they conduct each step, are given in the left half of Table 15.1. For context, also given are some of the overarching science questions associated with the extraterrestrial targets Mars, Titan, Europa, and some of the unique challenges posed by each of these environments. In this chapter, we describe *in situ* chemical analysis from the paradigm of an envisioned future state-of-the-art instrument system that contains the functionality of *all* the instruments considered in Table 15.1, and the means to perform each of the processing steps given. A system of this kind could take advantage of multiple detection methods, including laser-induced fluorescence (for ultra-high sensitivity to certain molecular classes) and mass spectrometry (for less sensitive broader coverage of all compound classes). However, it could also be modified to remove certain functionality as well, depending upon the nature of the desired chemistry investigation.

We focus our discussion largely on Steps 3–5 from the paradigm of the search for biochemical traces of past or present life on Mars. These analytical methods can be utilized for other *in situ* chemical investigations; a noteworthy example of this is the analysis of the chemical structure of Titan's organic aerosols (underway in our laboratory) [10,11]. Of course, for these and other *in situ* investigations, the instrument designer must first and foremost consider the inherent properties of the sample of interest. This becomes especially challenging when little is known about the chemical and physical properties of an extraterrestrial target sample, as is the case for Titan's organic aerosols and dune materials. Given the relative wealth of data we currently have from Mars orbit and surface measurements, the design of a robust instrument is less challenging for Martian regolith analyses. However, Mars almost certainly still poses unexpected challenges, as demonstrated by the cementing of Martian regolith within the Phoenix scoop upon extended exposure to sunlight [12]. Taking into account these and previously mentioned technical constraints, any spacecraft payload instrumentation intended for this type of study must generally meet the following criteria:

- low mass, power, and volume;
- ability to analyze multiple samples in an automated manner;
- extreme sensitivity for analytes of interest (i.e., organic and biological molecules).

Additionally, programmability designed into the instrumentation—and the adaptability this enables—provides risk mitigation and robustness against potential unexpected sample properties and analyte compositions.

These requirements are met by the technique of μCE, which can utilize miniaturized power supplies (<1 W; MS typically requires 50–60 W) to drive fluid flows (in contrast to chromatographic techniques such as high performance liquid chromatography that require large hydrostatic pumps). By utilizing multiwafer stack geometries containing microfluidic valving for fluidic pumping and manipulation, automation of all sample handling is possible. This includes mixing, dilution, delivery to and from electrophoresis wells, and fluorescent derivatization of samples, and enables extremely sensitive (subparts per trillion) analyses. In this chapter, we address the state-of-the-art in this burgeoning field of investigation.

15.2 INSTRUMENT DESIGN

In the design of a portable μCE instrument, such as the Mars Organic Analyzer (MOA, discussed in detail in Section 15.3), it must first be determined which steps in a desired end-to-end wet-chemical total organic analysis will be performed by the μCE instrument. Performance of these steps must be divided between the microdevice and the supporting hardware external to the microdevice. Figure 15.1a shows an example processing stream for a μCE instrument. For these discussions we consider an instrument where Steps 3–5 (Table 15.1) are conducted by the μCE instrument, while Steps 1 and 2 are conducted by other systems on the rover/lander platform.[2]

The above design decisions do not adequately differentiate the μCE instrument tasks between the microdevice and the compact external hardware that supports its operation. Ultimately, it is the nature of the sample and the chemical analysis desired that determines which operations take place on the chip and which operations take place elsewhere. In MOA technology, all wet chemistry occurs within the microdevice, including microfluidic processing and microcapillary electrophoretic separation (Figure 15.1b). The microdevice requires the support of solenoids for pneumatic actuation of microvalves during microfluidic processing, high-voltage power supplies to drive electrophoretic separations, an optical system for laser-induced fluorescence detection (or alternative hardware for other detection methods), and the electronic hardware to control these systems. These support subsystems are integrated into a single instrument external to the microdevice, and the microdevice conducts all wet chemical analysis in symbiotic interface with this instrument. For this reason, we discuss the instrumentation external to the microdevice next, and the microdevice itself in the subsequent sections.

[2] Sample extraction (Step 2) on-chip is also possible and early prototypes are under development at JPL.

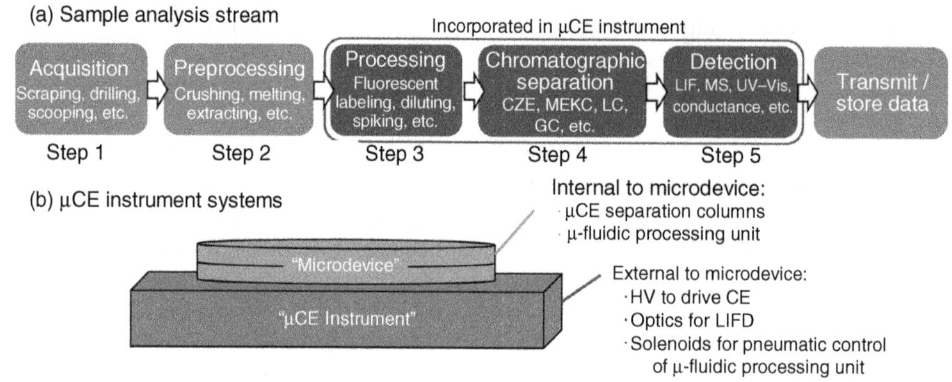

FIGURE 15.1 Subsystems required for a completely autonomous μCE sample analysis. (a) A general sample analysis scheme, with systems indicated that can easily be integrated into a single μCE instrument. Here, we focus our attention on Steps 3–5. (b) Schematic of an integrated μCE system with microdevice, illustrating the partition of functions between the microdevice and external systems.

15.3 INSTRUMENTATION EXTERNAL TO THE MICRODEVICE

Examples of the support instrumentation external to the microdevice can be found in the MOA microcapillary electrophoresis instruments (Figure 15.2) developed at UC Berkeley [13–15] and the lab-on-a-robot developed at UT San Antonio [16]. The single-channel MOA (Figure 15.2) [14] is a portable lab-on-a-chip microcapillary electrophoresis (μCE) instrument that contains all the necessary optics for laser-induced fluorescence detection (Step 5), pneumatics for control of on-chip fluidic manipulation (Step 3), and high-voltage power supplies for microchip capillary electrophoretic separations (Step 4). The optical system uses a 405 nm laser, which is passed through a dichroic and focused to a 10 μm spot inside the electrophoresis channel. Fluorescence is collected by the objective, reflected off the dichroic, passed through a long-pass filter, and collected by a photomultiplier tube (PMT). High-voltage power supplies can provide a maximum of −15 kV at the cathode, and a maximum of −2.5 kV to the

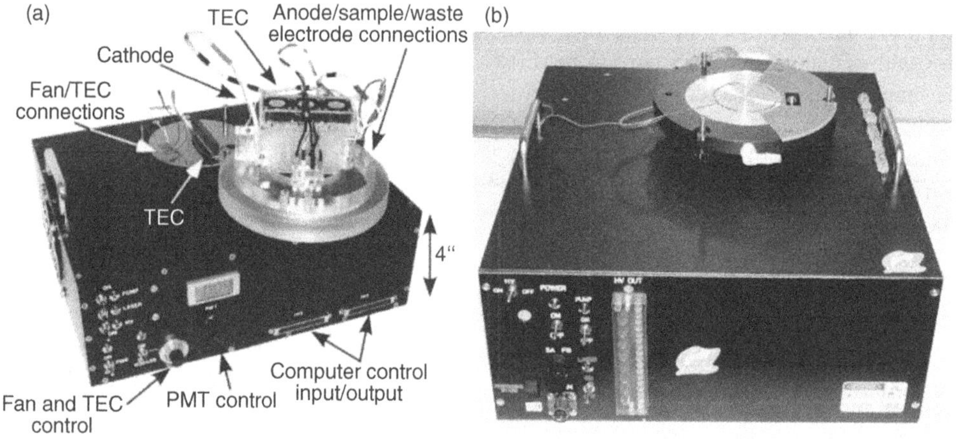

FIGURE 15.2 Mars Organic Analyzer (MOA) portable microchip capillary electrophoresis instruments. The original single-channel MOA (a) houses a bank of 16 solenoids for pneumatic control of microfluidic devices and a stationary optical system with a PMT for laser-induced fluorescence detection from a single channel in a microdevice [13,14,17–21,23]. A next-generation multichannel MOA (McMOA, b) houses a bank of 32 solenoids for pneumatic control of the microfluidics for multiple channels. The optical system contains an objective mounted on a translating stage for detection of laser-induced fluorescence from multiple channels and a CCD spectrometer for real-time fluorescence spectral determination [15,38]. (a) Reprinted with permission from Ref. 4, copyright 2005 and (b) Ref. 5, copyright 2010.

sample and waste. Additionally, the sample, waste, and anode can be either grounded or floated. The pneumatic system houses 16 solenoids that can control up to 16 independent pneumatic outputs, enabling the control of multilayer hybrid devices for on-chip fluidic manipulation. A LabVIEW program sends commands to the instrument and collects resulting data via a laptop interface [14]. The MOA has been used to detect primary amines, amino acids [14,17,18], nucleobases [19], polycyclic aromatic hydrocarbons (PAHs) [20], aldehydes [21], ketones [21], and carboxylic acids [22,23]. It has also been used to determine the chirality of amino acids via cyclodextrin-assisted CE [19] and micellar electrokinetic chromatography [17].

The MOA's capability for high-resolution separations with ultra-sensitive detection of these analytes has been demonstrated by a number of successful analyses both in the laboratory and in the field. In the laboratory, the MOA was used to analyze organic molecules present in samples from the Yungay Hills region of the Atacama Desert [14], the Murchison meteorite [17], highly acidic and saline Rio Tinto deposits [18], saturated brines from the Saline Valley, CA [18], sediments from lava tube caves in the Pisgah lava flows in the Mojave Desert, CA [23], and effluent from the Bumpass Hell geothermal area in Lassen National Park, CA (Figure 15.3) [23]. In 2004, the MOA was field tested in the jarosite-rich Panoche Valley, successfully analyzing amino acids sublimed from a jarosite sample down to subparts per billion (ppb) levels [14].[3] The MOA was also successfully tested in the Atacama Desert, the oldest and driest terrestrial desert and hence a possible analog to Martian conditions, in 2005. Subcritical water extracts of samples taken from the Yungay Hills region were analyzed by the MOA, detecting amino acids down to 4 ppb and identifying the age of the organics in the sample as 2×10^3–1×10^5 years, consistent with known El Nino precipitation events 1×10^4–1.6×10^4 years ago [13]. These analyses represent a $>10^3$ improvement in sensitivity over Viking's organic detection GC–MS instrument [17,18,20,21,24].

The multichannel Mars organic analyzer (McMOA, Figure 15.2) is similar in design to the MOA, but possesses additional functionality. The detection optics are designed so that the objective is mounted on a linearly translating stage, allowing for programmable detection in any one of several channel locations. The detection optics also use a charge-coupled device (CCD) spectrometer for fluorescence spectral measurement as opposed to the MOA's PMT, allowing for multiwavelength detection with sensitivity similar to the single-channel MOA. The high-voltage power supplies are capable of independently driving electrophoresis in up to eight individual channels. Finally, the McMOA is equipped with pneumatic controls for up to 32 individually controlled microfluidic valves,[4] enhancing the complexity and functionality of microfluidic devices controllable by the McMOA as compared to the MOA [15].

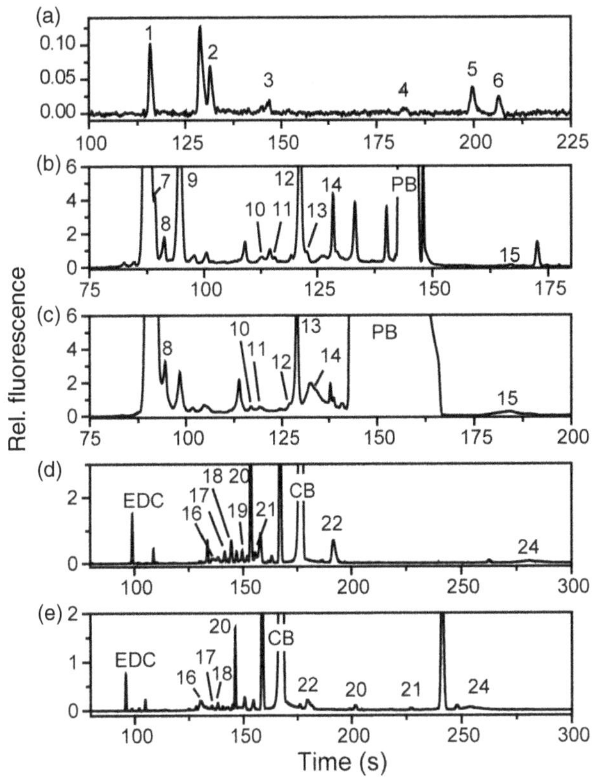

FIGURE 15.3 Electropherograms of samples analyzed on the Mars Organic Analyzer. (a) Analysis of polycyclic aromatic hydrocarbons (PAHs) in a subcritical water extract (SCWE) of a sample from the Yungay Hills region of the Atacama Desert, Chile. Sample contains: 1. 9,10-diphenylanthracene, 2. anthracene, 3. anthanthrene, 4. fluoranthene, 5. perylene, and 6. benzo[*ghi*]fluoranthene. (b and c) Analysis of Pacific Blue derivatized amines and amino acids in a (b) SCWE of a sediment from the Rio Tinto, Spain, and (c) in a saturated brine from the Saline Valley, CA. Samples contain: 7, ammonia; 8, methylamine; 9, ethylamine; 10, leucine and isoleucine; 11, valine; 12, serine; 13, alanine; 14, glycine; 15, and aspartic acid. (d and e) Analysis of Cascade Blue derivatized carboxylic acids in (d) a sample from the Pisgah lava tube caves in the Mojave Desert, CA, and (e) the effluent from the Bumpass Hell geothermal area in Lassen National Park, CA. Samples contain: 16, hexanoic, heptanoic, and octanoic acids; 17, pentanoic acid; 18, butyric acid; 19, propionic acid; 20, pyruvic acid; 21, acetic acid; 22, formic acid; 23, citric acid; and 24, α-ketoglutaric acid.

[3] Concentrations determined as weight percentages for the dry weight of analyte in the original soil sample.

[4] It should be mentioned that a given number of individual off-chip solenoid valves can be multiplexed to address a greater number of on-chip microfluidic valves (or effective valving operations). However, this can result in systems that are especially prone to catastrophic single point failures, in which a single solenoid failure disrupts a large number of valving operations. Therefore, we do not give a detailed description of multiplexing, but refer the reader to the work of Grover et al. [24].

The McMOA has been fully demonstrated in the laboratory. The analysis of samples on eight independent channels in a single microdevice has been demonstrated with consistent peak heights, peak areas, resolutions, and peak efficiencies [15]. On-chip fluidic manipulation has been demonstrated with addressing to any individual channel of the eight available [15]. The CCD spectrometer allows for spectrographic analysis of individual peaks, enabling a second dimension of analysis of species with unique fluorescent spectra, such as PAHs [15].

The MOA systems inherently require low power (<15 W) and reagent consumption (~90 μL buffer/analysis). Even laboratory prototypes using off-the-shelf parts and standard electronics boards have low mass (~11 kg) and volume (<14,000 cm^3). With an effort toward further miniaturization and optimization, these instruments could easily be made much smaller, lighter, and potentially require even less power by using integrated electronic circuits off chip and custom manifolds with integrated pneumatic circuits [25]. These instruments clearly demonstrate the utility of the MOA technology for autonomous rover or lander-based extraterrestrial exploration platforms.

The only published autonomous integration of μCE systems with a rover platform, the "lab-on-a-robot" was developed utilizing pulsed amperometric detection (Figure 15.4) [16]. The lab-on-a-robot is a wireless mobile unit with robotics controlled by an on-board global positioning system (GPS) and a user-operated LabVIEW program on a remote laptop computer. The mobile unit also houses a three-channel high-voltage power supply to drive electrophoresis, an electrochemical detector utilizing signal-conditioning analog circuitry and a digital feedback range controller, an air pump, and a CE microchip fabricated in poly(dimethylsiloxane) (PDMS). The entire device is controlled by digital hardware implemented on a field programmable gate array development board. The lab-on-a-robot successfully demonstrated the ability to navigate to a GPS location, acquire an air sample (Step 1), perform μCE separation (Step 4) with electrochemical detection (Step 5), and transmit acquired data to a remote station without requiring manual intervention that would expose the human operator to the testing environment. This lab-on-a-robot device was successfully demonstrated via the detection of a phenol source of air contamination [16].

FIGURE 15.4 A lab-on-a-robot containing a PDMS microdevice, HV power supplies to drive electrophoresis, electrochemical detector, mobile platform, field programmable gate array, RF modem, compass, and GPS [16]. Reprinted with permission from Ref. 6, copyright 2008, Wiley-VCH Verlag GmbH & Co.

15.4 MICRODEVICE BASICS

MOA technology utilizes glass–polymer hybrid wafer stacks with microfabricated capillaries and pneumatically actuated monolithic membrane valves. As mentioned earlier, off-chip components are used to operate and monitor these microdevices to perform the necessary wet chemical steps of sample dilution, derivatization, standard spiking, and electrophoretic separation. We first describe the microchannels for separation, then the microvalve fluidic processors, and finally the integration of the components into a single stack.

15.4.1 All-Glass Devices for Microchip Capillary Electrophoresis

Capillary electrophoresis has traditionally been conducted in long (~30–50 cm), narrow-bore (~50 μm inner diameter), fused silica capillaries. First successfully conducted in microfabricated amorphous glass in 1992 [26], μCE typically uses short length (~20 cm), narrow (~100 μm), shallow (~35 μm) microfabricated channels, and thus achieves high-resolution separations with shorter analysis times and reduced reagent and sample consumption than traditional CE [26,27]. An example μCE channel is presented in Figure 15.5a. This device has been used for μCE separations of multiple targets by multiple separation techniques, including amino acids by capillary zone electrophoresis [28], and PAHs by cyclodextrin-assisted CE [29]. Microchip capillaries and conventional capillaries differ in a number of key ways, particularly in sample injection, coupling of the electrophoresis device to the detection system, and the ease of coupling the electrophoresis device to a fully automated sample handling system.

Traditional capillaries rely on one of two injection methods. Electrokinetic injection relies on a field that is applied for several seconds with the end of the capillary submerged in the sample to introduce sample. In hydrodynamic injection, the capillary is placed in the sample and then elevated relative to the detector to introduce a sample plug into the end of the column. During injection, analytes diffuse and broaden the injection plug, making precise control of plug

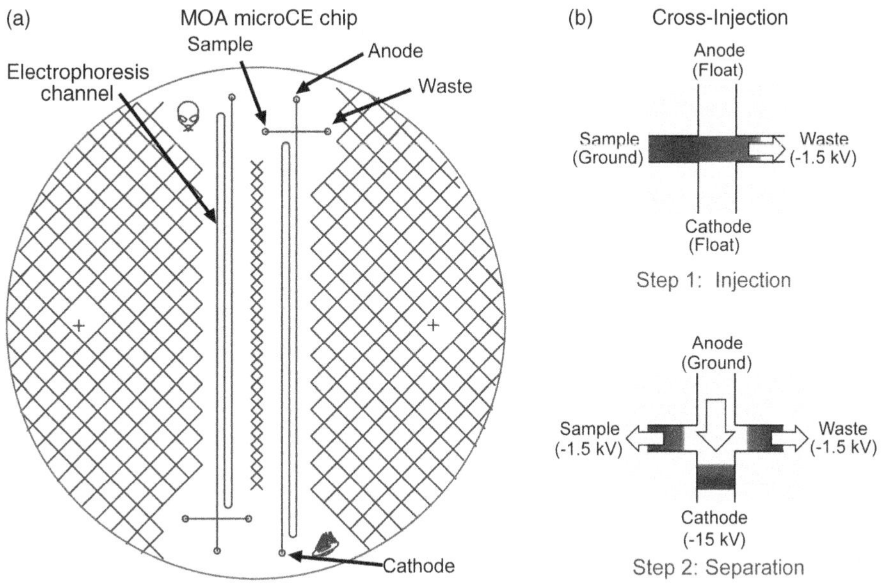

FIGURE 15.5 A standard microdevice for electrophoretic separations with cross-injection of sample [17,18,21,23]. (a) A schematic of a standard folded μCE separation channel. This 22.4 cm long electrophoresis channel is typically etched to 35 μm depth and 200 μm width. The cross-channel 1.5 cm from the anode is typically etched to 35 μm depth and 80 μm width. The cross-hatching represents etched "bonding lines" that facilitate glass–glass thermal bonding of otherwise featureless areas by allowing trapped air an escape route. (b) Depiction of a standard cross-injection of sample into the electrophoresis column [22]. In the first step, a field is applied so that EOF drives sample into the entire cross-channel. When the fields are switched to separation mode, a plug of sample is injected into the separation channel. Backbiasing, driving some EOF into the sample and waste wells, prevents sample from leaking into the channel during separation and reducing separation quality.

size challenging. In a microdevice, electrokinetic injection is the most attractive option, and the cross-injection structure (Figure 15.5b) allows for the most facile control of small sample plug sizes. Injection is achieved by applying a voltage (~1.5 kV) across the sample/waste arm, inducing electro-osmotic flow (EOF) from the sample to waste and driving sample into the cross channel. After a brief (~10–30 s) equilibration time, the voltages are switched so that the main flow is from the anode toward the detector and a small, repeatable sample plug is introduced into the separation column. Generally, "back-biasing" is performed to prevent leakage of sample from the cross channel into the separation channel. Back-biasing involves applying a low voltage (~1.5 kV) during separation to drive flow from the anode toward the sample and waste wells.

Microfabricated devices for electrophoresis are readily coupled to laser-induced fluorescence detectors. In confocal laser-induced fluorescence detection, the laser is focused through an objective to a small spot (~20 μm) in the separation channel. The objective collects emitted fluorescence and passes it to the rest of the detection system, typically involving a dichroic, filter, and PMT, charge-coupled device (CDD), or other detector. In optical detection coupled to conventional capillaries, the laser beam and the emitted fluorescence must pass through the curved surface of the capillary, which distorts the beam waist. While sheath flow cuvettes with flat windows have been used to make detection in conventional capillaries simpler [30], microfabricated CE devices avoid the problem altogether with their flat air–glass interface at the detection side and a flat glass–fluid interface on the analyte side.

The most advantageous aspect to microfabricated CE is the ability to integrate multiple components into a single chip with relative ease, which is discussed in detail in the following sections. In the UC Berkeley-based Mathies laboratory alone, examples of integrated microfluidic components implemented have included multiple μCE channels [15,31], pH-sensitive electrodes [32], PCR reactors [31,33–37], monolithic membrane valves and self-priming pumps based on these valves [15,31,37–40], dilution circuits [38–40], microfabricated heaters [31,33–37], and resistive temperature detectors [31,33–37]. These components have been integrated into devices for many applications, including DNA sequencing [33,35,41], short tandem repeat (STR) typing for forensics analysis [35,41], and droplet generation for stochastic studies of cellular populations [42]. The remainder of this chapter focuses on the automation of on-chip μCE microfluidic processing required for

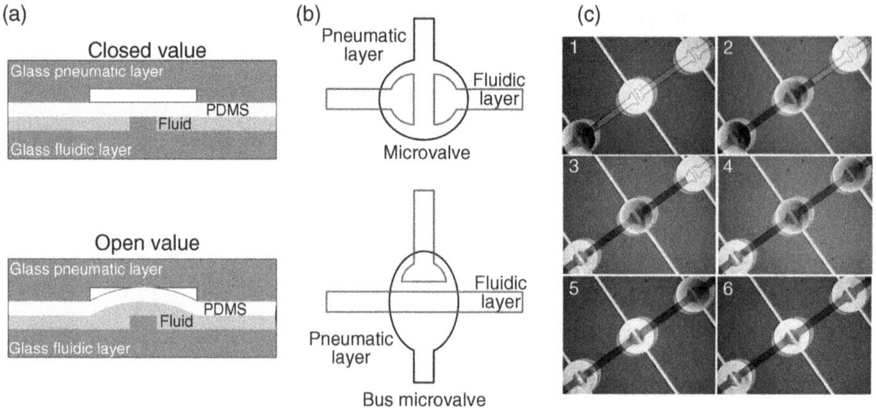

FIGURE 15.6 Monolithic membrane microvalve actuation (a), top views of a standard microvalve and a bus valve (b) [4,40], and the peristaltic pump created by operating three microvalves in series (c) [33]. The microvalve is created by sandwiching a flexible membrane (e.g., PDMS) between a glass pneumatic layer with an etched displacement chamber and a glass fluidic layer with an etched discontinuous channel. When a vacuum is applied to the pneumatic layer, the membrane deflects to allow fluidic flow across the discontinuity in the fluidic channel. Sequential actuation of three microvalves in series creates a peristaltic pump, driving fluid along a channel. A closed bus valve allows fluidic flow along the main (bus) channel while blocking access to the side channel it controls. Actuation of a bus valve enables fluidic access of the side channel via the bus channel.

extraterrestrial exploration, including fluorescent derivatization, spiking with standards, serial dilutions, μCE channel loading, and device cleaning.

15.4.2 Three-Layer Hybrid Substrate Glass–PDMS Devices for Fluidic Manipulation

A fully autonomous microdevice for extraterrestrial exploration capable of the necessary sample processing steps will require precise microfluidic manipulation, including mixing and fluidic transfer. Monolithic membrane microvalves fabricated in hybrid three-layer devices have been used to control fluids on the nanoliter to microliter volume scale [14,39,43], and thus provide the technological backbone for the microfluidic processors required for fully autonomous *in situ* analysis. Microvalve structures are produced by sandwiching a flexible elastomer membrane between two glass layers, each etched with features corresponding to pneumatic inputs and fluidic channels (Figure 15.6a and b) [14,39,43,44]. Within a microvalve, the fluidic feature is an etched discontinuous fluidic channel, and the pneumatic feature an etched displacement chamber. When a vacuum is applied to the pneumatic feature, the membrane deflects into the displacement chamber, enabling fluidic flow across the channel discontinuity. Microfabricated valves have small dead volumes ($<20\,nL$) and remain sealed against fluidic pressure up to 75 kPa when pressure is applied to the pneumatic feature [43]. Three valves in series can be operated in sequence to form a peristaltic pump that can transfer fluid at a rate of nanoliters to microliters per cycle with typical rates between 1 and $100\,nL\,s^{-1}$ (Figure 15.6c) [43,44].

While monolithic membrane microvalves enable fluidic manipulation of small (nanoliter–microliter scale) volumes on-chip, each open actuation draws fluid from both valve inputs. This provides a mechanism for potential contamination of fluids at the site of the microvalve. Potential contamination is mitigated by the "bus" valve (in analogy to an electronic bus), which continuously allows fluidic connection along the bus channel and regulates the fluidic connection between the bus channel and a separate input/output channel (Figure 15.6b, bottom). Bus valves thus enable the cleaning of the bus channel with buffer prior to accessing contamination-sensitive samples. The first publication of bus valves was their utilization by Skelley et al. in μCE development work for extraterrestrial analysis [14]. These valves have permitted essentially contamination-free fluidic manipulation on-chip for numerous chip designs, which are discussed further in Section 15.5 [14–16].

The most commonly utilized polymer for monolithic membrane microvalves has been PDMS, which exhibits good elasticity and easily adheres to glass [14,15,45]. PDMS therefore serves as a good membrane for rapid device prototyping and design evaluation, but it has properties that make it unsuitable for spaceflight applications, including chemical and gas permeability, susceptibility to damage by repeated freeze-thaw cycles, and the tendency to permanently bond to either the discontinuity in the fluidic layer or the pneumatic feature when not actuated for long time periods (months). Other membrane options have been explored, with Teflon® [44,46] and Fluorocur™

perfluoropolyether (PFPE) [47] successfully demonstrated as membranes in multiwafer stack fluidic manipulation devices. These fluorinated membrane options maintain performance after repeated thermal cycling, exhibit good chemical resistance, and enable flight-ready devices [44,46,47].

15.4.3 Integrating Fluidic Manipulation with Electrophoresis

Of the multiple approaches to integrating three-layer fluidic manipulation systems with μCE separation channels, the most commonly utilized approach has been the full integration of the components into a single four-layer wafer stack. The functionalities of these systems are described in detail in the following sections, and a brief description of their fabrication and operation is presented here. Briefly, the μCE channels and fluidic structures are etched on opposite faces of a single glass wafer. Holes are drilled in this wafer to permit fluidic access between the functionalities represented on each face. The μCE face is then thermally bonded to a blank glass "backing wafer." A polymer elastomer (e.g., PDMS) layer, with holes punched for access to the fluidic and μCE layers, is then bonded to the fluidic side of the wafer. A separate glass wafer is etched with pneumatic features, holes are drilled in this wafer for access to the μCE, fluidic, and pneumatic layers, and its pneumatic layer is bonded to the elastomer gasket layer. Therefore, this device incorporates the two-layer glass μCE device in the "bottom" two layers and the three-layer fluidic processor in the "upper" three layers. Because one glass layer contains both μCE channel and fluidic features, the entire stack is only four layers thick [14,15,44,45].

15.5 MICRODEVICES AND THEIR APPLICATIONS

For full functionality of a microfluidic device for extraterrestrial exploration, the device must be capable of microfluidic manipulations, particularly the successful and rapid mixing of fluidic inputs for labeling reactions and dilutions. However, since microfluidic devices operate at low Reynolds number, flow profiles are normally laminar and the turbulent flow conducive to rapid (i.e., not diffusion limited) mixing is challenging to access. Additionally, high Schmidt numbers are encountered by these systems, with analytes having low molecular diffusivity compared to viscous momentum. Therefore, micromixers must be utilized for efficient on-chip mixing of fluids. A number of micromixers utilizing monolithic membrane valves have been demonstrated, including continuous flow micromixers [38,48], rotary pump micromixers that accomplish fixed-volume mixing via pumping in a circular loop [39], and mixers that utilize the repeated transfer of combined fluid between microvalves and/or fluidic reservoirs [15,45,49]. Reagent and sample storage must also be considered, as should the individual addressing of μCE channel wells.

15.5.1 Microdevices with Bus-Valve Control of Microfluidic Manipulation

One of the first devices developed for rapid on-chip microfluidic manipulation was a bus-valve enabled dilution circuit [39] (Figure 15.7) fabricated in a three-layer glass–PDMS–glass device. The 400 nL serial dilution circuit contains five

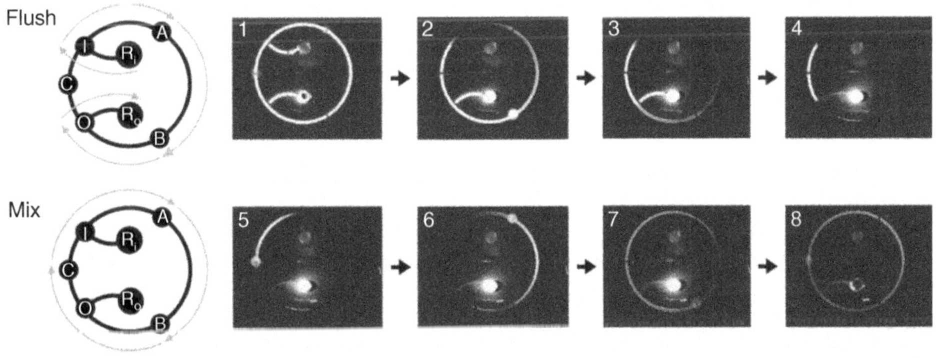

FIGURE 15.7 Serial dilution circuit pumping program schematics and epifluorescence still-frames with a circuit primed with fluorescein dye. Fluid flow paths are indicated with gray arrows overlaid on the circuit schematics. Valves A, B, and C are standard microvalves, while I and O are bus valves where the circular channel serves as the bus channel. Buffer is pumped from R_i to R_o, clearing the right side of the mixing loop while isolating the carryover aliquot on the left side (frames 1–4). Mixing is achieved by serially actuating A, B, and C while I and O are kept closed (frames 5–8). The output reservoir, R_o, was manually evacuated between frame 7 and frame 8 for the purpose of visualizing the fully mixed sample. This device achieves dilution ratios of 0.04–0.2 depending on fabricated device geometry, and can produce 400 nL fully mixed solution in 22 s [39]. Reprinted with permission from Ref. 29, copyright 2006.

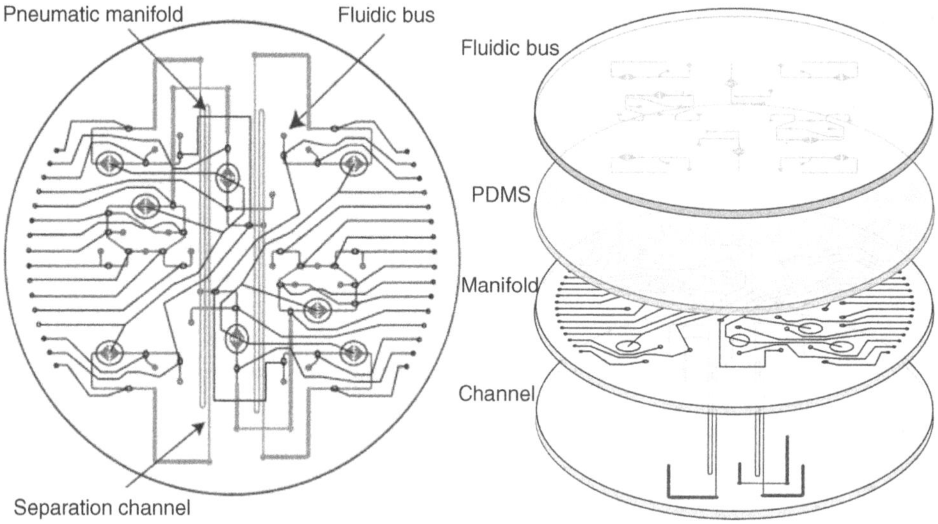

FIGURE 15.8 A single-channel MOA-compatible four-layer microdevice capable of pumping buffer from an on-chip reservoir to an off-chip desiccated prelabeled sample, pumping rehydrated sample to the sample reservoir of a μCE separation channel, and performing electrophoretic separation. After separation, the device autonomously evacuates the sample well of the electrophoresis channel, replaces it with buffer, and conducts an electrophoretic channel cleaning step before proceeding to the next sample. This chip was successfully field tested in the jarositic Panoche Valley, detecting amino acids to parts per billion levels. Reprinted with permission from Ref. 4, copyright 2005.

monolithic membrane valves: three standard microvalves arranged in a loop to drive cyclic mixing of the diluent and carryover, and two bus valves to control fluidic access to the circuit by input and output channels. By varying the valve placement in the circuit, carryover fractions from 0.04 to 0.2 were obtained. Each dilution process, which is composed of a diluent flush cycle followed by a mixing cycle, is carried out with no pipetting, and a sample volume of 400 nL is sufficient for conducting an arbitrary number of serial dilutions. Mixing is precisely controlled by changing the cyclic pumping rate, with a minimum mixing time of 22 s [39].

The first multilayer wafer stack demonstrated for use with the MOA instrument system is shown in Figure 15.8 [14]. This design routes buffer to the cathode of a channel, where it is allowed to wick via capillary action throughout the entire channel. After the channel is filled, the anode and waste wells are also filled with buffer. For field-testing in the Panoche Valley, the device pumped buffer from an on-chip reservoir through a polyether ether ketone (PEEK) tubing sipper to a puck that had been coated with the amine-reactive fluorogenic probe fluorescamine and amino acids sublimed from a jarositic sample acquired in the region. The device then pumped the dissolved fluorescamine-labeled amino acids to the sample well and performed electrophoresis, detecting amines at parts per billion levels and amino acids at sub parts per billion levels [14].[5]

[5] Concentrations determined as weight percent of dry analyte in sample.

A fluidic routing chip for use with the McMOA instrument is shown in Figure 15.9 [15]. This chip is similar to the single-channel MOA chip (Figure 15.8) [14], but possesses additional functionality [15]. This chip has eight channels that can be independently operated by the McMOA instrument. The microdevice has a fluidic bus structure that allows for individual addressing of a microchannel. Once a channel is addressed by actuation of the bus valves that control access to the channel wells, buffer is pumped to the cathode and fluid is routed individually to the waste, sample, and anode wells. Each of the wells can select from two potential fluidic inputs. This device was successfully demonstrated by the routing and analysis of Pacific Blue succinimidyl ester (PB) labeled amino acids, exhibiting a channel-to-channel variation less than the run-to-run variation and a carryover/contamination between sequential analytes of less than the measurable 1%. Additionally, this device permitted the enhanced compositional analysis of the PAH content of an environmental contamination standard sediment of Lake Erie [15].

While the above chip designs enable successful sample routing, the sample input must be prelabeled and diluted prior to injection on these microdevices for μCE analysis. For spaceflight applications, the fluidic sample processing steps of labeling, dilution, and spiking with a standard must also be performed autonomously. Figure 15.10 shows a bus valve-based design that enables the required fluidic sample processing steps in addition to fluidic routing to the μCE channel and μCE analysis [45]. In this design, bus valves

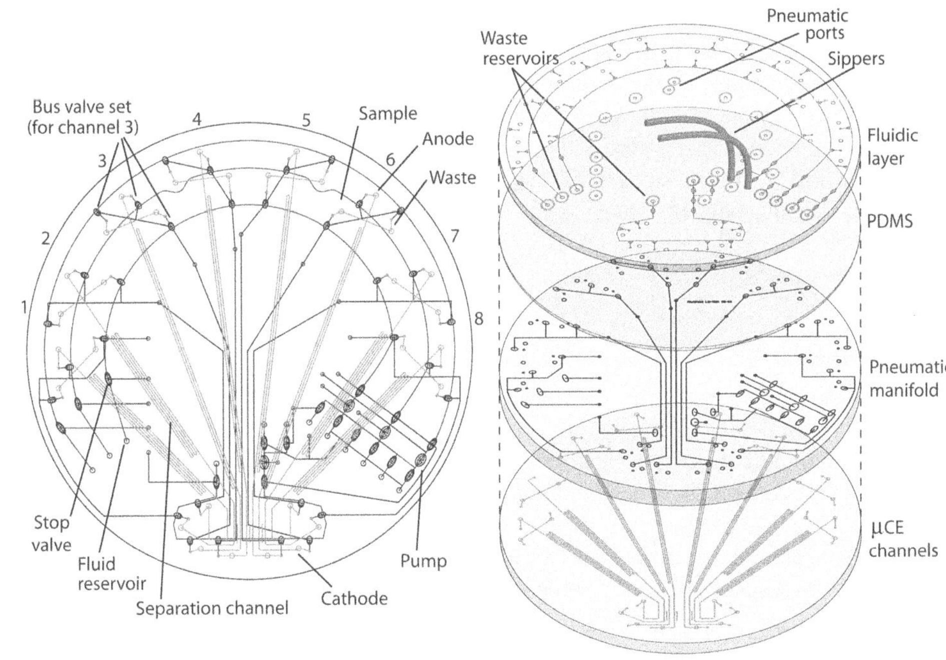

FIGURE 15.9 A McMOA-compatible four-layer microdevice capable of routing multiple buffers or prelabeled samples to any one of the eight separation channels of the microdevice. Bus valves are utilized to select the desired channel and three standard microvalves in series drive fluidic flow. Separations exhibited lower channel-to-channel variation than run-to-run variation, and routing of alternate samples to a single channel exhibited a carryover/contamination of less than the detectable 1%. Reprinted with permission from Ref. 5, copyright 2010.

around the perimeter of the structure control access to drilled fluidic reservoirs in the center of the structure. Mixing for labeling, dilution, or standard spiking is accomplished by pumping appropriate ratios of the fluids from their fluidic reservoir into a new reservoir. Contamination is mitigated by flushing the perimeter bus with buffer between operations. Once the sample is prepared for analysis, the μCE channel is autonomously filled similarly to previously described

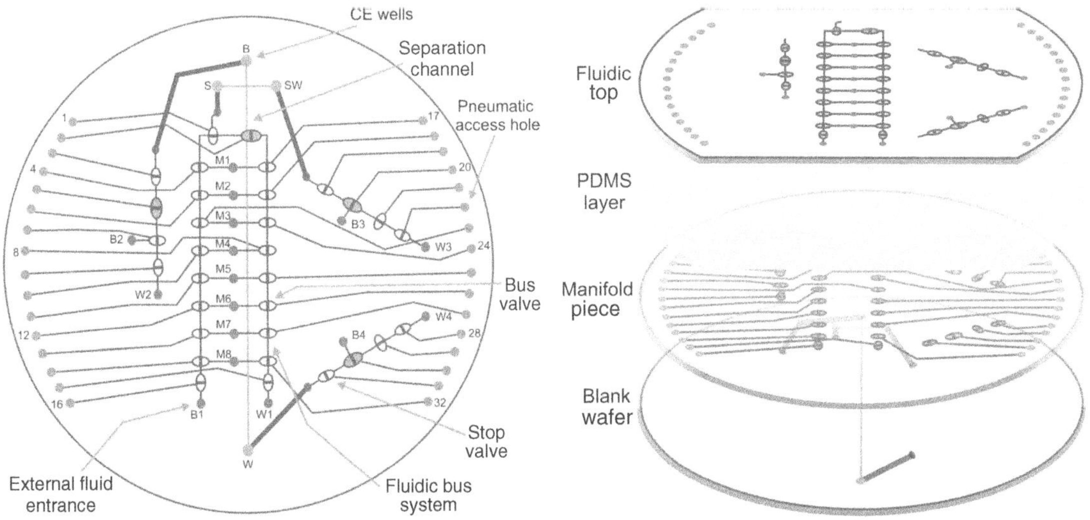

FIGURE 15.10 A single-channel MOA-compatible four-layer microdevice capable of performing on-chip Pacific Blue succinimidyl ester (PB) labeling of amines, dilution, and combination with a standard (spiking) with combination ratios of 1:1 to 1:10. This device enables end-to-end fully autonomous PB-labeled amino acid sample preparation and analysis. Reprinted with permission from Ref. 35, copyright 2011.

designs and the sample routed to the sample reservoir of the μCE channel. All operations are performed completely autonomously after a manual priming of the reservoirs with buffer, unlabeled aqueous sample, standard, and dye solutions; however, it would be trivial to access these solutions autonomously via PEEK tubing sippers similar to the single-channel MOA autosampler design (Figure 15.8). This device enabled the end-to-end analysis of laboratory amino acid standards [45].

15.5.2 Automaton Devices for Programmable Microfluidic Manipulation

The ideal MOA system for extraterrestrial analyses will possess a programmable microfluidic processor capable of implementing multiple assay procedures, thus enabling not only diverse routine analyses but also a degree of programmable flexibility to address unanticipated sample matrices and analyte compositions. The automaton (Figure 15.11) is a programmable microfluidic platform based on an 8×8 rectilinear array of pneumatically actuated monolithic membrane microvalves [40,49]. Basic operations are achieved by digital transfer of fluid between microvalves in the array. These operations can be implemented combinatorially, enabling the automation of diverse assay protocols. The automaton has a high degree of flexibility and programmability with 2^{64} possible valve configurations available for each step in an operation.

This device was originally developed for automated nanoliter-scale multistep quantitative biomolecular assays. Precise metering of small volumes (~14 nL) and serial dilution protocols were demonstrated with linearity over a 1000-fold dilution range [40]. Initially demonstrated on a small-volume hydrogen peroxide assay with a sub micromolar detection limit [40], the automaton has more recently been characterized for the automated processing of larger volumes [48]. Because the automaton can process large sample volumes (10–30 μL) on a microfluidic format compatible with

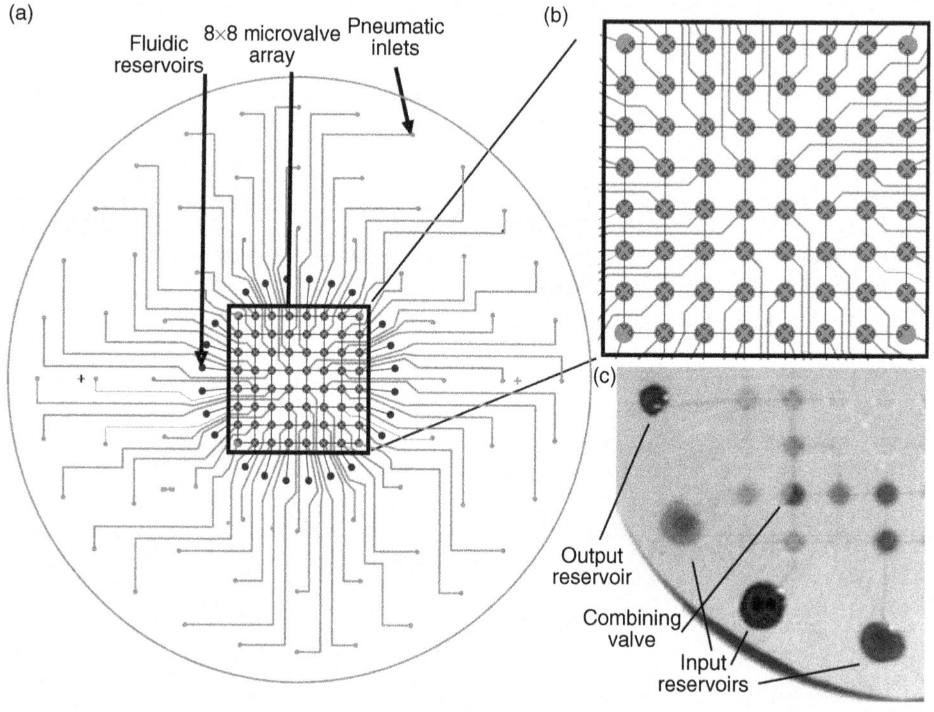

FIGURE 15.11 The automaton array of microvalves for programmable microfluidic manipulation (a), a close-up of the schematic (b), and a photograph of an example operation (c) [40]. Although designed for the precise metering and rapid mixing of nanoliter-scale volumes, the automaton was recently demonstrated for the processing of up to milliliter-scale volumes. In the example photographed here, fluid from reservoirs containing red (light gray), blue (dark gray), and yellow (white) dye is combined in a single "combining" valve and transferred to an output reservoir. A demonstration of EDC-activated Cascade Blue (CB) labeling of carboxylic acids for MOA analysis required an initial 1:1:1 combination of EDC, CB, and carboxylic acids to a total volume of 10 μL, followed by a 1/3 dilution of the reaction mixture with buffer. The autonomously generated results exhibited peak efficiencies within the run-to-run deviation of manually generated results, and peak areas of manual and autonomous manipulation were comparable [48].

MOA μCE technology, it can provide the microfluidic sample processing required for autonomous MOA analysis, and proof-of-concept experiments utilizing it for the labeling and dilution of carboxylic acids have been demonstrated [48].

The automaton requires no bus valves for contamination control if "stop" valves are utilized to control fluidic access to the rectilinear array. These valves have no connectivity except along the channel connecting the drilled fluidic reservoir to the array. By closing these stop valves, the array can be purged rapidly with buffer between operations by simply opening all the valves and then closing them sequentially to drive fluid to a waste well. Utilizing the rectilinear array structure over legacy bus-valve-based structures (1) reduces on-chip real-estate required for microfluidic sample processors, (2) enables a high degree of assay programmability, and (3) reduces the number of valves and thus solenoid controllers required to perform processing. Further iterations of the Automaton design are currently being developed for contamination-free rapid sample processing interfaced to MOA μCE devices.

15.6 CONCLUSIONS

Microcapillary electrophoresis systems have reached an impressive level of technological maturity over the past decade and are finally ready for spaceflight. These systems inherently provide the wet-chemical sample processing required for many extraterrestrial targets, and do so with ultra-high sensitivity in a compact, low-mass and low-power package. The supporting hardware and the chemical analyses have been demonstrated in field tests in the Panoche Valley and the Atacama desert, validating their performance in relevant environments. The automation of all steps including acquiring a sample from an off-chip source, fluorescent derivatization, dilution, and μCE analyses on microfluidic chips has been demonstrated, enabling completely independent end-to-end sample analysis. The chemistry has been optimized to enable the most highly sensitive (sub parts-per-trillion) analyses of small organic molecules currently feasible. All the necessary developments required for autonomous extraterrestrial μCE analysis have been accomplished; all that remains is building the flight-ready system.

A model of a flight-ready μCE instrument that could be built today with existing technology is shown in Figure 15.12. The instrument is very small, housed within a box of volume approximately 11.5 cm × 11.5 cm × 9 cm that maintains the microdevice at elevated temperature and pressure for electrophoresis. Samples are transferred to the microdevice via sippers driven by on-chip microvalve pumps. The instrument houses all the necessary optics for laser-induced fluorescence detection, high-voltage power supplies to drive electrophoresis, and

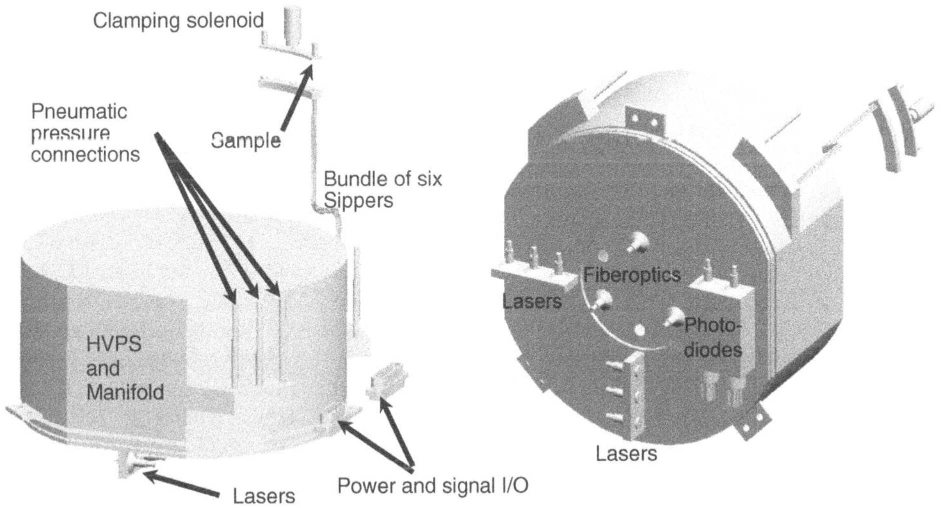

FIGURE 15.12 A conception of a future flight-ready μCE instrument based on MOA technology. In this design, the preprocessing (i.e., Steps 1–2 from Table 15.1) will take place outside the system, delivering either a liquid sample to the sippers or a solid sample that can be rehydrated in buffer pumped from an on-chip reservoir as in the single-channel MOA chip. The support hardware is indicated in this conception, including multiple lasers for redundancy and potentially multiple excitation wavelengths, high-voltage power supplies, and interface with the electronic control system. The chip would be housed on the interior of this system, and would be capable of performing completely autonomous fluidic manipulation and analysis of samples for amines, amino acids, aldehydes, ketones, carboxylic acids, and polycyclic aromatic hydrocarbons. The system dimensions are approximately 11.5 cm × 11.5 cm × 9 cm with a mass of 3 kg and a power use of ~1 W.

pneumatic controls to power the programmable automation in a 3 kg package that would require only 0.5 W (time average, 2 W peak) power during operation.

This flight-ready instrument meets the goals of robotic chemical exploration we described at the beginning of this chapter: it is capable of end-to-end analyses of multiple organic compound classes without any direct human intervention. This end-to-end analysis begins with sample acquisition via the sipper inputs. This is accomplished either by pumping a liquid sample from an extraction subsystem into an on-chip sample reservoir, or by pumping buffer from an on-chip reservoir to a sample to solubilize/dilute it before pumping the dissolved sample back to an on-chip sample reservoir. The programmable microfluidic sample processor labels the sample with an amine reactive fluorescent probe, a carbonyl reactive probe, and a carboxylic acid reactive probe, providing highly sensitive (subparts-per-trillion) coverage of amines, amino acids, aldehydes, ketones, and carboxylic acids. PAHs are solubilized by a buffer system containing cyclodextrins and are detected by their native fluorescence. Combinations of samples with standards are conducted, as is serial dilution of samples with high concentrations of the target molecules. Samples are then sent to the sample well of the electrophoresis channel for μCE separation with laser-induced fluorescence (LIF) detection. The total analysis conducted on a sample yields highly sensitive information about multiple organic compound classes and thus paints a complete picture of the organic chemistry of the extraterrestrial target.

On a final note, although our interests in microcapillary electrophoresis instrumentation are motivated by the fundamental scientific goals of astrobiology and planetary exploration, we are also committed to developing this technology to a level of maturity that will enable commonplace usage here on Earth in the coming decade. In that larger context, the technological story we have told in this chapter is essentially one of integration, automation, and packaging of a decades-old technology: capillary electrophoresis. It is our belief that it is this process of packaging that will ultimately enable the average person to use a handheld unit with the essential features we outline here without having to understand the intricacies associated with the technology itself (just as the average person is not required to understand the technologies inside a cellular telephone to make a phone call). All that these systems would require is the input of a small amount of liquid or solid sample in order to output a list of reagents present in the sample and their concentrations. We are highly enthusiastic about integrating multiple tools for chemical analysis into these future handheld chemistry laboratories as other technologies reach an analogous level of technological maturity, including mass spectrometers, electrochemical detectors, and so on. With the technological developments described here and in other chemical analytical arenas, we are finally approaching our long-held dream of a powerful handheld laboratory for personalized chemical analysis in addition to a powerful analytical tool for extraterrestrial exploration.

ACKNOWLEDGMENTS

We would like to acknowledge an ongoing collaboration with Rich Mathies' group at the University of California. We thank Erik Jensen for help with Figure 15.11. We thank Morgan Cable for her editing assistance. A. Stockton is supported by a NASA Postdoctoral Program fellowship. The writing of this chapter was carried out at the Jet Propulsion Laboratory, California Institute of Technology, under a contract with the National Aeronautics and Space Administration. Financial support was provided by NASA's Astrobiology Science and Technology Instrument Development (ASTID) Program (Project No. 104320).

REFERENCES

1. Des Marais, D.J., Nuth, J.A., Allamandola, L.J., Boss, A.P., Farmer, J.D., Hoehler, T.M., Jakosky, B.M., Meadows, V.S., Pohorille, A., Runnegar, B., and Spormann, A.M. (2008) *Astrobiology 8*, 715.
2. Shortt, B.J., Darrach, M.R., Holland, P.M., and Chutjian, A. (2005) *J. Mass Spectrom. 40*, 549.
3. Schlappi, B., Altwegg, K., Balsiger, H., Hassig, M., Jackel, A., Wurz, P., Fiethe, B., Rubin, M., Fuselier, S.A., Berthelier, J.J., De Keyser, J., Reme, H., and Mall, U. (2010) *J. Geophys. Res-Space Phys. 115.*
4. Balsiger, H., Altwegg, K., Bochsler, P., Eberhardt, P., Fischer, J., Graf, S., Jackel, A., Kopp, E., Langer, U., Mildner, M., Muller, J., Riesen, T., Rubin, M., Scherer, S., Wurz, P., Wuthrich, S., Arijs, E., Delanoye, S., De Keyser, J., Neefs, E., Nevejans, D., Reme, H., Aoustin, C., Mazelle, C., Medale, J.L., Sauvaud, J.A., Berthelier, J.J., Bertaux, J.L., Duvet, L., Illiano, J.M., Fuselier, S.A., Ghielmetti, A.G., Magoncelli, T., Shelley, E.G., Korth, A., Heerlein, K., Lauche, H., Livi, S., Loose, A., Mall, U., Wilken, B., Gliem, F., Fiethe, B., Gombosi, T.I., Block, B., Carignan, G.R., Fisk, L.A., Waite, J.H., Young, D.T., and Wollnik, H. 2007 *Space Sci. Rev. 128*, 745.
5. Becker, L., Richardson, C., Chaicharoen, K., Vanamerom, F., Cornish, T., Antoine, M., Pinnick, V., Cotter, R., Goesmann, F., Raulin, F., and Ehrenfreund, P. (2010) *Geochim. Cosmochim. Acta 74*, A67.
6. Mahaffy, P. (2008) *Space Sci. Rev. 135*, 255.
7. Klein, H.P. (1977) *J. Geophys. Res. 82*, 4677.
8. Biemann, K., Oro, J. and Toulmin, P. III, Orgel, L.E., Nier, A.O., Anderson, D.M., Simmonds, P.G., Flory, D., Diaz, A.V., Rushneck, D.R., and Biller, J.A. (1976) *Science 194*, 72.
9. Bada, J.L., Ehrenfreund, P., Grunthaner, F., Blaney, D., Coleman, M., Farrington, A., Yen, A., Mathies, R., Amudson, R., Quinn, R., Zent, A., Ride, S., Barron, L., Botta, O., Clark, B.,

Glavin, D., Hofmann, B., Josset, J.L., Rettberg, P., Robert, F., and Sephton, M. 2008 *Space Sci. Rev. 135*, 269.

10. Cable, M.L., Stockton, A.M., Mora, M.F., and Willis, P.A. (2013) *Anal. Chem. 85*, 1124–1131.
11. Mora, M.F., Stockton, A.M., and Willis, P.A. (2012) *Electrophoresis 34*, 1–8.
12. Smith, P.H., Tamppari, L.K., Arvidson, R.E., Bass, D., Blaney, D., Boynton, W.V., Carswell, A., Catling, D.C., Clark, B.C., Duck, T., DeJong, E., Fisher, D., Goetz, W., Gunnlaugsson, H.P., Hecht, M.H., Hipkin, V., Hoffman, J., Hviid, S.F., Keller, H.U., Kounaves, S.P., Lange, C.F., Lemmon, M.T., Madsen, M.B., Markiewicz, W.J., Marshall, J., McKay, C.P., Mellon, M.T., Ming, D.W., Morris, R.V., Pike, W.T., Renno, N., Staufer, U., Stoker, C., Taylor, P., Whiteway, J.A., and Zent, A.P. 2009 *Science 325*, 58.
13. Skelley, A.M., Aubrey, A.D., Willis, P.A., Amashukeli, X., Ehrenfreund, P., Bada, J.L., Grunthaner, F.J., and Mathies, R.A. (2007) *J. Geophys. Res. Biogeosci. 112*, G04S11.
14. Skelley, A.M., Scherer, J.R., Aubrey, A.D., Grover, W.H., Ivester, R.H.C., Ehrenfreund, P., Grunthaner, F.J., Bada, J.L., and Mathies, R.A. (2005) *Proc. Natl. Acad. Sci. U.S.A. 102*, 1041.
15. Benhabib, M., Chiesl, T.N., Stockton, A.M., Scherer, J.R., and Mathies, R.A. (2010) *Anal. Chem. 82*, 2372.
16. Berg, C., Valdez, D.C., Bergeron, P., Mora, M.F., Garcia, C.D., and Ayon, A. (2008) *Electrophoresis 29*, 4914.
17. Chiesl, T.N., Chu, W.K., Stockton, A.M., Amashukeli, X., Grunthaner, F., and Mathies, R.A. (2009) *Anal. Chem. 81*, 2537.
18. Stockton, A.M., Chiesl, T.N., Lowenstein, T.K., Amashukeli, X., Grunthaner, F., and Mathies, R. (2009) *Astrobiology 9*, 823.
19. Skelley, A.M., Cleaves, H.J., Jayarajah, C.N., Bada, J.L., and Mathies, R.A. (2006) *Astrobiology 6*, 824.
20. Stockton, A.M., Chiesl, T.N., Scherer, J.R., and Mathies, R.A. (2009) *Anal. Chem. 81*, 790.
21. Stockton, A.M., Tjin, C.C., Huang, G.L., Benhabib, M., Chiesl, T.N., and Mathies, R.A. (2010) *Electrophoresis 31*, 3642.
22. Stockton, A.M. (2010) *Analysis of Organic Molecules Using the Mars Organic Analyzer, a Portable, Automated Microfabricated Capillary Electrophoresis Instrument*, University of California, Berkeley.
23. Stockton, A.M., Chandra Tjin, C., Chiesl, T., and Mathies, R. (2010) *Electrophoresis 31*, 3642.
24. Stockton, A.M., Chandra Tjin, C., Chiesl, T., and Mathies, R. (2011) *Astrobiology 11*, 519.
25. Behnam, M., Kaigala, G.V., Khorasani, M., Marshall, P., Backhouse, C.J., and Elliott, D.G. (2008) *Lab Chip 8*, 1524.
26. Manz, A., Harrison, D.J., Verpoorte, E.M.J., Fettinger, J.C., Paulus, A., Lüdi, H., and Widmer, H.M. (1992) *J. Chromatogr. A 593*, 253.
27. Reyes, D.R., Iossifidis, D., Auroux, P.A., and Manz, A. (2002) *Anal. Chem. 74*, 2623.
28. Hutt, L.D., Glavin, D.P., Bada, J.L., and Mathies, R.A. (1999) *Anal. Chem. 71*, 4000.
29. Nguyen, A.-L. and Luong, J.H.T. (1997) *Anal. Chem. 69*, 1726.
30. Zarrin, F. and Dovichi, N.J. (1985) *Anal. Chem. 57*, 2690.
31. Liu, C.N., Toriello, N.M., and Mathies, R.A. (2006) *Anal. Chem. 78*, 5474.
32. Douglas, E.S., Hsiao, S.C., Onoe, H., Bertozzi, C.R., Francis, M.B., and Mathies, R.A. (2009) *Lab Chip 9*, 2010.
33. Beyor, N., Yi, L.N., Seo, T.S., and Mathies, R.A. (2009) *Anal. Chem. 81*, 3523.
34. Thaitrong, N., Toriello, N.M., Del Bueno, N., and Mathies, R.A. (2009) *Anal. Chem. 81*, 1371.
35. Yeung, S.H.I., Liu, P., Del Bueno, N., Greenspoon, S.A., and Mathies, R.A. (2009) *Anal. Chem. 81*, 210.
36. Toriello, N.M., Liu, C.N., Blazej, R.G., Thaitrong, N., and Mathies, R.A. (2007) *Anal. Chem.*, *79*, 8549.
37. Liu, P., Seo, T.S., Beyor, N., Shin, K.J., Scherer, J.R., and Mathies, R.A. (2007) *Anal. Chem. 79*, 1881.
38. Benhabib, M. (2009) *Multichannel Mars Organic Analyzer for Planetary Analysis of Organic Molecules and Biomarkers*, University of California, Berkeley.
39. Paegel, B.M., Grover, W.H., Skelley, A.M., Mathies, R.A. and Joyce, G.F. (2006) *Anal. Chem. 78*, 7522.
40. Jensen, E.C., Bhat, B.P., and Mathies, R.A. (2010) *Lab Chip 10*, 685.
41. Liu, P. and Mathies, R.A. (2009) *Trends Biotechnol. 27*, 572.
42. Zeng, Y., Novak, R., Shuga, J., Smith, M.T., and Mathies, R.A. (2010) *Anal. Chem. 82*, 3183.
43. Grover, W.H., Skelley, A.M., Liu, C.N., Lagally, E.T., and Mathies, R.A. (2003) *Sens. Actuator B Chem. 89*, 315.
44. Willis, P.A., Hunt, B.D., White, V.E., Lee, M.C., Ikeda, M., Bae, S., Pelletier, M.J., and Grunthaner, F.J. (2007) *Lab Chip 7*, 1469.
45. Mora, M.F., Greer, H.F., Bryant, S., and Willis, P.A. (2011) *Anal. Chem. 83*, 8636.
46. Grover, W.H., von Muhlen, M.G., and Manalis, S.R. (2008) *Lab Chip 8*, 913.
47. Willis, P.A., Greer, F., Lee, M.C., Smith, J.A., White, V.E., Grunthaner, F.J., Sprague, J.J., and Rolland, J.P. (2008) *Lab Chip 8*, 1024.
48. Jensen, E.C., Chiesl, T., Stockton, A.M., Bera, A., and Mathies, R. (2012) *Lab Chip* DOI: 10.1039/C2LC40861A.
49. Jensen, E.C., Zheng, C.G., Kim, J., and Mathies, R. (2010) *J. Lab. Autom. 15*, 455.
50. Grover, W.H. and Mathies, R.A. (2005) *Lab Chip 5*, 1033.

16

RAPID ANALYSIS OF CHARGE HETEROGENEITY OF MONOCLONAL ANTIBODIES BY CAPILLARY ZONE ELECTROPHORESIS AND IMAGED CAPILLARY ISOELECTRIC FOCUSING

YAN HE, JIM MO, XIAOPING HE, AND MARGARET RUESCH

16.1 INTRODUCTION

As progress in biotechnology continues, a large number of therapeutic proteins have been developed using recombinant DNA technology. Monoclonal antibodies (mAbs) have become the most rapidly growing class of biotherapeutics in development for many different disease indications in the last decade [1–3]. Novel mAb molecules are entering clinical studies at a rate of nearly 40 per year, and the research pipeline includes about 250 therapeutic mAbs in clinical studies [2]. Concomitantly, there is an increasing need for robust analytical methods that can be used for high-throughput analysis of a large number of samples to support bioprocess development, formulation screening, and quality control testing.

Similar to other biotherapeutics, mAbs are complex molecules, and a variety of methods are required to monitor product heterogeneity to ensure product quality and consistency [4]. IgG mAbs, with molecular weights of approximately 150 kDa, comprise most of the immunoglobulin (Ig) in human serum as well as most mAb therapeutics currently marketed or in development. An IgG molecule consists of four polypeptide chains, two heavy and two light chains, assembled with intrachain and interchain disulfide bonds. Each IgG (Figure 16.1) is composed of two identical Fab regions (fragment-antigen binding) and an Fc region (fragment crystallizable). In IgGs, the unique complementarity-determining regions (CDRs) that reside in the variable fragment portion of the Fab define antigen specificity. The hinge region provides flexibility for the two Fab domains relative to the Fc domain, and affects bivalent antigen binding and activation of Fc effector functions. Meanwhile, the Fc portion of IgG participates in pH-dependent electrostatic interactions with Fc receptors, which activate immune effector functions and protect the mAb from degradation [5–10].

Monoclonal antibodies are heterogeneous, containing microheterogeneity in the Fc carbohydrate moiety and often in the Fab or CDR regions. Most of the heterogeneity results in changes in the surface charge of the antibody, either directly as a change in the number of charged residues, or indirectly as a chemical or physical alteration that changes surface-charge distribution [11–13]. Table 16.1 presents an overview of the sources of charge-related heterogeneity in mAbs. The charge variants can be classified into two categories: acidic and basic species. For example, deamidation, sialylation, or glycation result in an increase in the net negative charge and therefore a decrease in p*I* values, leading to the formation of acidic species. On the other hand, the presence of heavy chain C-terminal lysine residues, succinimide formation, or removal of sialic acid introduces additional positive charges or removes negative charges, thereby causing an increase in p*I* values to form basic species. Antibody charge heterogeneity has gained considerable

Capillary Electrophoresis and Microchip Capillary Electrophoresis: Principles, Applications, and Limitations, First Edition.
Edited by Carlos D. García, Karin Y. Chumbimuni-Torres, and Emanuel Carrilho.

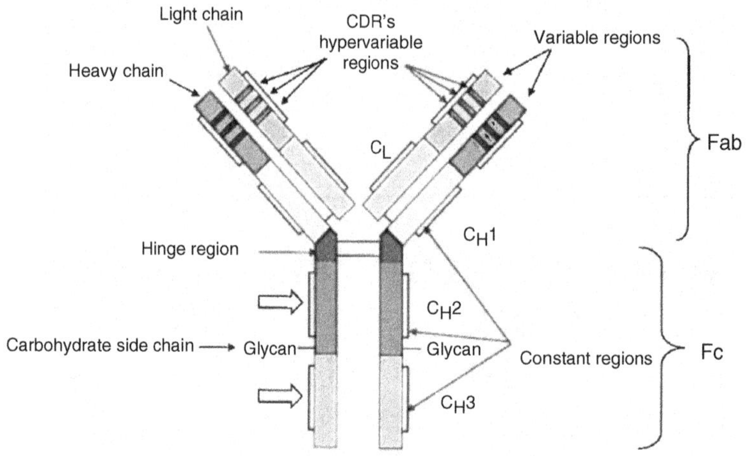

FIGURE 16.1 Schematic illustration of an IgG antibody structure.

TABLE 16.1 Sources of Charge Heterogeneities of IgGmAbs

S. No.	Degradation Pathways	Effect	Charged Variants	References
1	Succinimide formation	Loss of COOH	Basic	14,16
2	Asialylation	Loss of COOH	Basic	17
3	C-terminal proamidation	Loss of COOH	Basic	18
4	Oxidation	Conformational	Basic	19
5	Disulfide-mediated	Conformational	Basic	20,21
6	Deamidation	New COOH	Acidic	22,23
7	Sialylation	New COOH	Acidic	23
8	C-terminal Lys cleavage	Loss of NH_2	Acidic	23,24
10	Cysteinylation	Conformational	Acidic	25
11	N-terminal glutamine cyclization	Loss of NH_2	Acidic	13,26
12	Glycation	Loss of NH_2	Acidic	27,28

attention in the biotechnology industry due to its potential influence on stability and biological activity [12,14,15]. An assessment of the charge heterogeneity of mAbs is important in understanding lot-to-lot product consistency at the end of manufacture and on stability.

Charge-related heterogeneity in mAbs is usually examined by ion-exchange chromatography (IEC) [12,15,29–32] or isoelectric focusing (IEF) [33,34]. The development of a robust IEC method requires the optimization of many parameters including column, mobile-phase composition, pH, temperature, and gradients. IEC method development can be time-consuming, and each molecule usually requires a unique set of separation conditions. An advantage of IEC is that it is amenable to characterization of different charge species, since isolated peaks can be collected and subsequently analyzed by orthogonal techniques. Another traditional method for analyzing charge heterogeneity of proteins is slab gel IEF, which was introduced by Svensson and Vesterberg in 1960s [33,34]. While slab gel IEF provides high-resolution separation of charge variants of proteins, it is a time-consuming and labor-intensive method that often results in high assay variability in quantitation.

In the last three decades, several different capillary electrophoresis (CE) techniques (Table 16.2) have been developed and applied to the separation and quantitation of mAb charge species. Capillary isoelectric focusing (CIEF) and imaged capillary isoelectric focusing (iCE)

TABLE 16.2 CE Techniques for the Analysis of Protein Charge Heterogeneity

Techniques	Principle of Separation	References
CIEF (capillary isoelectric focusing)	Isoelectric point	35
iCE (CIEF with whole column imaging detection)	Isoelectric point	36,37
CZE (capillary zone electrophoresis)	Charge to size	38,39

utilize the same separation, based on isoelectric point. The difference between these techniques is the mobilization and detection steps, with iCE utilizing a unique whole-column imaging technique that eliminates the need for mobilization of the focused proteins past the detector. While capillary isoelectric focusing (CIEF) using traditional mobilization and detection will not be discussed in detail in this chapter, the development and application of iCE separations is applicable to CIEF method development and application. Capillary zone electrophoresis (CZE) is a CE separation based on a molecule's charge and size. In this chapter, we focus on rapid charge separations of mAbs by CE, using CZE and iCE methods.

16.2 CAPILLARY ZONE ELECTROPHORESIS

Separation by CZE is based on both the charge and size of molecules, which is clearly indicated in the following equation [40].

$$\mu_{ep} = \frac{q}{6\pi\eta r} \tag{16.1}$$

In this equation, q is the net charge of the analyte, r is the Stokes radius, and η is the viscosity of the separation medium. Equation 16.1 can be made more useful, but more approximate, by expressing the size dependency of mobility using protein molecular weight rather than radius as shown in Equation 16.2 [41].

$$\mu_{ep} = \frac{K(1)q}{K(2)M^{1/3} + K(3)M^{2/3}} \tag{16.2}$$

Equation 16.2 indicates that a protein's charge is its most important attribute with regard to determining its mobility. Molecular weight exhibits an inverse influence that varies continuously from $M^{1/3}$ to $M^{2/3}$ depending on the magnitude of the molecular weight and the ionic strength range under examination. For most mAb charge isoforms (e.g., deamidation, lysine variants, glycation), it is reasonable to assume that they have similar molecular weights. Hence, the separation selectivity of charge isoforms by CZE is predominantly governed by charge differences.

16.2.1 Separation and Detection Strategy

Routine application of CZE for the analysis of mAb charge heterogeneity requires the rapid development of an optimized separation as well as reproducible method performance. The crucial separation variables include capillary construction and temperature, buffer type, pH, and concentration.

16.2.1.1 Capillary Construction One of the greatest challenges in the analysis of proteins by CE is the adsorption of proteins to the walls of the uncoated capillary. Basic proteins, such as most mAbs with p*I*s in the range of 7.5–9.5, tend to interact electrostatically with the negatively charged silanols on the inner surface of the capillary. Adsorption may cause many problems: low efficiency, asymmetric peak shapes, and a drifting EOF and thus migration times. Dynamic or permanent coatings on the capillary inner surface can efficiently prevent wall interactions of most proteins and significantly reduce protein adsorption. Separation of IgG mAbs by CZE using both coated and uncoated capillaries have been investigated [41–43]. Using a diluted buffer (50 mM) of e-amino-caproic acid (EACA), charge isoforms were shown to be well-separated using coated capillaries [42], but very broad peaks were obtained with an uncoated capillary [43]. Buffer conditions that improve CZE separations in uncoated capillaries are discussed further in this chapter. The capillary internal diameter (i.d.) and length also have a significant impact on separation resolution and speed. Capillaries with i.d.s in the range of 25–75 μm are typically used. Faster separations can be achieved using capillaries with narrower i.d.s, which provide better heat dissipation under higher field strength, although at the cost of detection sensitivity. Improved separations can be attained with longer capillaries, but at the cost of separation speed. In practice, the selection of capillary length and i.d. is made to balance the requirements of separation resolution, speed, and detection sensitivity.

16.2.1.2 Buffer Composition The separation of protein charge variants is strongly affected by the buffer composition, including buffer pH, concentration, and type, as well as the presence of additives. Protein charge is the most sensitive parameter governing mobility, and it is directly controlled through the adjustment of buffer pH. The strong effect of buffer pH on the ability to resolve similarly sized mAb charge variants is illustrated in Figure 16.2. Increasing the pH from 4.7 to 5.2 resulted in a significant improvement in the separation of both acidic and basic species, with a slight increase in migration time. A further increase in the pH from 5.2 to 5.7 results in an additional improvement in the separation of acidic and main species, but a deteriorated separation of the basic and main species. Moreover, the peaks become broadened, especially for the more basic species. Peak broadening becomes more significant at pH 6.0, bringing about a total loss in the separation of the basic species [44].

The vast majority of published protein separations by CZE include the use of additives in the buffer. These additives are used to alter selectivity and/or decrease adsorption tendencies, thereby improving efficiency. The additives include salts, zwitterions, cations, organic solvents, and surfactants. In addition to permanent coatings, the use of concentrated zwitterionic buffers and/or diluted

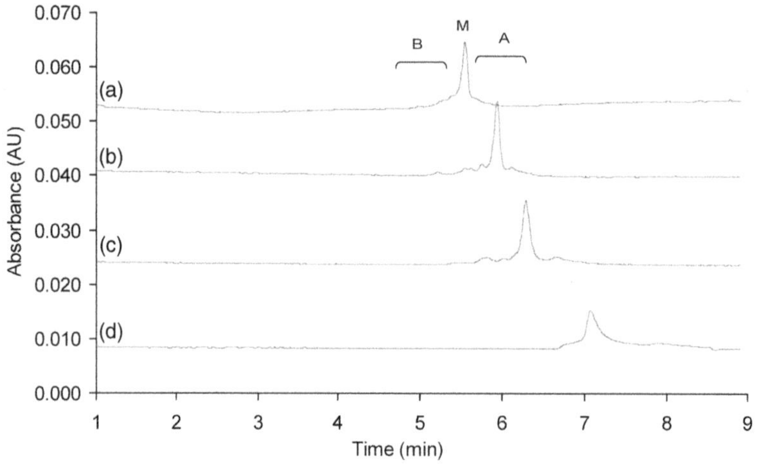

FIGURE 16.2 CZE separation of mAb charge variants at different pH conditions. Conditions for bare fused-silica capillary: total length, 30 cm; effective length, 10 cm; i.d., 40 μm; capillary temperature, 25 °C; voltage, 20 kV; injection, 2.1 Pa × 5 s; detection, 214 nm. Separation buffer: 400 mM EACA–acetic acid, 0.05 % (m/v) HPMC. pH, (a) 4.7; (b) 5.2; (c) 5.7; (d) 6.0. Beckman Coulter PA800: IgG concentration, 1 mg mL^{-1}. Peak identity: A, acidic; M, main; B, basic. From Ref. 44.

cationic amines can efficiently eliminate protein adsorption to the inner surface of bare capillaries. Figure 16.3 shows the separation of different mAbs and their charge variants by CZE using different concentrations of the zwitterionic buffer, EACA at pH 5.5, in an uncoated capillary [43,44]. Peak shape and separation resolution improve as the concentration of EACA buffer is increased from 50 to 600 mM.

Adsorption of proteins to the inner surface of bare capillaries can also be reduced by the addition of a dynamic coating to the separation buffer. A divalent amine (1,3-diaminopropane) has been used to efficiently reduce the adsorption of basic proteins to the capillary and improve the separation [45]. Increasing the 1,3-diaminopropane concentration from 10 to 50 mM in the buffer resulted in a significant improvement in the peak shape and separation resolution.

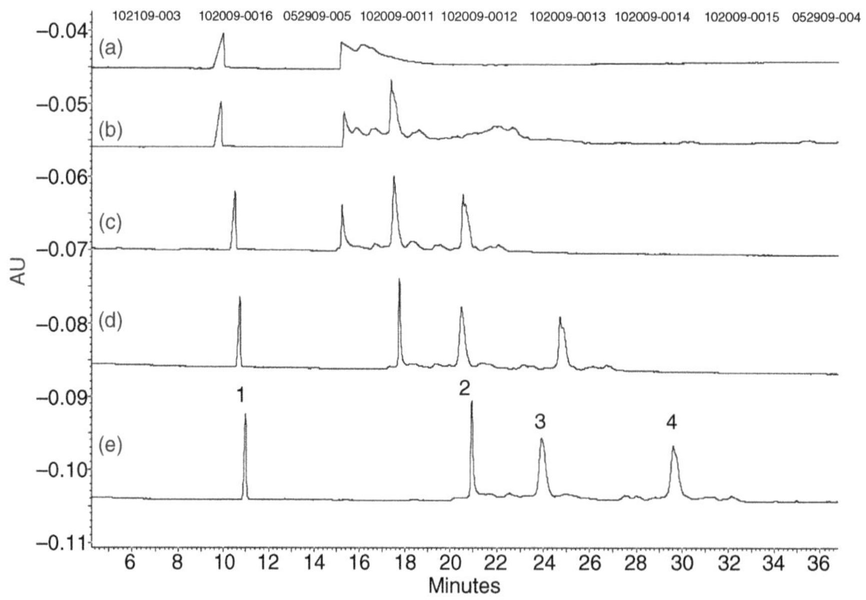

FIGURE 16.3 Effect of EACA concentration on the separation of IgG1 and IgG2 mAbs in a bare fused-silica capillary. Conditions for bare fused-silica capillary: total length, 50 cm; effective length, 40 cm; i.d., 50 μm; buffer: (a) 50 mM; (b) 100 mM; (c) 200 mM; (d) 400 mM; (e) 600 mM EACA–acetic acid, 0.1% HPMC; pH, 5.4. Peak identities: (1) histidine; (2) IgG1 mAb; (3) and (4) IgG2 mAbs. From Ref. 43.

16.2.1.3 Separation Voltage and Field Strength Using conventional buffers and 50 μm capillaries, CE can be performed at field strengths as high as 300–500 V cm^{-1} with acceptably low thermal zone deformation. Higher field strengths (up to 2000 V cm^{-1}) can be applied to greatly increase the separation speed without a decrease in resolution, if the experiments are performed in buffers with very low conductivity and/or capillaries with smaller i.d.s [46]. An example of this class of buffers is polyoxyethylene bis(3-amino-2-hydroxypropyl)/polyoxyethylene bis(acetic acid), at a pH of 8.6. The concentration of the buffer constituents was 0.3 % w/v and 0.025 % w/v, respectively. A rapid separation of albumin and its charge variants were completed within 30 s. The combined use of high electric field strength and a short capillary has also been applied for the high-speed separation of mAb charge variants (Figure 16.4). The maximum allowable voltage using the existing commercial CE instruments, 30 kV, was applied across a 30 cm (effective length 10 cm) long, 40 μm i.d. bare fused-silica capillary. The separation buffer consisted of 400 mM EACA–acetic acid, 2 mM TETA, and 0.05% HPMC, pH 5.7. A rapid separation of charge variants of two IgG mAbs was achieved within 3 min.

16.2.1.4 Detection CZE methods can utilize detection by UV at 214 nm, a universal absorption wavelength for the peptide bond present in proteins. However, a drawback of UV absorption detection in CE is that the concentration sensitivity is low due to the short optical path length limited by the internal diameter of capillary. For the sensitive analysis of trace proteins, other detection methods such aslaser-induced fluorescence (LIF) detection or native fluorescence must be used. For LIF detection, proteins are derivatized before the electrophoretic run. Using *o*-phthaldialdehyde-2-mercaptoethanol as the label, Lee et al. demonstrated a detection limit for carbonic anhydrase of 0.73 nM, and the detection response was linear over three orders of magnitude [47]. Finally, the integration of CE with MS is a promising approach that can overcome one of the major challenges of CE separations: characterization of peaks. Successful CE–MS combines the fast, biologically compatible separation of CE with MS capabilities for obtaining structural information of different charge variants [48,49].

16.2.2 Applications

CZE has been widely used to assess the charge heterogeneity of a wide variety of recombinant proteins, including rtPA [50], IgF-I [51], human erythropoietin [52], interleukin-6 [53], and IgG antibodies [38,42–44,54–56]. One of the earliest applications of CZE for the assessment of IgG mAb charge heterogeneity was investigated by Wenisch et al. in 1990 [38]. A side-by-side comparison of CZE and IEF in immobilized pH gradients (IPG) was performed for the analysis of the purity of a mAb. The mAb was isolated by IEF in a multicompartment electrolyzer with isoelectric immobiline membranes. The CZE separation was carried out in a capillary coated with hydrophilic polymer using UV detection at 200 nm and a separation buffer containing 100 mM phosphate (pH 5.8). While IPG IEF provided better resolving power in this study, CZE was superior in terms of speed of analysis and quantification. The separation time with CZE was about 20 min compared with 2–3 h for IPG IEF. In addition, CZE using UV 200 nm in the absorption region of the peptide bond provides a more accurate assessment of charge isoform distribution.

Ma and Nashabeh compared CZE and IEC methods for the assessment of charge distributions of mAbs [42]. The

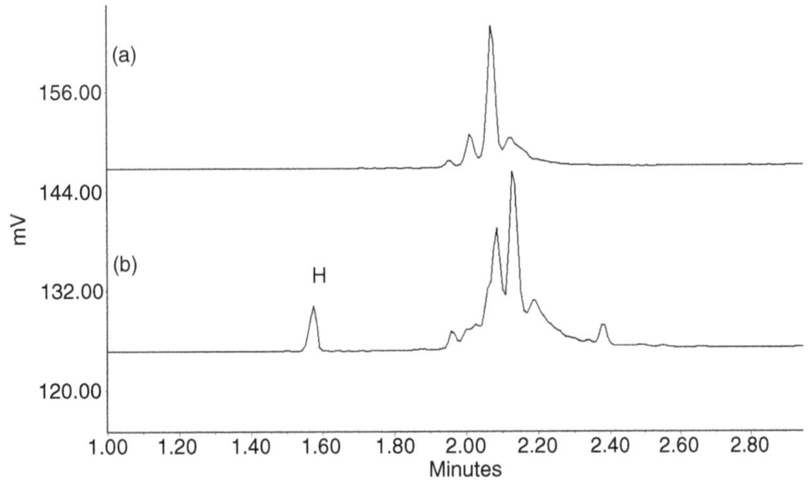

FIGURE 16.4 Rapid CZE separation of two IgG mAbs using a short capillary and high electric field strength. H indicates histidine in formulation buffer.

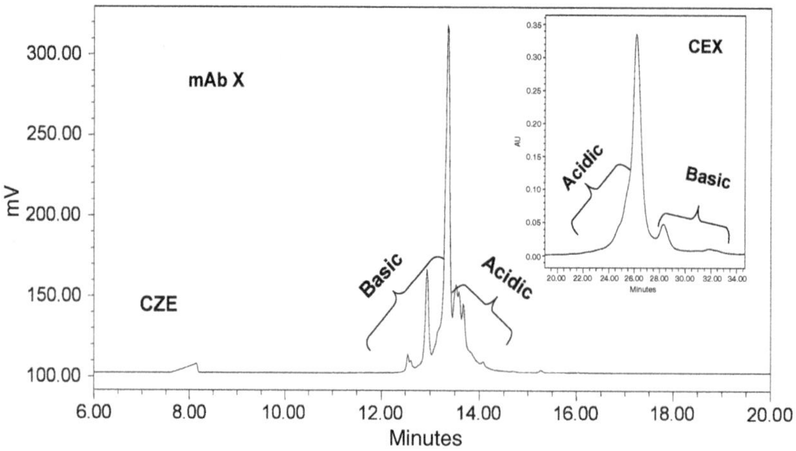

FIGURE 16.5 Monitoring charge heterogeneity of an IgG1 mAb by CZE and CEX. From Ref. 55.

separation was conducted in a coated capillary. The run buffer consisted of 45 mM EACA–acetic acid (pH 4.5) and 0.2 % HPMC to reduce protein interactions with the capillary wall. Notably, the CZE separation was very rapid (<10 min) when compared with IEC, with similar separation of the C-terminal lysine basic charge variants. A similar CZE method was applied to the analysis of an IgG1 mAb and resulted in an improved separation of charged species as compared with cation exchange chromatography (CEX) (Figure 16.5) [55]. The CZE separation was rapid and reproducible with RSDs of 1.2% and 0.8% for the percent acidic and basic species, respectively.

During the development of therapeutic mAbs, various cell lines, cell culture and bioreactor conditions, purification process conditions, and final formulations are evaluated to optimize product consistency and stability. One attribute that is often monitored during these evaluations is charge heterogeneity. Particularly when changes in the overall charge profile are observed during processing or on stability, further characterization of the sources of this heterogeneity may be warranted [12,14,18]. CZE is a technique suitable for the rapid screening of a large number of in-process and stability samples.

For example, Figure 16.6 shows the comparison of IEX (cation exchange chromatography indicated by CEX) and CZE for the monitoring of basic species of mAbs generated under different conditions in a bioreactor. Although the levels of basic species reported are different, the CZE and IEX show similar trending for the different samples. For this analysis, CZE has the advantage of a very short separation time, 3 min, compared with 50 min by IEX. However, IEX offers a means to more rapidly collect and characterize fractions. The CZE method used for the analysis in Figure 16.6 utilizes a bare fused-silica capillary with relatively harsh flushing conditions, resulting in robust performance of the method in the analysis of samples of varying sample matrices.

CZE can also be used for the high-throughput screening of large number of stability samples to support the selection of a formulation that yields the best stability profile. This is illustrated in Figure 16.7; acidic species of an IgG mAb stored in six different formulations for three months at different temperatures were monitored. Acidic species of the IgG mAb in all formulations change little when temperature rises from 5 to 25 °C, but increase significantly when the temperature rises from 25 °C to the stressed condition of 40 °C in this accelerated stability study. Differences in the rate of increase in various formulations are helpful in guiding formulation decisions.

As compared to IEX, CZE offers the advantages of faster analysis times and a set of conditions that is generally applicable to multiple products without the need for product-specific method development. Figure 16.8 shows the rapid separation of multiple mAbs by CZE using a bare

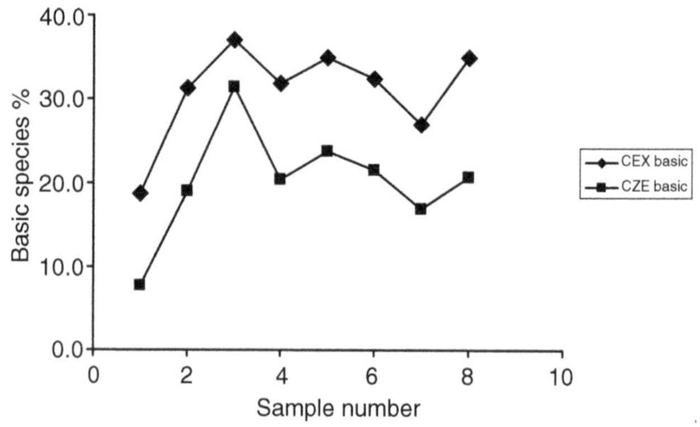

FIGURE 16.6 Comparison of CZE and CEX for the analysis of basic charge variants (reported as percent basic species) of an IgG mAb produced under different conditions in a small-scale bioreactor. Conditions for bare capillary: total and effective length, 30 and 10 cm; i.d., 40 μm; capillary temperature, 25°C; voltage, −20 kV; detection, 214 nm. Separation buffer: 400 mM EACA–acetic acid, 2 mM triethylenetetramine, 0.05% HPMC; pH 5.7.

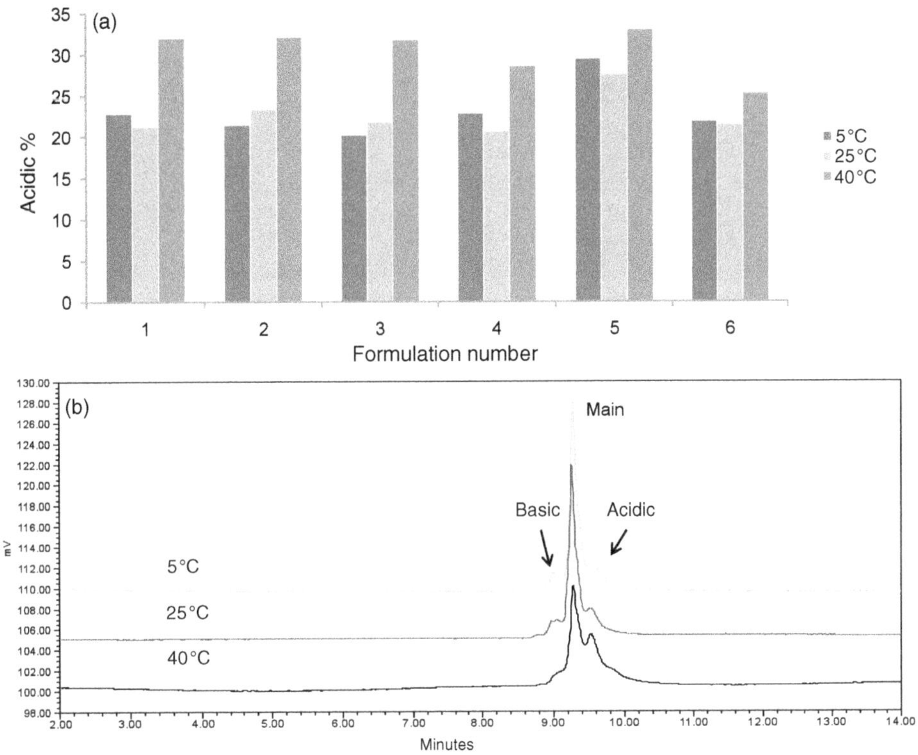

FIGURE 16.7 (a) Changes in acidic species ("acidic %") of an IgG mAb stored in various formulations at different temperatures. (b) Representative electropherograms of IgG formulation 3 at different temperatures, illustrating an overall increase in acidic species with increasing temperature. Separation conditions for BioCAP XL coated capillary: effective length, 40.2 cm; total length, 50.2 cm; i.d., 50 μm; buffer, 45 mM EACA–acetic acid; pH, 5.0; 0.2% HPMC; detection, 214 nm; voltage, 30 kV.

capillary. The migration times of each individual mAb are reproducible, with RSDs in the range of 0.1–0.15%. For biotherapeutics, identity of the material manufactured must be confirmed on release testing; the unique p*I* values determined by IEF or CIEF are one way to confirm identity by comparing with a known reference. Similarly, the unique and reproducible migration times of the main peaks of different mAbs can be used for demonstrating identity by CZE, as has been previously reported [42,56].

16.3 IMAGED CAPILLARY ISOELECTRIC FOCUSING

Imaged capillary isoelectric focusing (Imaged cIEF or iCE) is a one-step mode CIEF method that separates protein species based on their charge differences (isoelectric point, or p*I*) in a pH gradient. In CIEF separations, proteins are mixed with carrier ampholytes; following the application of voltage, the ampholytes form a pH gradient and the proteins or their isoforms are focused at their p*I*. While traditional CIEF requires subsequent mobilization of the separated proteins past a detector, iCE does not require a mobilization step prior to their detection [36,37]. The iCE technology focuses protein species according to their p*I* values within a short (ca. 5 cm) capillary column and monitors them in a real-time mode using whole column imaging detection (WCID). After the focusing is complete, the protein species with different charges appear as electrophoretic peaks that are measured using an imaged UV detector at a wavelength of 280 nm. A key advantage of iCE over conventional CIEF is the absence of a mobilization step. The iCE separation requires reduced analysis time due to the use of short capillary, and the ability to monitor the focusing process in real-time leads to faster method development [57,58].

The iCE technology is versatile in that it can be applied to proteins with a wide range of complexity, and it has been effective in providing data for a variety of purposes such as bioprocess development, protein formulation development, release, and stability testing for clinical or commercial biotherapeutics.

16.3.1 Method Development and Optimization

Separation of mAb charge variants in iCE is affected by a number of parameters including carrier ampholytes,

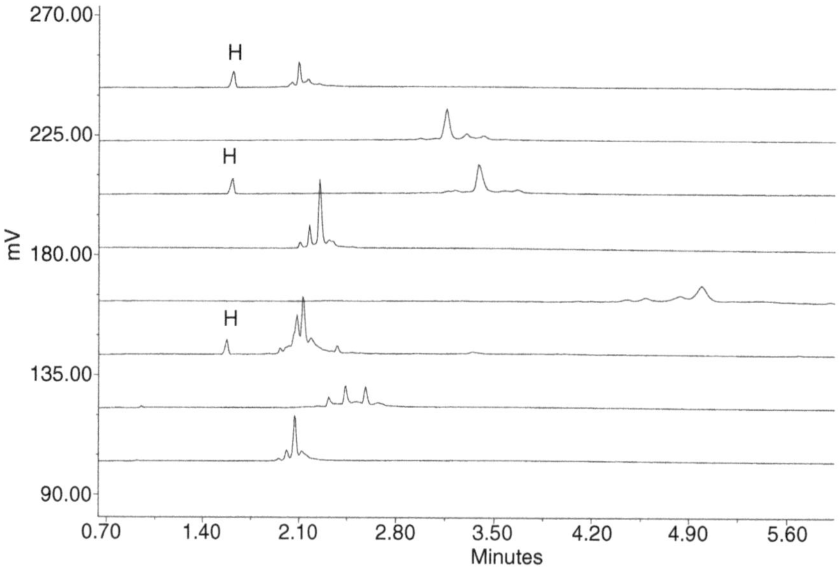

FIGURE 16.8 Rapid separation of multiple mAbs by CZE. H indicates histidine in the sample matrix. Conditions for bare fused-silica capillary: total length, 30 cm; effective length, 10 cm; i.d., 40 μm; buffer, 400 mM EACA–acetic acid; 2 mM, tirethylenetetramine; pH, 5.7; voltage, 30 kV; detection, 214 nm.

additives, focusing time, salt concentration, and sample protein concentration [59–61].

16.3.1.1 Carrier Ampholyte The choice of carrier ampholytes is critical, since ampholytes establish the pH gradient that directly affects resolution of the protein in the capillary. Commercial ampholytes from different vendors may perform differently for a given protein when there is an interaction between protein and ampholyte. When optimizing the analysis for a specific protein, the ampholytes that will be used must be evaluated according to a variety of criteria, including viscosity, conductivity, and transparency at 280 nm. There are five major brands of ampholytes: Pharmalyte, Servalyte, Ampholine Zoom, and Bio-lyte. A comparison of iCE profiles using three of the commercial ampholytes is shown in Figure 16.9. Differences such as baseline noise (Figure 16.9a) and isoform resolution (Figure 16.9b) are apparent in the electropherograms. For this separation, Pharmalyte resulted in the lowest baseline noise.

Most iCE separations employ 3–10 ampholytes mixed with the sample to establish a pH gradient across the pH 3–10 range. If higher resolution is desired in a more specific pH range, narrow-range ampholytes may be added to the mixture to improve the separation as shown in Figure 16.10.

Separation of mAb charge variants is also affected by ampholyte concentration. The concentration of ampholytes should be high enough to generate a stable pH gradient during the focusing process. However, too high a concentration of ampholytes will lead to increase of focusing time and baseline noise [59]. The optimal concentration is different for different brands of ampholytes. A concentration of 4–8% is optimal for Pharmalyte, while for 1–4% is the optimal for Servalyte. The optimal ampholyte concentration may vary depending on the individual mAb.

16.3.1.2 Additives Proteins generally have low solubility and a strong tendency to aggregate and precipitate near their p*I*s. Precipitation of proteins often manifests as spikes and irreproducible peaks in the electropherogram. Additives that can increase protein solubility can mitigate the problem of protein precipitation. Urea is a powerful protein denaturant, and is frequently added to the separation medium to eliminate protein precipitation and improve peak efficiency. Sucrose is a known osmolyte able to stabilize proteins against aggregations. A combination of urea and sucrose has been used as an additive in the separation matrix to prevent precipitation during the iCE focusing process. Figure 16.11 shows the impact of urea and sucrose on the separation of an IgG mAb. Without additives, the protein precipitates due to low solubility during the focusing process, resulting in spikes in the electropherogram (Figure 16.11a). Incorporation of 1 M urea and 3% sucrose into the iCE separation buffer results in an improved and reproducible separation (Figure 16.11b). It should be noted that urea addition also increases the viscosity of the solution and decreases the mobility of the proteins, leading to increase in focusing time.

16.3.1.3 Focusing Time and Voltage Focusing time is expected to be proportional to the length of a capillary and to the reciprocal of field strength. Rapid focusing can be

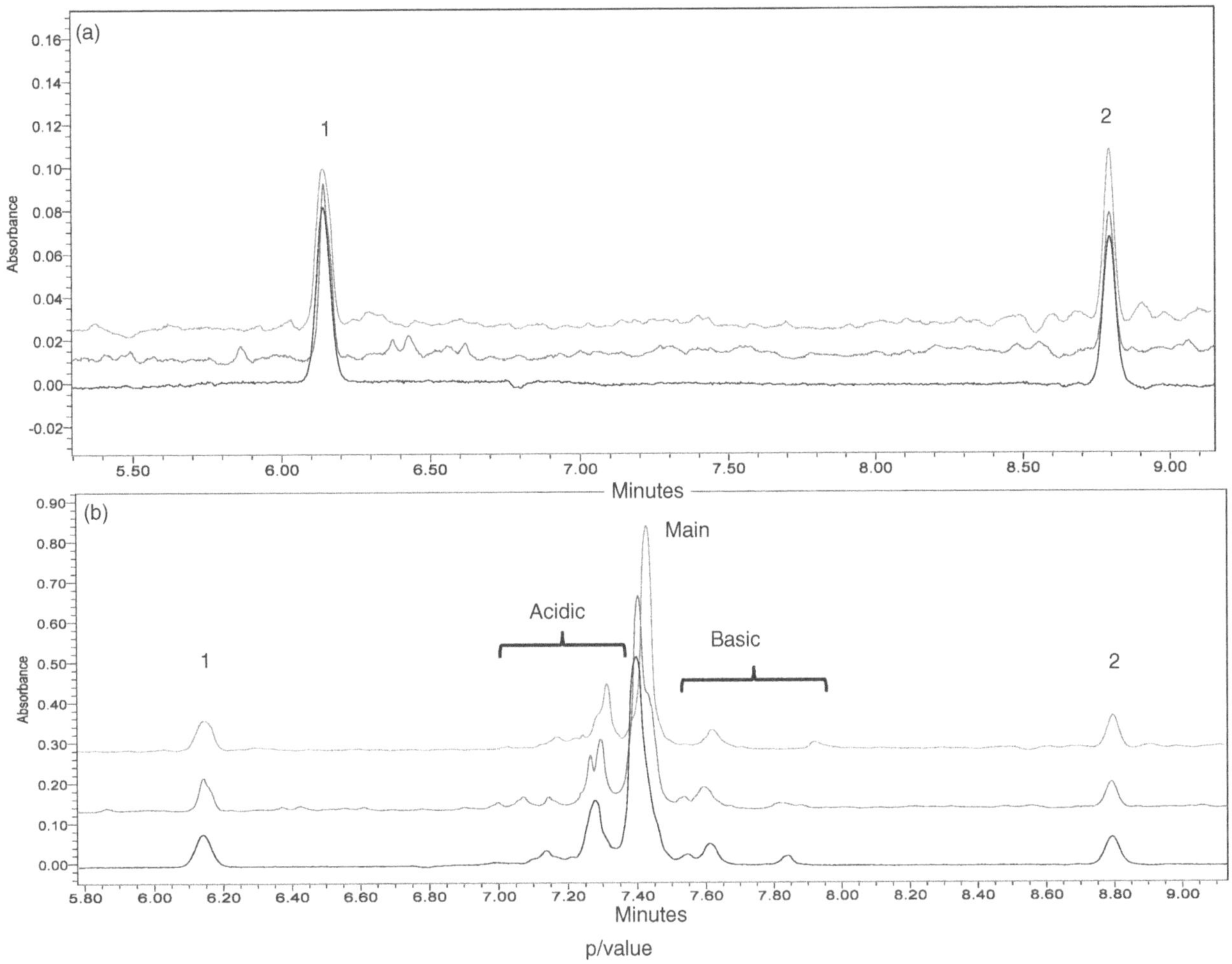

FIGURE 16.9 Comparison of commercial ampholytes in the separation of (a) p*I* markers and (b) an IgG mAb. Peak identity: (i) low p*I* marker, (ii) high p*I* marker. Top trace is 4% Servalyte 3–10; middle trace is 4% Bio-lyte 3–10; bottom trace is 4% Pharmalyte 3–10.

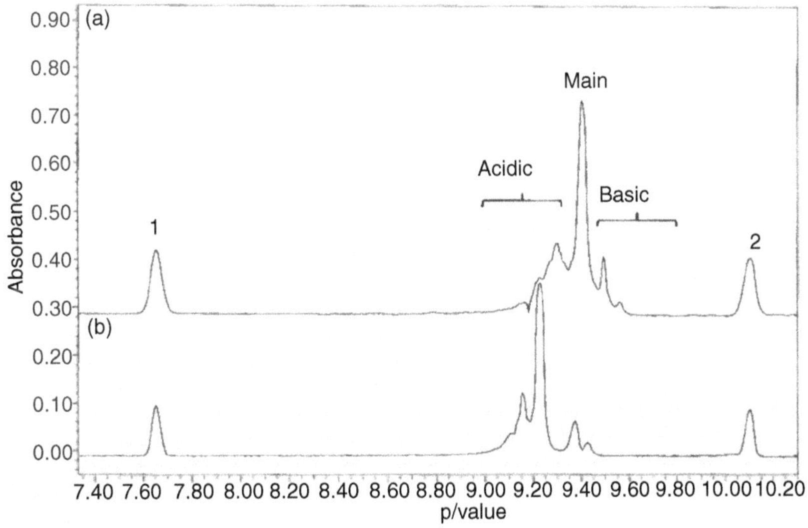

FIGURE 16.10 Separation of an IgG mAb by (a) broad-range Pharmalyte 3–10 and (b) narrow range Pharmalyte 8–10.5. Peak identity: (1) low p*I* marker; (2) high p*I* marker.

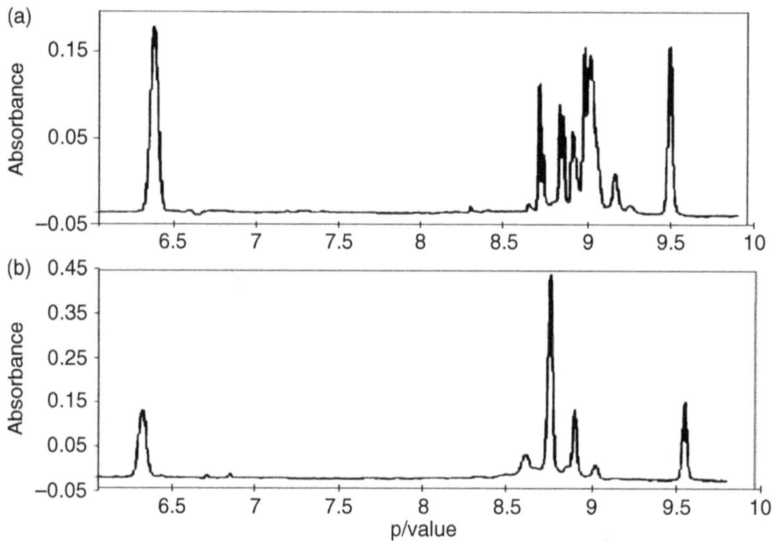

FIGURE 16.11 The effect of additives on an iCE separation for mAbs. Separation in the (a) absence of additives and (b) presence of additives (1 M urea and 3% sucrose). From Ref. 59

attained using a short capillary and high voltage. However, if the focusing voltage is too high, it will generate high current and heat inside the capillary that may lead to peak broadening, protein degradation, or precipitation, and even shutdown of current. Focusing time is also affected by the carrier ampholyte concentration and p*I* range, as well as the concentration of additives such as urea and sucrose. Inadequate focusing may not only lead to poor separation, but in some cases may also lead to artificial split peaks. However, focusing too long may increase the risk of protein precipitation, which may lead to spikes in the electropherogram. Therefore, the optimal focusing time should preferably be determined for each protein individually.

Capillaries used in the existing iCE instrument are stripped outside from the outer polyimide layer to allow the whole-column imaging. A snapshot of the focusing process can be taken every 30 s, enabling real-time monitoring of the focusing status. This significantly reduces the time needed for optimization of focusing time. As shown in Figure 16.12, a mAb molecule was analyzed by iCE and 12 snapshots were overlaid taken to recapture the focusing process from 0 to 20 min. The charge profile was completely resolved by 302 s, or 5 min.

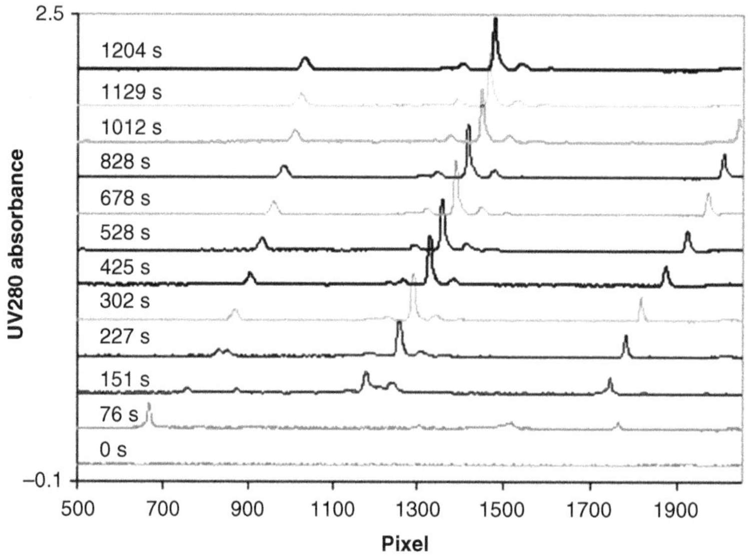

FIGURE 16.12 Effect of focusing time on iCE separation. Focusing time increases from 0 to 1204 s (bottom trace to top trace).

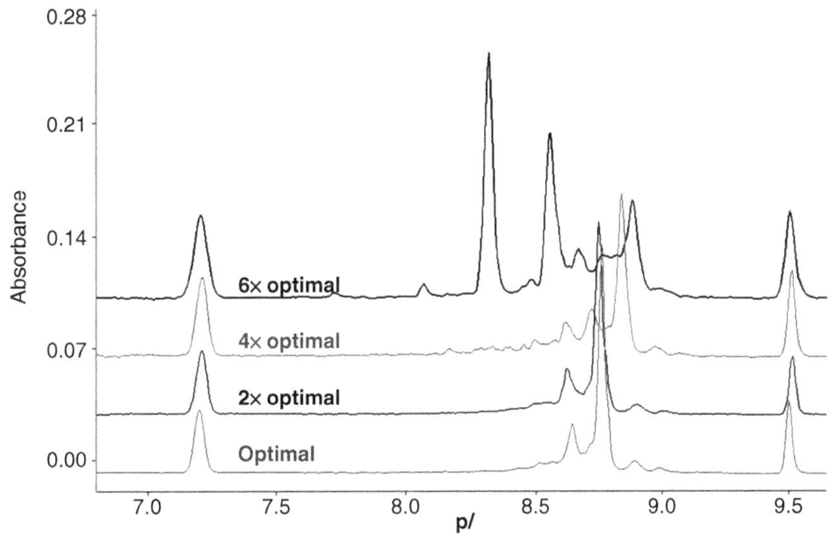

FIGURE 16.13 The effect of increasing salt concentration (two, four, and six times the optimal concentration) on the separation of a mAb. From Ref. 59

16.3.1.4 Salt Concentration Similar to other CE techniques, iCE analysis is sensitive to the presence of salts in sample matrix. A high salt concentration results in a poor separation of protein charge variants. There are many reasons for the poor performance caused by high salt concentration. Distortion of the pH gradient can be caused by salts in protein samples, resulting in alterations in the separation patterns. Salts also affect the iCE separation by altering the viscosity, EOF, and the interaction between the protein and the capillary walls. Sodium chloride (NaCl), sodium acetate (NaAc), and sodium citrate are common salts used in mAb formulations. Figure 16.13 shows the effect of NaCl and NaAc concentration on the separation of an IgG mAb. The optimal concentration of salt is 3.8 mM NaCl and 0.5 mM NaAc. Interference of the separation was observed when the salt concentration reached four times (4×) optimal. The separation was completely distorted with an increase in the salt concentration to six times (6×) optimal.

16.3.1.5 Protein Concentration Under the same sample matrix and focusing conditions, various mAbs have different solubilities in the separation matrix. When the mAb concentration is too high, mAbs with low solubility may precipitate, resulting in irregular spikes in the electropherogram. For proteins that are highly soluble, overloading may also cause deterioration of separation resolution. On the other hand, protein working concentrations should be kept high enough so that the sensitivity is adequate for detecting and quantifying minor charge variants. The iCE separation of two mAbs at different working concentrations has been tested, and the separation was affected by protein concentration for only one of the mAbs [59]. At 0.6 mg mL^{-1}, the separation was acceptable for one mAb, but spikes were apparent when the concentration was 0.6 mg mL^{-1} for a second mAb. An optimal protein working concentration for most mAb molecules is 0.3 mg mL^{-1}, resulting in an acceptable separation, sensitivity, and reproducibility.

An optimized general iCE method for mAbs consists of the following major parameters: 4% broad range ampholytes pH 3–10, 0.35% methylcellulose; two internal p*I* markers flanking the mAb; 0.3 mg mL^{-1} mAb injection concentration; 100 s for sample transfer time; 0 s for transfer delay time; 1–4 M urea and 3% sucrose as additives as necessary; 1 min prefocusing at 1500 V; and 5 min focusing at 3000 V [59]. The optimized iCE conditions can be applied to multiple mAbs as shown in Figure 16.14. The mAbs evaluated cover a wide range of p*I* values (p*I* = 7.0–9.4) and diverse charge compositions (10–50% acidic species).

16.3.2 iCE Method Validation

iCE methods have been employed for release and stability testing for clinical materials. GMP methods need to be qualified or validated for their intended use. Janini et al. have described an iCE method that was validated for the analysis of charge variants of MU-B3 mAb [60].

At Pfizer, a platform iCE method has been qualified for the analysis of monoclonal antibodies to determine charge heterogeneity profile. The method can be used for monitoring the acidic and basic species of monoclonal antibody drug substance and drug product, and is suitable for release, stability, and in-process samples. Specificity was assessed by evaluating the charged species composition of each mAb as well as the actual p*I* value of each iCE peak. Each of the mAb samples was analyzed according to the generic iCE

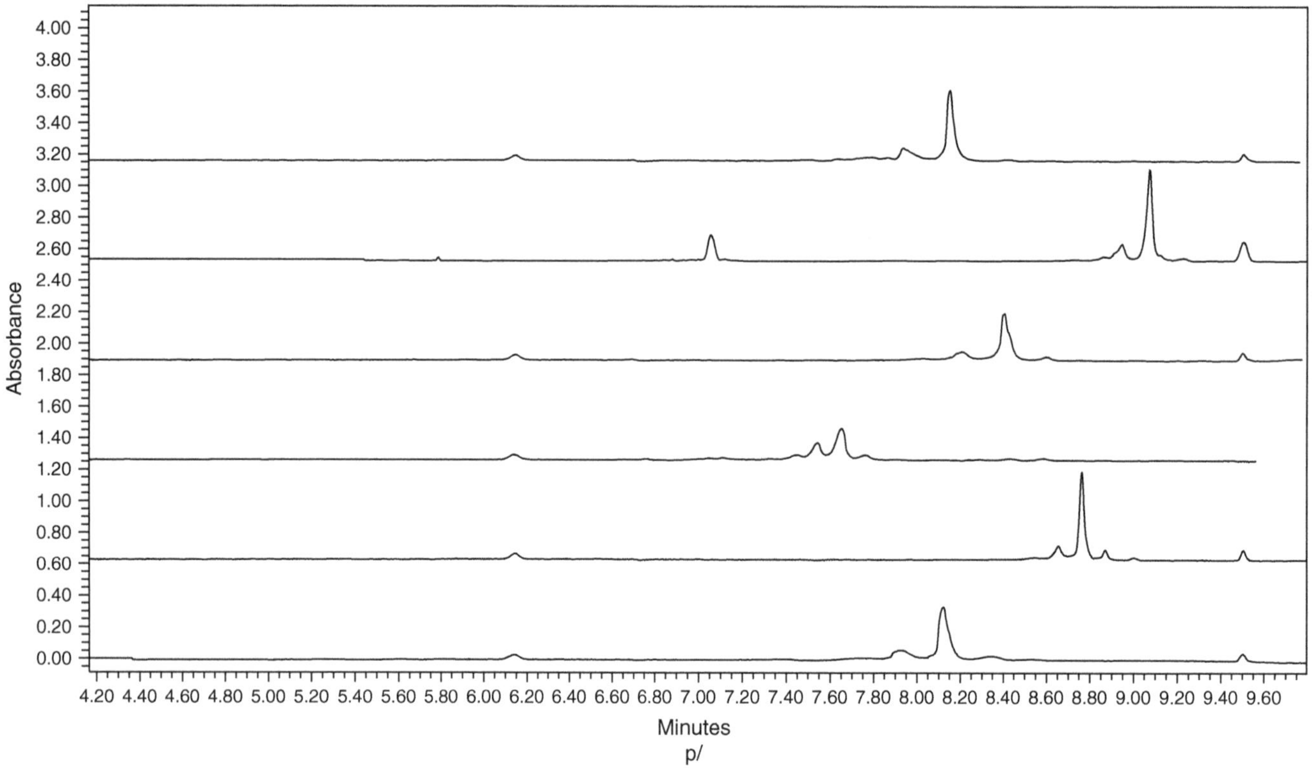

FIGURE 16.14 iCE separation of charge variants of multiple mAbs using the same method conditions.

method. The generic iCE method demonstrated was able to resolve IgG molecules with charge difference of 0.1 p*I* unit. Linearity of the method was demonstrated over a range of 3.0–420 μg mL^{-1} with correlation coefficients $\geq$0.99. The Quantitation Limit (QL) for the method was found to be 1% of the nominal concentration of 0.3 mg mL^{-1}. Accuracy of the method was demonstrated by evaluating recovery at 1–5, 10, 20–50, and 100–140% of the nominal injection concentration (0.3 mg mL^{-1}), and were found to be within 75–125, 80–120, 85–115, and 90–110%, respectively. Repeatability (interinjection) and intermediate precision (different sample preparations and different columns) had relative standard deviations (RSD) for peak area and percent peak area of not more than 15.0% and 5.0%, respectively, at the nominal injection level. The precision and robustness of the method demonstrate iCE can be used in Quality Control (QC) laboratories as a technique to monitor charge heterogeneity of mAbs.

16.3.3 Applications

iCE methods have been widely employed to support every aspect of bioprocess development for mAbs, including cell line selection, cell culture and purification development, and formulation screening, as well as general characterization of charged species.

16.3.3.1 Cell Line Development Support Monitoring charge heterogeneity is an important tool for the characterization of molecules and processes, as well as to aid in the assessment of comparability of cell line or process changes. For example, the iCE technique was used to evaluate the comparability of a mAb molecule produced by a new set of potential cell lines and an existing cell line (Figure 16.15). The iCE method was used to monitor the overall charge heterogeneity profile of the mAb produced by the new cell lines compared with the previous cell line. Figure 16.15 illustrates the charge heterogeneity profiles of materials generated by different cell lines for this mAb molecule. The acidic and basic species shown in the separations in Figure 16.15 are quantified in Table 16.3 along with the p*I* of the parent (main) peak. Parameters evaluated for cell line selection may include titer, cell viability, cell line stability, aggregate content, glycosylation profile, and charge heterogeneity profile. The iCE method is well suited to rapidly determine the charge heterogeneity profiles of a set of mAbs from various cell lines, making it useful in guiding cell line selection.

16.3.3.2 Formulation Screening Similar to most proteins, the IgG mAb is intrinsically unstable, requiring lyophilized and/or low-temperature storage when formulated as drug product. A large number of formulations are routinely

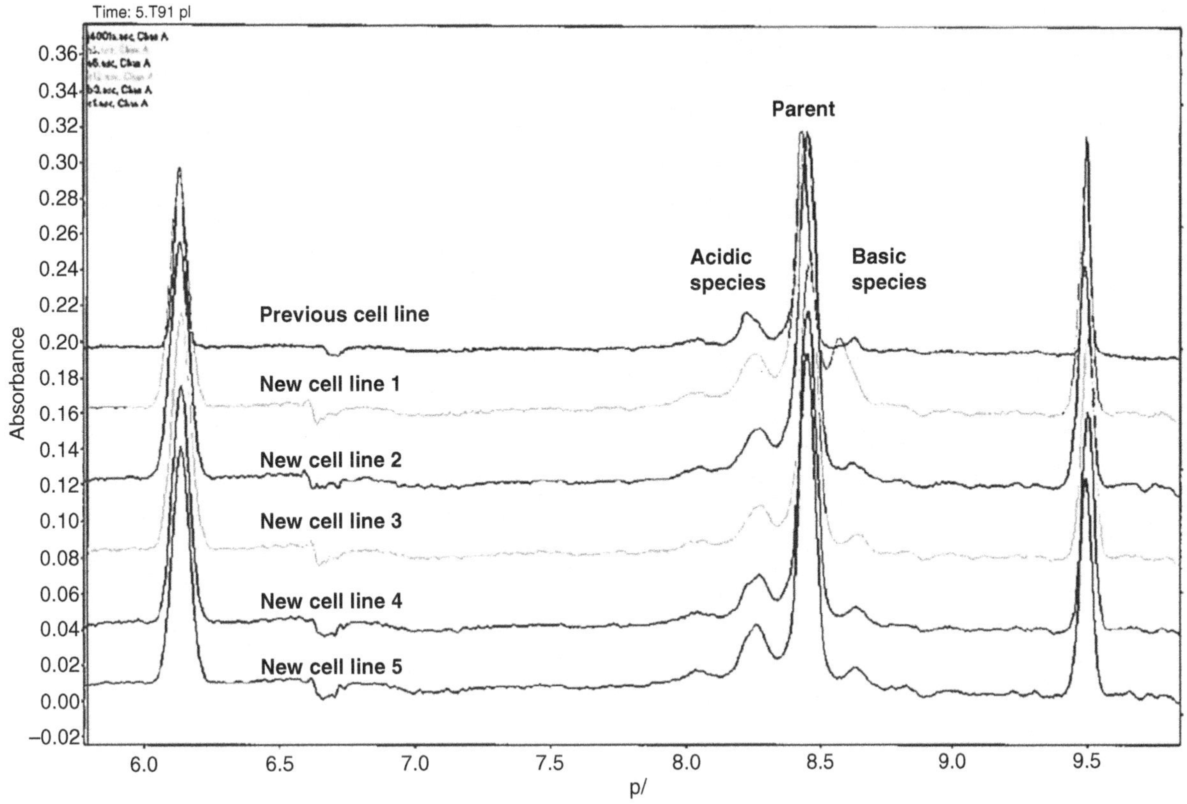

FIGURE 16.15 Use of iCE for comparing the charge heterogeneity of monoclonal antibodies derived from various cell lines under development. As a tool for screening cell lines, charge separations for new candidate cell lines are compared against the previous cell line.

screened to identify a composition that maximizes drug product stability. The iCE technology has been applied as high-throughput tool for formulation development of drug product [61]. Figure 16.16 shows the comparison of iCE and cation exchange HPLC (CEX) for the analysis of an IgG2 mAb in different formulations stored at 40 °C for 4 weeks. Although the percent value of acidic species reported by CEX and iCE is different, the relative ranking of formulations was similar. Notably, iCE took about 13 min per sample to analyze whereas CEX took 30 min, although as described earlier CEX is most amenable to the characterization of collected fractions.

TABLE 16.3 Charge Heterogeneity and Cell Line Selection for a mAb Molecule

Cell Line	Parent p*I*	% Acidic Species	% Parent	% Basic Species
Previous	8.43	27.0	67.7	5.0
New 1	8.45	25.0	56.1	19.0
New 2	8.45	23.5	70.6	5.8
New 3	8.47	23.3	71.5	5.3
New 4	8.46	22.1	72.8	5.2
New 5	8.46	25.7	68.4	6.0

16.3.3.3 Characterization of Acidic Species As has been described, screening of samples for overall charge profile is an important tool for developing product and process understanding. One useful way to characterize the mAb charge variants that are separated by iCE is to treat the mAbs with enzymes or stressed conditions before analysis by iCE to obtain the parent, acidic, and basic species composition.

For example, Table 16.4 and Figure 16.17 illustrate further characterization of a mAb that had 45–50% acidic species based on iCE. To elucidate the nature of the acidic species, the mAb was subject to treatment by N-glycosidase F (PNGase F) to remove all the N-linked glycans from the mAb molecules. As shown in Table 16.4, there is a significant decrease in the acidic species following deglycosylation, suggesting that glycans may be impacting the overall charge profile. On removal of the N-linked glycans by PNGase F, the asparagine is converted into an aspartic acid remain and therefore results in a lower parent (main) peak p*I*. The material was then treated with sialidase A to remove the sialic acid, if present, from the glycans. The iCE

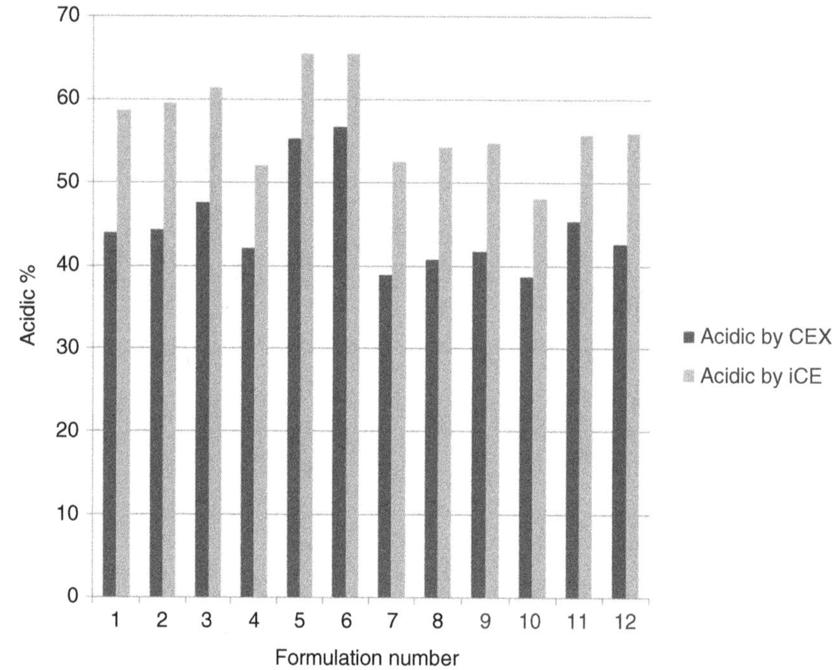

FIGURE 16.16 Comparison of iCE and CEX in screening of mAb stability in different formulations. As a tool for analyzing monoclonal antibody stability, iCE provides a rapid alternative for determining the percent acidic species that may change on stability especially at stressed storage conditions.

results of both the PNGase F and Sialidase A treated samples had significantly reduced acidic species but increased parent species. Figure 16.17 compares the iCE profiles of the mAb before and after the Sialidase A treatment that show a significant portion of the acidic species of the mAb reverted to parent molecule after sialic acid is removed, suggesting that a significant portion of the acidic species is related to sialylated glycan. This finding can be further analyzed by orthogonal methods for glycan characterization to confirm the findings suggested by the iCE analysis.

TABLE 16.4 Charged Species Composition for a mAb Molecule—Before and After Deglycosylation

Batch	Parent Peak p*I* Value	% Acidic Species	% Parent	% Basic Species
		Before deglycosylation		
1	8.50	48.9 ± 0.4	47.3 ± 0.3	3.8 ± 0.1
2	8.50	45.7 ± 0.8	50.9 ± 1.0	3.4 ± 0.2
3	8.49	50.0 ± 0.3	46.7 ± 0.3	3.3 ± 0.1
4	8.50	45.5 ± 0.3	50.5 ± 0.2	4.0 ± 0.1
		After deglycosylation		
1	8.22	28.6 ± 0.3	65.1 ± 0.4	6.3 ± 0.1
2	8.22	26.5 ± 0.9	67.6 ± 1.4	5.9 ± 0.3
3	8.23	29.2 ± 0.2	66.1 ± 0.2	4.7 ± 0.2
4	8.22	25.0 ± 0.3	69.3 ± 0.5	5.6 ± 0.4

16.4 SUMMARY

The CE techniques described in this chapter provide viable alternative approaches to IEC and IEF for the assessment of charge heterogeneity of mAbs in development as biotherapeutics. While these examples have focused on rapid methods for the analysis of mAbs in particular, these techniques are useful for other protein biotherapeutics as well. Indeed, many of the same separation conditions are useful for proteins other than mAbs. The methods can be used in all the stages of development. The methods can be validated for drug substance and drug product release and stability testing in a QC laboratory setting. Table 16.5 compares the performance of various charge-related separations typically used for the analysis of mAb biotherapeutics.

Use of shorter and narrower capillaries and higher field strength has furthered the development of CZE and iCE as high-resolution and high-speed techniques for the analysis of mAbs. Future directions include the further exploration of microchip-based IEF [62–64] and ultrahigh voltage CZE [65–67] for even faster separation speeds. Additionally, the development of both efficient and MS friendly separation conditions will lead to better characterization of separated charge variants by online CE–MS. CZE and iCE analyses of charge heterogeneity, taken together with data from orthogonal techniques, are valuable and efficient tools that support the development of mAb biotherapeutics.

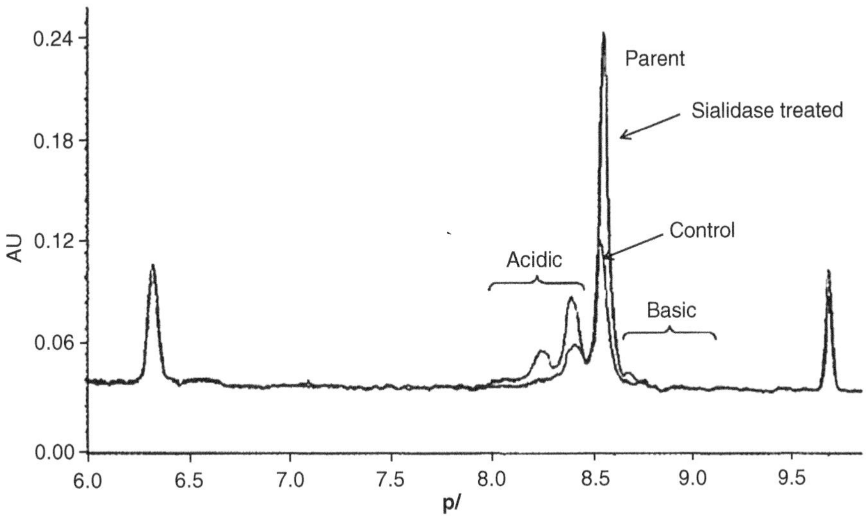

FIGURE 16.17 iCE profiles of a mAb before and after sialidase treatment.

TABLE 16.5 Performance Comparison of Charge-Sensitive Methods

	Gel IEF	HPLC–IEC	CZE	CIEF	iCE
Separation time	4–8 h	20–60 min	3–10 min	10–60 min	6–10 min
Sample preparation	Tedious				
Gel handling	Simple	Simple	Medium	Medium	
Quantitation/automation	Semi/no	Yes	Yes	Yes	Yes
Resolution	+++	++	++/+++	+++	+++
Capability to couple with MS	MS of collected fractions	MS of collected fractions	Sometimes	Sometimes	No
Throughput	Low	Medium	High	Medium/high	High
Platform method	Yes	No	Yes/no	Yes/no	Yes

REFERENCES

1. Ziegelbauer, K. and Light, D.R. (2007) *J. Commerc. Biotechnol. 14*(1), 65–72.
2. Reichert, J.M. (2010) *MAbs 2*(1), 423–430.
3. Reichert, J.M. (2008) *Curr. Pharm. Biotechnol. 9*, 423–430.
4. ICH ICoH. (Q6B) Specifications: Test Procedures and Acceptance Criteria for Biotechnological/Biological Products1999.
5. Zola, H. (1987) *Monoclonal Antibodies: A Manual of Techniques*, CRC Press, Boca Raton, FL.
6. Price C.P. and Newman D.J. (eds.) (1991) *Principles and Practice of Immunoassay*, Stockton Press, New York.
7. Jiang, X.R., Song, A., and Bergelson, S. et al. (2011) *Nat. Rev. Drug Disc. 10*, 101–110.
8. Goding, J.W. (1983) *Monoclonal Antibodies: Principles and Practice*, Academic Press, New York, NY.
9. Chan, A.C. and Carter, P.J. (2010) *Nat. Rev. Immunol. 10*, 301-316.
10. Abbas, A.K., Lichtman, A.H., and Pober, J.S. (1991) *Cellular and Molecular Immunology*, W.B. Saunders Company, Philadelphia.
11. Wang, W., Singh, S., Zeng, DL., King, K., and Nema, S. (2007) *J. Pharm. Sci., 96*(1), 1–26.
12. Vlasak, J. and Ionescu, R. (2008) *Curr. Pharm. Biotechnol. 9*, 468–481.
13. Santora, L.C., Krull, I.S., and Grant, K. (1999) *Anal. Biochem. 275*(1), 98–108.
14. Harris, R.J., Kabakoff, B., and Macchi, F.D. et al. (2001) *J. Chromatogr. B Biomed. Sci. Appl. 752*(2), 233–245.
15. Farnan, D. and Moreno, G.T. (2009) *Anal. Chem. 81*(21), 8846–8857.
16. Chu, G., Chelius, D., Xiao, G., Khor, H., Coulibaly, S., and Bondarenko, P. (2007) *Pharm. Res, 24*(6), 1145–1156.
17. Wright, A. and Morrison, S.L. (1998) *J. Immunol. 160*(7), 3393–3402.
18. Johnson, K.A., Paisley-Flango, K., Tangarone, B.S., Porter, T. J., and Rouse, J.C. (2007) *Anal. Biochem. 360*(1), 75–83.
19. Chumsae, C., Gaza-Bulseco, G., Sun, J., and Liu, H. (2007) *J. Chromatogr. B. 850*(1–2), 285–294.
20. Wypych, J. et al. (2008) *J. Bio. Chem. 283*(23), 16194–16205.
21. Dillon, T.M., et al. (2008) *J. BioChem. 283*(23), 16206–16215.

22. Zheng, J.Y. and Janis, L.J. (2006) *Int. J. Pharm. 308*(1–2), 46–51.
23. Lyubarskaya, Y., Houde, D., Woodard, J., Murphy, D., and Mhatre, R. (2006) *Anal. Biochem. 348*(1), 24–39.
24. Antes, B. et al. (2007) *J. Chromatogr. B. 852*(1–2), 250–256.
25. Banks, D.D. et al. (2008) *J. Pharm. Sci. 97*, 764–779.
26. Moorhouse, K.G., Nashabeh, W., Deveney, J., Bjork, N.S., Mulkerrin, M.G., and Ryskamp, T. (1997) *J. Pharm. Biomed. Anal. 16*(4), 593–603.
27. Quan, C. et al. (2008) *Anal. Biochem. 373*(2), 179–191.
28. Fischer, S., Hoernschemeyer, J., and Mahler, H.-C. (2008) *Eur. J. Pharm. Biopharm. 70*(1), 42–50.
29. Shan, L. and Anderson, D.J. (2001) *J. Chromatogr. A 909*(2), 191–205.
30. Liu, Y. and Anderson, D.J. (1997) *J. Chromatogr. A 762*(1–2), 207–217.
31. Bates, R.C. and Frey, D.D. (1998) *J. Chromatogr. A 814*(1–2), 43–54.
32. Andersen, T., Pepaj, M., Trones, R., Lundanes, E., and Greibrokk, T. (2004) *J. Chromatogr. A 1025*(2), 217–226.
33. Vesterberg, O. (1967) *Acta Chem. Scand. 21*, 206–216.
34. Svensson, H. (1961) *Acta. Chem. Scand. 15*, 325–341.
35. Hjertén, S. and Zhu, M.-D. (1985) *J. Chromatogr. A 346*, 265–270.
36. Wu, J. and Pawliszyn, J. (1992) *Anal. Chem. 64*(23), 2934–2941.
37. Wu, J. and Pawliszyn, J. (1994) *Anal. Chem. 66*(6), 867–873.
38. Wenisch, E., Tauer, C., Jungbauer, A., Katinger, H., Faupel, M., and Righetti, P.G. (1990) *J. Chromatogr. A 516*(1), 133–146.
39. Jorgenson, J.W. and Lukacs, K.D. (1981) *Anal. Chem. 53*(8), 1298–1302.
40. Skoog, D.A., Holler, F.J., and Crouch S.R. (2007) *Principles of Instrumental Analysis*, Thomson Brooks/Cole Publishing, Belmont, CA.
41. Compton, B.J. (1991) *J. Chromatogr. A 559*(1–2), 357–366.
42. Ma, S. and Nashabeh, W. (2001) *Chromatographia 53*, S75–S89.
43. He, Y. et al. (2010) *Anal. Chem. 82*(8), 3222–3230.
44. He, Y., Isele, C., Hou, W., and Ruesch, M. (2011) *J. Sep. Sci. 34*(5), 548–555.
45. Bullock, J.A. and Yuan, L.-C. (1991) *J. Microcol. Sep. 3*(3), 241–248.
46. Hjertén, S., Valtcheva, L., Elenbring, K., and Liao, J.-L. (1995) *Electrophoresis 16*(1), 584–594.
47. Lee, I.H., Pinto, D., Arriaga, E.A., Zhang, Z., and Dovichi, N.J. (1998) *Anal. Chem. 70*(21), 4546–4548.
48. Liu, T., Shao, X.-X., Zeng, R., and Xia, Q.-C. (1999) *J. Chromatogr. A 855*(2), 695–707.
49. Gennaro, L.A. and Salas-Solano, O. (2008) *Anal. Chem. 80* (10), 3838–3845.
50. Thorne, J.M., Goetzinger, W.K., Chen, A.B., Moorhouse, K. G., and Karger, B.L. (1996) *J. Chromatogr. A 744*(1–2), 155–165.
51. Lüdi, H., Gassmann, E., Grossenbacher, H., and Märki, W. (1988) *Anal. Chim. Acta 213*, 215–219.
52. Chen F.-T. and Evangelista, R.A. (1998) *Electrophoresis 19*(15), 2639–2644.
53. Bergmann, J., Jaehde, U., and Schunack, W. (1998) *Electrophoresis 19*(2), 305–310.
54. Wu S.-L., Teshima, G., Cacia, J., and Hancock, W.S. (1990) *J. Chromatogr. A 516*(1), 115–122.
55. Tangarone, B. Development of a CE charge heterogeneity method for a challenging protein. Paper presented at IBC 12th Well Characterized Biologicals 2008; Reston, VA.
56. Ma, S. Analysis of protein therapeutics by capillary electrophoresis: applications and challenges. In: Mire-Sluis, A.R. (ed.), *State of the Art Analytical Methods for the Characterization of Biological Products and Assessment of Comparability*, Basel, pp. 49–68.
57. Wu, J., Tragas, C., Watson, A., and Pawlizyn, J. (1999) *Anal. Chim. Acta 383*, 67–78.
58. Salas-Solano, O., Gennaro, L., and Felten, C. (2008) *LCGC Europe* 615–622.
59. He, X.Z., Que, A.H., and Mo, J.J. (2009) *Electrophoresis 30*(5), 714–722.
60. Janini, G., Saptharishi, N., Waselus, M., and Soman, G. (2002) *Electrophoresis 23*(11), 1605–1611.
61. Li, N., Kessler, K., Bass, l, and Zeng, D. (2007) *J. Pharm. Biomed. Anal. 43*, 963–972.
62. Vlcková, M., Kalman, F., and Schwarz, M.A. (2008) *J. Chromatogr. A 1181*(1–2), 145–152.
63. Tan, W., Fan, Z.H., Qiu, C.X., Ricco, A.J., and Gibbons, I. (2002) *Electrophoresis 23*(20), 3638–3645.
64. Shimura, K. (2009) *Electrophoresis 30*(1), 11–28.
65. Hutterer, K.M. and Jorgenson, J.W. (1999) *Anal. Chem. 71*(7), 1293–1297.
66. Palonen, S., Jussila, M., Porras, S.P., Hyötyläinen, T., and Riekkola, M.-L. (2001) *J. Chromatogr. A 916*(1–2), 89–99.
67. Hutterer, K.M. and Jorgenson, J.W. (2005) *Electrophoresis 26*(10), 2027–2033.

17

APPLICATION OF CAPILLARY ELECTROPHORESIS FOR HIGH-THROUGHPUT SCREENING OF DRUG METABOLISM

Roman Řemínek, Jochen Pauwels, Xu Wang, Jos Hoogmartens, Zdeněk Glatz, and Ann Van Schepdael

17.1 INTRODUCTION

Recent advances in the branches of chemistry, computer modeling, and molecular biology have transformed the incidental finding of active natural molecules into a targeted drug development process [1]. Although computer modeling allows the design of a molecule structure that fits into the desired target, finding a compound with drug-like properties still remains a complicated, time- and resources-consuming practice. Taken together, the entire development of a new drug (Figure 17.1) lasts 10–15 years and costs more than US$ 800 million, on average [2]. The main portion of these expenses is spent on extensive clinical testing. Unfortunately, there is a high failure rate in the preclinical and clinical stages. Major efforts of pharmaceutical companies are directed toward removing compounds with unsuitable properties from further testing as early as possible within the development process, before more time and resources have been unnecessarily invested. The biological activity, that is, the ability to cause a desired pharmaceutical effect, is fundamental for a promising drug-like compound. Nevertheless, an acceptable drug candidate cannot be selected only on the basis of high efficacy. All compounds are required to pass the ADME/Tox (absorption, distribution, metabolism, excretion, and toxicity) testing and comply with several additional criteria (preparation costs, odor, taste, and so on) as well [3–5]. Although advances in combinatorial chemistry have enabled the creation of large and diverse series of brand-new compounds, the technology used in developing screens for pharmacokinetic properties has fallen behind, causing a bottleneck in the drug development process. At the present time, truly high-throughput screening methods, such as those used to screen for early leads in combinatorial libraries, may achieve rates of 100,000 compounds per week. In contrast, routine methods used in ADME/Tox testing do not exceed 100 compounds per week and a rate of ten tested compounds per week is considered high for *in vivo* studies [6].

This chapter deals with the metabolism part of ADME/Tox tests, more specifically with the metabolic stability of a new drug candidate. Metabolic stability refers to the susceptibility of compounds to biotransformation. Results of these tests are usually reported in terms of intrinsic clearance, from which secondary pharmacokinetic parameters such as half-life can be calculated when other data on distribution volume and fraction absorbed are available [7]. Cytochromes P450 (CYP) constitute the major group of enzymes involved in drug metabolism [8]. This superfamily of membrane-bound mixed-function oxidases located in the smooth endoplasmic reticulum of the liver and other extrahepatic tissues has been shown to be active in the metabolism of approximately 60% of commonly prescribed drugs [4]. Estimated contributions of individual human CYP isoforms to drug metabolism are depicted in Figure 17.2 [9]. CYP oxidation, reduction, or hydrolysis of drugs generally results in their deactivation, and many drugs are simultaneously rendered more hydrophilic, so that they are amenable to further modification or excreted easily from the body [10]. *In vitro* human liver-related test systems such as recombinant enzymes, microsomes, hepatocytes, and liver

Capillary Electrophoresis and Microchip Capillary Electrophoresis: Principles, Applications, and Limitations, First Edition.
Edited by Carlos D. García, Karin Y. Chumbimuni-Torres, and Emanuel Carrilho.

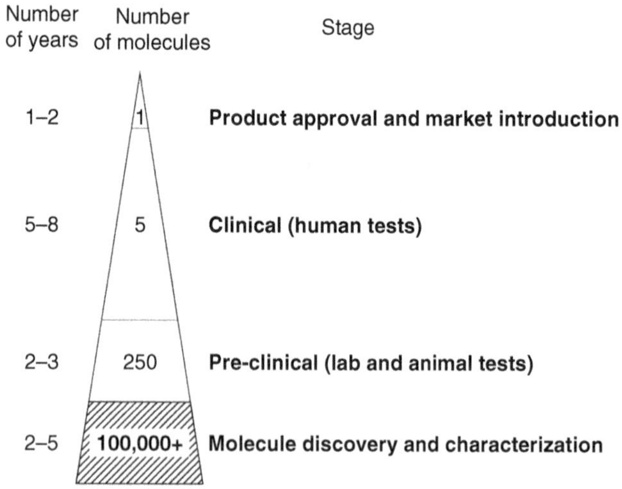

FIGURE 17.1 Schematic illustration of the drug development process. The first preclinical stage is particularly characterized by enormous numbers of routinely tested compounds. Based on data from Ref. 2.

slices can provide important information on CYP activity. Although utilization of a mixture of individual CYP isoforms enables fast analyses suitable for high-throughput screening, its predictive abilities with respect to *in vivo* drug processing are rather low [11]. In contrast, the use of whole hepatocytes or liver slices for drug stability tests offers the advantage of a realistic simulation of conditions inside the liver. However, these approaches are more laborious and time-consuming and less apt for high-throughput screening. Human liver microsomes (HLM), on the other hand, combine a good simulation of the native status inside the human liver with the suitability for high-throughput

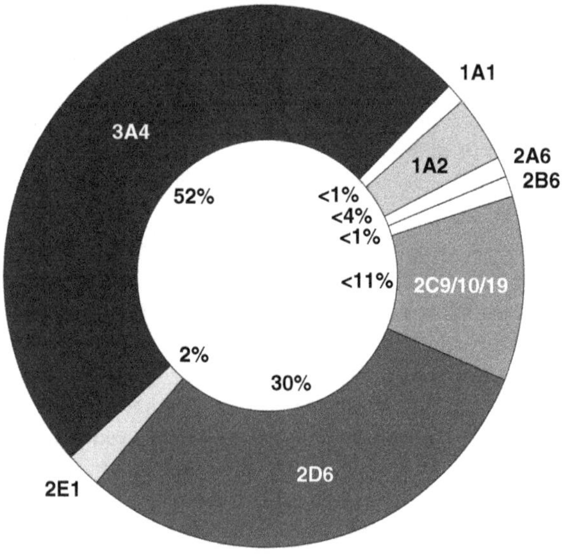

FIGURE 17.2 Estimated fractions of drugs metabolized by the main CYP isoforms. Based on data found in Ref. 9.

screening. Therefore, HLM have been chosen as the enzymatic system in the following study.

In view of the importance of CYP, a number of analytical approaches have been introduced to study their activity. These were mostly based on colorimetric, fluorimetric, high-performance liquid chromatography–mass spectrometry (HPLC–MS), and radiometric methods [12–15]. Traditional analytical formats do not allow the rapid determination of the metabolism of thousands of drug candidates. Consequently, new formats of *in vitro* assays have been developed from traditional setups through protocol simplification and technology advancement [16]. Although progress in miniaturization and robotics, in combination with protocol simplification, has synergistically increased the rates of drug candidate screening [17], the experimental capacity still remains low. In addition to the above-mentioned, highly recognized methodologies, capillary electrophoresis (CE) has recently emerged as a powerful and versatile technique for enzymatic studies [18–20]. CYP reactions often result in only minor structural modifications of the parent compound [21], making separation of these molecules by common HPLC methods a challenge. On the contrary, CE methodology is based on a completely different selectivity process and, therefore, even complex molecules varying only in, for example, one additional charge can be separated with reasonable resolution. Efficient separation of structurally similar compounds together with other attractive features such as the low sample and reagents volume requirement, the short analysis times, the possibility of chiral separations without the need for special, expensive columns, and the speed and ease of method development represent advantages of CE over HPLC for drug metabolism studies. Besides rapid and efficient analyses, CE also offers a high degree of automation and the option to use a multi-capillary format. All these features predetermine CE as a promising tool for high-throughput screening.

The analysis of incubation mixtures involving HLM can only be performed properly if two issues are addressed, namely the presence of proteins in the sample and the possibly low sensitivity of the analytical method. Solutions to these issues will first be described in general. Then, a practical example will be given as an illustration.

17.2 SAMPLE DEPROTEINIZATION

The high protein content in biological material such as HLM is a burden, complicating both HPLC and CE analyses, since these molecules tend to adsorb to the stationary phase or the inner wall of the separation capillary. Adsorbed proteins then gradually change the chromatographic or electrophoretic conditions of the system leading to low repeatability, peak broadening, or even to column obstruction. Although CE methods for direct analysis of biological media without any sample pretreatment have been published, and adsorbed molecules can usually be removed by application of rinsing procedures, including strong hydroxide, acid, or sodium dodecyl sulfate (SDS) solution [22], certain sample pretreatment is generally imperative. The prevailing sample preparation techniques are centrifugation, protein precipitation, ultrafiltration, liquid–liquid extraction (LLE), solid-phase extraction (SPE), liquid-phase microextraction (LPME), and solid-phase microextraction (SPME).

Centrifugation represents the basic technique suitable for fast displacement of insoluble membrane-bound proteins. Soluble proteins are partly removed during this process as well, while remaining molecules and fragments are still able to negatively affect the separation conditions. Therefore, it is necessary to couple centrifugation with additional precautions such as strong rinsing or micellar electrokinetic chromatography (MEKC) where protein molecules are denatured by the surfactant present in the background electrolyte (BGE) and then removed from the capillary during rinsing [23,24].

During ultrafiltration, a sample is processed by centrifugal force through a semipermeable membrane with a certain cutoff limit allowing separation of molecules according to their weight and structure [25]. This simple and fast technique is suitable for pretreatment of samples containing higher concentrations of analytes, since a significant portion of lipophilic analyte molecules can be withheld by the separated sample matrix.

The aforementioned issue regarding lipophilic analytes can be solved with protein precipitation using an organic solvent. The organic agent disrupts the solvation layers of the proteins, leading to their aggregation by attractive electrostatic and dipole forces into an insoluble form that can be removed by filtration or centrifugation. Moreover, the solvent releases the bound lipophilic analytes and increases their solubility [26,27]. Usually, a considerable amount of solvent needs to be added to achieve the desired effect. Hence, an important disadvantage of this method is dilution of the sample making precipitation more appropriate for HPLC than for CE.

LLE, also known as solvent extraction or partitioning, separates the analytes from the sample matrix on the basis of their higher relative solubility in different immiscible liquids added to the sample solution. The most commonly used system of water or aqueous buffer and organic solvent allows the extraction of lipophilic components [28]. In SPE, the sample is processed through a disposable mini-column filled with an adsorbent phase chosen depending on the chemical properties of the selected compounds. Reversed-phase SPE separates the analytes based on their polarity. Weakly adsorbed compounds can be eluted by a polar solvent, whereas lipophilic analytes require a nonpolar solvent [29]. Contrary to precipitation, extracts obtained by LLE and SPE may be evaporated and thus preconcentrated. However, the entire procedures are laborious and time-consuming and thus not suitable for high-throughput analyses. Furthermore, LLE and SPE are not very efficient in handling microvolume samples while the low sample volume requirement is one of the most important CE advantages over HPLC.

The issue of dealing with small sample volume can be overcome by utilization of LPME or SPME. Although these techniques may look like LLE and SPE transferred to the microscale, there are significant differences in their implementation. In LPME, organic phase is used as part of a separation fiber segregating the aqueous solution and acceptor phase. After adjustment of the sample pH to neutralize the analytes of interest, they migrate through the organic solvent into the acceptor phase. Since these molecules regain a charge in this compartment, a preconcentration of cleared analytes is taking place in the acceptor phase. SPME is a much faster procedure compared to SPE. Equilibrium extraction of analyte molecules happens directly on a fused-silica fiber previously coated with liquid polymer film, providing integration of sample clean-up and enrichment into one step [29,30]. Although microscale extraction techniques allow handling of small sample volumes and fast procedures, prior knowledge of the pK_a values of the analytes, or the distribution constants in LPME or SPME, respectively, is a prerequisite, thus making their implementation in high-throughput applications more complex.

As pointed out earlier in this chapter, every deproteinization technique has its own benefits and drawbacks; the final selection thus primarily depends on the composition of a given sample. Application of fast and simple procedures is advisable in screening studies of extensive chemical libraries; therefore, the combination of CE with centrifugation and a strong rinsing procedure is presented in this chapter.

17.3 ON-LINE PRECONCENTRATION

In spite of all benefits mentioned earlier, two decades after the introduction of the first commercially available device, CE is still only a complementary analytical technique to HPLC. This is among other reasons due to the low concentration sensitivity, one of the most significant CE drawbacks. The small sample volumes injected, the short UV absorption path length—determined by the inner capillary diameter—and the low concentrations of therapeutically relevant drugs, all contribute to this poor sensitivity. Apart from the application of high-sensitivity detectors or extended–path length capillaries, the main focus is now oriented toward sample preconcentration techniques especially in an on-capillary setup [31–33]. These simple procedures can be carried out without any modifications to a commercial CE instrument. Field-amplified sample stacking (FASS), transient isotachophoresis (ITP), and sweeping represent the most widely used techniques in this context, and are described more in detail in Chapter 2.

On-line preconcentration techniques are in principle based on injection of a long plug of sample and subsequent focusing of analytes into a narrow zone that is afterwards separated by means of CE. In FASS, the sample is prepared in a solvent with lower conductivity than the BGE. The ions in the injected sample region thus experience a higher electric field strength on application of the separation voltage and migrate faster than the ions in the BGE region. The sample ions then concentrate as they cross the region boundary and their velocity drops. Besides a difference in conductivity, other mechanisms such as pH [34] or viscosity [35] changes have also been used to acquire regions with a difference in ion velocity.

In ITP, the sample is introduced between a region of leading electrolyte (LE) and terminating electrolyte (TE) containing ions with higher and lower electrophoretic mobility, respectively, than the sample components. After the application of voltage, analytes form narrow zones according to their electrophoretic mobilities and move with uniform velocity. The preconcentration within each separated zone depends on the concentration of LE, which is typically 1000-fold higher than the predicted analytes' concentration in the initial sample, allowing an increase in concentration by a factor of 1000 or more [36].

Sweeping is the most extensively used on-line preconcentration technique with MEKC [37–39]. In principle, the sample prepared in a solvent without micelles is injected into the capillary previously filled with a BGE containing a micellar pseudostationary phase. On the application of voltage, micelles from the inlet vial penetrate the sample region, then pick up and accumulate analyte molecules into a narrow zone. This procedure results from chromatographic partitioning and other interactions between the micelles and analyte molecules. Sweeping has a great potential in CYP metabolism studies, since it is favorable for preconcentration of lipophilic compounds. Moreover, sweeping is also probably the only on-line preconcentration technique suitable for samples with a relatively high ionic strength such as incubation mixtures. The variant known as sweeping in a heterogeneous electric field even requires the sample to have a higher ionic strength than the BGE [40]. Nevertheless, until all processes taking place on concentration boundaries are completely understood, the optimization of sweeping methods will stay a laborious and time-consuming task.

In Section 17.4, a method is introduced allowing fast stability screening of new drug candidates against HLM. It is highlighted how the various challenges can be approached during method development. The chosen deproteinization technique needed to be compatible with screening of chemical libraries. The sensitivity, on the other hand, was sufficient, and there was no need to perform on-line preconcentration techniques.

17.4 METHOD DEVELOPMENT

As mentioned earlier, the goal of this study was to introduce a method based on CE, allowing high-throughput screening of drug candidates' metabolism. Since the compound libraries contain many chemicals with different electrophoretic properties, it is impractical to perform the screening on the basis of product formation or substrate consumption. However, all CYP use reduced nicotinamide adenine dinucleotide phosphate (NADPH) as a source of electrons for oxygen molecule cleavage in the oxygenase reaction [41]:

$$\mathrm{RH + O_2 + NADPH + H^+ \rightarrow ROH + NADP^+ + H_2O}$$

For this reason, it should be possible to determine a compound's stability against HLM by monitoring the formation of nicotinamide adenine dinucleotide phosphate ($NADP^+$). $NADP^+$ as an analyte thus ensures not only the generic applicability of the method but also good sensitivity even for analyses of substrates at therapeutically relevant concentrations. It has to be taken into consideration though that CYP consume NADPH in nonproductive reactions as well [42–44]:

$$\mathrm{O_2 + NADPH + H^+ \rightarrow NADP^+ + H_2O_2}$$

$$\mathrm{O_2 + 2\,NADPH + 2\,H^+ \rightarrow 2\,NADP^+ + 2\,H_2O}$$

Therefore, the amount of $NADP^+$ produced in connection with metabolism of the tested compound should be determined from the difference between two samples: the blank devoid of tested compound and the full incubation mixture.

Initial analyses were performed with the following separation conditions—capillary: 40 cm effective length; BGE: 25 mM sodium phosphate buffer, pH: 6.6; applied voltage: 18 kV positive polarity; capillary temperature: 37°C; detection: 260 nm. The rinsing procedure, consisting of 1 min with 0.1 M NaOH, 1 min with Milli-Q® water, and 3 min with BGE, showed very poor repeatability of $NADP^+$ migration times with a relative standard deviation (RSD) of 13%. The observed gradual increase of the migration times of $NADP^+$ was expected because the HLM preparation represents a very complex mixture of polypeptides, proteins, phospholipids, and other cell-building material. This is a result of the HLM production procedure where a homogenate of human livers is centrifuged at 9,000–12,000×g [45]. The obtained supernatant is very rich in microsomes, that is, in small vesicles derived from endoplasmic reticulum with a mean diameter of about 150 nm. However, it still contains a variety of compounds as mentioned earlier. Three precautions were introduced to improve the repeatability: (i) dynamic coating of the inner capillary wall, (ii) short-end injection CE mode, and (iii) strong between-run rinsing procedure.

17.4.1 Dynamic Coating of Inner Capillary Wall

The addition of the water-soluble polymer linear polyacrylamide (LPA) to the BGE, protecting the inner capillary surface against protein adsorption, represents the first precaution [46]. The repeatability improvement in the presence of LPA was observed during previous research on in-capillary dextromethorphan metabolism in the presence of HLM [47]. Hence, different volume proportions of LPA and concentrations of disodium hydrogen phosphate buffer were tested to determine the most suitable BGE composition. A content of 10% (v/v) LPA (average FW 10,000) in the BGE was found to be the most appropriate, since higher proportions caused practical problems due to the high viscosity of the solution. The protective layer is stabilized by the formation of hydrogen bonds between the silanol groups of the silica surface and the oxygen atom of the carbonyl groups of LPA. Since the coating is refreshed before every analysis, issues such as capillary fouling, coating inhomogeneity, or limited lifetime of a covalent coating can be overcome. Moreover, LPA reduces the electroosmotic flow (EOF) and thus enables to take the second precaution: the short-end injection mode of CE.

17.4.2 Short-End Injection

In CE instrumental design, the minimal available capillary length is given by the distance between the detection window and the outlet end of the capillary. To employ this part of the capillary for separation, a sample can be injected from the outlet end. This procedure is known as short-end injection and reduces analysis time by at least two-thirds compared to conventional injections [48]. In addition, it is well known that the extent of peak broadening caused by diffusion processes is directly dependent on the analysis time. As a result, this injection mode leads to sharper peaks and approximately two- to threefold enhancement of the method sensitivity [48,49]. The setting used in this study is depicted in Figure 17.3. The incubation mixture was injected hydrodynamically from the outlet capillary end. The LPA contained in the BGE sufficiently reduces the EOF, allowing negatively charged analytes to migrate toward the anode and reach the detection window after the application of a voltage with positive polarity. The short-end injection procedure not only diminishes the analysis time but also reduces the possibility of protein adsorption and thus contributes to the improvement of method repeatability.

17.4.3 Strong Rinsing Procedure

The introduction of a strong rinsing procedure between the analyses represents the third precaution. Several procedures combining rinsing steps with SDS, ethanolic sodium hydroxide solution and increasing concentrations of sodium hydroxide and phosphoric acid were tested. The lowest RSD values of migration times were achieved by rinsing with 1 M phosphoric acid for 2 min, with Milli-Q® water for 1 min, with 1 M NaOH for 2 min, with Milli-Q® water for 1 min, and with BGE for 3 min. This procedure was used in the optimized method.

17.4.4 Optimized Method

The optimized method made use of the following chemicals and reagents. Pooled human liver microsomes were obtained from BD Biosciences (Woburn, MA). Ciclosporin, dextromethorphan hydrobromide, erythromycin dihydrate, lidocaine hydrochloride, and quinidine sulfate were purchased from European Pharmacopoeia Laboratory (Strasbourg, France). Bufuralol hydrochloride, cisapride monohydrate and tolbutamide were kindly provided by Ron Gilissen (Johnson & Johnson Pharmaceutical Research and Development, Beerse, Belgium). Bupivacaine hydrochloride was a gift from Professor Y. Vander Heyden (Laboratory of Analytical Chemistry and Pharmaceutical Technology, Free University of Brussels, Brussels, Belgium). Propranolol hydrochloride was kindly donated by Professor J. Crommen

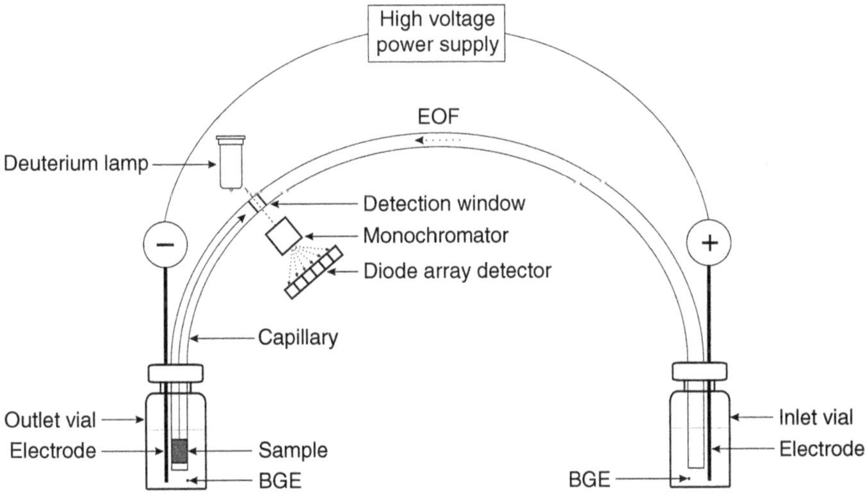

FIGURE 17.3 Schematic illustration of the short-end injection mode in CE.

(Department of Analytical Pharmaceutical Chemistry, University of Liège, Liège, Belgium). (*R*)-warfarin was from Ultrafine (Manchester, UK). Nifedipine, verapamil hydrochloride, LPA 50% solution (average FW 10,000), $NADP^+$, and NADPH were supplied by Sigma-Aldrich (Steinheim, Germany). Disodium hydrogen phosphate was purchased from Ferak (Berlin, Germany) and sodium dihydrogen phosphate and SDS from Acros Organics (Geel, Belgium). Methanol, ethanol, and phosphoric acid were received from VWR (Leuven, Belgium). Phthalic acid was supplied from UCB (Brussels, Belgium). Sodium hydroxide was obtained from Fisher Scientific (Loughborough, UK).

All solutions were prepared with water from a Millipore Milli-Q® Gradient system (Milford, MA). The incubation buffer (consisting of 10 mM phosphate buffer, pH 7.4) and the phosphate buffer for the BGE (see the last paragraph of this section) were prepared freshly every 3 days and stored at 4 °C. The stock solutions of the probe drugs were prepared in an appropriate solvent: methanol (ciclosporin, cisapride, erythromycin, nifedipine, quinidine, and testosterone); 0.1 M NaOH ((*R*)-warfarin); water (bufuralol, bupivacaine, lidocaine, propanolol, and verapamil) and kept at 4 °C. Working solutions were then obtained by dilution of the stock solution with incubation buffer (Table 17.1). The content of primary solvent in the incubation mixture should be kept below 1% (v/v) so that it has no effect on enzymatic activity. Working solutions of HLM, NADPH, $NADP^+$, probe drugs, and phthalic acid (internal standard I.S.) were prepared freshly every day and stored at 4 °C until use.

Enzyme reaction monitoring was performed as follows. A reaction mixture without NADPH was preincubated in an Eppendorf tube at 37 °C for 5 min. The enzymatic reaction was initiated by rapid addition and mixing of a given amount of NADPH. As a result, the reaction mixture contained certain concentrations of a probe drug, 1 mg mL^{-1} HLM (mg protein mL^{-1}), 0.3 mM NADPH in 10 mM phosphate buffer, pH 7.4. The incubation was performed in an Eppendorf Thermomixer (Hamburg, Germany) at a mixing frequency of 450 rpm and a temperature of 37 °C for 30 min. Phthalic acid (I.S.) was added at a final concentration of 1.2 μM, and the sample was immediately centrifuged at 12,100×g for 10 min. The supernatant was cooled at −25°C for 5 min and stored at 4 °C before use. All samples were analyzed within 12 h after the incubation with the use of CE.

All analyses were carried out on a P/ACE™ MDQ CE System (Beckman Coulter, Fullerton, CA) equipped with a 190–600 nm diode array detector. Data collection and report analysis were performed by 32 Karat™ Software v. 5.0. A 75 μm fused-silica capillary (10 cm effective length, 50 cm total length) from Polymicro Technologies (Phoenix, AZ) was used for all separations. The capillary was rinsed with 0.1 M NaOH for 10 min, with Milli-Q® water for 10 min and with BGE for 10 min at the beginning of each day, and the optimized washing procedure was applied between analyses to ensure repeatable separations.

The BGE was prepared by stirring LPA solution with 50 mM disodium hydrogen phosphate (1:9; v/v) for 20 min. The resulting mixture was at pH 8.0. Samples were injected by application of a pressure of 0.5 psi (3.45 kPa) onto the outlet vial for 3 s. Analyses were performed at 8 kV (positive polarity) in a capillary thermostated to 37 °C. The analytes were detected at 260 nm. The results were evaluated using SigmaPlot v. 8.02 (SPSS, Chicago, IL) and Statistica v. 6.0 (StatSoft, Tulsa, OK) software.

TABLE 17.1 List of Probe Drugs Used for Stability Screening

Probe Drug	c_{fin} (μM)	K_m (μM)	References
Bufuralol	100	10[1]	50
Bupivacaine	600	125[1]	51
Cisapride	50	4.3[b]	52
Ciclosporin	10	1[b]	53
Erythromycin	250	33 ± 4[b]	54
Lidocaine	5000	1032[b]	55
Nifedipine	100	15[1]	56
Propranolol	200	4.1–31.9[1]	57
Quinidine	500	47.1 ± 5.9[c]	58
(*R*)-warfarin	1000	166 ± 12[b]	59
Testosterone	300	47[1]	56
Verapamil	200	21.8[b]	60

Substrate concentrations in the final incubation mixtures (c_{fin}) were five to ten times their respective Michaelis–Menten (K_m) values based on tests with:
[a]HLM.
[b]recombinant human CYP3A4.
[c]rat liver microsomes reported in the literature.

17.5 METHOD VALIDATION

Although the above-mentioned measures led to an improvement in repeatability of migration time, LPA caused an increase in the viscosity of the BGE with consequent deviations in injected sample volume and peak area. This phenomenon likely arises from subtle differences in composition (viscosity) of the BGE used for capillary pretreatment, since detrimental effects such as siphoning, spontaneous injection, and loss of sample by evaporation or diffusion are limited by utilization of a viscous BGE [61]. The issue was overcome by using phthalic acid as an internal standard. This compound is stable against HLM activity [62]. A typical electropherogram acquired with the optimized method is shown in Figure 17.4. Repeatability of the method was tested by analyzing the complete reaction mixture containing microsomes, 100 μM bufuralol as a probe drug, and 200 μM standards of $NADP^+$ and NADPH. The sample was stored at 4°C in the sample chamber inside the CE device to minimize $NADP^+$ formation

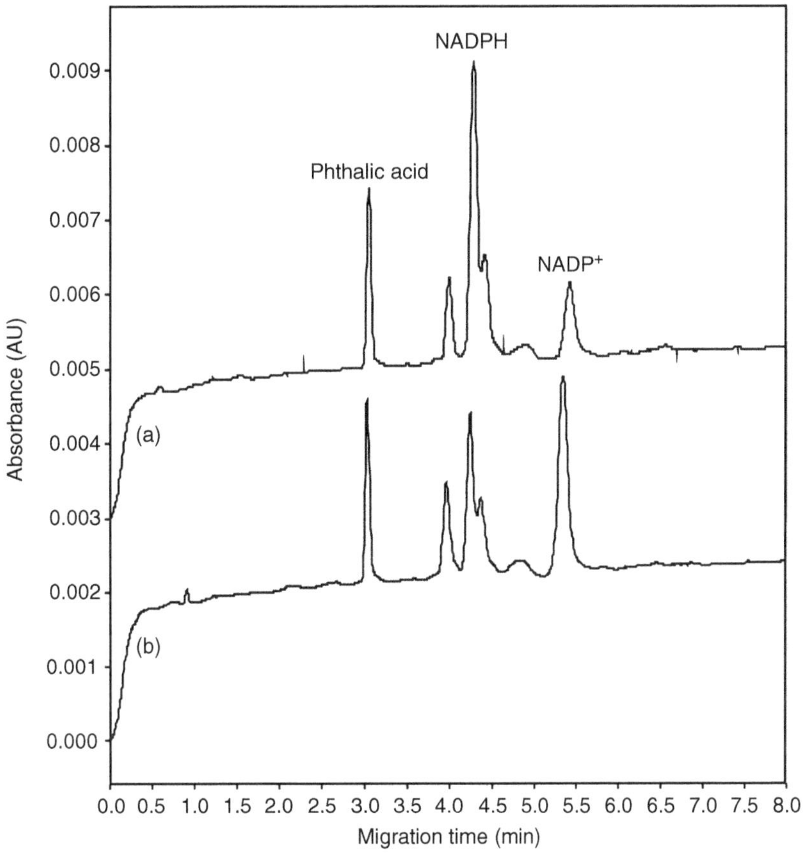

FIGURE 17.4 Electropherograms acquired with the optimized method. Comparison of blank (a) and full incubation mixture (b) from the stability assay of lidocaine. Both samples consisted of 1 mg mL^{-1} HLM, 1.2 μM phthalic acid (I.S.), and 0.3 mM NADPH prepared in 10 mM sodium phosphate buffer, pH 7.4; the full incubation mixture also contained 5 mM lidocaine; 30 min incubation. Separation conditions: 75 μm fused-silica capillary (10 cm effective length), temperature of capillary 37 °C, detection at 260 nm, 10% (v/v) LPA in 50 mM disodium hydrogen phosphate as BGE, injection at 0.5 psi for 3 s, separation voltage 8 kV (positive polarity).

during the run sequence. Also, the calibration curve, limit of detection, limit of quantification, and recovery were determined by analyzing the reaction mixtures without probe drug. The results are summarized in Table 17.2. RSD values of 1.8% and 2.0% for migration time and relative peak area of NADP$^+$, respectively, confirm the good method repeatability. The calibration graph was linear over the range 25–400 μM of NADP$^+$ with a coefficient of determination (r^2) of 0.996.

TABLE 17.2 Validation Parameters of the Optimized Method

Validation Parameters	NADP$^+$
Migration time repeatability (RSD, $n = 6$)	1.8%
Relative peak area repeatability (RSD, $n = 6$)	2.0%
Linearity range	25–400 μM
Determination coefficient	0.996
Recovery ($n = 6$)	100.6–103.9%
Limit of quantification (S/N = 10)	20 μM
Limit of detection (S/N = 3)	6.5 μM

17.6 METHOD APPLICATIONS

17.6.1 Drug Stability Screening

After the validation, the fast stability screening of 12 drugs was performed. These drugs are all substrates of individual CYP isoforms. The probe drugs were analyzed at one concentration point corresponding to five to ten times their Michaelis–Menten values (K_m) described in the literature (see Table 17.1). Samples of each probe drug were incubated in three separate vials; each vial was analyzed in duplicate. Blank incubations were done in the absence of substrate; blank samples were analyzed in triplicate. The amounts of NADP$^+$ produced in the samples were determined from the calibration curve. For each substrate, the difference in $NADP^+$ formation between blank and full incubation was statistically evaluated. Values of $NADP^+$ concentration in blank samples constitute the first group of results; the values of $NADP^+$ concentration in full incubation mixtures form the second group. First, the mean and variance of both groups were calculated for each tested drug. Since the variances of the first and second group were not equal via F-test in case of several probe drugs, a Welch's modified *t*-test, that is, independent two-sample *t*-test was carried out to determine the statistical significance of the difference between the means of both groups. The statistical *t* value was calculated as follows:

$$t = \left| \frac{\overline{X}_1 - \overline{X}_2}{\sqrt{(S_1^2/n_1) - (S_2^2/n_2)}} \right|$$

where X_i, S_i^2, n_i, $i=1$, and $i=2$ are the sample mean, sample variance, and number of analyses of the group of full incubation mixture analyses and blank analyses, respectively [63].

For the significance testing, the distribution of the test statistic was approximated as being an ordinary Student's *t* distribution with degrees of freedom (ν) calculated using [64]:

$$\nu = \frac{\left[(S_1^2/n_1) - (S_2^2/n_2)\right]^2}{\left[(S_1^2/n_1)^2/(n_1-1)\right] + \left[(S_2^2/n_2)^2/(n_2-1)\right]}$$

The drug was considered unstable (metabolized) when the difference between $NADP^+$ production in blank and full incubation mixture was statistically significant. Table 17.3 indicates the four probe drugs showing significant microsomal metabolism. Testosterone was marked as a stable substrate in spite of the fact that most of its metabolism is located in hepatic tissue [65–67]. In this respect, it is necessary to stress that a compound's stability against HLM highly depends on its ability to pass through the vesicle membrane [68], which is particularly determined by the polarity of the molecule, and the proportional content of CYP isoform responsible for the compound's metabolism in the human liver. In addition, the criterion for significant difference had an important influence on the final result.

TABLE 17.3 Results of the Stability Screening

Bufuralol	–	Nifedipine	–
Bupivacaine	+	Propranolol	–
Ciclosporin	+	Quinidine	–
Cisapride	–	(*R*)-warfarin	–
Erythromycin	+	Testosterone	–
Lidocaine	+	Verapamil	–

The drugs showing a significant difference in $NADP^+$ production between blank and full incubation mixture ($p=0.05$) were considered unstable (metabolized) (+) within the frame of the testing criteria.

17.6.2 Kinetic Study

Since the amount of $NADP^+$ coming on top of the nonproductive reaction directly corresponds to the substrate reaction rate, the apparent kinetic parameters for metabolism of a compound can be established using the present method. In a kinetic study, apparent Michaelis–Menten constants (K'_m) can be obtained, providing deeper insight into the prospective dosing frame and compound applicability. A kinetic parameters determination was carried out with different concentrations of lidocaine in the reaction mixture (from 0.1 to 5 mM). The incubation was performed for 30 min with 1 mg mL^{-1} HLM. The blank and each concentration were analyzed in triplicate. The resulting K'_m and apparent maximum reaction velocity (V'_{max}) values estimated by nonlinear regression for the Michaelis–Menten model were 353.5 ± 140.3 μM and 2.2 ± 0.2 nmol min^{-1} mg^{-1} of protein, respectively. Performing a kinetic study with the aforementioned method could experience higher experimental error compared to ordinary approaches based on product formation, since errors of both blank and full incubation mixture analyses are reflected in the determined $NADP^+$ amount. CYP3A4 represents the main CYP isoform responsible for lidocaine metabolism in the high-substrate-concentration region [69]. Since CYP3A4 constitutes approximately 30% of all liver CYP activity [70], the resulting K'_m value is in good agreement with the findings of Wang et al. determining a K_m of 1032 μM for lidocaine *N*-deethylation by recombinant human CYP3A4 [55].

17.7 CONCLUSIONS

The presented study is based on measuring the production of $NADP^+$ during microsomal metabolism of drugs. As long as the drug and its metabolites are separated from $NADP^+$ and the internal standard, the method is applicable and thus constitutes a rather universal approach. Good method repeatability is achieved through the utilization of soluble polymeric gel, the short-end injection mode, and a strong rinsing procedure. The versatile substrate application, minimal sample and running solutions consumption, and simple sample preparation procedures predetermine this methodology as a promising tool for high-throughput screening in the early stages of drug development, especially if it is applied in a multicapillary format. Compared to commonly used HPLC, the presented CE methodology is more economically and environmentally friendly, thus meeting the demands of both the public and the pharmaceutical industry.

ACKNOWLEDGMENTS

This work was supported by research scholarships from the Flemish Ministry of Education and the Flemish Fund for Scientific Research (Research project G.0681.08), by grants No. P206/10/0057 and GAP206/12/G014 from The Czech Science Foundation, and by "CEITEC—Central European Institute of Technology" (CZ.1.05/1.1.00/02.0068) from European Regional Development Fund.

REFERENCES

1. Ansede, J.H. and Thakker, D.R. (2004) *J. Pharm. Sci. 93*, 239–255.
2. DiMasi, J.A., Hansen, R.W., and Grabowski, H.G. (2003) *J. Health Econ. 22*, 151–185.
3. Fecik, R.A., Frank, K.E., Gentry, E.J., Menon, S.R., Mitscher, L.A., and Telikepalli, H. (1998) *Med. Res. Rev. 18*, 149–185.
4. Venkatakrishnan, K., von Moltke, L.L., and Greenblatt, D.J. (2001) *J. Clin. Pharmacol. 41*, 1149–1179.
5. Lin, J.H. and Lu, A.Y.H. (1997) *Pharmacol. Rev. 49*, 403–449.
6. White, R.E. (2000) *Ann. Rev. Pharmacol. Toxicol. 40*, 133–157.
7. Masimirembwa, C.M., Bredberg, U., and Andersson, T.B. (2003) *Clin. Pharmacokin. 42*, 515–528.
8. Ioannides, C. (2002) In: Ioannides, C. (ed.), *Enzyme Systems that Metabolise Drugs and Other Xenobiotics*, Wiley, Chichester.
9. Anzerbacher, A. and Anzerbacherová, E. (2001) *Cell. Mol. Life Sci. 58*, 737–747.
10. Guengerich, F.P. (2002) In: Ioannides, C. (ed.), *Enzyme Systems that Metabolise Drugs and Other Xenobiotics*, Wiley, Chichester.
11. Pelkonen, O., Tolanen, A., Turpeinen, M., Uusitalo, J. (2008) In: Gad, S.C. (ed.), *Preclinical Development Handbook: ADME and Biopharmaceutical Properties*, Wiley, Hoboken, NJ.
12. Weaver, R., Graham, K.S., Beattie, I.G., and Riley, R.J. (2003) *Drug Metab. Dispos. 31*, 955–966.
13. Kassel, D.B. (2004) *Curr. Opin. Chem. Biol. 8*, 339–345.
14. Di Marco, A., Cellucci, A., Chaudhary, A., Fonsi, M., and Laufer, R. (2007) *Drug Metab. Dispos. 35*, 1737–1743.
15. Kumar, R.A. and Clark, D.S. (2006) *Curr. Opin. Chem. Biol. 10*, 162–168.
16. Yu, H. and Adedoyin, A. (2003) *Drug Discov. Today 8*, 852–861.
17. Kerns, E.H. and Di, L. (2003) *Drug Discov. Today 8*, 316–323.
18. Glatz, Z. (2006) *J. Chromatogr. B. 841*, 23–37.
19. Zhang, J., Konecny, J., Glatz, Z., Hoogmartens, J., and Van Schepdael, A. (2007) *Curr. Anal. Chem. 3*, 197–217.
20. Saevels, J., Van Schepdael, A., and Hoogmartens, J. (1999) *Anal. Biochem. 266*, 93–101.
21. Pratt, W.B. (1990) *Principles of Drug Action: The Basis of Pharmacology*, 3rd edn., Churchill Livingstone, New York.
22. Lloyd, D.K. and Wätzig, H. (1995) *J. Chromatogr. B 663*, 400–405.
23. Konečný, J., Mičíková, I., Řemínek, R., and Glatz, Z. (2008) *J. Chromatogr. A 1189*, 274–277.
24. Konečný, J., Juřica, J., Tomandl, J., and Glatz, Z. (2007) *Electrophoresis 28*, 1229–1234.
25. Huang, Z., Timerbaev, A.R., Keppler, B.K., and Hirokawa, T. (2006) *J. Chromatogr. A. 1106*, 75–79.
26. Alnouti, Y., White, C.A., and Bartlett, M.G. (2004) *Biomed. Chromatogr. 18*, 523–531.
27. Soetebeer, U.B., Schierenberg, M.O., Schulz, H., Hempel, G., Andresen, P., and Blaschke, G. (2001) *Anal. Chem. 73*, 2178–2182.
28. Sun, J., Xu, X., Wang, C., and You, T. (2008) *Electrophoresis 29*, 3999–4007.
29. Martínez, D., Cugat, M.J., Borrull, F., and Calull, M. *J. Chromatogr. A* (2000) *902*, 65–89.
30. Xie, W., Mullett, W.M., Miller-Stein, C.M., and Pawliszyn, J. (2008) *J. Chromatogr. B 877*, 415–420.
31. Silva, M. (2009) *Electrophoresis 30*, 50–64.
32. Kartsova, L.A. and Bessonova, E.A. (2009) *J. Anal. Chem. 64*, 326–337.
33. Aranas, A.T., Guidote, A.M., and Quirino, J.P. (2010) *Anal. Bioanal. Chem. 394*, 175–185.
34. Ward, E.M., Smyth, M.R., O'Kennedy, R., and Lunte, C.E. (2003) *J. Pharm. Biomed. Anal. 32*, 813–822.
35. Guillouzic, S., McCormick, L.C., and Slater, G.W. (2002) *Electrophoresis 23*, 1822–1832.
36. Kvasnička, F. (2000) *Electrophoresis 21*, 2780–2787.
37. Huang, C.-W., Jen, H.-P., Wang, R.-D., and Hsieh, Y.-Z. (2006) *J. Chromatogr. A 1110*, 240–244.
38. Tsai, I.-L., Sun, S.-W., Liao, H.-W., Lin, S.-C., and Kuo, C.-H. (2009) *J. Chromatogr. A 1216*, 8296–8303.
39. Zhao, Y., Kong, Y., Wang, B., Wu, Y., and Wu, H. (2007) *J. Chromatogr. A 1146*, 131–135.
40. Quirino, J.P., Kim, J.-B., and Terabe, S. (2002) *J. Chromatogr. A. 965*, 357–373.
41. Paine, M.J.I., Scrutton, N.S., Munro, A.W., Gutierrez, A., Roberts, G.C.K., and Wolf, C.R. (2005) In: Ortiz de Montellano, P.R., (ed.), *Cytochrome P450: Structure, Mechanism, and Biochemistry*, 3rd edn., Kluwer Academic/Plenum Publishers, New York.
42. Rashba-Step, J. and Cederbaum, A.I. (1994) *Mol. Pharmacol. 45*, 150–157.
43. Gorsky, L.D., Koop, D.R., and Coon, M.J. (1984) *J. Biol. Chem. 259*, 6812–6817.
44. Kuthan, H. and Ullrich, V. (1982) *Eur. J. Biochem. 126*, 583–588.
45. Nelson, A.C., Huang, W., and Moody, D.E. (2001) *Drug Metab. Dispos. 29*, 319–325.
46. Kim, H.S. and Wainer, I. (2006) *Anal. Chem. 78*, 7071–7077.
47. Moisés, E. C. D., Duarte, L. de B., Cavalli, R. de C., Marques, M. P., Lanchote, V. L., Duarte, G., and da Cunha, C. P., (2008) *Eur. J. Clin. Pharmacol. 64*, 1189–1196.

48. Altria, K.D., Kelly, M.A., and Clark, B.J. (1996) *Chromatographia 43*, 153–158.
49. Němec, T. and Glatz, Z. (2007) *J. Chromatogr. A 1155*, 206–213.
50. Jones, B.C., Hyland, R., Ackland, M., Tyman, C.A., and Smith, D.A. (1998) *Drug Metab. Dispos. 26*, 875–882.
51. Gantenbein, M., Attolini, L., Bruguerolle, B., Villard, P.-H., Puyoou, F., Durand, A., Lacarelle, B., Hardwigsen, J., and Le-Treut, Y.-P. (2000) *Drug Metab. Dispos. 28*, 383–385.
52. Pearce, R.E., Gotschall, R.R., Kearns, G.L., and Leeder, J.S. (2001) *Drug Metab. Dispos. 29*, 1548–1554.
53. Kenworthy, K.E., Bloomer, J.C., Clarke, S.E., and Houston, J.B. (1999) *J. Clin. Pharmacol. 48*, 716–727.
54. Riley, R.J. and Howbrook, D. (1997) *J. Pharmacol. Toxicol. Methods 38*, 189–193.
55. Wang, J.-S., Backman, J.T., Taavitsainen, P., Neuvonen, P.J., and Kivistö, K.T. (2000) *Drug Metab. Dispos. 28*, 959–965.
56. Pelkonen, O., Mäenpää, J., Taavitsainen, P., Rautio, A., and Raunio, H. (1998) *Xenobiotica 28*, 1203–1253.
57. Johnson, J.A., Herring, V.L., Wolfe, M.S., and Relling, M.V. (2000) *J. Pharmacol. Exp. Ther. 294*, 1099–1105.
58. Bhoopathy, S., Sarkar, M.A., and Karnes, H.T. (2002) *Eur. J. Pharm. Sci. 16*, 265–272.
59. Ngui, J.S., Chen, Q., Shou, M., Wang, R.W., Stearns, R.A., Baillie, T.A., and Tang, W. (2001) *Drug Metab. Dispos. 29*, 877–886.
60. Von Richter, O., Burk, O., Fromm, M.F., Thon, K.P., Eichelbaum, M., and Kivistö, K.T. (2004) *Clin. Pharmacol. Ther. 75*, 172–183.
61. Mayer, B.X. (2001) *J. Chromatogr. A 907*, 21–37.
62. Kato, R., Takanaka, A., and Shoji, H. (1969) *Jpn. J. Pharmacol. 19*, 315–322.
63. Welsh, B.L. (1947) *Biometrica 34*, 28–35.
64. Satterthwaite, F.E. (1946) *Biometrics Bull. 2*, 110–114.
65. Waxman, D.J. and Chang, T.K.H. (2005) In: Ortiz de Montellano, P.R., (ed.), *Cytochrome P450: Structure, Mechanism, and Biochemistry*, 3rd edn., Kluwer Academic/Plenum Publishers, New York.
66. Sivertsson, L., Ek, M., Darnell, M., Edebert, I., Ingelman-Sundberg, M., and Neve, E.P.A. (2010) *Drug Metab. Dispos. 38*, 995–1002.
67. Mäenpää, J., Pelkonen, O., Cresteil, T., and Rane, A.J. (1993) *Steroid Biochem. Mol. Biol. 44*, 61–67.
68. Panuganti, S.D. and Svensson, C.K. (2008) In: Gad, S.C. (ed.), *Preclinical Development Handbook: ADME and Biopharmaceutical Properties*, Wiley, Hoboken, NJ.
69. Bargetzi, N.J., Aoyama, T., Gonzales, F.J., and Meyer, U.A. (1989) *Clin. Pharmacol. Ther. 46*, 521–527.
70. Shimada, T., Yamazaki, H., Mimura, M., Inui, Y., and Guengerich, F.P. (1994) *J. Pharmacol. Exp. Ther. 270*, 414–423.

18

ELECTROKINETIC TRANSPORT OF MICROPARTICLES IN THE MICROFLUIDIC ENCLOSURE DOMAIN

QIAN LIANG, CHUN YANG, AND JIANMIN MIAO

18.1 INTRODUCTION

Electrokinetic phenomena including electro-osmosis, electrophoresis, and dielectrophoresis are widely used for pumping liquid buffer solutions and separating particles in lab-on-a-chip systems. Generally, the electrokinetic transport velocity of a microparticle suspended in an electrolyte solution is described by the well-known Smoluchowski equation. Such equation is applicable to the transport of particles in an unbounded domain [1]. However, when the particle size is comparable with the dimension of the fluidic channels, and/or the particle motion is close to (either perpendicular to or parallel to) the channel wall, the Smoluchowski equation is no longer valid because the local electrical potential distribution and the flow field are significantly affected by the presence of the channel wall. Thus, the present study aims at examining the solid boundary effects on the electrokinetic transport of microparticles in a microfluidic confined domain.

Numerous studies have been reported to examine the solid boundary effects on electrokinetic transport of microparticles in finite microfluidic domains. The pioneer studies [2–7] analyzed the boundary effects on electrophoretic motion of a microsphere moving in parallel and perpendicular to a plane surface and along the axis of a circular orifice. With the successive reflections method, the corrections for the classic Smoluchowski equation were derived, which are valid at relatively large particle–wall separations. At small particle–wall separations, however, the net wall effect was predicted to reinforce the particle electrophoretic mobility [8,9]. Compared to theoretical analyses, very few experimental studies of the wall effects on particle electrokinetic transport have been reported. Although the key parameter, that is, the particle–wall separation, could not be well controlled and is still unknown in these studies [10,11], their results, to some extent, have confirmed the prior theoretical predictions.

In addition to most investigations of the boundary effects on electrokinetic transport of microparticles in flow-through channels, the electrokinetic transport of particles in a microfluidic enclosure domain is of great significance to many applications such as the development of electrophoretic displays [12–15] and electrophoretic deposition technology [16,17]. The electrophoretic display (EPD) is a technology that operates under ambient light illumination conditions based on the transport of charged pigment particles in a suspending medium. The pigment particles with the suspending medium are in a closed EPD cell. On the application of an electrical stimulus, different charged particles can move toward respective electrodes of counter-polarity. The phenomenon of particle transport in a microfluidic confined domain is a complex process simultaneously involving electrophoresis, electro-osmosis, and dielectrophoresis. Different from the electrokinetic transport of particles in infinite domains, on the one hand the presence of solid boundaries plays an important role in particle motion because of the hydrodynamic and electrical interactions among the particle–fluid–solid walls. On the other hand, the presence of electrical charges on the solid wall surfaces can also produce electroosmotic flow (EOF) under

Capillary Electrophoresis and Microchip Capillary Electrophoresis: Principles, Applications, and Limitations, First Edition.
Edited by Carlos D. García, Karin Y. Chumbimuni-Torres, and Emanuel Carrilho.

applied electrical field. The development of EOF in a confined domain can then induce vortices, which in turn have significant impact on the particle transport.

This chapter is to examine the characteristics of electrokinetic transport of a microparticle in a microfluidic confined domain. The domain boundary effects on the particle motion are studied numerically.

18.2 NUMERICAL MODEL

18.2.1 Problem Description

Both theoretical and experimental investigations [18,19] have shown that 2D models can be sufficient to describe the essential physical process of the electrokinetic particle transport. Therefore, a 2D mathematical model is presented in this study as shown in Figure 18.1. Owing to the symmetry, only half geometry is chosen here. Consider a circular particle of radius a in a microfluidic enclosure domain with $2h$ in height and $2w$ in width. The dimension of the domain is assumed 100 times of the particle radius, say, $h = w = 100a$. Therefore, the particle rotation is not considered in this study. The distances from the particle center to the bottom wall and the horizontal central line of the domain are defined as b and c, respectively. w_1 and w_2 denote the distance from the particle center to the symmetric line and the sidewall, namely, $w = w_1 + w_2$. The upper and bottom walls of the confined chamber are set as electrodes; thus, the electrical field is imposed vertically. The negatively charged particle is located at the horizontal center line of the chamber, and it moves toward the anode once the electrical field is turned on. The eccentricity of particle motion, e, is defined as $e = w_1/w$. The following assumptions are made in the model development: (i) The electrical double-layer (EDL) thickness of the particle and domain boundary is assumed to be comparable to the particle size. (ii) The particle and side boundaries of the domain are rigid and nonconducting. (iii) Both the particle and solid walls bear uniform surface charges. (iv) The two electrodes, located on the upper and lower walls of the domain, are set as perfect conductor. (v) The fluid inside the confined domain is Newtonian and incompressible. (vi) The Brownian motion of the particle is negligible.

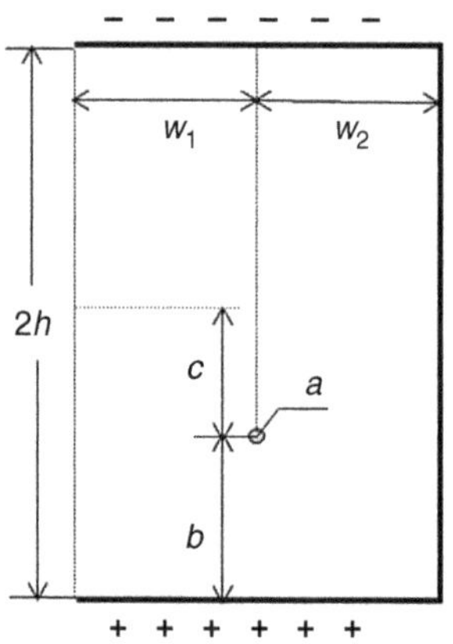

FIGURE 18.1 Schematic illustration for the modeling of electrokinetic transport of a particle in a confined microfluidic chamber.

18.2.2 Mathematical Model

For a charged particle, the electrical potential φ is related to the ionic concentration through Poisson's equation,

$$\nabla \cdot (\varepsilon\varepsilon_0 \nabla\varphi) = -\rho_e = -\sum_i n_i z_i e_0 \tag{18.1}$$

where $\varepsilon\varepsilon_0$, ρ_e, n_i, z_i, and e_0 are the dielectric permittivity of the solution medium, the free charge density, the ionic concentration for the ith species, the valence of type i ion, and the fundamental electrical charge, respectively. The ion distributions are assumed to satisfy the Boltzmann distribution,

$$n_i = n_0 \exp\left(\frac{-z_v e_0}{k_b T}\varphi\right) \tag{18.2}$$

where k_b is the Boltzmann constant, T is the absolute temperature and n_0 is the number concentration of the type i ions in the bulk solution. It is possible to treat most electrolytes as though they were symmetric ($z_+ = z_- = z_v$), with the valence z_v, equal to the valence of the counterions. In that case, the Poisson–Boltzmann equation is expressed as

$$\nabla^2\varphi = \frac{2z_v e_0 n_0}{\varepsilon\varepsilon_0} \sin h\left(\frac{z_v e_0}{k_b T}\varphi\right) \tag{18.3}$$

which is subject to the following boundary conditions,

$$\mathbf{n} \cdot (\varepsilon\varepsilon_0 \nabla\varphi) = -\sigma_p, \text{ at the particle surface} \tag{18.4}$$

$$\mathbf{n} \cdot (\varepsilon\varepsilon_0 \nabla\varphi) = -\sigma_w, \text{ at the boundaries of the confined domain} \tag{18.5}$$

where $\mathbf{n}$ is the unit normal vector pointing toward the liquid phase, and σ_p and σ_w are the surface charge density of the particle and the chamber wall, respectively. The electrical potential ψ of applied electric field is described by the Laplace equation,

$$\nabla^2\psi = 0 \tag{18.6}$$

which is subject to the boundary conditions expressed as

$$\mathbf{n} \cdot \nabla\psi = 0, \text{ at the side walls and the particle surface} \tag{18.7}$$

$$\psi = 0, \text{ at the upper wall} \tag{18.8}$$

$$\psi = \psi_0, \text{ at the lower wall} \tag{18.9}$$

With the linear superposition assumption for the electrical potentials contributed by EDL and applied electric field, we can obtain an expression for the total electrical field **E**,

$$\mathbf{E} = -\nabla\phi = -\nabla(\varphi + \psi) \tag{18.10}$$

The flow of incompressible electrolyte in the closed chamber is governed by the Stokes equation and the continuity equation, respectively, expressed as

$$-\nabla p + \eta\nabla^2\mathbf{v} - \rho_e\nabla\phi = 0 \tag{18.11}$$

$$\nabla\cdot\mathbf{v} = 0 \tag{18.12}$$

where $\mathbf{v}$ is the fluid velocity, η is the fluid dynamic viscosity, and p is the pressure. The term $-\rho_e\nabla\phi$ accounts for the electrical body force without consideration of the electrostriction and gradient of permittivity. Thus, the flow boundary conditions are

$$\mathbf{v} = 0, \text{ at solid walls of the confined domain} \tag{18.13}$$

$$\mathbf{v} = \mathbf{U}, \text{ at the particle surface} \tag{18.14}$$

where $\mathbf{U}$ is the transport velocity of the particle. The particle motion can be described by Newton's second law,

$$m\frac{d\mathbf{U}}{dt} = \mathbf{F}_{net} \tag{18.15}$$

$$\frac{d\mathbf{X}_p}{dt} = \mathbf{U} \tag{18.16}$$

where m, t, $\mathbf{F}_{net}$, $\mathbf{X}_p$ are, respectively, the mass of the particle, time, the net force, and the position vector of the particle center. By ignoring the gravity force, we can obtain the net force acting on the particle as

$$\mathbf{F}_{net} = \mathbf{F}_h + \mathbf{F}_e \tag{18.17}$$

where $\mathbf{F}_h$ and $\mathbf{F}_e$ are the hydrodynamic force and the electrostatic force, respectively. They are determined by integrating the hydrodynamic stress tensor and the Maxwell stress tensor with regard to the particle surface, respectively,

$$\mathbf{F}_h = \int_S \left\{-p\mathbf{I} + \eta[\nabla\mathbf{v} + (\nabla\mathbf{v})^T]\right\}\cdot\mathbf{n}ds \tag{18.18}$$

$$\mathbf{F}_e = \int_S \left[\varepsilon\varepsilon_0\mathbf{E}\mathbf{E} - \frac{1}{2}\varepsilon\varepsilon_0\mathbf{E}^2\mathbf{I}\right]\cdot\mathbf{n}ds \tag{18.19}$$

where $\mathbf{I}$ is the unit dyadic.

To nondimensionalize the model, we choose the particle radius $l = a$ and the thermal potential $\phi_0 = k_bT/z_v e_0$ as the characteristic length and the characteristic potential. The velocity scale, pressure scale, charge density scale, and concentration scale are set as $\mathbf{U}_0 = \varepsilon\varepsilon_0\phi_0{}^2/\eta l$, $p_0 = \eta\mathbf{U}_0/l$, $\sigma_0 = \varepsilon\varepsilon_0\phi_0/l$, and $c_0 = \varepsilon\varepsilon_0\phi_0/Fl^2$, respectively.

The EDL-induced electrical potential distribution φ^* is given by the Poisson–Boltzmann equation in dimensionless form,

$$\nabla^{*2}\varphi^* = (\kappa a)^2\sinh\varphi^* \tag{18.20}$$

where $\kappa^{-1} = \sqrt{\varepsilon\varepsilon_0 RT/(2F^2z_v^2n_0)}$ is the EDL thickness for a symmetric electrolyte, R is the universal gas constant, and F is the Faraday constant. The boundary conditions in dimensionless form are

$$\mathbf{n}\cdot(\nabla^*\varphi^*) = -\sigma_p^*, \text{ at the particle surface} \tag{18.21}$$

$$\mathbf{n}\cdot(\nabla^*\varphi^*) = -\sigma_w^* \tag{18.22}$$

The electrical potential ψ^* of the applied electric field is given by

$$\nabla^{*2}\psi^* = 0 \tag{18.23}$$

with the corresponding boundary conditions,

$$\mathbf{n}\cdot(\nabla^*\psi^*) = 0 \tag{18.24}$$

$$\psi^* = 0, \text{ at the upper wall} \tag{18.25}$$

$$\psi^* = \psi^*, \text{ at the lower wall} \tag{18.26}$$

and the total electrical field is given by

$$\mathbf{E}^* = -\nabla^*\phi^* = -\nabla^*(\varphi^* + \psi^*) \tag{18.27}$$

The flow field

$$-\nabla^*p^* + \nabla^{*2}\mathbf{v}^* + (\kappa a)^2\sinh\varphi^*\nabla^*\phi^* = 0 \tag{18.28}$$

$$\nabla^*\cdot\mathbf{v}^* = 0 \tag{18.29}$$

with boundary conditions:

$$\mathbf{v}^* = 0, \text{ at solid walls of the confined domain} \quad (18.30)$$

$$\mathbf{v}^* = \mathbf{U}^*, \text{ at the particle surface} \quad (18.31)$$

The motion of the particles is described by Newton's second law:

$$m^* \frac{d\mathbf{U}^*}{dt^*} = \mathbf{F}^*_{\text{net}} \quad (18.32)$$

$$\frac{d\mathbf{X}^*_{\text{p}}}{dt^*} = \mathbf{U}^* \quad (18.33)$$

where $t^* = \rho l^2/\eta$ and $m^* = \rho_{\text{p}} l^3/m$ are dimensionless time and dimensionless mass of the particle. ρ, ρ_{p}, and m are the density of the fluid, the density of the particle, and the mass of the particle, respectively. The net force is

$$\mathbf{F}^*_{\text{net}} = \mathbf{F}^*_{\text{h}} + \mathbf{F}^*_{\text{e}} \quad (18.34)$$

18.3 NUMERICAL SIMULATION

The phenomenon of particle transport in a microfluidic confined domain is a complex process simultaneously involving electrophoresis, electro-osmosis, and dielectrophoresis. The velocity and trajectory of the particle are dependent on external forces imposed, which in turn are determined by the electrical potential distribution and flow field. When the particle approaches a solid boundary, the presence of the boundary can affect the electrical and flow field as well, and such effect is significant especially at small particle–wall separations. Therefore, all equations presented in the mathematical model section have to be solved simultaneously. The finite element method based commercial software COMSOL Multiphysics 3.4 was used to solve this coupling problem. In particular, the flow field and electrical potential distributions associated with particle motion were obtained by using the arbitrary Lagrangian–Eulerian (ALE) technique, and the particle motion was tracked by using a moving mesh approach [20,21]. As the mesh deforms gradually with the particle motion within the computational domain, to ensure the accuracy a termination condition is predefined to stop the computation when the mesh quality degrades to a certain designed level. Subsequently, a new geometry is created based on the preceding deformed grid, and thus a new mesh is generated to continue the computation until the next mesh degradation. To facilitate the ALE algorithm, quadrilateral mesh elements were employed during the entire computation so that the "inverted mesh elements" problem was minimized [22]. In addition, mesh independent tests were performed to ensure that the numerical solutions are not dependent on grid size.

In order to verify the numerical simulation, we put the particle along the symmetric line of the confined chamber employing the 2D axial symmetry module. After setting the upper and lower boundaries as inlet and outlet, it actually represents a 3D test case involving electrokinetic transport of a spherical particle with thick EDL in an open-end cylindrical microchannel under an applied electrical field. The results acquired from this test example are compared with those calculated using the well-known Huckel equation [1], and a very good agreement is obtained.

18.4 RESULTS AND DISCUSSION

As mentioned earlier, two important factors affect the electrokinetic transport of particle in a confined microfluidic domain: one is the deformation of the electrical potential distribution and flow field between the particle–wall gap, and the other is the vortices induced by the EOF along the sidewalls of the microfluidic confined chamber. When the particle moves in bulk flow at large particle–bottom wall separations, the vortex effect dominates, since the bottom boundary has negligible effect in this circumstance. However, the vertical component of vortex streams reduces dramatically with decreasing particle–bottom wall gap, and the bottom wall starts playing a vital role at small particle–bottom wall separations. In order to address the two factors clearly, this section consists of two parts: the particle transport in bulk flow and the particle transport near the bottom electrode.

18.4.1 Particle Transport in the Bulk Flow

To explore the electrokinetic transport of single particles in bulk flow within the confined chamber, a particle is initially put at different positions along the horizontal center line of the computational domain, and these locations are indicated by the eccentricity e as illustrated in Figure 18.2. The solid lines and arrows represent streamlines and directions of the flow field, respectively, inside the confined domain at $\gamma = -0.5$, where $\gamma = \zeta_w/\phi_0$ denotes the dimensionless zeta potential of the sidewall. We are interested in the particle motion within a range of $0 < c/h < 0.9$ so as to exclude the bottom wall effect.

18.4.1.1 The Particle Velocity in the Confined Domain As clearly shown in Figure 18.3, the particle translational velocity varies pronouncedly with regards to its initial

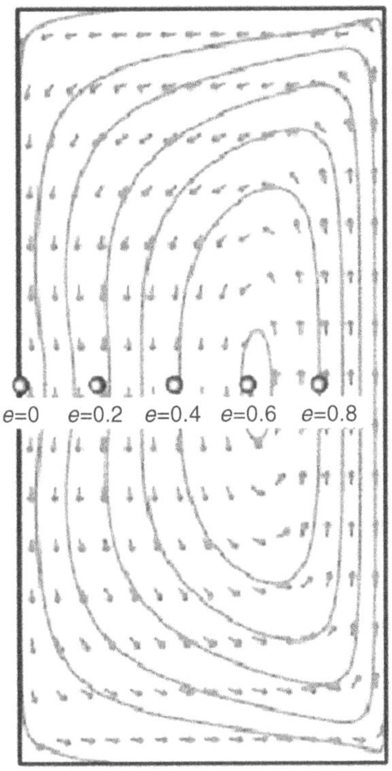

FIGURE 18.2 The initial positions of single particle within the computational domain.

locations. $\mathbf{U}_\infty$ and $\mathbf{U}_\mathrm{v}$ are the respective translational velocity in an infinite domain and the vertical component of the translational velocity in the confined chamber. At smaller eccentricities (e.g., $e=0$, 0.2, and 0.4), the particle moves with greater translational velocities at the beginning along the migration direction due to vortex acceleration, and followed with the deceleration stage when the particle approaches the bottom boundary because of the reduction of the local flow along the vertical direction. At the final stage shown in the figure, the values of these curves drop to unit one approximately, suggesting negligible vortex effect on the particle transport. When placed at $e=0.6$, the particle moves with a nearly constant velocity close to $\mathbf{U}_\infty$.

The reason is that, in this case, the initial position of the particle coincides at the center area of the vortices (as shown in Figure 18.2), at which the magnitude of the flow field is very small. The particle consequently is only accelerated near the bottom area within the confined domain. However, such velocity increment is very limited because of the small magnitude of the vertical component of the local vortices streams at that region. It is noted that the particle velocity development corresponding to $e=0.8$ exhibits entirely different from other situations (the solid arrow above the curve indicates the direction of the velocity evolution as time elapses). As we know, the magnitude of the vortex stream generated by the EOF along the sidewall increases as moving close to the side boundary. Owing to the opposite direction of the local flow, the particle experiences a quite low initial velocity ($\mathbf{U}_\mathrm{v}/\mathbf{U}_\infty=0.34$), and is further slowed down by the vortex until the particle moves oppositely. In this situation, the particle horizontal motion plays a significant role, and it pushes the particle approach to the sidewall, thus yielding deceleration of the particle velocity. As a result, the particle moves back with the local vortex stream. The magnitude of the opposite velocity at which the particle arrives at the horizontal center line ($\mathbf{U}_\mathrm{v}/\mathbf{U}_\infty=-0.27$, when $c/h=0$) is observed less than that of its initial velocity. This is also attributed to the horizontal motion which causes the particle to move closer to the sidewall.

18.4.1.2 The Trajectory of Particle Transport within the Confined Domain Not only the vertical transport is significantly affected by vortices, but also the horizontal motion of the particle is affected by the hydrodynamic effect. As indicated in Figure 18.4, the trajectories of the particle

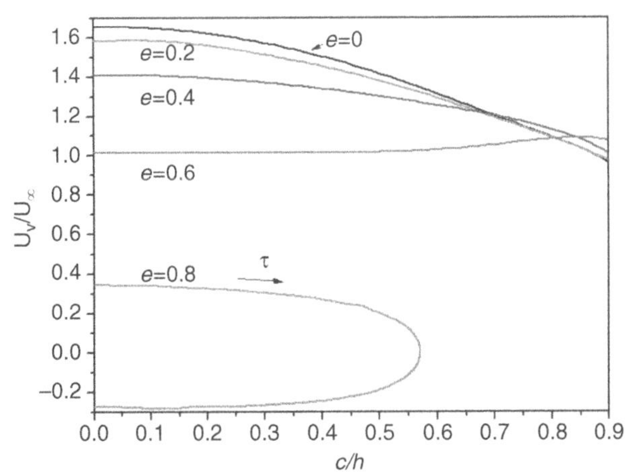

FIGURE 18.3 The particle velocity evolution with respect to various initial positions ($\gamma=-0.5$).

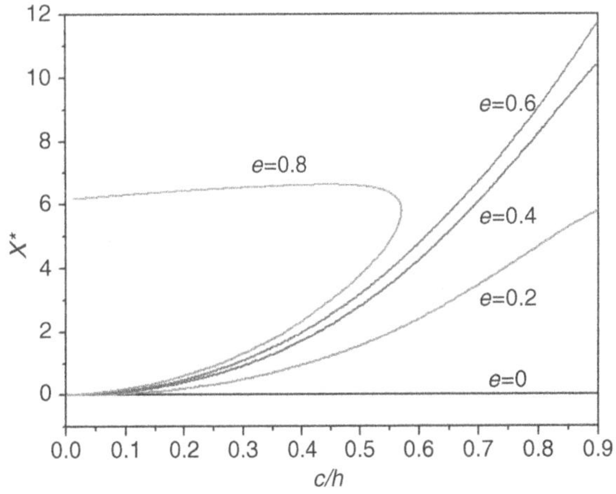

FIGURE 18.4 The trajectories of the particle motion with respect to initial positions ($\gamma=-0.5$). (a) $e=0.4$ and (b) $e=0.8$.

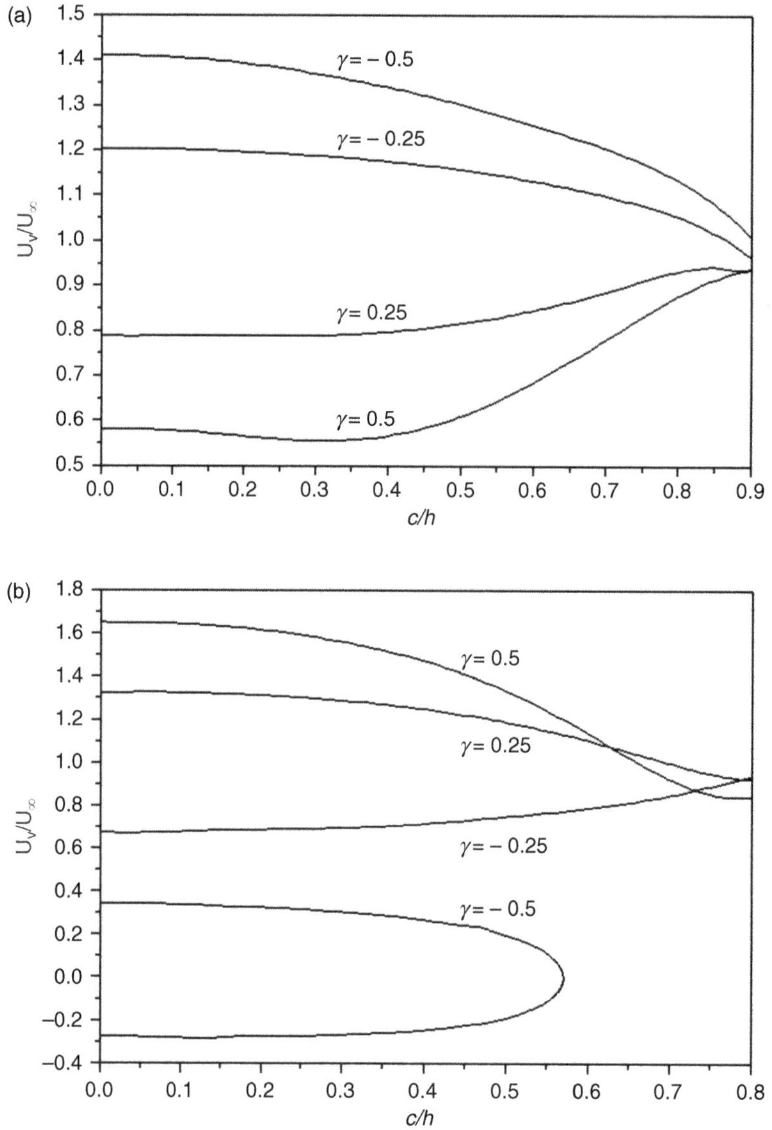

FIGURE 18.5 Effect of sidewall zeta potential on the particle motion. (a) $e = 0.4$ and (b) $e = 0.8$.

motion are strongly dependent on the vortex as well. In this figure, the vertical coordinate is the dimensionless horizontal displacement of the particle motion, and the horizontal coordinate is c/h denoting the dimensionless displacement along length direction of the domain. It is noted that particle motion closer to the sidewall normally experiences larger horizontal displacements.

18.4.1.3 The Effect of Sidewall Zeta Potential on the Particle Motion Since both the transport velocity and trajectory of the particle are significantly influenced by sidewall induced EOF vortex, the zeta potential of the sidewall is important to this situation. Here, the effect of the sidewall zeta potential on the particle motion is examined for various values including −0.5, −0.25, 0.25, and 0.5. Our consideration here is for a particle located at $e = 0.4$ and $e = 0.8$, at which the local stream directions around the particle are opposite.

Figure 18.5a and b show the effect of sidewall zeta potential on the particle motion for a particle positioned at $e = 0.4$ and $e = 0.8$, respectively. As discussed earlier, at $\gamma = -0.5$ the particle located initially at $e = 0.8$ is found to move oppositely. This phenomenon however vanishes when the side boundary carries less surface charge (e.g., $\gamma = -0.25$), indicating weak vortex streams along the sidewall. At this situation, the particle is speeded up when approaching the bottom wall. This trend also explains why the particle shows significant horizontal displacements at the final stage near the bottom surface. Because the particle initially placed at $e = 0.4$ and $e = 0.8$ is located on the left- and right-hand sides of the vortex center, respectively, the influence of the vortex on the particle motion in these two cases is competing with each other. For the clockwise vortex (i.e., $\gamma > 0$), the vortex effects cause the particle at $e = 0.4$ to move with lower velocities ($\mathbf{U}_v/\mathbf{U}_\infty < 1$) at early stage, followed by the acceleration stage. At $e = 0.8$, however, higher initial velocities ($\mathbf{U}_v/\mathbf{U}_\infty > 1$) are observed because of the vortex enhancement, and the velocities drop to unit one approximately at the final stage with decreasing vortex effects.

18.4.2 Particle Transport Near the Bottom Surface

When the particle moves close to the bottom wall electrode, the particle motion is dominated by the hydrodynamic and electrical fields that resulted from the particle–fluid–solid wall interactions. In addition, the vertical component of vortex stream velocity is very small in this situation, and thus its effect on the particle transport along the vertical direction is negligible. To emphasize the bottom electrode surface effect, we restrict our analysis within a near wall region, namely, $0.2 < a/b < 1$.

18.4.2.1 The Effect of the EDL Thickness on the Near Wall Motion of the Particle The effect of particle EDL thickness on the particle motion for various κa is presented in Figure 18.6. The denominator $\mathbf{U}_i$ in the plot represents the particle vertical velocity at $a/b = 0.2$. It is obvious that the particle velocity drops sharply when approaching the bottom wall. This is due to the intensive hydrodynamic interaction within such narrow particle–wall gap, which tends to push the particle away from the bottom surface. In addition, the results demonstrate that the particle vertical velocity reduces more rapidly with increasing the EDL thickness of the particle. As we known, the overlapped EDLs of two objectives can result in repulsive force if the particle and the bottom wall carry the

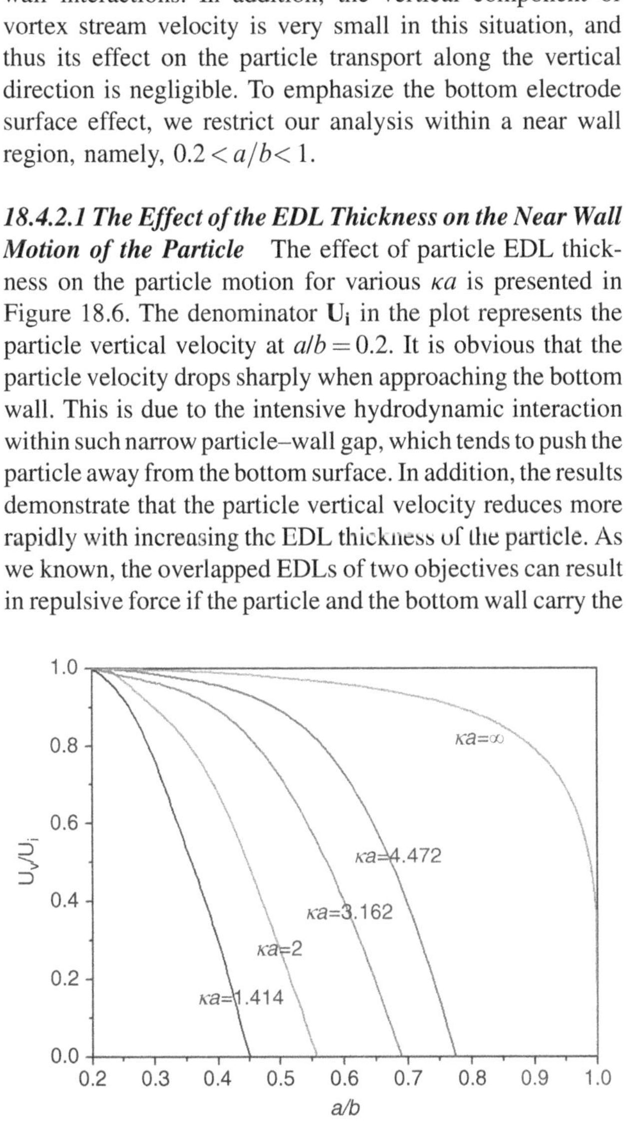

FIGURE 18.6 The effect of EDL thickness on particle vertical velocity.

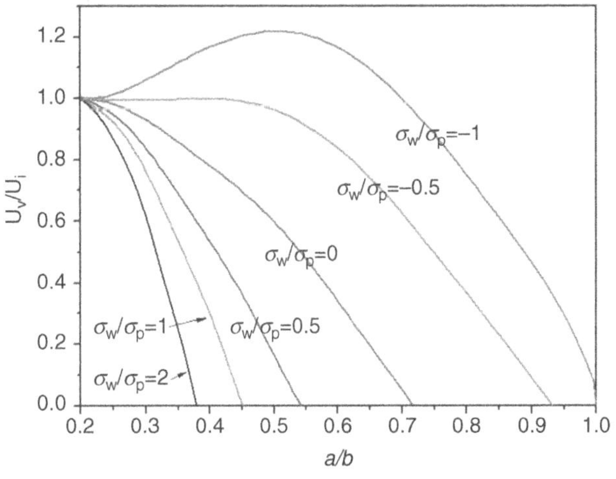

FIGURE 18.7 The effect of surface charge on particle vertical velocity.

same type of charges. For a certain separation between the particle and the bottom wall, the EDL interaction between these two objectives is more significant with increasing EDL thickness, thus yielding stronger repulsive force. The figure also shows the particle equilibrium height (along the vertical coordinate) at which the electrophoretic force is balanced against the EDL repulsive force. Obviously, the particle with thicker EDL can have a higher equilibrium height. The curve with $\kappa a = \infty$ implies the limiting case of the thin EDL condition, in which the EDL repulsive force vanishes due to the negligible EDL thickness.

18.4.2.2 The Effect of Surface Charge on the Near Wall Transport of the Particle Figure 18.7 presents the effect of surface charge on particle vertical velocity close to the bottom electrode; such effect is denoted by using the charge density ratio of the bottom wall to the particle surface. It is seen that with increasing the σ_w/σ_p value, the particle equilibrium height increases, and the particle experiences faster velocity drop. This is attributed to the increasing repulsive force with increasing the charge density of the bottom wall. However, the particle is observed to accelerate a bit before its velocity decreases at the negative σ_w/σ_p cases, indicating an attractive force between the particle and the bottom wall. As a result, the particle can move closer to the bottom wall with small separation gaps in comparison with the positive σ_w/σ_p cases.

18.5 MODEL APPLICATION

Nowadays electrophoretic display (EPD) becomes an important technology used in electronic papers due to paper-like appearance and low power consumption. A typical EPD cell consists of a thin layer of particle suspensions sandwiched by two electrode surfaces. A schematic illustration of the

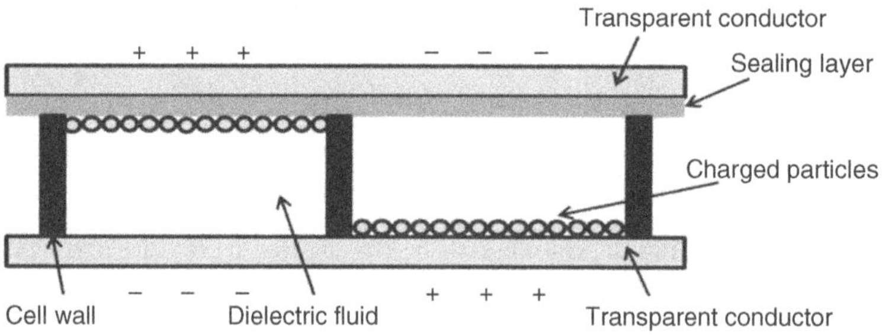

FIGURE 18.8 Schematic illustration of a typical electrophoretic display (EPD) cell.

cross section of an EPD cell is provided in Figure 18.8. Depending on applications, the dimension of the EPD cell may vary in a range of 60–200 μm (width) × 30–60 μm (height) [23]. The motion of particles in such EPD cell and thus the display information shown is mainly determined by the operation of electrical field applied. For example, when the electrical voltage is applied, the particles will be driven to the anode electrode if the particles are negatively charged. The current study presents a numerical model to examine the characteristics of the electrokinetic transport of particles in a microfluidic confined domain. In particular, the present work addresses the effects of wall and induced EOF flow on particle transport. Therefore, the results of this work are expected to provide a better understanding of the complex physical processes related to particle motion in EPD cells.

18.6 CONCLUSIONS

In this chapter, a numerical model for electrokinetic transport of particles in microfluidic confined domains is presented. The velocity and trajectory of a particle are analyzed, and they are affected not only by the disturbance of local electrical and flow fields due to the presence of solid wall but also by the vortices induced by the EOF along the sidewall of the domain. The particle closer to the sidewall moves at a lower velocity and experiences a significant horizontal migration because of the EOF-induced vortices. When the particle approaches the lower surface, the particle velocity is reduced dramatically due to the hydrodynamic effect in the particle–lower wall gap. The equilibrium heights of particle motion above the lower electrode are found to decrease on increasing the EDL thickness and surface charge of the particle and the surface.

REFERENCES

1. Masliyah, J.H., and Subir, B. (2006) *Electrokinetic and Colloid Transport Phenomena*, Wiley.
2. Keh, H.J., and Anderson, J.L. (1985) *J. Fluid Mech. 153*, 417–439.
3. Keh, H.J., and Chen, S.B. (1988) *J. Fluid Mech. 194*, 377–390.
4. Keh, H.J., and Lien, L.C. (1991) *J. Fluid Mech. 224*, 305–333.
5. Zydney, A.L. (1995) *J. Colloid Interface Sci. 169*, 476–485.
6. Ennis, J. and Anderson, J.L., (1997) *J. Colloid Interface Sci. 185*, 497–514.
7. Shugai, A.A. and Carnie, S.L. (1999) *J. Colloid Interface Sci. 213*, 298–315.
8. Yariv, E. and Brenner, H. (2003) *J. Fluid Mech. 484*, 85–111.
9. Unni, H.N., Keh, H.J., and Yang, C. (2007) *Electrophoresis 28*, 658–664.
10. Xuan, X., Raghibizadeh, S., and Li, D. (2006) *J. Colloid Interface Sci. 296*, 743–748.
11. Xuan, X., Ye, C., and Li, D. (2005) *J. Colloid Interface Sci. 289*, 286–290.
12. Ota, I., Ohnishi, J., and Yoshiyama, M. (1973) *Proc. IEEE 61*, 832–836.
13. Dalisa, A.L. (1977) *IEEE Trans. Elect. Dev. ED-24*, 827–834.
14. Hopper, M.A., and Novotny, V. (1979) *IEEE Trans. Electr. Dev. ED-26*, 1148–1152.
15. Comiskey, B., Albert, J.D., Yoshizawa, H., and Jacobson, J. (1998) *Nature 394*, 253–255.
16. Besra, L., and Liu, M. (2007) *Progr. Mater. Sci. 52*, 1–61.
17. Sarkar, P., and Nicholson, P.S. (1996) *J. Am. Ceram. Soc. 79*, 1987–2002.
18. Davison, S.M., and Sharp, K.V. (2006) *J. Colloid Interface Sci. 303*, 288–297.
19. Ai, Y., Park, S., Zhu, J., Xuan, X., Beskok, A., and Qian, S. (2009) *Langmuir 26*, 2937–2944.
20. Takashi, N. and Hughes, T.J.R., (1992) *Comput. Meth. Appl. Mech. Eng. 95*, 115–138.
21. Hu, H.H. (1996) *Int. J. Multiphase Flow 22*, 335–352.
22. *Comsol Multiphysics User's Guide*; COMSOL Inc., 2007.
23. Liang, R.C., Hou, J., Zang, H., and Chung, J. (2003) *IDMC 2007*, Fr- 17–15.

19

INTEGRATION OF NANOMATERIALS IN CAPILLARY AND MICROCHIP ELECTROPHORESIS AS A FLEXIBLE TOOL

Germán A. Messina, Roberto A. Olsina, and Patricia W. Stege

19.1 INTRODUCTION

"Good things come in small packages." The unique properties of nanomaterials and structures on the nanometer scale have sparked the attention of material developers. Nowadays, everybody is talking about nanotechnology and nanomaterials applied to different fields of investigation (e.g., Chemistry, Physics, Medicine, Electronics, and Materials Sciences). Also, a large number of books, journals, and publications are devoted to this subject. Even in the area of industry, nanomaterials are having an important impact. Advancing nanoscience from the laboratory into a technology suitable for product applications has been the goal of many new companies. Introducing nanomaterials into products and in the commercial field is today the objective of major chemical and materials firms. Incremental shifts in product performance using nanomaterials (for example, as fillers in plastics, as coatings on surfaces, and as UV protectives in cosmetics) are already occurring. Although technology holds more promise for the future, it is expected to bring more disruptive changes to both products and markets, because nanotechnology could pervade almost any industry, including medicine, plastics, energy, electronics, and even aerospace. Figure 19.1 illustrates the sciences that contribute to the synthesis and application of nanomaterials.

The objective of this chapter is to introduce the implementation of nanomaterials in capillary electrophoresis (CE). We describe, through published studies, the advantages and disadvantages of the use of different nanoparticles (NPs) in this separation technique. Moreover, our intention is to bring readers interested in nanotechnology into this field that is almost unexplored and just beginning to be developed. However, the papers cited in this chapter represent only a small selection of the current literature in this new field of research.

19.1.1 Historical Overview of Nanotechnology

Fifty years ago, Physics Nobel Laureate Richard Feynman gave a lecture entitled "There is Plenty of Room at the Bottom" [1]. Feynman made reference to the apparently wild idea of trying to make an ant do what we want it to do; he even suggested the possibility of introducing a mechanical surgeon within the blood to "look around." Feynman could not imagine how important that lecture would be, not only for the scientists present in the room that day, but also for the development of nanotechnology. Twenty years later, Drexler described a new approach involving molecular manipulation and molecular engineering to build molecular machines and molecular devices with atomic precision [2]. It was Drexler who finally popularized the term nanotechnology. There are two main entries into the nanoworld: molecular manufacturing, involving the manipulation of individual atoms (working from the bottom up), and molecular miniaturization, resulting in smaller and smaller devices (a top-down approach). In the "bottom-up" approach, the materials are formed by self-assembly with an atom-by-atom or molecule-by-molecule motif creating complex structures. For this approach, the limitations are an incomplete understanding of the synthesis processes, including how to control them to obtain desirable materials, and

Capillary Electrophoresis and Microchip Capillary Electrophoresis: Principles, Applications, and Limitations, First Edition.
Edited by Carlos D. García, Karin Y. Chumbimuni-Torres, and Emanuel Carrilho.

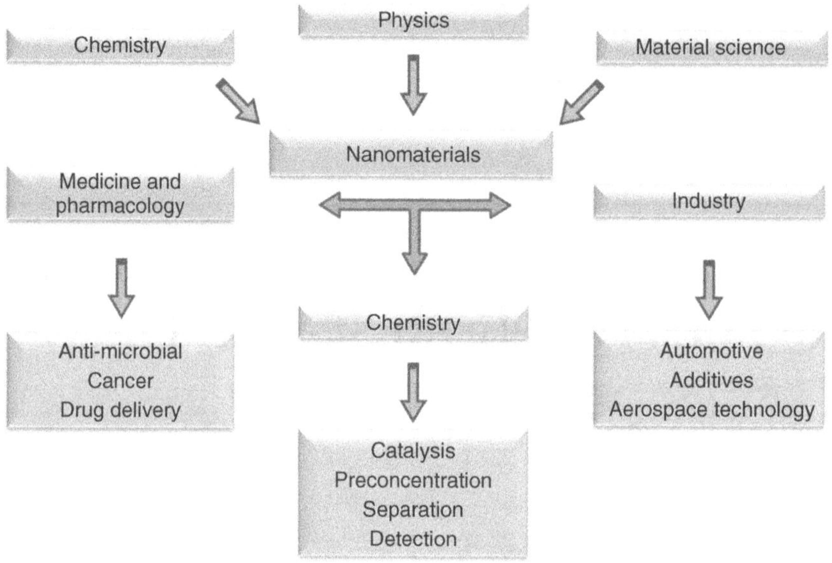

FIGURE 19.1 Schematic representation of the interaction between the different sciences in the synthesis of nanoparticles and their application.

the types of structures created. The "top-down" approach starts with bulk materials and ends with the formation of nanomaterials through a series of degradation processes. The limitations are the size, the shape, and the poor reproducibility of the materials formed [3].

Nanoscience and nanotechnology include the areas of synthesis, characterization, exploration, and application of nanostructured and nanosized materials. The application of nanomaterials can be historically traced back to even before the generation of modern science and technology. NPs were used as dyeing materials in ceramics by ancient people [4]; colloidal gold (Au) was used in medical treatment to cure dipsomania, arthritis, and so on, as early as the nineteenth century. Systemic experiments conducted on nanomaterials had already been started from Faraday's experiments in 1857 [5]. However, the real burst of nanotechnology did not come until the early 1990s. In recent decades, sophisticated instruments for characterization and manipulation, such as scanning electron microscopy, transmission electron microscopy, and scanning probe microscopy, have become more available for researchers to approach the nanoworld.

The term "nano" is used as a prefix for any unit such as a second or a meter, and it means one billionth of that unit. Hence a nanometer (nm) is a billionth of a meter, or 10^{-9} m. The scheme in Figure 19.2 gives a perspective of the scale of a nanometer.

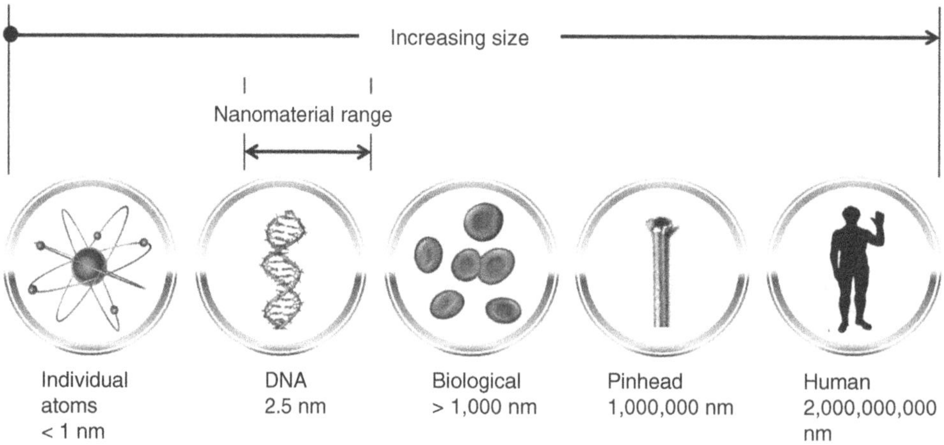

FIGURE 19.2 Sequence of images showing the various levels of scale. Adapted from Interagency Working Group on Nanoscience, Engineering and Technology Council Committee on Technology, "Nanotechnology: Shaping the World Atom by Atom," September 1999.

In spite of the wide use of the term "nanotechnology," it has been misleading in many instances, because some of the technology deals with systems in the micrometer and not the nanometer range (1–100 nm). Furthermore, research frequently involves basic and applied science of nanostructures and not basic and applied nanotechnology. Also, nanomaterials are not discovered materials, but are nanoscale forms of well-known materials, such as gold (Au), silver (Ag), platinum (Pt), and iron (Fe). It is important to bear in mind that some past technology would currently be considered a form of nanotechnology, for instance carbon nanoparticles (CNPs) used to reinforce tires, and Nature's photosynthesis.

19.1.2 Nanomaterials

The variety of existing nanomaterials is vast, and their range of properties and possible applications appear to be enormous, from extraordinarily tiny electronic devices, including miniature batteries, to biomedical uses, and further packaging films, superabsorbants, components of armor, and parts of automobiles. It is clear that researchers are merely on the threshold of understanding and development, and that a great deal of fundamental work remains to be done.

Nanomaterials are extremely small, featuring sizes on the same scale as the critical size for fundamental physical phenomena—electronic, magnetic, optical, and chemical. While proteins are 10–1000 nm in size, and cell walls are 1–100 nm thick, their behavior on encountering a nanomaterial may be quite different from that seen in relation to larger-scale materials. Nanocapsules and nanodevices may present new possibilities for drug delivery, gene therapy, and medical diagnostics.

Surfaces and interfaces are also important in explaining nanomaterial behavior. In bulk materials, only a relatively small percentage of atoms are at or near a surface or interface (such as a crystal grain boundary). In nanomaterials, the small feature size ensures that many atoms, perhaps half or more in some cases, are near interfaces. Surface properties such as energy levels, electronic structure, and reactivity can be completely different from interior states, and give rise to very different material properties.

There are many types of intentionally produced nanomaterials, and a variety of others are expected to appear in the future. For the purpose of this chapter, most current nanomaterials are organized into four types: (i) carbon-based materials, (ii) metal-based materials, (iii) dendrimers, and (iv) composites.

19.1.2.1 Carbon-Based Nanomaterials Carbon nanotubes (CNTs) are molecular-scale tubes of graphitic carbon with outstanding properties. They are among the stiffest and strongest fibers known, and have remarkable electronic properties and many other unique characteristics. For these reasons, they have attracted huge academic and industrial interest; thousands of papers on NTs are being published every year. The current growing interest in CNTs is a direct consequence of the synthesis of buckminsterfullerene, C_{60}, and other fullerenes, in 1985 [6]. The discovery that carbon could form stable, ordered structures other than graphite and diamond has stimulated researchers worldwide to search for other new forms of carbon. The search received a new impetus when it was shown in 1990 that C_{60} could be produced in a simple arc-evaporation apparatus readily available in most laboratories. It was using such an evaporator that the Japanese scientist Iijima discovered fullerene-related CNTs in 1991 [7].

CNTs can be described as a graphite sheet rolled up into a nanoscale tube; these CNTs have diameters in the range between fractions of nanometers and tens of nanometers and lengths up to several centimeters with or without their ends capped by fullerene-like structures. The multi-walled CNTs contained at least two layers, often many more, and ranged in outer diameter from about 3 to 30 nm. They were invariably closed at both ends. A transmission electron micrograph of some multiwalled CNTs is shown in Figure 19.3a. In 1993, a new class of CNTs was discovered with just a single layer. These single-walled CNTs are generally narrower than the multiwalled CNTs, with diameters typically in the range 1–2 nm, and tend to be curved rather than straight. Figure 19.3b shows some typical single-walled tubes. It was soon established that these new fibers had a range of exceptional properties, and this sparked an explosion of research into CNTs. However, it is important to note that nanoscale tubes of carbon, produced catalytically, had been known for many years before Iijima's discovery. The main reason why these early tubes had not excited wide interest was the fact that they were structurally rather imperfect, so they did not have particularly interesting properties. Recent research has focused on improving the quality of catalytically produced NTs. CNTs have been extensively used in many analytical applications. Although the use of nanodiamonds, nano-onions, peapods, nanofibers, nanorings, and nanotubules has been reported, most applications have focused on fullerenes and CNTs [8].

19.1.2.2 Metal-Based Nanomaterials Metal-based nanomaterials include quantum dots (QDs), gold (AuNPs), silver (AgNPs), and metal oxides, such as titanium dioxide (TiO_2 NPs), and aluminum trioxide (Al_2O_3NPs). Owing to their fascinating optical and electronic properties, semiconductor nanocrystals have attracted considerable attention in recent years [9]. The three-dimensional quantum confinement in QDs induces a discretization of electronic states, very much like in atoms or molecules. The resulting optical properties depend on the crystal size and can therefore be tuned by varying the dimensions of the nanocrystals. QDs are considered promising candidates for optoelectronic

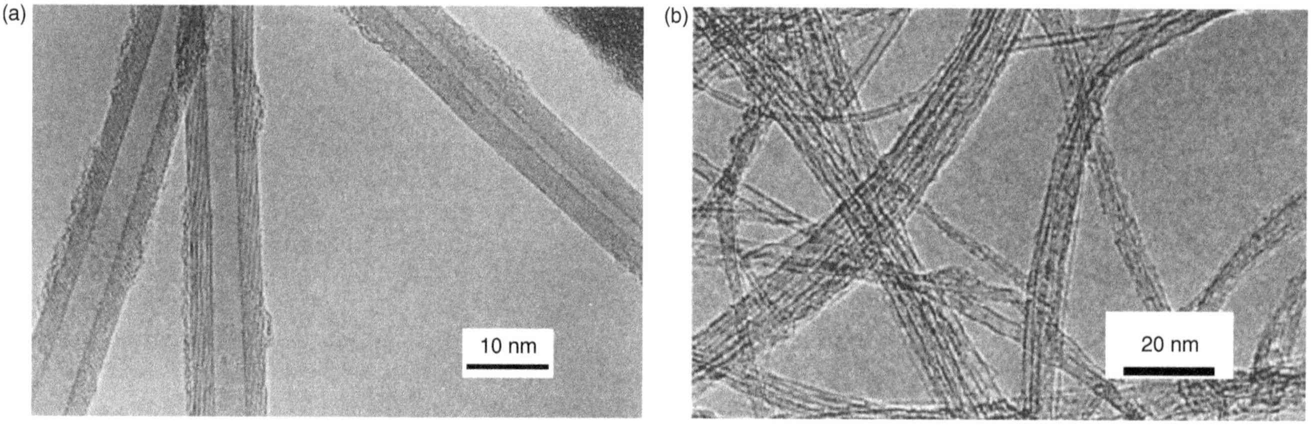

FIGURE 19.3 Transmission electron micrograph and scheme of multiwalled CNTs (a). Transmission electron micrograph and scheme of single-walled CNTs (b). Reprinted with permission from www.personal.reading.ac.uk/~scsharip/tubes.htm.

applications, including light-emitting diodes [10] and QD lasers [11]. The synthesis of QDs can be achieved by growth in colloidal solutions [12,13], on surfaces in vacuum deposition processes [14–16], by electron beam lithography [17], and by sol–gel techniques [18,19]. Further, AuNPs exhibit several interesting physical and chemical properties, making them an integral part of research in nanoscience. AuNP use has found the greatest application in the improvement of detection systems. Owing to their high surface-to-volume ratio and high surface energy, nanomaterials provide stable immobilization of a large number of biomolecules which, even after the immobilization, retain their bioactivity. Typically, colloidal solutions of spherical AuNPs are red with the surface plasmon resonance band centered at ~ 520 nm. This band is weakly dependent on the size of the particle and the refractive index of the surrounding media [20–22], but strongly affected by shape [20,22,23] and inter-particle distance [24]. The morphology and surface chemistry of NPs are usually controlled by using appropriate synthesis methods. Owing to its simplicity and high yield, the citrate reduction method of Turkevich et al. [25] is the most commonly used method for the preparation of spherical AuNPs in the range 9–120 nm [26]. The simultaneous use of AuNPs absorbing at different wavelengths might allow multiplexed analysis.

As a typical noble metal nanomaterial, AgNPs are easy to synthesize, and for this reason, they have attracted considerable attention because of their quantum characteristics, such as small granule diameter, large specific surface area, and high catalytic activity; in addition, they also show excellent biocompatibility and conductivity [27]. Nowadays, AgNPs are added to many common household products [28,29]. In addition, the medical use [30–32] of these nanomaterials as broad-spectrum antimicrobial agents against more than 650 different types of disease-causing organisms, including viruses [33–36], is widely known. There are many processes for synthesizing AgNPs, but chemical reduction is the most frequently applied method for the preparation of stable colloidal dispersions of AgNPs in water or organic solvents [37–42]. The unprecedented freedom to design and modify NPs to accomplish very specific tasks is currently being realized. Nanometer-sized metal particles or metal nanoparticles (MNPs) are known to have unique physical and chemical properties (e.g., electrical, magnetic, optical, and ionization potentials), which are distinct from those of both bulk metals and isolated atoms and molecules. TiO_2 is one of the most recent nanomaterials that has attracted great attention because of its unique properties. Its powder possesses interesting optical, dielectric, and catalytic properties, which result in industrial applications such as pigments, fillers, catalyst supports, and photocatalysts [43–48]. The most common synthesis procedures were based on the hydrolysis of acidic solutions of Ti(IV) salts [49]. Alternatively, gas-phase oxidation reactions of $TiCl_4$ and hydrolysis reactions of titanium alkoxides were used to generate finely separated, high-purity TiO_2 powders. Al_2O_3 has been considered as one of the most promising advanced materials for a variety of applications because of its distinctive chemical, mechanical, and thermal properties. The high surface-to-volume ratio in NPs can lead to modifications in structure and properties. There is great interest in the study of the surfaces of Al_2O_3 NPs as their catalytic activity is closely related to the structure of exposed crystalline faces [50]. Silicon dioxide (silica) NPs have great practical importance in industrial applications, such as the fabrication of electric and thermal insulators [51]. They are also widely used in biomedical applications [52] and gene delivery [53]. Colloidal silica crystals with periodicity within the optical wavelength range also have a photonic band gap, which makes them well suited for electronic applications ranging from microwave to optical devices [54].

In recent years, with the rapid development of nanostructured materials, MNPs have attracted considerable attention for their novel properties [55,56]. MNPs are a class of NPs (i.e., engineered particulate materials of size $<100\,nm$) that can be manipulated under the influence of an external magnetic field. MNPs are commonly composed of magnetic elements, such as iron, nickel, cobalt, and their oxides. To date, the application of MNPs has extended to numerous areas, including magnetic recording, magnetic fluids [57], immobilization of proteins [58], detection of DNA hybridization [59], and drug delivery [56]. Recently, MNPs have also been studied by electrochemistry [60]. The excellent physical properties of the NPs, such as high surface-to-volume ratio, unique absorption spectrum, and electric conductivity, make them a highly beneficial and suitable matrix for applications in immobilization. The use of MNPs as versatile and efficient substrates for the immobilization of antibodies or antigens not only shows enhancements of the amount of antibodies or antigens immobilized on its surface, but also preserves the activity of the immobilized biomolecules [61,62].

19.1.2.3 Dendrimers The dendritic structure is a widespread motif in Nature, often utilized where a particular function needs to be exposed or enhanced. Dendrimers are symmetrical, highly branched polymers with a compact spherical structure [diameter ranging from 1.1 nm for the first generation (G1) to 9 nm for G8] [63] and unique behavior [64,65]. The first successful attempt to create and design dendritic structures by organic synthesis was carried out by Vögtle and co-workers in 1978 [66]. However, after this first report, several years passed before the field was taken up by Tomalia's group [67]. They developed a new class of amide-containing cascade polymers, which brought these hitherto fairly small molecular motifs into well-defined macromolecular dendritic structures. Moreover, Newkome et al. [68] discussed the huge opportunity for nanotechnological research. Dendrimers are monodispersed if prepared correctly. They are routinely synthesized from a central polyfunctional core by repeated addition of monomers. The core is characterized by a number of functional groups. Dendrimer generation is created by adding monomers to each functional group, in turn leaving end groups that are able to react again [69,70]. The structure of the polymer is determined by the number of reactive groups of the core, the length of the branch, and the dimensions of the surface group [71,72]. The maximum size is limited to the generation at which the dendrimer becomes tightly packed, looking like a ball (see Figure 19.1). Their physicochemical properties are intrinsically dependent on generation number and surface functionalities [63]. Dendrimer syntheses involve divergent or convergent hierarchical assembly strategies [73,74].

19.1.2.4 Composites The increased interest in multifunctional materials and structures is driven by the need for the development of new materials and structures that simultaneously perform:

(i) multiple structural functions, which would require a composite structure that has high strength, high stiffness, high fracture toughness, and high damping;
(ii) combined nonstructural and structural functions, which would require a load-bearing structure that has the capability of providing its own noise and vibration control, self-repair, thermal insulation, and energy harvesting/storage; or
(iii) both, which would require a structure combining the functions of both types (i) and (ii).

Most of the recent developments in multifunctional materials and structures tend to be of type (ii). Multifunctional materials are necessarily composite materials, and the strong growth in the use of composites has been greatly influenced by multifunctional design requirements. The traditional approach to the development of structures is to address the load-carrying function and other functional requirements separately, resulting in a suboptimal load-bearing structure with add-on attachments that perform the nonstructural functions with the penalty of added weight. Recently, however, there has been increased interest in the development of load-bearing materials and structures that have integral non-load-bearing functions, guided by recent discoveries about how multifunctional biological systems work.

Metal Matrix Composites (MMCs). Metal matrix composites (MMCs) are composed of a metallic matrix (aluminum, magnesium, iron, cobalt, and copper) and a dispersed ceramic (oxides and carbides) or metallic (lead, tungsten, and molybdenum) phase. These composites are produced by combining two pre-existing constituents (a metal and a ceramic fiber). This commonly used process includes powder metallurgy, diffusion bonding, liquid-phase sintering, squeeze infiltration, and stir-casting. Alternatively, the typically high reactivity of metals at processing temperatures can be exploited to form the reinforcement and/or the matrix *in situ*, that is, by chemical reaction within a precursor of the composite.

The performance of these materials and their physical and mechanical characteristics depend on the nature of two components: the volume fraction of the adopted reinforcement and the production technology. The MMCs utilize the properties of the matrix (light weight and good thermal conductivity and ductility) and of the reinforcement, usually ceramic (high stiffness, high wear resistance, and low coefficient of thermal expansion). The

MMCs are classified according to the type of reinforcement used and their geometric characteristics. The main classification groups of composites are continuous reinforcement composites (constituted by continuous fibers or filaments) and discontinuous reinforced composites (containing short fibers, whiskers, or particles).

Ceramic Matrix Composites (CMCs). Ceramic matrix composites (CMCs) are composed of a ceramic matrix and embedded fibers of another ceramic material (dispersed phase). There is a wide range of CMCs depending on the chemical composition of the matrix and the reinforcement. The reinforcement of ceramic materials consists of polycrystalline or amorphous inorganic fibers or carbon fibers. The term "ceramic fibers" covers all nonmetallic inorganic fibers (oxide or nonoxide), with the exception of fibers manufactured via solidification of glass melts. Nonoxide CMCs have been the most studied composites. Between them, carbon and silicon carbide fibers are the highest-performance fibers in terms of stiffness, failure strength, refractoriness, and density. Moreover, nonoxide fibers avoid compatibility problems, which are crucial at high temperatures. Carbon fibers are extensively used in the production of polymer–matrix composites because they are cheaper than other fibers (except glass fibers). The oxide CMCs have long-term applications in oxidizing atmospheres. These oxide fibers, although refractory, tend to undergo grain growth at high temperatures (which results in degradation of fiber strength) and exhibit poor creep resistance. They display much higher densities than, say, carbon fibers.

The traditional processing techniques of CMCs are hot pressing and high temperature. Recently, ceramic and CMC manufacturing from pyrolytic conversion of preceramic polymers has attracted considerable attention due to their unique combination of low-temperature processing, applicability of versatile plastic shaping technologies, and microstructural control capabilities.

Polymer Matrix Composites (PMCs). Polymer matrix composites (PMCs) are composed of a matrix from thermosetting plastics [unsaturated polyester (UP), epoxy (EP)] or thermoplastic (polycarbonate (PC), poly(vinyl chloride), nylon, polystyrene). Embedded glass, carbon, steel, or Kevlar fibers (dispersed phase) are used as reinforcement. Thermoplastic polymers are distinguished by their ability to be reshaped on the addition of heat (above the glass transition temperature of the amorphous phase or the melting temperature of the crystalline phase). This cycle can be carried out repeatedly. Thermosetting polymers, on the other hand, undergo chemical reactions during curing, which cross-linked the polymer molecules. Once cross-linked, thermosets become permanently hard and simply undergo chemical decomposition under the application of excessive heat. Thermosetting polymers typically have greater abrasion resistance and dimensional stability than thermoplastic polymers, which typically have better flexural and impact properties. The traditional processing techniques of PMCs are solution impregnation, powder impregnation/pultrusion, and film stacking.

19.2 NANOMATERIALS IN ANALYTICAL CHEMISTRY

The aim of analytical nanotechnology is to use the excellent properties of NPs to improve well-established analytical methods or to develop others for new analytes or matrices. In addition to the typical advantages of NPs, their use should lead to improved selectivity, sensitivity, rapidity, miniaturizability, or portability of the analytical systems. As nanotechnology progresses, new nanomaterials with properties that can be unequivocally ascribed to their nanometric dimensions are bound to emerge as powerful analytical chemical tools. These materials are employed in, for example, sensing, separations, and extraction applications. These materials have attracted much attention because of their potential for interesting applications, broad applicability, and tunable properties according to application needs. These interesting materials include hybrids, nanomaterials, and biomolecular materials. Moreover, these materials offer unique properties (thermal, mechanical, electronic, biological, etc.) [75–79] that are not found in conventional materials. The combination of the remarkable recognition capabilities of biomolecules with the unique properties of nanomaterials have resulted in systems with significantly improved performance [80] and novel applications across physics, chemistry, biology, engineering, and computer science. Apart from mechanical strength and light weight, most of the extraordinary biological properties of nanomaterials are linked to their unique surface properties (surface area, surface roughness, energetics, and altered electron distributions) [81], which allow improved interactions between material surfaces and many biological entities. Such interactions depend not only on the methods of fabrication, but also on their size and specific geometry [82]. As expected, these factors together with the ability to form hydrogen bonds, π–π stacking, dispersion forces, dative bonds, and hydrophobic effects influence the stability and selectivity of the nanomaterials [83]. Consequently, different nanomaterials have been used to develop unique applications in sensing [84–89], separation [90–93], and extraction [94–97].

Considering analytical chemistry, nanomaterials have played an important role in improving the fields of sensing and separation. There are clear differences in the criteria employed for selecting appropriate materials in sensing and separation applications. For example, although it is desirable to employ materials with analyte selectivity in sensing applications, it is, however, an essential criterion in separation applications. Sensing applications require materials with transducer properties, whereas separation applications require materials with the ability to influence mass transfer behaviors and/or phase partitioning of analytes. Sensing applications continually seek appropriate materials to achieve highly sensitive and predictable responses over wide analyte concentration ranges, whereas materials with the ability to achieve near complete resolution of analyte components within the shortest possible time are ideal for separation applications. As regards analytical tools for analyte separation, NPs (particularly CNTs) have been used to boost the resolution in chromatographic and electrochromatographic applications [98].

NPs for gas chromatography are usually packed into a tubular column. These properties can be used to prepare highly selective, thermally stable chromatographic sorbent columns. As an alternative to packing, CNTs can be self-assembled into a gas chromatographic column [99]. NPs have also been used in liquid chromatography (LC) [i.e., by inserting them into a monolithic column of vinylbenzyl chloride–ethylene dimethacrylate (VBC–EDMA) polymer [100]]. Recently, they have been used in CE as a pseudostationary phase (PSP) in electrokinetic chromatography (EKC); because it does not interfere with the ionization process, this technique is more compatible with mass spectrometry (MS) detection than classical micellar electrokinetic chromatography (MEKC).

The modification of the surface of some of the NPs with functional groups and/or molecules is a key during the eventual procedure to prevent aggregation and to control the particle size of the NPs. Such modification also provides extra selectivity when the NPs are used as separation media. NPs are mostly stabilized by steric exclusion and electrostatic repulsion. On the other hand, it is interesting that particle aggregation can be utilized in the sensitive detection of particular molecules that can induce NP aggregation, indicated by solution color change and monitored by electron microscopy [101].

NPs also play an important role in electrochemical and optical detection. CNTs and AuNPs are the types of NPs most widely used for developing electrochemical (bio) sensors, as are AgNPs and QDs for optical systems. By virtue of their excellent electrochemical properties, NPs are often added as mediators to composites in order to facilitate electron-transfer reactions with electroactive species in solution [102]. The NPs minimize fouling of the electrode surface, enhance electrocatalytic activity, and simplify the immobilization process of biomolecules such as enzymes or antibodies on the electrode surface [103–105]. In addition to classical optical sensors for surface recognition between analytes and NPs, salient applications in this field include those of QDs for the *in vivo* imaging of, for example, tumoral cells [106]. Specifically, silica NPs mostly applied to large-scale electronic devices have found a great potential for use in nanoscale devices with the possibility of creating versatile chip-based biosensors [107]. Silicon-based arrays made of antibody-conjugated nanowires coupled with transistors have also been multiplexed to detect simultaneously single copies of multiple analytes [108].

Another field in analytical chemistry that was deeply modified by the development of nanotechnology was the purification and preconcentration of analytes from the sample matrix. Nanomolecular imprinted polymers (nanoMIPs) [109] and CNTs [110,111] have been widely used as sorbents for solid-phase extraction and membrane filtration [112,113].

19.3 NANOPARTICLES IN CAPILLARY ELECTROPHORESIS

Extensive efforts devoted to miniaturization of analytical instruments have significantly promoted the development of this field; therefore, it could be said that separation science has been progressing with size. The goal in CE was to find a strategy to increase the surface area and/or the interaction during the running using the nanomaterials as either a stationary phase or PSP, in order to enhance sample capacity. The application of nanosized novel materials as separation media has offered more and more opportunities to solve separation puzzles. The potential of nanostructured materials in chromatographic and electrophoretic techniques has gradually been discovered in recent years.

Basic requirements for NPs suitable for use in CE are as follows [114]: (1) form stable suspensions in several types of electrolytes; (2) provide selectivity in the interactions with analytes; (3) be charged, in order not to co-elute with electroosmotic flow (EOF), thereby allowing separations of neutral analytes; (4) show equal mobility, in order not to contribute to band broadening; (5) show minimal mass-transfer resistance; (6) undisturbed detection; and (7) be small and porous, to provide a large surface area. It is important to keep these rules of thumb in mind in order to be able to achieve high separation efficiencies.

Among the existing NPs, the most often used in CE [115,116] are fullerenes, CNTs, silica, precious metals, and metal oxides (e.g., polymers, molecular micelles, molecularly imprinted polymers). CE has several modifications that result in the different modes of this methodology. Nanomaterials have found application in capillary electrochromatography (CEC) and EKC. This last technique can be developed in two different ways, using the NPs in partial filling or continuous filling form. The variation between CEC and EKC is the role that the nanomaterials play during the separation. Probably due to the cost, the most commonly used nanomaterials in CE as both PSPs and stationary phases have been the carbon-based NPs.

19.3.1 Nanoparticles in Capillary Electrochromatography

CEC is a separation technique that combines the high efficiency of CE with the selectivity offered by LC. It offers rapid separations, high efficiency, high selectivity, and low consumption of samples and reagents. The observed mobility of analytes in CEC depends on the EOF, the electrophoretic mobility of the analytes, and the chromatographic retention due to partitioning of the analytes between the electrolyte and the stationary phase. Three different types of CEC using nanomaterials are generally distinguished based on the capillary column used: packed-column CEC [117,118], open-tubular CEC [119,120] and monolithic-column CEC [121–123]. In open-tubular CEC, the materials that constitute the stationary phase are attached to the inner wall of the capillary and the mobile phase is driven by the EOF. The inner diameter of the capillary should be small to achieve efficient diffusion of the analytes to the stationary phase. However, the use of too narrow capillaries is not convenient due to decreased sensitivity in UV absorbance detection, high flow resistance, and problems with blockage. In addition, the sample capacity is relatively poor in open-tubular CEC, although there are techniques to improve it by surface enlargement, that is, by etching or sol–gel chemistry [120]. This section mostly describes the use of NPs as coating materials in open-tubular CEC.

19.3.1.1 Organic Nanoparticles *Carbon-Based Nanoparticles.* CNTs possess a large surface area, high electrical conductivity, good chemical and thermal stability, and significant mechanical strength. CNTs are normally difficult to disperse in pure aqueous buffers and in several organic solvents because of their hydrophobicity. However, CNTs can be oxidized with a strongly acidic medium to produce CNTs with carboxylic groups on their surface. These carboxylic groups make possible the attachment of the CNTs to a large variety of surfaces. Among the groups who have worked with single-walled CNT coated capillaries, Luong et al. [124] separated seven aniline derivatives within 10 min. The phenomenon that allowed the separation was the ionic interactions between the analytes and single-walled CNTs. The silica surface was first treated with positively charged poly(diallydimethylammonium chloride) (PDDA) on which the negatively charged CNTs were adsorbed. Later, Sombra et al. [125] developed a simple procedure to attach the CNT to the wall of the capillary through a covalent interaction between the carboxylic group of the NTs and the activated silica of the capillary. This procedure allowed the separation of eight nonsteroidal anti-inflammatory drugs, β-lactams antibiotics, and chloramphenicol with high resolution, without band broadening, and without distortion of the baseline. Moreover, the same procedure was utilized by Stege et al. [126]. They implemented this mode of CE to determine melatonin in complex food matrices, such as plant and grape extracts and wine. The results indicated high electrochromatographic resolution, good capillary efficiencies, and improved sensitivity with respect to those obtained with conventional capillaries.

Carbon nanohorns (CNHs) were discovered by Iijima et al. [127]. Single-walled CNHs are a new class of material related to single-walled CNTs but with a unique structural feature composed by a typical diameter of 2 nm and a length between 30 and 50 nm. The end of these tubes is closed by a conical cap with an average diameter cone angle of 20°. Single-walled CNHs easily form stable aggregates that have been identified as dahlia flower-like structures [128] in which the tubule walls are attracted to each other by van der Waals forces. This CNH structure provides high porosity and large surface area, and therefore it presents high affinity for organic compounds showing a large adsorbing capacity [129]. Single-walled CNHs have been used as the stationary phase in CEC for the separation of five water-soluble vitamins [130]. The electrochromatographic features obtained when the nanohorns were immobilized on the capillary showed good separation efficiencies and higher retention factors than those obtained with a bare fused-silica capillary. Furthermore, the results were comparable to those obtained with single-walled CNTs immobilized on the capillary and showed significant improvements in resolution. A neutral monolithic stationary phase with single-walled CNTs was also obtained by *in situ* copolymerization of VBC and EDMA in the presence of 2-propanol with CNTs and formamide as the porogen (Figure 19.4). The authors separated a mixture of peptides that contained Gly–Tyr (G), Val–Tyr–Val (V), methionine enkephalin (M), leucine enkephalin (L), and angiotensin II (A) to evaluate the column. The mobile phase contained acetonitrile and phosphate buffer at pH 2.5. The conclusion was that not only the retention but also the separation efficiency were enhanced by incorporation of CNTs into the stationary phase [131].

Polymer Nanoparticles. Kleindienst et al. [132] described a coated capillary with a permanently adsorbed monolayer of chloromethylated polystyrene NPs (40–140 nm) for the separation of proteins and peptides using an open-tubular CEC with UV detection with high plate numbers (up to 1,900,000 m^{-1}). These NPs were coated with ethylenediamine and 2,3-epoxy-1-propanol to obtain a stable hydrophilic, positively charged surface. The same coating was used by Huber et al. [133], who applied electrospray ionization (ESI) MS detection, obtaining a high number of theoretical plates (above 1,000,000 m^{-1}). To fit this procedure to an "MS-friendly" system, they changed some

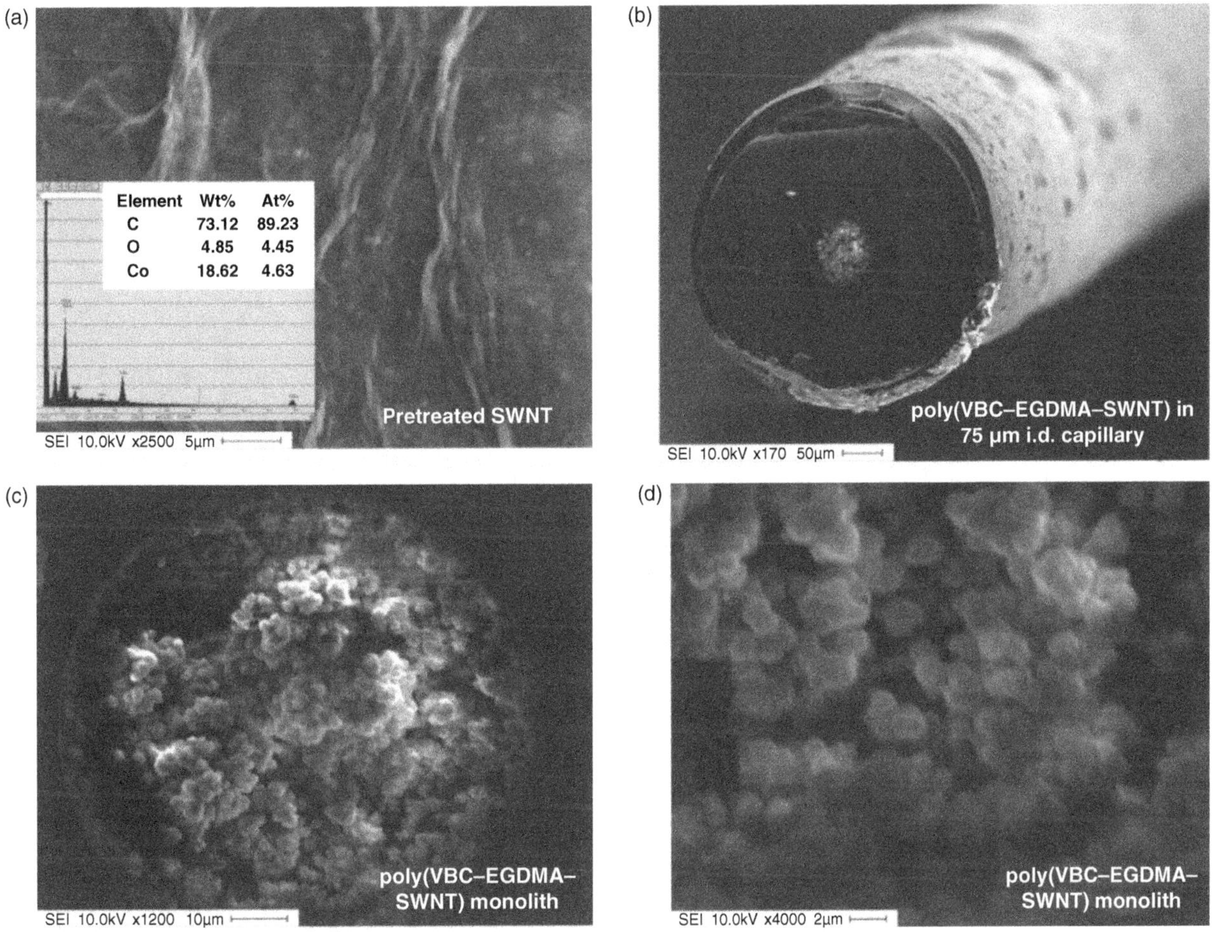

FIGURE 19.4 SEM images of H_2SO_4/H_2O_2-pretreated SWNT (a) and poly(VBC–EDMA–SWNT) monolith (b–d). Reprinted with permission from Ref. 131.

characteristics. The most important change was the background electrolyte (BGE) that, in this case, was ammonium formate, pH ~3, an acidic, volatile buffer. An ion-exchange stationary phase attached to the capillary wall using 70 nm quaternary ammonium-functionalized latex NPs used to preconcentrate two ions was proposed by Breadmore et al. [134]. The NPs were electrostatically adsorbed on the capillary wall. Moreover, they introduced a new electroseparation concept, namely using an isotachophoretic gradient. In this method, the sample was injected and then the capillary was filled with a BGE containing a weak ion-exchange (WE) competing anion of low mobility. A stronger anion-exchange (SE) competing anion with higher-mobility electrolyte was injected following the WE. On applying the separation voltage, some of the more mobile stronger competing anions from SE will migrate ahead of the less mobile weaker competing anions from WE, establishing a gradient boundary according to isotachophoretic principles. First, a low-mobility, WE competing anion (F^-) was used for preconcentration. Second, a high-mobility, strong ion-exchange competing ion (ClO_4^-) was used for elution of the analytes. In a deeper analysis, Breadmore et al. [135] studied the characteristics of the isotachophoretic gradient for elution after preconcentration. The maximum focusing efficiency and the strongest electrolyte should have the highest possible eluotropic strength for a given weak electrolyte concentration. The most suitable BGE was 1,5-naphthalene disulfonate (NDS) for the 77 anions studied as an SE. This system was later optimized by Breadmore et al. [136] for the preconcentration/separation of UV-absorbing anions using NDS as the SE anion. The approach was used to determine nitrate in Antarctic ice cores with detection limits in the range 2.2–11.6 ppb with a total analysis time of 25 min. The same group published three more studies using aminated latex NPs (Figure 19.5) as an online preconcentrator [137–139] and they concluded that the main contribution to band broadening came from mass transfer in the mobile phase, which could be improved by decreasing the capillary inner diameter.

Nanomolecular Imprinted Polymers. Achiral nanoMIPs were developed using (*S*)-ibuprofen as the template molecule,

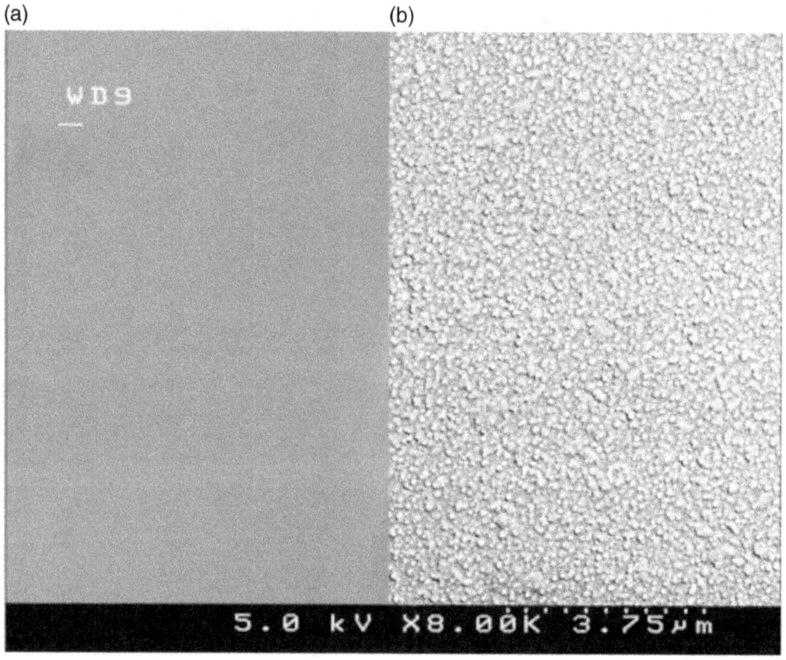

FIGURE 19.5 SEM images of inner surface of (a) bare fused-silica capillary and (b) AS5A latex-coated capillary. Reprinted with permission from Ref. 138.

γ-methacryloxypropyltrimethoxysilane (γ-MPTMS) as the ionizable precursor, 4-vinylpyridine (4-VPY) as the functional monomer, and ethylene glycol dimethacrylate as the cross-linker (Figure 19.6) [140]. By incorporating γ-MPTMS into the prepolymerization mixture, the charged groups were successfully coupled to the surface of 4-VPY nanoMIPSs to generate a stable EOF for CEC separation. The nanoMIPs made without γ-MPTMS showed a very weak EOF that could not be determined, and the nanoMIPs made with 4-VPY yielded better separation than those made with methacrylic acid. The resulting nanoMIP stationary phases showed a good chiral discrimination toward ibuprofen in aqueous and nonaqueous media. The separation was governed by both hydrophobic and hydrogen bonding interactions. In aqueous media, hydrophobic interactions played an important role in the retention and enantioselectivity, whereas in nonaqueous media, polar interactions (hydrogen bonding, ionic interactions, etc.) and also the mobile phase elution ability were mainly responsible for binding and recognition.

The synthesis of nanoMIPs using microwave irradiation can provide a viable alternative to the traditional means such as UV irradiation and thermal initiation for the preparation of monolithic capillary columns. Polystyrene-based monolithic stationary phases were prepared *in situ* in fused-silica capillaries and simultaneously in vials, and then they were evaluated using pressure-assisted capillary electrochromatography (pCEC), CEC, and low-pressure liquid chromatography (LPLC) (Figure 19.7). Under microwave irradiation,

FIGURE 19.6 Scanning electron micrographs of the ends of the nanoMIPs3, nanoMIP4, and nanoMIPs. Reprinted with permission from Ref. 140.

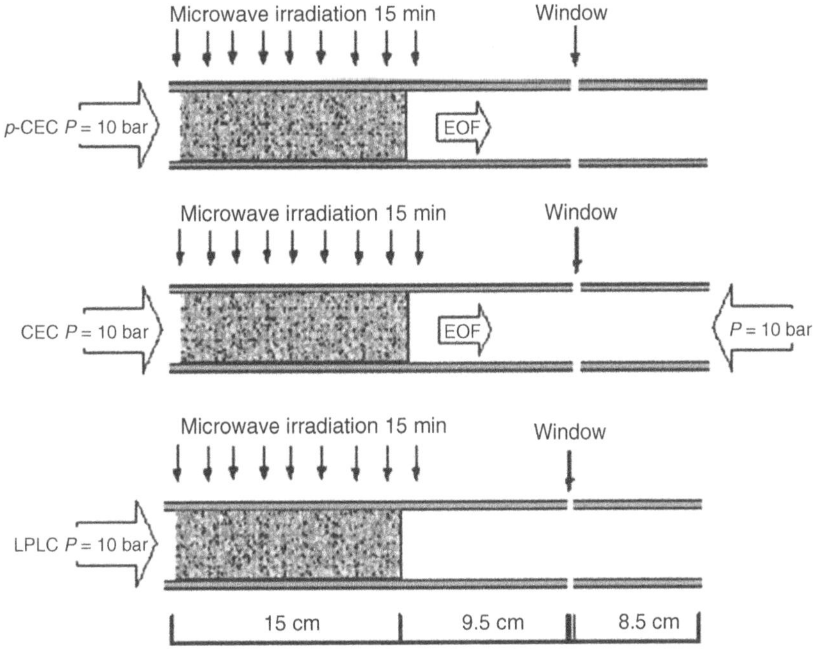

FIGURE 19.7 Schematic diagram of pCEC, CEC, and LPLC operation modes. Reprinted with permission from Ref. 141.

the polymerization time was reduced from several hours to 5 min in spite of an identical recipe, including template, monomers, and porogen. The molecular recognition of 4-HBA was achieved, and the column efficiency of the template was 35,000 plates m^{-1} on the resulting monolithic nanoMIPs columns [141].

A method for the preparation of nanomolecularly imprinted polymers (nanoMIPs) derivatized on to the surface of a monolithic silica capillary column was successfully developed. The vinyl groups were first introduced on to the silica monolith by immobilization of γ-methacryloxypropyltrimethoxysilane. Then, the nanoMIP coating was copolymerized and anchored on the surface of the silica monolith. Acetonitrile was selected as the porogen (solvent). The obtained nanoMIP-derivatized silica monolith using L-tetrahydropalmatine (l-THP) and (5*S*,11*S*)-(−)-Tröger's base (S-TB) as the imprinted template was characterized in terms of the retention behavior of thiourea and toluene. Under the optimized CEC conditions, baseline enantioseparations of THP and TB were achieved in 4 min. This result indicates that there were sufficient recognition sites on the surface of the silica monolith, resulting in strong recognition ability. Compared with a nanoMIP organic monolith, the nanoMIP-derivatized silica monolith exhibits better column efficiency and stability in CEC [142]. A room-temperature ionic liquid (RTIL)-mediated nonhydrolytic sol–gel (NHSG) protocol was explored for the fabrication of new nanoMIP silica-based hybrid monoliths for the chiral separation of a basic template, zolmitriptan, by CEC. The RTIL-mediated NHSG protocol involved free-radical copolymerization and the NHSG process. Three carboxylic acids (trifluoromethylacrylic acid, cinnamic acid, and MAA) were examined as both functional monomers and catalysts for the NHSG condensation of methacryloxypropyltrimethoxysilane (MPTMS) to form a silica-based framework. The RTIL was incorporated to reduce gel shrinkage and also to act as the pore template. The synergism of the free radical copolymerization of the C=C bond of carboxylic acids and MPTMS with the NHSG condensation of MPTMS catalyzed by carboxylic acids was demonstrated. The incorporation of RTIL increased the porosity, and hence improved the selectivity of the prepared hybrid monoliths. The suitability of the combination of nanoMIPs and CEC for use as a powerful tool in environmental or food analysis was studied for the first time and successfully demonstrated [143]. An MIP monolith column was synthesized and evaluated as stationary phase for the selective determination of the fungicide thiabendazole (TBZ) in citrus samples by nonaqueous CEC. The influence of the mobile phase composition, the voltage of the power supply, and the separation temperature on the recognition of TBZ by MIPs was evaluated, and the imprint effect in the column was clearly demonstrated. The high selectivity achieved by the developed nanoMIP–CEC procedure allowed the unambiguous detection and quantification of TBZ in citrus samples by direct injection of the crude sample extracts, without any previous clean-up, in less than 6 min [144].

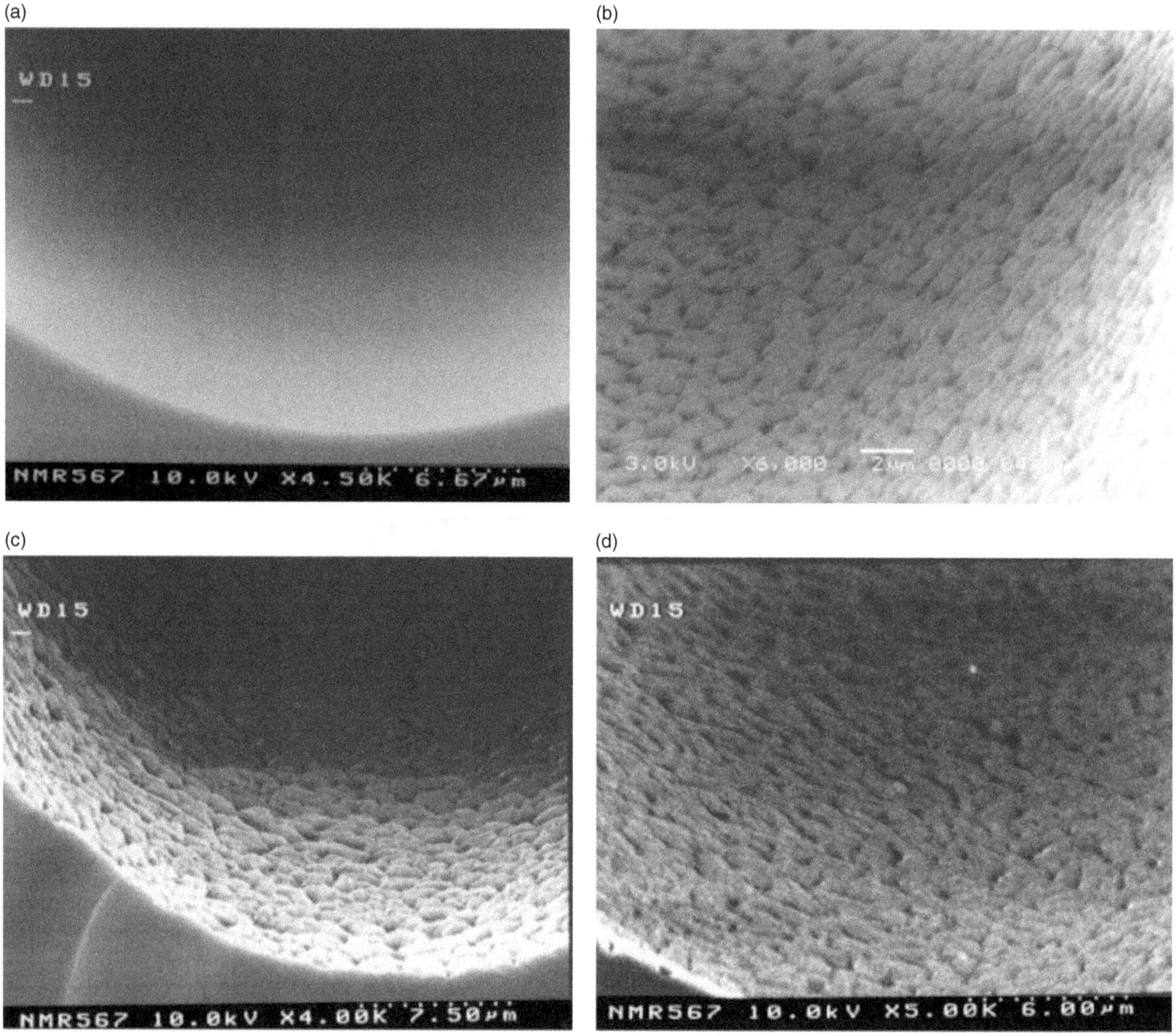

FIGURE 19.8 SEM images of (a) bare capillary, (b, c) etched bare capillary, and (d) etched Au nanoparticle capillary. Reprinted with permission from Ref. 147.

19.3.1.2 Inorganic Particles

Gold Nanoparticles. The first article describing the application of AuNPs for CEC separation was reported by Neiman et al. [145]. They used modified AuNPs to improve selectivity and control the EOF into capillaries to separate aromatic acids and bases. Two different Au-based NPs were added to the run buffer. In one case, citrate-stabilized AuNPs (18 nm) were used in conjunction with capillaries treated with poly(diallyldimethylammonium chloride) (PDDA). The positively charged PDDA layer on the capillary walls adsorbed the negatively charged AuNPs. In another case, thiol-stabilized AuNPs (5 nm) were used without PDDA in the capillary. Additionally, O'Mahony et al. [146] developed an open-tubular column for CEC by immobilizing dodecanethiol AuNPs on previously derivatized fused-silica capillaries with 3-aminopropyltrimethoxysilane (APTMS) or 3-mercaptopropyltrimethoxysilane (MPTMS) (Figure 19.8). The fused-silica capillaries were etched with ammonium hydrogen difluoride, followed by prederivatization of the new surface with MPTMS for the immobilization of dodecanethiol AuNPs (2–3 nm). The electrochromatography of a "reversed phase" was used to test a mixture (thiourea, naphthalene, and biphenyl) and four polycyclic aromatic hydrocarbons (PAHs) with high efficiency (up to 200,000 plates m^{-1}). The electroosmotic flow

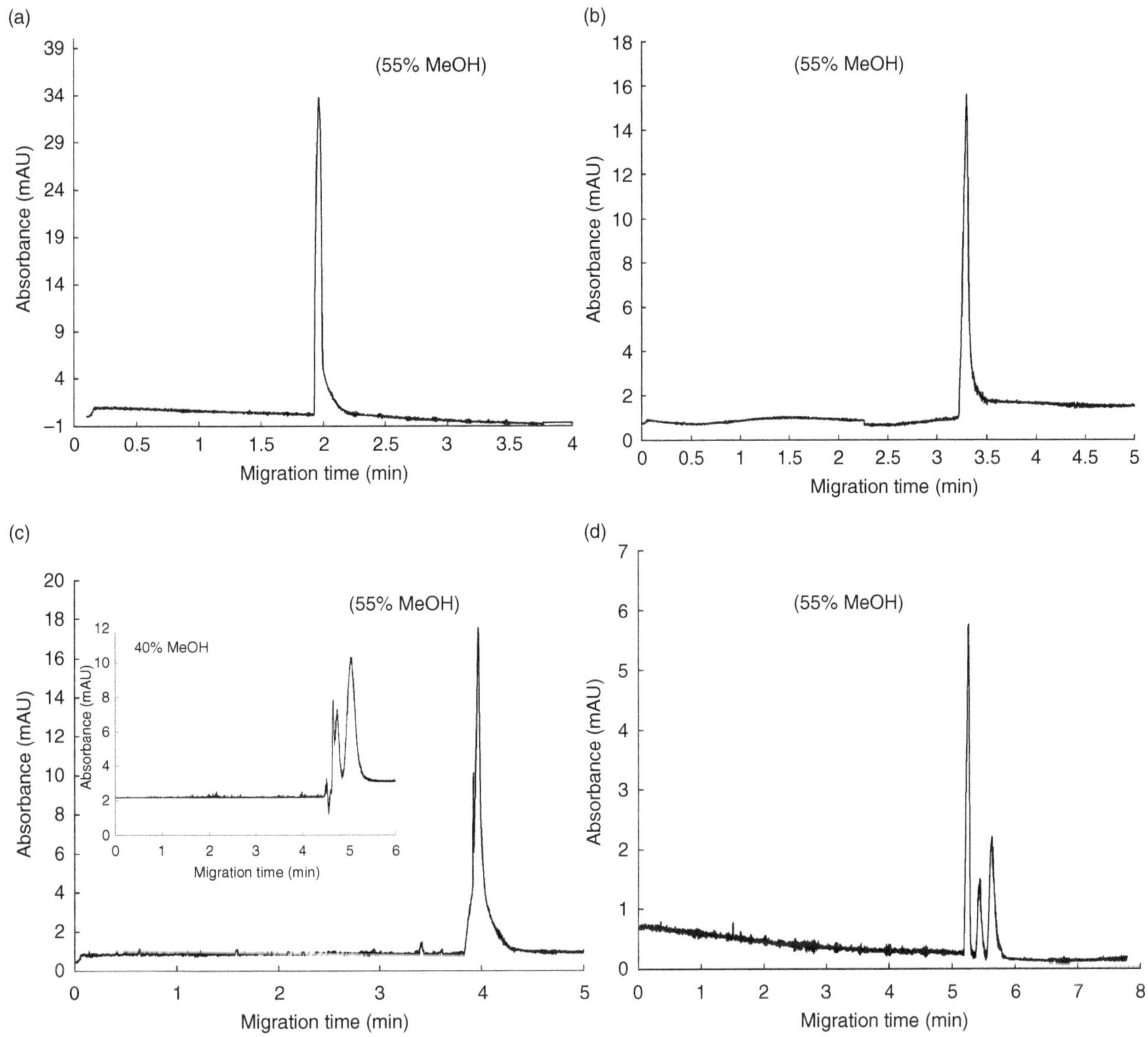

FIGURE 19.9 Electrochromatograms of the test mixture on (a) bare capillary, (b) sol–gel capillary, (c) alkanethiol Au nanoparticle capillary, and (d) sol–gel based alkanethiol Au nanoparticle capillary. Reprinted with permission from Ref. 148.

characteristics of the etched AuNP capillary, unetched AuNP capillary, bare capillary, and etched bare capillary were studied by varying the percentage of organic modifier in the buffer, buffer pH, and separation voltage. An etching process was used to increase the surface area of the capillary inner wall by a factor of up to 1000. This methodology produces high separation efficiencies, although the etching process is time consuming and complicated [147].

Yang et al. [148] also developed an open-tubular CEC column that was prepared by immobilizing dodecanethiol AuNPs on prederivatized fused-silica capillary columns with sol–gel technology. MPTMS was selected as a sol–gel precursor to develop a sol–gel layer on the inner wall of the capillary, prior to assembly of dodecanethiol AuNPs on the generated sol–gel layer through specific interaction between the AuNPs and surface terminating thiol groups. This open-tubular CEC method was tested for the separation of thiourea, naphthalene, and biphenyl and three PAHs, and also three selected drug substances (propiophenone, benzoin, and warfarin). Efficient separations were obtained with theoretical plate numbers of 168,000 plates m^{-1} for naphthalene and 51,000 plates m^{-1} for biphenyl (Figure 19.9). The results confirmed that dodecanethiol AuNPs, bound on the sol–gel-based inner layer of the fused-silica capillary can provide sufficient solute-bonded phase interactions for open-tubular CEC with reproducible retention and characteristic reversed-phase behavior.

Alkanethiol self-assembly and dithiol layer-by-layer (LBL) self-assembly processes to prepare an AuNP (~18 nm)-coated open-tubular CEC column for the separation of three neutral steroid drugs (testosterone, progesterone, and testosterone propionate) were performed by Liu et al. [149]. The capillary was prederivatized with 3-aminopropyltrimethoxysilane and the AuNPs were functionalized through self-assembly of alkanethiols. Alkanedithiols were used as a cross-linker to form multiple layers of AuNPs by an LBL method. Multiple layers of AuNPs were required to achieve sufficient separation resolution, as it was not possible to separate the steroids using a monolayer of AuNPs. In addition, the separation performance did not deteriorate over 100 injections, which indicated that the capillary columns coated with alkanethiols self-assembled on multilayer AuNP films were fairly stable [150]. Newly developed porous polymer monolithic capillary columns, which were modified with AuNPs coated with exchangeable functionalities, allowed easy switching of separation modes by a simple ligand-exchange process [151]. These columns were prepared from a poly(glycidyl methacrylate-*co*-ethylene dimethacrylate) monolith through reaction of its epoxide moieties with cysteamine to afford a monolith rich in surface thiol groups. Functionalization of the surface of the bound AuNPs was then carried out using low molecular weight thiol-containing surface ligands (Figure 19.10). It was clear

(a)

(b)

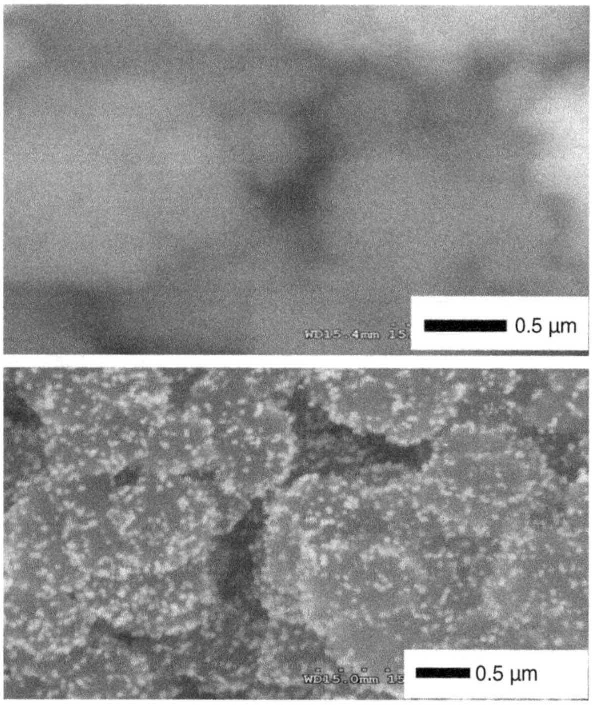

FIGURE 19.10 SEM images of the internal structure of original poly(glycidyl methacrylate-*co*-ethylene dimethacrylate) monolith (a) and its counterpart modified with AuNPs prepared *in situ* (b). The white spots in (b) are the AuNPs [151].

that changing the chemistry of the surface had a drastic effect on the separation process with the fastest separation (<2 min) being achieved using the column with carboxylic groups that contributed to generating a strong EOF, thus enhancing the native electrophoretic migration of peptides. In contrast, although the column with hydroxyl functionalities also separated the peptides, the speed was lower because hydroxyl groups do not afford any appreciable EOF and the driving force was just electromigration. The amine-functionalized column produced a reversal in the elution order with higher retention of the peptides. In addition to the chemical differences of the column, this change in the separation profile appears also to be affected by reversing the polarity of the driving voltage. This novel approach extends the application range of monoliths, since a single column can be used in different separation modes such as CE and nano-LC separations of peptides and proteins.

Field-amplified sample stacking using a fused-silica capillary coated with AuNPs embedded in PDDA was investigated for the electrophoretic separation of indoxyl sulfate, homovanillic acid (HVA), and vanillylmandelic acid (VMA). AuNPs (27 nm) exhibit ionic and hydrophobic interactions and also hydrogen bonding with the PDDA network to form a stable layer on the internal wall of the capillary. This approach reversed the EOF, allowing fast migration of the analytes while retarding other endogenous compounds including ascorbic acid, uric acid, catecholamines, and indoleamines. The detection limit obtained with this modification using a boron-doped diamond electrode was ~75 nM [152]. AuNPs allowed the development of a new strategy to coat fused-silica capillaries through ionic adsorption of AuNPs on a polyelectrolyte multilayer (PEM)-modified capillary wall. The coating was constructed *in situ* by alternating rinses with positively charged poly (diallydimethylammonium chloride), negatively charged poly(sodium-4-styrenesulfonate), and positively charged AuNPs. After self-assembly of *n*-octadecanethiol on the surface of AuNPs, the modified capillary was investigated as a new medium for the separation of neutral analytes and proteins in open-tubular CEC. The theoretical plate numbers per meter under these conditions were 46,900 for naphthalene and 18,700 for biphenyl, respectively. The separation efficiency and detection sensitivity of this new column were nearly the same as those for the column prepared using traditional coating methods. However, there were two advantages of this new coating method: first, the coating process is easy, although four steps were needed; second, the coating obtained was very stable, since the AuNPs can be adsorbed tightly on the inner surface of the fused-silica capillary column modified with polyelectrolyte. This modification was made through electrostatic interaction and enhanced interaction of van der Waals forces between AuNPs and the polyelectrolyte. The AuNP-coated PEM-modified capillary column not only showed good

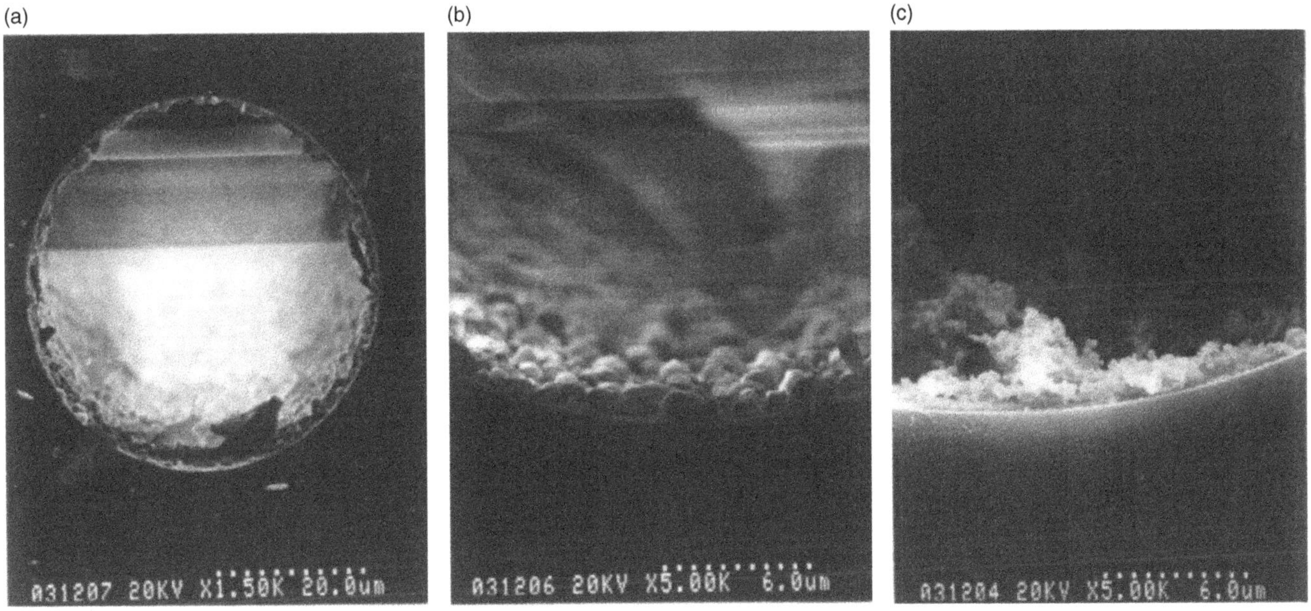

FIGURE 19.11 SEM images of TiO_2 NP-coated column. (a) Cross-section of the column prepared by two-cycle coating procedures. (b) Edge of the column prepared by two-cycle coating procedures. (c) Edge of the column prepared by a single coating procedure. Reprinted with permission from Ref. 154.

performance for neutral analytes but was also suitable for the analysis of both basic and acidic proteins [153].

Metal Oxide NPs. TiO_2 started to be applied in separation science with the modification of octadecylsilane (ODS) for the high-performance liquid chromatographic (HPLC) separation of biomolecules also to compare the results with those of the traditional reversed-phase system. Since 1994, however, several groups started to attach NPs of different metallic oxides (i.e., ZrO_2, TiO_2, and Al_2O_3) and their composites to chromatographic columns, and also the inner wall of capillaries. The latter approach was developed to manipulate the EOF, minimize the interaction of basic molecules with the capillary, and improve the separation efficiency. Hsieh et al. exploited the function of TiO_2 NPs as the stationary phases in open-tubular CEC. Column preparation by the sol–gel approach was an efficient approach, since it allowed direct stationary phase bonding to capillary walls. Therefore, the possibilities of preparing an open-tubular CEC column by condensation reaction from the TiO_2 NPs directly with the silanol groups on the surface of the inner wall of a fused-silica capillary were described (Figure 19.11). To characterize the prepared column further, angiotensin-type oligopeptides were employed as model compounds to evaluate the separation performance. The average separation efficiency was 31,000 plates m^{-1} [154].

Later, the same group reported the separation of cobalamin, apo-transferrin, ovalbumin, and bovine serum albumin using the same coating [155]. The TiO_2 NP-coated open-tubular column exhibited an exceptional surface property that was dependent on the buffer composition and pH. In the presence of phosphate buffer (40 mM, pH 8.0), the TiO_2 NP-coated column formed a stable surface for the CEC separation of the mixture of glycoisoforms. Ligand exchange seemed to be the main basis for the separation. Egg-white proteins (lysozyme, conalbumin, and ovalbumin a and b) were separated with an efficiency of 10,000 plates m^{-1} for ovalbumin and five different proteins. Moreover, another approach was described by Lia et al. [156]. TiO_2 NPs were deposited on a column by the liquid-phase deposition (LPD) method, which was used as an open-tubular CEC separation of proteins and peptides.

Over the past few decades, MNPs have also attracted increasing attention because of their unique advantages such as high surface-to-volume ratio, minimum disturbance of the structure of attached biomolecules, fast binding rates, high miscibility, and high specificity. An interesting application of MNPs is their use as a support in immunoaffinity assay capillary electrophoresis (IA-CE) [157]. In the case depicted in Figure 19.12, MNPs were used as a support to perform an online enzyme-linked immunosorbent assay (ELISA) to determine antibodies against *Helicobacter pylori* in human serum. The high surface area of the MNP allowed the binding of a great amount of antigen, requiring the reactor to be only 4 cm long. Moreover, the automation of the CE allowed the entire analysis to be performed without incubation steps. Also, the increased surface area of the MNP and

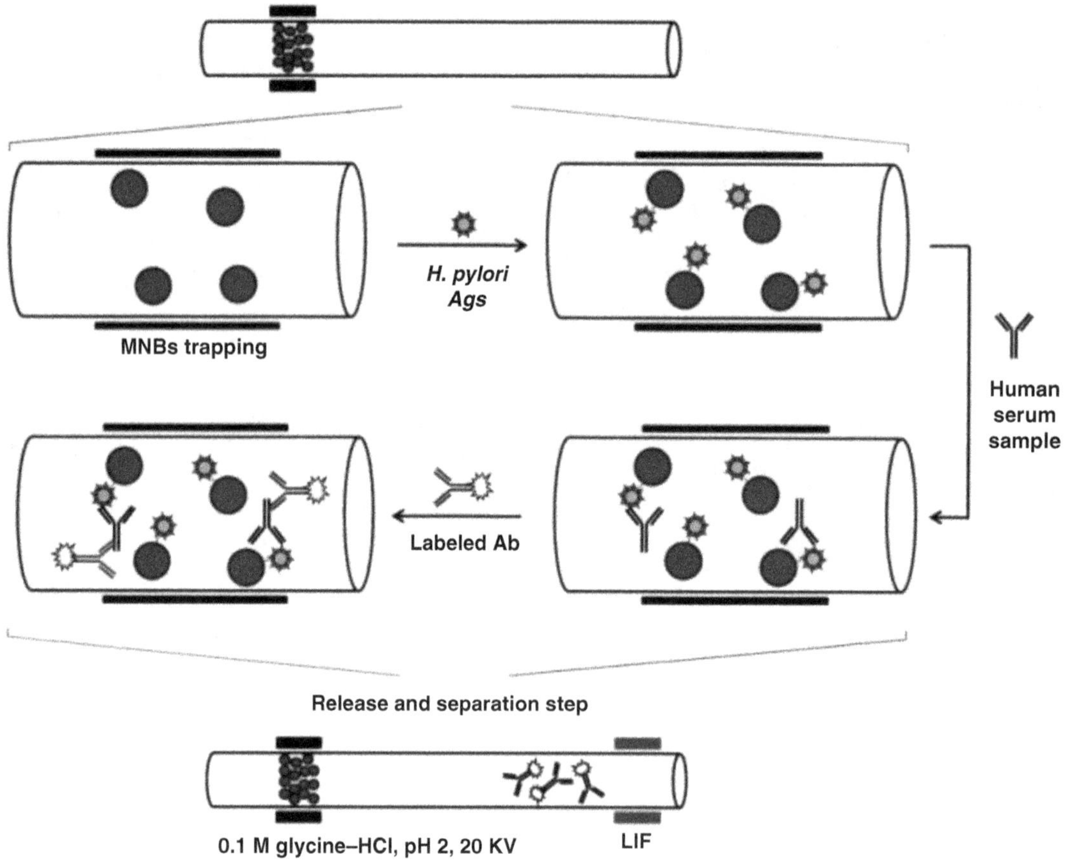

FIGURE 19.12 Schematic representation of the immunoaffinity assay–capillary electrophoresis. Reprinted with permission from Ref. 157.

the reduced diffusion distances permitted a reduction in the time of analysis to only 35 min at clinically relevant concentrations.

19.3.2 Nanoparticles in Electrokinetic Chromatography

The use of NPs as part of the running buffer provides the BGE with new sites of interaction with the analyte, which allow high resolution and, in some cases, high sensitivity to the separation and detection. Using NPs as a PSP, the analytes can be separated not only because of their charge but also based on their capacity to interact with the NPs following their own differential partitioning between the solution buffer and the NPs. This additional partitioning effect increases the number of separation degrees of freedom and makes this system capable of separating charged and neutral compounds. Along with the infinite possibilities of sensitizing different NPs with different compositions, the possibility arises of controlling the nature of the interactions and the opportunity to adapt the system to a specific analyte. Moreover, the chances do not end here: the advances in the modification of several surfaces in fact provide other alternatives for the additives and the subsequent interactions.

The following requirements should be met for the NPs to be used as a PSP in EKC:

- Their size should fall in the nanometer range.
- Their surface should possess residual charge in order to facilitate the formation of stable suspensions and electrophoretic mobility of the PSP. In fact, NPs move through the capillary at an apparent rate resulting from the effects of the EOF and their own native electrophoretic mobility.
- They should be able to interact with the target analytes. Therefore, interaction or adsorption of compounds on NP surfaces should be weak in order to ensure a low resistance to mass transfer. However, there should be substantial partitioning differences between the analytes. NPs differing in chemical composition can obviously also be expected to differ in properties.
- They should be compatible with the detection system in use.

From our point of view, the use of NPs as a PSP includes this methodology in the EKC category, but other authors prefer to include it in the CEC category. PSPs are interaction phases that move with (or against) the mobile phase. Among the benefits of these systems is the absence of stationary phase carry-over effects, which simplifies sample analysis in complex matrices, that is, of biological origin. Depending on the particular NP solubilization method, EKC can be used in two different operational modes that will be further described. The most important disadvantage in the application of nanomaterials as a PSP is the detection, due to the large amount of optical absorption that the nanomaterials have. A way to solve this problem is to use the partial filling technique [158,159]. This interesting approach allows the analytes and buffer additives to enter the detector at different times and hence minimizes interference from the additives during detection of the analytes. However, a continuous full filling technique [160,161] enables more demanding separations to be performed (as more NPs can be introduced into the capillary) and is much easier to be optimized.

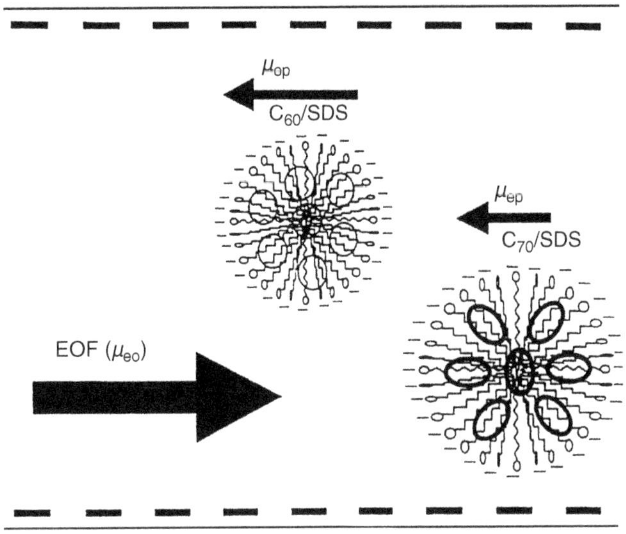

FIGURE 19.13 Schematic representation of the separation of C_{70}–SDS and C_{60}–SDS by CE using buffer. EOF, electroosmotic flow; μ, electrophoretic mobility. Reprinted with permission from Ref. 172.

19.3.2.1 Organic Nanoparticles

Fullerenes. The most commonly used nanomaterials as a PSP in CE have been the carbon-based NPs. They include fullerenes; buckminsterfullerenes are closed-cage carbon molecules containing pentagonal and hexagonal rings arranged in such a way that they have the formula C_{20+m}, with m being an integer. They are the third allotropic form of carbon, the others being graphite and diamond. Fullerenes comprise a wide range of isomers and homologous series, both the most studied, C_{60} or C_{70}, and the so-called higher fullerenes, C_{240}, C_{540}, and C_{720}. C_{60} was discovered in 1985 by Kroto et al. [162] through spectrometric measurements on interstellar dust. However, their structure was confirmed later in the laboratory. Kroto, Smalley, and Curl received the Nobel Prize in 1996 for this discovery. Because of their solubility, fullerenes have been limited to nonaqueous solvents, such as hexane, toluene, and dichloromethane; as a consequence, fullerenes could not be used directly dissolved in the BGE. In spite of this, researchers have developed some ways to dissolve fullerenes in aqueous environments. It has been found that fullerenes; for example, C_{60} and C_{70}, can be dissolved in aqueous media through derivatization [163–167] or complexation with liposomes, micelles, or other aggregate systems [168–171].

The first article describing the use of fullerenes as a PSP was published in 2000. Treubig and Brown [172] investigated the use of two different fullerenes for the separation of PAHs after the evaluation of C_{60} and C_{70} fullerenes (Figure 19.13). They compared the results with those of an MEKC procedure. One difference between the two conditions is a higher retention for the group of PAHs in C_{60}–SDS buffer. A higher retention was expected since the C_{60} fullerenes, which are encapsulated within the region of the micelles, enhanced the hydrophobic retention of the PAHs through π–π interactions. Compared with an MEKC system, and having the same SDS concentration, the hydrophobic retention, capacity factors, and resolution improved with the addition of fullerenes.

Following Treubig and Brown's work, Moliner-Martínez et al. used fullerenes (C_{60}) coated with SDS as a PSP, but in this case they optimized the method using half the concentration of the surfactant in the BGE [173,174]. The use of this additive in the BGE improved the separation of β-lactam antibiotics, amphenicols, and anti-inflammatory drugs. Moreover, they compared the separation effect of some phenolic compounds using not only fullerenes but also multi- and single-walled CNTs. They concluded that the presence of C_{60} fullerenes as a PSP increased the time of migration of the selected analytes but without significant improvements in sensitivity. They inferred that the improvement of the resolution was mostly due to the interaction of the π-clouds of both the analytes and the C_{60} fullerenes.

Carbon Nanotubes. The literature on the use of multi- and single-walled CNTs as buffer additives in CE is extensive. They possess a large surface area, high electrical conductivity, good chemical stability, and significant mechanical strength. CNTs are normally difficult to disperse in pure aqueous buffers and in various organic solvents because of their hydrophobicity. However, they can be dissolved in aqueous solution when they have been previously oxidized. The oxidation of the CNT is possible due to their open structure. In addition, CNTs can interfere with UV detection when continuously present in the electrolyte during EKC

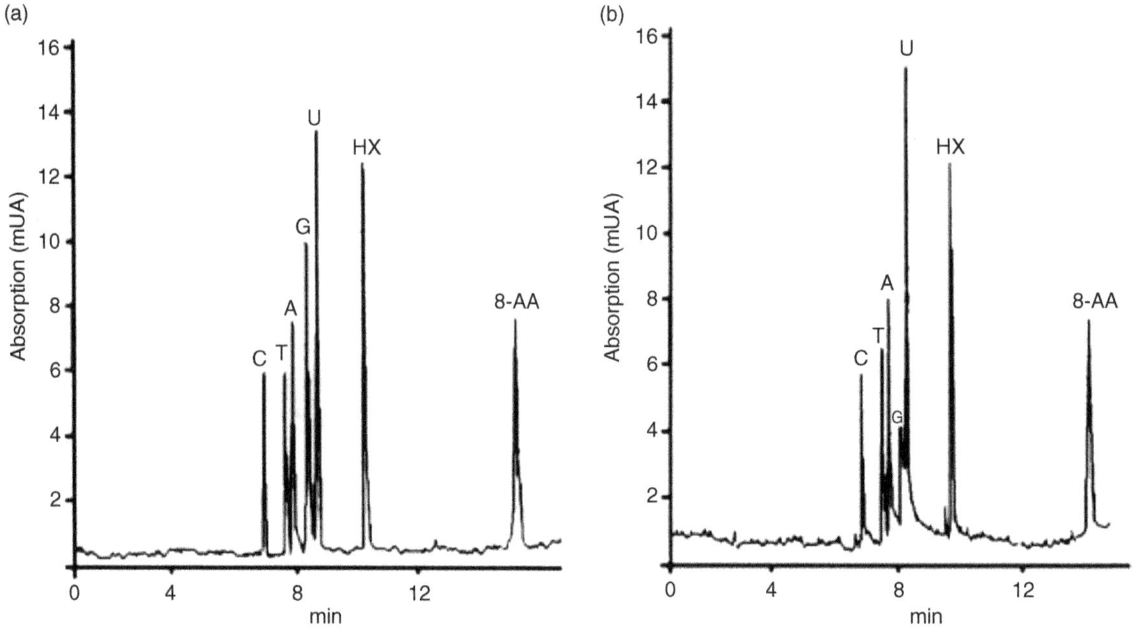

FIGURE 19.14 Electropherograms of a standard mixture solution (a) containing 8.06 g mL^{-1} *c*-MWNTs in the running buffer; (b) containing 1.061024 g mL^{-1} *c*-MWNTs in the running buffer. Reprinted with permission from Ref. 177.

separations. The first use of CNTs as a PSP was described by Wang et al. [175], who used oxidized single-walled CNTs with a diameter of 1.2–1.4 nm as running buffer additives with the aim of improving the resolution and peak shape in the separation of caffeine and theobromine, and structural isomers of catechol and hydroquinone (HQ). In the first case, the authors suggested that a defined concentration of oxidized single-walled CNTs could form a network in the running buffer based on the unique tubule structure. The migration time of the analytes was not affected at low concentrations of single-walled CNTs, supposedly due to alignment of these CNTs in the electric field. In the second case, the separation of the structural isomers was mainly due to the functional groups on the oxidized single-walled CNTs that had an effect on the electrophoretic behaviors by forming intermolecular hydrogen bonds with the analytes. Finally, the presence of charged oxidized single-walled CNTs suppressed the electrodiffusion and decreased the adsorption between the capillary wall and solutes, which led to better peak shapes for the isomer separation.

Moreover, Na et al. [176] compared the use of four different kinds of β-CD-modified NPs, that is, multiwalled CNTs, polystyrene NPs, TiO_2 NPs, and Al_2O_3 NPs, and achieved an effective method to separate the two enantiomers of clenbuterol using multiwalled CNTs coated with β-CD. They stated that the tubular structure and the largest surface area enhanced the resolution in the case of CNTs, allowing a better attachment of β-CD to the surface. Also in this work they proposed TX-100 as the surfactant with best dispersing ability in promoting the dispersion of modified-NPs, which is due to its benzene ring and alkyl chain length. These organic NPs were capable of improving the separation of purine and pyrimidine bases in yeast [177]. The multiwalled CNTs, with diameters in the range 8–15 nm, were treated with strong acid and sonication to become negatively charged and more soluble. The separation of the bases was achieved in less than 16 min (Figure 19.14).

A network of multiwalled CNTs with molecular sieving properties was proposed to be responsible for the separation. It was suggested that the presence of multiwalled CNTs decreased analyte–capillary wall interactions and stacked the analytes into narrower zones, due to their large surface area and abundance of functional groups on the surface. The use of such a system could be favorable for protein separation where capillary wall interactions are an issue, as in the case of RNA molecules (Figure 19.15). The results of this work showed that the CNTs can provide great advantages related to the resolution compared with the resolution obtained using only TX-100.

Xu and Li [178] improved the separation of DNA fragment by adding a mixture of polyvinylpyrrolidone (PVP) and multiwalled CNTs to the running buffer. They hypothesized that a synergic effect between the network generated by PVP and that generated by the CNTs was responsible for the improvements in the separation. The multiwalled CNT concentration increased mobility and decreased resolution; the authors suggested that this was due to an increase in the electrostatic repulsion of the negatively charged DNA. The

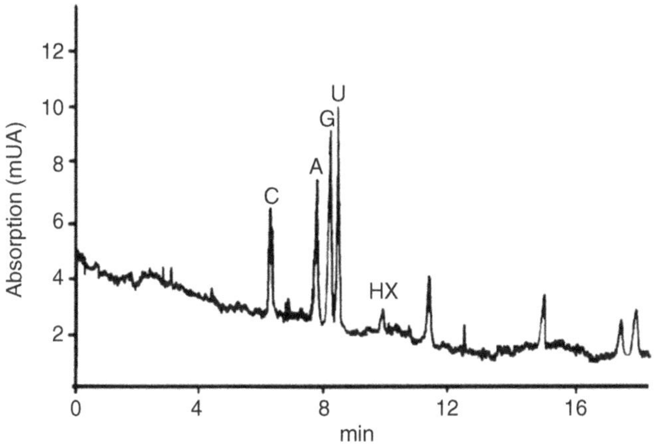

FIGURE 19.15 Typical electropherogram for the diluted hydrolyzates of yeast RNA. Reprinted with permission from Ref. 177.

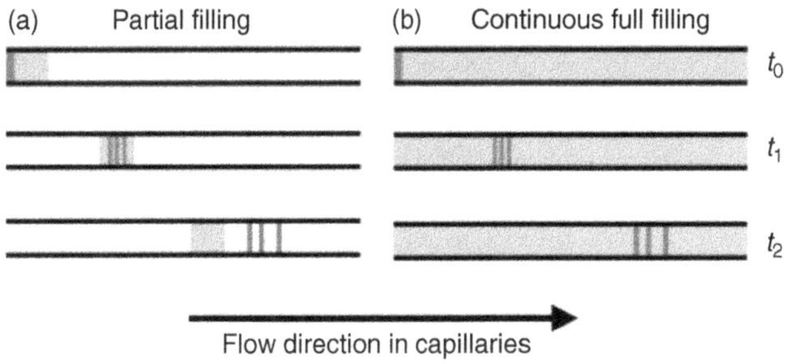

FIGURE 19.16 Schematic illustrations of an EKC during (a) the partial filling technique and (b) the continuous full filling technique. The capillary (black) is illustrated during three subsequent times (t_0, t_1, t_2) for the two techniques. Samples are illustrated by dark gray bars, slurry is illustrated by light gray fields, and background electrolyte is illustrated by white fields (confined by the capillary). By the time the sample molecules reached the interface on the detector (not included in the illustration), they had already passed the slurry plug in (a) whereas in (b) a mix of the samples and the slurry had reached the interface. Reprinted with permission from Ref. 181.

repulsive effect dominated the sieving effect at high multi-walled CNTs concentration, thus yielding faster separations but with reduced resolution. The separations took ~30 min and it was the first study on this theme. The detection method used in this work was contactless conductivity.

Polymer Nanoparticles. To the best of our knowledge, the first work using sulfonated polymer NPs of 20 nm size as PSPs was published in 1989 [179]. The separation of five catecholamines was not totally optimized, but this application opened up the possibility of developing other kinds of PSP instead of surfactants in EKC. However, it took almost a decade until other researchers entered this new area. The viability of using ion-exchange particles as a PSP in EKC for the separation of monovalent inorganic cations was investigated by Breadmore et al. [180]. Using sulfonated polymeric particles (average diameter 225 nm) as the PSP, the separation selectivity for alkali metals and ammonium was examined using indirect absorbance detection. The monovalent cations were successfully separated, but indirect absorbance detection gave poor sensitivity when polymer NPs were continuously present in the electrolyte, because of their light scattering effect. Although the EKC technique has several advantages, the use of additives in the buffer, such as surfactants, cyclodextrins, or proteins, suffers from increased background noise that may introduce problems in trace analysis. For instance, sodium dodecyl sulfate (SDS) is known to decrease the signal response in ESI-MS due to suppression of the analyte ions in the electrospray ionization (ESI) and to contaminate the mass spectrometer. Continuous full filling and partial filling techniques (Figure 19.16) with methacrylic acid-based polymer NPs (160 nm) as PSPs were evaluated by Viberg et al. [181] by coupling the capillary to the mass spectrometer via an orthogonal ESI interface to prevent NPs from entering the mass spectrometer. Figure 19.16 shows the separation of three analytes (nortriptyline, salbutamol, and diphenhydramine) using both techniques.

It was concluded that the continuous full filling technique was easier to optimize, it did not require time-consuming particle packing or retaining frits, and had the advantage of an unused interaction phase for every new analysis. Using the above-mentioned NPs in a continuous full filling technique, Nilsson et al. [182] performed the reversed-phase separation of four neutral analytes (dialkyl phthalates). The NPs (600 nm) were synthesized by polymerization of methacrylic acid and trimethylolpropane trimethacrylate and coated with dextran. These particles formed stable suspensions in electrolytes having broad ranges of polarities. Highly efficient separations were achieved with plate numbers as high as 700,000 m^{-1} for a retained analyte and they optimized this system to make it work with a low amount of organic modifier. Recently, a novel method using poly (*N*-isopropylacrylamide) to separate double-stranded DNA and identify single-nucleotide polymorphisms was proposed by Song et al. [183]. These NPs have an interesting temperature-sensitive behavior; they were water-soluble polymers at a lower critical solution temperature (LCST) of 32 °C. Above the LCST, the particles are phase separated and precipitate out from water. The phase-separated poly(*N*-isopropylacrylamide) particles, which have a narrow size distribution, were used as both a dynamic coating and a PSP, thus avoiding the need to immobilize the capillary coating. This study represents an interesting approach to performing EKC and CEC at the same time, taking advantage of both methods. The polystyrene NPs in combination with hydroxypropyl-β-cyclodextrin (β-CD) were a good alternative to improve the separation of the two isomers of propranolol [184], a chiral compound. These chiral selector NPs (15 nm)

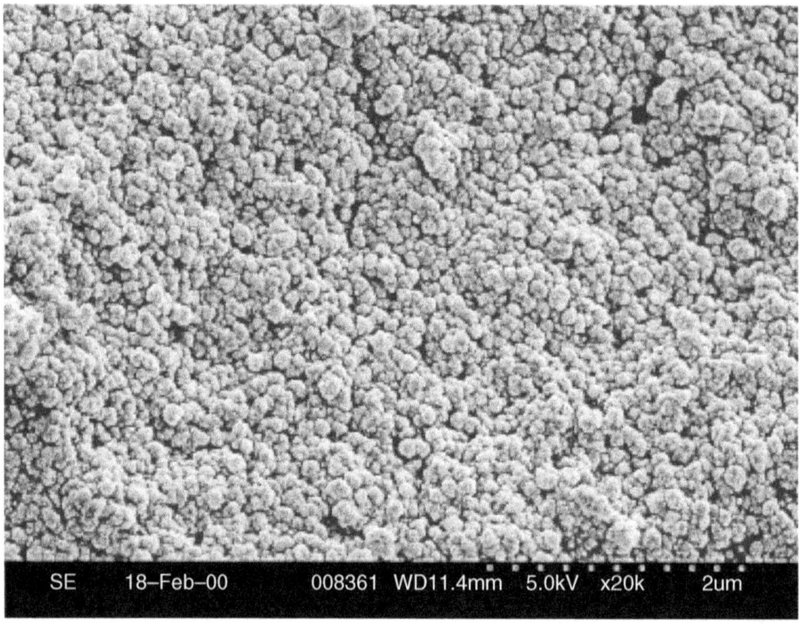

FIGURE 19.17 SEM image of a nanoMIP. Reprinted with permission from Ref. 188.

were used as a PSP; the hydroxypropyl-β-CD interacted differently with the two enantiomers, allowing the separation of the chiral compound at a low concentration of the additive. Selectivity improvements were reflected by changes in the observed mobility as a result of interactions governing the absorption process on the NP surface. Moreover, it was observed that the capillary wall was saturated with polymer NPs already at low concentration, which suggested that these absorbed NPs could play an important role during the separation.

Molecularly Imprinted Polymer NPs. A nanoMIP or plastic antibody is a polymer that is synthesized in the presence of a molecule that is subsequently extracted, leaving complementary cavities behind. Functional monomers are oriented around a template (i.e., analyte), and subsequently polymerized in a solvent in the presence of an initiator and a cross-linking monomer to form a well-defined chemically complementary 3D structure around the template. These polymeric materials that are chemically and mechanically stable selectors show selectivity toward a certain ligand, that is, the template molecule [185]. NanoMIPs are formed in a solvent where the polymer particles have poor solubility and the monomers have good solubility. The particles precipitate as they reach a certain size and their growth is interrupted. The first nanoMIP was synthesized and used as a PSP chiral selector by Schweitz et al. in 2000 [186]. They studied the capacity for chiral separation of propranolol enantiomers, using the partial filling technique. Following this work, Spégel et al. [187] used the same NPs synthesized by precipitation/polymerization and the capillary was derivatized with (methacryloxy)propyltrimethoxysilane to reduce the EOF in order to allow easier optimization of the partial filling technique. They demonstrated that electrostatic interactions were mainly responsible for the separation and that the use of a weak functional monomer increased the efficiency. Later, monodisperse microspheres were imprinted with (+)-ephedrine as the template molecule. de Boer et al. [188] synthesized the above-mentioned nanoMIPs (~100 nm in diameter) by precipitation/polymerization and used them in the partial filling technique for the chiral separation of ephedrine and salbutamol enantiomers (Figure 19.17). The best separation was obtained with an acidic buffer (pH 2.5) containing 40% acetonitrile, which helped to stabilize the NP suspensions. The EOF was considerably decreased due to the low pH used.

Spégel et al. developed two different approaches using nanoMIPs to achieve selectivity toward multiple predetermined targets [189]. NanoMIPs that utilized (*S*)-ropivacaine and (*S*)-propranolol as template were synthesized and used for the separation of enantiomers of ropivacaine and propranolol in a single run. In the first approach, they used a mixture of two different types of nanoMIPs with selectivity against two different templates. An advantage of the mixed-particle approach is the fast and easy optimization of the preparation and the small amount of nanoMIPs needed for each of the analytes studied. The second approach used another type of nanoMIPs with selectivity toward both templates simultaneously; they incorporated two different templates during the preparation of the NPs. The presence of two templates in the preparation of the nanoMIPs caused a multifaceted situation during complexation by the introduction of additional equilibrium to the basic template-to-monomer ratio. Owing to competition between the two different templates for the finite amount of functional monomers, the template showing the strongest interactions with the functional monomers could affect the possibility for the second template to form stable complexes. The amount of template used in the preparation of the nanoMIPs most likely affected the number of affinity sites present in the resulting polymer.

Priego-Capote et al. [190] developed the first nanoMIPs with "monoclonal antibody"-like behavior based on miniemulsion polymerization. The synthesis of nanoMIPs (30–150 nm) was developed for the analysis of *rac*-propranolol, where (*S*)-propranolol was used as a template. Baseline separation with a plate number of 25,000–60,000 was achieved. These effects were attributed to the reduction of the nanoMIP site heterogeneity by means of peripheral location of the core cross-linked NPs and to nanoMIP binding sites with the same ordered radial orientation. This new nanoMIP approach was based on the replacement of the functional monomers with a surfactant monomer (sodium *N*-undecenoyl glycinate (SUG) to improve the inclusion in the nanoMIP structure and to use a miniemulsion in the nanoMIP synthesis.

19.3.2.2 Inorganic Particles

Silica Nanoparticles. Silica NPs have also been used as PSPs in CE. The first study using silica NPs as a PSP in a full filling system was reported by Göttlicher and Bächmann [191]. A suspension of reversed-phase NPs (500 nm) was used for the separation of PAHs with fluorescence detection. Silica NPs were coated with SDS and stabilized in a suspension containing β-CD to increase the solubility of the PAHs. The retention was dependent on the distribution of analytes between the SDS-coated particles and the electrolyte containing β-CD, and the separation was achieved with an efficiency of 20,000–100,000 plates m^{-1}. Unfortunately, the lack of stability of the particles used could not be overcome. Improving on their previous study, Bächmann and Göttlicher [192] developed reversed-phase silica NPs, modifying the surface by fixing chemically negative functional groups. Consequently, the charge of the silica-modified NPs was changed, so it was possible to use a large amount of organic modifier. PAHs and naphthalene derivatives were separated and detected with fluorescence and UV detection using 500 nm diameter NPs. It was concluded that the plate numbers (100,000 plates m^{-1}) were reduced mainly due to mass transfer resistance at the particle surface, which could be solved with the application of a higher field strength, and due to the different particle velocities, which could be overcome by classifying the particles according to their velocities before including them in the PSP. By using very small silica NPs, Fujimoto and Muranaka [193] separated phthalates and steroids. Monodisperse NPs (14.2 nm) were used as a PSP in a continuous full filling system with no significant noise and long-term fluctuations in the UV signal. It was suggested that the separation was based on hydrogen bonding between the analytes and NP surfaces. An interesting development and comparison between several organically modified silicates used as PSPs were described by Neiman et al. [194]. One neutrally charged and two positively charged groups of NPs were synthesized by the sol–gel process and used for the separations of aromatic acids and also structural isomers of aromatic acids. Methyltrimethoxysilane (MTMOS), aminopropyltrimethoxysilane (APTMS), and *N*-[3-(trimethoxysilyl)propyl]ethylenediamine (EDAS) were the alkoxysilane precursors used for the synthesized of the NPs. All of these alkoxysilanes had three hydrolyzable methoxy groups and an additional functional group that was covalently attached to the cross-linked siloxane backbone after sol formation. The MTMOS-based sol was found to improve the selectivities between various aromatic acids. The APTMS sol interacted differently from MTMOS with structural isomers of aromatic acids. At low pH, with APTMS sol in the run buffer, neutral solutes could also be separated. The separation of the neutral solutes seemed to be facilitated by the formation of hydrogen bonds between the solutes and the APTMS sol. APTMS and EDAS affected the separation of the same compounds differently, thus indicating that even small changes of the functional groups of the sol had a pronounced effect on the interactions between the sols and the solutes. Migration times were longer compared with CZE but the separation selectivity, resolution, and peak shape were considerably improved due to the addition of NPs to the buffer. The EOF was reversed due to adsorption of positively charged NPs on the capillary wall. The wide particle size distribution, that is, particle velocity distribution, was the main drawback due to its band broadening effect. Another study was developed using silica NPs in the BGE as a PSP to separate biologically active amines. The EOF-minimized interactions between the capillary wall and analytes were suppressed through a dynamic coated using PVP and poly(ethylene oxide) (PEO). The adsorption of the silica NPs on the capillary wall suppressed the unspecific interactions of the analytes with the inner wall of the capillary, which resulted in sharper peaks, and higher plate numbers were therefore achieved. Moreover, silica NPs increased the EOF, producing a enhancement of separation speed [195]. Wang et al. used silica NPs as an additive to the BGE to enhance the separation of seven quinolones, improving the resolution and the selectivity of the separation. The addition of the silica NPs to the running buffer resulted in an EOF decrease and permitted full interaction between silica NPs and analytes [196].

Gold Nanoparticles. AuNPs possess a large surface area and have been used mostly to separate biological molecules due to the special affinity that molecules have for NPs.

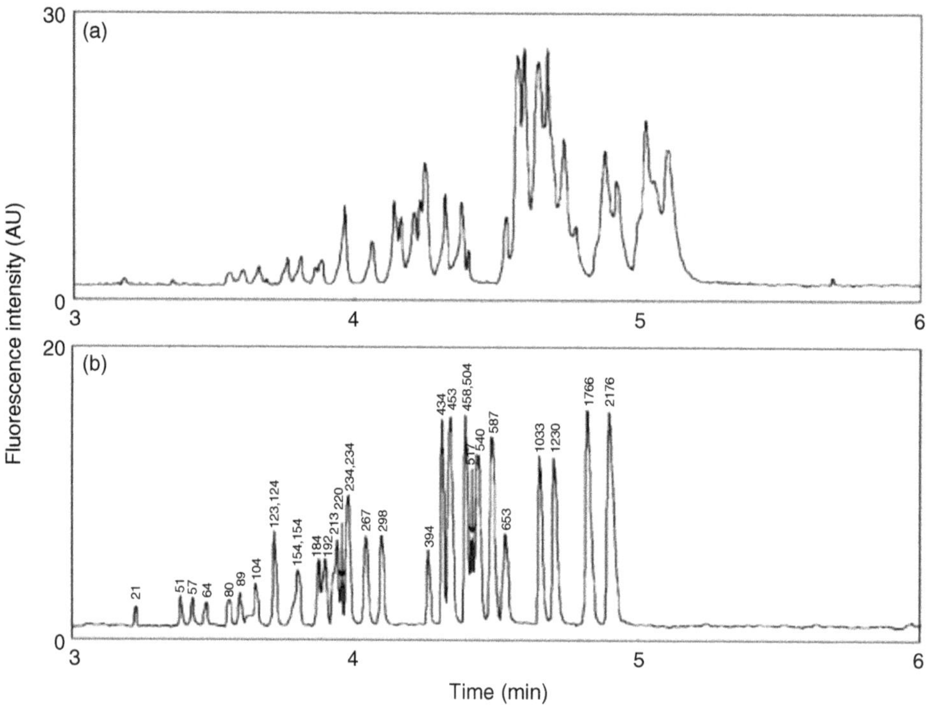

FIGURE 19.18 Separations of 10 mg mL^{-1} DNA markers V and VI using (a) 0.2% PEO and (b) 0.2% PEO–AuNPs. Reprinted with permission from Ref. 199.

Several approaches have been developed, especially by Chang's group, to improve the separation of double-stranded DNA (dsDNA) [197–200] using PEO-coated AuNPs in the continuous full filling technique with fluorescence detection. PEO interacts with the AuNPs via their hydrophobic regions, leaving the hydrophilic regions available to interact with the DNA in the surrounding medium. The function of the PEO molecule is to prevent aggregation and agglomeration of the AuNPs, and to enhance interactions between the DNA and PEO-coated NPs. Direct laser-induced fluorescence (LIF) detection via intercalation of ethidium bromide was used to detect DNA in the presence of NPs. AuNP suspensions possess low viscosity and are more easily injected into capillaries than conventional polymer solutions, which are normally used for DNA separations. It was suggested that the separation mechanism meant that DNA temporarily intertwines with the PEO adsorbed on the AuNPs. The capillary was dynamically coated with PVP, to suppress the EOF and to minimize unspecific interactions between the DNA and capillary wall. The first study reported by Huang et al. [197] using PEO–AuNPs was aimed at evaluating the effect of the size of the AuNPs on the resolution and speed of the separation. They evaluated AuNPs in the size range from 3.5 to 56 nm and separated dsDNA from 8 to 2176 bp in 5 min using 56 nm AuNPs, while dsDNA from 5 to 40 kbp were separated within 25 min using 3.5 and 32 nm AuNPs. Following this work, they performed the separation of long dsDNA (0.12–23.1 kbp) in 3 min using PEO-coated AuNPs (32 nm) [198]. Moreover, dsDNA (8.27–48.5 kbp) was separated with plate numbers $>10^6$ within 7 min. The amount of salt was an important parameter because DNA becomes more compact and the interactions with the capillary wall become less pronounced; then the separation efficiency is greater with small amounts of salt present in the buffer. The possibility of separating long-stranded DNA molecules makes this method a very promising alternative for the analysis of longer DNA molecules such as chromosomes. Later, the same group [199] reported that 56 nm PEO–AuNPs as a PSP were used to separate dsDNA and the capillary was dynamically coated with 5.0% PVP. The separation reproducibility increased on increasing the NP concentration, due to the adsorption of NPs on the wall of the capillary decreasing the unspecific DNA interactions with the capillary wall. DNA separation (21 bp to 23 kbp) was achieved within 7 min (Figure 19.18). One of the most important parameters optimized in this work was the concentration of ethidium bromide, due to the complexes that DNAs are capable of forming with this compound. Thus, DNA become longer and stiffer and the net charge decreases, which makes the complexes migrate more slowly than free DNA. It is important to control the concentration of ethidium bromide to prevent it from disturbing the DNA separations.

The separation of different sized DNA fragments was developed also by Tseng et al. [200]. μ-DNA fragments and high molecular weight DNA fragment mixtures were resolved in the presence of PEO-coated AuNPs (32 nm) as a PSP within 5 min with up to 350,000–2,200,000 theoretical plate numbers. These highly efficient separations were achieved using an electric field and a hydrodynamic flow simultaneously. It was suggested that the presence of AuNPs decreased the degree of auto cleavage and aggregation in the electric field. An AuNP-filled CE method combined with three multiplex polymerase chain reactions (PCRs) was established for the simultaneous diagnosis of five common α-thalassemia deletions, including the $-\alpha^{3.7}$ deletion, $-\alpha^{4.2}$ deletion, Southeast Asian (– –SEA), Filipino (– –FIL), and Thai (– –THAI) deletions. AuNPs were used as a PSP to improve the resolution between DNA fragments in a low-viscosity polymer. The relative standard deviations (RSDs) of the migration times were less than 2.81%. A total of 21 patients with α-thalassemia deletions were analyzed using this method, and all results showed good agreement with those obtained by gel electrophoresis. According to the authors, the addition of AuNPs improved the interactions between DNA and the polymer adsorbed on the AuNP surface, without markedly changing the viscosity [201]. The first important approach made using AuNPs to separate acidic and basic proteins was presented by Yu et al.. The separation was performed at low pH (3), achieving high plate numbers and high run-to-run reproducibility [202]. By using PEO–AuNPs, the theoretical plate number increased to over 1,000,000 for lysozyme. The AuNPs were first protected by a bilayer of didodecyldimethylammonium bromide (DDAB). The DDAB-capped AuNPs were noncovalently coated with PEO and dispersed in the electrolyte to enhance interactions with proteins (Figure 19.19). The inner wall of the capillary was then coated with these NPs to prevent protein adsorption; the adsorbed positively charged NPs in the wall of the capillary also produced a reversal of the EOF. The authors demonstrated the applicability of this method by analyzing real sample matrices such as saliva, red blood cells, and plasma.

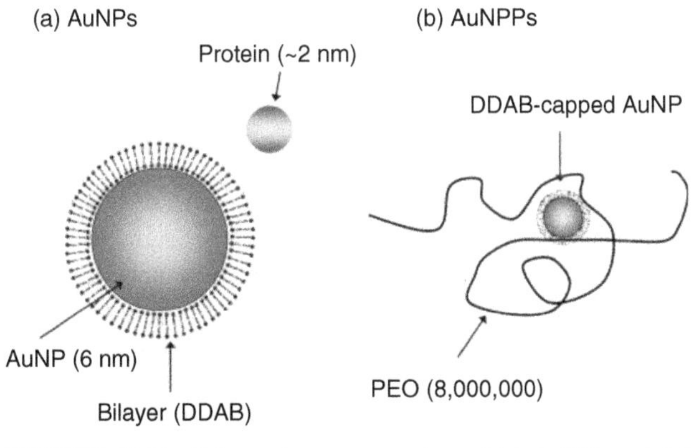

FIGURE 19.19 Schematic representation of (a) AuNPs and (b) AuNPPs. Schematic illustration of a capped AuNP showing the relative size of the particle core, bilayer, protein, and PEO. Reprinted with permission from Ref. 202.

Recently, Ivanov et al. used AuNPs with either primarily covalently functionalized carboxylic acid (Au–COOH) or amine (Au–NH_2) surface groups to separate three neurotransmitters [dopamine (DA), epinephrine (EP), and pyrocatechol] [203]. They described three NP-specific mechanisms for the effective separation. First, the degree of NP interactions was quantified using a new parameter termed the critical nanoparticle concentration (CNC). The CNC is defined as the lowest concentration of NPs that induces predominant NP aggregation under specific buffer conditions and it is determined by using dual-wavelength photodiode array detection. Second, NP analyte interactions were dictated by electrostatic interactions, which depend on the pK_a of the analyte and the surface charge of the NP. Finally, NP capillary interactions occurred in a surface chemistry-dependent manner. An important optimized factor was the viscosity of the running buffer, which was influenced by the formation of an NP steady-state PSP along the capillary wall. As a result of these three NP-specific interactions, Au–NH_2 NPs increased the mobility of the neurotransmitters while a smaller opposite effect was observed for Au–COOH NPs. The presented approach of exploiting NP behavior in the presence of an electric field has significant impacts in separation science where NPs are employed. In 2010, Yang et al. [204] developed the application of thiolated β-CD-modified AuNPs as chiral selectors for the enantioseparation of four pairs of dinitrophenyl-labeled amino acid enantiomers (DL-Val, Leu, Glu, and Asp) and three pairs of drug enantiomers [(*RS*)-chlorpheniramine, zopiclone, and carvedilol[. This was the first approach to analyze the potential of modified AuNPs as chiral selectors in a full filled EKC separation. These particles had an average diameter of 9.5 nm and they were soluble in water and stable in both acidic and basic conditions. Efficient baseline separation was achieved with very low concentrations of thiolated β-CD-modified AuNPs (0.8–1.4 mg mL^{-1}); the corresponding β-CD concentration in the buffer was only 0.30–0.53 mM, which was much lower than the optimum concentration of 15mM if pure β-CD was used as chiral selector. The results showed that thiolated β-CD-modified AuNPs can provide sufficient interaction with the analytes, leading to enhancement of enantioseparation. In conclusion, AuNPs have great potential and prospective use in enantioseparation as PSPs with advantages of high separation efficiency at low concentration of chiral selector and easy modification with various organic molecules through covalent bonding.

19.3.3 Nanoparticles in Microchip Electrochromatography

Nowadays, microchip capillary electrophoresis (MCE) is certainly the most rapidly advancing and successful branch of miniaturized separation systems. The reasons are mainly the inherent simplicity of its fabrication and operation, combined with its unique features of separation speed, sample injection, and reagent consumption. Early developments and applications of CE microchips employed glass or quartz as chip substrates because they were optically transparent and exhibited EOF properties similar to those of fused silica. Furthermore, a number of other materials including polymers/plastics were investigated. The implementation of nanomaterials to improve the separation in microchips allowed the new technology to grow very fast. As has already been discussed, the presence of NPs improves the precision of the analysis and increases the separation efficiency. The use of NPs for DNA separation in MCE has so far shown great promise and was summarized by Lin et al. [205]. The application of a BSA immobilized on the surface of short single-walled CNTs as a stationary phase was developed for the chiral separation of tryptophan enantiomers in a poly(methyl methacrylate) (PMMA) chip by Weng et al. [206]. With this achievement, a successful separation of tryptophan enantiomers was obtained in less than 70 s with a resolution factor of 1.35. The stability of the stationary phase in the channel was examined and the EOF was shown to be more stable and reproducible compared with the use of an unmodified chip. The single-walled CNTs used as stationary phases were cut by chemical oxidation under ultrasonication to gain more carboxylic acid groups on the surface so as to provide more binding sites for proteins and, as a consequence, more interactions with the analytes. Different types of NPs could be applied to improve the separation in MCE. Polymer NPs were introduced by Tabuchi et al. [207], who reported an MCE separation of DNA fragments in a wide size range (1–15 kbp) within 100 s. The channels used were 100 μm wide and 30 μm deep with an effective separation length of 30 mm. The goal of this study was to use an uncharged NP medium, core–shell-type nanospheres with a size of 30 nm, in conjunction with a pressurization technique during MCE. The NPs, which were composed of a cross-linked hydrophobic core of poly(lactic acid) and a hydrophilic shell of poly(ethylene glycol), had low viscosity compared with conventional polymers due to their globular structure. The NPs were used as a packing material and as a sieving matrix in microchip capillary electrochromatography (MCEC). DNA fragments up to 15 kbp were successfully analyzed within 100 s without observing any saturation in migration rates. DNA fragments migrated in the medium while maintaining their characteristic molecular structure. To guarantee effective DNA loading and electrofocusing in the nanosphere solution, the authors developed a double pressurization technique. Optimal pressure conditions and concentrations of packed nanospheres are critical to achieve improved DNA separations. An electrofocusing process increased the sample concentration due to the stacking effect of the packed NPs. The introduced DNA fragments migrated and became focused in the NP plug under an electric field. The DNA showed minimal stretching during separation. Moreover, the same group [208] presented an advanced MCE using an NP-doped polymer solution that allowed greater separation of DNA. PEG-coated polystyrene latex NPs with different sizes (80, 110, and 193 nm) were mixed with a conventional solution for the separation of dsDNA (10 bp to 3 kbp). When a 0.49 wt% hydroxypropylmethylcellulose (HPMC) buffer solution was mixed with a 2.25 wt% 80 nm PEGylated latex, a higher separation efficiency and a higher mobility of a wider molecular range of dsDNA (10 bp to 2 kbp) were achieved. A polymer solution in combination with NPs was used to separate large and small DNA simultaneously. The PEG-coated NPs were prepared by emulsion polymerization of styrene in the presence of PEG macromonomers. The separation performance was dependent on the particle size nd concentration. Furthermore, the effectiveness of the larger PEGylated latex (193 nm) was not as great as that of the smaller one (80–110 nm). The observed separation improvement using the polymer solution with latex NPs seemed to derive from the balance between the wider polymer mesh size and the structural obstacles of particles in the buffer. Fast filling of the separation medium was possible due to its low viscosity. Even nanoMIPs have been investigated in microchips. An MCEC using nanoMIPs as a stationary phase was designed for the rapid enantioseparation of ofloxacin by Qu et al. [209]. The NPs were synthesized by the copolymerization of methacrylic acid and ethylene glycol dimethacrylate on 3-(methacryloyloxy)propyltrimethoxysilane-functionalized magnetic NPs (25 nm diameter) in the presence of template molecules. The imprinted NPs (200 nm diameter) were used like a stationary phase in the microchannel of the microfluidic device with the help of an external magnetic field. The authors analytical performance was improved by varying the packing length of the NP zone, showing an advantage over the conventional packed CEC. Also, monodispersed silica NPs were added to an electrophoresis buffer solution to obtain high mobility and high separation of dsDNA (100–1500 bp) within 150 s [210]. Various particle sizes (20–140 nm) of monodispersed silica NP solutions were mixed with conventional 0.7% HPMC buffer solution to achieve MCE. The speed and resolution were affected by the addition of silica NPs to the buffer. It was suggested that the DNA was accelerated by the repulsive forces of the silica NPs and thereby the speed and resolution were enhanced. The mobility and separation showed a strong dependence on the silica NP concentration because of change in diffusion distances, that is, mass transfer. An ingenious MCEC methodology was reported by Pumera et al. [211], who described the use of AuNPs in conjunction with MCEC to improve the selectivity between solutes and to increase the efficiency of the separation. They coated the microchannel wall of a microfluidic device with a layer of PDDA and then collected on it citrate-stabilized AuNPs (10 nm). The resolutions and the plate numbers of the solutes were doubled in the presence of the AuNPs. Such selectivity improvements reflected changes in the observed mobility accrued from interactions of solutes with the particle surface. The electrochemical detection and the quantification of the solutes were not affected by the PDDA and the AuNPs. The impact of these NPs as a PSP and stationary phase in MCE was also described by Lin et al. [212], who developed the separation of dsDNA (8–2176 bp) using PEO with LIF detection. A PMMA microchip with a 75 μm separation channel and an effective length of 3 cm was coated using the LBL technique in sequence with PVP, PEO, and 13 nm AuNPs, improving the reproducibility and resolution of the separation. The analysis of the PCR products of UGT1A7 was completed in 7 min, with RSDs of the peak heights and migration times of less than 2.3% and 2.0%, respectively. The peak profile was sharper and the resolution was improved by the three-layer coating. Citrate was added to stabilize the PEO–AuNPs that were added to the electrolyte during separation; urea was also added to improve the resolution further. In conjunction with stepwise changes of the concentrations of ethidium bromide (0.5 and 5 μg mL^{-1}), this method resulted in improved resolution and sensitivity for DNA markers V and VI, which were normally separated on physical gels. In addition, DNA was separated in MCEC on PMMA chips with efficiencies up to 1,700,000 plates m^{-1} by the same group [213]. After testing a number of polymers, they found that coating of the PMMA microchannels with PEO ($M_r = 6.0 \times 10^5$ g mol^{-1}) on the first layer was essential to minimize the interaction of DNA with the PMMA surface. To achieve high efficiency, multilayer coating of PMMA chips with PEO, PVP, and PEO–AuNPs were used. A 2-(PEO–PVP)–PEO–AuNPs PMMA chip, which was repeatedly coated with 1.0% PEO and 5.0% PVP twice, and then coated with 0.75% PEO–AuNPs each for 30 min, provided a high efficiency (up to 1.7×10^6 plates m^{-1}) for the separation of DNA markers V (pBR 322/*Hae*III digest) and VI (pBR 328/*Bgi*I digest and pBR 328/*Hin*fI digest) when using 0.75% PEO–AuNPs. The authors demonstrated the separation of hsp65 gene fragments of *Mycobacterium Hae*III digests by MCE within 90 s. The multilayer coating was important to minimize unspecific DNA adsorption and to accomplish very high efficiency even when the multilayer process was complicated. The separation of DA and EP in MCEC was performed by Wang et al. [214], who used polydimethylsiloxane (PDMS) microfluidic channels modified by citrate-stabilized AuNPs after coating a layer of linear polyethylenimine (LPEI), which were successfully used to separate DA and EP from the baseline in native and hybrid PDMS microchannels. The PDMS was first coated with a layer of polyelectrolyte and then by the citrate-stabilized AuNPs, using an electrostatic LBL method. Amperometric detection was used with a single carbon fiber cylindrical electrode. The analytes were well separated within 100 s in a 3.7 cm long separation channel at a separation voltage of +800 V using a 30 mM phosphate buffer solution (pH 7.0). Linear responses of the analytes were obtained from 25 to 600 μM with detection limits of 2 μM for DA and 5 μM for EP. The separation efficiency was enhanced and better control of the EOF was achieved by modification of the PDMS with AuNPs. The modified PDMS channels showed long-term stability and excellent separation reproducibility. In a later study, Wang et al. [215] reported a separation of neurotransmitters and environmental pollutants using protein-coated film modification of PDMS microfluidic chips based on the LBL assembly technique for enhancing separation efficiency. Two kinds of protein-coated films were prepared. One was obtained by successively immobilizing the cationic polyelectrolyte chitosan, AuNPs (13 nm), and albumin on the PDMS microfluidic channel surface. The other was achieved by sequentially coating lysozyme and albumin. Neurotransmitters (DA, EP) and environmental pollutants (*p*-phenylenediamine, *p*-PDA; 4-aminophenol, 4-AP; hydroquinone, HQ) as two groups of separation models were studied to evaluate the effect of the functional PDMS microfluidic chips. The results clearly showed that these analytes were efficiently separated within 140 s in a 3.7 cm long separation channel and successfully detected with in-channel amperometric detection using a single-fiber cylindrical electrode at the end of the separation channel. Through modification, the more homogeneous channel surface displayed higher reproducibility and better stability. The coating was used to improve the stability of the EOF, make the surface more hydrophilic, and eliminate adsorption. The implementation of metal-based NPs in MCE has been used to separate amino acids. Qiu et al. [216] determined four amino acids using a modified PDMS microchip with TiO_2 NPs as a functional film on the PDMS microfluidic channel surface. It was formed by sequentially immobilizing PDDA and TiO_2 NPs on the PDMS surface by the LBL assembly technique. The modified PDMS microchip was used for the determination of arginine, phenylalanine, serine, and threonine. The four amino acids were efficiently separated within 100 s in a 3.7 cm long separation channel and successfully detected on the carbon fiber electrode in conjunction with in-channel indirect amperometry. The resolutions and theoretical plate numbers of the analytes were considerably enhanced in the presence of TiO_2 NPs. When a bilayer of PDDA/TiO_2 NPs was coated on the PDMS surface, the EOF was decreased at pH 9.2 compared with the native microchip, which means that it exhibited a pH-dependent tendency. The decreased EOF was favorable for the separation of biomolecules with similar properties. The RSD of the EOF on the coated microchip was 0.5% at pH 9.2, exactly one-seventh of that on the native microchip (RSD 3.5%). In other words, a suppressed and more stable EOF was obtained on the modified microchip compared with the native microchip. Continuing with the use of metal-based NPs, Doyle et al. [217] used MNPs retained in the separation channel by applying an external magnetic field and achieving the separation of different dsDNAs (15, 33.5, and 48.5 kbp) in a wider size range than with a normal separation gel. When a constant homogeneous magnetic field was applied to the NP suspension, they self-organized into a fixed, quasi-regular array of columns. The superparamagnetic NPs showed low viscosity, allowing easy filling in the absence of the magnetic field.

19.4 CONCLUSIONS

Nanotechnology has become a real interdisciplinary research field in which many physicists, chemists, biologists, materials scientists, and other specialists are involved. Nano investigations, being widely interdisciplinary by their very nature, promote the joining and merging of various sciences and technology fields such as powder technology, colloid chemistry, surface chemistry and physics, clusters and aerosols, tribology, catalysis, simulation and modeling, and computer techniques. During the last decade, several amounts of different nanomaterials have been synthesized and applied to almost every analytical method with a variety of advantages and disadvantages. Nanomaterials open up possibilities to develop new analytical methodologies that could never have been carried out before. As we have established in this chapter, nanomaterials present different properties than those of larger sized particles. These unique physical and chemical properties have been found to be a very valuable tool for integrating them into a broad variety of systems to improve their performance and versatility. They have improved the performance of sensors, detection systems, extraction/preconcentration, catalysis, and separation systems.

Especially in CE, nanomaterials have played an important role related to the improvement of the resolution in the separation. Nanotechnology has redefined the capillary electrochromatography and electrokinetic chromatography concepts. Miniaturized NP-based systems have been shown to have many obvious advantages. The main and most important advantage of the NPs is the size–volume relation. The great surface area of these new materials allows an important increase in interactions with the analytes. These interactions depend not only on the chemical structure of the analyte but also on the exposed groups on the surface of the NPs. The fact that the surface of the NPs can be modified permits the interaction to be improved and this interaction to be focused in order to obtain better selectivity based on the chemical structure of the analytes, and also to manage the EOF direction and velocity. These advantages, combined with a flat flow velocity profile, should be able to offer an optimal separation system with respect to separation efficiency and selectivity.

Several dynamic and covalent coating techniques have been developed to allow stable NP suspensions to be formed in a variety of electrolytes. It is important to have stable suspensions, although NP–analyte interaction kinetics must be considered, that is, both criteria should be fulfilled simultaneously. The major problems in using NPs as a PSP are connected with detection of analytes and suspension stability of the NPs. A partial filling technique was used to allow detection of the analytes. In CEC, the main problems are the poor mass transfer, the poor phase ratio, and problems with the column itself.

The advantage related to the implementation of NPs compared with their implementation in HPLC and micro-LC is probably the difficulties encountered in developing frits for nanomaterials. Packed particle beds require a retaining frit, often prepared by sintering particles of silica or its derivatives, or *in situ* polymerization. However, these frits have the disadvantage of introducing band-broadening effects. Owing to this problem, most of the early reports described the use of micrometer-sized particles in these separation methods. Another problem with stationary phases in chromatography is that they may be contaminated by substances present in the sample and electrolyte and by the analytes themselves. Contamination is of specific concern in the analysis of samples in complex matrixes, such as blood plasma or cell lysates. Moreover, the simplicity of the process to use NPs in CE compared with the other chromatographic methods makes CE the best alternative to investigate the properties of new materials in separation techniques.

Finally, future studies on NPs and their applications in CE should cover the following topics:

(1) Synthesis of smaller, more porous, and more monodisperse NPs for faster mass transfer, higher sample capacity, and narrower band broadening.
(2) Design of highly selective interaction sites on the surface of NPs for on-line preconcentration and screening of target compounds such as biomacromolecules and small drug molecules. Immobilization of aptamers and antibodies on the particle surface and preparation of surface-imprinted core–shell NPs are promising strategies for separation with high selectivity.
(3) Further application of NPs in chip-based electrophoresis for faster separation. *In situ* synthesis of NPs in chips will eliminate tedious solution-transfer procedures.
(4) Analysis of nanomaterials by NP-based CE techniques. This topic involves classification and characterization of NPs of different sizes, separation and

purification of nanomaterials of different shapes (e.g., nanorod, nanotube, nanobelt, and nanocage), and determination of some important properties of interesting nanomaterials.

(5) Application of NPs in highly sensitive detection of a variety of analytes. As mentioned earlier, fluorescence and electrochemical detection are favorable choices in NP-based CE analysis. Utilizing the fluorescence-quenching effect of NPs toward specific molecules and/or fluorescence resonance energy transfer between NPs and analytes may offer sensitive online detection strategies. Some NPs with electrocatalytic activity could be used as electrode modifiers so as to improve detection sensitivity. With the development of materials science, more and more novel NPs will be proposed and employed in separation science.

REFERENCES

1. Feynman, R. (1960) *Eng. Sci. 23*, 22.
2. Drexler, K.E. (1986) Anchor Press/Doubleday, New York.
3. European Commission (2002) *Nanotechnology, Small Science with a Big Potential. European Research in Action.* Office for Official Publications of the European Communities, Luxembourg.
4. Zhao, H., and Ning, Y. (2000) *Gold Bull. 33*, 103.
5. Faraday, M. (1857) *Philos. Trans. R. Soc. Lond. 147*, 145.
6. Kroto, H.W., Heath, J.R., O'Brien, S.C., Curl, R.F., and Smalley, R.E. (1985) *Nature 318*, 162.
7. Iijima, S. (1991) *Nature 354*, 56.
8. Valcarcel, M., Cardenas, S., Simonet, B.M., Moliner-Martinez, Y., and Lucena, R. (2008) *Trends Anal. Chem. 27*, 34.
9. Nirmal, M., Dabbousi, B.O., Bawendi, M.G., Macklin, J.J., Trautman, J.K., Harris, T.D., and Brus L.E. (1996) *Nature 383*, 802.
10. Tasco, V., Todaro, M.T., De Vittorio, M., De Giorgi, M., Cingolani, R., Passaseo, P., Ratajczak, J., and Katcki, J.W. (2004) *Appl. Phys. Lett 84*, 4155.
11. Darugar, Q., Qian, W., and El-Sayed, M.A. (2006) *Appl. Phys. Lett. 88*, 261108.
12. Murray, C.B., Norris, D.J., and Bawendi, M.G. (1993) *J. Am. Chem. Soc. 115*, 8706.
13. Crouch, D., Norager, S., O'Brien, P., Park, J.H., and Pickett, N. (2003) *Philos. Trans. R. Soc. Lond. A 361*, 297.
14. Lee, S., Shin, D.W., Kim, W.M., Cheong, B., Lee, T.S., Lee, K.S., and Cho, S. (2006) *Thin Solid Films 514*, 296.
15. Schüler, A., Gampp, R., and Oelhafen, P. (1999) *Phys. Rev. B 60*, 16164.
16. Schüler, A., Videnovic, I.R., Oelhafen, P., and Brunold, S. (2001) *Sol. Energy Mater. Sol. Cell 69*, 271.
17. Ishikawa, T., Nishimura, T., Kohmoto, S., and Asakawa, K. (2000) *Appl. Phys. Lett. 76*, 167.
18. Parvathy, N.N., Pajonk, G.M., and Venkateswara Rao, A. (1997) *J. Cryst. Growth 179*, 249.
19. Lifshitz, E., Dag, I., Litvin, I., Hodes, G., Gorer, S., Reisfeld, R., Zelner, M., and Minti, H. (1998) *Chem. Phys. Lett. 288*, 188.
20. Jain, P.K., Lee, K.S., El-Sayed, I.H., and El-Sayed, M.A. (2006) *J. Phys. Chem. B 110*, 7238.
21. Khlebtsov, N.G., Trachuk, L.A., and Melnikov, A.G. (2005) *Opt. Spectrosc. 98*, 77.
22. Lee, K.S., and El-Sayed, M.A. (2005) *J. Phys. Chem. B 109*, 20331.
23. Perez-Juste, J., Pastoriza-Santos, I., Liz-Marzan, L.M., and Mulvaney, P. (2005) *Coord. Chem. Rev. 249*, 1870.
24. Sonnichsen, C., Reinhard, B.M., Liphardt, J., and Alivisatos, A.P. (2005) *Nat. Biotechnol. 23*, 741.
25. Turkevich, J., Stevenson, P.C., and Hillier, J. (1951) *Discuss Faraday Soc. 11*, 55.
26. Kimling, J., Maier, M., Okenve, B., Kotaidis, V., Ballot, H., and Plech, A. (2006) *J. Phys. Chem. B 110*, 15700.
27. Ren, C.B., Song, Y.H., Li, Z., and Zhu, G.Y. (2005) *Anal. Bioanal. Chem. 381*, 1179.
28. Percival, S., Bowler, P.G., and Dolman, J. (2007) *Int. Wound J. 4*, 186.
29. Vigneshwaran, N., Kathe, A., Varadarajan, P., Nachane, R., and Balasubramanya, R. (2007) *J. Nanosci. Nanotechnol. 7*, 1893.
30. Chen, X., and Schluesener, H.J. (2008) *Toxicol. Lett. 176*, 1.
31. Heggers, J., Goodheart, R., Washington, J., Mccoy, L., Carino, E., Dang, T., Edgar, P., Maness, C., and Chinkes, D. (2005) *J. Burn Care Rehabil. 26*, 53.
32. Davenport, K., and Keeley, F. (2005) *J. Hosp. Infect. 60*, 298.
33. Hardman, S., Cope, A., Swann, A., Bell, P., Naylor, A., and Hayes, P. (2004) *Ann. Vasc. Surg. 18*, 308.
34. Lansdown, A. (2006) *Curr. Probl. Dermatol. 33*, 17.
35. Elechiguerra, J.L., Burt, J., Morones, J., Camacho-Bragado, A., Gao, X., Lara, H., and Yacaman, M. (2005) *J. Nanobiotechnol. 3*, 6.
36. Raffi, M., Hussain, F., Bhatti, T., Akhter, J.I., Hameed, A., and Hasan, M.M. (2008) *Sci. Technol. 24*, 192.
37. Sergeev, M.B., Kasaikin, A.V., and Litmanovich, A.E. (1999) *Mendeleev Commun. 9*, 130.
38. Sun, Y., and Xia, Y. (2002) *Science 298*, 2176.
39. Zhang, Y., Chen, F., and Zhuang, J. (2002) *Chem. Commun. (Camb.) 7*, 2814.
40. Pyatenko, A., Shimokawa, K., and Yamaguchi, M. (2004) *J. Appl. Phys. A Mater. Sci. Process. A 79*, 803.
41. Bogle, K.A., Dhole, S.D., and Bhoraskar, V.N. (2006) *Nanotechnology 17*, 3204.
42. Chen, X., and Schluesener, H.J. (2007) *Toxicol. Lett. 176*, 1.
43. Barbe, C.J., Arendse, F., and Grätzel, M. (1997) *J. Am. Ceram. Soc. 80*, 3157.

44. Ahmed, M., and Attia, Y.A. (1995) *J. Non-Cryst. Solids 186*, 402.
45. Ferroni, M., Guidi, V., Martinelli, G., and Nelli, P. (1996) *Nanostruct. Mater. 7*, 709.
46. Rekoske, J.E., and Barteau, M.A. (1997) *J. Phys. Chem. B 101*, 1113.
47. Yeh, Y.C., Tseng, T.Y., and Chang, D.A. (1990) *J. Am. Ceram. Soc. 73*, 1992.
48. Atta, A.K., Biswas, P.K., and Ganguli, D. (1992) *Mater. Lett. 15*, 99.
49. Barringer, E.A. and Bowen, H.K. (1985) *Langmuir 1*, 414.
50. Rozita, Y., Brydson, R., and Scott, A.J. (2010) *J. Phys. Conf. Ser. 241*, 012096.
51. Hoet, P.H.M., Brueske-Hohlfeld, I., and Salata, O. (2004) *J. Nanotoxicol. 2*, 1.
52. Roy, I., Ohulchanskyy, T., Bharali, D., Pudavar, H., Mistretta, R., Kaur, N., and Prasad, P. (2005) *Proc. Natl. Acad. Sci. U.S. A 102*, 279.
53. Csogor, Z.S., Nacken, M., Sameti, M., Lehr, C.M., and Schmidt, H. (2003) *Mater. Sci. Eng. C 23*, 93.
54. Wang, W., Gu, B., Liang, L., and Hamilton, W. (2003) *J. Phys. Chem. B 107*, 12113.
55. Josephson, L., Perez, J.M., and Weissleder, R. (2001) *Angew. Chem. Int. Ed. 40*, 3204.
56. Huang, S.H., Liao, M.H., and Chen, D.H. (2003) *Biotechnol. Prog. 19*, 1095.
57. Wang, Z.H., Choi, C.J., Kim, B.K., Kim, J.C., and Zhang, Z.D. (2003) *Carbon 41*, 1751.
58. Rossi, L.M., Quach, A.D., and Rosenzweig, Z. (2004) *Anal. Bioanal. Chem. 380*, 606.
59. Pumera, M., Castañeda, M.T., Pividori, M.I., Eritja, R., Merkoc,, A., and Alegret, S. (2005) *Langmuir 21*, 9625.
60. Zhao, G., Feng, J.J., Zhang, Q.L., Li, S.P., and Chen, H.Y. (2005) *Chem. Mater. 17*, 3154.
61. Xiao, Y., Ju, H.X., and Chen, H.Y. (1999) *Anal. Chim. Acta 391*, 73.
62. Xu, C.H., He, C.P., and Fang, Y.Z. (2001) *J. Electroanal. Chem. 510*, 78.
63. Shi, X., Bányai, I., and Rodriguez, K. (2006) *Electrophoresis 27*, 1758.
64. Venuganti, V.V.K., and Perumal, O.P. (2008) *J. Pharm. Sci.*
65. Klajnert, B., and Bryszewska, M. (2001) *Acta Biochim. Pol. 48*(1), 199.
66. Buhleier, E., Wehner, W., and Vögtle, F., (1978) *Synthesis, 2*, 155.
67. Tomalia, D.A., Baker, H., and Dewald, J. (1985) *Polym. J. 17*(1), 117.
68. Newkome, G.R., Behera, R.K., Moorefield, C.N., and Baker, G.R. (1991) *J. Org. Chem. 56*, 7162.
69. Malik, N., Wiwattanapatapee, R., and Klopsch, R. (2000) *J. Control. Release 65*, 133.
70. Delong, R., Stephenson, K., and Loftus, T. (1997) *J. Pharm. Sci. 86*, 762.
71. Stöckigt, D., Lohmer, G., and Belder, D. (1996) *Rapid Commun. Mass Spectrom. 10*, 521.
72. Klajnert, B., and Bryszewska, M. (2002) *Cell. Mol. Biol. Lett. 7*, 1087.
73. Nourse, A., Millar, D.B., and Minton, A.P. (2000) *Biopolymers 53*, 316.
74. Krämer, M. (2004) Polymeric Nanocarriers with Dendritic Core–Shell Architectures. Inaugural Dissertation, Breisgau.
75. Tomalia, D.A. (2005) *Prog. Polym. Sci. 30*, 294.
76. Dufès, Ch., Uchegbu, I.F., and Schätzlein, A.G. (2005) *Adv. Drug Deliv. Rev. 57*, 2177.
77. Cheng, M.-D. (2005) *J. Environ. Sci. Health 39*, 2691.
78. Webster, T.J., Ergun, C., Doremus, R.H., Siegel, R.W., and Bizios, R. (2000) *J. Biomed. Mater. Res. 51*, 475.
79. Cheng, F.-Y., Wang, S.P.-H., Su, C.-H., Tsai, T.-L., Wu, P.-C., Shieh, D.-B., Chen, J.-H., Hsieh, P.C.-H., and Yeh, C.-S. (2008) *Biomaterials 29*, 2104.
80. Chung, Y.-C., Chen, I.H., and Chen, C.-J. (2008) *Biomaterials 29*, 1807.
81. Deng, C., Chen, J., Chen, X., Xiao, C., Nie, L., and Yao, S. (2008) *Biosens. Bioelectron. 23*, 1272.
82. Engel, E., Michiardi, A., Navarro, M., Lacroix, D., and Planell, J.A. (2008) *Trends Biotechnol. 26*, 39.
83. Murray, R.W. (2009) *Anal. Chem. 81*, 1723.
84. Liu, H., and Webster, T. (2007) *J. Biomater. 28*, 354.
85. Murray, R.W. (2008) *Chem. Rev. 108*, 2688.
86. Valcarcel, M., Simonet, B.M., and Cardenas, S. (2008) *Anal. Bioanal. Chem. 391*, 1881.
87. Wang, J., Chen, G., Chatrathi, M.P., and Musameh, M. (2004) *Anal. Chem. 76*, 298.
88. Sotiropoulou, S., Gavalas, V., Vamvakaki, V., and Chaniotakis, N.A. (2003) *Biosens. Bioelectron. 18*, 211.
89. Glynou, K., Ioannou, P.C., Christopoulos, T.K., and Syriopoulou, V. (2003) *Anal. Chem. 75*, 4155.
90. Park, J., Kurosawa, S., Watanabe, J., and Ishihara, K. (2004) *Anal. Chem. 76*, 2649.
91. Bavastrello, V., Stura, E., Carrara, S., Erokhin, V., and Nicolini, C. (2004) *Sens. Actuators B 98*, 247.
92. Haick, H. (2007) *J. Phys. D Appl. Phys. 40*, 7173.
93. Kiyokatsu, J., Kunihiko, Y., John, C.F., and Wilton, R.B. (1992) *J. Microcol. Sep. 4*, 187.
94. Huang, M.-F., Huang, C.-C., and Chang, H.-T. (2003) *Electrophoresis 24*, 2896.
95. Mari, T., Katsuyama, Y., Nogami, K., Nagata, H., Wakuda, K., Fujimoto, M., Nagasaki, Y., Yoshikawa, K., Kataoka, K., and Baba, Y. (2005) *Lab Chip 5*, 199.
96. Gomes, D., Nunes, S.P., and Peinemann, K.-V. (2005) *J. Membr. Sci. 246*, 13.
97. Cai, Y.-q., Cai, Y.-e., Mou, S.-f., and Lu, Y.-q. (2005) *J. Chromatogr. A 1081*, 245.
98. Xiao, C., Han, S., Wang, Z., Xing, J., and Wu, C. (2001) *J. Chromatogr. A 927*, 121.

99. Sudhir, P.-R., Wu, H.-F., and Zhou, Z.-C. (2005) *Anal. Chem. 77*, 7380.
100. Zhang, Z., Yan, B., Liao, Y., and Liu, H. (2008) *Anal. Bioanal. Chem. 391*, 925.
101. Saridara, C., and Mitra, S. (2005) *Anal. Chem. 77*, 7094.
102. Li, Y., Chen, Y., Xiang, R., Ciuparu, D., Pfefferle, L.D., Horvath, C., and Wilkins, J.A. (2005) *Anal. Chem. 77*, 1398.
103. Lu, C., Zu, Y. B., and Yam, V.W.W. (2007) *Anal. Chem. 79*, 666.
104. Trojanowicz, M. (2006) *Trends Anal. Chem. 25*, 480.
105. Wang, J. (2005) *Analyst 130*, 421.
106. Panini, V., Messina, G.A., Salinas, E., Fernández, H., and Raba, J. (2008) *Biosens. Bioelectron. 23*, 1145.
107. Martinez, N.A., Messina, G.A., Bertolino, F.A., Salinas, E., and Raba, J. (2008) *Sens. Actuators B 133*, 256.
108. Smith, A.M., and Nie, S. (2004) *Analyst 129*, 672.
109. Liu, J., Liu, J., Yang, L., Chen, X., Zhang, M., Meng, F., Luo, T., and Li, M. (2009) *Sensors 9*, 7343.
110. Arter, J.A., Taggart, D.K., McIntire, T.M., Penner, R.M., and Weiss, G.A. (2010) *Nano Lett. 10*, 4858.
111. Urraca, J.L., Moreno-Bondi, M.C., Orellana, G., Sellergren, B., and Hall, A.J. (2007) *Anal. Chem. 79*, 4915.
112. Suárez, B., Simonet, B.M., Cárdenas, S., and Valcárcel, M. (2007) *J. Chromatogr. A 1159*, 203.
113. López-Lorente, A.I., Simonet, B.M., and Valcárcel, M. (2010) *Anal. Chem. 82*, 5399.
114. Suárez, B., Santos, B., Simonet, B.M., Cárdenas, S., and Valcárcel, M. (2007) *J. Chromatogr. A 1175*, 127.
115. Majumder, M., Chopra, N., and Hinds, B.J. (2005) *J. Am. Chem. Soc. 127*, 9062.
116. Göttlicher, B., and Bächmann, K. (1997) *J. Chromatogr. A 780*, 63.
117. Pretorius, V., Hopkins, B.J., and Schieke, J.D. (1974) *J. Chromatogr. 99*, 23.
118. Jorgenson, J.W., and Lukacs, K.D. (1981) *J. Chromatogr. 218*, 209.
119. Tsuda, T., Nomura, K., and Nakagawa, G. (1982) *J. Chromatogr. 248*, 241.
120. Guihen, E., and Glennon, J.D. (2004) *J. Chromatogr. A 1044*, 67.
121. Svec, F. (2005) *J. Sep. Sci. 28*, 729.
122. Birnbaum, S., and Nilsson, S. (1992) *Anal. Chem. 64*, 2872.
123. Eeltink, S., and Svec, F. (2007) *Electrophoresis 28*, 137.
124. Luong, J.H.T., Bouvrette, P., Liu, Y.L., Yang, D.Q., and Sacher, E. (2005) *J. Chromatogr. A 1074*, 187.
125. Sombra, L., Moliner-Martínez, Y., Cárdenas, S., and Valcárcel, M. (2008) *Electrophoresis 29*, 3850.
126. Stege, P.W., Sombra, L.L., Messina, G., Martinez, L.D., and Silva, M.F. (2010) *Electrophoresis 31*, 2242.
127. Iijima, S., Yudasaka, M., Yamada, R., Bandow, S., Suenaga, K., Kokai, F., and Takahashi, K. (1999) *Chem. Phys. Lett. 309*, 165.
128. Berliocchi, M., Orlanducci, S., Reale, A., Regoliosi, P., Di Carlo, A., Lugli, P., Terranova, M.L., Brunetti, F., Bruni, G., and Cirillo, M. (2004) *Synth. Met. 145*, 171.
129. Pagona, G., Sandabayaka, A.S.D., Maigne, A., Fan, J., Papavassiliou, G.C., Petsalakis, I.D., Steele, B.R., Yudasaka, M., Iijima, S., Tagmatarchis, N., and Ito, O. (2007) *Chem. Eur. J. 13*, 7600.
130. Jiménez-Soto, J.M., Moliner-Martínez, Y., Cárdenas, S., and Valcárcel, M. (2010) *Electrophoresis 31*, 1681.
131. Li, Y., Chen, Y., Xiang, R., Ciuparu, D., Pfefferle, L.D., Horvath, C., and Wilkins, J.A. (2005) *Anal. Chem. 77*, 1398.
132. Kleindienst, G., Huber, C.G., Gjerde, D.T., Yengoyan, L., and Bonn, G.K. (1998) *Electrophoresis 19*, 262.
133. Huber, C.G., Premstaller, A., and Kleindienst, G. (1999) *J. Chromatogr. A 849*, 175.
134. Breadmore, M.C., Boyce, M.C., Macka, M., Avdalovic, N., and Haddad, P.R. (2000) *Analyst 125*, 799.
135. Breadmore, M.C., Macka, M., Avdalovic, N., and Haddad, P.R. (2001) *Anal. Chem. 73*, 820.
136. Breadmore, M.C., Palmer, A.S., Curran, M., Macka, M., Avdalovic, N., and Haddad, P.R. (2002) *Anal. Chem. 74*, 2112.
137. Boyce, M.C., Breadmore, M., Macka, M., Doble, P., and Haddad, P.R. (2000) *Electrophoresis 21*, 3073.
138. Breadmore, M.C., Macka, M., Avdalovic, N., and Haddad, P.R. (2000) *Analyst 125*, 1235.
139. Breadmore, M.C., Boyce, M., Macka, M., Avdalovic, N., and Haddad, P.R. (2000) *J. Chromatogr. A 892*, 303.
140. Deng, Q., Lun, Z., Gao, R., Zhang, L., Zhang, W., and Zhang, Y. (2006) *Electrophoresis 27*, 4351.
141. Zhang, Y.P., Ye, X.W., Tian, M.K., and Qu, L.B. (2008) *J. Chromatogr. A 1188*, 43.
142. Ou, J., Li, X., Feng, S., and Dong, J. (2007) *Anal. Chem. 79*, 639.
143. Wang, H., Zhu, Y., Yan, X., Gao, R., and Zheng, J. (2006) *Adv. Mater. 18*, 3266.
144. Cacho, C., Schweitz, L., Turiel, E., and Pérez-Conde, C. (2008) *J. Chromatogr. A 1179*, 216.
145. Neiman, B., Grushka, E., and Lev, O. (2001) *Anal. Chem. 73*, 5220–5227.
146. O'Mahony, T., Owens, V.P., Murrihy, J.P., Guihen, E., Holmes, J.D., and Glennon, J.D. (2003) *J. Chromatogr. A 1004*, 181.
147. Yang, L., Guihen, E., Holmes, J.D., Loughran, M., O'Sullivan, G.P., and Glennon, J.D. (2005) *Anal. Chem. 77*, 1840.
148. Yang, L., Guihen, E., and Glennon, J.D. (2005) *J. Sep. Sci. 28*, 757.
149. Liu, F.K., Hsu, Y.T., and Wu, C.H. (2005) *J. Chromatogr. A 1083*, 205.
150. Liu, F., Hsu, Y., and Wu, C. (2005) *J. Chromatogr. A 1083*, 205.
151. Cao, Q., Xu, Y., Liu, F., Svec, F., and Fréchet, J.M.J. (2010) *Anal. Chem. 82*, 7416.

152. Zhou, L., Glennon, J.D., and Luong, J.H.T. (2010) *Anal. Chem. 82*, 6895.
153. Qu, Q., Liu, D., Mangelings, D., Yang, C., and Hu, X. (2010) *J. Chromatogr. A 1217*, 6588.
154. Hsieh, Y.L., Chen, T.H., Liu, C.P., and Liu, C.Y. (2005) *Electrophoresis 26*, 4089.
155. Hsieh, Y.L., Chen, T.H., and Liu, C.Y. (2006) *Electrophoresis 27*, 4288.
156. Lia, T., Xua, Y., and Fenga, Y.Q. (2009) *J. Liq. Chromatogr. Relat. Technol. 32*, 2484.
157. Stege, P.W., Raba, J., and Messina, G.A. (2010) *Electrophoresis 31*, 3475.
158. Valtcheva, L., Mohammad, J., Pettersson, G., and Hjertén, S. (1993) *J. Chromatogr. 638*, 263.
159. Schweitz, L., Spegél, P., and Nilsson, S. (2000) *Analyst 125*, 1899.
160. Göttlicher, B., and Bächmann, K. (1997) *J. Chromatogr. A 768*, 320.
161. Viberg, P., Jornten-Karlsson, M., Petersson, P., Spégel, P., and Nilsson, S. (2002) *Anal. Chem. 74*, 4595.
162. Kroto, H.W., Heath, J.R., O'Brien, S.C., Curl, R.F., and Smalley, R.E. (1985) *Nature 318*, 162.
163. Chiang, L.Y., Bhonsle, J.B., and Hwu, J.R. (1996) *Tetrahedron 52*, 4963.
164. Sun, Y.P., Liu, B., and Moton, D.K. (1996) *Chem. Commun. 24*, 2698.
165. Guidi, D., Hungerbuhler, H., and Asmus, K.D. (1997) *J. Phys. Chem. 101*, 1783.
166. Guidi, D.M. (1997) *J. Phys. Chem. A 101*, 3895.
167. Tatiana, D.R., Prato, M., Novello, F., Maggini, M., and Banfi, E. (1996) *J. Org. Chem. 61*, 9070.
168. Bensasson, R.V., Bienvenue, E., Dellinger, M., Leach, S., and Seta, P. (1994) *J. Phys. Chem. 98*, 3492.
169. Andersson, T.K., Nilsson, K., Sundahl, M., Westman, G., and Wennerstrom, O. (1992) *J. Chem. Soc. Chem. Commun. 8*, 604.
170. Hungerbuhler, H., Guldi, H., and Asmus, K.D. (1993) *J. Am. Chem. Soc. 115*, 3386.
171. Andersson, T., Sundah, M., Westman, G., and Wennerstrom, O. (1994) *Tetrahedron Lett. 35*, 7103.
172. Treubig, J.M., and Brown, P.R. (2000) *J. Chromatogr. A 873*, 257.
173. Moliner-Martínez, Y., Cárdenas, S., and Valcárcel, M. (2007) *J. Chromatogr. A 1167*, 210.
174. Moliner-Martínez, Y., Barrios, M., Cárdenas, S., and Valcárcel, M. (2008) *J. Chromatogr. A 1194*, 128.
175. Wang, Z.H., Luo, G.A., Chen, J.F., Xiao, S.F., and Wang, Y.M. (2003) *Electrophoresis 24*, 4181.
176. Na, N., Hu, Y.P., Jin, O.Y., Baeyens, W.R.G., Delanghe, J.R., Taes, Y.E.C., Xie, M.X., Chen, H.Y., and Yang, Y.P. (2006) *Talanta 69*, 866.
177. Xiong, X., Ouyang, J., Baeyens, W.R.G., Delanghe, J.R., Shen, X.M., and Yang, Y.P. (2006) *Electrophoresis 27*, 3243.
178. Xu, Y., and Li, S.F.Y. (2006) *Electrophoresis 27*, 4025.
179. Wallingford, R.A., and Ewing, A.G. (1989) *Anal. Chem. 61*, 98.
180. Breadmore, M.C., Macka, M., and Haddad, P.R. (1999) *Electrophoresis 20*, 1987.
181. Viberg, P., Jornten-Karlsson, M., Petersson, P., Spegel, P., and Nilsson, S. (2002) *Anal. Chem. 74*, 4595.
182. Nilsson, C., Viberg, P., Spegel, P., Jornten-Karlsson, M., Petersson, P., and Nilsson, S. (2006) *Anal. Chem. 78*, 6088.
183. Song, J.M., Asthana, A., and Kim, D.P. (2006) *Talanta 68*, 940.
184. Na, N., Hu, Y.P., Ouyang, J., Baeyens, W.R.G., Delanghe, J.R., and De Beer, T. (2004) *Anal. Chim. Acta 527*, 139.
185. Vlatakis, G., Andersson, L.I., Muller, R., and Mosbach, K. (1993) *Nature 361*, 645.
186. Schweitz, L., Spégel, P., and Nilsson, S. (2000) *Analyst 125*, 1899.
187. Spégel, P., Schweitz, L., and Nilsson, S. (2001) *Electrophoresis 22*, 3833.
188. de Boer, T., Mol, R., de Zeeuw, R.A., de Jong, G.J., Sherrington, D.C., Cormack, P.A.G., and Ensing, K. (2002) *Electrophoresis 23*, 1296.
189. Spégel, P., Schweitz, L., and Nilsson, S. (2003) *Anal. Chem. 75*, 6608.
190. Priego-Capote, F., Ye, L., Shakil, S., Shamsi, S.A., and Nilsson, S. (2008) *Anal. Chem. 80*, 2881.
191. Göttlicher, B., and Bächmann, K. (1997) *J. Chromatogr. A 768*, 320.
192. Bächmann, K., and Göttlicher, B. (1997) *Chromatographia 45*, 249.
193. Fujimoto, C., and Muranaka, Y. (1997) *J. High Resolut. Chromatogr. 20*, 400.
194. Neiman, B., Grushka, E., Gun, J., and Lev, O. (2002) *Anal. Chem. 74*, 3484.
195. Kuo, I., Huang, Y. and Chang, H. (2005) *Electrophoresis 26*, 2643.
196. Wang, Y., Baeyens, W.R.G., Huang, C., Fei, G., He, L., and Ouyang, J. (2009) *Talanta 77*, 1667.
197. Huang, M.F., Huang, C.C., and Chang, H.T. (2003) *Electrophoresis 24*, 2896.
198. Huang, M.F., Kuo, Y.C., Huang, C.C., and Chang, H.T. (2004) *Anal. Chem. 76*, 192.
199. Chiou, S.H., Huang, M.F., and Chang, H.T. (2004) *Electrophoresis 25*, 2186.
200. Tseng, W.L., Huang, M.F., Huang, Y.F., and Chang, H.T. (2005) *Electrophoresis 26*, 3069.
201. Chen, Y.L., Shih, C.J., Ferrance, J., Chang, Y.S., Chang, J.G., and Wu, S.M. (2009) *J. Chromatogr. A 1216*, 1206.
202. Yu, C.-J., Su, C.L., and Tseng, W.L. (2006) *Anal. Chem. 78*, 8004.
203. Ivanov, M.R., Bednar, H.R., and Haes, A.J. (2009) *ACS Nano 3*, 386.
204. Yang, L., Chen, C., Liu, X., Shi, J., Wang, G., Zhu, L., Guo, L., Glennon, J., Scully, N., and Doherty, D. (2010) *Electrophoresis 31*, 1697.

205. Lin, Y.W., Huang, M.F., and Chang, H.T. (2005) *Electrophoresis 26*, 320.

206. Weng, X.X., Bi, H.Y., Liu, B.H., and Kong, J.L. (2006) *Electrophoresis 27*, 3129.

207. Tabuchi, M., Ueda, M., Kaji, N., Yamasaki, Y., Nagasaki, Y., Yoshikawa, K., Kataoka, K., and Baba, Y. (2004) *Nat. Biotechnol. 22*, 337.

208. Tabuchi, M., Katsuyama, Y., Nogami, K., Nagata, H., Wakuda, K., Fujimoto, M., Nagasaki, Y., Yoshikawa, K., Kataoka, K., and Baba, Y. (2005) *Lab Chip 5*, 199.

209. Qu, P., Lei, J., Zhang, L., Ouyang, L., and Ju, H. (2010) *J. Chromatogr. A 1217*, 6115.

210. Ohno, T., Matsuda, T., Suzuki, H., and Fujimoto, M. (2006) *Adv. Powder Technol. 17*, 167.

211. Pumera, M., Wang, J., Grushka, E., and Polsky, R. (2001) *Anal. Chem. 73*, 5625.

212. Lin, Y.W., Huang, M.J., and Chang, H.T. (2003) *J. Chromatogr. A 1014*, 47.

213. Lin, Y.W., and Chang, H.T. (2005) *J. Chromatogr. A 1073*, 191.

214. Wang, A.J., Xu, J.J., Zhang, Q., and Chen, H.Y. (2006) *Talanta 69*, 210.

215. Wang, A.J., Xu, J.J., and Chen, H.Y. (2006) *J. Chromatogr. A 1107*, 257.

216. Qiu, J.-D., Wang, L., Liang, R.-P., and Wang, J.-W. (2009) *Electrophoresis 30*, 3472.

217. Doyle, P., Bibette, J., Bancaud, A., and Viovy, J. (2002) *Science 295*, 5563.

20

MICROCHIP CAPILLARY ELECTROPHORESIS TO STUDY THE BINDING OF LIGANDS TO TEICOPLANIN DERIVATIZED ON MAGNETIC BEADS

Toni Ann Riveros, Roger Lo, Xiaojun Liu, Marisol Salgado, Hector Carmona, and Frank A. Gomez

20.1 INTRODUCTION

Microfluidics has revolutionized many areas of analytical chemistry allowing for rapid, efficient analyses with minuscule reagent and sample consumption over conventional analytical techniques [1–3]. Applications incorporating microfluidics include cell sorting, microreactor technologies, the polymerase chain reaction (PCR), protein crystallization, and organic synthesis and enzymatic digestion [4–11].

The use of magnetic polymeric particles with surface functional groups has been applied to many areas of chemistry, biology, and medicine. Recent work has detailed the use of magnetic microspheres in thermotherapy for cancer cells, immunoassays, DNA hybridization, cell separation, protein and peptide separation, and pathogen detection [12–17]. Their simple separation and conjugation with a wide array of biomolecules has led to their integration in a variety of detection methods. Microbeads have a high density of surface functional groups making them easy to modify, thereby making it possible to use higher concentrations of samples leading to an increase in limits of detection (LOD). Various surface chemistry modifications are available for magnetic microbeads. Furthermore, microbeads are available in micron- and smaller-sized diameters appropriate for microfluidics applications and other micro total analysis systems (μTAS). The most important advantage of magnetic microsphere-based analysis is easy separation of bound complexes from unbound detectable species without the use of sophisticated and expensive techniques and instrumentation.

Herein, we detail several novel microfluidic bead-based assays to assess the binding of teicoplanin (Teic) from *Actinoplanes teichomyceticus*, immobilized on magnetic microspheres, to D-Ala–D-Ala terminus peptides [18–20]. In the first technique, frontal analysis microchip CE (FAMCE) is used to examine the binding of Teic, covalently attached to magnetic microbeads, to small peptides. Direct and competitive assays are demonstrated. The binding constants obtained from the FAMCE technique are comparable to those values using affinity CE, fluorescence spectrometer, and flow cytometry. In the second technique called microchip frontal affinity chromatography (MFAC), an electrokinetically controlled microfluidic system containing two magnetic microbead columns fabricated into the microchannels is used to estimate the binding of a small peptide to magnetic microbeads derivatized with Teic. One column is filled with Teic-modified beads and is used as the affinity column. The second column contains nonderivatized beads and is used as the control. The extent of interaction between the peptide and the two types of beads in either microchannel resulted in differences in migration time of the ligand, and this is used in the Scatchard analysis to obtain a value for the binding constant.

20.2 EXPERIMENTAL SECTION

20.2.1 Materials and Methods

5-Carboxyfluorescein–D-Ala–D-Ala–D-Ala (DA3) was custom synthesized by Anaspec (San Jose, CA). Glycidyl

Capillary Electrophoresis and Microchip Capillary Electrophoresis: Principles, Applications, and Limitations, First Edition.
Edited by Carlos D. García, Karin Y. Chumbimuni-Torres, and Emanuel Carrilho.

methacrylate (GMA), dimethylacrylamide (DMA), tetramethylethylenediamine (TMED), N_α,N_ε-diacetyl-Lys-D-Ala–D-Ala (DA2), *N*-(2-hydroxyethyl)piperazine-*N*′-(2-ethanesulfonic acid)) (HEPES), Tween-20, 2-(*N*-morpholino)ethanesulfonoic acid (MES), resazurin sodium salt, bovine serum albumin (BSA), *N*-(3-dimethylaminopropyl)-*N*′-ethylcarbodiimide hydrochloride (EDC), dimethylacrylamide (DMA), tetramethylethylenediamine (TMED), and *N*-(3-dimethylaminopropyl)-*N*′-ethylcarbodiimide hydrochloride (EDC) were purchased from Sigma (St. Louis, MO). Zinc chloride, 2-(*N*-morpholino)-ethanesulfonic acid (MES), potassium chloride and dialysis tubing with 3500 molecular weight cutoff was purchased from Fisher (Pittsburgh, PA). Negative type photoresist SU-8 2025 and its developer were obtained from Microchem (Newton, MA). Sylgard 184 PDMS oligomer and its cross-linking agent were purchased from Dow Corning (Midland, MI). Carboxylic group terminated Dynal® magnetic beads M-270 (⊥2.8 μm diameter) were purchased from Invitrogen (Carlsbad, CA). Teicoplanin (Teic) was obtained from Advanced Separation Technologies (Whippany, NJ).

20.2.1.1 Equipment and Fabrication of the Microchips MCE studies were performed on a homemade microfluidic system consisting of two high voltage power supplies and a confocal laser induced fluorescence (LIF) detector. Briefly, a laser beam was horizontally focused into the microchannel. In the vertical direction, fluorescence was detected by a photomultiplier tube (PMT) after passing through two filters and two objectives.

The PDMS chips were prepared by using a mold created by soft photolithography. The pattern (Figure 20.1a) was designed using AutoCAD software (San Rafael, CA) and printed as a high resolution (20,000 dpi) photomask (CAD/Art Services, Inc., OR). Negative-type photoresist was spin-coated onto a 3 in. silicon wafer at 1200 rpm for 30 s. Sylgard 184 PDMS prepolymer was mixed thoroughly with its curing agent at 10:1 (v/v), degassed and then baked for 2 h at 75°C. The PDMS replicas were peeled off from the mold. Holes (3 mm diameter) were made for the buffer and sample wells by punching through the PDMS chip. The channels were 25 μm high and 50 μm wide at the top. The distances from the wells 1, 2, 4, and 5 to the well 3 were 5.0, 4.8, 4.0, and 5.0 cm, respectively (Figure 20.1a). The PDMS replica was irreversibly sealed onto a clean substrate glass of 2 mm thickness after treated with oxygen plasma.

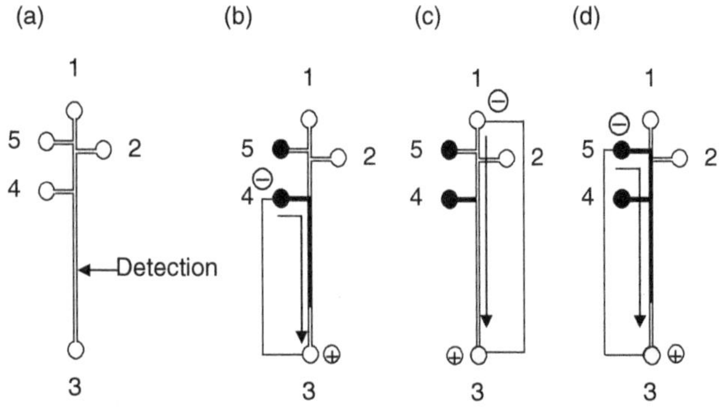

FIGURE 20.1 (a) Schematic of microchip and (b–d) MCE protocol. b–d represents the process for injecting the supernatant solution, rinsing the channel, and injecting the external standard. Arrows indicated the solution movement. 1 and 2, buffer wells; 3, waste well; 4, the supernatant well; 5, the external standard well.

20.2.1.2 Surface Coating The channel surface was modified with epoxy-modified hydrophilic polymer following the method described by Wu et al. [21]. In brief, 5.0% DMA (w/v) and 0.1% GMA (v/v) were mixed in distilled water, and thoroughly degassed for 10 min. To initiate the polymerization reaction, 0.1% TMED (v/v) and 0.05% potassium persulfate (w/v) were added into the mixture and the reaction was allowed to run for 30 min at room temperature. Then, to remove unpolymerized monomers and small polymer molecules, the polymer solution was extensively dialyzed by 3500 molecular weight cutoff dialysis membranes for 2 days with distilled water changed every 12 h. The polymer solution was introduced into the just sealed microchannel, and incubated for 15 min at room temperature. The solution was completely pumped out and PDMS microchip was directly heated at 110°C for 10 min.

20.2.1.3 Teic Immobilization on Magnetic Microbeads A solution (100 μL) of carboxylic acid terminated magnetic beads from a stock solution (2×10^9 beads mL^{-1}) was washed (5×) with MES buffer (25 mM, pH 5). A solution (60 μL) of Teic (10 mg mL^{-1}) in MES (25 mM, pH 5) was added to the beads and mixed well. A solution (30 μL) of freshly prepared EDC solution (100 mg mL^{-1} in cold water) and MES buffer (10 μL) were added to the microbead suspension. The mixture was incubated for 6 h at room temperature with slight shaking. The supernatant solution was discarded and microbeads were suspended in Tris (50 mM, pH 7.4) buffer for 15 min to quench nonreacted carboxylic acid groups. Microbeads were washed with phosphate buffer (100 mM, pH 7.4, 150 mM NaCl, 0.1% (w/v) bovine serum albumin) (5×) and then resuspended in 1 mL phosphate buffer (100 mM, pH 7.4, 150 mM NaCl).

20.2.2 Procedures

20.2.2.1 FAMCE Studies A solution of 20 mM HEPES (pH 7.40) containing 0.10% Tween-20 was used as running buffer. DA3 was dissolved in the running buffer and DA2 in a solution of DA3. For estimating the binding constant of DA3 to Teic, a constant concentration of a solution of DA3 was incubated with various concentrations of Teic immobilized on microbeads (Teic-beads) for 30 min. For DA2, a constant concentration of Teic-beads and DA3 was then incubated with various concentrations of DA2 for 30 min. The microbeads were then separated from the supernatant by use of an external magnet. The amount of free DA3 in the supernatant was measured with FAMCE using the original solution as the external standard. Figure 20.1b–d showed the voltage application sequence for the process. A voltage of 1.0 kV was applied to the supernatant and waste wells (Figure 20.1b) followed by application of a voltage of 1.3 kV to the buffer and waste wells (Figure 20.1c), and finally a voltage of 1.3 kV to the external standard and waste wells (Figure 20.1d), respectively, for 50 s. After each run, the channel was rinsed with the running buffer for 90 s.

20.2.2.2 MFAC Studies The binding study was performed on a homemade microfluidic system consisting of two high voltage power supplies and a confocal LIF detector described previously. A three-step procedure was used to fabricate the microchips for MFAC: (1) magnetic microbead derivatization with Teic; (2) fabrication of a hydrophilic microchip, and; (3) packing of magnetic beads into the microchannels. A solution (100 μL) of carboxylic acid terminated magnetic beads from a stock solution (2×10^9 beads mL^{-1}) was washed (5×) with 2-(*N*-morpholino)-ethanesulfonic acid (MES) buffer (25 mM, pH 5). A solution (60 μL) of Teic (10 mg mL^{-1}) in MES (25 mM, pH 5) was added to the beads and mixed. A solution (30 μL) of freshly prepared EDC solution (100 mg mL^{-1} in cold water) and MES buffer (10 μL) were added to the bead solution and the mixture was incubated for 6 h at room temperature with slight shaking. The solution was then discarded and beads were suspended in Tris (50 mM, pH 7.4) buffer for 15 min to quench nonreacted carboxylic acid groups. The beads were then washed with PBS (100 mM, pH 7.4, 150 mM NaCl, 0.1% BSA) (5×) and resuspended in PBS (1 mL) (100 mM, pH 7.4, 150 mM NaCl) buffer. With the help of a "holding" magnet, two 2 mm diameter rare earth magnets were placed above the side channels where the magnetic microbeads were initially packed.

20.3 RESULTS AND DISCUSSION

20.3.1 FAMCE Studies

The goal of this work is to develop the FAMCE technique to estimate the binding parameters of ligands to the receptors. The binding of Teic to terminal D-Ala–D-Ala peptides was chosen as the model system. As a glycopeptide antibiotic, Teic belongs to the vancomycin (Van) family. Van-group antibiotics kill bacterial cells by inhibiting peptidoglycan biosynthesis [22–25]. They function by binding to the terminal D-Ala–D-Ala dipeptide (binding also occurs to D-Tyr–D-Ala and D-Tyr–D-Ser terminus peptides) of bacterial cell wall precursors, thereby, impeding further processing of these intermediates into peptidoglycan. Prevention of peptidoglycan formation makes bacterial cells prone to lysis and represents one of the three main mechanisms of antibiotic action the others being inhibition of protein synthesis and inhibitionof RNA/DNA synthesis. Besides Van, Teic has been used to treat methicillin-resistant *Staphylococcus aureus* infections [24,25]. Though it is not routinely used in hospitals, Teic is more active against *Staphylococcus* species than van including methicillin-resistant strains and has fewer side effects and exhibits a longer half-life in humans.

20.3.1.1 Nonspecific Adsorption Resistance

Magnetic Microbeads. The magnetic microbeads were made from highly cross-linked polystyrene with magnetic material precipitated in pores evenly distributed throughout the particles, and further coated with a hydrophilic layer of glycidyl ether for concealing the iron oxide inside the beads. Carboxylic acid groups are then introduced on the surface. Spectrofluorometric measurements were performed to monitor the nonspecific adsorption of DA3 on the magnetic microbeads. An aliquot of DA3 was mixed with nonderivatized microbeads and Teic-beads, respectively. Using fluorescence (excitation and emission at 488 and 520 nm, respectively), the amount of free DA3 was quantified. The relative amount of free DA3 in the supernatant solution to the original was calculated to be 98.2% and 57.0% for nonderivatized and Teic-beads, respectively, proving that DA3 had no nonspecific adsorption to the magnetic microbeads and that the binding of DA3 to the Teic-beads was specific.

Microchannel Surface. The surface of PDMS is hydrophobic in its natural state, which causes nonspecific adsorption. Oxygen plasma treatment oxidized the surface of PDMS making it temporarily hydrophilic by converting —OSi $(CH_3)_2O$— groups to silanol, carboxyl, and hydroxyl groups. These groups on the surface of the PDMS make it possible for irreversible chip bonding and permanent modification. In our work, the surface was modified to a permanently hydrophilic one via the active groups on the microchannel surface and the epoxy groups of synthesized polymers. Fluorescence microscopy was used to monitor the native and modified microchannel surface adsorption of peptides. Figure 20.2 shows the fluorescence images of the microchannels after a solution of DA3 (0.10 mg mL^{-1}) was washed through the microchannel for 20.0 min followed by a buffer rinse for 1.5 min. The native channel showed nonspecific adsorption of DA3 while the modified channel did not.

Surface modification with the epoxy-modified hydrophilic polymer afforded stable EOF. As electrolysis and evaporation of the buffer would result in an increase in current, the time of analysis was limited to 500 s. There was

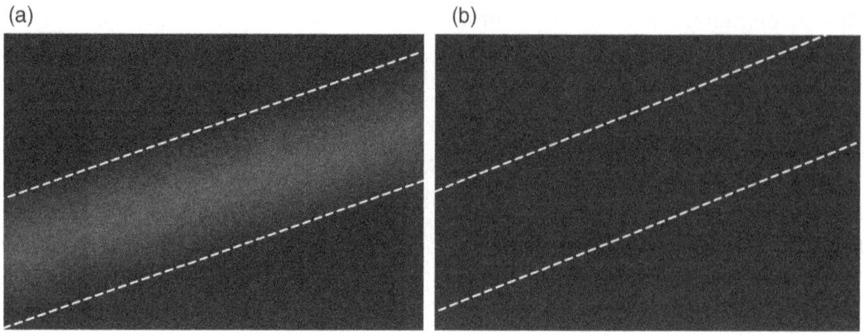

FIGURE 20.2 Fluorescence graphs of the different microchannels after incubation with DA3 followed by a rinse with the running buffer (a) Native microchannel and (b) Modified microchannel.

no change in current up to 500 s, and EOF was calculated to be much less than $0.2 \times 10^{-4}\,\mathrm{cm^2\,V^{-1}\,s^{-1}}$.

20.3.1.2 The Binding of DA3 to Teic-Beads We first examined the binding of DA3 to Teic-beads. Figure 20.3 is a representative electropherogram for a single run. The total run time was 150 s. The injection volume was estimated to be approximately 4.0 nL based on the dimensions of the microchannels and the injection length, while fluorescence spectrometer requires at least 10 μL samples. The peak for DA3 was greatly reduced in the supernatant compared to the original solution, indicating that DA3 was bound to Teic-beads. The original was used as the external standard in this work.

The concentration of free DA3 was calculated using Equation 20.1:

$$[\mathrm{DA3}] = 25.0\left(\frac{H_{\mathrm{DA3}}}{H_{\mathrm{E}}}\right) \tag{20.1}$$

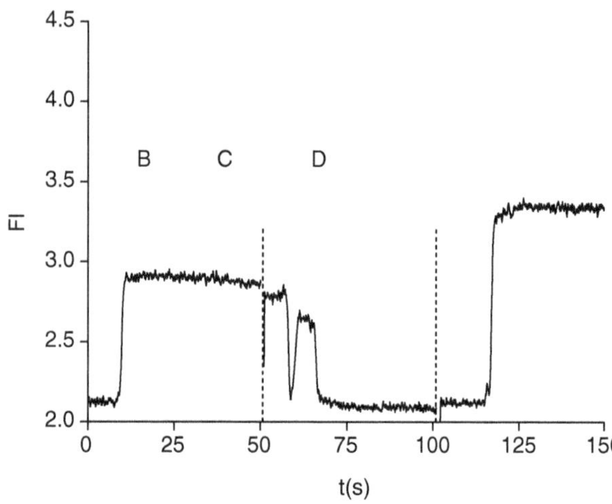

FIGURE 20.3 Representative electropherogram of a single measurement: (B) the supernatant solution, (C) rinsing the channel, and (D) the external standard solution.

Here [DA3] and H_{DA3} were the concentration and the step height of free DA3 in the supernatant, and 25.0 and H_{E} were the concentration (μM) and step height of the external standard, respectively. Using the original solution as the external standard saved us the need for a calibration curve greatly saving time in the binding assay.

Scatchard analysis is a common way to linearize the binding data. In the present experiments, the model can be expressed as Equation 20.2.

$$\frac{r}{C_{\mathrm{f}}} = -Kr + nK \tag{20.2}$$

Here, r is the concentration ratio of the bound ligand to total receptor, C_{f} is the concentration of free ligand, K is the apparent binding constant between ligand and receptor, and n is the number of binding sites. In our study, since Teic binds to D-Ala–D-Ala terminus peptides in a 1:1 stoichiometry and the total amount of peptide was kept constant (25.0 μM), Scatchard analysis could be transformed into Equation 20.3:

$$\frac{(C_{\mathrm{DA3}} - [\mathrm{DA3}])\nu_{\mathrm{DA3}}}{\nu_{\mathrm{Teic}}[\mathrm{DA3}]} = -K_{\mathrm{DA3}}(C_{\mathrm{DA3}} - [\mathrm{DA3}])\nu_{\mathrm{DA3}}/\nu_{\mathrm{Teic}} + C_{\mathrm{Teic}}K_{\mathrm{DA3}} \tag{20.3}$$

Here, [DA3] is the concentration of the free peptide, ν_{Teic} and ν_{DA3} are the volumes of Teic-beads suspension and the peptide solution, respectively, K_{DA3} is the binding constant between DA3 and Teic, and C_{DA3} and C_{Teic} are the total concentration of DA3 and the Teic concentration of Teic-bead suspension, respectively. Figure 20.4a showed the electropherograms for increasing amount of Teic-beads. With increasing Teic-beads, the step height of free DA3 was decreased, indicating an increasing amount of DA3 was being bound to Teic-beads. Figure 20.4b was the transformative Scatchard plot using the data obtained from the earlier experiments. A K_{DA3} of

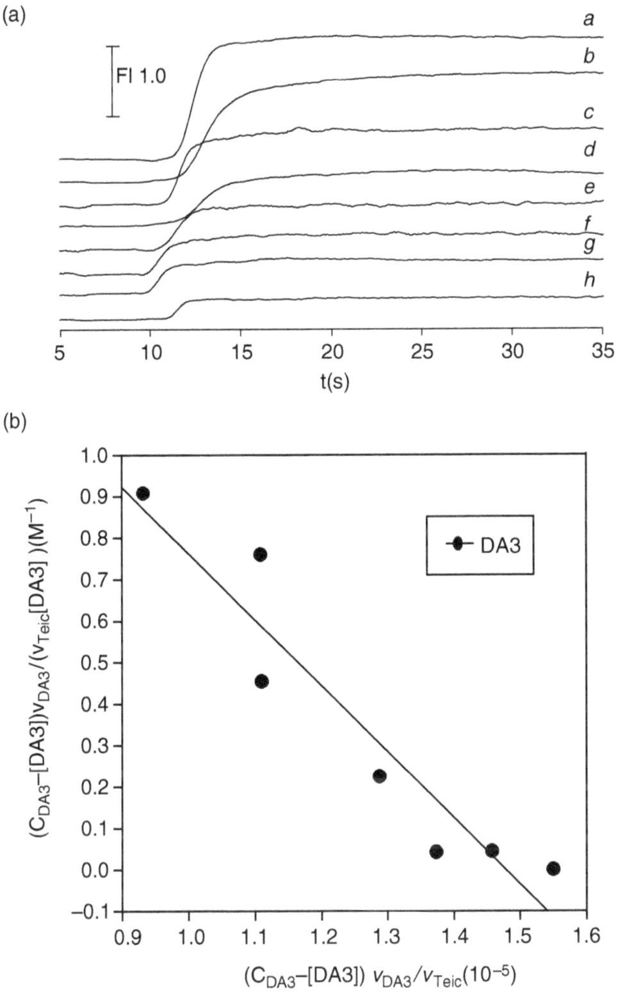

FIGURE 20.4 (a) Representative electropherograms used for the measurement of the binding constant between DA3 and Teic-beads and (b) the transformative Scatchard plot. 60 μL 25 μM DA3 was incubated with *(a)* 0 μL, *(b)* 0.5 μL, *(c)* 1.0 μL, *(d)* 1.5 μL, *(e)* 2.0 μL, *(f)* 3.0 μL, *(g)* 4.0 μL, and *(h)* 5.0 μL Teic-beads.

$1.6\pm0.2\times10^5\ M^{-1}$ was obtained from the slope. This value agreed well with our previous values—$K_{DA3}=3.3\times10^5\ M^{-1}$ using flow cytometry and $1.5\times10^5\ M^{-1}$ using traditional fluorescence [19]. This technique required approximately 0.4% of sample compared to traditional fluorescence techniques. A competitive binding assay using the ligand DA2 was also developed to demonstrate the versatility of the technique. The binding constant was found to be $(3.6\pm0.6)\times10^5\ M^{-1}$ for this assay comparable to our previous value $(2.6\times10^5\ M^{-1})$ using affinity CE [25].

20.3.2 MFAC Studies

For the MFAC technique, Figure 20.5b–d shows the voltage application sequence for the FAC process. A voltage of 1.0 kV was applied to the Teic column (Figure 20.5b) for 300 s

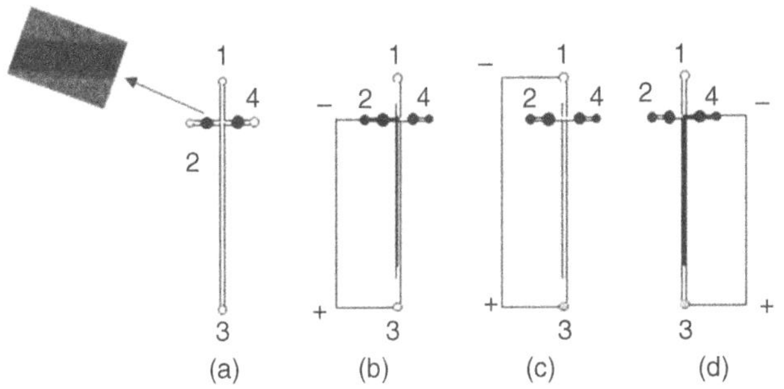

FIGURE 20.5 (a) Microchip and representative microchannel. (b–d) Microchip frontal analysis chromatography (MFAC) protocol (1, buffer well; 2, sample well; 3, waste well; 4, control well).

followed by application of a voltage of 1.0 kV to the buffer and waste wells (Figure 20.5c), for 100 s, and finally a voltage of 1.0 kV to the control column (Figure 20.5d) for 300 s.

Electroosmotic flow (EOF) was measured with running buffer using the current monitoring method described by Huang [26]. Both the affinity and control columns were found to have the same EOF. Figure 20.6 shows the fluorescence images of the affinity and control columns after a solution of DA3 ($0.10\ mg\ mL^{-1}$) was washed through the microchannel for 6 min followed by a buffer rinse for 3 min. The control column showed no adsorption of DA3 while the affinity column did, proving that the binding of DA3 with the Teic-beads was specific.

In this study, we first examined the binding of 80 μM DA3 to the columns. Figure 20.7 is a representative electropherogram for a single run. The migration time and peak shape of DA3 in the affinity column was different from those realized in the control column under the same conditions. The migration time and slope of the curve for DA3 in the affinity column is greater and smaller, respectively, than that found in the control column. We attribute these differences to the fact that DA3 binds to the Teic-beads and has no affinity for unmodified beads.

The amount of bound peptide can be calculated from Equation 20.4:

$$q=\frac{L}{t_0}(t_1-t_0)AC \tag{20.4}$$

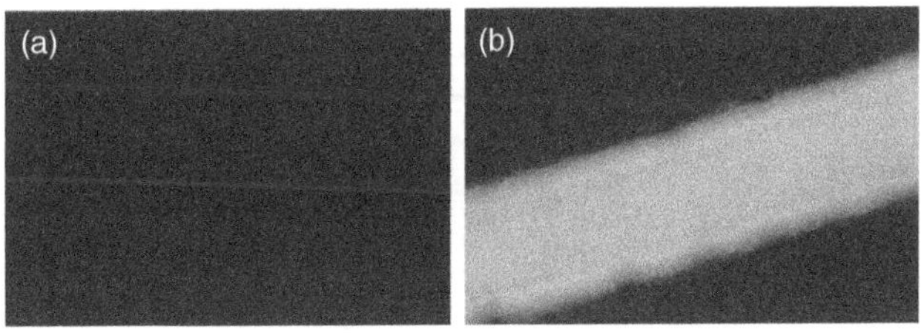

FIGURE 20.6 Fluorescence micrographs of (a) control channel and (b) affinity channel after incubation with DA3 and after rinsing with the running buffer.

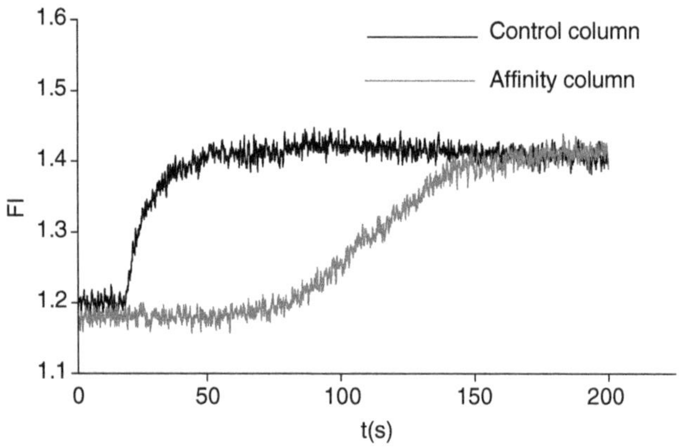

FIGURE 20.7 Representative electropherogram from a single binding measurement.

Here, q, L, t_0, t_1, A, and C were the amount of bound DA3, effective separation length, the migration time in the control column, the migration time in the affinity column, the intersection area of the channel and the sample concentration, respectively. For a single run, the sample consumption was estimated to be approximately 0.4 μL, while conventional FAC usually requires several milliliters for the binding assay.

To estimate the binding constant, six different sample concentrations were used (Figure 20.8a).

Scatchard analysis was used to linearize the binding data and is expressed in Equation 20.5:

$$\frac{r}{C} = -kr + nk \tag{20.5}$$

Here, r is the ratio of bound DA3 to total Teic-beads, C is the concentration of DA3, K is the apparent binding constant, and n is the number of binding sites present on a Teic-bead. Substitution of Equation 20.7 into 20.8 yields, after simplification, Equation 20.6:

$$\frac{t_1 - t_0}{t_0} = -kC\left(\frac{t_1 - t_0}{t_0}\right) + \left(\frac{nk}{ALq_T}\right) \tag{20.6}$$

Here, q_T stands for the total amount of Teic-beads used in this study. Figure 20.8b is a plot of the binding of DA3 to the Teic-beads according to Equation 20.9. The binding constant was calculated to be $5.4 \times 10^4\,M^{-1}$.

20.4 CONCLUSIONS

FAMCE and MFAC were developed to examine the binding of Teic-derivatized magnetic beads to D-Ala–D-Ala terminus peptides. In the FAMCE technique, a solution of peptide was exposed to increasing concentrations of Teic-beads. The amount of bound peptide was quantified by frontal analysis MCE and used in the Scatchard analysis to obtain a binding constant between receptor–ligand pair. In the MFAC technique, a cross-chip was used holding derivatized and underivatized Teic-beads. Application of a voltage through the columns resulted in differences in migration time of the ligand as detected by fluorescence. The obtained binding constants for the FAMCE technique are comparable to the literature values (this study, $K_{DA3} = 1.6 \times 10^5\,M^{-1}$; previous studies, $K_{DA3} = 3.3 \times 10^5\,M^{-1}$ using flow cytometry and $1.5 \times 10^5\,M^{-1}$ using fluorescence spectrometer; this study, $K_{DA2} = 3.6 \times 10^5\,M^{-1}$; previous study, $K_{DA2} = 2.6 \times 10^5$, using affinity CE). For the MFAC technique, the $K_{DA3} = 5.4 \times 10^4\,M^{-1}$. Compared to other MCE modes, there is no requirement on electrophoretic mobility of the reacting species and bound complex. At present, work

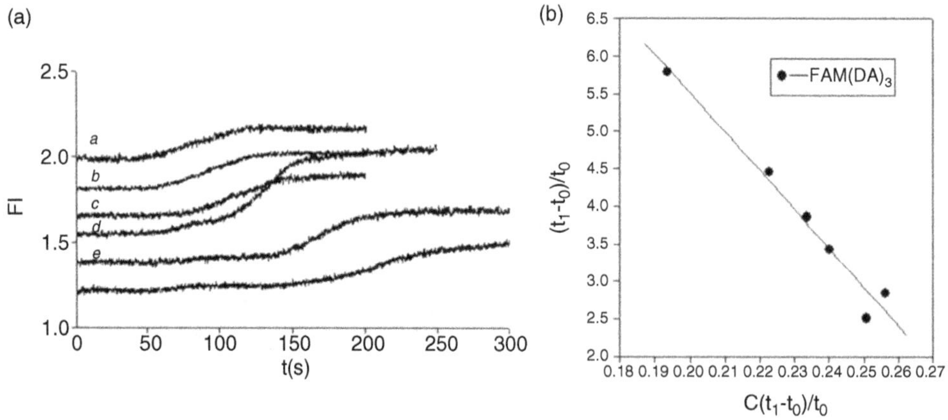

FIGURE 20.8 (a) Representative electropherograms used for the measurement of the binding constant between DA3 and Teic-beads. ([DA3] = *a*, 100 μM; *b*, 90 μM; *c*, 80 μM; *d*, 70 μM; *e*, 60 μM; *f*, 50 μM). (b) Scatchard plot of the data for DA3 and Teic-beads.

is focused on integrating the mixing of the ligand and receptor immobilized on magnetic microbeads into the microchip so the assay will be fully chip-based as well as examining tighter receptor–ligand interactions. Compared to other techniques, the FAMCE and MFAC techniques offer several advantages including extremely low sample consumption, ease of packing and reproducibility.

ACKNOWLEDGMENTS

The authors gratefully acknowledge financial support for this research by grants from the National Science Foundation (CHE-0515363, DMR-0351848, MCB-0448676, and CBET-0723271).

REFERENCES

1. Gomez, F.A. (2008) *Biological Applications of Microfluidics*, Wiley, Hoboken, NJ.
2. Martinez, A.W., Phillips, S.T., Whitesides, G.M., and Carrilho, E. (2010) *Anal. Chem. 82*, 3–10.
3. Yang, H., Luk, V.N., Abelgawad, M., Barbulovic-Nad, I., and Wheeler, A.R. (2009) *Anal. Chem. 81*, 1061–1067.Mukhopadhyay, R. (2009) *Anal. Chem. 81*, 4169–4173.
4. Applegate, R.W., Squier, J., Vestad, T., Oakey, J., Marr, D.W. M., Bado, P., Dugan, M.A., and Said, A.A. (2006) *Lab Chip 6*, 422–426.
5. Chen, C.C., Zappe, S., Sahin, O., Zhang, X.J., Fish, M., Scott, M., and Solgaard, O. (2004) *Sens. Actuators B. 102*, 59–66.
6. Pamme, N. and Wilhelm, C. (2006) *Lab Chip 6*, 974–980.
7. MacDonald, M., Spalding, G., and Dholakia, K. (2003) *Nature 426*, 421–424.
8. Sundararajan, N., Kim, D., and Berlin, A.A. (2005) *Lab Chip 5*, 350–354.
9. Johnson Hou, C., Milovic, N., Godin, M., Russo, P.R., Chakrabarti, R., and Manalis, S.R. (2006) *Anal. Chem. 78*, 2526–2531.
10. Khandurina, J., McKnight, T.E., Jacobson, S.C., Waters, L.C., Foote, R.S., and Ramsey, J.M. (2000) *Anal. Chem. 72*, 2995–3000.
11. Easley, C.J., Karlinsey, J.M., and Landers, J.P. (2006) *Lab Chip 6*, 601–610.
12. Jordan, A., Scholz, R., Maier-Hauff, K., van Landeghem, F., Waldoefner, N., Teichgraeber, U., Pinkernelle, J., Bruhn, H., Neumann, F., Thiesen, B., von Deimling, A., and Felix, R. (2006) *J. Neurooncol. 78*, 7–14.
13. Richardson, J., Hawkins, P., and Luxton, R. (2001) *Biosens. Bioelec. 16*, 989–993.
14. Hayes, M.A., Polson, N.A., Phayre, A.N., and Garcia, A.A. (2001) *Anal. Chem. 73*, 5896–5902.
15. Choi, K.W.O. Thomas, J.H., Heineman, W.R., Halsall, H.B., Nevin, J.H., Helmicki, A.J., Henderson, T., and Ahn, C.H. (2002) *Lab Chip 2*, 27–30.
16. Liu, X., Lo, R., and Gomez, F.A. (2009) *Electrophoresis 30*, 2129–2133.
17. Lin, Y.S., Tsai, P.J., Weng, M.F., and Chen, Y.C. (2005) *Anal. Chem. 77*, 1753–1760.
18. Liu, X. and Gomez, F.A. (2009) *Electrophoresis 30*, 1194–1197.
19. Piyasena, M.E., Real, L.J., Diamond, R.A., Xu, H., and Gomez, F.A. (2008) *Anal. Bioanal. Chem. 392*, 877–886.
20. Sarakhanikhorami, M., Lo, R.C., and Gomez, F.A. (2009) *Bioanalysis 1*, 715–719.
21. Wu, D.P., Zhao, B.X., Dai, Z.P., Qin, J.H., and Lin, B.C. (2006) *Lab Chip 6*, 942–947.
22. Kahne, D., Leimkuhler, C., Lu, W., and Walsh, C. (2005) *Chem. Rev. 105*, 425–448.
23. Liu, X., and Gomez, F.A. (2009) *Anal. Bioanal. Chem. 393*, 615–621.
24. Soriano, A., et al. (2006) *Euro. J. Clin. Microb. Infect. Dis 25*, 35–38.
25. Silverio, C.F., Plazas, A., Moran, J., and Gomez, F.A. (2002) *J. Liq. Chromatogr. Rel. Tech. 25*, 1677–1691.
26. Lin, Y.H., Lee, G.B., Li, C.W., and Huang, G.R. (2001) *J. Chromatogr. A 937*, 115–125.

21

GLYCOMIC PROFILING THROUGH CAPILLARY ELECTROPHORESIS AND MICROCHIP CAPILLARY ELECTROPHORESIS

YEHIA MECHREF

21.1 INTRODUCTION

The most common and structurally diverse post-translational modification of proteins is glycosylation. Multiple glycosylation sites of proteins and their associated microheterogeneity contribute to the common complexity of glycosylation patterns and the frequent difficulties associated with resolving such fine structural differences in large biopolymers. The function and activity of a glycoprotein are ultimately defined by the resulting structural variants that are known as protein glycoforms [1].

Key biochemical processes are defined by the glycosylation of proteins, including protein folding, stability, and localization [2]. Moreover, specific interactions between a glycan and its target protein(s) have been implicated in cellular communication, such as cell–cell, cell–matrix, protein–protein, and sugar–sugar interactions [3–6]. Therefore, alterations in the glycosylation of proteins either through various site occupancy changes on the polypeptide chain, or in the variation of the oligosaccharide structures occupying a particular site on a protein change the biological activity of these proteins (glycoproteins). Mammalian diseases, such as hereditary disorders, immune deficiencies, cardiovascular disease, and cancer, have been linked to aberrant glycosylation of glycoconjugates [7,8]. Recently, many biomedical research efforts have been focused on devising approaches that allow monitoring the subtle, yet biologically significant, glycosylation changes that correlate with these afflictions. The ultimate goal of these activities is to aid in the diagnosis and treatment of these diseases as well as permitting a better understanding of the molecular attributes associated with these diseases [8].

Currently, the characterization of biomolecules, including glycan structures released from a glycoprotein, is commonly attained using mass spectrometry (MS) [9–13]. Although MS and tandem MS techniques provide a plethora of information, isomeric glycan structures are not readily characterized by either technique. The analysis of glycan pools containing many isomeric structures is only attained using various separation methods. It is thus natural to consider combining them with the MS-based technologies. Because carbohydrates are relatively polar molecules, hydrophilic-interaction chromatography (HILIC) has been recognized as a powerful method for the separation of oligosaccharides [14,15]. The retention of different oligosaccharides on silica- or amide-based stationary phases appears sufficient for the identification of core glycan structures present in a typical glycan mixture. However, the resulting separations are typically time-consuming, and, in many incidents, lack the resolution necessary to distinguish closely related isomers. As an alternative, high-performance anion-exchange chromatography with pulsed amperometric detection (HPAEC–PAD) allows glycan analysis at better resolution [16]. Nevertheless, HPAEC–PAD analyses also suffer from often-long analysis time (>60 min), relatively poor reproducibility, and relatively low sensitivity [17].

Capillary electrophoresis (CE) and microfluidics capillary electrophoresis (MCE) offers inherently high

Capillary Electrophoresis and Microchip Capillary Electrophoresis: Principles, Applications, and Limitations, First Edition.
Edited by Carlos D. García, Karin Y. Chumbimuni-Torres, and Emanuel Carrilho.

efficiencies and selectivity, allowing the effective resolution of numerous glycan isomers associated with glycoproteins [18–20]. Various capillary electromigration techniques, including capillary zone electrophoresis (CZE) [21,22], capillary gel electrophoresis (CGE) [23], micellar electrokinetic chromatography (MEKC) [24], and capillary electrochromatography (CEC) [25–28], have been utilized to achieve the separation of different glycans.

CEC, which is a hybrid separation technique combining the best of CE and LC, is based on the separation of neutral molecules that are transported through a column containing a stationary phase by a mobile phase, the movement of which is driven by an electroosmotic flow (EOF). Several structurally similar components in a mixture, in addition to certain structural isomers, have been resolved by CEC [26–28]. However, CGE and CZE thus far have offered superior separation efficiency and resolution of glycans to those observed in CEC.

The separation of glycans in CGE results from the sieving of glycans as they migrate through the capillary filled with a physical gel. The mechanism of separation in CGE is based on the molecular sieving of glycans that possess different hydrodynamic volumes in solution. A distinct resolution of certain structural isomers has been demonstrated using CGE [23,29]. The simplicity and versatility of CZE have prompted it to be considered the most widely used mode of CE. The separation in CZE occurs as solutes migrate in discrete zones at different nobilities in a capillary filled with a buffer, thus attributing to its simplicity [21,30].

The high potential of merging the most effective isomer separation tool (capillary electromigration, CE, and MCE) with the most powerful, informative, and sensitive measurement technique (MS) remains very relevant. In this chapter, the utility of different CE and MCE approaches to the structural characterization of glycoproteins are highlighted, while different means to interface these techniques to mass spectrometry are discussed. This chapter also briefly describes the method commonly employed to release glycans, and briefly discusses the general aspects of CE-LIF.

21.1.1 Release of *N*-Glycans from Glycoproteins

Although *N*-glycans can be released through chemical approaches, the release of *N*-glycans is routinely attained using enzymes, which permit efficient release and subsequently higher detection sensitivity. This is mainly due to the fact that enzymatic release does not involve excessive sample treatments and preparation steps commonly associated with chemical release. Sample losses increase substantially as the number of steps needed to prepare the sample increases.

21.1.1.1 Chemical Release Hydrazine chemically cleaves *N*- and *O*-glycans from glycoproteins [31,32]. The release of *N*-glycan is only achieved at 95°C, while a milder temperature (60°C) is required to release *O*-glycans [32]. Despite hydrazinolysis earlier success, this approach suffers from several major disadvantages, including loss of information pertaining to (i) the glycosylation sites, (ii) the modification of monosaccharide residues such as sialic acid acylation or glycolation, (iii) modification of indigenous *N*- or *O*-glycans, and (iv) the loss of terminal *N*-acetylglucosamine or *N*-acetylgalactosamine. This approach induces excessive site reactions in the absence of strictly anhydrous conditions, which is very common. Hydrazinolysis has now become secondary to the more favored enzymatic release.

21.1.1.2 Enzymatic Release Peptide *N*-glycosidase F (PNGase F) is the most widely used enzyme for the release of *N*-glycans from glycoproteins, while other enzymes, which are commercially available, have also been used [33]. PNGase F is an endoglycosidase enzyme that cleaves off the intact glycans as glycosylamines, which is readily converted to reducing glycans in the phosphate reaction buffer. This enzymatic release of glycans also results in the conversion of asparagine (the amino acid to which glycans are attached) to aspartic acid. This chemical change is accompanied by an m/z shift that is often used to determine the site of glycosylation using a mass spectrometer with high mass resolution capabilities. Moreover, PNGase F has a very wide specificity, cleaving all glycans linked to asparagine except those possessing fucose residues that are α(1–6) linked to the reducing-terminal GlcNAc. Glycoproteins possessing glycan structures with such linkages are commonly found in plants and insects. These *N*-glycan structures are enzymatically cleaved using peptide *N*-glycosidase A [33].

21.1.2 Release of *O*-Glycans from Glycoproteins

The analysis of protein *O*-glycosylation is still substantially less sensitive than that of *N*-glycosylation mainly because of the chemical approaches used that involve several steps, resulting in substantial sample loss that subsequently limits the sensitivity of *O*-glycan analysis.

21.1.2.1 Chemical Release Hydrazinolysis [31,32] and alkaline β-elimination [34] are the two chemical approaches commonly used to release *O*-glycans of proteins. Although hydrazinolysis yields reducing glycans, its tedious procedure with many needed precautions have limited its use [32]. An improved version involving the use of 70% (w/v) aqueous ethylamine was recently reported [35,36]. However, the overall reaction yields were low, as the *O*-glycans suffered significant peeling reaction and/or other forms of degradation.

On the other hand, β-elimination approach is currently viewed as the most reliable method for the effective release of *O*-glycans. A strong reducing agent, which converts glycans to their respective alditols, is necessary to minimize the "peeling reactions" caused by the needed alkaline medium for the effective release [37]. Although β-elimination is generally effective, the formation of alditols limits its utility, especially since glycans in general lack a chromophore. Therefore, a derivatization allowing the inclusion of appropriate structural moieties can be beneficial in enhancing CE, LC, or MS sensitivities of MS investigations [38–42]. This chemical approach is also not microscale compatible, as the minute quantities of released glycans are easily lost during the cleaning procedure needed to reduce/eliminate the excessive amounts of salts needed for this procedure.

The aforementioned limitations of β-elimination method was alleviated by the so-called ammonia-based β-elimination method introduced by Huang et al. [43], which yields free reducing end *N*- and *O*-glycans amenable to MS analysis. This procedure provides alternative to the hitherto used hydrazinolysis and the Carlson β-elimination methods. The method involves minimum sample handling, and lacks peeling reactions and deacetylation by-products. The reactants used in this method are easily removed through methanol evaporation of excess ammonium–borane complex, thus rendering this method very effective in releasing glycans from low-microgram quantities of glycoproteins.

21.1.2.2 Enzymatic Release Unlike *N*-glycans, there is no endoglycosidases capable of efficient release of *O*-glycans. The only endoglycosidase that has been somewhat employed for the release of *O*-glycans is *endo*-α-*N*-acetylgalactosaminidase, which only releases unsubstituted Core-1 *O*-glycans [44–46]. At this time, chemical release methods provide the only effective means for the release of *O*-glycans.

Mechref and coworker recently introduced a sensitive alternative to enzymatic digestion, involving the solid-phase permethylation of pronase E-treated glycoproteins [47]. The high pH associated with solid-phase permethylation allows β-elimination to proceed before methylation of hydroxyl and amino groups. The use of pronase E was necessary to expose the *O*-glycosylation sides to the sodium hydroxide beads used. This new method appeared to be at least 50 times more sensitive than all the other chemical methods. However, this method will only produce permethylated *O*-glycans, thus limiting their analysis to only MS [47].

21.2 GENERAL CONSIDERATIONS OF CAPILLARY ELECTROPHORESIS AND MICROCHIP CAPILLARY ELECTROPHORESIS OF GLYCANS

Although matrix-assisted laser desorption/ionization (MALDI)–MS permits the analysis of glycans derived from model glycoproteins, purified glycoproteins, and glycoproteins present in biological mixtures, complementary information is provided through the analysis of the same samples by CE–laser-induced fluorescence (LIF) after fluorescence derivatization. This is mainly true because of the ability of different forms of CE to resolve structural isomers, which are otherwise indistinguishable by MS alone. Glycomic analysis by CE has been achieved by different capillary electromigration techniques, including the CZE mode employing free-buffer media or gels, or through MEKC. However, employing sensitive detectors such as LIF or MS is a must.

Spectroscopic detection of carbohydrates is not trivial, since most native carbohydrates lack a chromophore or fluorophore in their molecules that would enable their sensitive detection. Although direct or indirect detection modes have been employed for the analysis of carbohydrates, these techniques failed to offer the necessary sensitivity to analyze glycan mixtures from biological samples [48]. Electrochemical techniques have proven more useful for the detection of underivatized carbohydrates than spectroscopic detection other than LIF. Amperometric and pulsed-amperometric detection are two approaches that have been demonstrated for the detection of carbohydrates; however, they lack reproducibility and the necessary selectivity in the presence of interfering compounds. Detection of carbohydrates at high sensitivity is only attained through LIF detection, which is recommended if MS is not available, after attaching a fluorophore to the reducing end of the carbohydrate molecules [13]. Since not all *N*-glycans are electrically charged, the derivatization procedure is also expected to introduce ionic groups needed for electromigration. However, some neutral derivatizing reagents have also been utilized in conjunction with the MEKC mode or in CZE mode using borate buffer. The latter introduces a charge through the formation of a charged complex.

Unlike nano-LC, CZE does not permit the injection of a large aliquots derived from biological samples. While some approaches such as sample stacking and sample preconcentration can somewhat reduce this problem, they adversely influence the performance of CZE, which is more suited for very small solute concentrations and peak volumes and, correspondingly, small column diameters. The detection challenges of CZE, and to a various degree also the other capillary electromigration techniques, can be offset through the use of the highly sensitive LIF and MS techniques.

21.2.1 Capillary Electrophoresis–Laser-Induced Fluorescence (CE–LIF) Analysis of Glycans

LIF detection of glycans is based on the use of fluorescence-tagging reagents suited to optimally absorb the output wavelengths of reliable lasers, such as argon-ion or helium–cadmium lasers. In 1991, the first carbohydrate analysis by CE–LIF was demonstrated, using 3-(4-carboxybenzoyl)-2-quinolinecarboxaldehyde (CBQCA) fluorescence-tagging reagent, for amino sugars [30], and

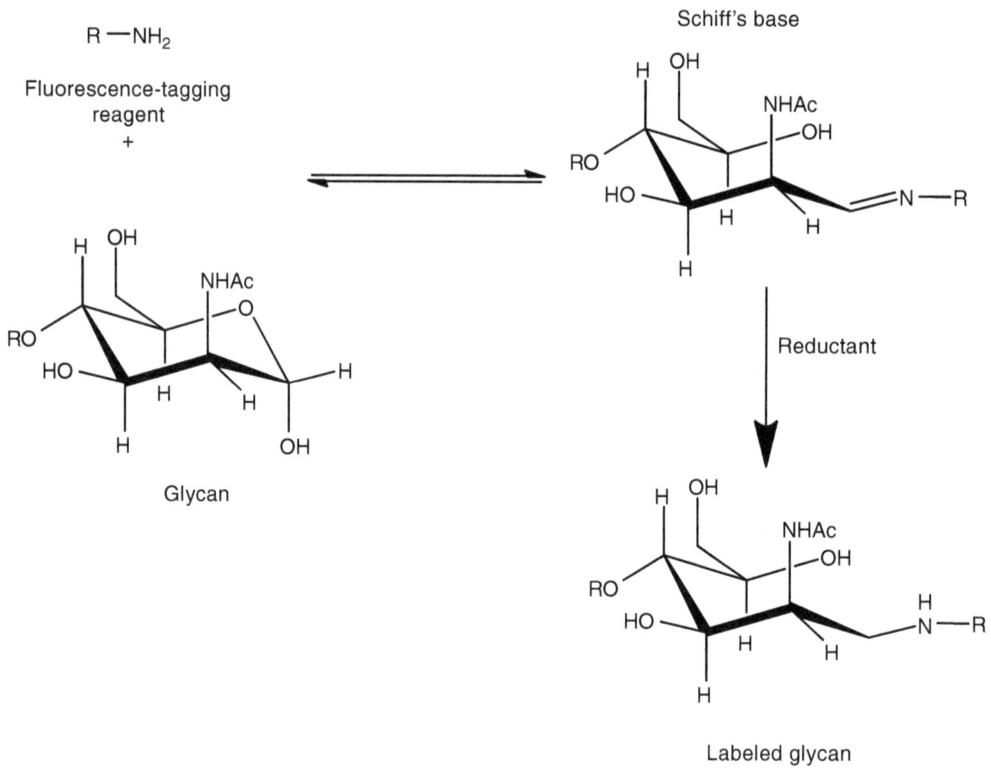

FIGURE 21.1 Reductive amination derivatization approach.

N-glycans derived from glycoproteins [22,49,50]. The described procedure permitted the analysis of *N*-glycans derived from fetuin at detection limits reaching subattomole levels. Since then, additional derivatizing reagents employed in conjunction with CE–LIF for sensitive analysis and detection of glycans have been introduced.

Several fluorophores, including 8-aminonaphthalene-1,3,6-trisulfonic acid (ANTS), 7-aminonaphthalene-1,3-disulfonic acid (ANDS), and 2-aminonaphthalene-1-sulfonic acid (ANS), have been attached to glycans through reductive amination chemistry (Figure 21.1) before CE–LIF analysis [51–53]. Among those fluorescence-tagging reagents, ANTS was demonstrated to be the most effective for the CE analysis of *N*-glycans derived from glycoproteins because of its greater charge, thus allowing faster analyses and offering higher resolution. Accordingly, this reagent was successfully employed for characterization of glycans derived from various glycoproteins, including human immunoglobulin G [54], ovalbumin [55,56], fetuin [29,55], recombinant HIV gp-120 [57], and α_1-acid glycoprotein [57].

However, the use of ANTS was hampered by the instability and high cost of the required He–Cd laser. This prompted the need to explore alternative fluorophores, requiring a more convenient light source such as the highly reliable argon-ion laser. Today, the most commonly used fluorescence reagents for CE–LIF is 1-aminopyrene-3,6,8-trisulfonic acid (APTS) [58–60]. Using this reagent, Chen and Evangelista [58] devised a nearly comprehensive method, allowing the analysis of *N*-glycans by combining specific chemical and enzymatic conversions and CE–LIF. First, reductive amination conditions needed to preserve sialic acid and fucose residues were employed to derivatize enzymatically released *N*-glycans with APTS. The reductive amination conditions employed in this study were effective in the successful profiling of heavily sialylated *N*-glycans. Many investigators used this method for the analyses of *N*-glycans derived from various glycoproteins, including ribonuclease B [23,29,58,59,61], fetuin [58,59], recombinant human erythropoietin [58], kallikrein [58], and a chimeric recombinant monoclonal antibody [21].

Guttman used APTS in conjunction with CGE and exoglycosidase digestions for multistructure sequencing of *N*-glycans [18]. Carefully selected exoglycosidases and a subsequent comparison of the positions of the separated exoglycosidase digest fragments to maltooligosaccharides of a known size in terms of relative migration shifts of CGE separation allowed the deduction of glycan linkages and compositions. This method for the first time allowed comprehensive characterization of glycans through CE–LIF analysis.

A modified form of Guttman method was later adopted by Ma and Nashabeh [21] to monitor certain variations in the glycosylation of rituximba, a chimeric recombinant monoclonal antibody, during production. However, in this case,

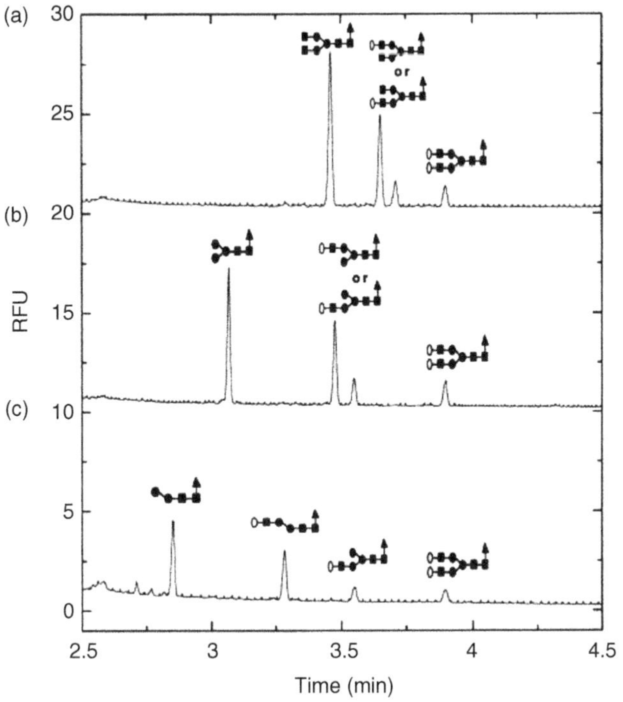

FIGURE 21.2 CE profile of APTS-labeled glycans derived from a monoclonal antibody. (a) Represents standard core-fucosylated biantennary/disialylated, monosialylated, and asialylated glycans. Conditions: column, polyacrylamide-coated 50/365 μm i.d./o.d.; length, 50.5 cm total, 40.5 cm effective length; temperature, 257°C; injection pressure, 0.5 psi for 5.0 s; voltage, 15 kV anodic electro-osmotic flow; λex 488 nm, λem 520. Incubation of released *N*-glycans with β-*N*-acetylglucosaminidase removed the terminal GlcNAc residues on the G0 and G1 isomers (b). The migration time of G2 remained unchanged, since it does not possess terminal GlcNAc. On the other hand, G0 exhibited a larger change in migration time than G1 isomers because of the removal of two GlcNAc residues from G0, relative to one from G1. The use of α1-2,3) mannosidase, which specifically cleaves terminal α(1–3) linkage, in conjunction with a change in the CE profile, indicated the predominance of G1 with a galactose residue on the mannose α(1–6) arm (c). Symbols: black rectangle, *N*-acetylglucosamine; black circle, mannose; black triangle, fucose; white circle, galactose. Reproduced from Ref. 62 with permission.

CZE was used. The *N*-glycans derived from rituximba are neutral, complex biantennary oligosaccharides with zero, one, and two terminal galactose residues (G0, G1, and G2, respectively). Like the method described by Guttman [18], this method was based on releasing *N*-glycans from the glycoprotein via PNGase F, then derivatizing them with APTS before a CE mapping. All observed glycans were fully resolved, including the positional isomers of G1 (Figure 21.2). The two G1 positional isomers were identified by comparing CE profiles obtained from sequential enzymatic digestions. The method has several advantages over other schemes, including simplicity, accuracy, precision, high throughput, and robustness.

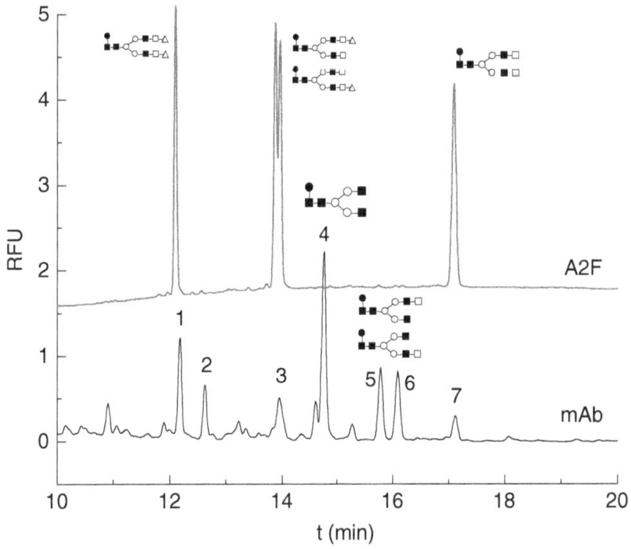

FIGURE 21.3 CE profile of APTS-labeled glycans derived from a monoclonal antibody. The upper trace represents standard core-fucosylated biantennary/disialylated, monosialylated, and asialylated glycans. Conditions: column, polyacrylamide-coated 50/365 μm i.d./o.d.; length, 50.5 cm total, 40.5 cm effective length; temperature, 25°C; injection pressure, 0.5 psi for 5.0 s; voltage, 15 kV anodic electroosmotic flow; λex 488 nm, λem 520. Symbols: black rectangle, *N*-acetylglucosamine; white circle, mannose; black circle, fucose; white rectangle, galactose; white triangle, *N*-acetylneuraminic acid. Reproduced from Ref. 62 with permission.

In an another study by Mechref et al. [62], closely related *N*-glycan structural isomers were baseline resolved using CZE–LIF analysis of APTS labeled *N*-glycans. The CE–LIF trace (Figure 21.3) of the fluorescently labeled standards (core-fucosylated biantennary, asialylated, and mono- and disialylated glycans) clearly illustrates the ability of this technique to separate structural isomers. Monosialylated structures,with terminal sialic acid residue attached to either the 1–6 or 1–3 antenna, were partially resolved (Figure 21.3). Monofucosylated biantennary glycan structures lacking one terminal galactose residue are baseline resolved (see Figure 21.3), further suggesting the ability of CZE to resolve closely related structural isomers.

CE–LIF method was also devised to attain carbohydrate analysis at ultramicroscale [63]. This method is based on initial conversion of carbohydrates to *N*-methylglycamines before their tagging with 7-nitro-2,1,3-benzoxadiazole (NBD) and subsequent CE–LIF analysis. Conversion of *N*-glycans to *N*-methylglycamines was achieved by incubating the sample with dimethylamine–borane complex at 40°C. Next, *N*-methylglycamines were coupled to NBD in 5 min at 70°C. Although the derivatization involves two steps, the procedure is carried out simultaneously in a "one pot" arrangement without any loss in derivatization efficiency. This method allowed a high-resolution profiling of *N*-glycans derived from fetuin through the use of borate buffer [63].

Callewaert et al. were the first to demonstrate the clinical applicability of CE–LIF for the profiling of serum *N*-glycans using capillary electrophoresis gel-based DNA sequencer for assessing liver cirrhosis [64–66]. The quantification of 14 peaks that were detectable in all samples was feasible using this method. Although five peaks demonstrated a decrease or increase in abundance that could be correlated with disease progression, quantitative assessment of liver cirrhosis was achievable using the ratio of only two glycan structures identified in the desialylated *N*-glycans derived from blood serum samples as a triantennary and fucosylated GlcNAc bisecting glycan structures. Although this CE-based *N*-glycan profiling method distinguished cirrhotic from noncirrhotic chronic liver disease patients, with 79% sensitivity and 86% specificity, it was rather complicated and hard to use.

The same group recently modified and simplified the same approach that was designated as GlycoFibro test [66]. It involved the use of 96-well plate-based serum *N*-glycomics sample preparation in conjunction with incubation in a polymerase chain reaction (PCR) thermocycler. This configuration yielded APTS-labeled *N*-glycans, which were subsequently analyzed by the capillary electrophoresis-based DNA sequencer. This approach was employed to analysis 376 consecutive chronic hepatitis C virus patients for which liver biopsies were available. Quantification was based on the ratio between a bisecting GlcNAc-modified agalactose glycan and a triantennary glycan. The ratio of these two glycans correlated with the histological fibrosis stage data generated through FibroTest ($p = 0.4 - 0.5$ [66]). The developed method appears to be effective in assessing liver fibrosis in chronic hepatitis patients, thus offering an alternative to liver biopsy, which is routinely hampered by sampling error and up to 20% interlaboratory variance [66].

Although CE–LIF allows the analysis of *N*-glycans derived from both purified glycoproteins and biological samples at high sensitivity and resolution, positive identification of structures is only achieved using standards or a battery of exoglycosidase to identify glycan structures. The use of MS rather than LIF somewhat eliminates the need for standards. However, optimum electrophoretic separation of glycans is commonly attained using buffers that are not MS compatible. This factor has been the driving force behind the development of CE analysis for *N*-glycans using buffers that are compatible with mass spectrometry without jeopardizing resolution and separation efficiencies.

21.2.2 Interfacing Capillary Electrophoresis and Capillary Electrochromatography to Mass Spectrometry

Over the past several years, the popularity of CE–MS interfacing has substantially increased due to the enhanced MS sensitivity and improved CE–MS reproducibility. This interfacing takes a full advantage of the high sensitivity of mass spectrometers and the unmatched separation efficiency of CE. Interfacing CE to MS technique also offers high mass accuracy and tandem MS, thus generating a wealth of information. Obviously, MS and tandem MS allow the positive identification of glycans separated by CE, including isobaric structures that could be resolved by CE and subsequently detected by MS.

The hyphenation of liquid chromatography and MS has paved the road for interfacing of CE and capillary electrochromatography to MS. In the late 1980s, CE–MS couplings to a continuous-flow fast atom bombardment mass spectrometer (cfFAB–MS) [67,68], electrospray mass spectrometer (ESI–MS) [69,70], and ion spray mass spectrometer (ISP–MS) [71] were reported. Overviews of interfaces available for coupling CE to MS have been comprehensively reviewed [72–76]. Coupling CE to MALDI–MS was initially reported by Chang and Yeung [77], and shortly after by Karger et al. [78]. CEC coupling to MALDI–MS was later reported by Tegeler et al. [79].

21.2.2.1 ESI Interfaces for Capillary Electrophoresis
Flow-rate compatibility and proper grounding of the CE electrical circuit are the two obstacles that need to be overcome to successful couple electromigration techniques to MS through ESI. CE usual flow rates are 1–100 nL min^{-1}, while that for cf-FAB, electrospray, ion spray, and MALDI–MS interfaces is 1–200 μL min^{-1}. This entails the addition of sheath liquid, thus prompting analyte dilution and "chemical noise." However, micro- and nano-electrospray systems have provided a solution to the flow rate incompatibility, thus allowing higher efficiency and ion sampling rates at extremely low flow rates compatible with CE. On the other hand, improper grounding of the CE electrical circuit associated with ESI interface leads to poor sensitivity arising from the reduced ionization of the analyte or arcing between the ESI sprayer needle and the electrode of the CE. This problem is easily rectified using a resistor of 40–100 MΩ to ground CE column outlet and the ESI sprayer needle. There are three means by which CE or CEC is interfaced to MS through electrospray: (i) sheathless, (ii) (coaxial) sheath flow, and (iii) liquid-junction devices.

21.2.2.2 Sheathless-Flow Interface The use of stainless-steel capillary sheath for electrical contact was described in 1987 by Smith et al. [69]. The incorporation of silver on the stainless-steel capillary sheath was described later [70]; however, the effective use of this arrangement was hampered by the lack of stability of this interface originating from the flaking of the silver and gold coatings of the capillary. The direct application of the high voltage to the inlet of the separation capillary with a tapered outlet to form a microspray tip was devised by Mazereeuw et al. [80] This arrangement eliminated the need for the electrode or conductive coating of the spray capillary and the sheath liquid. On the other hand, Zare et al. [81] established the electrical potential needed for ESI through the insertion of a thin gold wire about 2 mm into the capillary outlet of the separation capillary, which then acted as the ESI needle. In this arrangement, a stable electrospray current was sustained by the CE electroosmotic flow; however, others questioned this aspect of the interface design [80]. Although other arrangements based on the sheathless-flow concept were reported, the instability and high day-to-day variations associated with this design have limited the use of this type of interface.

21.2.2.3 Sheath-Flow Interface A coaxial delivery of sheath liquid to make up for the discrepancy between CE flow rate and ESI required flow rate was first described by Smith et al. [82]. Coupling CE to cfFAB–MS [68] and ISP–MS [83] was subsequently achieved using similar designs. Generally, this type of interface requires a sheath liquid that mixes with the CE outlet flow at the tip of the separation capillary to generate a flow rate needed to sustain the electrospray. The design consists of three concentric capillaries. The innermost capillary is the CE separation capillary, while the sheath liquid (make-up flow) is provided by a second stainless-steel capillary through which the ESI potential is applied. A nebulizing gas is supplied through the third (outer) capillary to sustain ESI and suppresses the effects of corona discharge. The dilution and the "chemical noise" introduced through the make-up liquid render this interface insensitive. Moreover, several operational parameters, including capillary position and flow rate as well as composition of sheath liquid, need to be optimized.

21.2.2.4 Liquid Junction Interface A third design for CE–MS interface involved partial disconnection of the CE separation column and the ESI emitter, thus allowing for a choice of appropriate make-up solvent and flow rate for optimized ESI performance. This decoupling of the CE and ESI emitter allows individual optimization for each of the two components. Henion et al. [71] were the first to use this interface to couple CE to MS. In their reported arrangement, a stainless-steel tee was used to provide a 10–25 μm gap between the ends of the CE separation column and the electrospray needle. The required high degree of precision in the alignment and the spacing of the CE and sprayer capillaries, the poor mixing of the CE electrolyte and the make-up liquid, and the increased band broadening associated with this arrangement have contributed to its limited use. However, these issues were later rectified by a controlled flow-rate delivery of the make-up fluid and the implementation of a self-alignment system [84]. Other improvements of the original arrangement include the use of a porous glass joint-μESI immersed in a reservoir of 1% acetic acid [85] and a microdialysis tubing [86]. In the case of the later, the limited mass transfer of the microdialysis tubing substantially reduced/eliminated the loss of sensitivity caused by the make-up liquid introduced in the sheath flow or liquid junction interface. The liquid flow in this arrangement is due to the column electroosmosis, the pressure resulting from the capillary effluent, and the flow induced by ESI. A liquid junction interface using a removable electrospray tip enclosed in a subatmospheric chamber and placed in front of the sampling orifice of the mass spectrometer was described by Karger et al. [87]. This arrangement allowed a good control of the flow from the liquid reservoir that feeds the liquid junction by the chamber pressure.

CE was recently interfaced to MS using a pressurized liquid junction nanoflow electrospray interface and surface-coated capillaries [88]. In this configuration, the separation capillary and the spray tip were positioned in an electrode vessel containing the appropriate spray liquid. Inside the liquid junction, an electrode spray potential was applied, thus sustaining a stable electrospray at nanoliter per minute flow rates without an external pump. This interface provided high durability of the spray tip and allowed independent optimization of the CE separation and ESI conditions, thus permitting a nanogram per milliliter detection of β-blocker drugs (β-adrenergic receptor inhibitors) [88].

21.2.2.5 MALDI Interfaces for Capillary Electrophoresis The coupling of CE to MALDI–MS was first described by Chang and Yeung [77]. Their arrangement consisted of a UV laser that vaporizes and ionizes the effluents from a CE separation capillary, while calcium chloride solution (0.5 mM) acted as both a CE electrolyte and a matrix. This arrangement produced pulsed ions analyzed by a TOF mass spectrometer. Karger et al. [78] introduced a vacuum deposition interface for coupling CE to MALDI-TOF–MS. Liquid samples consisting of the analyte and matrix were deposited on a moving tape in the evacuated source chamber of a TOF mass spectrometer. The vacuum deposition procedure generated significantly more reproducible signal intensity, eliminating the need for a "sweet spot" searching relative to the conventional dried-droplet method.

Murray et al. [89] online CE–MALDI-TOF–MS interface involves a rotating stainless steel ball that transports samples from atmospheric pressure to the high vacuum of the mass spectrometer for desorption and ionization. The ball is rotating at 0.03–0.3 revolutions min^{-1}. The sample is deposited directly from the separation capillary on a 19 mm ball on which a mixing of the matrix and the sample takes place using a capillary delivering matrix solution. The sample and matrix mixture is dried before passing a polymer gasket into the MALDI-TOF–MS ionization chamber. The ball is cleaned after it rotates out of the ionization chamber, using a solvent-saturated felt.

The above-mentioned MALDI–MS interfaces were developed mainly for CE–MS of peptides. A sample deposition device for interfacing CEC to MALDI–MS for the analysis of glycans was recently described by Tegeler et al. [79] The devise consisted of an inlet and outlet buffer reservoirs connected to a matrix reservoir through a connection sleeve. The matrix reservoir is connected to a deposition capillary via another connection sleeve. CEC eluent was transported to the matrix reservoir via a capillary that was connected to the deposition capillary by the connection sleeve inside the matrix reservoir. This connection sleeve also acted as a mixing chamber, allowing the CEC eluent to be mixed with the matrix before deposition. The utility of this arrangement was demonstrated for complex glycan mixtures separated by CEC using hydrophilic-phase monolithic columns, with the capillary flow being deposited on a standard MALDI plate along with a suitable matrix solution [79].

21.2.2.6 CE–MS Analysis of Glycans Several reviews have recently highlighted the application of CE–MS and tandem MS to glycoscreening in the biomedical field [13,90–94]. Suzuki et al. were the first to describe the offline coupling of CE to MALDI–MS for the structural characterization of APTS-labeled oligosaccharides and *N*-glycans derived from a glycoprotein [61]. An automated high-resolution fraction collector was utilized to collect the CE-resolved components. On-probe sample cleanup with a cation-exchange resin and a matrix mixture of 6-hydroxypicolinic and 3-hydroxypicolinic acids (1:1 ratio) was employed to acquire negative-ion mass spectra of the APTS-labeled *N*-glycans derived from glycoproteins. This approach allowed the detection of singly deprotonated ions with a detection limit of approximately 30 fmol for APTS-labeled maltoheptose. The detection sensitivity of highly sialylated structures (tri-, tetra-, pentasialylated and higher) was not at par with that of asialylated or mono- and disialylated structures. This is attributed to the in-source fragmentation of the highly labile sialic acid. This effect is more pronounced especially in the case of heavily sialylated glycans.

Mannooligosaccharide caps in mycobacterial lipoarabinomannan (ManLAMs), which are key molecules in the immunopathogenesis of tuberculosis, were characterized using CE–ESI–MS [95] and CE in conjunction with MALDI–MS [96]. In the case of CE–ESI, the APTS-labeled glycans were detected as $[M-2H]^{2-}$ molecular ions in the negative ion mode using triethylammonium formate buffer. Unequivocal structural characterization at the picomole level was attained, since some sequence fragment ions were detected in the spectra. Mannooligosaccharide caps were released by a mild acid hydrolysis from 85 μg of cellular ManLAMs of *Mycobacterium bovis*. Preferential ManLAMs cleavages occurring in the arabinan domain were achieved as a result of this treatment, thus leading to the formation of arabinose, mannooligosaccharide caps, and the mannan core. This was unequivocally confirmed through CE–ESI–MS in which arabinose, mannose, mannose–arabinose, mannose–mannose–arabinose, and mannose–mannose–arabinose–arabinose were detected [95]. The last two structures were previously misassigned using CE–UV and CE–LIF analysis in which the structural assignment was based on matching the migration times to standards [97]. CE–MALDI–MS allowed the analysis of only 5 μg (300 pmol) of *Mycobacterium tuberculosis* ManLAMs subjected to a mild acid hydrolysis before APTS labeling and CE separation. Offline MALDI–MS analysis using (9:1) mixture of 2,5-dihydroxybenzoic acid (DHB) and 5-methoxysalicylic acid (MSA) as the matrix and the on-probe sample cleanup procedure with cation-exchange resin was performed on the fractions collected from the CE separation. Mass characterization of the deprotonated molecular ions $[M-H]^{-}$ and the y-type ion fragments obtained in post-source decay experiments allowed the successful assignment of the mannooligosaccharide cap structures from Ref. 96. These two studies confirmed that offline CE–MALDI–MS, in this case, was more sensitive than CE–ESI–MS as suggested by the amount of ManLAMs consumed in each study. This better sensitivity of the offline CE–MALDI–MS was also confirmed through the ability of this approach to detect the minor compounds observed in CE–LIF analysis, which were not identified in the CE–ESI–MS study [96].

Offline CE–ESI–quadrupole time-of-flight (QTOF) MS configuration was also utilized for glycoscreening of *O*-glycosylated peptides derived from the urine of patients with a hereditary *N*-acetylhexosaminidase deficiency (Schindler's disease) [98]. The sensitive identification of the different glycopeptides present in urine samples was achieved using the data-dependent acquisition capability of the QTOF instrument, which allowed on-the-fly switching between the negative-ion mode MS scans and low-energy collision-induced dissociation tandem MS on selected precursor ions. The low detection levels of glycans typically derived from biological matrixes was attained using this approach. Sialylated *O*-glycosylated peptides, undetectable in complex mixtures by direct ESI–MS, were identified using this offline CE–ESI–MS configuration.

CE–ESI sheath-flow coupled to an ion-trap mass spectrometer for the analysis of negatively charged labeled and native glycans was recently described by Harvey et al. [99]. A stable spray was sustained 60–100% methanol solution as a coaxial sheath liquid and nitrogen sheath gas. 8-Aminonaphthalene-1,3,6-trisulfonic acid (ANTS) labeled *N*-glycans derived from model glycoproteins such as ribonuclease B and fetuin as well as native *N*-glycans derived from fetuin were analyzed using this configuration. A CE separation attained in this study using 20 mM 6-aminocaproic acid buffer was comparable to the ones usually achieved using

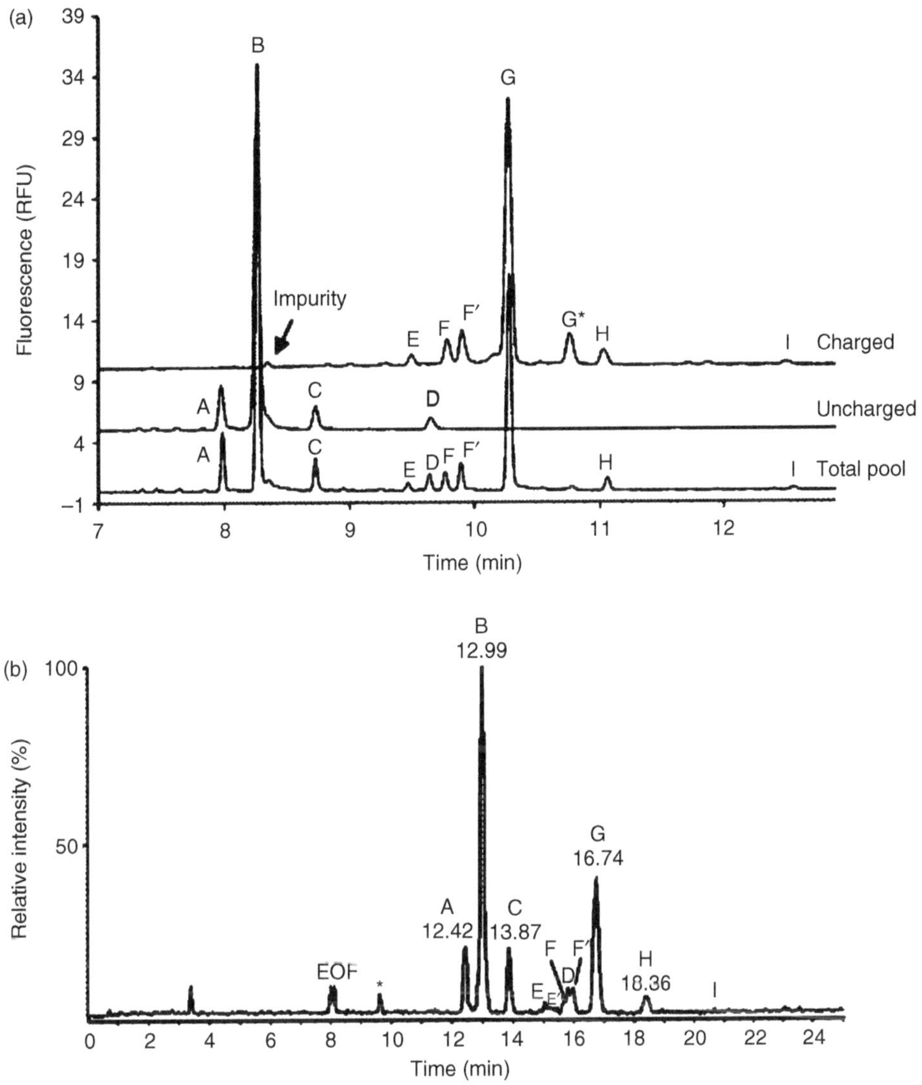

FIGURE 21.4 CE–LIF electropherograms of the APTS-derivatized total CBH I *N*-glycan pool and of the uncharged and charged glycans (a). CE–MS base peak electropherogram of the total CBH I *N*-glycan pool (b). The neutral glycans are labeled A–D, the charged ones E–I. Unlabeled neutral glycans are detected in the EOF; unlabeled phosphorylated glycans migrate more slowly and are indicated by an asterisk. Reproduced from Ref. 101 with permission.

10 mM citric acid, which is commonly utilized in the CE–LIF analyses of labeled glycans. The ionization efficiency of ANTS over a wide pH range and its resistance to fragmentation during tandem MS experiments were the driving forces behind using this labeling reagent [99].

A Q-Trap mass spectrometer was recently utilized in conjunction with CE to characterize *N*-glycans derived from cellobiohydrolase I, with a special focus on phosphorylated residues [100]. A tandem MS was acquired through a modified triple quadrupole, where the Q3 region was employed as a conventional quadrupole mass filter or as a linear ion trap with axial ion ejection [101]. APTS labeling of the glycans, which imparted negative charges, allowed the simultaneous negative ion detection of both neutral and charged glycans. Generally, APTS labeling of these oligosaccharides permits high-resolution CE, a better ionization in the negative ion mode, a simultaneous detection of neutral and charged oligosaccharides, and generation of predictable tandem MS spectra (Figure 21.4). This configuration allowed the distinguishing of phosphorylated glycan isomers, while tandem MS data furnished additional structural information [100].

Labeling of *N*-glycans associated with glycoproteins attained through the transglycosylation activity of *endo*-β-*N*-acetylglucosaminidase (*endo*-M) was recently reported [102]. This endoglycosidase possesses both a hydrolytic

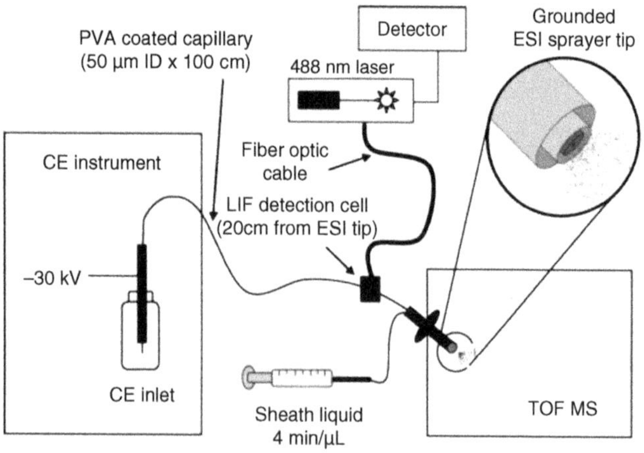

FIGURE 21.5 Schematic of the CE–MS system with online LIF detection. Reproduced from Ref. 103 with permission.

activity toward the glycosidic bond in the *N,N'*-diacetylchitobiose moiety of the *N*-glycans of glycopeptides, and a transglycosylation activity to transfer both the complex-type and high-mannose type oligosaccharides of the N-linked sugar chains from glycopeptides to suitable acceptors having an *N*-acetylglucosamine residue. A suitable fluorescent acceptor such as NDA–Asn–GlcNAc was used in conjunction with this enzyme to fluorescently label *N*-glycans derived from glycoproteins before CE–TOF–MS analysis. This approach was initially validated using *N*-glycans derived from ovalbumin. Pronase digestion of glycoproteins, which was required before treatment with *endo*-M, made this derivatization approach lengthy. However, the derivatization approach alone reaches completion in 30 min [102].

Online CE–LIF–MS configuration was first described by Gennaro and Salas-Solano, which permitted the direct characterization of N-linked glycans from therapeutic antibodies [103]. This online CE–LIF–MS configuration is schematically shown in Figure 21.5. The system comprised a CE system with a CE–MS capillary cartridge allowing for external detection. Fitting a PVA-coated capillary with an ellipsoid from Picometrics (Toulouse, France) and using a separate Picometrics Zeta-LIF Discovery detection system consisting of a 488 nm laser unit and an LIF detection cell connected to the detector via a fiber-optic cable permitted online LIF detection, as illustrated in Figure 21.5. The detection cell is fitted with an adjustable arm, approximately 20 cm from the ESI tip. MS was achieved using a MicrOTOF mass spectrometer equipped with an orthogonal, grounded ESI source. A sheath liquid consisting of a 50:50 isopropanol/water (0.2% ammonia) was added coaxially to the separation capillary at 4 μL min^{-1} flow rate. The identification of major and minor glycan species was permitted through accurate masses furnished by the mass spectrometer.

This study demonstrated for the first time the ability to attain CE–MS separation efficiency comparable to that commonly acquired through CE–LIF [103]. Figure 21.6 depicts expanded-scale views of (a) a conventional CE–LIF electropherogram using a 60 cm capillary, (b) a CE–LIF trace obtained online with MS detection, and (c) a CE–MS base-peak electropherogram. The four early migrating species are seen clearly in both the online LIF and MS ion electropherograms. However, the MS-based electropherograms (b and c) suffer from a shift in migration time and some loss in separation efficiency, which is due to dead volumes originating from the addition of the MS detector.

21.2.2.7 Glycomic Analysis by CEC–MS Generally, CEC is commonly referred to as a technique combining the desirable features of both capillary electromigration (flow-induced radial solute mass transfer) and chromatography (a wide choice of suitable stationary phases). Moreover, preconcentration of dilute samples at the capillary inlet is easily attainable through CEC. Although a recent progress in CEC features prominently monolithic columns, which principally offer a wide range of retention selectivity together with the separation conditions that appear compatible with MS operation, stationary phases designed specifically for the separation of glycans have not been in abundance.

Examples of carbohydrate applications of CEC include the separation of aminobenzamide derivatives [104], using a hydrophobic monolithic stationary phase. Other approaches to separating native glycans using "MS-friendly" mobile phases lead to the use of columns featuring hydrophilic interactions. CEC monolithic columns suited for the separation of very complex pools of glycans have been demonstrated [26–28]. However, the production of these CEC monolithic columns has been limited to one research group, suggesting the difficulty associated with the assembly of such columns.

Monolithic CEC columns based on hydrophilic interactions have been recently configured to FT–MS for the analysis of closely related glycan structures [28]. The mixture of *O*-glycans chemically released from mucin was resolved by the CEC–ESI–FTMS, as shown in Figure 21.7a. This type of mass spectrometer offers 3.9 ppm average mass measurement accuracy with external calibration. The mass resolution and accuracy of FT–MS for saccharides was demonstrated in this study in the characterization of two glycan structures that differ from each other by 1 *m/z* unit. These were an acidic glycan with *m/z* 756 (dot at 10.5 min in Figure 21.7b) and a neutral glycan with *m/z* 757 (dot at 12 min in Figure 21.7b). The average resolution of the spectrum illustrated in Figure 21.7a is 30,000. Previously, these structures could not be distinguished by MALDI-TOF–MS because of its inability to isolate either ions for further postsource decay or collision-induced dissociation studies. These two oligosaccharides, however, could be electrochromatographically separated from each other, while the high mass accuracy of FT–MS allowed

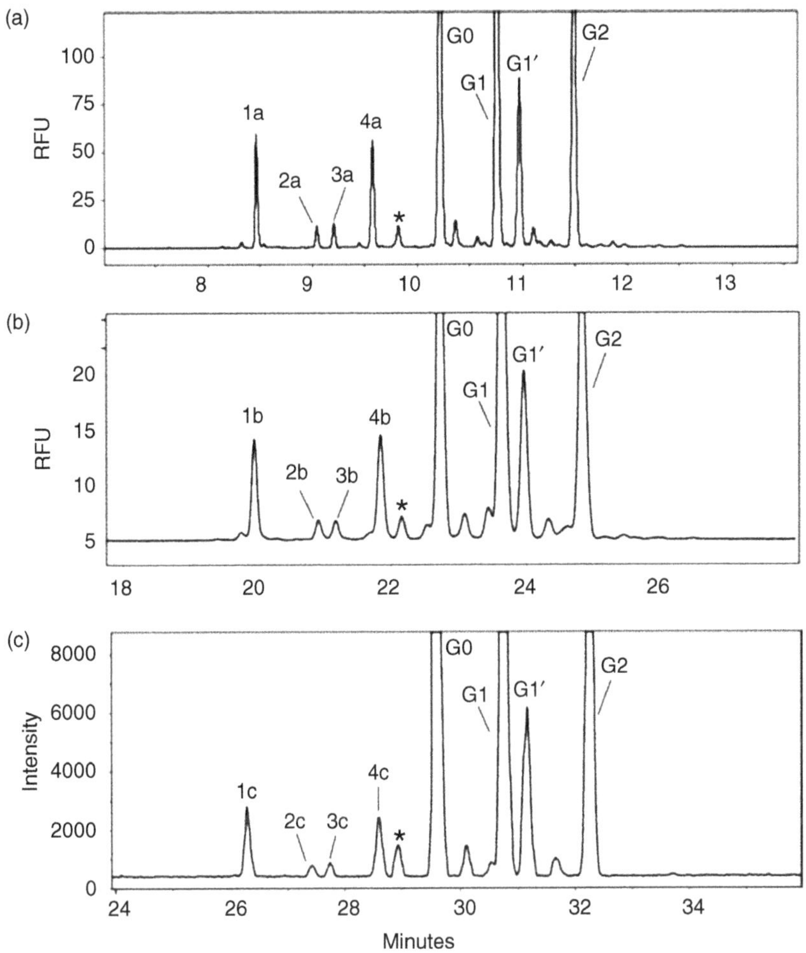

FIGURE 21.6 Expanded-scale electropherograms of APTS labeled *N*-glycans derived from rMAb 1: (a) standard CE–LIF electropherogram using a 60 cm capillary. (b) CE–LIF trace obtained online with MS detection. (c) CE MS base peak electropherogram. Reproduced from Ref. 103 with permission.

their accurate mass determination (756.2683 *m/z* and 757.2851 *m/z*, Figure 21.7c, d, respectively). These structures were further characterized by tandem MS. On collisional activation, the acidic glycan with *m/z* value of 756 and GlcNAc(NeuGc)GalNAc-ol composition easily lost its acidic residue at the nonreducing end to form the fragment at *m/z* value of 449 (Figure 21.7e). Sialic acid residues are very labile in collision-induced dissociation (CID); therefore, they are readily lost. Conversely, the neutral glycan with *m/z* value of 757 produced two product ions with *m/z* values of 449 and 611. These fragments originate from the loss of a fucose residue and fucose and hexose residues from the nonreducing end of the structures (Figure 21.7f), thus suggesting FucHexHexNAcGalNAc-ol to be the composition of this glycan.

Employing the CEC–MALDI–MS interface described earlier [79], the analysis of *N*- and *O*-glycans derived from bile salt-stimulated lipase (BSSL) from human breast milk, which is a relatively large glycoprotein consisting of 722 amino acid residues with numerous *O*-glycosylation sites near the C-terminus [105], was achieved. The high microheterogeneity of the glycan structures derived from this glycoprotein is evident in Figure 21.8, which depicts 3D and 2D CEC–MALDI–MS electropherograms. Approximately, 50 distinct peaks are displayed in this figure. Many of the depicted peaks differ by a few mass units, while some peaks appear to represent structural isomers.

21.3 MICROCHIP CAPILLARY ELECTROPHORESIS

Since glycoproteins are commonly encountered at the submicrogram levels, miniaturization of the previously described separation methodologies represents an excellent fit to such predicament. Although modern instrumental

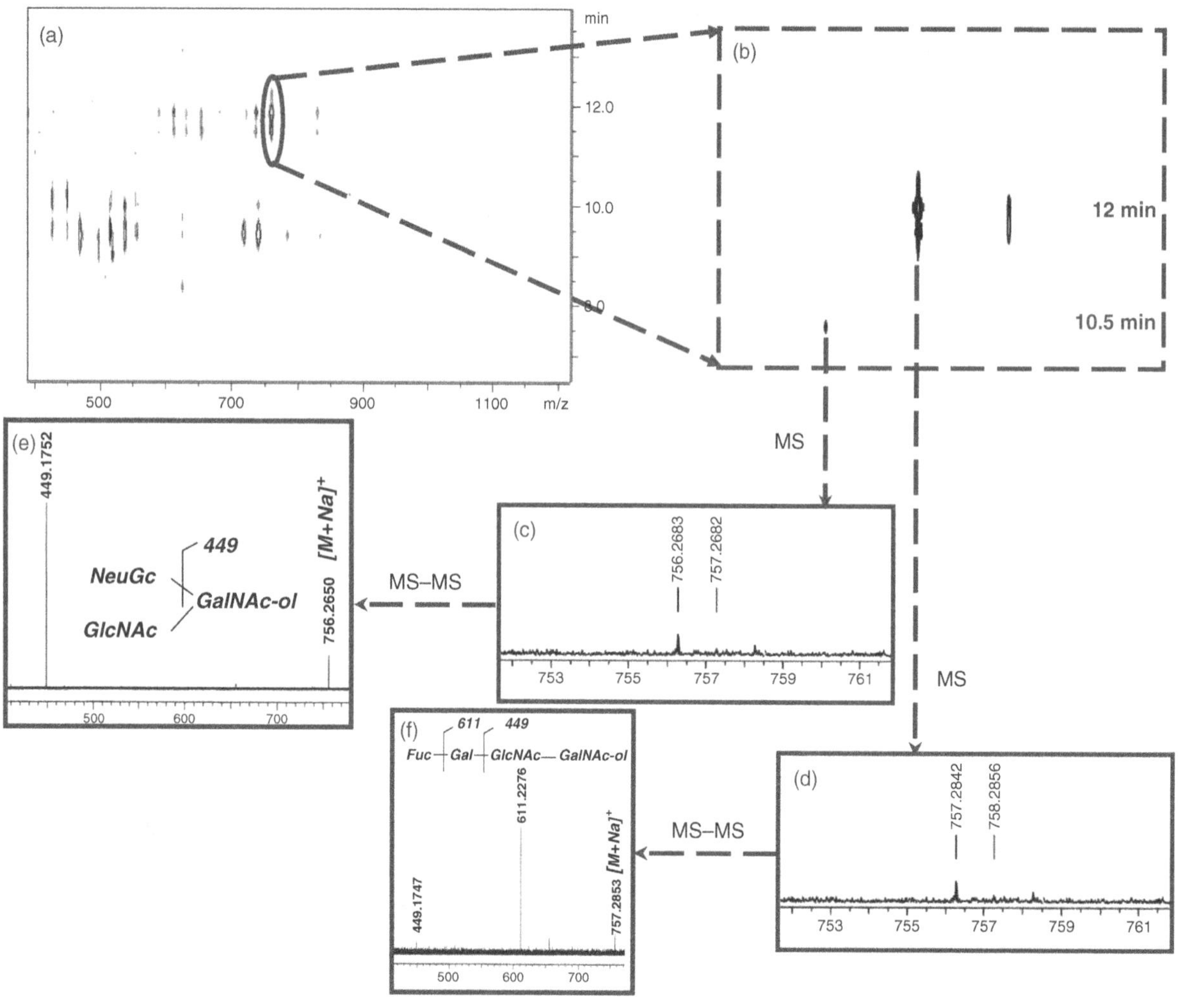

FIGURE 21.7 (a) 2D contour plot of a mixture of *O*-glycans cleaved from mucin. (b) zoomed 2D contour plot, (c and d) spectra of glycans with *m/z* values of 756 and 757 and their corresponding tandem mass spectra (e and f). An average resolution of ~30,000 is demonstrated. Experimental conditions: cyano column, 26 cm; mobile phase, 2.4 mM ammonium formate buffer (240 mM, pH 3) and 0.2 mM sodium acetate in the mixture of acetonitrile/water (71:29, v/v); field strength, 600 V cm^{-1}; injection, 12 kV, 30 s. Reproduced from Ref. 28 with permission.

techniques allow accurate measurements of such small amounts, sample loss to the surfaces of devices being used can occur before such measurements. A sample loss during ultrafiltration, dialysis, lyophilization, and so on, can easily become a bottleneck of the entire analysis. Another problem with working at such a reduced scale is contamination (dust, solvent, reagent impurities, and so on). It is thus crucial to minimize the number of handling and transfer steps during the analysis. Miniaturizing separation, in terms of reduced column diameters, solvent flow rates, and the overall surface area that a glycan sample may encounter during analysis, is thus becoming significant in high-sensitivity work. The ability to integrate multiple steps such as sample preparation, purification, separation, and detection to a small analytical unit, such as a microchip, is invaluable to minimizing sample losses and consequently enhancing analysis sensitivity.

Thus far, chip-based approaches employed in glycomic analysis involve devices that incorporate microchannels, allowing a sample injection, preconcentration, and separation. Both electrophoretic and chromatographic modes of separation in microchips have been performed. The former is commonly referred to as microchip capillary electrophoresis (MCE), which commonly uses different detection principles such as refractometry [106], electrochemical detection [107–110], UV [111], and LIF [112–114].

An interesting application of MCE involving a total serum protein *N*-glycome profiling was recently reported [65,115]. This method was an extension of the method described earlier [64]. Alumina–silicate glass chips with a

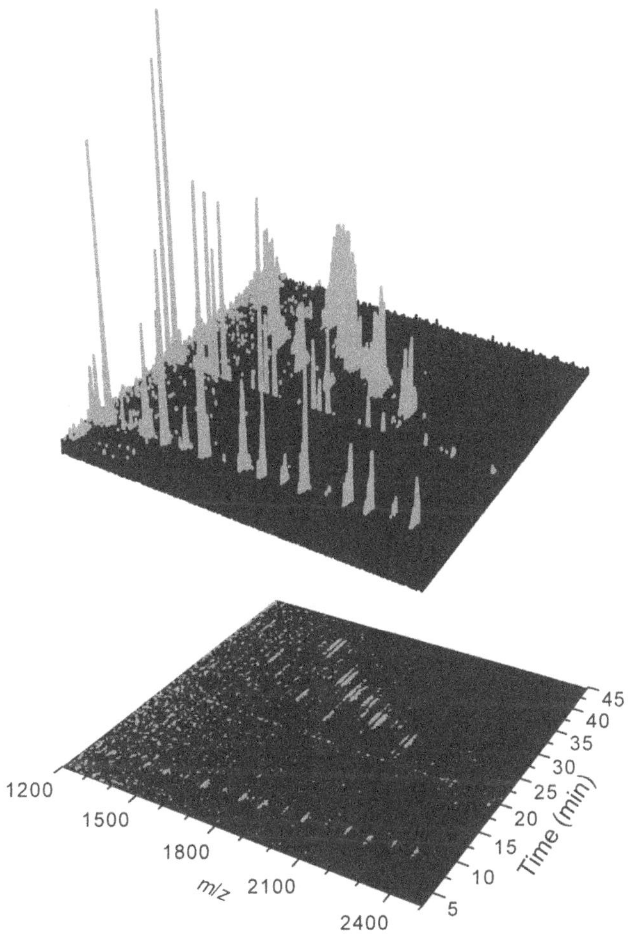

FIGURE 21.8 3D electrochromatogram of the mixture of N- and O-linked glycans derived from human bile salt-stimulated lipase. Experimental conditions: cyano capillary column, 28 cm; mobile phase, 2.4 mM ammonium formate buffer in a 60:40 mixture of acetonitrile/water; field strength, 535 V cm^{-1}; injection, 10 kV, 10 s; matrix, 20 mg mL^{-1} DHB in 80:20 methanol/water. Reproduced from Ref. 79 with permission.

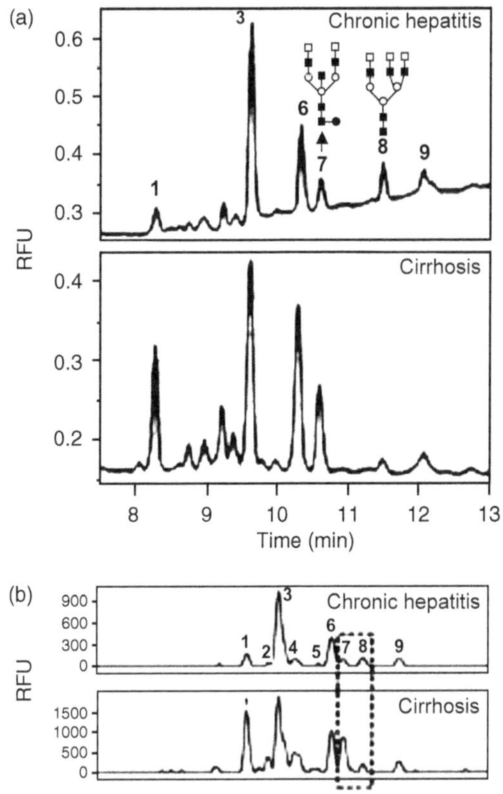

FIGURE 21.9 (a) MCE profiling of serum samples from a noncirrhotic chronic hepatitis patient (upper trace) and cirrhotic patient (lower trace). (b) Profiling of the same samples using the ABI377 gel-based DNA-sequencer. Symbols: ■, *N*-acetylglucosamine; ○, mannose; □, galactose; ●, fucose. Reproduced from Ref. 65 with permission.

double-T injector were fabricated. A uniform introduction of all molecular types without a bias commonly associated with an electrokinetic injection in CE was achieved through the use of this type of injector. The profiling of the major *N*-glycans in human serum was attained in 12 min with a resolution comparable to what has been achieved through gel-based DNA sequencers using MCE with 11.5 cm effective length filled with 4% linear polyacrylamide permitted. Profiling of two serum samples from a noncirrhotic chronic hepatitis patient and a cirrhotic patient demonstrated the potential of ME in clinical studies (Figure 21.9). The change in the ratio of two glycan structures (namely, complex triantennary and fucosylated bisecting biantennary) was determined by the same group to be diagnostic for liver cirrhosis in a previous study using gel-based DNA sequencer (Figure 21.9b) [65]. MCE generated profiles (Figure 21.9a) supported these findings. The two peaks associated with glycan structures were well resolved in MCE, while the ratio between the peaks representative of the two structures was comparable to the gel-based DNA sequencer results. Accordingly, this study demonstrated the practical aspects of MCE analysis, allowing rapid, reproducible, and quantitative comparison between the different liver cohorts.

Although the same group reported the development of a new method that is substantially shorter than the aforementioned methods, the separation efficiency demonstrated by the new method was substantially lower [116]. Moreover, the new method involved only 10 min incubation with PNGase F, which is substantially shorter than what is commonly used. This short incubation time was not justified or supported by any data. The new method was supposed to be an improvement of the other methods that were based on comparing the ratio of two peaks corresponding to fucosylated GlcNAc bisecting and triantennary glycan structures; however, these two peaks are not

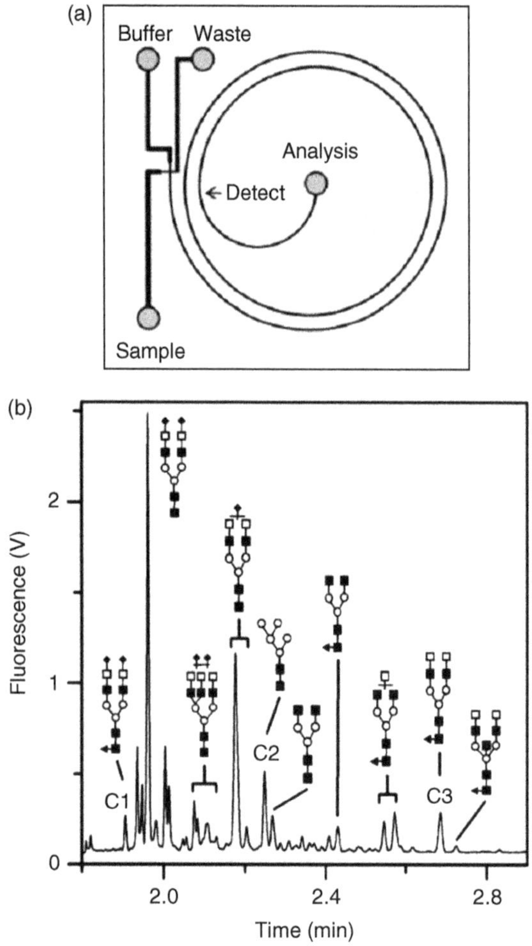

FIGURE 21.10 (a) Schematic of the microfluidic device with the spiral channel design used for glycan analysis. The effective separation length was 22 cm from the cross intersection to the detection point indicated by the arrow. The reservoirs and corresponding channels are labeled with their functions. The microchip substrate was 5×5 cm. The narrow and wide channels were 60 and 240 μm wide at the top, respectively. (b) Electropherogram of *N*-glycans released from a blood serum sample from a stage IV breast cancer patient separated on the microfluidic device. The separation length was 22 cm, and the separation field strength was 750 V cm^{-1}. The separation efficiencies for components C1, C2, and C3 are listed in Table 2. The symbols are as in Figure 21.9. Reproduced from Ref. 65 with permission.

base-line resolved, thus making their quantification partially nonpractical.

Jacobson et al. recently reported the efficient electrophoretic separations of *N*-glycans on MCE with analysis times less than 3 min [117]. A spiral channel design with an effective separation length of 22 cm was employed (Figure 21.10a). MCE performance was comparable to that of CE, thus confirming the absence of any negative effects arising from the noncircular cross-section of the microchannel. The microfluidic device was employed to analyze APTS-labeled glycans derived from model glycoproteins and blood serum. The separation efficiency provided by MCE was adequate to resolve *N*-glycan positional isomers derived from ribonuclease B and linkage isomers derived from asialofetuin. The separation efficiencies attained for the complex APTS-labeled *N*-glycans derived from blood serum (Figure 21.10b) were highly comparable to those observed in the case of APTS-labeled *N*-glycans derived from model glycoproteins. The same group recently enhanced the separation efficiency of APTS-labeled glycan using serpentine separation channels and asymmetrically tapered turn [125]. The tapered turns were designed to minimize band broadening originating from the "racetrack" effect. This serpentine design permitted an increase in the length of the separation channels beyond what was attainable using spiral channel design described earlier without sacrificing separation efficiency.

21.4 CONCLUSIONS

The unmatched CE spatial resolution in relatively short separation times has been demonstrated in this chapter. This technique is not yet widely used in glycomics because of the discrepancy between the amount detected and the needed sample concentration. Although LIF is permitting sensitive analysis of glycans in conjunction with CE definitive identification of glycans is only attained when standards are available, which is not very common. Therefore, MS is still considered to be the method for the reliable sensitive identification of glycans. A reliable and easy interfacing of CE to MS is expected to prompt a greater use of CE in glycomics.

Over the past three decades, the efforts of many research groups interested in the identification of glycans isomers have resulted in the development of many new technologies enabling effective and reliable interfacing of CE to MS. Nevertheless, many of the buffers that are routinely employed for CE of glycans are not MS-friendly. An improvement in this regard is still warranted. Derivatization of glycans is still considered as a means to enhance detection, even in the case of MS. The suitability of these derivatization methods for MS detection is currently being emphasized.

Although the efficiencies of MCE separations have been shown to be very comparable to those of CE, at much shorter analysis time (second vs. minute), The misconception that complex samples are not amenable to separation by MCE has limited the use of this technique in glycomics. This misconception is starting to dissipate as evident by the increase in MCE publications and the acceptance of the ability of this approach to accommodate the separation of complex systems.

REFERENCES

1. Fukuda, M.N. (1989) *Blood 79*, 84–89.
2. Varki, A. (1993) *Glycobiology 3*, 97–130.
3. Dwek, R.A. (1996) *Chem. Rev. 96*, 683–720.
4. Helenius, A. and Aebi, M. (2001) *Science 291*, 2364–2369.
5. Rudd, P.M., Woods, R.J., Wormald, M.R., Opdenakker, G., Downing, A.K., Campbell, I.D., and Dwek, R.A. (1995) *Biochem. Biophys. Acta 1248*, 1–10.
6. Rudd, P.M. et al. (1999) *J. Mol. Biol. 293*, 351–366.
7. Dennis, J.W. and Granovsky, M. (1999) *Bioassays 21*, 412–421.
8. Lowe, J.B. and Marth, J.D. (2003) *Annu. Rev. Biochem. 72*, 643–691.
9. Dell, A. and Morris, H.R. (2001) *Science 291*(5512), 2351–2356.
10. Geiser, H., Silvescu, C., and Reinhold, V. (2006) *Sep. Methods Proteomics* 321–343.
11. Harvey, D.J. (2003) *Int. J. Mass Spectrom. 226*, 1–35.
12. Mechref, Y. and Novotny, M. (2002) *Chem. Rev. 102*(2), 321–369.
13. Novotny, M.V. and Mechref, Y. (2005) *J. Sep. Sci. 28*, 1956–1968.
14. Churms, S.C. (1996) *J. Chromatogr. A 720*, 75–91.
15. Hemström, P. and Irgum, K. (2006) *J. Sep. Sci. 29*, 1784–1821.
16. Townsend, R.R., Hardy, M.R., Cumming, D.A., Carver, J.P., and Bendiak, B. (1989) *Anal. Biochem. 182*, 1–8.
17. Lee, Y.C. (1996) *J. Chromatogr. A 720*, 137–149.
18. Guttman, A. (1997) *Electrophoresis 18*, 1136–1141.
19. Hermentin, P., Doenges, R., Witzel, R., Hokke, C.H., Vliegenthart, J.F.G., Kamerling, J.P., Conradt, H., Nimtz, M., and Brazel, D. (1994) *Anal. Biochem. 221*, 29–41.
20. Honda, S., Makino, A., Suzuki, S., and Kakehi, K. (1990) *Anal. Biochem. 191*, 228–234.
21. Ma, S. and Nashabeh, W. (1999) *Anal. Chem. 71*, 5158–5192.
22. Liu, J., Shirota, O., and Novotny, M. (1991) *J. Chromatogr. A 559*, 223–35.
23. Guttman, A. and Pritchett, T. (1995) *Electrophoresis 16*, 1906–1911.
24. Camilleri, P., Harland, G.B., and Okafo, G. (1995) *Anal. Biochem. 230*, 115–122.
25. Palm, A. and Novotny, M. (1997) *Anal. Chem. 69*(22), 4499–4507.
26. Que, A. and Novotny, M.V. (2003) *Anal. Bioanal. Chem. 375*, 599–608.
27. Que, A.H. and Novotny, M.V. (2002) *Anal. Chem. 74*, 5184–5191.
28. Que, A.H., Mechref, Y., Huang, Y., Taraszka, J.A., Clemmer, D.E., and Novotny, M.V. (2003) *Anal. Chem. 75*, 1684–1690.
29. Guttman, A., Chen, F.-T.A., Evangelista, R.A., and Cooke, N. (1996) *Anal. Biochem. 233*, 234–242.
30. Liu, J., Shirota, O., Wiesler, D., and Novotny, M.V. (1991) *Proc. Natl. Acad. Sci. USA 88*, 2302–2306.
31. Takasa, S., Misuochi, T., and Kobata, A. (1982) *Methods Enzymol. 83*, 263–268.
32. Patel, T., Bruce, J., Merry, A., Bigge, C., Wormald, M., Jaques, A., and Parekh, R. (1993) *Biochemistry 32*, 679–693.
33. O'Neill, R.A.J. (1996) *J. Chromatogr. A 720*, 201–215.
34. Carlson, D.M. (1968) *J. Biol. Chem. 243*, 616–626.
35. Hanisch, F.G., Jovanovic, M., and Peter-Katalinic, J. (2001) *Anal. Biochem. 290*, 47–59.
36. Chai, W., Feizi, T., Yuen, C., and Lawson, A.M. (1997) *Glycobiology 7*, 861–872.
37. White, C.C. and Kennedy, J.F. (1988) *An Introduction to the Chemistry of Carbohydrates*, Clarendon Press, Oxford.
38. Feizi, T. and Childs, R.A. (1994) *Methods Enzymol.* 205–217.
39. Yoshino, K., Takao, T., Murata, H., and Shimonishi, Y.A. (1995) *Anal. Chem. 67*, 4028–4031.
40. Whittal, R.M., Palcic, M.M., Hindsgaul, O., and Li, L. (2001) *Anal. Chem. 67*, 3508–3514.
41. Bardelmeijer, H.A., Waterval, J.C.M., Lingeman, H., van't Hof, R., Bult, A., and Underberg, W.J.M. (1997) *Electrophoresis 18*, 2214–2227.
42. Paulus, A. and Klockow, A.J. (1996) *J. Chromatogr. A 720*, 353–376.
43. Huang, Y., Mechref, Y., and Novotny, M.V. (2001) *Anal. Chem. 73*, 6063–6069.
44. Brooks, M.M. and Savage, A.V. (1997) *Glycoconj J. 14*, 183–190.
45. Endo, Y., and Kobata, A. (1976) *J. Biochem. 80*, 1–8.
46. Fan, J.-Q., Kadowaki, S., Yamamamoto, K., Kumagai, H., and Tochikura, T. (1988) *Agric. Biol. Chem. 52*, 1715–1723.
47. Goetz, J.A., Novotny, M.V., and Mechref, Y. (2009) *Anal. Chem. 81*, 9546–9552.
48. Hoffstetter-Kuhn, S., Paulus, A., Gassmann, E., and Widmer, H.M. (1991) *Anal. Chem. 63*, 1541–1547.
49. Liu, J., Dolnik, V., Hsieh, Y.-Z., and Novotny, M.V. (1992) *Anal. Chem. 64*, 1328–1336.
50. Liu, J., Shirota, O., and Novotny, M.V. (1992) *Anal. Chem. 64*, 973–975.
51. Chiesa, C. and O'Neill, R.A. (1994) *Electrophoresis 15*, 1132–1140.
52. Stefansson, M. and Novotny, M. (1994) *Carbohydr. Res. 258*, 1–9.
53. Stefansson, M. and Novotny, M.V. (1994) *Anal. Chem. 66*, 1134–1140.
54. Klockow, A., Widmer, H.M., Amado, R., and Paulus, A. (1994) *Fresenius J. Anal. Chem. 350*, 415–425.
55. Klockow, A., Amado, R., Widmer, H.M., and Paulus, A. (1995) *J. Chromatogr. A 716*, 241–257.
56. Klockow, A., Amado, R., Widmer, H.M., and Paulus, A. (1996) *Electrophoresis 17*, 110–119.
57. Guttman, A. and Starr, C.M. (1995) *Electrophoresis 16*, 993–997.
58. Chen, F.-T. and Evangelista, R.A. (1998) *Electrophoresis 19*, 2639–2644.

59. Evangelista, R.A., Chen, F.-T., and Guttman, A. (1996) *J. Chromatogr. A 745*, 273–280.
60. Guttman, A., Chen, F.-T., and Evangelista, R.A. (1996) *Electrophoresis 17*, 412–417.
61. Suzuki, H., Mueller, O., Guttman, A., and Karger, B.L. (1997) *Anal. Chem. 69*, 4554–4559.
62. Mechref, Y., Muzikar, J., and Novotny, M.V. (2005) *Electrophoresis 26*, 2034–2046.
63. Honda, S., Okeda, J., Iwanaga, H., Kawakami, S., Taga, A., Suzuki, S., and Imai, K. (2000) *Anal. Biochem. 286*, 99–111.
64. Callewaert, N., Geysens, S., Molemans, F., and Contreras, R. (2001) *Glycobiology 11*, 275–281.
65. Callewaert, N., van Vlierberghe, H., van Hecke, A., Laroy, W., Delanghe, J., and Contreas, R. (2004) *Nat. Med. 10*, 429–434.
66. Vanderschaeghe, D., Laroy, W., Sablon, E., Halfon, P., van Hecke, A., Delanghe, J., and Callewaert, N. (2009) *Mol. Cell. Proteomics 8*, 986–994.
67. Minard, R.D., Luckenbill, D., Curry, P.J., and Ewing, A.G. (1989) *Adv. Mass Spectrom. 11*, 436–437.
68. Moseley, M.A., Deterding, L.J., Tomer, K.B., and Jorgenson, J.B. (1989) *Rapid. Commun. Mass Spectrom. 3*, 87–93.
69. Olivares, J.A., Nguyen, N.T., Yonker, C.R., and Smith, R.D. (1987) *Anal. Chem. 59*, 1230–1232.
70. Smith, R.D., Barinaga, C.J., and Udseth, H.R. (1988) *Anal. Chem. 60*, 1948–1952.
71. Lee, E.D., Muck, W., Henion, J.D., and Covey, T.R. (1989) *Biomed. Environ. Mass Spectrom. 18*, 844–850.
72. Cai, J. and Henion, J. (1995) *J. Chromatogr. A 703*, 667–692.
73. Ding, J. and Vouros, P. (1999) *Anal. Chem. 71*, 378–385.
74. Gelpi, E. (2002) *J. Mass Spectrom. 37*, 241–253.
75. Niessen, W.M.A. (2000) *J. Chromatogr. A 794*, 407–435.
76. Niessen, W.M.A., Tjaden, U.R., and van der Greef, J. (1993) *J. Chromatogr. 636*, 3–19.
77. Chang, S.Y. and Yeung, E.S. (1997) *Anal. Chem. 69*, 2251–2257.
78. Preisler, J., Hu, P., Rejtar, T., and Karger, B.L. (2000) *Anal. Chem. 72*, 4785–4795.
79. Tegeler, T.J., Mechref, Y., Boraas, K., Reilly, J.P., and Novotny, M.V. (2004) *Anal. Chem. 76*, 6698–6706.
80. Mazereeuw, M., Hofte, A.J.P., Tjaden, U.R., and van der Greef, J. (1997) *Rapid Commun. Mass Spectrom. 11*, 981–986.
81. Fang, L., Zhang, R., Williams, E.R., and Zare, R.N. (1994) *Anal. Chem. 66*, 3696–3701.
82. Smith, R.D., Olivares, J.A., Nguyen, N.T., and Udseth, H.R. (1988) *Anal. Chem. 60*, 436–441.
83. Thibault, P., Pleasance, S., and Laycock, M.V. (1991) *J. Chromatogr. 542*, 483–501.
84. Wachs, T., Sheppard, R.L., and Henion, J. (1996) *J. Chromatogr. B 685*, 335–342.
85. Settlage, R.E., Russo, P.S., Shabanowitz, J., and Hunt, D.F. (1998) *J. Microcol. Sep. 10*, 281–285.
86. Severs, J.C. and Smith, R.D. (1997) *Anal. Chem. 69*, 2154–2158.
87. Foret, F., Zhou, H., Gangi, E., and Karger, B.L. (2000) *Electrophoresis 21*, 1363–1371.
88. Fanali, S., D'Orazio, G., Foret, F., Kleparnik, K., and Aturki, Z. (2006) *Electrophoresis 27*, 4666–4673.
89. Musyimi, H.K., Narcisse, D.A., Zhang, X., Stryjewski, W., Soper, S.A., and Murray, K.K. (2004) *Anal. Chem. 76*, 5968–5973.
90. Campa, C., Coslovi, A., Flamigni, A., and Rossi, M. (2006) *Electrophoresis 27*, 2027–2050.
91. Mechref, Y. and Novotny, M.V. (2002) *Chem. Rev. 102*, 321–369.
92. Mechref, Y. and Novotny, M.V. (2006) *J. Chromatogr. B 841*, 65–78.
93. Novotny, M.V., Soini, H.A., and Mechref, Y. (2008) *J. Chromatogr. B 866*, 26–47.
94. Zamfir, A. and Peter-Katalinic, J. (2004) *Electrophoresis 25*, 1949–1963.
95. Monsarrat, B., Brando, T., Condouret, P., Nigou, J., and Puzo, G. (1999) *Glycobiology 19*, 335–342.
96. Ludwiczak, P., Brando, T., Monsarrat, B., and Puzo, G. (2001) *Anal. Chem. 73*, 2323–2330.
97. Venisse, A., Fournie, J.J., and Puzo, G. (1995) *Eur. J. Biochem. 231*, 440–447.
98. Bindila, L., Almeida, R., Sterling, A., Allen, M., Peter-Katalinić, J., and Zamfirl A. (2004) *J. Mass Spectrom. 39*, 1190–1201.
99. Gennaro, L.A., Delaney, J., Vouros, P., Harvey, D.J., and Domon, B. (2002) *Rapid Commun. Mass Spectrom. 16*, 192–200.
100. Sandra, K., Devreese, B., Stals, I., Claeyssens, M., and van Beeumen, J. (2004) *J. Am. Soc. Mass Spectrom. 15*, 413–423.
101. Sandra, K., van Beeumen, J., Stals, I., Sandra, P., Claeyssens, M., and Devreese, B. (2004) *Anal. Chem. 76*, 5878–5886.
102. Min, J.Z., Toyo'oka, T., Kato, M., and Fukushima, T. (2005) *Chem. Commun.* 3484–3486.
103. Gennaro, L.A. and Salas-Solano, O. (2008) *Anal. Chem. 80*, 3838–3845.
104. Palm, A. and Novotny, M.V. (1997) *Anal. Chem. 69*, 4499–4507.
105. Mechref, Y., Chen, P., and Novotny, M.V. (1999) *Glycobiology 9*, 227–234.
106. Buggraf, N., Krattiger, B., de Mello, A.J., de Rooij, N.F., and Manz, A. (1998) *Analyst 123*, 1443–1447.
107. Fanguuy, J.C. and Henry, C.S. (2002) *Analyst 127*, 1021–1023.
108. Fu, C.-G. and Fang, Z.-L. (2000) *Anal. Chim. Acta 422*, 71–79.
109. Hebert, N.E., Kuhr, W.G., and Brazill, S.A. (2002) *Electrophoresis 23*, 3750–3759.
110. Schwarz, M.A., Galliker, B., Fluri, K., Kappes, T., and Hauser, P.C. (2001) *Analyst 126*, 147–151.
111. Suzuki, S., Ishida, Y., Taga, A., Arai, A., Nakanishi, H., and Honda, S. (2003) *Electrophoresis 24*, 3828–3833.
112. Dang, F., Zhang, L., Hagiwara, H., Mishina, Y., and Baba, Y. (2003) *Electrophoresis 24*, 714–721.

113. Dang, F., Zhang, L., Jabasini, M., Kaji, N., and Baba, Y. (2003) *Anal. Chem.* *75*, 2433–2439.

114. Suzuki, S., Shimotsu, N., Honda, S., Arai, A., and Nakanishi, H. (2001) *Electrophoresis* *22*, 4023–4031.

115. Callewaert, N., Contreas, R., Mitnik-Gankin, L., Carey, L., Matsudaira, P., and Ehrlich, D. (2004) *Electrophoresis* *25*, 3128–3131.

116. Vanderschaeghe, D., Szekrenyes, A., Wenz, C., Gassmann, M., Naik, N., Bynum, M., Yin, H., Delanghe, J., Guttman, A., and Callewaert, N. (2010) *Anal. Chem.* *82*, 7408–7415.

117. Zhuang, Z., Starkey, J., Mechref, Y., Novotny, M.V., and Jacobson, S.C. (2007) *Anal. Chem.* *79*, 7170–7175.

INDEX

Capillary Electrophoresis and Microchip Capillary Electrophoresis: Principles, Applications, and Limitations, First Edition.
Edited by Carlos D. García, Karin Y. Chumbimuni-Torres, and Emanuel Carrilho.

www.ingramcontent.com/pod-product-compliance
Lightning Source LLC
LaVergne TN
LVHW062140200926
848484LV00007B/1836